REVIEWS in MINERALOGY

Volume 37   1998

ULTRAHIGH-PRESSURE MINERALOGY

Physics and Chemistry of the Earth's Deep Interior

Russell J. Hemley, *Editor*

Geophysical Laboratory
Carnegie Institution of Washington
Washington, DC

This volume was supported in part by

The Center for High Pressure Research

A National Science Foundation
Science and Technology Center

Paul H. Ribbe, *Series Editor*

Department of Geological Sciences
Virginia Tech, Blacksburg, Virginia

MINERALOGICAL SOCIETY of AMERICA

Washington, DC

D0721662

REVIEWS IN MINERALOGY

(Formerly: SHORT COURSE NOTES)

ISSN 0275-0279

Volume 37

ULTRAHIGH-PRESSURE MINERALOGY:
Physics and Chemistry of the Earth's Deep Interior

ISBN 0-939950-48-0

ADDITIONAL COPIES of this volume as well as others in this series may be obtained at moderate cost from:

THE MINERALOGICAL SOCIETY OF AMERICA
1015 EIGHTEENTH STREET, NW, SUITE 601
WASHINGTON, DC 20036 U.S.A.

ULTRAHIGH-PRESSURE MINERALOGY:

Physics and Chemistry of the Earth's Deep Interior

FOREWORD

This volume was edited by Russell J. Hemley in preparation for a short course by the same title, organized together with Ho-kwang Mao and sponsored by the Mineralogical Society of America, December 4-6, 1998 on the campus of the University of California at Davis. This is number 37 in the *Reviews in Mineralogy* series, begun in 1974 and continuing with vigor, at least into the near future, with volumes on "Uranium Mineralogy and Geochemistry," on "Sulfate Mineralogy," and on "Natural Zeolites" planned for 1999 and 2000. Volume 36 (1056 pages!), "Planetary Materials," edited by J.J. Papike, appeared earlier this year, and it is certainly not unrelated to the subject of this volume on the physics and chemistry of the deep interior of one of the planets. Rus Hemley has highlighted *Reviews in Mineralogy* Volumes 14, 18, 20 and 29 as particularly relevant to the subject matter of "Ultrahigh-Pressure Mineralogy."

Paul H. Ribbe
Department of Geological Sciences
Virginia Tech
Blacksburg, VA   November 1998

PREFACE

High-pressure mineralogy has historically been a vital part of the geosciences, but it is only in the last few years that the field has emerged as a distinct discipline as a result of extraordinary recent developments in high-pressure techniques. The domain of mineralogy is now no less than the whole Earth, from the deep crust to the inner core—the entire range of pressures and temperatures under which the planet's constituents were formed or now exist. The primary goal of this field is to determine the physical and chemical properties of materials that underlie and control the structural and thermal state, processes, and evolution of the planet. New techniques that have come 'on-line' within the last couple of years make it possible to determine such properties under extreme pressures and temperatures with an accuracy and precision that rival measurements under ambient conditions. These investigations of the behavior of minerals under extreme conditions link the scale of electrons and nuclei with global processes of the Earth and other planets in the solar system. It is in this broad sense that the term 'Ultrahigh-Pressure Mineralogy' is used for the title of this volume of *Reviews in Mineralogy*.

This volume sets out to summarize, in a tutorial fashion, knowledge in this rapidly developing area of physical science, the tools for obtaining that knowledge, and the prospects for future research. The book, divided into three sections, begins with an overview (Chapter 1) of the remarkable advances in the ability to subject minerals—not only as pristine single-crystal samples but also complex, natural mineral assemblages—to extreme pressure-temperature conditions in the laboratory. These advances parallel the development of an arsenal of analytical methods for measuring mineral behavior under those conditions. This sets the stage for section two (Chapters 2-8) which focuses on high-pressure minerals in their geological setting as a function of depth. This top-down approach begins with what we know from direct sampling of high-pressure minerals and rocks brought to the surface to detailed geophysical observations of the vast interior. The third section (Chapters 9-19) presents the material fundamentals, starting from properties of a chemical nature, such as crystal chemistry, thermochemistry, element partitioning, and melting, and moving toward the domain of mineral physics such as melt properties, equations of state, elasticity, rheology, vibrational dynamics, bonding, electronic structure, and magnetism.

The Review thus moves from the complexity of rocks to their mineral components and finally to fundamental properties arising directly from the play of electrons and nuclei. The following themes crosscut its chapters.

Composition of the mantle and core. Our knowledge of the composition of the Earth in part is rooted in information on cosmochemical abundances of the elements and observations from the geological record. But an additional and essential part of this enterprise is the utilization of the growing information supplied by mineral physics and chemistry in detailed comparison with geophysical (e.g. seismological) observations for the bulk of the planet. There is now detailed information from a variety of sources concerning crust-mantle interactions in subduction (Liou et al., Chapter 2; Mysen et al., Chapter 3). Petrological, geochemical, and isotope studies indicate a mantle having significant lateral variability (McDonough and Rudnick, Chapter 4). The extent of chemical homogeneity versus layering with depth in the mantle, a question as old as the recognition of the mantle itself, is a first-order issue that threads its way throughout the book. Agee (Chapter 5) analyzes competing models in terms of mineral physics, focusing on the origin of seismic discontinuities in the upper mantle. Bina (Chapter 6) examines the constraints for the lower mantle, with particular emphasis given to the variation of the density and bulk sound velocity with depth through to the core-mantle boundary region (Jeanloz and Williams, Chapter 7). Stixrude and Brown (Chapter 8) examine bounds on the composition of the core.

Mineral elasticity and the link to seismology. The advent of new techniques is raising questions of the mineralogy and composition of the deep interior to a new level. As a result of recent advances in seismology, the depth-dependence of seismic velocities and acoustic discontinuities have been determined with high precision, lateral heterogeneities in the planet have been resolved, and directional anisotropy has been determined (Chapters 6 and 7). The first-order problem of constraining the composition and temperature as a function of depth alone is being redefined by high-resolution velocity determinations that define lateral chemical or thermal variations. As discussed by Liebermann and Li (Chapter 15), measurements of acoustic velocities can now be carried out simultaneously at pressures that are an order of magnitude higher, and at temperatures that are a factor of two higher, than those possible just a few years ago. The tools are in hand to extend such studies to related properties of silicate melts (Dingwell, Chapter 13). Remarkably, the solid inner core is elastically anisotropic (Chapter 8); with developments in computational methods, condensed-matter theory now provides robust and surprising predictions for this effect (Stixrude et al., Chapter 19), and with very recent experimental advances, elasticity measurements of core material at core pressures can be performed directly (Chapters 1 and 15).

Mantle dynamics. The Earth is a dynamic planet: the rheological properties of minerals define the dynamic flow and texture of material within the Earth. Measurement of rheological properties at mantle pressures is a significant challenge that can now be addressed (Weidner, Chapter 16). Deviatoric stresses down to 0.1 GPa to pressures approaching 300 GPa can be quantified in high-pressure cells using synchrotron radiation (Chapter 1). The stress levels are an appropriate scale for understanding earthquake genesis, including the nature of earthquakes that occur at great depth in subducted slabs (deep-focus earthquakes) as these slabs travel through the Earth's mantle. Newly developed high-pressure, high-precision x-ray tools such as monochromatic radiation with modern detectors with short time resolution and employing long duration times are now possible with third-generation synchrotron sources to study the rheology of deep Earth materials under pressure (Chapter 1).

Fate of subducting slabs. One of the principal interactions between the Earth's interior and surface is subduction of lithosphere into the mantle, resulting in arc volcanoes, chemical heterogeneity in the mantle, as well as deep-focus earthquakes (Chapters 2 and 3). Among the key chemical processes associated with subduction is the role of water in the recycling process (Prewitt and Downs, Chapter 9), which at shallower levels is essential for understanding arc volcanism. Mass and energy transport processes govern global recycling of organic and inorganic materials, integration of these constituents in the Earth's interior, the evolution (chemically and physically) of descending slabs near convergent plate boundaries, and the fate of materials below and above the descending slab. Chapters 5 and 6 discuss the evidence for entrainment and passage of slabs through the 670 km discontinuity, and the possibility of remnant slabs in the anomalous D'' region near the core-mantle boundary (Chapter 7). The ultimate fate of the materials cycled to such depths may affect interactions at the core-mantle boundary and may also hold clues to the initiation of diapiric rise. The evolution and fate of a subducting slab can now be addressed by experimental simulation of slab conditions, including in situ monitoring of a simulated slab in high-pressure apparatus in situ x-ray and spectroscopic techniques. The chemistry of volatiles changes appreciably under deep Earth conditions: they can be structurally bound under pressure (Prewitt and Downs, Chapter 9).

Melting. Understanding pressure-induced changes in viscosity and other physical properties of melts is crucial for chemical differentiation processes ranging from models of the magma ocean in the Earth's early history to the formation of magmatic ore deposits. (Chapter 13). Recent evidence suggests that melting may take place at great depth in the mantle. Seismic observations of a low-velocity zone and seismic anisotropy at the base of the mantle have given rise to debate about the existence of regions of partial melt deep in the mantle (Chapter 7). Deep melting is also important for mantle convection from subduction of the lithosphere to the rising of hot mantle plumes. Very recent advances in determination of melting relations of mantle and core materials with laser-heating techniques are beginning to provide accurate constraints (Shen and Heinz, Chapter 12). Sometimes lost in the debate on melting curves is the fact that a decade ago, there simply were no data for most Earth materials, only guesses and (at best) approximate models. Moreover, it is now possible to carry out in situ melting studies on multi-component systems, including natural assemblages, to deep mantle conditions. These results address whether or not partial melting is responsible for the observed seismic anomalies at the base of the mantle and provide constraints for mantle convection models (Chapter 7).

The enigma of the Earth's core. The composition, structure, formation, evolution, and current dynamic state of the Earth's core is an area of tremendous excitement (Chapter 8). The keys to understanding the available geophysical data are the material properties of liquid and crystalline iron under core conditions. New synchrotron-based methods and new developments in theory are being applied to determine all of the pertinent physical properties, and in conjunction with seismological and geodynamic data, to develop a full understanding of the core and its interactions with the mantle (Chapter 7). There has been considerable progress in determining the melting and phase relations of iron into the megabar range with new techniques (Chapter 12). Constraints are also obtained from theory (Chapter 19). These results feed into geophysical models for the outer and inner core flow, structural state, evolution, and the geodynamo. Moreover, there is remarkable evidence that the Earth's inner core rotates at a different rate than the rest of the Earth. This evidence in turn rests on the observation that the inner core is elastically anisotropic, a subject of current experimental and theoretical study from the standpoint of mineral physics, as described above.

The thermodynamic framework. Whole Earth processes must be grounded in accurate thermodynamic descriptions of phase equilibria in multi-component systems, as discussed by Navrotsky (Chapter 10). New developments in this area include increasingly accurate equations of state (Duffy and Wang, Chapter 14) required for modeling of phase equilibria as well as for direct comparison with seismic density profiles through the planet. Recent developments in in situ vibrational spectroscopy and theoretical models provide a means for independently testing available thermochemical data and a means for extending those data to high pressures and temperatures (Gillet et al., Chapter 17). Accurate determinations of crystal structures provide a basis for understanding thermochemical trends (Chapter 9). Systematics for understanding solid-solution behavior and element partitioning are now available, at least to the uppermost regions of the lower mantle (Fei, Chapter 11). New measurements for dense hydrous phases are beginning to provide answers to fundamental questions regarding their stability of hydrous phases in the mantle (Chapters 3 and 9) and the partitioning of hydrogen and oxygen between the mantle and core (Chapter 8).

Novel physical phenomena at ultrahigh pressures. One of the key recent findings in high-pressure research is the remarkable effect of pressure on the chemistry of the elements, at conditions ranging from deep metamorphism of crustal minerals (Chapter 2) to "contact metamorphism" at the core-mantle boundary (Chapter 7). Pressure-induced changes in Earth materials represent forefront problems in condensed-matter physics. New crystal structures appear and the chemistry of volatiles changes (Chapter 9). Pressure-induced electronic transitions and magnetic collapse in transition metal ions strongly affect mineral properties and partitioning of major, minor, and trace elements (Chapter 11). Evidence for these transitions from experiment (Chapter 18) and theory (Chapter 19) is important for developing models for Earth formation and chemical differentiation. The conventional view of structurally and chemically complex minerals of the crust giving way to simple, close-packed structures of the deep mantle and a simple iron core is being replaced by a new chemical picture wherein dense silicates, oxides, and metals exhibit unusual electronic and magnetic properties and chemistry. In the end, this framework must dovetail with seismological observations indicating an interior of considerable regional variability, both radially and laterally depending on depth (e.g. Chapters 6 and 7).

New classes of global models. Information concerning the chemical and physical properties of Earth materials at high pressures and temperatures is being integrated with geophysical and geochemical data to create a more comprehensive global view of the state, processes, and history of the Earth. In particular, models of the Earth's interior are being developed that reflect the details contained in the seismic record but are bounded by laboratory information on the physics and chemistry of the constituent materials. Such "Reference Earth Models" includes the development of reference data sets and modeling codes. Tools that produce seismological profiles from hypothesized mineralogies (Chapters 4 and 5) are now possible, as are tools for testing these models against 'reference' seismological data sets (Chapter 6). These models incorporate the known properties of the Earth, such as crust and lithosphere structure, and thus have both an Earth-materials and seismological orientation.

Other planets. The Earth cannot be understood without considering the rest of the solar system. The terrestrial planets of our solar system share a common origin, and our understanding of the formation of the Earth is tied to our understanding of the formation of its terrestrial neighbors, particularly with respect to evaluating the roles of homogeneous and heterogeneous processes during accretion. As a result of recent developments in space exploration, as well as in the scope of future planetary missions, we have new geophysical and geochemical data for the other terrestrial planets. Models for the accretion history of the Earth can now be reevaluated in relation to this new data. Experiments on known Earth materials provide the thermodynamic data necessary to calculate the high-pressure mineralogy of model compositions for the interior of Mars and Venus. Notably, the outer planets have the same volatile components as the Earth, just different abundances. Studies of the outer planets provide both an additional perspective on our own planet as well as a vast area of opportunity for application of these newly developed experimental techniques (Chapter 1 and 17).

New techniques in the geosciences. The utility of synchrotron radiation techniques in mineralogy has exceeded the expectations of even the most optimistic. New spectroscopic methods developed for high-pressure mineralogy are now available for characterizing small samples from other types of experiments. For example, the same techniques developed for in situ studies at high pressures and temperatures are being used to investigate microscopic inclusions such as coesite in high-pressure metamorphic rocks (Chapter 2) and deep-mantle samples as inclusions in diamond (Chapter 3). With the availability of a new generation of synchrotron radiation sources (Chapter 1) and spectroscopic techniques (Chapter 17), a systematic application of new methods, including microtomographic x-ray analysis of whole rock samples, is now becoming routinely possible.

Contributions in technology. Finally, there are implications beyond the geosciences. Mineralogy has historically has led many to conceptual and technical developments used in other fields, including metallurgy and materials science, and the new area of ultrahigh pressure mineralogy continues this tradition. As pointed out in Chapter 1, many high-pressure techniques have their origins in geoscience laboratories, and in many respects, geoscience leads development of high-pressure techniques in physics, chemistry, and materials science. New developments include the application of synthetic diamond for new classes of 'large-volume' high-pressure cells. Interestingly, information on diamond stability, including its metastable growth, feeds back directly on efforts to grow large diamonds for the next generation of such high-pressure devices (Chapter 1). Micro-analytical techniques, such as micro-spectroscopy and x-ray diffraction, developed for high-pressure research are now used outside of this field of research as well. The study of minerals and mineral analogs under pressure is leading to new materials. As in the synthesis of diamond itself, these same scientific approaches promise the development of novel, technological materials.

A number of books have reviewed certain aspects of the material covered in this volume. Although many are cited in the chapters, some of the key publications are listed here. Among the most recent are the proceedings of the US-Japan seminar on high-pressure mineral physics, *Properties of Earth and Planetary Materials at High Pressures and Temperatures,* edited by M.H. Manghnani and T. Yagi [American Geophysical Union, Washington, D.C., 1998], which contains numerous recent developments, and earlier volumes in the series contain landmark papers in the field. Very recent developments in high-pressure research are given in *Review of High-Pressure Science and Technology, Vol. 7,* edited by M. Nakahara [Japan Society for High-Pressure Science and Technology, Kyoto, 1998]. T.J. Ahrens' *Mineral Physics and Crystallography* [AGU Reference Shelf, Vol. 2, American Geophysical Union, Washington, D.C., 1995] contains useful tabulations of data as well as introductions to both properties and techniques. *Equations of State of Solids for Geophysics and Ceramic Science,* by O.L. Anderson [Oxford University Press, New York, 1995] is an important recent resource.

Earlier books include the following. Previous volumes of *Reviews of Mineralogy* provide useful pedagogic background, particularly Volume 14, *Microscopic to*

Macroscopic, edited by S.W. Kieffer and A. Navrotsky; Volume 18, *Spectroscopic Methods*, edited by F.C. Hawthorne; Volume 20, *Modern Powder Diffraction*, edited by D.L. Bish and J.E. Post; and Volume 29, *Silica*, edited by P.J. Heaney et al. L.G. Liu and W.A. Bassett [*Elements, Oxides, Silicates,* Oxford University Press, New York, 1986] provided an encyclopedic overview of the high *P-T* properties of mantle and core materials generally in the sub-megabar range (mostly <30 GPa), summarizing data to 1986. D.L. Anderson's *Theory of the Earth* [Blackwell, Boston, 1989] and A.E. Ringwood's earlier *Composition and Petrology of the Earth's Mantle* [McGraw-Hill, New York, 1975] defined integrated views of the deep interior with information available at that time. The classic, early papers in the field have been compiled by T.J. Shankland and J.D. Bass in *Elastic Properties and Equations of State* [American Geophysical Union, Washington, D.C., 1988].

Finally, I thank the following people for their help with this project. As is evident, this is the product of 34 authors, all very busy people who put aside other tasks to complete their chapters, sometimes with ridiculous deadlines. I am indebted to Dave Mao for much help with the organization and planning. Gordon Brown suggested the idea, and was persistent in his encouragement. The effort was supported by the NSF Center for High Pressure Research; its staff and especially the members of its Executive Committee, Don Weidner (Director), Charlie Prewitt, Bob Liebermann, and Alex Navrotsky, contributed to many aspects of the project. Dave Mao, Martin Wilding, and Alex Navrotsky helped with the short course. A number of additional people contributed immeasurably in completing the volume: Steve Gramsch, James Badro, Viktor Struzhkin, Maddury Somayazulu, Merri Wolf, and Amanda Davis. And last, but actually first, I thank Paul Ribbe, who manages to put together these volumes with seeming effortlessness, uniform excellence, and ever good cheer, year after year after year.

Russell J. Hemley

Geophysical Laboratory
Center for High Pressure Research
Carnegie Institution of Washington
Washington, DC   November 1998

Table of Contents

3   THE UPPER MANTLE NEAR CONVERGENT PLATE BOUNDARIES

B. O. Mysen, P. Ulmer, J. Konzett, M. W. Schmidt

8 THE EARTH'S CORE

L. Stixrude, J. M. Brown

9 HIGH-PRESSURE CRYSTAL CHEMISTRY

C. T. Prewitt, R. T. Downs

10 THERMODYNAMICS OF HIGH PRESSURE PHASES

Alexandra Navrotsky

11   SOLID SOLUTIONS AND ELEMENT PARTITIONING AT HIGH PRESSURES AND TEMPERATURES

Yingwei Fei

12   HIGH-PRESSURE MELTING OF DEEP MANTLE AND CORE MATERIALS

Guoyin Shen, D. L. Heinz

18 HIGH-PRESSURE ELECTRONIC AND MAGNETIC PROPERTIES

R. J. Hemley, Ho-kwang Mao, R. E. Cohen

19 THEORY OF MINERALS AT HIGH PRESSURE

L. Stixrude, R. E. Cohen, R. J. Hemley

Chapter 1

NEW WINDOWS ON THE EARTH'S DEEP INTERIOR

Ho-kwang Mao and Russell J. Hemley

Geophysical Laboratory and Center for High Pressure Research
Carnegie Institution of Washington
5251 Broad Branch Road, NW
Washington, DC 20015

INTRODUCTION

Nearly all minerals have high-pressure origins. Crustal minerals may originally come from high-pressure sources such as rising plumes or meteorite impacts. They may lose part or all of their high-pressure signatures through melting, recrystallization, assimilation, and other geological processes, thus becoming "low-pressure minerals." Some may preserve high-pressure forms, thus directly revealing their genealogy. By far the major fraction of minerals in the solid Earth is still hidden at great depth under pressures. They control the formation and evolution, and therefore the present and future state, of the solid Earth. Studies of high-pressure physical and chemical properties of minerals are the key for understanding global geophysics and geochemistry. For example, plate tectonics is a surface manifestation of the deep-rooted dynamics of minerals and rocks at very high pressures. Only very recently have we begun to gain a firm grasp of how to accurately characterize most of the crucial high-pressure properties throughout the *P-T* range of the Earth. High-pressure technology is now ripe for major advances: the next decade will be a harvesting period for high-pressure mineralogists to make major impact on the geosciences. The present chapter presents an overview of these technical developments, together with a brief history of the field and an assessment of prospects for the future. Opportunities for new research directions are emphasized.

In his classic 1952 paper, Francis Birch, the father of modern high-pressure mineralogy, showed how the constitution of the Earth's deep interior could be understood from knowledge available at that time for the high-pressure density and elasticity of minerals (Birch 1952). His important paper also included the following footnote:

Unwary readers should take warning that ordinary language undergoes modification to a high-pressure form when applied to the interior of the Earth; a few examples of equivalents follow:

High-pressure form:	*Ordinary meaning:*
certain	*dubious*
undoubtedly	*perhaps*
positive proof	*vague suggestion*
unanswerable argument	*trivial objection*
pure iron	*uncertain mixture of all the elements*

Though tongue-in-cheek, the footnote carries a sober message that theories and models of the Earth's interior are only as good as the high-pressure data behind them. Acquisition of definitive high-pressure data depends upon three basic prerequisites:

(1) reaching *P-T* conditions to study the stable phases,
(2) measuring material properties *in situ* at high *P-T,*
(3) achieving necessary accuracy.

0275-0279/98/0037-0001$05.00

The progress in high-pressure mineralogy in the past half-century has been the result of following precisely these goals.

The reconnaissance period

In 1953, Lawrence Coes of Norton Company reported (Coes 1953) a new dense crystalline silica (subsequently named coesite) synthesized at 3 GPa (1 gigapascal = 10^9 Newton m^{-2}). This discovery was a rude awakening that our understanding of mineralogy was only skin deep: quartz, a rock-forming mineral, so common and so familiar, was actually unstable under moderate pressures. In 1954, General Electric Company announced the high-pressure synthesis of diamond (at 5 GPa). This success quickly grew into a multi-billion dollar industry, and further demonstrated the enormous potential of high-pressure mineralogy in materials applications. Coesite and diamond are now used extensively as index minerals for "ultrahigh pressure metamorphism" indicating deep upper mantle origin (McCammon et al. 1997, Wang et al. 1989; Liou et al., Chapter 2).

In 1959, Ted Ringwood converted Fe_2SiO_4 fayalite to the spinel structure (subsequently named ringwoodite; Ringwood 1959). This is the precursor of the magnesium-rich olivine-spinel transformation at the top of the transition zone (15 GPa). In 1961, a young Russian student Sergei Stishov and his associate Svetlana Popova synthesized yet another new dense crystalline silica (subsequently named stishovite) above 8 GPa (Stishov and Popova 1961, Stishov 1995). This is not simply a new phase: the silicon tetrahedra (silicon in four-fold coordination by oxygen), the building blocks of all rock-forming silicate minerals, convert to silicon octahedra (six-fold coordination by oxygen). The foundation of the old mineral kingdom began to crumble as it became apparent that common rocks changed to entirely new assemblages at pressures equivalent to the mantle transition zone—that is, less than 1/10 the distance to Earth's center!

The Earth's core is mainly made of iron. Unlike the high-pressure silicates and diamond that are quenchable and can be studied under ambient conditions after quench (releasing P-T), the dense form of iron observed by shock-wave compression is non-quenchable, and must be studied in situ at high pressure. In 1964, Taro Takahashi and Bill Bassett developed an x-ray diffraction technique to probe samples in a newly invented diamond cell and identified the hexagonal-close-packed (ε) phase of iron above 13 GPa (Takahashi and Bassett 1964). The phase is likely to be the main constituent of the Earth's core.

In the late 1960s to 1970s, attention shifted to the quest for "post-spinel" phases in the lower mantle, which required pressures higher than 25 GPa. Mao (1971) and Bassett and Ming (1972) observed that Fe_2SiO_4 spinel indeed breaks down to closed-packed oxides at the lower mantle conditions as predicted by Birch (1952). Simultaneous heating to very high temperatures at high pressure is essential for re-creating the conditions of the geotherm at great depth. Ming and Bassett (1974) used a focused infrared laser to heat samples in a diamond cell and discovered a series of post-spinel transitions in olivines having variable Fe/Mg ratio. X-ray diffraction patterns of post-spinel phase(s) of forsterite above 30 GPa were rather complicated, and was at first thought to indicate mixed oxides (periclase plus stishovite, as presaged by Birch) with additional unidentified lines. Lin-gun (John) Liu, a student of Bassett, showed that the patterns actually belong to periclase plus a new $MgSiO_3$ phase with perovskite structure (Liu 1975, 1976). The $(Mg,Fe)SiO_3$-perovskite is now believed to be the most abundant mineral in the Earth. In the next four years, Liu conducted a combinatorial-like study of mantle minerals with the Ming-Bassett laser-heating technique (Ming and Bassett 1974) and found a plethora of new transitions (Liu 1974, 1977, 1978; Liu and Ringwood 1975). Under the intense P-T conditions that prevail in the lower mantle, nearly all common rock-forming minerals convert, often through several transitions, to new phases or new assemblages.

Period of exploration

The multi-anvil high-pressure apparatus developed in Japan (Kawai and Endo 1970) and later utilized or adapted in high-pressure geoscience centers around the world provided a powerful tool for petrological studies and crystal growth at conditions equivalent to the top of the lower mantle (25-30 GPa). The diamond cell, on the other hand, went beyond 50 GPa (Mao and Bell 1975), broke the megabar (100 GPa) barrier (Mao and Bell 1976), and proceeded to cover the entire pressure range of the Earth (Xu et al. 1986). In the 1980s, high-pressure techniques developed symbiotically with many microanalytical probes, especially those using laser (Hemley et al. 1987) or synchrotron radiation (Bassett 1980)

These developments enabled the study of a great variety of physical and chemical properties of minerals under high *P-T* conditions. Important chemical properties include multi-component phase equilibrium, crystal structure, liquid-state and amorphous structure, site occupancy, melting, solution, phase separation, major and minor element partitioning, thermochemical parameters, diffusivity, 'crystal-field' energy, and reaction kinetics. Important physical properties include *P-T-V* equations of state, single-crystal and aggregate elasticity, compressional and shear acoustic velocity, molecular and lattice vibrational frequencies, anelasticity, viscosity, strength, thermal conductivity, electrical conductivity, dielectric parameter, optical parameters, electronic structure, and magnetism. Chemical and physical transformations include oxidation-reduction, hydration-dehydration, amorphization-crystallization, order-disorder, high-low electron spin, insulator-metal, and magnetic transitions. Each of these may be profoundly different under high *P-T* conditions as a result of fundamental alterations in bonding and interatomic interactions induced by these extreme conditions.

Meanwhile, theoretical studies advanced from early empirical and speculative approaches to recent highly accurate first-principles methods. Throughout, these have provided guidance, inspiration, and understanding for high-pressure experimentation. For example, Birch cited in the acknowledgements to his 1952 paper Jim Thompson's unpublished prediction of rutile-type silica; Stishov's reading of this led him to conduct his high-pressure experiments on quartz. Ringwood's (1959) discovery of olivine-spinel transition was originally predicted by Bernal (1936) on the basis of crystal chemistry systematics. Liu's (1974) identification of perovskite was inspired by Ringwood's crystal chemical prediction of perovskite as a post-spinel phase (Ringwood 1969). Steady advances in condensed-matter theory now provide quantitative predictions of mineral stability and numerous physical properties (Stixrude and Cohen 1995, Cohen et al. 1997, Stixrude et al, Chapter 19). These calculations have led to the discovery of new high-pressure behavior, including the transformation of Stishov's silica to a new, higher pressure phase (Kingma et al. 1995).

New opportunities and beyond

These advances are at best still described as reconnaissance or exploratory. High-pressure research has often been dictated by available technology rather than guided by key Earth problems. High-pressure data were often obtained with a severe trade-off in accuracy, thereby precluding the solution of problems with certainty. This situation, however, has changed drastically in the past several years. With the arrival of "third-generation" synchrotron sources and development of numerous novel in situ micro-analytical techniques, high-pressure mineralogy is finally reaching maturity. The greatly improved measurement capability and accuracy is a silent revolution, one that does not attract as much attention as the breaking of new pressure records or the discovery of new minerals, but is more crucial for providing answers to key questions about the deep Earth. With data obtained as a result of this technical revolution, old high-pressure data have been questioned and modified; Earth models have been revised, sometimes radically. Answers to such basic questions as "Is there a compositional boundary between upper and lower mantles?" or "How hot is the core?" swing back-and-forth with new determinations of elasticity or melting curves (e.g. Anderson 1997). To *some unwary readers*, this may seem to be another validation of Birch's footnote on language and high-pressure science. In fact, with the steady improvement in data quality, definitive answers are within reach, if not in hand.

Turning our attention to more distant realms, high-pressure mineralogy also includes the study of the deep interiors of other planets (Hemley and Mao 1998). The equivalent of upper and lower mantles in giant planets are the dense gaseous (H_2 and inert gases) and icy H_2O, NH_3, CH_4, CO, and CO_2) layers (Zharkov and Gudkova 1992, Duffy et al. 1994). Properties of these planetary high-pressure "minerals"—i.e. solidified gases and ices and their compounds—have been surprisingly counter-intuitive. For example, noble gases form stoichiometric compounds at moderate pressures, and are no longer "inert." The discovery of new high-pressure crystalline compounds in the H_2-H_2O system (Vos et al. 1993) and five new high-pressure compounds in the H_2-CH_4 system (Somayazulu et al. 1996) reveal a rich unexpected high-pressure chemistry in these systems that may also have applications as energetic materials. Even noble gases form compounds, such as $He(N_2)_{11}$, $NeHe_2$, and $Ar(H_2)_2$ (Loubeyre et al. 1993, 1994; Vos et al. 1992); they assume intriguing new bonding character under pressure. Argon and xenon are considered "volatile" at ambient pressure, but their melting points increase steeply with pressure and appear to exceed the melting point of iron at 30 GPa (Jephcoat 1998): i.e. iron will melt while argon and xenon remain as crystalline solids. Shall we consider iron "volatile" and argon and xenon "refractory" in the deep interior? Reactivity and chemical affinity are altered with changes in ionic radii, electronic structure, and bonding character under pressure (Caldwell et al. 1997). Associated elements in crustal minerals (e.g. Fe^{2+} and Mg^{2+}) may dissociate at depth (Bell et al. 1979, Mao et al. 1997); conversely, incompatible elements, e.g. Fe and K, may form alloys (Parker et al. 1996). To paraphrase Birch, the words "siderophile," "lithophile," "volatile," "compatible," and "noble," *undergo modification to a high-pressure form*. In effect we are dealing with a new order of the Periodic Table.

Beyond the applications to the deep interiors of Earth and planets, the study of minerals and mineral-like compounds under pressure is a fascinating field in its own right. Pressure is an ideal variable to tune properties for understanding fundamental chemistry and physics of materials. Historically, progress in most branches of the geosciences has been made by following advances in pure chemistry and physics. High-pressure mineralogy, on the other hand, as a result of its tangible objectives concerning the deep Earth, has in many respects led the field of high-pressure science and technology as a whole, particularly in the extension high *P-T* capabilities and the development of microanalysis, laser spectroscopy, and the application of synchrotron techniques

PRESSURE AND TEMPERATURE

High-pressure techniques

Reaching high P-T conditions with dynamic and static compression. Accurate experimental constraints on Earth and planetary deep interiors require knowledge of a great variety of physical and chemical properties of minerals obtained under pertinent *P-T* conditions (Fig. 1). These include conditions corresponding to *P-T* profiles of planetary bodies for direct comparison with the properties and processes the planet's

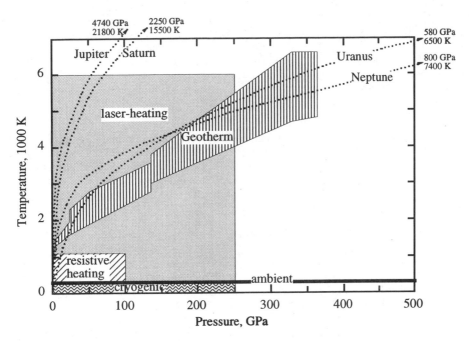

Figure 1. *P-T* range possible for static high-pressure experiments along with estimated temperature profiles for the Earth (geotherm) and other planets (Duffy and Hemley 1995, Zharkov and Gudkova 1992).

interior, as well as a wide range of high-pressure, and high-and-low temperature conditions for fundamental understanding of minerals. The range is covered by complementary dynamic and static compression techniques.

In shock-wave compression, pressure increases as a wave front passes through the medium as a result of high-velocity impact. Extremely high pressures can be generated with these dynamic processes since the peak pressure is only controlled by the intensity of the shock but not limited by the strength of the materials. Pressures as high as 50 TPa (1 terapascal = 10^{12} Newton m^{-2}) have been reported. Temperature rises steeply with pressure in such dynamic compression experiments. The dynamic compression can provide simultaneous high *P-T* conditions, but not high-pressure, low-temperature conditions. In situ observations must be compatible with the microsecond duration of the shock. Among the techniques that satisfy this timing requirement, measurements of high *P-T* acoustic velocity, density, and equation of state are well developed (Ahrens 1987), but other time-resolved techniques including x-ray diffraction and vibrational spectroscopy have come on line (see Nakahara 1998).

For static compression, the sample is confined in a pressure vessel. The high-pressure condition is stable for hours to years. The maximum pressure is limited by the strength and design of the vessel. In situ observations depend upon the ability of probing the sample through the wall of the vessel. Commonly used designs are given as follows.

Piston-cylinder and anvil devices. Pressures of the lower crust to upper mantle conditions (Chapters 2-5) can be generated by pushing a piston against sample cell assembly in a cylindrical chamber (Boyd and England 1960) (Fig. 2). Pressure is measured directly from force divided by the area of the cross section of the piston. The piston is under compression, and the cylinder is under tension. Because the compressional strength

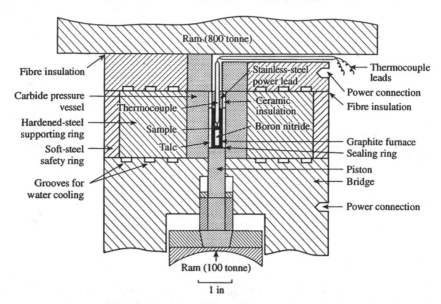

Figure 2. Piston-cylinder apparatus. The device consists of a double ram with an axially supported tungsten carbide vessel (after Boyd and England 1960).

of tungsten carbide, or other hard materials used for the piston and cylinder, is significantly higher than the tensile strength, the attainable pressure is limited by the fracturing point of the cylinder: 2 GPa for unsupported tungsten carbide core, 4-6 GPa for tungsten carbide core supported by a pre-stressed (interference) steel ring. The sample chamber is "very large," ranging in dimension from several centimeters in the Boyd-England piston-cylinder apparatus used in research (Boyd and England 1960) to tens of centimeters in belt apparatus used for industrial diamond synthesis. The very large chamber is ideal for synthesis of large samples or accommodating internal probes for measurements of such properties as acoustic velocity, viscosity, and electrical conductivity, but it is inaccessible for optical or x-ray probes.

Beyond the limit of the piston-cylinder apparatus, anvil devices can be used for reaching pressures of the mantle transition zone (>13 GPa) and higher. The anvil consists of a small pressure-bearing tip on one side and larger areas on opposite sides. Low pressure applied to the large side intensified at the small high-pressure tip by the area ratio of the two sides. At high pressures, anvils deform elastically and remain unyielding while the gasket, another crucial component, deforms plastically to allow the advance of the anvils. The plastic deformation stops after each pressure increment when the gasket reaches frictional equilibrium and builds up a pressure gradient to support the anvil tips and the high-pressure sample chamber. Achievable pressure depends upon the geometrical design of the device, frictional strength of the gasket, and compressional strength of anvils. Improvements in these areas have led to numerous designs of high-pressure apparatus. Most of them are variations of two basic types: multiple symmetrical anvils and Bridgman opposing anvils (see Besson 1997). A schematic of a large volume multi-anvil press is shown in Figure 3. Examples of other multi-anvil devices and their applications are discussed by Leibermann and Li (Chapter 15) and by Weidner (Chapter 16). Varieties of the symmetric opposed anvil design are depicted in Figure 4; these include the basic Bridgman anvil design, the supported anvil Drickamer cell, the Troitsk toroid cell, Paris-Edinburgh cell, and the diamond cell.

Hydraulic Press

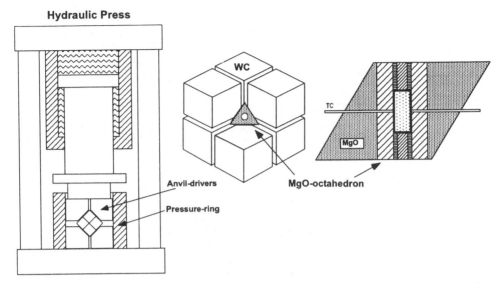

Figure 3. Schematic of a multi-anvil apparatus, which consists of a retaining ring which houses six removable push wedges. The wedges transmit the uniaxial compressive force of the hydraulic ram onto the faces of a cube that is assembled from eight separate tungsten carbide cubes. Each of the eight cubes has a truncated corner that rests against the face of an MgO octahedral pressure medium. The truncated cubes, which converge on the octahedron, are separated from one another by compressible pyrophyllite gaskets. Sample material is placed inside a furnace assembly that fits into a hole in the center of the octahedron (Bertka and Fei 1997).

Strength and transparency of anvils. Various high-strength materials, including steel, tungsten carbide, boron carbide, sapphire, cubic zirconia (Xu and Huang 1994), sintered diamond (Bundy 1975), and single-crystal diamond, are chosen as anvil materials for their strength, available sizes and shapes, optical clarity, x-ray transmission, and other mechanical, thermal, electric, and magnetic properties. For instance, with an anvil base-to-tip area ratio of 100, 30 GPa can be reached with tungsten carbide multi-anvils or Bridgman anvils, 60 GPa can be reached with sintered-diamond Bridgman anvils, and 140 GPa can be reached with single-crystal diamond anvils. Pressures of three hundred GPa are routinely reached with beveled-culet single-crystal diamond anvils with an area ratio of 1000. Single-crystal diamond, the strongest material known, allows the attainment of pressures far greater than any other materials. Even more significantly, single-crystal diamond is essentially transparent to radiation below 5 eV (i.e. UV-visible-IR) and x-radiation above 10 keV, thus providing a wide window for probing sample in situ at high *P-T*. As a non-magnetic insulator, diamond anvils are also suitable for electrical conductivity and magnetic susceptibility studies.

Large-volume and miniature anvil devices. High-pressure anvil devices can be scaled up or down; the sample volume is proportional to the size of anvil and press. To gain pressure intensification in the ultrahigh pressure range, a moderate sample size requires large anvils and even larger presses. Above 15 GPa, a "large sample chamber" refers to millimeter to centimeter size samples, which requires inch to meter size anvils and 200 to 50,000 ton presses. The sample diameter is limited by the size of the anvil tip and the pressure intensification ratio. Bridgman anvils produce a thin disc-shaped sample with thickness to diameter ratio of 1/10. Various types of multi-anvil apparatus have sample chambers of equal dimensions. The Paris-Edinburgh cell (Besson et al. 1992), as well as

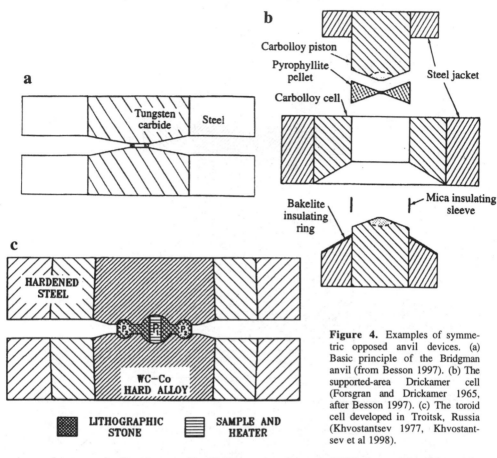

Figure 4. Examples of symmetric opposed anvil devices. (a) Basic principle of the Bridgman anvil (from Besson 1997). (b) The supported-area Drickamer cell (Forsgran and Drickamer 1965, after Besson 1997). (c) The toroid cell developed in Troitsk, Russia (Khvostantsev 1977, Khvostantsev et al 1998).

its predecessor, the Russian toroid cell (Khvostantsev et al. 1977), have spheroid sample chambers. The additional thickness increases the volume for the same anvil size, but it places a geometric limit on the maximum pressures.

The large-volume apparatus is superior for synthesizing larger mineral and rock specimens at high pressure. The large sample chamber also provides ample space for internal heaters, probes, thermocouples, and electrical leads. It provides sufficient sample size for certain in situ measurements that rely on large samples; e.g. ultrasonic interferometry (Chen et al. 1996) and neutron scattering (Klotz et al. 1997).

Miniature cells were originally developed because of the small size of available diamond anvils. With typical diamond anvils of 60 mg (0.3 carat) in weight and 2.5 mm in thickness (Fig. 5), the sample chamber is limited to 300 to 10 μm diameter for the pressure range of 30 to 300 GPa, respectively. It turns out, however, that not only is the miniature cell complementary to large volume apparatus in its pressure range, the small size of miniature cells imparts a number of major advantages even at the same pressures. Transport-dependent chemical and thermal equilibria are reached in small samples at shorter time (because these are proportional to the square of the sample dimension). The miniature high-pressure cell (1 to 5 inch) is highly versatile for combining high pressure with other extreme conditions, for example millikelvin temperatures (Haselwimmer et al. 1998) or very high magnetic fields (Tozer 1993).

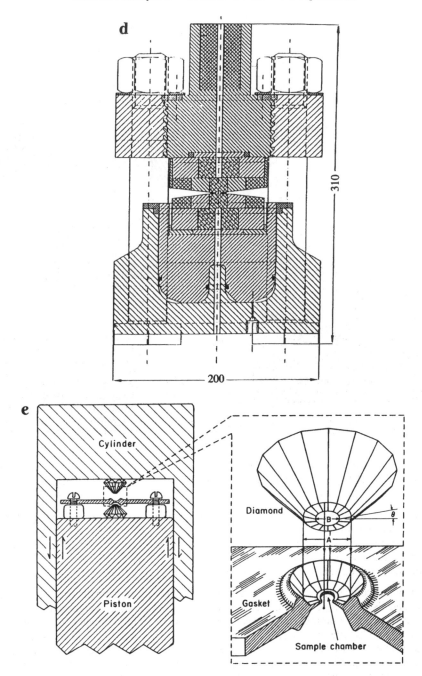

Figure 4 (cont.). Examples of symmetric opposed anvil devices. (d) Paris-Edinburgh cell (Besson et al. 1992), which uses the basic principles of the toroid-cell design. (e) Megabar-type diamond-anvil cell showing the piston-cylinder assembly and beveled anvil design. Tip diameter B and the bevel angle θ are the critical parameters for reaching ultrahigh static pressures; A is culet diameter (Mao and Bell 1976, 1978).

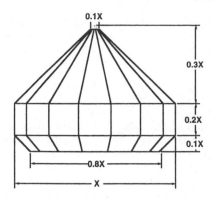

Figure 5. Typical dimensions of a single-crystal diamond anvil.

These devices are also highly maneuverable, allowing rotations along two or more axes for access of all single-crystal orientations for x-ray and optical measurements (Mao and Hemley 1996a,b). With improvements in x-ray, optical, electrical, and magnetic microprobes, size requirements have been greatly reduced to match diamond cell samples. In comparison to large volume devices with thick gaskets, which absorb probe radiation exponentially, miniature cells with thin and transparent anvils and gaskets can access a much wider energy range with minimal loss of signal. For regions of strong diamond absorption, further miniaturizing the anvil may even be desirable. Diamonds of less than 2 mg (1 mm in thickness) have been used successfully for infrared spectroscopy near the diamond second-order absorption (~2000 cm^{-1}; Goncharov et al. 1996) and x-ray spectroscopy below 12 keV.

Another class of high-pressure experiments (e.g. neutron scattering, nuclear magnetic resonance, and ultrasonic measurements at megabar pressures) may still require inch-size diamond anvils. Large natural single-crystal diamonds are rare and prohibitively expensive; the cost increases with the square of the weight. Perfect, inch-size, natural diamonds are simply unavailable at any cost. With the recent development of high-pressure single-crystal growth techniques, large synthetic diamond crystals of gram size are becoming available commercially. The chemically-pure, low-defect, synthetic diamonds have proven superior than natural diamonds as lower background window and higher strength anvil. The cost of synthetic diamonds increases only linearly with size. Further increase in size could be realized by homoepitaxial growth by chemical vapor deposition (CVD) of diamond on top of large synthetic diamonds (McCauley and Vohra 1995). It is conceivable, therefore, that inch-size single-crystal diamonds for Bridgman anvils or multi-anvils may be a reality in the near future.

Diamond cell. The use of single-crystal diamonds as Bridgman anvils started in 1959 and was soon applied to high-pressure mineralogy. With decades of development, the diamond cell has emerged as uniquely providing the capability of a wide range of in situ measurements over the entire *P-T* range of the Earth (Fig. 1). Its earlier limitations, include-ing pressure uncertainty, stress anisotropy, temperature gradients, lack of equilibrium, and small sample size, have been gradually eliminated or turned into advantages.

In 1976, Mao and Bell reported a mechanical design of the cell with which pressures above 100 GPa were reached (Mao and Bell 1976). This design and its variations have been used widely for ultrahigh-pressure research. The small culet face at the tip of the single-crystal, brilliant-cut diamond anvil is polished to be parallel to the large table facet within 0.5 mrad (Fig. 5). The ratio of the table area to the culet area gives the pressure intensification. Two opposing anvils compress a metal gasket which contains the sample chamber at the center (Fig. 4e). Modification of the diamond by adding a bevel to the culet led to another major advance (Mao and Bell 1978). The beveled diamond formed the configuration of the second-stage anvils that allowed ultrahigh pressures beyond 150 GPa to be reached routinely (Vohra et al. 1988, Mao et al. 1990).

The main improvements in mechanical design were features that allowed precise and stable diamond anvil alignment at very high pressure. The alignment was achieved by two half-cylindrical diamond seats (rockers); alignment was maintained at high-loading conditions by the rigidity of rockers (tungsten carbide) and the long, closely-fitted piston and cylinder. In a later improvement, the rockers were replaced by flat tungsten carbide seats with parallelism within 0.5-1 mrad (Mao et al. 1994). The hardened steel piston and cylinder are each made of a single piece to assure rigidity. Pressures are raised by compressing the rigid piston-cylinder assembly with various mechanisms, including lever arms, screws, or hydraulic diaphragms. Many varieties of the diamond cell having different forms, shapes, and sizes optimized for different measurements are derived from the basic piston-cylinder arrangement (Fig. 6).

Gasket. The gasket in anvil devices serves three critical functions: (1) encapsulating the sample, (2) building a gradient from ambient to the peak pressure, and (3) supporting the tip of anvils. In diamond cells, the gasket outside the flat culet area (Fig. 4e) forms a thick ring that supports the anvils like a belt apparatus; without this belt the anvils do not survive above 40 GPa. The effectiveness of the support depends on the tensile strength and initial thickness of the gasket. On the other hand, the thin, flat portion of the gasket between two parallel culets sustains large pressure gradient within the gasket while reducing the pressure gradient in the sample which it confines. Its thickness, t, does not depend upon the initial thickness but is a function of shear strength, σ, and pressure gradient, $\Delta P/\Delta r$ (Sung et al. 1977):

$$t = \sigma/(\Delta P/\Delta r)$$

Hardened steel has been used as an all-purpose gasket material. The high strength rhenium can be used for experiments requiring large thickness or high temperature. Composite gaskets can be constructed to optimize different functions at different parts of the gasket; e.g. insulating inserts (MgO or Al_2O_3) can be inserted in metallic gasket for introduction of electrical leads into the high-pressure region. Recent developments in diamond coating of the central flat region of the gasket (Boehler et al. 1997) greatly increases the shear strength σ. For the same pressure range and anvil size, the diamond coated gasket doubles the thickness of the sample chamber and effectively doubles the sample volume.

Conventional diamond cells with opaque metal gaskets are restricted geometrically by 'tunnel vision' through the diamond anvils (Fig. 7). Furthermore, they are limited by the absorption of the diamond at photon energies above 10 (1% transmission, T) to 12 keV (10% T). High-pressure cells with opposing diamond anvils and beryllium or amorphous boron gaskets have been developed to free these limitations (Mao et al. 1996). Beryllium provides a side window for x-ray studies in the radial direction, which was inaccessible previously (Fig. 8). With the new x-ray diffraction geometry, single-crystal elasticity tensors, shear strength, and preferred orientation of materials have been obtained up to a pressure of 290 GPa (Hemley et al. 1997a) with gaskets made of high-strength beryllium (ultimate tensile strength, 517 MPa; micro-yield strength 68.9 MPa); similar experiments have been carried out to 30 GPa with amorphous boron gaskets (Mao et al. 1996). The low attenuation of beryllium also allows high-pressure x-ray studies to be extended down to 4 (1% T) to 5 (10% T) keV. The new range includes all transition element K edges and rare earth element L edges (Fig. 7), and represents fertile ground for characterizing electronic and magnetic properties of minerals by x-ray absorption and emission spectroscopies.

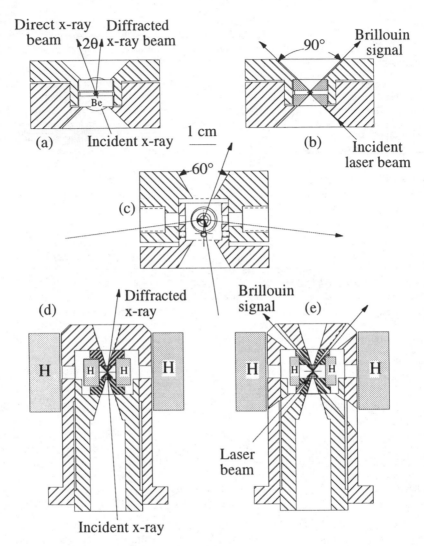

Figure 6. Recent diamond-anvil cell designs. (a) Symmetrical (mini) piston-cylinder cell (1:2 length-to-diameter ratio) for single-crystal x-ray diffraction. (b) Large opening mini piston-cylinder cell (1:2) for Brillouin scattering and other spectroscopic measurements. (c) Symmetrical piston-cylinder cell with longer length-to-diameter ratio of 1:1 for radial x-ray diffraction and spectroscopy at megabar pressures. (d) Double furnace, externally heated long piston-cylinder cell for x-ray diffraction. (e) Double furnace cell modified with additional access ports for Brillouin scattering spectroscopy.

High pressure

Measurement of pressure. Pressure in the piston-cylinder apparatus is determined by dividing the applied force by the cross sectional area of the piston (the definition of pressure). Beyond the limit of the piston cylinder, *P-V* equations of state are used as pressure calibration (Anderson et al. 1989, Heinz and Jeanloz 1994, Ming et al. 1983). Experimentally the *P-V* relations can be determined by a pair of independent variables related to the equation of state. For instance in shock-wave measurements, the paired

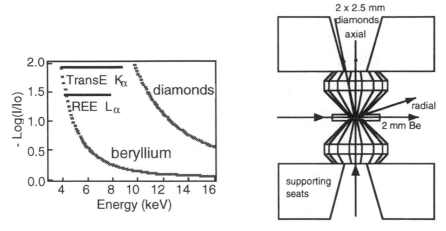

Figure 7. Schematic of the radial diffraction and spectroscopy technique showing x-ray transmission through the gasket. The plot at the top shows the x-ray transmission through 5.0 mm diamond anvils as compared to 4.0 mm of beryllium gasket. The low absorption between 4 and 10 keV allows absorption and resonance studies near the K and L edges of numerous transition and rare earth elements to be measured.

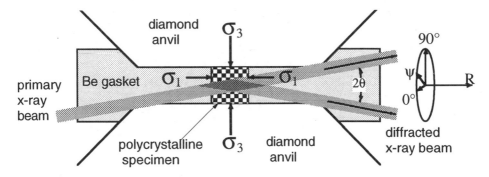

Figure 8. Geometry for radial x-ray diffraction of the uniaxially compressed sample in a diamond cell. (Singh et al. 1998a,b). σ_3 and σ_1 are axial and radial stresses. The diffraction from the sample is measured through the gasket as a function of ψ by rotating the cell. 2θ is the diffraction angle.

variables are particle velocity (U_P) and shock velocity (U_S), from which and with Hugoniot equations of state, the pressure-density relations are calculated. Such direct determinations of pressure are called primary pressure calibrations.

Once calibrated against a primary standard, any other pressure-dependent variable can be used as a secondary standard for pressure determination. Secondary standards are chosen for their accessibility and resolution in a specific high-pressure device. "Fixed point" secondary standards are commonly used for large-volume apparatus in which pressure-induced phase transitions of metals can be readily detected as a discontinuous change in electrical conductivity occurs (Drickamer 1970). The continuous "ruby scale" is commonly used for diamond cells; here tiny ruby grains are added in the sample chamber with minimum disturbance of the sample (Piermarini et al. 1975), and the pressure-shift of ruby fluorescence wavelength can be easily probed with a laser beam through the diamond window (Mao et al. 1978). The ruby scale has been calibrated up to 180 GPa against primary shock-wave standards (Bell et al. 1986).

With improvements in spectroscopic techniques, the precision of ruby measurements and other secondary scale can now resolve 0.2% to 1% in pressure, but the shock-wave primary calibration that they rely on still carry 5% uncertainty pressure and become the limit in accuracy. Recently, a new pair of variables has been chosen to improve the accuracy of primary calibration for anvil devices. They are the density (ρ) measured with x-ray diffraction and the acoustic velocity (V_ϕ) measured with ultrasonic method or Brillouin scattering on the same sample under the same compression (Mao and Hemley 1996b). Pressure is derived directly from

$$P = \int V_\varphi^2 d\rho$$

The resultant P-ρ relation is a primary pressure standard. Such measurements have been carried out to 60 GPa at 300 K (Zha et al. 1997) and to 8 GPa at 1600 K (Chen et al. 1998) for MgO (see Chapters 15 and 17).

Hydrostaticity and homogeneity of pressure. Fluid media transmits hydrostatic pressure to a sample; i.e. the stress is uniform in all directions. However, all fluids solidify above 11 GPa at room temperature, requiring the use of solid media to transmit pressure under these conditions. Solids have finite strength and cause pressure anisotropy or hydrostaticity, which refers to directional deviatoric stress at a point and pressure gradients or homogeneity, which refers to the variation of stress conditions at different points in the sample. They depend upon the mode of compression and the strength of the pressure-transmitting medium.

It is desirable to eliminate or to reduce the pressure anisotropy and inhomogeneity in some high-pressure experiments. This can be achieved by choosing a soft pressure transmitting media. Methanol-ethanol mixtures are commonly used in diamond cells as a fluid medium to 10 GPa (Piermarini et al. 1973). Helium solidifies at 11 GPa but remains as a very weak solid up to 120 GPa (Loubeyre et al. 1996). Microprobing techniques provide the capability of selecting a minute sample volume under homogeneous pressure conditions. Multi-anvil devices apply pressure more uniformly (e.g. from eight symmetrical directions) but are not easily used with gas pressure transmitting media. Solid pressure media such as NaCl soften and reduce pressure anisotropy and inhomogeneity when heated to melting temperatures.

Deviatoric stress. When quantified, experiments under deviatoric stress provide rich additional information about strength, elasticity and rheology that are unavailable with hydrostatic experiments. Deviatoric stress can be measured in multi-anvil apparatus from the broadening of x-ray diffraction lines (Weidner et al. 1994; Weidner, Chapter 16). The uniaxial compression in the diamond cell is ideal for quantitative study of deviatoric stress at ultrahigh pressures (Singh et al. 1998a, b). With radial x-ray diffraction (Fig. 8), the difference in d-spacings obtained from the $\psi = 0°$ and $\psi = 90°$ x-ray patterns give the deviatoric strains, $Q = (d_{0°} - d_{90°})/3d_P$, where $d_P = (d_{0°} + 2d_{90°})/3$ is the d-spacing under hydrostatic pressure $\sigma_P = (\sigma_3 + 2\sigma_1)/3$ where σ_3 and σ_1 are axial and radial stresses, respectively. Together with the hydrostatic equation of state, these data can be used to a measure of the deviatoric stress, $t = \sigma_1 - \sigma_3$, also called the uniaxial component. The maximum t supported by a material is determined by its strength; i.e. $t = \sigma_y = 2\tau$, where σ_y and τ are the yield and shear strengths of the material, respectively. It is often assumed in high-pressure experiments that $t = \sigma_y$; however, in general t varies with sample environment and the equality (von Mises condition) obtains only if the sample is observed to plastically deform under pressure.

Maximum pressure. Future extension of the maximum pressure requires further improvements in the diamond-cell design, including the diamond anvil quality, diamond culet configuration, gasket materials, diamond support, cell alignment, and loading devices. Ultrahigh static pressures are achieved concentrating high stresses over a very small area. In situ microprobing capabilities are essential for characterizing the pressure distribution and improving the pressure vessel itself. In the past, without ultrahigh pressure stress-strain data on the pressure vessel and gasket, the design of diamond cells was mainly empirical, consisting of trying out different designs until breakage occurred. As a result, only minor adjustments have been possible, with the basic design of the diamond cell hardly changing during the past two decades.

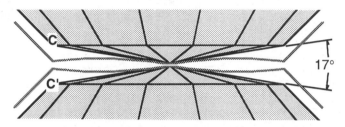

Figure 9. Deformation of diamond observed at a peak pressure of 300 GPa by synchrotron x-ray imaging techniques (Hemley et al. 1997a). *C* and *C'* mark the culet edges of the anvils. The lighter lines trace the edge of the diamonds against the rhenium gasket (not shown) at the peak load, illustrating the elastic collapse of the two bevels. Beveled diamond anvils with small central flats of 10 μm and outer diameter of 300 μm were used. Initially at zero pressure, the combined bevel angle was 17°. The central sample thickness was only 3 μm after gasket indentation, while the edges of the culets of the two diamonds were separated by 45 μm of Re. Because the Re gasket strongly absorbs x-rays, the magnitude of the transmission at a point measured with the 5 x 5 μm x-ray beam gives the thickness of the gasket as determined by plastic flow of the metal and the concomitant elastic deformation of the anvil. With increasing load, the 17° combined angle at the anvil tip begins to decrease; that is, the anvil tip begins to flatten. At the highest loads, however, the originally straight slope of the bevel transforms to a cup, with the bevel angle reversed at the edge.

Recent synchrotron x-ray imaging techniques have been developed to directly measure the macroscopic strain of diamond anvils at ultrahigh pressure (Hemley et al. 1997a). In this method, the transmission intensity of a fine x-ray beam perpendicular to the diamond culet is monitored while the culet is scanned across the beam. Because of the great contrast in absorption coefficients of diamond and gasket materials (e.g. Re or stainless steel), the transmission intensity profile reflects the thickness of the gasket and the shape of the diamond tips. The results show that a huge amount of macroscopic strain can be accommodated by the cupped anvil tip (Fig. 9). With the edges of the two anvils approaching, further loading only increases pressure at the edge and causes catastrophic diamond failure. The cupping in the transmission radiograph can thus be used as a signal to stop further loading and save the diamonds. Using focused synchrotron microbeam to probe the elastic and plastic deformation of anvils and gasket for feedback comparison with analytical calculations (e.g. combined first-principles and finite element), various designs can be tested systematically for extending the maximum pressure to the next level.

Variable temperature

Cryogenic conditions. Understanding the fundamental physical properties of minerals requires low-temperature measurements. In principle, the use of high-pressure instrumentation does not affect the lowest temperatures possible for simultaneous high-pressure, low-temperature studies. A variety of lower pressure devices, including the piston-cylinder apparatus, have been used to liquid helium temperatures (4.2 K). For practical purposes, however, again the small size and therefore low thermal load of miniature anvil devices are ideally suited for low-temperature studies with flow and bath cryostats as well as refrigerators. Diamond-cell experiments into the megabar range (>100 GPa) have been performed with cryostats at liquid helium temperatures (Moshary et al. 1993, Goncharov et al. 1998). There have also been a few studies at lower pressures (~30 GPa) to temperatures of 0.03 K (Haselwimmer et al. 1998).

External resistive heating. With electrical resistance coils external to the diamond anvils (Fig. 6d,e), 1100 K and 100 GPa have been reached for in situ x-ray diffraction measurements (Fei and Mao 1994). The *P-T* conditions with external resistance heating are well defined. The singularly high thermal conductivity of diamond causes the temperature variation within the sample chamber to be typically less than ±1 K. Because stress-bearing components, including the sample gasket, diamonds, diamond seats and the cell body, are also heated, the maximum temperature is limited to 1500 K due to softening of these components at high temperatures. Parameters for *P-V-T* equations of state applicable over a wider range of conditions can be obtained by accurate x-ray diffraction measurements. Systematic studies of *P-V-T* equations of state and phase transitions of lower mantle phases have been carried out with external resistive heating and simultaneous x-ray diffraction.

Internal resistive heating. The large sample volumes in the piston-cylinder apparatus, multi-anvil apparatus, Drickamer cell, and Paris-Edinburgh cell are suitable for internal resistive heating. For example, millimeter-size sample capsules are placed in the middle of a heater in the high-pressure cell assembly. Thermocouples are inserted for temperature measurements. Under pressure, electric current is passing through the heater to provide uniform heating of the sample up to thousands of kelvins while the pressure vessel is cooled by flowing water. The large sample volume is ideal for petrologic studies to 30 GPa. For higher pressures in the diamond cell, the internal heater, which is often the sample itself, must be significantly smaller. Iron wire or foil of 5-20 μm size has been heated by passing electric current to its melting temperatures (Dubrovinsky et al. 1997, Liu and Bassett 1975).

Laser heating. Temperatures in excess of 5000 K can be achieved for samples under pressure in diamond cells by heating with high-power infrared lasers (Ming and Bassett 1974). The temperature is measured from the black-body (or grey-body) radiation, which at high *P-T* conditions is intrinsically superior than the use of a thermocouple, which is a secondary calibration and is affected by the stress effects on EMF as well as by possible sample contamination effects (Mao et al. 1971). In addition, the spatially resolved black-body measurements provide a real-time non-intrusive probe of the temperature profile while thermocouples provide only an intrusive point measurements. Due to the transient nature of laser heating and the steep temperature gradient (the temperature varies from the maximum value to ambient over 20 μm in radial distance and is steeper in the axial direction), contradictory results have been reported by different laboratories in the past, causing confusion about the reliability of the laser-heating technique. Significant progress in shaping and defining the temperature distribution in laser-heated diamond cells has been made recently (Boehler 1996, Jeanloz and Kavner 1996, Jephcoat and Besedin 1996, Sweeney and Heinz 1998).

These advances include the development of the double-sided laser heating technique (**Fig. 10**). Here, the heating laser is split into two beams that pass through the opposing diamond anvils to heat the high-pressure sample simultaneously from both sides. A multimode YAG laser provides a flat-top power distribution at the focal spot (Shen et al. 1996, Mao et al. 1998). The method has been further improved by precise mixing two separate YLF lasers in TEM_{00} and two in TEM_{01} modes to create a flat temperature

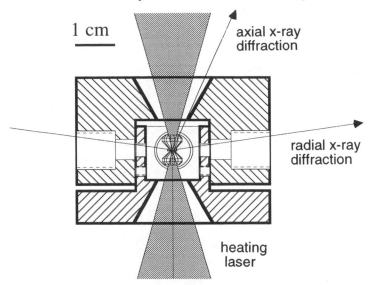

Figure 10. Double-sided laser heating in the megabar symmetrical diamond cell, which allows both radial and axial x-ray measurements at high *P-T* conditions.

distribution with a deviation of less than 1% (see Shen and Heinz, Chapter 12). A double 'hot-plate' configuration is created where the heat generation and temperature measurement are concentrated at the planar interface of an opaque sample layer and transparent medium. The axial temperature gradient in the sample layer in the diamond cell is eliminated within the cavity of the two parallel hot plates. Temperatures on the two sides are measured separately with an imaging spectrograph and CCD and equalized by controlling the beam splitting. Uniform temperatures of 3000 K (±20) K have been achieved in a high-pressure sample of 15-μm diameter x 10-μm thickness. With this technique, the laser-heating method is approaching the accuracy of other high *P-T* techniques using internal heaters.

The technique is readily combined with in situ x-ray diffraction characterization of the sample at these *P-T* conditions. A significantly smaller x-ray microprobe beam is used for in situ. For example, phase relations and melting of a growing number of materials, including Fe and FeO, have been studied at high *P-T* conditions of the Earth's geotherm (see Chapter 12 by Shen and Heinz, and Chapter 18 by Hemley et al.). An extension of the technique involves the coupling of doubled-sided laser-heating with radial diffraction; this is most readily achieved by EDXD with multi-element solid state detectors (Fig. 11). Preliminary results have been obtained on FeO and Au at high *P-T* conditions.

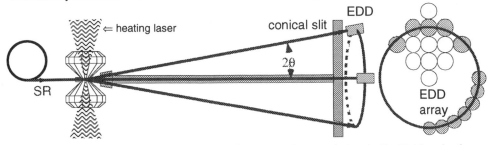

Figure 11. Schematic of high *P-T* radial x-ray diffraction technique employing double-sided laser heating and a multi-element solid state detector (EDD array).

HIGH-PRESSURE MINERAL PHYSICS AND CHEMISTRY

Quenching high P-T samples

Minerals and mineral assemblages synthesized at high P-T conditions are often quenchable; they may be studied by analyzing the run products at ambient temperature while at high pressures in the high-pressure vessel (temperature-quench) (Knittle and Jeanloz 1987) or after releasing pressure (P-T quench) (Campbell et al. 1992, Knittle and Jeanloz 1991). Some minerals may metastably preserve their high pressure form (*e.g.* diamond and $MgSiO_3$-perovskite); others may revert to the ambient pressure phase but preserve the chemical and textural signatures acquired at high P-T conditions. Large-volume devices are ideal for P-T quench studies and synthesis of large specimens. New minerals synthesized at high pressures and temperature can be brought to ambient conditions for detailed characterization (Agee et al. 1995, Katsura and Ito 1996, Li and Agee 1996). The frozen positions of dropping or floating calibration spheres in quenched silicate melt mark the melt viscosity at high P-T (see Dingwell, Chapter 13). Element partitioning, crystal chemistry and electronic structure of quenched specimens removed from the high pressure vessel can be studied with low-pressure surface probes, including SEM, TEM, electron energy loss, SIMS, and photoemission spectroscopy, for which the pressure vessel is impenetrable by electrons, ions, and VUV to soft x-ray radiation. Large metastable crystals can also be used as starting materials for other in situ investigations at high P-T conditions (Kudoh et al. 1987, Yeganeh-Haeri et al. 1989).

The diamond cell is suitable for studying temperature-quenched samples at high pressure and provides a means for P-T quench studies above 30 GPa. With recent improvements in temperature uniformity and stability during laser heating, combined with the spatial resolution (~5 μm) of synchrotron x-ray and IR microprobes, a growing body of information about multicomponent phase relations and major element partitioning in important geochemical systems can be obtained, as demonstrated by studies of Fe/Mg partitioning in lower mantle perovskite and magnesiowüstite phases (Mao et al. 1997; Fei, Chapter 11).

In situ probes at high pressure and variable temperature

The history of high-pressure mineralogy is synonymous with the history of advances in analytical probes. To measure accurately physical and chemical properties in situ at ultrahigh pressures, powerful micro-sampling probes must be developed to reach minute samples through the strong anvils and gasket of the pressure vessel and to separate the weak sample signal from that of massive surrounding vessel materials. Without such probes, all properties, including the pressure itself, cannot be quantified. Indeed, although pioneering multimegabar experiments were reported in early 1970s, they were sometimes referred as "trick ways of producing record high pressure, such as pressing the point of a diamond indenter against the flat face of a large diamond ... they are not of much use to science because specimens of test material cannot be contained in the region of high pressure in a manner allowing reliable measurement" (Bundy 1977). What motivated this sentiment was the lack of accurate microprobe techniques that could be applied. Attainment of megabar pressures was later established by laser microprobe of ruby fluorescence, which in turn was calibrated by x-ray diffraction microprobes. The arrival of a battery of laser, synchrotron, and other microprobing techniques has completely changed the scene.

Electrical and magnetic measurements. With increasing compression, electrons in a mineral will become increasingly unstable localized in bonds or near atomic cores. From the early days of quantum theory it was predicted that above a critical density electrons in any material must delocalize into conduction states thus forming a metal

(Wigner and Huntington 1935). However, before reaching the ultimate metallic state, the conductivity varies, depending in part on factors such as crystal structure changes and charge-transfer processes; in some cases such as graphite, the material may change from metal to insulator (Li and Mao 1994). Conductivity measurement techniques are well developed for large volume presses (Balchan and Drickamer 1961, Kawai and Mochizuki 1971) and diamond cells (Block et al. 1977, Kavner et al. 1995, Mao and Bell 1977). For ultrahigh pressures, a new method that uses a lithographic layout of electrical leads covered by epitaxial CVD diamond overgrowth provide "designer diamonds" for micron-size conductivity measurements.

There are important implications of these high-pressure electrical and magnetic properties for mineralogy (see Hemley et al., Chapter 18). Elemental sulfur, an insulating mineral at ambient conditions, transforms to a body-centered orthorhombic (bco) metallic phase at 90 GPa. Magnetic susceptibility measurements in diamond cells show that the new phase is not only a metal, but is a superconductor with T_c of 10 K at 90 GPa. At 160 GPa, sulfur further transform to the β-Po-type phase (Luo et al. 1993) with T_c of 17 K (Struzhkin et al. 1997), a record high T_c for an elemental superconductor. Modifications of this technique allow its extension to general magnetic susceptibility measurements of minerals at high pressure, as demonstrated by feasibility studies of iron (Timofeev et al., unpublished). Finally, we point out that the magnetic character of minerals and their analogys can be obtained by Mössbauer spectroscopy to pressures in the megabar range, as demonstrated in recent studies using both a conventional radioactive source (Pasternak et al. 1997) and new synchrotron radiation methods (Lübbers et al. 1997, Zhang 1997).

Optical observations. In the ultraviolet to infrared region, "colorful" minerals reveal a wealth of physical and chemical information. Their optical appearance changes as the underlying properties are altered with pressure. The changes can be viewed through the clear, colorless, single-crystal diamond windows with modified transmission and reflectance polarizing microscopes, with spectrometers, or with the naked eye. High-pressure melting, crystallization, and phase transitions are often visually identifiable by movements of grain boundaries and changes in texture, morphology, and refractive indices. High-pressure dielectric properties can be determined from measurements of refractive indices with optical interferometry (Hemley et al. 1991). Pressure-induced changes in crystal symmetry are revealed by distortion of interfacial angles or birefringence under crossed polarization. Pressure effects on optical absorption spectra of transition elements in minerals reveal the change of crystal-field stabilization (or orbital hybridization) energy, oxidation state, site occupancy, *d*- or *f*-electron spin state, or charge-transfer processes; these in turn affect radiative heat transfer, electrical conductivity and oxidation-reduction processes in the Earth's deep interior (Mao and Bell 1972). High-pressure metal-insulator transitions also show conductivity changes in optical range; e.g. at the aforementioned pressure-induced transition of graphite to an insulating state, the graphite becomes transparent (Hanfland and Syassen 1989, Utsumi and Yagi 1991).

Laser-excited spectroscopy. Highly collimated, coherent, monochromatic laser beams can be readily focused through diamond windows of the high-pressure cell to provide a versatile microprobe with μm spatial resolution to multimegabar pressures (Hemley, et al. 1987). The use of these laser microprobes has opened new areas for high-pressure Raman, luminescence, and Brillouin spectroscopies. In optical fluorescence spectroscopy, laser photons are used to excite electrons in the sample to a higher energy level. Energy levels in minerals are revealed as the excited state falls back to the ground state and emitting a fluorescence photon at energy equal to the energy difference of the two states (ΔE). Electronic states are tuned continuously with pressure, or changed abruptly at transitions. By monitoring the frequency and lifetime of the fluorescence as a function of pressure, the changes of states due to continuous compression or abrupt transition can be understood. The pressure tuning of fluorescence spectra also provide a scale for calibrating the pressure itself at ambient and high temperatures (Hess and Schiferl 1992).

Raman spectroscopy is widely used for characterizing the properties of minerals, glasses, and melts at high pressure (McMillan et al. 1996; Gillet et al., Chapter 17). Unlike fluorescence peaks, which appear at the frequencies $v = \Delta E$ independent of the excitation laser frequency, Raman peaks appear at frequencies shifted from the laser frequency, v_L, by subtracting or adding ΔE, i.e. $v = v_L - \Delta E$ (Stokes) or $v = v_L + \Delta E$ (anti-Stokes). Although it is most commonly used for probing vibrational states, it can also be used very successfully to study magnetic and electronic exciations. The appearance of Raman peaks follows selection rules governed by the symmetry of the system. The shift of Raman peaks indicates the change of bonding or electronic energy levels. Developments in Raman technology have taken several significant leaps in the past two decades: from the scanning double-monochromator system in the 1970s to the subtractive-grating filter stage and dispersive spectrograph with a diode array detector in the 1980s, to the single-grating imaging spectrometers with notch filters and CCD detectors in the 1990s (Asher and Stein 1996). New Raman spectrometers are much more sensitive and compact, and thus easily coupled with diamond cells for measurements at ultrahigh pressures. Moreover, recent work has shown that ultra-pure synthetic diamond anvils provide a window of unprecedented clarity for optical studies such as Raman spectroscopy at ultrahigh pressures (Goncharov et al. 1998).

Detailed information on the structure of the deep Earth is mostly derived from seismic observations in the form of wave velocities (e.g. Dziewonski and Anderson 1981; Bina, Chapter 6, and Liebermann and Li, Chapter 15—both in this volume). At the highest pressures, velocities of acoustic waves in a crystal can be obtained at high pressure by Brillouin spectroscopy (Bassett and Brody 1977); in this technique an optical beam is scattered by an acoustic wave and the Doppler shift of the laser frequency reveals the wave velocity (Fig. 6b,e). Unlike Raman shifts, which are usually tens to thousands wave numbers (cm^{-1}) from v_L, Brillouin shifts are usually less than 1 cm^{-1}. A multiple (3- to 6-) pass Fabry-Perot interferometer is typically used for gaining extremely high resolution and rejecting the background caused by Rayleigh scattering of the laser line. By measuring velocities of a primary (longitudinal) wave, V_P, and two secondary (shear) waves, V_S, as functions of orientation in a single crystal, the elasticity tensor (compliance S_{ij} or stiffness C_{ij}) can be determined. Refractive indices and phase transitions at high pressure can also be investigated by Brillouin spectroscopy as a by-product (Brody et al. 1981). This method has been applied to major mantle phases (e.g. Zha et al. 1998) and glasses up to 60 GPa (Zha et al. 1994). The results have been used to constrain composition models and the velocity anisotropy of the deep Earth.

Single-crystal wave velocities have also been obtained up to 20 GPa with a related laser technique, impulsive stimulated scattering (ISS; Chai et al. 1997; see also Gillet et al., Chapter 17). In ISS, a laser beam is split into two beams and recombined to form a standing wave grating in a material; a second laser is scattered from the grating. The data are obtained in the time domain, rather than the frequency domain as in conventional Brillouin spectroscopy. In addition to acoustic velocities, thermal diffusivity at high pressure can be obtained with ISS (Chai et al. 1996). Recent developments have combined the two techniques for simultaneous measurements in both frequency and time domains.

Acoustic interferometry. Acoustic wave velocities and elasticity are central to geophysical research. At pressures up to 10 GPa, these properties can be obtained by direct measurements of wave travel time or echo interference (Jackson et al. 1990). Unlike

Brillouin scattering, which generally requires pristine, colorless, unstrained single crystals, direct acoustic measurements are much more general and can be applied equally well to opaque and/or polycrystalline samples. Recent breakthroughs have extended acoustic interferometry to high *P-T* conditions in multi-anvil devices (Chen et al. 1998, Li et al. 1998, Sinelnikov et al. 1998). Preliminary measurements of this type in diamond cells have been promising. Because of the limitation that the interfering wavelength must be significantly shorter than the small sample thickness dictated by the diamond cell, further development depends on the advances in GHz interferometry (Bassett et al. 1998).

New opportunities with synchrotron radiation

A vast new frontier has opened for applications of synchrotron radiation in high-pressure mineralogy. Concurrent with the development of ultrahigh pressure capabilities, the use of synchrotron radiation as a light source has grown to become the ultimate probe in high-pressure research. Synchrotron radiation started as a nuisance in electron accelerators in which magnets around the ring apply centripetal forces to bend the electron beam to form a circular orbit. The centripetal acceleration of charged particles also causes transverse emission of electromagnetic radiation—the synchrotron radiation—in the tangential direction of the ring. With the progress in higher energy accelerators, synchrotron radiation have become ever more powerful (Fig. 12). Scientists began to build beamlines to tap into the unwanted by-product emissions and used the unprecedented brightness of these light sources for scientific investigations in a parasitic mode to the main research of particle physics: the first generation synchrotron radiation research was started. In the 1980s, dedicated accelerators were build and optimized to be used primarily as light sources (second generation sources). In the 1990s, the third generation sources took the next leap by being optimized for 'insertion devices' (undulators and wigglers) at the straight sections between bending magnets as the main radiation sources. Insertion devices enhance the brilliance and energy of the radiation by orders of magnitude. Third-generation sources also boast larger rings, higher energies, and much more user-friendly environments.

Synchrotron radiation is a perfect match for the study of materials under extreme pressures and temperatures: problems in high-pressure mineralogy that were previously considered completely unapproachable can now being addressed. Synchrotron radiation provides a powerful high-energy beam for penetrating the strong wall of pressure vessels. The high-brilliance, low-emittance beams are ideal for focusing to micron-sized dimensions to probe minute samples at ultrahigh pressures. Early high-pressure x-ray measurements using conventional sources (Bassett and Takahashi 1964) were limited to lower pressures, lower spatial resolution (50 μm), and very long measurement times (days to weeks). With the arrival of first- and second-generation synchrotron sources, samples have been characterized above 300 GPa, which are as small as 3-10 μm (Mao et al. 1989, Liu and Vohra 1996), and new transformations have been discovered routinely by x-ray diffraction. The symbiotic evolution of high-pressure and synchrotron radiation is now at a take-off point for exploiting new developments. Already, two beamlines at the Consortium for Advanced Radiation Sources (GSECARS) have dedicated instrumentation for high-pressure mineralogical studies at the Advanced Photon Source (APS), the third-generation synchrotron facility in the U.S.

Synchrotron radiation has many extraordinary properties. First, its energy range extends continuously from the far infrared (10^{-4} eV) to hard x-ray (10^5 eV), thereby providing probes over an exceptionally wide portion of the electromagnetic spectrum. Second, the highly parallel radiation originates from a very small source (0.01-0.1 mm). It can produce a very small focal spot (submicron in x-ray region) with exceptional brightness or spatial resolution. Third, it has a well-defined time structure of 10 to 100 ps per pulse.

Fourth, it is linearly polarized in the plane of the electron orbit and elliptically polarized above and below the plane. All these properties have implications for high-pressure applications. Previously, some of the most fundamental properties of minerals, including crystal structure, elasticity, electronic density of states, and band structure, simply could not be measured to deep mantle and core pressures. High-energy synchrotron radiation enjoys a unique advantage, as conventional methods are mostly unsuitable; i.e. the pressure vessel is impenetrable by electron or vacuum ultraviolet (VUV) to soft x-ray probes for electron energy loss or photoemission spectroscopy.

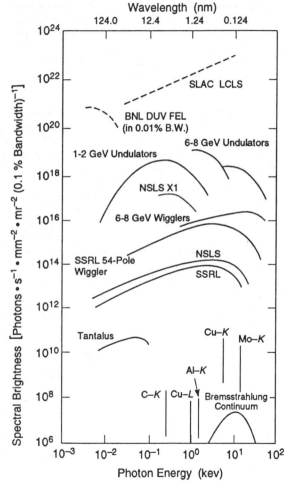

Figure 12. Spectral brightness from various radiation sources. The conventional sealed tube and rotating anode sources are at the bottom. The higher energy spectra correspond to different synchrotron sources having the following abbreviations: National Synchrotron Light Source (NSLS), Brookhaven National Laboratory (BNL), Standord Synchrotron Radiation Laboratory (SSRL). The curves labelled 6-8 GeV undulators and wigglers refer to approximate spectra for insertion devices at the third generation facilities: European Synchrotron Radiation Facility (ESRF, France), Advanced Photon Source (APS, USA), and SPRing-8 (Japan). The dotted lines at the top are free electron laser proposals. From Winick (1994). Dashed lines (BNL DUV FEL and SLAC LCLS) refer to prospective free-electron lasers of the next (fourth) generation.

Infrared spectroscopy. Vibrational infrared spectroscopy provides detailed information on bonding properties of crystals, glasses, and melts, thereby yielding a microscopic description of thermochemical properties. Infrared measurements also provide information on electronic excitations, including 'crystal-field', charge-transfer, and excitonic spectra of insulators and semiconductors, interband and intraband transitions in metals, and novel transitions such as pressure-induced metallization of insulators. The recent advent of high-pressure synchrotron infrared spectroscopy has been a major advance in such investigations. Infrared radiation at the VUV ring of the National Synchrotron Light

Source (NSLS), for example, has up to ~10^4 times the brightness of a conventional thermal (lamp) source. Because of the significant enhancement in the ability to probe microscopic samples provided by this source, it is ideally suited to studies of minerals under extreme pressures. Recent experiments have shown the potential of the technique for study Earth and planetary materials at ultrahigh pressures (Hemley et al. 1998). Studies of dense hydrogen have led to the discovery of a number of unexpected phenomena, including a striking intensity enhancements of vibrational modes (Hanfland et al. 1993, Hemley et al. 1994), an unusually complex phase diagram (Goncharov et al. 1995, Mao and Hemley 1994, Mazin et al. 1997), new classes of excitations in the solid (Hemley et al. 1997b), and accurate bounds on metallization at megabar pressures (Hemley et al. 1996). The system is exceedingly rich in physical phenomena and continues to provide a database that is crucial for planetary modeling (Hemley and Mao 1998, Hemley et al. 1995). Compressed H_2O, another important planetary component, is another excellent example of the remarkable effects of pressure: synchrotron infrared and x-ray measurements show that ice transforms from a molecular system to a dense symmetric non-molecular oxide material (Goncharov et al. 1996).

Angle or energy dispersive x-ray diffraction. X-ray diffraction (XRD) is by far the predominant high-pressure technique involving synchrotron radiation. It provides a means for definitive identification of crystalline (e.g. Funamori and Jeanloz 1997) and amorphous transitions (Hemley et al. 1988), detailed atomic positions in a crystal or radial distribution functions in a glass (e.g. Meade et al. 1992), and yields precise unit cell and equation-of-state data for crystalline solids as a function of *P-T* conditions. High-energy synchrotron sources offer x-radiation capable of penetrating the strong pressure container to probe high-pressure samples. For the multi-anvil (Wang et al. 1998) or Paris-Edinburgh cells (Nelmes et al. 1992), hard x-rays above 20 keV can penetrate through low-absorption gaskets. Diamond anvils are transparent to hard x-rays above 12 keV. The newly developed high-strength Be gasket extends this limit down to 4 keV. Angle dispersive x-ray diffraction (ADXD) with a monochromatic beam, or energy dispersive x-ray diffraction (EDXD) with polychromatic or a "white" beam, can be obtained in situ at high pressure.

In ADXD the incident x-ray beam is monochromatized at a fixed wavelength, λ. In the Bragg Equation,

$$n\lambda = 2d\sin\theta \tag{1}$$

d-spacings are determined from the observed θ angles. The wavelength, λ in Å, is related to the x-ray photon energy, *E* in keV, by

$$\lambda E = 12.4 \text{ Å-keV} \tag{2}$$

A single-crystal monochromator is used to select a small segment of energy, typically several eV, for high-resolution ADXD. The highly collimated monochromatic x-ray beam impinges upon a polycrystalline mineral or mineral assemblage in the high-pressure vessel and produces powder diffraction rings of various 2θ angles which are collected by area detectors such as an image plate or CCD (Nelmes, et al. 1992). Two-dimensional data collection not only increases efficiency, but also provides more reliable peak intensities by integrating the intensity along each diffraction ring, thus reducing the effects of coarse crystallinity and preferred orientation. Intensity information from ADXD has been used extensively for high-pressure crystal structure determination with Rietveld and maximum entropy methods. The high resolution of ADXD (typically $\Delta d/d$ fwhm of 0.001) is crucial for the next generation of high-precision, high-pressure mineralogy: splitting of peaks in low-symmetry crystals can be resolved, and multiple phases can be separated; high-order parameters of P-V-T equations of state, such as the Anderson-Grüneisen parameter δ and dK/dT, can be determined unequivocally, and second-order phase transitions can be identified. Quasi-hydrostatic stress conditions, which can be achieved in a diamond cell with a helium pressure medium or in a large-volume apparatus with heating, is essential for minimizing the peak broadening due to deviatoric stress; in this way, the intrinsic instrumental resolution of ADXD can be realized.

EDXD uses the entire x-ray energy spectrum of synchrotron radiation and does not select a small segment with a monochromator (Mao and Hemley 1996a). The collimated polychromatic (white) x-ray beam passes through the high-pressure vessel and impinges on the mineral specimen. The polychromatic diffracted beam is collimated at a fixed 2θ angle and collected by a solid-state detector (e.g. intrinsic Ge detector) which disperses the diffracted photons in the energy spectrum. At a fixed Bragg angle 2θ, d-spacings are determined from the peak energies or wavelengths. The crucial advantages of EDXD with a point detector over ADXD with area detectors are a minimal requirement of accessible angle as well as high spatial resolution along the primary x-ray beam path. The small opening of the cell for x-ray access leaves maximum support of the anvils for reaching maximum pressures. Because both primary and diffracted beams are finely collimated, the detector only receives signals from the small sample volume where the two beams overlap, but not from the anvils or gasket. Minute samples at ultrahigh pressures and samples of low diffraction intensity (low atomic number) can be studied with minimal interference from scattered x-rays from the anvils; accurate intensities can thus be obtained. This is also crucial for diffraction of amorphous materials at high pressures (Meade et al. 1992), which yield very broad diffraction patterns comparable in some respects to the background signal from non-Bragg scattering of the diamonds or amorphous boron epoxy gasket. Utilizing the entire x-ray spectrum of polychromatic synchrotron radiation, EDXD has high efficiency and high counting statistics. The time resolution provided by the technique is an additional advantage for high-pressure laser-heating experiments and for kinetic studies of transient states.

Single-crystal x-ray diffraction. Single-crystal diffraction is the most definitive technique for high-pressure crystallography (Prewitt and Downs, Chapter 9, this volume). It yields considerably more information and higher resolution than polycrystalline diffraction, and in principle can provide accurate electron density distributions that may reveal the evolution of bonding under pressure (Hemley et al., Chapter 18, this volume). Uncertainties in peak positions and intensities caused by overlapping reflections in a polycrystalline pattern are resolved for single-crystal diffraction, in which reflections are well separated in reciprocal space. For each diffraction condition satisfied for a single crystal, only one peak (*hkl*) is present. When using a He pressure medium, for example, the range for meeting this condition is very small (rocking curves of <0.1°), so reflections can be resolved despite twinning, multiple crystals, mosaic spread, or continuous changes in symmetry caused by displacive phase transitions. Overtone (*nh, nk, nl*) and superlattice (*h/n, k/n, l/n*) diffraction peaks appear in the same EDXD spectrum, which provides a convenient and powerful way to detect and study superstructures. With high energies up to 100 keV, values of the reciprocal lattice Q ($Q = 4\pi\sin\theta/\lambda$) as high as 20 Å^{-1} are accessible. The large Q range can also be used for many novel measurements, including direct electron density determination of single crystals at high pressure.

To meet all diffraction conditions of a single crystal under pressure, the high-pressure vessel must have a large portion of its reciprocal space accessible by x-rays, and one must be able to turn the entire vessel around two axes. Both are difficult conditions to achieve with large-volume presses but easy with small diamond cells (Fig. 6a). Single-crystal x-ray diffraction measurements have been performed to 65 GPa in a study of stishovite at a

NSLS wiggler beamline. The reversible phase transition from the rutile to $CaCl_2$ structure was observed at 56 GPa on increasing pressure and 40 GPa on releasing pressure (Mao and Hemley 1996b). The observations of splitting of the *hkl* reflections and non-splitting of *hhl* reflections confirm that the transition involves a distortion of (tetragonal) stishovite to the (orthorhombic) $CaCl_2$–type structure. For planetary minerals, single-crystal diffraction of hydrogen studied to record pressures of 119 GPa at an ESRF wiggler beamline reveals directly the crystal structure and equation of state over this range (Loubeyre et al. 1996). These measurements provide important constraints for models of the "gas-ice" mineralogy of the planets of the outer solar system.

X-ray spectroscopy techniques. X-ray spectroscopy has been flourishing at synchrotron facilities because of its many applications, but its potential for high research has barely been tapped. The development of high-pressure x-ray spectroscopy was hindered in the past by insufficient intensity and opaqueness of high-pressure vessels below 10 keV. Recently, the third generation synchrotron radiation sources have greatly boosted the intensity, and the high-strength Be gasket has opened the 4-10 keV energy window (Fig. 7). The new x-ray energy range makes available techniques for investigating atomic coordination, structures, and electronic properties of minerals with a wide variety of x-ray spectroscopies. Studies carried out during the past year suggest numerous breakthroughs in this new frontier area.

X-ray absorption at the vicinity of absorption edges (e.g. *K*, *L*, or *M*) of an atom contains a wealth of information regarding the structural, electronic, and magnetic interactions of the atom with its neighboring atoms. X-ray absorption spectroscopy (XAS) is the most widely used spectroscopic technique in high-pressure synchrotron radiation research. Extended x-ray absorption fine structure (EXAFS) provides local structure information (Galoisy et al. 1995, Waychunas et al. 1986), whereas near edge x-ray absorption fine structure (XANES) provides information on electronic properties (Waychunas 1987). For magnetic samples, x-ray magnetic circular dichroism (XMCD) can be observed in both XANES and EXAFS by using circularly polarized x-rays. These measurements can be combined in different ways for new classes of experiments; for example, XMCD in XANES measures spin-resolved electronic densities of states, and XMCD in EXAFS provide local magnetic structural information. There have been major advances that exploit the temporal structure of synchrotron radiation to perform nuclear resonance spectroscopy in the time domain for high-pressure Mössbauer spectroscopy. This has been used successfully in high-pressure studies of Fe and Eu compounds at second-generation synchrotron sources (Takano et al. 1991, Nasu 1996, Zhang 1997). Another new method made possible by recent developments in synchrotron radiation is inelastic x-ray scattering (IXSS). Like inelastic optical scattering and neutron scattering, it probes elementary excitations, such as phonons and plasmons, which are essential for understanding thermal, optical, magnetic, and transport properties (Kao et al. 1996). It has great potential for high-pressure studies, particularly with the larger volume anvil-type cells (e.g. Fig. 4).

CONCLUDING PROLEGOMENON

Recent advances in high-pressure techniques present us with an extraordinary opportunity to solve major problems associated with the enigma of the deep Earth. The key to these problems lies in accurate *in situ* measurements of a wide range of physical and chemical properties at the appropriate *P-T* conditions that prevail within the planet. Many of these methods have just come 'on-line.' In fact, experimental studies of earth materials are experiencing a surge of breakthroughs that were deemed inconceivable only a few years ago. For example, recent progress at the second- and third-generation synchrotron sources

Table 1. Integration of high-pressure analytical techniques and applications.

Analytical techniques	*High-pressure properties*	*Materials*
P-T Quench	crystal structure	hydrogen
Calorimetry	amorphous state	noble gases
Optical absorption	amorphization	noble gas compounds
Optical reflection	metallization	volatiles
Luminescence	superconductivity	planetary gases and ices
Infrared	magnetism	clathrates
Raman scattering	ferroelectricity	glasses
Brillouin scattering	piezoelectricity	melts
Impulsive stimulated scattering	spin state	transition elements
Ultrasonic interferometry	electronic bands	lanthanides
Electrical conductivity	dielectric function	actinides
Magnetic susceptibility	'crystal-field' energy	thermoelectric materials
Neutron scattering	charge transfer	luminescent materials
Nuclear magnetic resonance (NMR)	radiative heat transfer	superhard materials
Electron paramagnetic resonance (EPR)	thermal conductivity	superconductors
	thermodynamic parameters	semiconductors
Mössbauer	phase equilibria	silicates
Nuclear resonance forward scattering (NRFS)	reaction kinetics	oxides
	reactivity	sulfides
Energy dispersive x-ray diffraction (EDXD)	solubility	metals
	diffusivity	metal alloys
Angle dispersive x-ray diffraction (ADXD)	element partitioning	mineral assemblages
	site occupancy	
X-ray absorption near edge structure (XANES)	order-disorder	
	oxidation states	
Extended x-ray absorption fine structure (EXAFS)	hydration	
	melting	
X-ray magnetic circular dichroism (XMCD)	differentiation	
	density	
X-ray emission spectroscopy (XES)	equation of state	
	wave velocity	
Inelastic x-ray scattering spectroscopy (IXSS)	elasticity	
	phonons	
	anisotropy	
	strength	
	rheology	
	viscosity	

has fundamentally altered high-pressure experimentation, from reconnaissance studies with limited capabilities to high-precision measurements with comprehensive material characterization over a wide *P-T* range. Some measurements still require larger volumes at high pressure, but with the development of synthesis methods for large diamond anvils (Israel and Vohra 1998), the potential exists for classes of 'large-volume' cells based on single-crystal diamond.

Moreover, in ultrahigh-pressure mineralogy, the power of an integrated approach is far greater than the sum of individual techniques. Comprehensive understanding of high-pressure phenomena relies on the combination of complementary measurements. Table 1 is a list (which is overlapping and by no means exhaustive) of techniques, properties, and classes of materials to be measured and correlated. As a result of fundamental alterations in bonding and interatomic interactions induced by extreme *P-T* conditions, these phenomena may be mutually related: transformations may involve changes in density, structure, elasticity, electronic properties, and magnetic state. Measurement by only one technique may reveal only a small footnote of a larger epic. Techniques now exist so many of these properties can be measured on the same sample over a *P-T* range, and with accuracy, precision, and sensitivity approaching that measured under ambient conditions. The remaining chapters describe applications of these and other emerging techniques in deep Earth mineralogy.

ACKNOWLEDGMENTS

We are grateful to our colleagues at the Geophysical Laboratory and the Center for High-Pressure Research for contributing to many of the results and ideas presented above. We are indebted to Steve Gramsch for critically reading the manuscript. This work was supported by NSF, NASA, DOE, and W.M. Keck Foundation.

REFERENCES

Agee CB, Li J, Shannon MC, Circone S (1995) Pressure-temperature phase diagram for the Allende meteorite. J Geophys Res 100:17725-17740

Ahrens, TJ (1987) Shock wave techniques for geophysics and planetary physics. *In:* CG Sammis and TL Henyey (eds) Methods of Experimental Physics, Vol 24. p 185-235, Academic, San Diego

Anderson OL (1997) Iron: beta phase frays. Science 278:821-822

Anderson OL, Isaak DG, Yamamoto S (1989) Anharmonicity and the equation of state for gold. J Appl Phys 65:1534-1543

Asher SA, Stein P (eds) (1996) Fifteenth International Conference on Raman Spectroscopy. Wiley, New York

Balchan AS, Drickamer HG (1961) High pressure electrical resistance cell, and calibration points above 100 kilobars. Rev Sci Instrum 32:308-313

Bassett WA (1980) Synchrotron radiation, an intense x-ray source for high pressure diffraction studies. Phys Earth Planet Inter 23:337-340

Bassett WA, Brody EM (1977) Brillouin scattering: a new way to measure elastic moduli at high pressures. *In:* MH Manghnani, S Akimoto (eds) High Pressure Research—Applications in Geophysics. p 519-532 Academic, New York

Bassett WA, Ming LC (1972) Disproportionation of Fe_2SiO_4 to $2FeO + SiO_2$ at pressure up to 250 kilobars and temperatures up to 3000°C. Phys Earth Planet Inter 6:154-160

Bassett WA, Spetzler H, Angel RJ, Chen GR, Shen AH, Reichmann RJ, Yoneda A (1998) Simultaneous gigahertz ultrasonic interferometry and x-ray diffraction in a new diamond anvil cell. *In:* M Nakahara (eds) Review of High Pressure Science and Technology 7:142-144. Japan Society of High Pressure Science and Technology, Kyoto

Bassett WA, Takahashi T (1964) Specific volume measurements of crystalline solids at pressures up to 200 kilobars by x-ray diffraction. *In:* Lloyd (eds) ASME 1964 Symposium on High-Pressure Technology

Bell PM, Xu J, Mao HK (1986) Static compression of gold and copper and calibration of the ruby pressure scale to pressures to 1.8 megabars. *In:* Y Gupta (ed) Shock Waves in Condensed Matter. p 125-130 Plenum, New York

Bell PM, Yagi T, Mao HK (1979) Iron-magnesium distribution coefficients between spinel [$(Mg,Fe)_2SiO_4$, magnesiowüstite [$(Mg,Fe)O$], and perovskite [$(Mg,Fe)SiO_3$]. Carnegie Inst Washington Yearb 78:618-621

Bernal JD (1936) Geophysical discussions. Observatory 59:268

Bertka CM, Fei Y (1997) Mineralogy of the Martian interior up to core-mantle boundary pressures, J Geophys Res 102:5251-5264

Besson JM (1997) Pressure generation. *In:* Holzapfel WB, Isaacs NS (eds) High-Pressure Techniques in Chemistry and Physics: A Practical Approach. p 1-45 Oxford University Press, Oxford

Besson JM, Hamel G, Grima T, Nelmes RJ, Loveday JS, Hull S, Häusermann D (1992) A large volume pressure cell for high temperatures. High Pressure Res 8:625-630

Bina CR, Silver PG (1990) Constraints on lower mantle composition and temperature from density and bulk sound velocity profile. Geophys Res Lett 17:1153-1156

Birch F (1952) Elasticity and constitution of the Earth's interior. J Geophys Res 57:227-286

Block S, Forman RA, Piermarini GJ (1977) Pressure and electrical resistance measurements in the diamond cell. *In:* M Manghnani, S Akimoto (eds) High Pressure Research—Applications in Geophysics. p 503-508 Academic, New York

Boehler R (1996) Melting of mantle and core materials at very high pressures. Phil Trans R Soc Lond A 354:1265-1278

Boehler R, Ross M, Boercker DB (1997) Melting of LiF and NaCl to 1 Mbar: systematics of ionic solids at extreme conditions. Phys Rev Lett 78:4589-4592

Boyd FR, England JL (1960) Apparatus for phase-equilibrium measurements at pressures up to 50 kilobars and temperatures up to 1750°C. J Geophys Res 65:741-748

Brody EM, Shimizu H, Mao HK, Bell PM, Bassett WA (1981) Acoustic velocity and refractive index of fluid hydrogen and deuterium at high pressures. J Appl Phys 52:3583-3585

Bundy FP (1975) Ultrahigh pressure apparatus using cemented tungsten carbide pistons with sintered diamond tips. Rev Sci Instrum 46:1318-1324

Bundy FP (1977) Designing tapered anvil apparatus for achieving higher pressures. Rev Sci Instrum 48:591-596

Caldwell WA, Nguyen JH, Pfrommer BG, Mauri F, Louie SG, Jeanloz R (1997) Structure, bonding and geochemistry of xenon at high pressures. Science 277:930-933

Campbell AJ, Heinz DL, Davis AM (1992) Material transport in laser-heated diamond anvil cell melting experiments. Geophys Res Lett 19:1061-1064

Chai M, Brown JM, Slutsky LJ (1996) Thermal diffusivity of mantle minerals. Phys Chem Minerals 23:470-475

Chai M, Brown M, Slutsky LJ (1997) The elastic constants of a pyrope-grossular-almandine garnet to 20 GPa. Geophys Res Lett 24:523-526

Chen G, Li B, Liebermann RC (1996) Selected elastic moduli of single-crystal olivines from ultrasonic experiments to mantle pressures. Science 272:979-980

Chen G, Liebermann RC, Weidner DJ (1998) Elasticity of single-crystal MgO to 8 gigapascals and 1600 kelvin. Science 280:1913-1916

Coes L (1953) A new dense crystalline silica. Science 118:131-132

Cohen RE, Mazin II, Isaak DE (1997) Magnetic collapse in transition metal oxides at high pressure: implications for the Earth. Science 275:654-657

Drickamer HG (1970) Revised calibration for high pressure electrical resistance cell. 1667-1668

Dubrovinsky LS, Saxena SK, Lazor P (1997) X-ray study of iron with in situ heating at ultrahigh pressure. Geophy Res Lett 24:1835-1838

Duffy TS, Ahrens TJ (1992a) Hugoniot sound velocities in metals with applications to the Earth's inner core. *In:* Y Syono, MH Manghnani (eds) High Pressure Research: Application to Earth and Planetary Sciences. p 353-361 Terra Scientific, Tokyo

Duffy TS, Ahrens TJ (1992b) Sound velocities at high pressure and temperature and their geophysical implications. J Geophys Res 97:4503-4520

Duffy TS, Hemley RJ (1995) Temperature structure of the Earth. Rev Geophys (US Nat Rep to IUGG) 5-9

Duffy TS, Vos WL, Zha CS, Hemley RJ, Mao HK (1994) Sound velocities in dense hydrogen and the interior of Jupiter. Science 263:1590-1593

Dziewonski A, Anderson DL (1981) Preliminary reference earth model. Phys Earth Planet Inter 25:297-356

Fei Y, Mao HK (1994) In situ determination of the NiAs phase of FeO at high pressure and high temperature. Science 266:1678-1680

Forsgran KF, Drickamer HG (1965) Design variables for a high pressure cell with supported taper pistons. Rev Sci Instrum 36:1709-1712

Funamori N, Jeanloz R (1997) High-pressure transformation of Al_2O_3. Science 278:1109-1111

Galoisy L, Calas G, Brown GE (1995) Intracrystalline distribution of Ni in San Carlos olivine: an EXAFS study. Am Mineral 80:1089-1092

Goncharov AF, Hemley RJ, Mao HK, Shu JF (1998) New high-pressure excitations in para-hydrogen. Phys Rev Lett 80:101-114

Goncharov AF, Mazin II, Eggert JH, Hemley RJ, Mao HK (1995) Invariant points and phase transitions in deuterium at megabar pressures. Phys Rev Lett 75:2514-2517

Goncharov AF, Struzhkin VV, Somayazulu M, Hemley RJ, Mao HK (1996) Compression of ice to 210 GPa: evidence for a symmetric hydrogen bonded phase. Science 273:218-220

Hanfland M, Hemley RJ, Mao HK (1993) Novel infrared vibron absorption in solid hydrogen at megabar pressures. Phys Rev Lett 70:3760-3763

Hanfland M, Syassen K (1989) Optical reflectivity of graphite under pressure. Phys Rev B 40:1951-1954

Haselwimmer RKW, Tyer AW, Pugh E (1998) Millikelvin diamond anvil cell for the study of quantum critical phenomena. *In:* M Nakahara (eds) Review of High Pressure Science and Technology 7:481-483. Japan Society of High Pressure Science and Technology, Kyoto

Hastings JB, Siddons DP, vanBürck U, Hollatz R, Bergmann U (1991) Mössbauer spectroscopy using synchrotron radiation. Phys Rev Lett 66:770-773

Heinz DL, Jeanloz R (1994) The equation of state of the gold calibration standard. J Appl Phys 55:885-893

Hemley RJ, Bell PM, Mao HK (1987) Laser techniques in high-pressure geophysics. Science 237:605-612

Hemley RJ, Goncharov AF, Lu R, Struzhkin VV, Li M, Mao HK (1998) High-pressure synchrotron infrared spectroscopy at the National Synchrotron Light Source. Nuovo Cimento 20:1-13 .

Hemley RJ, Hanfland M, Mao HK (1991) High-pressure dielectric measurements of hydrogen to 170 GPa. Nature 350:488-491

Hemley RJ, Jephcoat AP, Mao HK, Ming LC, Manghnani MH (1988) Pressure-induced amorphization of crystalline silica. Nature 334:52-54

Hemley RJ, Mao HK (1998) Static compression experiments on low-Z planetary materials. *In:* MH Manghnani, T Yagi (eds) Properties of Earth and Planetary Materials at High Pressure and Temperature. p 173-183 Am Geophys Union, Washington, DC

Hemley RJ, Mao HK, Duffy TS, Eggert JH, Goncharov AF, Hanfland M, Li M, Somayazulu M, Vos W, Zha CS (1995) Dense hydrogen in the outer solar system: implications from recent high-pressure experiments. *In:* KA Farley (eds) Volatiles in the Earth and Solar System. p 250-260 American Institute of Physics, New York

Hemley RJ, Mao HK, Goncharov AF, Hanfland M, Struzhkin VV (1996) Synchrotron infrared spectroscopy to 0.15 eV of H_2 and D_2 at megabar pressures. Phys Rev Lett 76:1667-1670

Hemley RJ, Mao HK, Shen G, Badro J, Gillet P, Hanfland M, Häusermann D (1997a) X-ray imaging of stress and strain of diamond, iron, and tungsten at megabar pressures. Science 276:1242-1245

Hemley RJ, Mazin II, Goncharov AF, Mao HK (1997b) Vibron effective charge in dense hydrogen. Europhys Lett 37:403-407

Hemley RJ, Soos ZG, Hanfland M, Mao HK (1994) Charge-transfer states in dense hydrogen. Nature 369:384-387

Hess NJ, Schiferl D (1992) Calibration of the pressure-induced frequency shift of Sm:YAG using the ruby and nitrogen vibron pressure scales from 6 to 900 K and 0 to 300 kbar. J Appl Phys 71:2082-2085

Israel A, Vohra YK (1998) Growth of diamond anvils for high-pressure research by chemical vapor deposition, *In:* RM Wentzcovitch, RJ Hemley, WB Nellis, P Yu (eds) High Pressure Materials Research. p 179-184 Materials Research Society, Warrendale, PA

Itie J, Baudelet F, Dartyge E, Fontaine A, Tolentino H, San-Miguel A (1992) X-ray absorption spectroscopy at high pressure. High Pressure Res 8:697-702

Jackson I, Khanna SK, Revcolevschi A, Berthon J (1990) Elasticity, shear-mode softening and high-pressure polymorphism of wüstite ($Fe_{1-x}O$). J Geophys Res 95:21671-21685

Jayaraman A (1983) Diamond-anvil cell and high-pressure physical investigations. Rev Mod Phys 55:65-108

Jeanloz R, Kavner A (1996) Melting criteria and imaging spectroradiometry in laser-heated diamond-cell experiments. Phil Trans R Soc Lond A 354:1279-1305

Jephcoat AP (1998) Rare-gas solids in the Earth's interior. Nature 393:355-358

Jephcoat AP, Besedin SP (1996) Temperature measurement and melting determination in the laser-heated diamond-anvil cell. Phil Trans R Soc Lond A 354:1333-1360

Kao CC, Caliebe WA, Hastings JB, Hämäläinen K, Krisch MH (1996) Inelastic x-ray scattering at the National Synchrotron Light Source. Rev Sci Instrum 67:1-5

Katsura T, Ito E (1996) Determination of Fe-Mg partitioning between perovskite and magnesiowüstite. Geophys Res Lett 23:2005-2008

Kavner A, Li X, Jeanloz R (1995) Electrical conductivity of a natural $(Mg,Fe)SiO_3$ majorite garnet. Geophys Res Lett 22:

Kawai N, Endo S (1970) The generation of ultrahigh hydrostatic pressures by a split sphere apparatus. Rev Sci Instrum 41:1178-1181

Kawai N, Mochizuki S (1971) Metallic states in three 3d transition metal oxides, Fe_2O_3, Cr_2O_3 and TiO_2 under static high pressures. Phys Lett 36A:54-55

Khvostantsev LG, Vereshchagin LF, Novikov AP (1977) Device of toroid type for high pressure generation. High Temp-High Press 9:637-639

Khvostantsev LG, Sidorov VA, Tsiok OB (1998) High pressure toroid cell: applications in planetary and materials sciences. *In:* MH Manghnani, T Yagi (eds) Properties of Earth and Planetary Materials at High Pressure and Temperature. p 89-96 Am Geophys Union, Washington, DC

Kingma KJ, Cohen RE, Hemley RJ, Mao HK (1995) Transformation of stishovite to a denser phase at lower-mantle pressures. Nature 374:243-245

Klotz S, Besson JM, Braden M, Karch K, Pavone P, Strauch D, Marshall WG (1997) Pressure induced frequency shifts of transverse acoustic phonons in germanium to 9.7 GPa. Phys Rev Lett 79:1313-1316

Knittle E, Jeanloz R (1987) Synthesis and equation of state of (Mg,Fe)SiO₃ perovskite to over 100 Gigapascals. Science 235:668-670

Knittle E, Jeanloz R (1991) Earth's core-mantle boundary: results of experiments at high pressures and temperatures. Science 251:1438-1443

Kudoh Y, Ito E, Takeda H (1987) Effect of pressure on the crystal structure of perovskite-type MgSiO₃. Phys Chem Minerals 14:350-354

Li B, Liebermann RC, Weidner DJ (1998) Elastic moduli of wadsleyite (β-Mg₂SiO₄) to 7 gigapascals and 873 kelvin. Science 281:675-677

Li J, Agee CB (1996) Geochemistry of mantle-core differentiation at high pressure. Nature 381:686-689

Li X, Mao HK (1994) Solid carbon at high pressure: electrical resistivity and phase transition. Phys Chem Minerals 21:1-5

Liu J, Vohra YK (1996) Florescence emission from high purity synthetic diamond anvil to 370 GPa. Appl Phys Lett 68:2049-2051

Liu LG (1974) Silicate perovskite from phase transformations of pyrope-garnet at high pressure and temperature. Geophys Res Lett 1:277-280

Liu LG (1975) Post-oxide phases of olivine and pyroxene and mineralogy of the mantle. Nature 258:510-512

Liu LG (1976) The high-pressure phases of MgSiO₃. Earth Planet Sci Lett 31:200-208

Liu LG (1977) The post-spinel phases of twelve silicates and germanates. In: MH Manghnani, S Akimoto (eds) High Pressure Research: Applications in Geophysics. p 245-253 Academic Press, New York

Liu LG (1978) High-pressure phase transformations of albite, jadeite and nepheline. Earth Planet Sci Lett 37:438-444

Liu LG, Bassett WA (1975) The melting of iron up to 200 kilobars. J Geophys Res 80:3777-3782

Liu LG, Ringwood AE (1975) Synthesis of a perovskite-type polymorph of CaSiO₃. Earth Planet Sci Lett 28:209-211

Loubeyre P, Jean-Louis M, Toullec RL, Charon-Gerard L (1993) High pressure measurements of the He-Ne binary phase diagram at 296K: Evidence for the stability of a stoichiometric Ne(He)₂ solid. Phys Rev Lett 70:178-181

Loubeyre P, LeToullec R, Hausermann D, Hanfland M, Hemley RJ, Mao HK, Finger LW (1996) X-ray diffraction and equation of state of hydrogen at megabar pressures. Nature 383:702-704

Loubeyre P, Letoullec R, Pinceaux JP (1994) Compression of Ar(H₂)₂ up to 175 GPa: a new path for the dissociation of molecular hydrogen? Phys Rev Lett 72:1360-1363

Lübbers R, Hesse HJ, Grünsteudel HF, Rüffer R, Zukrowski J, Wortmann G (1997) Probing magnetism in the Mbar range: NFS high-pressure studies of RFe₂ Laves phases (R = Y, Gd, Sc). Highlights in X-ray Synchrotron Radiation Research, p 38 ESRF, Grenoble, France

Luo H, Greene RG, Ruoff AL (1993) B-Po phase of sulfur at 162 GPa: x-ray diffraction study to 212 GPa. Phys Rev Lett 71:2943-2946

Mao HK, Bell PM (1971) High-pressure decomposition of spinel (Fe₂SiO₄). Carnegie Inst Washington Yearb 70:176-178

Mao HK, Bell PM (1972) Electrical conductivity and the red shift of absorption in olivine and spinel at high pressure. Science 176:403-406

Mao HK, Bell PM (1975) Design of a diamond-windowed, high-pressure cell for hydrostatic pressures in the range 1 bar to 0.5 Mbar. Carnegie Inst Washington Yearb 74:402-405

Mao HK, Bell PM (1976) High-pressure physics: the 1-megabar mark on the ruby R₁ static pressure scale. Science 191:851-852

Mao HK, Bell PM (1977) Techniques of electrical conductivity measurement to 300 kbar. In: MH Manghnani, S Akimoto (eds) High Pressure Research—Applications to Geophysics. p 493-502 Academic Press, New York

Mao HK, Bell PM (1978) High-pressure physics: sustained static generation to 1.36 to 1.72 megabars. Science 200:1145-1147

Mao HK, Bell PM, England JL (1971) Tensional errors and drift of thermocouple electromotive force in the single-stage, piston-cylinder apparatus. Carnegie Inst Washington Yearb 70:281-287

Mao HK, Bell PM, Shaner J, Steinberg D (1978) Specific volume measurements of Cu, Mo, Pd, and Ag and calibration of the ruby R₁ fluorescence pressure gauge from 0.06 to 1 Mbar. J Appl Phys 49:3276-3283

Mao HK, Hemley RJ (1994) Ultrahigh-pressure transitions in solid hydrogen. Rev Mod Phys 66:671-692

Mao HK, Hemley RJ (1996a) Energy dispersive x-ray diffraction of micro-crystals at ultrahigh pressures. High Pressure Res 14:257-267

Mao HK, Hemley RJ (1996b) Experimental studies of the Earth's deep interior: accuracy and versatility of diamond cells. Phil Trans R Soc Lond A 354:1315-1333

Mao HK, Hemley RJ, Mao AL (1994) Recent design of ultrahigh-pressure diamond cell. *In:* SC Schmidt, JW Shaner, GA Samara, M Ross (eds) High Pressure Science and Technology—1993. p 1613-1616 American Institute of Physics, New York

Mao HK, Shen G, Hemley RJ (1997) Multivariant dependence of Fe-Mg partitioning in the lower mantle. Science 278:2098-2100

Mao HK, Shen G, Hemley RJ, Duffy TS (1998) X-ray diffraction with a double hot-plate laser-heated diamond cell. *In:* MH Manghnani, T Yagi (eds) Properties of Earth and Planetary Materials at High Pressure and Temperature. p 27-34 Am Geophys Union, Washington, DC

Mao HK, Shu J, Fei Y, Hu J, Hemley RJ (1996) The wüstite enigma. Phys Earth Planet Inter 96:135-145

Mao HK, Wu Y, Chen LC, Shu JF, Jephcoat AP (1990) Static compression of iron to 300 GPa and $Fe_{0.8}Ni_{0.2}$ alloy to 260 GPa: implications for composition of the core. J Geophys Res 95:21,737-21,742

Mao HK, Wu Y, Hemley RJ, Chen LC, Shu JF, Finger LW (1989) X-ray diffraction to 302 gigapascals: high-pressure crystal structure of cesium iodide. Science 246:649-651

Mazin II, Hemley RJ, Goncharov AF, Hanfland M, Mao HK (1997) Quantum and classical orientational ordering in solid hydrogen. Phys Rev Lett 78:1066-1069

McCammon C, Hutchison M, Harris J (1997) Ferric iron content of mineral inclusions in diamonds from Sao Luiz: a view into the lower mantle. Science 278:434-436

McCauley TS, Vohra YK (1995) Homoepitaxial diamond film deposition on a brilliant cut diamond anvil. Appl Phys Lett 66:1486-1488

McMillan PF, Hemley RJ, Gillet P (1996) Vibrational spectroscopy of mantle minerals. *In:* MD Dyar, C McCammon, MW Schaefer (eds) Mineral Spectroscopy: A Tribute to Roger Burns. p 175-213 Geochemical Society, Houston, Texas

Meade C, Hemley RJ, Mao HK (1992) High pressure x-ray diffraction of SiO_2 glass. Phys Rev Lett 69:1387-1390

Ming LC, Bassett WA (1974) Laser heating in the diamond anvil press up to 2000°C sustained and 3000°C pulsed at pressures up to 260 kilobars. Rev Sci Instrum 45:1115-1118

Ming LC, Manghnani MH, Balogh J, Qadri SB, Skelton EF, Jamieson JC (1983) Gold as a reliable internal pressure calibrant at high temperatures. J Appl Phys 54:4390-4397

Moshary F, Chen NH, Silvera IF (1993) Remarkable high pressure phase line of orientational order in solid hydrogen deuteride. Phys Rev Lett 71:3814-3817

Nakahara M (ed) (1998) Review of High-Pressure Science and Technology, Vol 7. Japan Society High-Pressure Science and Technology, Kyoto

Nasu S (1996) High-pressure Mössbauer spectroscopy with nuclear forward scattering of synchrotron radiation. High Pressure Res 14:405-412

Nelmes RJ, McMahon MI, Hatton PD, Piltz RO, Crain J, Cernik RJ, Bushnell-Wye G (1992) Angle-dispersive powder diffraction at high pressure using an image-plate area detector. High Pressure Res 8:677-684

Parker LJ, Atou T, Badding JV (1996) Transition element-like chemistry for potassium under pressure. Science 273:95-97

Pasternak MP, Taylor RD, Jeanloz R, Li X, Nguyen JH, McCammon CA (1997) High pressure collapse of magnetism in $Fe_{0.94}O$: Mössbauer spectroscopy beyond 100 GPa. Phys Rev Lett 79:5046-5049

Piermarini GJ, Block S, Barnett JD (1973) Hydrostatic limets in liquids and solids to 100 kbar. J Appl Phys 44:5377-5382

Piermarini GJ, Block S, Barnett JD, Forman RA (1975) Calibration of pressure dependence of the R_1 ruby fluorescence line to 195 kbar. J Appl Phys 46:2774-2780

Ringwood AE (1959) The olivine-spinel inversion in fayalite. Am Mineral 44:659-661

Ringwood AE (1969) Phase transformations in the mantle. Earth Planet Sci Lett 5:401-412

Shen G, Mao HK, Hemley RJ (1996) Laser-heating diamond-anvil cell technique: double-sided heating with multimode Nd:YAG laser. *In:* M Akaishi et al.(eds) Advanced Materials '96—New Trends in High Pressure Research. p 149-152 Nat Inst Res Inorganic Mat, Tsukuba, Japan

Sinelnikov YD, Chen G, Neuville DR, Vaughan MT, Liebermann RC (1998) Ultrasonic shear wave velocities of $MgSiO_3$ perovskite at 8 GPa and 800 K and lower mantle composition. Science 281:677-679

Singh AK, Balasingh C, Mao HK, Hemley RJ, Shu J (1998a) Analysis of lattice strains measured under non-hydrostatic pressure. J Appl Phys 83:7567-7575

Singh AK, Mao HK, Shu J, Hemley RJ (1998b) Estimation of single-crystal elastic moduli from polycrystalline x-ray diffraction at high pressure: applications to FeO and iron. Phys Rev Lett 80:2157-2160

Somayazulu MS, Finger LW, Hemley RJ, Mao HK (1996) New high-pressure compounds in methane-hydrogen mixtures. Science 271:1400-1402

Stishov SM (1995) Memoir on the discovery of high-density silica. High Pressure Res 13:245-280

Stishov SM, Popova SV (1961) A new dense modification of silica. Geochemistry 10:923-926

Stixrude L, Cohen RE (1995) High-pressure elasticity of iron and anisotropy of Earth's inner core. Science 267:1972-1975

Struzhkin VV, Hemley RJ, Mao HK, Timofeev YA (1997) Superconductivity at 10 to 17 K in compressed sulfur. Nature 390:382-384

Sung C-M, Goetze C, Mao HK (1977) Pressure distribution in the diamond anvil press and the shear strength of fayalite. Rev Sci Instrum 48:1386-1391

Sweeney JS, Heinz DL (1998) Laser heating through a diamond anvil cell: melting at high pressure. *In:* MH Manghnani, T Yagi (eds) Properties of Earth and Planetary Materials at High Pressure and Temperature. p 197-213 Am Geophys Union, Washington, DC

Takahashi T, Bassett WA (1964) A high pressure polymorph of iron. Science 145:483-486

Takano M, Nasu S, Abe T, Yamamoto K, Endo S, Takeda Y, Goodenough JB (1991) Pressure-induced high-spin to low spin transition in $CaFeO_3$. Phys Rev Lett 67:3267-3270

Tozer SW (1993) Miniature diamond-anvil cell for electrical transport measurements in high magnetic fields. Rev Sci Instrum 64:2607-2611

Utsumi W, Yagi T (1991) Light-transparent phase from room temperature compression of graphite. Science 252:1542-1544

Vohra YK, Duclos SJ, Brister KE, Ruoff AL (1988) Static pressure of 255 GPa (2.55 Mbar) by x-ray diffraction: comparison with extrapolation of the ruby pressure scale. Phys Rev Lett 61:574-577

Vos WL, Finger LW, Hemley RJ, Mao HK (1993) Novel H_2-H_2O clathrates at high pressures. Phys Rev Lett 71:3150-3153

Vos WL, Finger LW, Hemley RJ, Hu J, Mao HK, Schouten JA (1992) A high-pressure van der Waals compound in solid nitrogen-helium mixtures. Nature 358:46-48

Wang FM, Ingalls R (1997) XAFS study of the iron bcc-hcp transition. Rev High Pressure Sci Tech 6:91

Wang X, Liou JG, Mao HK (1989) Coesite-bearing eclogite from the Dabie Mountains in central China. Geol 17:1085-1088

Wang Y, Weidner DJ, Meng Y (1998) Advances in equation-of-state measurements in SAM-85. *In:* MH Manghnani, T Yagi (eds) Properties of Earth and Planetary Materials at High Pressure and Temperature. p 365-372 AGU, Washington, D.C.

Waychunas GA (1987) Synchrotron radiation XANES sectroscopy of Ti in minerals: effects of Ti bonding distances Ti valence, and sitegeometry on absorption edge structure. Am Mineral 72:89-101

Waychunas GA, Gordon GE, Apted MJ (1986) X-ray K-edge absorption spectra of Fe minerals and model compounds. II. EXAFS. Phys Chem Minerals 13:31-47

Weidner DJ, Wang Y, Vaughan MT (1994) Strength of diamond. Science 266:419-422

Wigner E, Huntington HB (1935) On the possibility of a metallic modification of hydrogen. J Chem Phys 3:764-770

Winick H (1994) Introduction and overview. *In:* H. Winick (ed) Synchrotron Radiation Sources – A Primer. p 1-29 World Scientific, New Jersey.

Xu J, Mao HK, Bell PM (1986) High pressure ruby and diamond fluorescence: observations at 0.21 to 0.55 terapascal. Science 232:1404-1406

Xu JA, Huang E (1994) Graphite-diamond transition in gem anvil cells. Rev Sci Instrum 65:204-207

Yeganeh-Haeri A, Weidner DJ, Ito E (1989) Elasticity of $MgSiO_3$ in the perovskite structure. Science 243:787-789

Zha CS, Duffy TS, Downs RT, Mao HK, Hemley RJ, Weidner DJ (1998) Single-crystal elasticity of $\alpha-$ and $\beta-Mg_2SiO_4$ polymorphs at high pressure. *In:* MH Manghnani, T Yagi (eds) Properties of Earth and Planetary Materials at High Pressure and Temperature. p 9-16 Am Geophys Union, Washington, DC

Zha CS, Hemley RJ, Mao HK, Duffy TS, Meade C (1994) Acoustic velocities and refractive index of SiO_2 glass to 57.5 GPa by Brillouin scattering. Phys Rev B 50:13105-13112

Zha CS, Hemley RJ, Mao HK, Duffy TS (1997) Elasticity measurement and equation of state from MgO to 60 GPa. Eos Trans Am Geophys Union 78:F752

Zhang L (1997) High pressure Mössbauer spectroscopy using synchrotron radiation on earth materials. Crystallography at High Pressure Using Synchrotron Radiation: The Next Steps. ESRF, Grenoble, France

Zharkov VN, Gudkova TV (1992) Modern models of giant planets. *In:* Y Syono, MH Manghnani (eds) High Pressure Research in Mineral Physics: Application to Earth and Planetary Sciences. p 393-401 Terra Scientific, Tokyo

Chapter 2

HIGH-PRESSURE MINERALS
FROM DEEPLY SUBDUCTED METAMORPHIC ROCKS

J. G. Liou, R. Y. Zhang, W. G. Ernst

Department of Geological and Environmental Sciences
Stanford University, Stanford, CA 94305

Douglas Rumble, III

Geophysical Laboratory
Carnegie Institution of Washington
5251 Broad Branch Road, NW
Washington, DC 20015

Shigenori Maruyama

Earth and Planetary Sciences
Tokyo Institute of Technology
Tokyo 152, Japan

INTRODUCTION

The discovery of coesite (Chopin 1984, Smith 1984) and microdiamonds (Sobolev and Shatsky 1990) in very high-pressure (VHP)[1] crustal rocks is revolutionizing our understanding of continental collision zones and mantle dynamics attending subduction. The realization that segments of continental and oceanic crust ± mantle underpinnings have returned to the surface from depths of 100 km or more is, in itself, remarkable. But VHP rocks offer more than astonishment: they record a complete geodynamic pathway and represent constraints for testing hypotheses of crust/mantle interactions, and mechanisms combining subduction and tectonic exhumation. Problems relating to the descent of low-density continental crust to great depths, the genesis of magmatic arcs, the cause of intermediate-focus earthquakes, the scale of geochemical recycling of elements from the top of the crust to deep within the upper mantle, and the tectonic processes responsible for the formation + growth versus rifting + destruction of the continents must be revisited.

Garnet-bearing ultramafic rocks are volumetrically minor but widespread as significant components in VHP terranes such as the Western Gneiss Region, Norway (Medaris and Carswell 1990), the Bohemian Massif (Medaris et al. 1995), the Kokchetav Massif (Zhang et al. 1997), and the Dabie-Sulu terrane, east-central China (Zhang et al. 1994, 1995a; Liou and Zhang 1998, Zhang and Liou 1998). Some garnet peridotites represent fragments of the mantle wedge overlying the subduction zone, and provide new constraints into mantle processes including slab/mantle interactions, metasomatism, and mantle dynamics associated with convergent plate boundaries. Olivine, garnet, and pyroxene of these garnet peridotites contain unidentified micro-inclusions that may be VHP phases or dense hydrous magnesian silicates previously synthesized only in diamond-cell or multi-anvil experiments. Discovery of natural representatives of these synthetic phases

---

[1] For most petrologic studies, lithostatic pressures required for the production of coesite (i.e. 2.5-2.7 GPa) are denoted as ultrahigh-pressure (UHP) and the term UHP metamorphism is well established in the metamorphic literature (e.g. Liou et al. 1994, Coleman and Wang 1995). In the following, however, we use the nomenclature very high pressure (VHP) to allow distinction of these parageneses and conditions from those associated with deeper mantle and core materials.

0275-0279/98/0037-0002$05.00

would constitute a major advance in understanding the evolution of the mantle.

VHP regimes bridge the gap between upper mantle and crustal processes, and provide vital clues to understanding subduction and continental collision. Study of VHP rocks requires knowledge from a wide range of Earth science disciplines, including mineral physics. In this chapter, we present a petrochemical summary of recent studies of several Eurasian VHP terranes with a specific focus on the constituent mineralogy.

DEFINITION AND P-T REGIMES FOR VHP METAMORPHISM

In general, VHP and HP conditions can be conveniently separated by the P-T boundary for the quartz-coesite equilibrium. Maximum temperatures for VHP and HP metamorphism of most crustal rocks are about 750-800°C; above these temperatures in the presence of H_2O, granitic melt and/or migmatite is generated. At lower P and progressively lower temperatures, assemblages of granulite, amphibolite, epidote amphibolite, and greenschist successively appear. Below temperatures of about 400-450°C, blueschist-facies metamorphism occurs at pressures ranging from 0.5 to 1.6 GPa, whereas greenschist assemblages are stable at lower P (Fig. 1).

Liou et al. (1994) defined VHP metamorphism as mineralogical and structural readjustment of supracrustal rocks and associated ultramafic slices at pressures greater than ~ 2.5 GPa (~80-90 km). In cases of appropriate bulk chemistry, metamorphism at great depths produces coesite, microdiamond, and/or other characteristic VHP minerals, including talc in eclogite and magnesite + diopside ± Ti-clinohumite in garnet peridotite. Many of these assemblages have been described by Smith (1988), Liou et al. (1994), and by various authors in the volume edited by Coleman and Wang (1995). The relevant P-T conditions we used to define VHP, high-P (HP), and low-P (LP) metamorphic regimes are shown in Figure 1; in addition, geotherms of about 5°C/km (extreme high P/T) and 20°C/km (ancient cratons) are illustrated. Index minerals for identification of VHP metamorphism are described in numerous publications, including Smith (1988), Liou et al. (1994), Massonne (1995), and Chopin and Sobolev (1995).

Experimental results by Schreyer (1988a) in the model pelitic system K_2O-MgO-Al_2O_3-TiO_2-SiO_2-P_2O_5-H_2O indicate that the stabilities of VHP phases require abnormally low geothermal gradients—approximately 7°C/km—which can be attained only by the subduction of old, cold, oceanic crust-capped lithosphere ± pelagic sediments or an ancient continent. A petrogenetic grid for the MORB + H_2O system has been experimentally established using a multi-anvil apparatus by Okamoto and Maruyama (in press), combined with other experiments by Schmidt and Poli (1994), Pawley (1994), Poli and Schmidt (1995), and Liu et al. (1996). Except for the long-duration experiments by Liu et al. (1996), all other data represent synthesis runs only, and phase equilibrium has not been demonstrated. The results shown in Figure 1 suggest that lawsonite may persist up to at least 6 GPa if temperature is lower than 650°C; P-T regimes for the eclogite facies can be subdivided into those for amphibole eclogite, epidote eclogite, lawsonite eclogite, and dry eclogite. Depending on the geotherms of subducting oceanic slabs, these hydrous phases—although not experimentally proven—may be stable at various depths in subduction zones and could be important H_2O carriers to mantle depths. The significance of hydrous phases in VHP regimes and the role of H_2O for mantle/slab interactions will be discussed in a later section.

Depending on the P-T conditions, rocks of VHP and HP terranes are classified as intermediate- and low-T eclogites (Carswell 1990), or as "hot eclogite" formed at P greater than the coesite = quartz transition and "cold eclogite" at P less than the transition (e.g. Chopin et al. 1991; Avigad 1992, Okay 1993). In practice, differentiation between HP

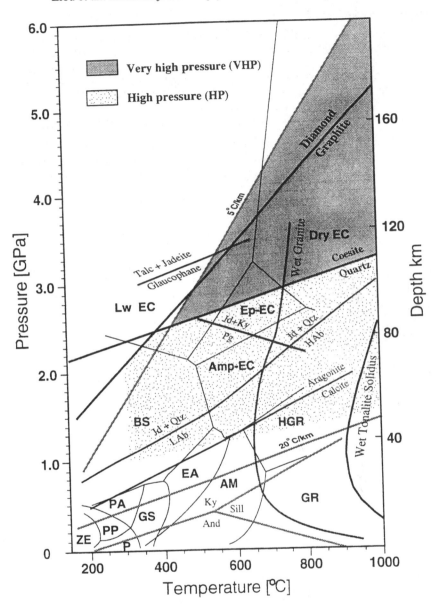

Figure 1. P-T regimes assigned to various metamorphic types: (1) Very high-P (VHP), (2) High-P, and (3) Low-P. Geotherms of 5°C /km and 20°C /km are indicated. Stabilities of diamond (Bundy 1980), coesite (Hemingway et al. 1998), glaucophane (Holland 1988), jadeite + quartz (Holland 1980), Al₂SiO₅ (Bohlen et al. 1991), paragonite (Holland 1979), aragonite (Hacker et al. 1992) and the minimum melting of granitic and tonalite solidus (Huang and Wyllie 1975) are shown. P-T boundaries of various metamorphic facies are from Spear (1993) and subdivision of the eclogite field into amphibole eclogite, epidote eclogite, lawsonite eclogite and dry eclogite are from Okamoto and Maruyama (in press).

"cold eclogite" and VHP "hot eclogite" is difficult, particularly for rocks in which VHP index minerals and their pseudomorphs are not preserved. In fact, in several VHP terranes described below, HP blueschist and "cold" eclogite belts occur in tectonic contact with the coesite-bearing VHP belt; these HP and VHP segments of a single terrane evidently were formed from a subduction-zone event prior to or during continent-continent collision.

Inclusions of quartz aggregates were reported more than 30 years ago as coesite pseudomorphs in eclogitic garnets with radial fractures in the HP Maksyutov Complex of the South Ural Mountains (Chesnokov and Popov 1965). However, this study was not widely publicized; except for the incompletely documented recognition of relict coesite in jadeite by Dobretsov and Dobretsova (1988), later studies of the Maksyutov metamorphics have not confirmed the occurrence of VHP phases (e.g. Beane et al. 1994; but see Lennykh et al. 1995, Leech and Ernst, 1998). However, the presence of relict coesite as micro-inclusions in garnet (± omphacite) was independently described from the Dora Maira Massif by Chopin (1984) and from the Western Gneiss Region of Norway by Smith (1984). This discovery caused a revision in our understanding of the metamorphism of crustal rocks. Subsequent findings of microdiamond inclusions in zircon and garnet in the Kokchetav Massif by Sobolev and Shatsky (1990), in garnet of the Dabie Mountains by Xu et al. (1992), and in two gneiss samples from Norway (Dobrzhinetskaya et al. 1993) further supported the concept of VHP metamorphism of supracrustal rocks.

Of course, diamond and coesite as VHP minerals of mantle origin have long been recognized in kimberlite pipes (e.g. Smyth and Hatton 1977, Ponomarenko and Spetsius 1978) and of shock phases in meteorite craters (e.g. Chao et al. 1960). Graphitized diamonds have been described in garnet clinopyroxenite from the Beni Bousera and Ronda peridotite massifs (Pearson et al. 1989, 1993, 1996). Some of these peridotites contain up to 15% carbon; several of the graphite octahedra contain faceted inclusions of garnet and clinopyroxene. Their occurrence constrains the depth of origin for pyroxenites to be in the diamond P-T stability field (>4.5 GPa and ~1100°C). Anomalously light carbon isotope values ($\delta^{13}C$ = -16 to -27.6 per mil) for the graphite suggest crystallization of diamond from subducted kerogenous carbon.

More than 100 diamond crystals ranging from 0.1 to 0.5 mm have been recovered from heavy mineral separates from two ophiolitic peridotite bodies in Tibet (Bai et al. 1993). Associated mineral separates include SiC (moissanite), octahedral olivine, and serpentine crystals believed to be pseudomorphs after β- or γ-olivine, native chromium, native nickel, and Cr^{+2}-bearing chromite. Although provocative, such unusual occurrences of VHP minerals require further study.

Abbreviations of minerals now to be discussed follow the format of Kretz (1983); they are listed in Table 1.

GLOBAL DISTRIBUTION OF VHP METAMORPHIC TERRANES

Occurrences of VHP rocks have been increasingly recognized and extensively reviewed (e.g. Liou et al. 1994, Coleman and Wang 1995, Ernst et al. 1995). Thus far, more than a dozen Eurasian VHP terranes have been documented; their locations and metamorphic ages are shown in Figure 2. These VHP terranes lie within major continental collision belts in Eurasia (and Africa), extending several hundred kms or more, and are confined to Alpine-type orogens. They share common structural and lithological characteristics: (1) VHP records occur mainly in eclogites and garnet peridotites included as pods and slabs within quartzofeldspathic gneissic units, a few of which contain coesite in garnet and omphacite, and microdiamonds in garnet and zircon as minute inclusions; (2) lithologies have continental and subcontinental geochemical and petrological characteristics; (3) exhumed VHP units are now present in the upper crust as thin subhorizontal slabs, bounded by normal faults on the top, and reverse faults on the bottom, and sandwiched in

Table 1. Abbreviations of minerals and other terms used in this paper.*

VHP - Very high-pressure	HP - high-pressure
AM - amphibolite facies	GR - granulite facies
BS - blueschist facies	GS - greenschist facies
EC - eclogite facies	PP - prehnite-pumpellyite facies
EA - epidote amphibolite facies	ZE - zeolite facies
Ab - albite	Kfs - K-feldspar
Act - actinolite	Ky - kyanite
Alm - almandine	Lw - lawsonite
Amp - amphibole	Mag - magnetite
And - andalusite	Mgs - magnesite
Ap - apatite	Mz - monazite
Arg - aragonite	Mrg - Margarite
Ap - apatite	Ms - muscovite
Cal - calcite	Ol - olivine
Chl - chlorite	Omp - omphacite
Chu - clinohumite	Opx - orthopyroxene
Coe - coesite	Pg - paragonite
Cpx - clinopyroxene	Phn - phengite
Cz - Clinozoisite	Pl - plagioclase
Di - diopside	Pm - pumpellyite
Dm - Diamond	Prp - pyrope
Do - dolomite	Qtz - quartz
Edn - Edenite	Rt - rutile
Elb - ellenbergerite	Sill - sillimanite
Ep - Epidote	Sps - spessartine
Fo - fosterite	Ti-chu - Ti-clinohumite
Gln - Glaucophane	Tlc - talc
Gr - graphite	Top - topaz-OH
Grs - grossular	Tr - tremolite
Grt - garnet	Ttn - titanite
Ilm - ilmenite	Zo - zoisite
Jd - jadeite	F - Fluid

* Mineral abbrevations are mainly after Kretz (1983).

amongst HP or lower grade units; and (4) coeval calc-alkali volcanic and plutonic rocks do not occur, whereas post-collisional or late-stage granitic plutons are common in some occurrences. In the Dabie-Sulu terrane, Proterozoic protoliths of the supracrustal rocks experienced Late Triassic subduction-zone metamorphism at mantle depths, followed by a retrograde amphibolite-granulite facies overprint during exhumation, and finally Cretaceous deformation and thermal recrystallization accompanying granitic intrusion. Trace VHP minerals are best preserved in strong containers such as zircon or garnet in eclogite and garnet peridotite. These occurrences are being increasingly recognized; for example, the latest one to be reported is in the eclogitic rocks from Sulawesi, Indonesia where minute coesite inclusions have been recently confirmed in zircons (Parkinson et al. 1998).

Among global VHP terranes, the Kokchetav Massif may be the oldest at 540-530 Ma. In order of descending geologic age, the VHP rocks are 480 Ma at Makbal, 425-408 Ma in the Western Gneiss Region, 375-380 Ma in the Maksyutov Complex, 374-324 Ma in the Bohemian Massif, 270 Ma in the Atbashy Complex, 240-210 Ma in the Dabie-Sulu belt, 48-41 Ma at Sifnos, and 35 Ma in the Dora Maira and Zermatt-Saas areas. VHP rocks from Mali have been dated as 1045-820 Ma. However, this date is suspect inasmuch as the VHP

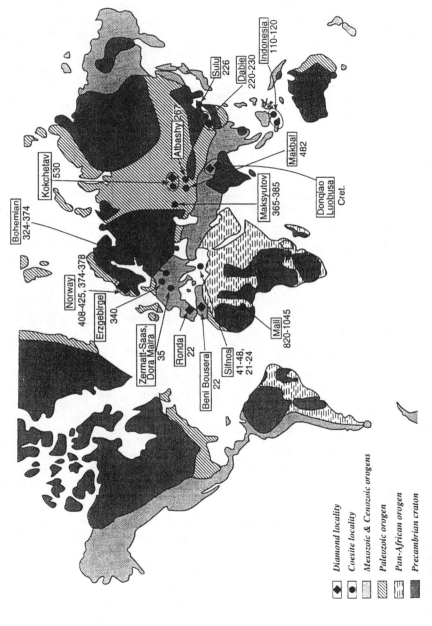

Figure 2. Distribution and age (in Ma) of recognized VHP terranes in the world (modified after Liou et al. 1994, Maruyama and Liou 1998).

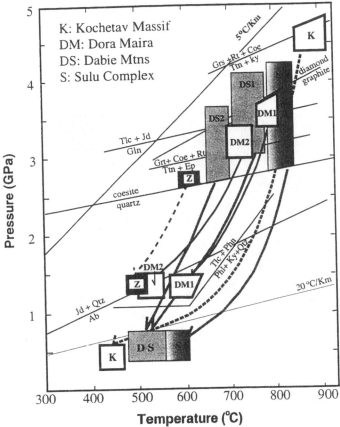

Figure 3. P-T estimates for both peak and retrograde metamorphism for major VHP terranes shown in Figure 2 (modified after Harvey and Carswell 1995).

terrane lies within the 750-550 Ma Pan-African collision zone between the West African and East Saharan cratons (Caby 1994). Apparently, VHP metamorphism occurred throughout the Phanerozoic; no clear indications of VHP metamorphism have survived from Archean or Proterozoic time (Maruyama and Liou 1998). Similarly, inclusions of microdiamond have been reported in garnet, kyanite and zircon from gneisses of the Saxonian Erzgeberge, Germany (Massonne 1998).

Well-investigated VHP terranes include the Dabie-Sulu belt of east-central China, the Kokchetav Massif of northern Kazakhstan, the Dora Maira Massif of the Western Alps, and the Western Gneiss Region of Norway. These four terranes contain trace of phases indicating metamorphism at mantle depths approaching or exceeding 100 km (Schreyer 1988b, Coleman and Wang 1995). Rocks are mainly quartzofeldspathic and pelitic gneisses, marbles, and quartzites, together with minor mafic eclogites and garnet peridotites. Thermobarometric measurements and phase-equilibrium constraints demonstrate that attending peak temperatures were about 700-900°C at confining pressures greater than 2.8-4.0 GPa as shown in Figure 3. These VHP conditions evidently reflect abnormally low geotherms of less than 7-8°C/km in subduction-zone environments. No other tectonic setting provides the appropriately high P and moderate T attending such solid-state recrystallization.

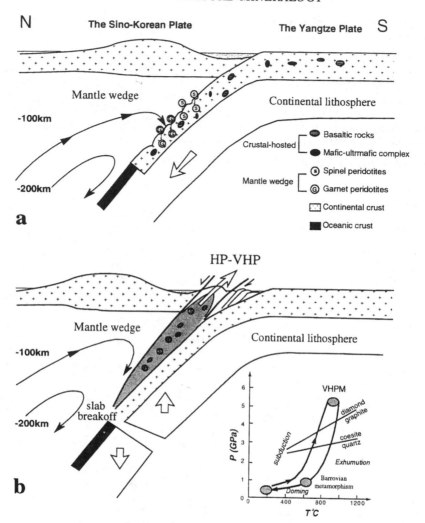

Figure 4. A tectonic model and P-T-time path for subduction and exhumation of crustal materials from continental lithosphere using the Dabie-Sulu collision zone as an example. The subducting slab includes crustal-hosted fragments and mantle wedge blocks. The VHP and HP slabs returned to shallow depths after recrystallization within coesite or diamond stability field at depths of over 100 km due to slab breakoff and buoyancy of a low-density continental sheet. Required time for subduction and exhumation is respectively at about 10 million years (modified after Zhang and Liou 1998).

Proposed tectonic settings for the formation of these VHP terranes are schematically shown in Figure 4. For VHP complexes, the profound descent of coherent tracts of ancient, cold sialic-crust-capped lithosphere is indicated (Ernst and Peacock 1996). Return to the surface is also qualitatively understandable, based on the buoyancy of VHP continental crust after it has decoupled from the downgoing slab (e.g. Ernst and Liou 1995). Because adiabatic decompression would result in the passage of decoupled subduction complexes through P-T regimes appropriate to the granulite and amphibolite facies (600-800°C, 0.3-1.0 GPa), significant overprinting of the earlier VHP assemblages should be the rule. The very rare occurrence of VHP relics suggests that these phases back reacted almost completely on return toward the Earth's surface due to later thermal overprinting.

Petrochemical characteristics of VHP minerals described below are mainly from the Dabei-Sulu terrane of east-central China, the largest recognized VHP belt in the world. This terrane is most familiar to us and is the target of the Chinese Deep Continental Drilling program beginning in 1998. Rare VHP minerals from the Dora Maira Massif, the Western Alps, the West Gneiss Region of Norway, and the Kokchetav Massif of northern Kazakhstan are also described. The geotectonic settings and general petrologic features of these four classic VHP terranes are summarized below.

The Dora Maira Massif, Western Alps, 35 Ma

The Dora Maira Massif of the Western Alps consists of Hercynian and older continental basement rocks metamorphosed under VHP and HP conditions as a result of the Mesozoic-Cenozoic convergence of European and African plates. The HP + VHP belt is a stack of four major units separated by low-angle faults; all have been affected by Alpine HP metamorphism and subsequent re-equilibration at lower P (Chopin et al. 1991, Kienast et al. 1991, Gebauer et al. 1997). The VHP unit consists of orthogneiss and intercalated metasediments with boudins of pyrope-quartzite containing VHP relics such as coesite, ellenbergerite, Mg-staurolite, and other VHP phases (Chopin 1984, 1987). This entity was subdivided by Chopin et al. (1991) into two metamorphic subunits: (a) a coesite-bearing, very-high-P subunit containing kyanite-eclogites with relict coesite and quartz pseudomorphs after coesite, with estimated peak P-T conditions of 3.7 GPa and 800°C, very close to the appearance of diamond (Schertl et al. 1991); and (b) a cold eclogite subunit metamorphosed at P lower than the coesite-quartz transition. In subunit (b), metapelites contain chloritoid, and eclogites carry paragonite, glaucophane, and epidote; P-T estimates for this subunit are 1.5 GPa and 500°C.

The time of VHP-metamorphism has long been considered to have culminated between 90 to 125 Ma, and retrograded between 35 to 41 Ma, judging from $^{39}Ar/^{40}Ar$ analyses of phengitic mica (e.g. Monié and Chopin 1991). However, Tilton et al. (1991) reported metamorphic ages of 30 to 55 Ma using U-Th-Pb methods on ellenbergerite and zircon, and Sm-Nd systematics on pyrope from pyrope-coesite rocks. These young age data are supported by recent SHRIMP analyses of zircons extracted from gneisses, schists, pyrope inclusions, and pyrope quartzites; results indicate the occurrence of 240-275 Ma-old zircon cores and newly formed 35 Ma rims for some zircons, with no data points matching the purported Eoalpine event at 100 Ma (e.g. Gebauer et al. 1997). Thus, VHP metamorphism must have taken place during Oligocene time.

The Kokchetav Massif, Northern Kazakhstan, 540-530 Ma

The Kokchetav Massif is a large, fault-bounded Proterozoic metamorphic complex surrounded by Caledonian rocks of the Ural-Mongolian foldbelt. The Early Cambrian metamorphism represent relict stages in what initially may have been a continuous P-T temporal series, but intense post-metamorphic deformation has resulted in a chaotically mixed sequence; several mapped units represent portions of a dismembered subduction-zone HP/VHP paragenesis (Dobretsov et al. 1995a,b; Ernst et al. 1995, Maruyama 1997). The VHP/HP unit extends NW-SE for at least 80 km; it is about 17 km wide, and is structurally overlain by a weak- to low-grade metamorphic unit on top and is underlain by a low-P facies unit. The HP/VHP unit is composed mainly of pelitic-psammitic gneiss with locally abundant eclogite lenses and minor garnet peridotite, orthogneiss, and metacarbonate. Diamonds occur as minute inclusions in garnet and zircon in pyroxene-carbonate-garnet rock, garnet-biotite gneiss, and schist. Inclusions of coesite pseudomorphs in garnet were recently discovered not only in diamond-bearing gneiss but also in diamond-free eclogite (Zhang et al. 1997). Coesite thus far is restricted to inclusions in zircon. The diamond-bearing metasediments and the associated lenses of eclogite, whiteschist, and garnet peridotite recrystallized under conditions at P > 4.0 GPa and T > 900°C. SHRIMP analysis of metamorphic zircons yields 530 ± 7 Ma (Claoué-Long et al. 1991) for VHP metamorphism, consistent with the Sm-Nd isochron age of 533 ± 20 Ma (Jagoutz et al. 1989) and 528 ± 7 to 505 ± 43 Ma for eclogite, biotite gneiss, and diamondiferous metasedimentary rocks (Shatsky and Jagoutz 1993).

The Sulu-Dabie Terrane, East-Central China, 210-240 Ma

The Qinling-Dabie-Sulu metamorphic complex, occupying the Triassic suture between the Sino-Korean and Yangtze cratons (Fig. 5), consists of three principal units: a northern migmatite high-T terrane (P< 2 GPa), a central VHP coesite- and diamond-bearing eclogite belt (2.6-4 GPa), and a southern HP blueschist-epidote amphibolite-eclogite belt (0.5-1.2 GPa). Inclusions of coesite and quartz pseudomorphs after coesite in garnet, omphacite, kyanite, epidote, zoisite, and dolomite are rare but widespread in both Sulu and Dabie blocks (Schertl and Okay 1994, Zhang et al. 1995a,b; Zhang and Liou 1996). Micro-diamond inclusions in garnet have been reported by Xu et al. (1992) in the Dabie, and by Xu (1997) in the Sulu terrane. Some garnet peridotites record much higher P (Yang et al. 1993, Zhang et al. 1994, 1995a; Hiramatsu et al. 1995) than the surrounding coesite-bearing eclogites, reaching 5-6 GPa. Zircons from VHP gneisses yielded U/Pb ages of ~209 to 227 Ma (Ames et al. 1993, 1996; Rowley et al. 1997). Nd/Sm isochron ages of garnet, omphacite and whole-rock ages of some eclogites give 210-240 Ma (Li et al. 1993). Employing $^{39}Ar/^{40}Ar$ techniques, phengitic mica from HP blueschist and eclogite record cooling ages through about 300°C of 190-230 Ma (Eide et al. 1992, Hacker and Wang 1995). A recent SHRIMP age for zircons gives 220-230 Ma (Hacker et al. 1998). Figure 5 is a simplified tectonic map of the Sulu-Dabie terrane and shows most localities described in this review.

The Dabie-Sulu VHP rocks are unique in the occurrence of: (1) abundant coesite and hydrous phases (such as talc, zoisite/epidote, nyböite, and phengite) in eclogites (Zhang et al. 1995a,c); (2) the world-record lowest $\delta^{18}O$ values (rutile is -15 per mil) for mineral separates from eclogites and metasediments (Yui et al. 1995, Zheng et al. 1996, 1998; Rumble and Yui 1998); (3) abundant garnet peridotites of mantle origin (Zhang et al. 1994, Zhang and Liou 1998); and (4) abundant exsolution textures in VHP minerals from garnet peridotite and eclogite (Hacker et al. 1997, Zhang and Liou 1998, Zhang et al. in press).

The Western Gneiss Region, Norway, 408-425 Ma

The Western Gneiss Region, Norway lies within the Caledonian collision zone extending from Trondjheimfjord in the north, to Songnefjord in the south. The gneissic unit, about 300 km long and 150 km wide, consists of interlayered pelite and migmatite, marble, quartzite, and amphibolite, with tectonic inclusions of gabbro, anorthosite, and peridotite. The gneissic unit exhibits mainly amphibolite-facies assemblages, but relics of high-P assemblages such as Grt + Cpx + Ky + Kfs + Qtz also occur. Eclogite boudins or lenses are widespread and consist mainly of garnet and omphacite with minor rutile, kyanite, phengite, and quartz. Coesite has been reported from Grytting in the Jlje district (Smith 1984, 1988); several new localities of coesite and quartz pseudomorphs after coesite have been recently recognized in both eclogites and the adjacent gneissic rocks (Wain 1997). Microdiamond grains 20-50 microns across have been described from residues separated from two gneisses (Dobrzhinetskaya et al. 1993); the associated kyanite-eclogites contain inclusions of coesite pseudomorphs in garnet. Garnet peridotite and associated Opx-bearing eclogites were metamorphosed at pressures less than 2.5-3.0 GPa at

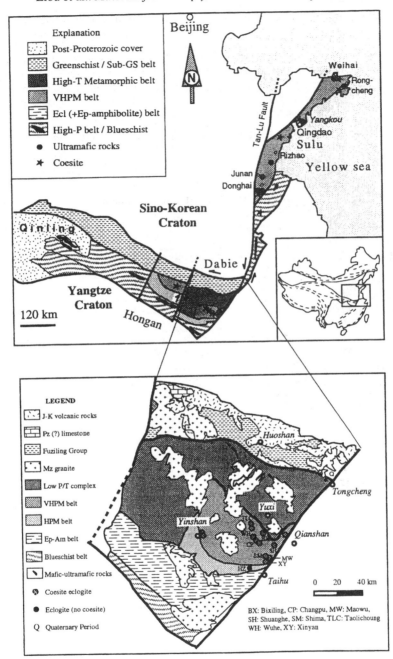

Figure 5. Regional tectonic map of the Qinling-Dabie collision zone between the Sino-Korean (SKC) and Yangtze cratons (YC) in east central China. Note the segmentation and displacement of the units by the Tan-Lu and other faults into several blocks including the Sulu, Dabie, Hongan, and Qinling blocks. The VHP belt is north of the high-P belt. Blueschist is discontinuously exposed in the HP belt for more than 2000 km. Subdivision of the Dabie block into various metamorphic belts is shown. Several reported localities of coesite and microdiamond are indicated. (modified after Zhang and Liou 1994).

T = 700-850°C. Sm-Nd isochron ages for the eclogites are 408-425 Ma for the peak metamorphism, and 374-378 Ma as the retrogressive mid-crustal Barrovian overprinting (Cuthbert et al. 1983, Griffin and Brueckner 1987).

DESCRIPTION OF VHP ROCKS

Eclogites from East-Central China

The end products of VHP metamorphism of gabbros, basaltic flows and tuffs, and diabasic sills + dikes are mafic eclogitic rocks. Metabasaltic eclogites containing Grt + Omp + Rt as essential phases occur in all known VHP terranes, and preserve the most complete record of both prograde and retrograde events. Therefore, mafic eclogitic rocks have received the most attention. Summary description of the Chinese eclogites can be found in several references (Wang et al. 1995, Liou et al. 1996, Zhang et al. 1996b, Cong 1996). Eclogites are classified into three types based on country-rock lithology: eclogite in gneiss, ultramafic rock, and marble. Eclogite in gneiss is most abundant; in this type, inclusions of coesite and quartz pseudomorphs after coesite are common in garnet and omphacite, rarely in kyanite, zoisite, or epidote (Fig. 6). Hydrous phases such as phengite, epidote/zoisite, talc, and nyböite occur as VHP minerals. Exsolution has been identified in many minerals, including titanite rods in olivine and clinopyroxene, quartz in omphacite, and monazite in apatite; details are discussed below.

Most eclogites within ultramafic rocks are bimineralic garnet-omphacite assemblage carrying minor rutile; they are less retrograded compared to eclogites in gneiss. A few contain minor quartz, kyanite, phengite, or amphibole. Round-to-oval shaped inclusions of polycrystalline quartz aggregates after coesite occur in garnet and omphacite with well-developed radial fractures in eclogites (Zhang et al. 1994, Zhang and Liou 1998). Minute K-feldspar inclusions were identified in eclogitic garnet (Zhang et al. 1994). Rare corundum-bearing garnetite and eclogite occur as lenses in the Zhimafang garnet lherzolite (Enami and Zang 1988).

Eclogite in marble contains a representative assemblage Grt + Omp ± Coe/Qtz ± Do ± Mgs ± Zo ± Amp ± high-Al Ttn. Inclusions of coesite/coesite pseudomorphs are common in garnets and occur also in dolomites (Schertl and Okay 1994, Zhang and Liou 1996). Some eclogites contain a large amount of carbonates including dolomite, magnesite, and late-stage calcite.

Several unusual Chinese coesite-bearing eclogites in gneissic rocks reflect the variation of protoliths of basaltic composition. These include glaucophane-bearing eclogite (Zhang and Liou 1994a), nyöbite-bearing eclogite (Hirajima et al. 1992), jadeite eclogite, and epidote eclogite (Zhang et al. 1995c), talc-bearing eclogite (Zhang and Liou 1994a, Zhang et al. 1995a,c), incomplete transformation of gabbro to coesite eclogite (Zhang and Liou 1997), and eclogite overprinted by a granulite assemblage (Zhang and Cong 1991, Wang et al. 1993, Zhang et al. 1995b).

Garnet peridotites

Garnet peridotites are widespread as an important component in VHP terranes. They include garnet-bearing lherzolite, harzburgite, wehrlite, websterite, and pyroxenite. These rocks are considered to have been derived from: (1) subducting lithosphere (Medaris and Bruckner personal com. 1997); (2) mantle-wedge peridotites that sank (Maruyama et al. 1996, Bruckner 1998) or were tectonically inserted (Zhang et al. 1994, Zhang and Liou 1998) into subducted continental lithosphere; or (3) products of subduction-zone metamorphism of peridotite previously emplaced in the crust (Evans and Trommsdorff 1978, Obata and Morten 1987, Zhang and Liou 1998). Garnet peridotites have experienced

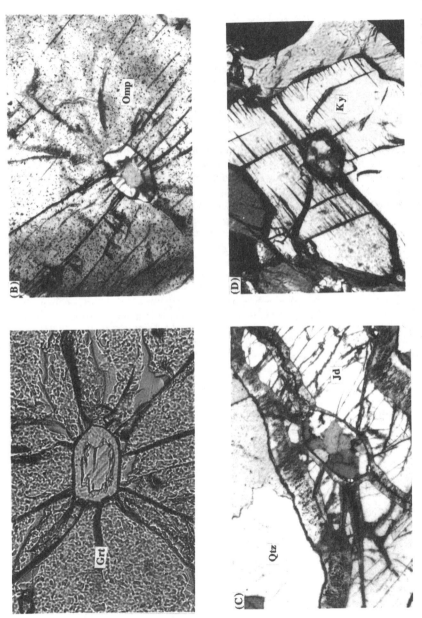

Figure 6. Photomicrographs of inclusions coesite and coesite pseudomorphs in VHP minerals from the Dabie-Sulu terrane: (A) Coesite with thin palisaded quartz aggregates in garnet (plane view; width of view = 0.85 mm). (B) Coesite with thin palisades quartz aggregates in omphacite (crossed polarizers; width of view = 1.12 mm) (Liou and Zhang 1998). (C) Quartz aggregates after coesite in jadeite (crossed polarizers; width of view = 1.12 mm). (D) Quartz aggregates after coesite in kyanite (crossed polarizers; width of view = 0.85 mm).

several distinct stages of recrystallization, including some of the following: (1) primary formation, (2) crustal hydration/metasomatism, (3) peak VHP metamorphism, (4) granulite and/or amphibolite facies retrogression during decompression, and (5) greenschist facies overprint. P-T estimates of VHP metamorphism are 800 to 1200°C and 3-5 GPa (Medaris and Carswell 1990, Yang et al. 1993, Zhang et al. 1994, 1995a; Krogh and Carswell 1995). The peak metamorphic conditions suggested that these Alpine-type peridotites originated from depths close to or exceeding ~140 km. Some garnet peridotites with titanate exsolution rods and magnetite lamellae in olivine may have been derived from the deep upper mantle (Dobrzhinetskaya et al. 1996, Green et al. 1997, Zhang et al., in press). This suggestion has been challenged (e.g. Hacker et al. 1997). Nevertheless, garnet peridotites in VHP terranes provide a new window into upper mantle processes.

Garnet peridotites from the Dabie-Sulu VHP terrane are characterized by occurrences of coesite inclusions in the enclosing eclogite or omphacitite, and ilmenite rods and magnetite plates in olivine. They have been recognized as of mantle-hosted (Type A) and crustal-hosted (Type B) origins. Type A garnet peridotites are fragments of the mantle wedge that have been sequestered in the subducted continental lithosphere, whereas Type B peridotites are portions of mafic-ultramafic igneous complexes intruded into continental crust prior to subduction. Type A lherzolites (Ol + Opx + Cpx + Grt ± Phl) are massive and have lower CaO, Al_2O_3 and TiO_2 suggesting residual mantle. Analyzed type A garnet lherzolites have total REE contents of about several ppm, and show a relatively flat pattern or moderately fractionated REE patterns with slight HREE depletion. Phlogopite-bearing garnet lherzolite from the Donghai area shows a U-shaped pattern with negative Eu anomaly; such a pattern for an anhydrous peridotite may reflect a complicated evolution that requires mantle enrichment or crustal contamination (Song and Frey 1989). The type A pyroxenites (Grt + Cpx ± Ol + Ilm) have higher total REE contents and show convex REE patterns; such convex patterns may be related to garnet and clinopyroxene cumulates. Pyroxenites preserve mantle O isotope values ($\delta^{18}O$: 5.5-5.6‰, Yang 1991) and $^{87}Sr/^{86}Sr$ ratios (0.7032-0.7036, Jahn 1998). In contrast, most type B peridotites show banded or layered structure and a wide range in CaO, Al_2O_3, TiO_2 and REE; peridotites from Bixiling also have lower total REE contents similar to the type A peridotites. However, most type B ortho- and clino-pyroxenites from Maowu are characterized by extremely high total REE contents and extremely (54C) to moderately fractionated, LREE-enriched REE patterns; some orthopyroxenites have concave REE patterns. These features imply that the Maowu pyroxenites were metasomatized by LREE-enriched fluid or melt prior to VHP metamorphism in the upper mantle (Zhang et al. 1998). Analyzed type B peridotites have relatively low mineral $\delta^{18}O$ values of +2.9 to +4.1‰ (Zhang et al. 1998), and high $^{87}Sr/^{86}Sr$ bulk rock ratios (0.707-0.708), and were subjected to crustal contamination and/or metasomatism prior to VHP metamorphism (Jahn 1998).

A schematic model for the dual origin of garnet peridotites from the Dabie-Sulu VHP terrane is shown in Figure 4A. This diagram illustrates the original settings of Type A and Type B peridotites before and during Triassic subduction of the Yangtze craton beneath the Sino-Korean craton, and the coeval nature of VHP and retrograde events with the surrounding gneissic rocks (Zhang and Liou 1998). Because of their different settings, these garnet peridotites possess different isotopic and metasomatic signatures. Figure 4B shows a scenario for exhumation of the VHP slab; the process mainly reflects buoyant return of continental crust after slab breakoff and later by a doming process (for details, see Maruyama et al. 1996, Ernst et al. 1997). These peridotites share similar retrograde overprints by granulite through amphibolite to greenschist-facies assemblages.

Among the various VHP lithologies, the garnet peridotites mentioned above yield the highest pressure estimates of the crustal sequence, up to 4.5-6 GPa based mainly on the Al-in-Opx geobarometers of Wood (1974), Nickel and Green (1985) and Brey and Köhler (1990). Some abnormally high-P peridotites may have been transported by mantle convection first from greater depths to the vicinity of the subduction zone and subsequently incorporated into subducted continental crust. P-T paths of garnet peridotites from Ronda and Beni Bousera exhibit a contrasting trajectory compared with the Alpe Arami body. Their path followed an upwelling mantle plume through a continental rift (see Fig. 29 of Maruyama et al. 1996). Adiabatically ascending mantle then became part of the lithosphere and thermal relaxation set in accompaning transportation toward the trench. The peridotite finally was incorporated into the subduction zone together with the continental crust.

Metasediments

VHP metasediments occur either as interlayers with eclogite or as host country rocks of both eclogites and garnet peridotites. They include mainly quartzofeldspathic gneiss and paraschist together with minor quartzite, pelite, marble, garnet-quartz jadeite rock, calc-silicate rock, and whiteschist. The classic locality of coesitite and whiteschist from the Dora Maria Massif of the Western Alps includes unusual minerals, including ellenbergerite (Chopin et al. 1986), bearthite (Chopin et al. 1993), and coarse-grained pyropic garnet (Chopin 1984) described in later sections. Inclusion of quartz pseudomorphs after coesite occurs in garnet, jadeite (Fig. 6C), epidote + kyanite in jadeite rock, and Ky-quartzite. Kyanite quartzite of the Sulu terrane is thickly banded, contains eclogite pods, and consists of quartz (>80 vol%), kyanite, zoisite, pyrite, and rutile. Quartzite interlayered with eclogite ranges from 1 to tens of meters in thickness, and consists of quartz (>60-70 vol%) and kyanite, subordinate Zo ± Phn ± Omp ± Ep, and minor garnet and rutile (Zhang et al. 1995c). Garnet-jadeite quartz rocks are composed of 50-65 vol% jadeite, 10-20 % garnet, 20-28 % quartz, minor rutile, and apatite with or without clinozoisite. Protoliths of Grt + Qtz + Jd rocks may be either sediments or intermediate igneous rock, subjected to Na metasomatism (Zhang et al. 1995c, Liou et al. 1997b). Typical metapelite (Qtz + Grt + Phn + Ky + Zo) associated with eclogite contains coarse-grained garnet and zoisite; abundant inclusions of mica, zoisite, quartz, and amphibole occur in garnet, and quartz and muscovite are included in zoisite.

Some eclogite-bearing marbles in the Dabie Mountains contain evidence of VHP conditions, such as inclusions of quartz pseudomorphs after coesite and calcite pseudomorphs after aragonite in garnet, and coesite inclusions in dolomite (Fig. 15D) (Wang and Liou 1992, Schertl and Okay 1994, Zhang and Liou 1996). Marble lenses and layers in gneiss range from centimeters to several tens of meters. The marbles are inhomogeneous, and range from rare, pure white marble to common, impure, gray marble in different bands; mineral parageneses of impure dolomitic marble provide conclusive assemblages suggesting VHP origin (e.g. Ogasawara et al. 1998, Omori et al. 1998). Some dolomitic marbles consist of dolomite, Mg-calcite, diopside, forsterite, and minor Ti-clinohumite and phlogopite. Mg-calcite occurs as a coarse-grained (3-5 mm in size) matrix mineral and as fine-grained inclusions in Ti-clinohumite (3 to 4 wt % TiO_2) and in diopside. Olivine and dolomite are finer grained than the calcite.

GEOCHEMICAL CHARACTERISTICS—EXAMPLES FROM CHINA

Petrological studies of various VHP rocks mentioned above have constrained the P-T-time paths for deep subduction of supracrustal and associated mafic-ultramafic rocks to mantle depths and their subsequent return to the surface. Figure 7 summarizes the process, timing, and nature of the protoliths, metamorphism, and recrystallization of VHP rocks described specifically for the Dabie-Sulu terrane. Precambrian protoliths, including granitic, pelitic, psammitic, carbonate, and minor mafic-ultramafic rocks, were subjected to VHP metamorphism at mantle depths prior to and during the Triassic collision of the

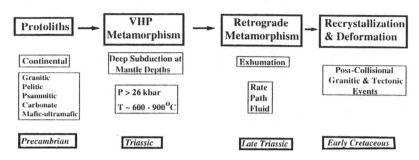

Figure 7. Flow chart showing the processes for formation and preservation of VHP assemblages in a collision zone of the Dabie-Sulu terrane, eastern central China (modified after Liou et al. 1996).

Yangtze and Sino-Korean cratons. The exhumation and preservation of minor VHP phases depended on the rate and path of return flow, the availability and infiltration of fluid, and the intensities of post-collisional deformation, recrystallization, and migmatization.

Because of the long return journey from mantle depths to the surface, and because of overprinting events during subsequent deformation and recrystallization, traces of VHP minerals are only preserved as inclusions in strong, refractory minerals such as zircon, garnet, and omphacite (e.g. Sobolev et al. 1994, Tabata et al. 1998). The effects of the return path (e.g. Hacker and Peacock 1995) and fluid infiltration (e.g. Austrheim 1998, Ernst et al. 1998) on the preservation and retrogression of VHP rocks have been described extensively. Stable isotope geochemistry is considered to be one of the most efficient tools in deciphering the source and the extent of fluid/rock interaction during metamorphic recrystallization. The preservation of remarkably low $\delta^{18}O$ values through VHP metamorphism and retrogression for the Sulu-Dabie VHP rocks (Yui et al. 1995, 1997; Zheng et al. 1996, 1998; Baker et al. 1997, Rumble 1998) is so unusual that it deserves special discussion.

Aqueous fluids play important roles in eclogitization and retrograde amphibolitization. Fluids participate in many metamorphic reactions, they act as catalysts in solid-solid reactions, lead to change in rock composition (metasomatism), and control rheology. In VHP regimes, deeply subducted packages of protoliths contain minor fluids in hydrous and carbonate phases; there is no evidence that H_2O has been present as a separate phase (e.g. Liou et al. 1997a). In such fluid-deficient environments, metamorphic reaction rates are severely retarded, and VHP phases and the geochemical signature of the primary tectonic setting can be preserved. For example, if garnet peridotites described above are indeed fragments of the mantle wedge above the subducting slab, they may provide the clearest indication of petrochemical processes of the deep lithosphere where the rocks have not been modified due to the infiltration of fluids.

The oxygen isotope compositions of surface waters together with common metamorphic and mantle rocks are plotted in Figure 8; also plotted are the $\delta^{18}O$ values from VHP eclogites and interlayered metasediments from Qinglongshan of the Sulu terrane. Crustal rocks range widely in composition but are generally greater than zero in $\delta^{18}O$. Mantle rocks are nearly homogeneous at +5 to +6‰ $\delta^{18}O$. Ocean water is zero in both $\delta^{18}O$ and δ D. Meteoric waters (rain, groundwater, snow, ice) have negative $\delta^{18}O$ and δD

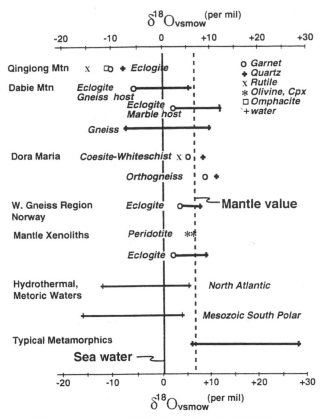

Figure 8. Variations of $\delta^{18}O$-isotope data for VHP minerals from the Sulu-Dabie terrane and other typical crustal metamorphic rocks and various waters (modified after Rumble 1998).

values. A comprehensive review of published $\delta^{18}O$ values of quartz from metamorphic rocks worldwide by Sharp et al. (1993) shows that the extreme lower limit of metamorphic quartz is +7‰, and the entire range was known to extend to +30‰. The discovery of $\delta^{18}O$ values of garnet and omphacite as low as -10‰ and quartz at -7‰ (Fig. 8) in coesite-bearing eclogites from the Sulu terrane was quite surprising because the values are so much lower than those previously measured (Yui et al. 1995, Zheng et al. 1996). Also remarkable were reports of eclogite garnets from the Dabie Shan with $\delta^{18}O$ of -5 to -7‰ (the equilibrium value for quartz at these conditions would be -2 to -4‰) (Baker et al. 1997, Yui et al. 1997). Unusually low δD values of -113 to -124‰ (VSMOW) have been found in phengite from eclogite and quartzite at Qinglongshan (Rumble and Yui 1998). The δD values are not as surprising as the low $\delta^{18}O$, but they are at the low end of the total range of natural variation in metamorphic micas (Fig. 8 of Sharp et al. 1993).

Stable isotope geochemistry for VHP rocks from the Dabie-Sulu terrane summarized in Figure 8 supports the following conclusions: (1) protoliths of the VHP rocks from this terrane were at the Earth's surface and were subjected to geothermal alteration involving cold-climate meteoric waters prior to Triassic subduction; (2) the presence of a pre-metamorphic, distinctive isotopic signature in both eclogites and their host rocks supports the hypothesis of *in situ* metamorphism for coesite-bearing eclogites; the length scale of structural coherence is at least 100 km; (3) the persistence of pre-metamorphic differences in oxygen isotopic abundance in different rock types argues against the existence of a pervasive fluid free to infiltrate across lithologic contacts during VHP metamorphism; and (4) the preservation of high-T oxygen isotope fractionations among the early-formed minerals support the absence of free fluid after the peak metamorphism during exhumation/cooling. Residence in the upper mantle evidently had no discernible metasomatic effect on the stable isotope composition of crustal rocks subducted during continental collision (Rumble 1998).

In essence, stable isotope results from the Dabie-Sulu terrane demonstrate that those VHP rocks have not been subjected to significant fluid/rock interactions since formation prior to the Triassic subduction. Primary records for VHP metamorphism at mantle depths and geochemical signatures of the mantle wedge as well as crustal rocks are well preserved. Therefore, the Dabie-Sulu terrane provides a natural laboratory to investigate geodynamic processes involving slab/mantle interactions. The plan to drill a 5-km scientific drill hole in the center of the oxygen and hydrogen isotope anomaly affords a unique opportunity to measure the depth and thickness of the Sulu VHP slab.

CHIEFLY DABIE-SULU VHP INDEX MINERALS

As stated above, VHP rocks of intracontinental suture zones have experienced complex and multi-stage metamorphic recrystallization; hence, most VHP minerals have been obliterated during their return to the surface. Strong, rigid, unreactive crystals including garnet, zircon, kyanite, and omphacite evidently function as pressure vessels, preserving some early-formed VHP mineral inclusions. Examination of zircons and garnets for included primary VHP phases such as coesite and diamond has proven to be a powerful technique, especially for highly retrograded metamorphic rocks. Characteristic minerals and mineral compositions indicative of VHP metamorphism are classified as (1) index minerals, (2) hydrous phases, (3) carbonate phases, (4) exsolution lamellae, and (5) other unusual phases.

Coesite and coesite pseudomorphs

Coesite is a high-P polymorph of SiO_2 stable at P > 2.6 GPa at 700°C (Fig. 1). Coesite is monoclinic, and distinguished from quartz by its higher birefringence, biaxial (+) character, lower birefringence, poor cleavage, and blue luminescence under the electron microprobe beam and Raman spectrum. Synthetic coesites at 7 GPa and 650°C contain about 100 ppm H_2O, documented by three infrared sharp absorption peaks at 3459, 3524, and 3572 cm^{-1} associated with OH stretching vibrations (Li et al. 1997). Reversal experiments and thermal calculations of the α-quartz = coesite phase boundary have been carried out (e.g. Bohlen and Boettcher 1982, Kuskov et al. 1991, Bose and Ganguly 1995, Mosenfelder et al. 1996, Hemingway et al. 1998); the results show significantly different P-T slopes and initial P-T values (e.g. see Fig. 1 of Hemingway et al. 1998) due to uncertainty in pressure calibration for the different apparatus (Mosenfelder et al. 1996). Hemingway et al. (1998) obtained thermodynamic properties of coesite to minimize the difference between calculated and experimental values and to relocate the equilibrium boundary of the α quartz = coesite transition. Results are quite comparable with the original boundary determined by Bohlen and Boettcher (1982).

SiO_2 inclusions in eclogitic rocks. Because quartz is one of the most common minerals in supracrustal rocks, positive identification of coesite, and quartz pseudomorphs after coesite, is considered to be most useful for the establishment of VHP metamorphism. Thus far, coesite and coesite pseudomorphs have been described in eclogitic and other rocks from the VHP terranes shown in Figure 2. Most coesite relics and pseudomorphs occur as minute oval to subrounded, 50 to 350 micron inclusions in zircon, garnet,

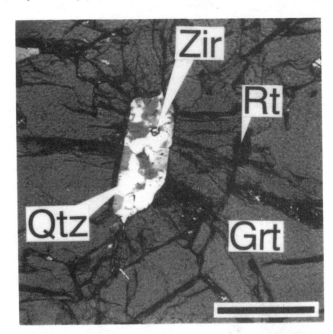

Figure 9. Photomicrograph of an inclusion of a coesite pseudomorph in garnet with radial fractures from eclogite enclosed in Dabie marble (crossed polarizers) (Omori et al. 1998).

omphacite, kyanite, and even in hydrous phases such as epidote, zoisite, and carbonate (Zhang et al. 1990, 1995a,c; Liou 1993, Zhang and Liou 1996). Coesite pseudomorphs consist of fine-grained polycrystalline quartz aggregates. Several examples of coesite from a Dabie-Sulu eclogites are shown in Figures 6 and 9. They are recognized by the occurrence of radial fractures in host minerals around the silica inclusion. The radial fractures developed because of the 10 % volume expansion involved in the transformation from coesite to quartz. Most reported coesites occur either as a single anhedral crystal invariably surrounded by a palisaded rim of fine polycrystalline quartz, or in an irregular form that is transacted by quartz veinlets. Depending on the extent of retrogression, some coesite grains are rimmed by a very thin layer of palisaded quartz aggregates (Fig. 6), whereas others are totally replaced by a mosaic of coarser grained quartz aggregates (Fig. 9). The sequence of retrogression from coesite through palisaded aggregates and mosaic intergrowth and granoblastic quartz has been described by Zhang et al. (1996a) and experimentally produced by Mosenfelder (1997).

SiO$_2$ inclusions in country rock gneisses. Coesite inclusions in zircon from country-rock orthogneiss of the Dabie VHP terrane were identified by Raman spectroscopy and electron microprobe microanalysis (Tabata et al. 1998). All coesite-bearing gneissic rocks contain amphibolite-facies assemblages in the matrix, indicating that these rocks experienced VHP metamorphism prior to amphibolite-facies retrogression. Quartz inclusions are also present in host zircons; some are accompanied by fractures and/or fluid inclusion arrays (Fig. 10). Coesite-bearing orthogneisses occur far from previously recognized VHP eclogites, indicating that cocsite-bearing lithologies occur throughout the southern Dabie unit, requiring subduction of the entire area to mantle depths. A similar conclusion was derived from the Sulu area characterized by strongly negative oxygen isotope fractionations described in the previous sections (e.g. see Rumble 1998).

Intergranular coesite. Several intergranular coesite crystals along garnet-omphacite grain boundaries in eclogites from the Sulu region have been described (Liou

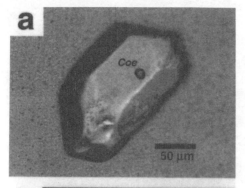

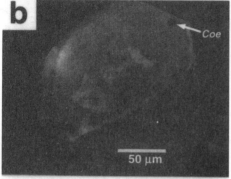

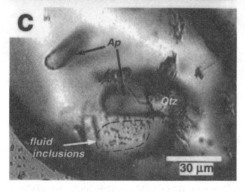

Figure 10. (a) Photomicrograph of coesite inclusion (Coe) and host zircon from Dabie quartzofeldspathic gneiss. This inclusion is not exposed on the polished surface. (b) Cathodoluminescence image of the same zircon grain as in (a), showing irregular-shaped interior surrounded by zoned rim. Exposed inclusion (coesite) is present at the outermost part of the grain. (c) Fluid inclusion array within the zircon grain separated from coesite-bearing gneiss (from Tabata et al. 1998).

and Zhang 1996, Ye et al. 1996, Zhang and Liou 1997, Banno et al. 1997). These grains exhibit variable extents of conversion of coesite to palisaded—through mosaic—to granoblastic-quartz aggregates (Fig. 11). This occurrence of intergranular matrix coesite and groundmass coesite pseudomorphs is unique. Metagabbroic and metagranitic rocks with preserved relict igneous minerals and textures also occur at this locality. The gabbroic rocks represent early stages of a progressive conversion of pre-existing lithologies to coesite-bearing eclogites (Zhang and Liou 1997). Most of the VHP eclogites show only incipient retrogression, with only minor late-stage amphibole. These petrologic features, together with the occurrence of intergranular coesite grains, suggest slow reaction kinetics. Laboratory experiments (Mosenfelder et al. 1997) indicate that the coesite = quartz reaction is sluggish in the absence of H_2O; the presence of only ~450 ppm H_2O is sufficient to

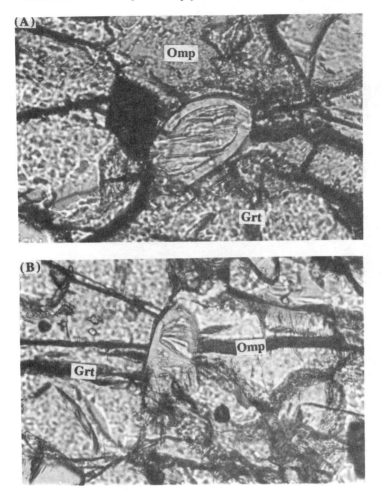

Figure 11. Photomicrographs of intergranular coesite grains at the contact between garnet (Grt) and omphacite (Omp) (plane polarized light; width of view = 0.37 mm) (modified after Liou and Zhang 1996).

transform coesite completely to quartz during exhumation at T > 300°C. Total lack of fluids during and after the VHP metamorphism may be the key for such sluggish prograde and retrograde reactions. This suggestion is consistent with the occurrence of similar intergranular coesite grains from a mantle-derived grospydite xenolith in a South Africa kimberlite pipe (Smyth 1977) and with the extremely negative $\delta^{18}O$ values for garnet, omphacite, phengite, and quartz from nearby eclogites and quartzites of the Sulu region (Yui et al. 1995, Zheng et al. 1996, 1998; Rumble and Yui 1998).

Diamond

Microdiamond of VHP origin was reported as inclusions in zircon and garnet of the Kokchetav Massif (Sobolev and Shatsky 1990, Zhang et al. 1997), in garnet, kyanite and zircon of the Erzgebirge gneisses (Massonne, 1998), in garnets from eclogite, garnet-pyroxenite and jadeitite in the Dabie Mountains (Xu et al. 1992), and in residues separated from Western Gneiss Region gneisses (Dobrzhinetskaya et al. 1993). Microdiamonds from

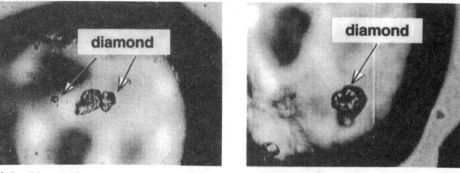

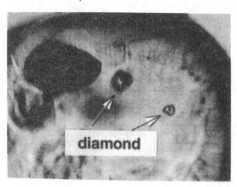

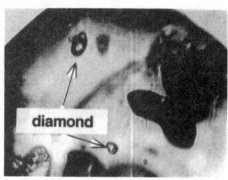

A8a8inc4,5   A8a12inc3,5   50µm

Figure 12. Photomicrographs of microdiamond inclusions in zircons from gneissic rocks of the Kokchetav Massif (courtesy of I. Katayama).

Norway have a grain size of 20-50 microns, are round and anhedral crystals; whether they are part of the primary assemblage or are due to contamination remains to be confirmed. The Kokchetav microdiamonds have cubo-octahedral faces, and average 12 µm in diameter; some grains have been partly or completely replaced by graphite. Several diamond inclusions in zircon are shown in Figure 12. The Kokchetav microdiamonds are characterized by (1) low $\delta^{13}C$ values, -10 to -19‰, suggestive of crustal biogenic carbon, (2) very high concentration of nitrogen impurities (~1470 ppm), and (3) unusual crystallographic habits (Finnie et al. 1994, Sobolev and Shatsky 1995). In the Dabie Mountains, more than 20 diamond crystals have been identified; they range in size from 150 to 700 µm, much coarser than those in the Kokchetav Massif. Positive identification of microdiamond was accomplished by X-ray diffraction of separated single crystals and by *in-situ* laser Raman microspectrometry. Since the discovery of microdiamond by Xu et al. (1992), other occurrences in both the Dabie Mountains and the Su-Lu region have been reported. In 1982, the No. 6 Geological Brigade of Jiangsu found two diamond grains in a bulk sample of the Zhimafang peridotite from Donghai; the largest one is 9.17 mg. In 1996, one diamond grain with a diameter of 0.7 x 0.5 mm was separated from a 2-ton eclogite sample from Donghai; this diamond is a transparent octahedron-cube glomerocryst, and was confirmed by laser Raman spectroscopy (Xu 1997).

The rare occurrence of microdiamond in the Dabie-Sulu terrane could be due to relatively higher f_{O_2} conditions. In such environments, diamond is not stable in most rocks, as evidenced by the abundance of carbonates including calcite, dolomite, and magnesite in the associated eclogitic rocks. As listed in the following reaction, at higher f_{O_2}, diamond reacts with the diopsidic component of pyroxene and the pyrope component of garnet to form magnesite, a minor phase in many eclogitic rocks from the Dabie-Sulu terrane:

$$Mg_3Al_2Si_3O_{12} + 3\ CaMgSi_2O_6 + 6\ C + 6\ O_2 = Ca_3Al_2Si_3O_{12} + 6\ MgCO_3 + 6\ SiO_2$$

Pyrope   Diopside   Diamond   Grossular   Magnesite Coesite

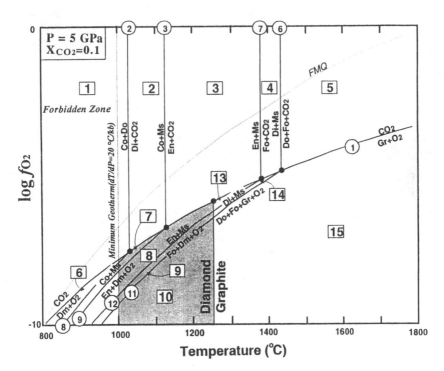

Figure 13. Log f_{O_2} - T diagram in the system $CaO-MgO-SiO_2-C-O_2-H_2O$ calculated at P = 5.0 GPa and $X_{CO_2} = 0.1$ showing the stability field of diamond-bearing assemblages as function of f_{O_2} (modified after Ogasawara et al. 1997).

In order to understand the paragenetic relations of diamond and associated phases in eclogitic and ultramafic rocks, f_{O_2}-T-P stabilities of diamond, coesite, enstatite, forsterite, graphite, magnesite, and dolomite in the system $CaO-MgO-SiO_2-C-O_2-H_2O$ have been calculated by Ogasawara et al. (1997). One example at P = 5 GPa and $X_{CO_2} = 0.1$ is illustrated in Figure 13. This and several other calculated diagrams suggest that the appearance of diamond depends on bulk composition, fluid composition in addition to f_{O_2}, and geothermal gradient. With X_{CO_2} lower than 0.1 and a geothermal gradient less than 5°C/km (20°C/kbar), diamond is stable at T < 1250°C. Figure 13 also indicates that: (1) the formation of diamond with coesite + diopside + enstatite in eclogitic rocks may be controlled by the reaction coesite + magnesite = enstatite + diamond + O_2; and (2) the formation of diamond with diopside + enstatite + forsterite in lherzolite may be controlled by the reaction enstatite + magnesite = forsterite + diamond + O_2. The former reaction occurs at slightly higher f_{O_2} conditions than the later reaction, consistent with previous suggestions (e.g. Luth 1993). Magnesite- and dolomite-rich carbonate rocks occur in several diamond-bearing assemblages, presumably due to low f_{O_2} conditions (Ogasawara et al. 1997).

VHP HYDROUS SILICATES

The role of H_2O and the stability of hydrous magnesium silicates in the Earth's upper mantle has attracted a great deal of attention (e.g. Ahrens 1989, Thompson 1992, Smyth 1994, Bose and Navrotsky 1998) because H_2O strongly affects solidus temperatures, and therefore magma genesis, the rheology of the subducting slab, and slab-mantle wedge interactions (Ernst et al. 1998). Because of observed continuous degassing and geochemical recycling in the upper mantle, determinations of the amount and nature of various volatile species have been attempted. Previous estimates of volatile concentrations have involved investigation of hydrous and carbonate phases in xenolith suites from the upper ~150 km of the mantle, and the determination of OH^- solubility in nominally anhydrous minerals. The recognition of VHP rocks derived from profound subduction-zone depths provides a unique opportunity to investigate the nature and source of mantle volatiles. Many of these rocks contain hydrous and carbonate phases, evidently produced within the P-T stability fields of coesite and diamond. VHP regimes thus bridge the gap between deep upper mantle and crustal processes. Recognized hydrous phases in VHP rocks are described below; representative compositions are listed in Table 2.

Phengitic mica

In the system K_2O-MgO-Al_2O_3-SiO_2-H_2O under conditions of excess SiO_2+H_2O, the number of Si atoms per formula unit (abbreviated p.f.u.) of phengite coexisting with K-feldspar + quartz + phlogopite, or at higher P, with Ky + Tlc + Qtz increases with elevated P; these relations have been experimentally calibrated by Massonne and Schreyer (1987, 1989). Figure 14 shows the systematic variation of the Si value of synthetic phengite mica

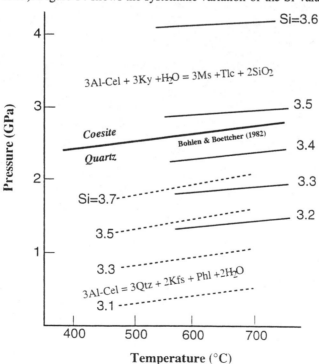

Figure 14. Experimentally calibrated isopleths of Si value per formula unit of phengitic mica in the system KMASH. Solid lines are for phengite coexisting with Tlc + Ky + Qtz and broken lines are for phengite coexisting with K-feldspar + phlogopite + Qtz (modified after Liou et al. 1994). The quartz-coesite transition is that of Bohlen and Boettcher (1982).

Table 2. Representative compositions of hydrous phases from Dabie-Sulu VHP eclogites.

Mineral:	Phn	Phn	Zo	Zo	Ep	Ep	Tlc	Tlc	Gln	Gln	Nyboite	Nyboite	Ti-Clinohumite	Ti-Clinohumite
SiO2	55.16	52.50	39.09	39.52	38.39	37.86	61.04	61.90	55.48	57.01	50.57	49.14	35.96	37.62
TiO2	0.36	0.30	0.05			0.30	0.00	0.01	0.09	0.01			4.75	3.89
Cr2O3	0.06	0.08	0.15				0.00	0.00	1.31	0.09			0.02	0.10
Al2O3	22.79	25.53	31.21	32.12	24.66	24.50	0.39	0.48	11.13	10.81	11.98	12.97	0.04	0.00
Fe2O3					12.20	12.01			1.92	1.48				
FeO	1.88	1.91	1.57	1.27		0.01	2.15	2.10	4.87	5.43	8.85	9.84	10.56	3.68
MnO	0.00	0.04	0.04				0.05	0.01	0.02	0.00	n.d	n.d	0.12	0.00
MgO	5.60	4.34	0.00	0.44	0.26	0.37	29.20	29.47	13.29	13.92	13.46	12.81	45.11	52.46
CaO	0.00	0.00	23.52	23.85	22.29	22.43	0.03	0.03	1.81	2.43	4.42	5.53	0.01	0.00
Na2O	0.20	0.47	0.03	0.00		0.02	0.03	0.02	7.25	7.10	6.30	5.79	0.00	0.00
K2O	9.94	10.08	0.00			0.01	0.01	0.00	0.11	0.00	0.25	0.27	0.00	0.05
Total	96.00	95.26	95.65	97.20	97.80	97.51	92.91	94.02	97.28	98.28	95.83	96.35	96.57	97.80
Si	3.62	3.49	6.10	6.04	6.06	5.95	3.99	4.00	7.62	7.73	7.18	7.01	3.94	4.00
Ti	0.02	0.01	0.01			0.04	0.00	0.00	0.01	0.00			0.39	0.31
Cr	0.00	0.00	0.02				0.00	0.00	0.14	0.01			0.00	0.01
Al	1.76	2.00	5.74	5.79	4.59	4.54	0.03	0.04	1.80	1.73	2.01	2.18	0.01	0.00
Fe3+					1.45	1.58			0.20	0.15	0.32	0.30		
Fe	0.10	0.11	0.20	0.15			0.12	0.11	0.56	0.62	0.73	0.88	0.97	0.33
Mn	0.00	0.00	0.00				0.00	0.00	0.00	0.00			0.01	0.00
Mg	0.55	0.43	0.00	0.10		0.09	2.85	2.84	2.72	2.81	2.85	2.72	7.36	8.32
Ca	0.00	0.00	3.93	3.91	3.77	3.78	0.00	0.00	0.27	0.35	0.67	0.85	0.00	0.00
Na	0.03	0.06	0.01				0.00	0.00	1.93	1.87	1.75	1.60	0.00	0.00
K	0.83	0.85	0.00				0.00	0.00	0.02	0.00	0.05	0.05	0.00	0.01
Total	6.91	6.96	16.02	15.99	15.92	15.96	7.00	6.99	15.27	15.26	15.56	15.58	12.67	12.97
Ref.	1	1	3	3	3	3	1	1	2	2	3	3	1	4

Ref.: 1. Zhang et al. (1995a); 2. Zhang and Liou (1994a); 3. Zhang et al. (1995c); 4. Ogasawara et al. (1998)

for these two assemblages. High-Si phengite is expected to be stable with Ky + Tlc in some pelitic and quartzofeldspathic rocks in VHP metamorphic terranes. Isopleths of Si p.f.u. of phengitic mica for a low-variance "buffered" assemblage are useful in order to delineate both prograde and retrograde P-T paths. However, white micas have continued to recrystallize at somewhat lower temperatures than the other index phases mentioned above, thus phengite compositions may generally reflect annealing during return towards the Earth's surface. The stability and composition of phengitic mica has been recently reinvestigated from 5.5 to 11 GPa, 700-1150°C in synthesis experiments performed in a multianvil apparatus. Domanik and Holloway (1996) concluded that phengitic mica systematically increases in Si p.f.u. with P from 5.5 to 11 GPa at 900°C; the maximum Si p.f.u. appears to be 3.80-3.85.

Phengitic mica is ubiquitous in VHP eclogite, and enclosing gneiss and marble; it formed in pre, syn- and post-VHP stages of metamorphism. Those considered as VHP phengitic micas occur as a minor matrix phase or as inclusions in garnet, have higher Si contents (>3.4 p.f.u.), coexist with talc and kyanite, and appear to have been stable in the coesite stability field. Some phengites display the compositional range of Ms_{60-70}, Pg_{00-08} and Cel_{26-38}; others show retrograde zoning with Si values varying from 3.5-3.6 in the core to 3.3 in the rim, compatible with a marked P drop (Massonne and Schreyer 1987, 1989). The analyzed phengites from VHP terranes are stable to at least 3.5 GPa in quartzofeldspathic rocks.

Zoisite and epidote

Zoisite/epidote minerals are rather common in VHP rocks; they occur as fine-grained inclusions or coarse-grained porphyroblasts in pre-, syn- and post-peak VHP metamorphic assemblages of coesite-bearing eclogites and associated rocks. Only those zoisite/epidote minerals with inclusions of coesite or coesite pseudomorphs are considered to be probable VHP phases. Zoisite coexisting with kyanite and talc in Mg-Al-rich coesite-bearing eclogites from Bixiling contains inclusions of coesite and coesite pseudomorphs (Fig. 15A), and has an $Fe^{3+}/(Fe^{3+}+Al)$ ratio of about 0.03 (Zhang et al. 1995a). Epidote occurs as a common matrix mineral or as porphyroblasts in eclogites from Donghai, and exhibits two distinct stages of recrystallization (Zhang et al. 1995c). Medium-grained, elongate epidote (Ps = 18) together with garnet, omphacite, kyanite, and rutile define an eclogitic foliation; the second-stage recrystallization resulted in the formation of coarse-grained epidote (Ps = 25) porphyroblasts that cut the rock foliation and include earlier minerals. Both medium-grained epidote and porphyroblastic epidote and kyanite contain inclusions of quartz pseudomorphs after coesite with radial fractures. Epidote porphyroblasts with inclusions of coesite pseudomorphs are associated with high-Al titanite in some Donghai eclogites (Fig. 15B). Thin beds of kyanite-bearing quartzite that locally contain abundant fine-grained neoblastic epidote (Ps = 23) coexisting with Ky, Grt, Phn, and Rt are interlayered with the coesite-bearing eclogites; both kyanite and epidote include coesite pseudomorphs. Some VHP zoisites and epidotes from several localities of the Sulu region contain up to 3 wt % SrO (Nagasaki and Enami 1998).

The Donghai occurrences indicate that epidote/zoisite minerals are almost certainly stable within the coesite stability field and have estimated peak temperature of 700-890°C at a minimum P of about 2.8 GPa for eclogites and kyanite-bearing quartzites (Zhang et al. 1995c). P-T estimates are consistent with the calculated stabilities of epidote/zoisite minerals (see Fig. 3 of Liou 1993) and the synthesis experiments of Poli and Schmidt (1995) for basaltic and andesitic compositions; zoisite/clinozoisite was produced at 600-800°C, 2.6-3.3 GPa. Liu et al (1996) however, demonstrated that epidote is unstable in MORB under such conditions.

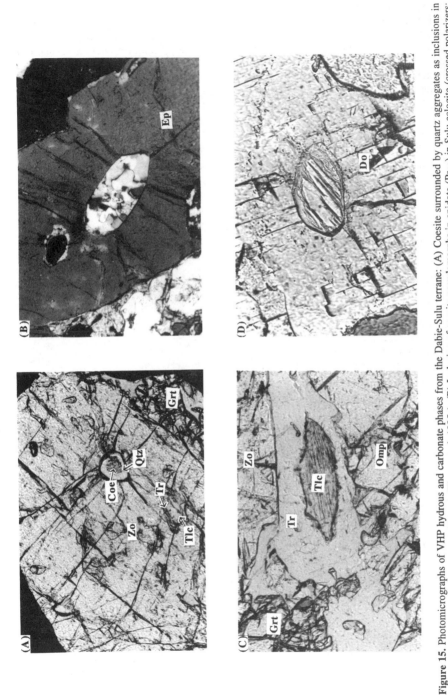

Figure 15. Photomicrographs of VHP hydrous and carbonate phases from the Dabie-Sulu terrane: (A) Coesite surrounded by quartz aggregates as inclusions in zoisite (plane light; width of view =1.44 mm) (Zhang et al. 1995a). (B) Inclusion of coesite pseudomorphs in epidote (Ps$_{23}$) in Sulu eclogite (crossed polarizers; width of view = 1.28 mm) (Zhang et al. 1995b). (C) Talc replaced by tremolite adjacent to garnet, omphacite and zoisite (plane light; width of view =1.44 mm) (Zhang et al. 1995a). (D) Coesite inclusion in dolomite from Dabie eclogite (plane polarized light; width of field view = 0.36 mm) (Zhang and Liou 1996).

Talc

In the VHP terrane of east-central China, talc is an eclogitic phase as inclusions in zoisite, omphacite, or garnet, and as discrete grains in the matrix of some coesite-bearing eclogites. Talc-bearing eclogites from Sulu, Dabie, and Hongan VHP regions contain similar assemblages of Grt + Omp ± Ky + Ep/Zo + Coe/Qtz + Rt + Tlc ± Phn. However, talc is partly replaced by lower P barroisite, tremolite, and cummingtonite in Hongan, Dabie, and Donghai respectively. Such alterations may be related to bulk composition and the eastward higher temperatures of retrograde metamorphism (Zhang et al. 1996a, Liou et al. 1996). Talc inclusions in a Bixiling eclogite are rimmed by a narrow band of tremolite at the contact with zoisite (Fig. 15C); relict coarse-grained matrix talc is surrounded by tremolite that is in direct contact with omphacite and other phases. These textural relations suggest that talc reacted with the Ca-bearing phases—omphacitic pyroxene or zoisite—to form tremolite, following a simple decompression reaction such as Tlc + Di = Tr. In garnet-bearing ultramafic rocks, most talc occurs as a secondary phase after enstatite; minor talc also occurs as inclusions within enstatite.

To illustrate distinct differences between talc parageneses in coexisting mafic and ultramafic rocks, we computed phase relations in the model system $CaO-MgO-SiO_2-H_2O$ (Al_2O_3, FeO) using the GE0-CALC program of Berman (1988). Phase relations for this system together with the stable assemblages for mafic eclogitic and ultramafic rocks are shown in Figure 16. Significant features are described below.

- (A). The stability of tremolite is limited by a dehydration reaction: Tr = Di + En + Qtz + H_2O at high T, and by a solid-solid reaction Tr = Di + Tlc at high P; both reactions have been experimentally determined (e.g. Jenkins 1987, Jenkins and Clare 1990) and thermodynamically calculated (e.g. Welch and Pawley 1991). The calculated end-member reaction, Tr = Tlc + Di, is located about 1 GPa lower than the coesite-quartz transition (Bohlen and Boettcher 1982) and is parallel to it. Together with another reaction, Tlc = En + Coe + H_2O, this restricts the stability of talc + coesite (in the presence of Di) to upper mantle pressures, whereas tremolite is limited to the quartz stability field.

- (B). The P-T positions of two invariant points in this system for Fo-absent and Qtz-absent equilibria were located. Several well-determined reactions involving tremolite and talc (± Fo) radiate from these invariant points. Stability relations in this simple system are well constrained.

- (C). Assemblages for ultramafic and mafic compositions of Figure 16 show systematic contrasts under similar P-T conditions. For example, for a basaltic protolith, the assemblage Coe + Di + Tlc is stable within the coesite field; in contrast, coexisting ultramafic rocks contain Fo + En + Di. During retrograde recrystallization along a clockwise P-T path, the mafic assemblage is replaced by Tr + Di + Qtz, whereas the ultramafic assemblage successively changes to (1) Tr + Fo + En, and then to (2) Tlc + Tr + Fo. Paragenetic sequences for mafic-ultramafic assemblages have been observed in the Bixiling mafic-ultramafic complex (Zhang et al. 1995c).

- (D). With introduction of other components such as jadeite in clinopyroxene, and tschermakite, glaucophane, and F substitution for tremolite, these simplified phase relations are significantly modified. For example, the solid-solid reaction Tr = Di+Tlc will be displaced toward lower P with the introduction of the Jd component in Cpx to form omphacite + talc. This assemblage is also observed in some Bixiling eclogites. Using the analyzed compositions of tremolite, omphacite (Di_{70}) and talc, and assuming ideal mixing models for amphibole and clinopyroxene, the position of this equilibrium was calculated at a P of about 1.5 GPa (Zhang et al. 1995a).

- (E). The occurrence of the eclogite assemblage Grt + Omp + Coe + Tlc ± Ky requires consideration of the additional components Al_2O_3 and FeO. Figure 17 shows calculated P-T relations for phases in the system $CaO-Al_2O_3-MgO-SiO_2-H_2O$, assuming a garnet composition of $Pyr_{75}Gr_{25}$ in the presence of excess SiO_2. Also shown are mineral assemblages for the Bixiling Tlc-bearing eclogite composition in various P-T fields. For this system, the assemblages Grt + Di + Tlc and Grt + Di + Ky are stable at higher P than that of the garnet-in univariant curve. At lower temperatures, kyanite + talc coexist with diopside + coesite, and the assemblage Ky + Tlc + Di occurs for the assumed composition. The garnet-in reaction, Tlc + Ky + Di = Grt, is located about 30°C lower than that for the end-member pyrope composition (dashed line in Fig. 17). The stability field of the assemblage Tlc + Grt + Di apparently is bounded by the garnet-in reaction and the upper-P limit of tremolite, and is restricted to Mg-rich bulk compositions (horizontally lined region of Fig. 17). As shown in this figure, the introduction of Jd component in clinopyroxene significantly displaces the upper-P limit of tremolite toward lower values. Thus, talc may appear in an eclogite assemblage at much lower P than that shown in Figure 17, so long as the bulk rock composition is appropriate (e.g. Liou et al. 1998b). Moreover, if FeO is considered, the phase relationships of Figure 17 will be significantly modified; eclogitic garnet will occur at lower T, and the assemblage Tlc + Ky + Omp + Grt + Coe, observed in the Bixiling eclogite, will be stable in a divariant P-T field.

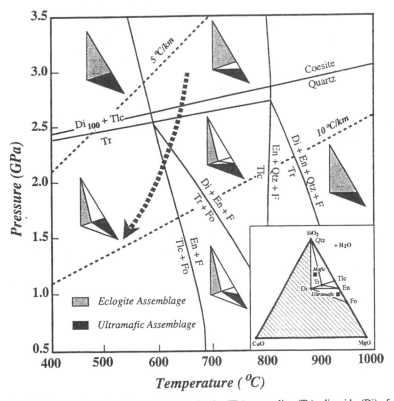

Figure 16. P-T diagram showing phase relations of talc (Tlc), tremolite (Tr), diopside (Di), forsterite (Fo), enstatite (En) and SiO_2 polymorphs calculated in the simple system $CaO-MgO-SiO_2-H_2O$ using GEO-CALC program of Berman (1988). Mineral parageneses for representative eclogite and ultramafic compositions together with geothermal gradients of 5°C /km and 10°C /km and estimated retrograded P-T path for Bixiling complex are also shown (after Liou and Zhang 1995).

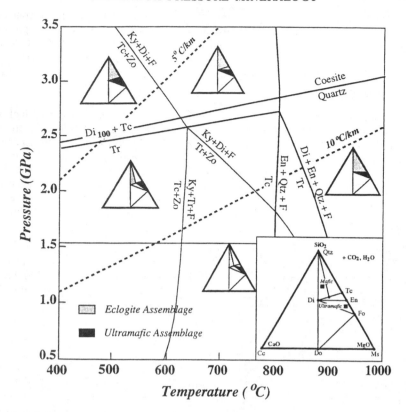

Figure 17. P-T diagram showing phase relations of talc (Tlc), tremolite (Tr), diopside (Di), zoisite (Zo), kyanite (Ky), garnet (Grt) and SiO$_2$ polymorphs calculated in the simple system CaO-MgO-Al$_2$O$_3$-SiO$_2$-H$_2$O using GE0-CALC program of Berman (1988) in the presence of excess SiO$_2$ phase. Garnet contains 25 mol % grossular and 75% pyrope. Univeriant lines involved pyrope end-member garnet are shown as dashed lines. Mineral assemblage Tlc + Di + Grt occurs only for Mg-rich composition as highlighted in a horizontally lined region. Mineral parageneses for calculated eclogitic bulk composition are also shown (after Liou and Zhang 1995).

Two reference geothermal gradients, 5°C/km and 10°C/km, are shown in Figure 16. The assemblage Tlc + Di + Coe in mafic rocks is stable, with Di + En + Fo in ultramafic rocks at VHP conditions, bounded by geothermal gradients of 5°C/km to 10°C/km. The talc-bearing ultramafic assemblage (Tlc + Fo + Di) occurs at upper mantle P at very low geothermal gradients less than 5°C/km, consistent with recent experiments by Bose and Ganguly (1993). The reaction Tlc = 3En + Coe + H$_2$O occurs at 830°C at ~3GPa and 750°C at 5 GPa, and talc is a VHP phase for ultramafic compositions.

Talc-bearing assemblages have also been described in other HP and VHP metamorphic terranes (e.g. Schreyer 1988b, Chopin et al. 1991, Zhang et al. 1997). The classic whiteschist assemblage Tlc + Ky from the Dora Maira Massif has a very broad stability field under H$_2$O-saturated conditions from 0.6 to probably 4 GPa (Schreyer 1988b). Tlc + Ky + Coe + pyrope from Mg-rich quartzites of the Dora Maria Massif formed at 3 GPa and 700-750°C (Chopin 1991). Whiteschist (Tlc + Ky + Grt + Amp) in the Kokchetav Massif shows a prograde P-T path defined by garnet compositional zoning and mineral inclusions. The inclusion assemblages, Qtz + Chl + Mrg, Ky + Zo + Chl, Chl + Mrg, and isolated zoisite occur in the mantle of the garnet, whereas Phl, Tlc, and Ky occur in the inner zone; garnet displays increasing pyrope and decreasing almandine content from core to rim. Talc inclusions as a pre-peak stage phase contain higher Al (0.17 p.f.u.) than matrix talc (0.09 p.f.u.). Its peak P probably falls in the diamond stability field at T > 720°C (Zhang et al. 1997). These VHP talc-bearing assemblages are mainly restricted to Mg-Al enriched mafic and pelitic rocks.

Ti-Clinohumite

Ti-clinohumite has been reported as a matrix mineral and as a fine-grained inclusion in garnet, pyroxene, olivine, and magnesite from marble and garnet peridotite in the Dabie-Sulu VHP terrane (Okay 1994, Zhang et al. 1994, Zhang and Liou 1994b, Zhang et al. 1995a, Ogasawara et al. 1998, Omori et al. 1998). TiO_2 and F contents of Ti-clinohumite range from 1.7 to 5.0 wt % and 0.05 to 1.05 wt %, respectively. In some Bixiling ultramafics, coarse-grained Ti-clinohumite was in textural equilibrium with magnesite and other ultramafic phases, but has partly broken down to ilmenite + olivine during exhumation. Ti-clinohumites also occur in garnet peridotites from the Kokchetav Massif (Zhang et al. 1997).

The published and observed parageneses and P-T estimates of Ti-clinohumite-bearing VHP rocks and the experimentally determined P-T field of this mineral (Engi and Lindsley 1980) all indicate that Ti-clinohumite is stable at upper mantle depths >100 km. This suggestion is consistent with the P-T estimate of ~740°C and >4 GPa for the Ti-clinohumite-bearing orthopyroxenite reported by Okay (1994).

The origin of Ti-clinohumite was reviewed by Dymek et al. (1988) and Gaspar (1992); this phase has been traditionally considered as a low-P mineral (Smith 1977, Mitchell 1979, Trommsdorff and Evans 1980). Dymek et al. (1988) claimed that Ti-chondrodite rather than Ti-clinohumite is a mantle phase inasmuch as Ti-clinohumite occurs as a metasomatic product during serpentinization of ultramafic rocks (Evans and Trommsdorff 1978, Medaris and Carswell 1990). However, Ti-clinohumite has been proposed as an important dense hydrous magnesium silicate in the Earth's mantle (Aoki et al. 1976, Thompson 1992) and may be stable at 150-300 km (5-10 GPa) and 700-1100°C (Ahrens 1989). It has been identified as an upper-mantle phase in ultramafic nodules from the Colorado Plateau (McGetchin and Besancon 1973). The maximum-P stability limit of Ti-clinohumite has not been determined; fluorine-free Ti-clinohumite breaks down to Ol + Ilm at 700°C and 2 GPa, and 800°C and 3 GPa (Engi and Lindsley 1980).

Nyböite and glaucophane

Occurrences of sodic amphibole such as nyböite and glaucophane as primary minerals in VHP rocks have been reported. For example, nyböite has been found in the Nybö eclogite from Norway (Smith 1988) and in coesite-bearing eclogite from the Sulu VHP terrane (Hirajima et al. 1992). Glaucophane coexists with pyrope, jadeite, phengite, and quartz (after coesite) in sodic whiteschist (Kienast et al. 1991), and occurs as inclusions in pyrope (Chopin 1987, Schertl et al. 1991), as secondary minerals along cracks in garnet, and along grain boundaries in garnet-jadeite-quartzite layers (Schertl et al. 1991) from the Dora Maira Massif. Glaucophane at the contact between talc and jadeite pseudomorphs has been proposed as a decompression product of the H_2O-conserving reaction Gln = Tlc + Jd (Chopin et al. 1991). In the Dabie Mountains, glaucophane porphyroblasts occurs in the matrix of coesite-bearing eclogites (Zhang and Liou 1994a) and exhibit clear compositional zoning; from core to rim, the Na^{M4} and Mg/(Mg+Fe) values decrease with concomitant increasing Ca, and the $Fe^{3+}/(Fe^{3+}+Al^{VI})$ ratio ranges from 0.09 to 0.12. The core composition (Na^{M4} = 1.68, Mg # = 0.83) represents that formed during the peak metamorphic stage within the Grt + Omp stability field, whereas the glaucophane rim

(Na^{M4} = 1.37, Mg# = 0.79) and barroisite porphyroblasts (Na^{M4} = 1.06-1.24 and Mg# = 0.87-0.89) may have formed during retrogression. Porphyroblastic nyböite in a Sulu jadeite eclogite contains higher Gln (Na^{M4} = 1.37-1.48) and Edn [$(Na + K)_A$ = 0.86-0.99] components in the core and has a taramitic rim composition (Zhang et al. 1995c). Nyböite is viewed as a stable phase during the peak metamorphism, because it is in textural equilibrium with garnet and omphacite; its high-P stability has been experimentally documented (Pawley 1992).

Compositions of some matrix glaucophanes in sodic whiteschists and glaucophane inclusions in pyrope from the Dora Maira Massif exhibit a calculated excess of cations in octahedral sites combined with a deficiency in the M^4 position (Chopin et al. 1991, Kienast et al. 1991). The maximum P-T condition for the sodic whiteschist is 680-750°C and >2.8 GPa, compatible with the P-T estimates of coesite eclogite of east-central China. Experimental studies and calculations using various thermochemical programs indicate that glaucophane is stable over a wide range of conditions. Holland (1988) calculated the reaction Gln = Tlc + Jd at 3.05 GPa and 600°C with a gentle positive P-T slope (Fig. 1). Similarly, the maximum P for the reaction, Jd + Grt + Tlc = Gln + Ky, is 3.3 GPa at about 640°C, and 3.1 GPa at about 750°C (Massonne 1995).

Discussion

The sources of volatiles and the extents of fluid-melt-rock interactions in subduction-zone metamorphism have been discussed extensively (e.g. Peacock 1990, 1991, 1993; Selverstone et al. 1992, Sharp et al. 1993, Philippot 1993, Getty and Selverstone 1994, Ernst et al. 1998). In subduction zones, substantial quantities of H_2O are released during progressive metamorphism, leading to eclogitization at depths of about 50-70 km. In especially cold subduction zones, the remaining volatile components may be transported to greater depths, up to 100-125 km, in hydrous minerals including phengite, glaucophane, epidote/zoisite and in less common, dense hydrous magnesian silicates (see Thompson 1992), especially if low temperatures promote P overstepping of the stability limits of those minerals. The scale of fluid flow in the subduction zone depends on the nature of the subducting lithosphere, the slab thermal structure, and other factors including rate of subduction. Cold subduction of sialic supracrustal rocks appear to be essential for the return of VHP metamorphosed rocks (Ernst and Liou 1995). In this section, we discuss the role of volatile-component-bearing phases during VHP metamorphism, implications for storage of fluid in the upper mantle, and the genesis of arc magmatism.

Low fluid contents of subducted supracrustal rocks. The initial supracrustal protoliths for VHP metamorphism may have been old, dry, and relatively cold (Coleman and Wang 1995). This suggestion is consistent with available geochronological data and with the low inferred thermobarometric gradients. Minor volatile contents of the protoliths would be in harmony with the cold subduction-zone occurrence of hydrous phases as described in previous sections.

Because VHP rocks from the Dabie-Sulu terrane experienced multistage deformation and recrystallization, it is difficult to estimate the amount of volatile components in the protoliths during the highest pressure metamorphism. Many VHP rocks have been analyzed and contain variable amounts of H_2O and CO_2. Some contain less than 0.5 wt % H_2O; this should be considered as the maximum H_2O content of the VHP rocks prior to their exhumation from mantle depths. Hydrous phases in aggregate constitute less than 5 vol% of most VHP rocks, and range in H_2O content from about 4 wt % for talc and phengitic mica through 3 wt % for epidote to 1-2 wt % for glaucophane. Our best estimate of the initial H_2O content of VHP rocks is between 0.1 to 0.3 wt %. This low H_2O content during VHP metamorphism in the Dabie-Sulu terrane is also supported by: (1) lack of coeval partial melting of supracrustal sialic rocks; (2) unusually low $\delta^{18}O$ values preserved in some VHP eclogites and associated quartzites; and (3) metastable preservation of protolithic textures and phase assemblages during prograde subduction-zone metamorphism. Similar incomplete forward reactions have been documented from the Bergen arcs by Austrheim (1987, 1998).

Storage of fluid in hydrous and dense, hydrous magnesium silicates. The occurrence of hydrous minerals in equilibrium with VHP eclogitic assemblages suggests that these lithologies retained minor volatile components during metamorphism. In a relatively low-T subduction zone, descent of mature, ~50 Ma old, oceanic lithosphere would result in the expulsion of fluids, initially abundant in the altered protoliths, at depths approaching 50-60 km (Peacock 1991); at deeper levels in the VHP environment, juicy supracrustal materials including metasediments may be partially melted to form pockets of melt of variable size, but only if aqueous fluid exists as a separate phase (Philipott 1993). Alternatively, in cold subduction zones characterized by rather dry, old continental crust, remaining volatiles may be stored in HP hydrous minerals such as zoisite/clinozoisite, glaucophane, phengite, talc, or possibly, lawsonite. These phases dehydrate at great depths; the expelled fluid migrates upward to shallow portions of the mantle wedge and would hydrate peridotites to form Ti-clinohumite, phlogopite, and clinoamphiboles. These garnet peridotites could be tectonically incorporated in the subducting slab; hydrous phases may transform into other dense hydrous magnesian silicates (DHMS) at greater depths. This process could result in direct transport of H_2O deep into the mantle (Thompson 1992, Maruyama and Okamoto, in press). However, subduction-zone geothermal gradients on the order of 1-4°C/km seems required to allow the preservation of DHMS.

Lack of partial melting and calc-alkaline magmatism. In subduction zones, the oceanic or continental crust-capped slab largely dehydrates before reaching melting conditions; partial melting may occur in the overlying mantle wedge as a result of regional-scale infiltration of an aqueous fluid derived from the devolatilization of clinoamphiboles in the downgoing slab (e.g. Peacock 1993, Philippot 1993, Ernst et al. 1998). However, the subduction of old, cold, dry supracrustal muscovite + biotite-bearing rocks provides only very limited devolatization and fluid flow at the proper depth (and temperature), and hence large-scale partial melting in the hanging wall is inhibited. In essence, the volatile-generating capacity of subducted supracrustal micaceous sialic materials at depth is too small to allow large-scale pervasive fluid flow into the overlying mantle wedge at subarc depths. Therefore, partial melting of the uppermost mantle and the overlying continental crust may not occur, and a calc-alkaline arc would not be produced.

The lack of a coeval magmatic arc along continental collision zones has long been recognized in the subduction of thick sialic crust. This feature is most apparent in the VHP terranes of the world shown in Figure 2. Evidently deep underflow of an ancient continental salient and storage of the subducted continental crust at great depths fails to provide enough fluid flux to generate melting of overlying mantle + continental crust.

Limited partial melting during VHP metamorphism has been suggested for the Dora Maira Complex (e.g. Massonne 1992) and the Kokchetav Massif (e.g. Shatsky et al. 1994). Estimated conditions of the Dabie-Sulu terrane are about 50 to 100°C higher than the granitic solidus (e.g. Huang and Wyllie 1975). In the presence of free H_2O, granitic gneiss will partially melt; this liquid could be trapped as melt inclusions or accumulate and migrate upwards as magma bodies. Minor inclusions of Kfs + Qtz ± Ab in garnet with radial fractures might represent melt inclusions; however, rare occurrences and different inclusion assemblages for a given sample disfavor this possibility. The negative $\delta^{18}O$ values for VHP minerals from both eclogitic rocks and the adjacent metasediments described above suggest extremely limited or no interaction with an oxygen-bearing fluid (Rumble 1998). Thus, an aqueous fluid phase is not ubiquitous or ever widespread in recognized VHP complexes.

CHIEFLY DABIE-SULU VHP CARBONATE PHASES

Experimental studies indicate that magnesite and dolomite should be the principal stable carbonates under mantle conditions, and may carry carbon in subducting plates into the deep upper mantle (Kushiro et al. 1975, Brey et al. 1983, Blundy et al. 1991, Kraft et al. 1991, Gillet 1993, Redfern et al. 1993). Reported occurrences of magnesite in mantle-derived kimberlites and VHP metamorphic ultramafics and eclogites support these experimental results (Lappin and Smith 1978, Okay 1993, Yang et al. 1993, Zhang and Liou 1994b, Zhang et al. 1995a). The discovery of coesite inclusions in dolomite from VHP rocks from the Dabie Mountains (Schertl and Okay 1994, Zhang and Liou 1996) provides additional evidence that some continental sediments have been subducted to depths of >100 km, and that dolomite is a stable phase at mantle depths. Inclusions of calcite pseudomorphs after aragonite have also been reported in coesite-bearing marbles from the Dabie Mountains (Wang and Liou 1993, Zhang and Liou 1996). Occurrence, parageneses, and compositions (Table 3) of these carbonates from the Dabie terrane, and calculated phase relations are discussed below.

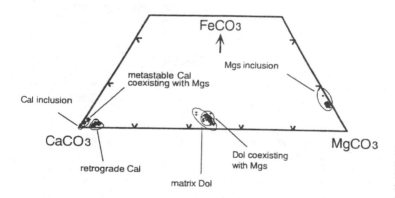

Figure 18. Compositions of carbonates in Dabie dolomitic marbles (after Omori et al. 1998).

Magnesite

Magnesite was identified as rounded to subhedral matrix crystals and as fine-grained inclusions in garnet and clinopyroxene in both VHP ultramafics and eclogites from the Dabie-Sulu terrane (Okay 1994, Yang et al. 1993, Zhang and Liou 1994b, Zhang et al. 1995a, Liou and Zhang 1998). Coarse-grained magnesite is in equilibrium with Fo, En, Grt, Di, and Ti-clinohumite in garnet peridotites, is nearly Ca free (0.12-0.13 wt %), and has a high Mg/(Mg + Fe) ratio of 0.93-0.96, as shown in Figure 18 (Zhang and Liou 1994b, Omori et al. 1998). Most magnesites from VHP ultramafics are rimmed by dolomite ± chlorite or periclase ± dolomite, and may have formed at P approaching 5-6 GPa and 800-900°C (Zhang and Liou 1994b). Magnesite in eclogite coexisting with Grt, Omp, Coe, and Rt contains substantial Fe and very little Ca ($Mg_{0.74}Fe_{0.25}Ca_{0.01}CO_3$), and is rimmed by dolomite. Euhedral garnet porphyroblasts show radial fractures around inclusions of magnesite. Magnesite inclusions in dolomite from dolomitic marble show concentric zoning from magnesite cores through dolomite to calcite rims; a few are rimmed by calcite aggregates, as shown in Figure 19.

Table 3. Representative compositions of VHP carbonates.*

UHP rock	Ecl	Ecl	Ecl	Ecl	Grt Peri	Grt Peri	Grt Peri	Ecl	Ecl	Ecl	Marble	Marble
Mineral:	Mgs	Mgs	Dol	Mgs	Mgs	Mgs	Mgs	Mgs	Dol	Cal	Mgs	Dol
SiO_2	0.00	0.07	0.00	0.00				0.00	0.00	0.00		
TiO_2	0.02	0.00	0.00	0.02				0.00	0.00	0.00		
Cr_2O_3	0.00	0.00	0.02	0.00								
Al_2O_3	0.00	0.00	0.02	0.00				0.00	0.00	0.00		
FeO	5.54	21.02	6.11	5.54	7.58	6.68	7.77	17.88	5.08	1.76	10.98	4.16
MnO	0.06	0.09	0.07	0.06				0.18	0.03	0.20	0.00	0.17
MgO	45.46	35.20	18.99	45.46	59.64	58.51	61.38	32.57	18.27	2.44	38.80	16.78
CaO	0.13	0.56	30.02	0.13	0.10	0.09	0.07	1.02	28.93	53.37	0.66	31.57
Na_2O	0.01	0.01	0.02	0.01				0.01	0.03	0.00		
K_2O	0.01	0.00	0.00	0.01								
Total	51.22	56.95	55.24	51.22	67.32	65.42	69.22	51.66	52.34	57.77	50.44	52.68
Si	0.00	0.00	0.00	0.00				0.00	0.00	0.00		
Ti	0.00	0.00	0.00	0.00				0.00	0.00	0.00		
Cr	0.00	0.00	0.00	0.00								
Al	0.00	0.00	0.00	0.00				0.00	0.00	0.00		
Fe	0.06	0.25	0.08	0.06	0.07	0.06	0.07	0.23	0.07	0.02	0.14	0.06
Mn	0.00	0.00	0.00	0.00				0.00	0.00	0.00	0.00	0.00
Mg	0.94	0.75	0.43	0.93	0.93	0.94	0.93	0.75	0.43	0.06	0.85	0.40
Ca	0.00	0.00	0.49	0.00				0.02	0.50	0.92	0.01	0.54
Na	0.00	0.00	0.00	0.00				0.00	0.00	0.00		
K	0.00	0.00	0.00	0.00				0.00	0.00	0.00		
Total	1.00	1.00	1.00	1.00	1.00	1.00	1.00	1.00	1.00	1.00	1.00	1.00
Ref.	1	1	1	1	3	3	3	2	2	2	4	4

* The formulae of magnesite and dolomite were calculated on the basis of 1 cation as no C analysis.

Ref.: 1. Zhang et al. (1995a); 2. Zhang and Liou (1996); 3. Zhang et al. (1994); Omori et al. (1998)

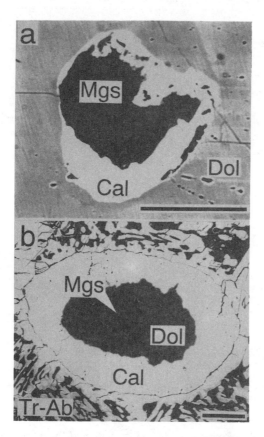

Figure 19. Back-scattered electron images of magnesite inclusions (a) in dolomite, and (b) in Tr-Ab symplectite after omphacite (scale bar = 50 μm) (after Omori et al. 1998). (a) Calcite in direct contact with magnesite, both minerals are surrounded by an almost homogeneous single grain of dolomite. (b) Concentric zoning of minerals around magnesite core. Dolomite exists between the magnesite cores and calcite rim.

Recent experimental studies of magnesite stability under mantle conditions indicate that $MgCO_3$ is a stable carbonate, acts as a major host for carbon storage in the Earth's upper (and lower) mantle, and can be a carrier for carbon in subducting plates (Kushiro et al. 1975, Brey et al. 1983, Blundy et al. 1991, Gillet 1993, Redfern et al. 1993). Coesite relics in magnesite-bearing eclogites, rare occurrences of magnesite in kimberlites (McGetchin and Besancon 1973) and in ultramafic diatremes (Smith 1979, 1987), together with other reported occurrences of magnesite in similar VHP rocks (e.g. Smith 1988), support experimental data for magnesite stability at mantle depths.

The stability field of Mgs + Di + En in the CaO-MgO-SiO_2-CO_2-H_2O system strongly depends on the X_{CO_2} of the fluid phase. Calculated phase relations of magnesite, diopside, enstatite, talc, tremolite, and olivine shown in Figure 20 indicate that the VHP metamorphism responsible for the Bixiling Complex took place under very low X_{CO_2} conditions (Ogasawara et al. 1994, 1995; Zhang et al. 1995a). For magnesite-bearing ultramafic rocks, thermobarometry yields a high P of 4.5-6.5 GPa at about 800°C (Zhang et al. 1995a).

Dolomite

Dolomite was identified as a retrograde phase after magnesite in VHP ultramafics, eclogites, and marbles from the Dabie-Sulu region. However, dolomite from a calc-silicate

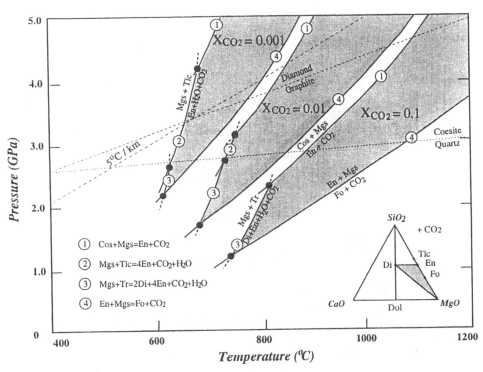

Figure 20. P-T diagram showing magnesite-enstatite-diopside stability fields at $X_{CO_2} = 0.1$, 0.01 and 0.001 for a model system $CaO-MgO-SiO_2-H_2O-CO_2$. Phase relations were calculated using non-ideal mixing of H_2O-CO_2 and the data set of Holland and Powell (1990) (modified after Ogasawara et al. 1992, Zhang et al. 1995).

rock interlayered with marble and coesite-bearing eclogites has recently been reported to contain inclusions of coesite partially reverted to quartz (Fig. 15D) (Schertl and Okay 1994, Zhang and Liou 1996). Dolomite has an average composition of $Ca_{47-52}Mg_{38-45}Fe_{07-09}$ and coexisting phengite has a maximum Si-content of 3.60 p.f.u. Coesite inclusions in dolomite indicate that the assemblage Do + Coe is stable at P > 2.8 GPa and at 800 ± 50°C, consistent with the experimentally determined high-P stability of dolomite (Kraft et al. 1991, Ross and Reeder 1992).

Coesite inclusions in dolomite occur in Dabie carbonate-bearing eclogite; the parent rock contains Grt (70%), Qtz (10%), carbonates (10%; magnesite, dolomite and calcite), Omp (5%), Rt (2-3%) and minor Phn, Ap, and zircon (Zhang and Liou 1996). Abundant inclusions of coesite and coesite pseudomorphs occur in the garnet; estimated P-T conditions for the peak metamorphism are 760°C and >2.8 GPa, with X_{CO_2} between 0.01 and 0.1. Dolomite is the major carbonate phase in the matrix; some grains of dolomite are rimmed by calcite. Minor magnesite occurs as either inclusions in dolomite or in the matrix. The inclusions in dolomite lack radial fractures; this may be related to the good cleavage of dolomite (Schertl and Okay 1994).

Dolomite occurs as a major phase in Kokchetav diamond-bearing dolomitic marbles which contain garnet and diopside characterized by Kfs lamellae. The original diopside evidently contained high K_2O (up to 1.56 wt %, Zhang et al. 1997). Dolomitic marbles in the Dabie-Sulu terrane range from silicate-free white marble through impure dolomite to calc-silicate rocks; these lithologies are interlayered at various scales and contain eclogite boudins. Dolomitic marbles within a single outcrop have assemblages of the major constituent minerals: (a) Mg-Cal + Do + Ol, (b) Mg-Cal + Do + Di + Ol, (c) Mg-Cal + Do + Di, (d) Mg-Cal + Do + Ti-Chu, (e) Mg-Cal + Ti-Chu + Di, and (f) Mg-Cal + Do + Ti-Chu + Di + Ol (Ogasawara et al. 1994, 1998, Omori et al. 1998). Such variation in assemblages suggests an irregular distribution of X_{CO_2} within individual outcrops.

Calcite pseudomorphs after aragonite

Wang and Liou (1993) described inclusions of calcite pseudomorphs in garnet and omphacite in Dabie marbles, and discussed the stability of aragonite under VHP conditions. Composite inclusions of Cc + Qtz aggregates after coesite in garnets from Dabie dolomite-bearing eclogites and Kokchetav garnet-zoisite gneisses were recognized by Zhang and Liou (1996) and Zhang et al. (1997); inclusions in the Dabie eclogite have well developed radial fractures, and the quartz pseudomorphs show palisade texture. Calcite aggregates are inferred to be an inversion product after aragonite, inasmuch as it is well known that this polymorph readily inverts to calcite during decompression (e.g. Carlson and Rosenfeld 1981, Ernst 1988). Inclusions of the inverted calcite aggregates after aragonite in marbles have very similar textures to quartz pseudomorphs after coesite inclusions in garnet and omphacite in eclogite.

Phase relations

Experimental studies have shown that aragonite, dolomite, and magnesite are stable under HP and VHP conditions. Five observations summarized below confirm that these carbonate phases are present in the coesite and perhaps even in the diamond P-T fields: (1) the presence of the magnesite + diopside assemblage in VHP ultramafic rocks (Zhang and Liou 1995a); (2) the coexistence of dolomite with magnesite, coesite, garnet, and omphacite; (3) inclusions of calcite ± quartz pseudomorphs after inferred aragonite ± coesite in garnet (Wang and Liou 1993, Zhang and Liou 1996, Zhang et al. 1997); (4) inclusions of coesite within dolomite (Schertl and Okay 1994, Zhang and Liou 1996); and (5) coexistence of diamond with dolomite as inclusions in garnet from marble of the Kokchetav Massif (e.g. Zhang et al. 1997). Magnesite is stable at even higher P than dolomite inasmuch as dolomite breaks down in the presence of $MgSiO_3$ to form magnesite and diopside at high P (e.g. Redfern et al. 1993).

A petrogenetic grid for VHP metamorphism in the model system $MgO-CaO-SiO_2-H_2O-CO_2$ has been proposed by Ogasawara et al. (1998); the occurrence of dolomite or magnesite is highly dependent on X_{CO_2} as shown in Figure 20. The coexistence of dolomite + coesite is limited in the model system by the reaction $Do + 2Coe = Di + 2CO_2$. Using the estimated P-T conditions of peak metamorphism for the Dabie eclogite of 760°C and >2.8 GPa, the stable coexistence of Coe + Do + Omp ($X_{Di} = 0.5$) requires $0.01 < X_{CO_2} < 0.1$. For this range of estimated P-T-X_{CO_2} conditions, magnesite is stable with coesite. Based on the analyzed compositions of dolomite and omphacite, and unit activity of coesite, T-X_{CO_2} relations for the reaction $Do + 2Coe = Di + 2CO_2$ were calculated for the coesite-bearing marble by Wang and Liou (1993). At T = 630-760°C and P = 3.0 GPa, they estimated the X_{CO_2} for this reaction at < 0.03, consistent with the calculation shown in Figure 20.

Independent study of calc-silicate rocks, eclogites, and impure marbles from several outcrops in the Dabie Mountains yields consistent mineral parageneses and P-T estimates (Wang et al. 1992, Omori et al. 1998). Quartz aggregates in the matrix of the calc-silicate rock are similar to the retrograde products of coarse-grained matrix coesite from the Dora Maira Massif (Chopin et al. 1991). Field relations and petrologic data suggest that eclogites, marbles, and country rocks were subjected to coeval VHP metamorphism: supracrustal rocks of the continental crust were subducted to mantle depths of >100 km, and dolomite and magnesite were stable under such conditions.

EXSOLUTION LAMELLAE IN DABIE-SULU AND OTHER ROCKS

Exsolution textures are common in VHP minerals from both eclogitic and ultramafic rocks; these include (a) ilmenite rods in olivine and clinopyroxene, (b) magnetite plates or rods in olivine and clinohumite, (c) quartz rods in omphacite, (d) monazite lamellae in apatite (Zhang et al., in press), and (e) rutile needles in garnet (Zhang et al. 1996a, 1997, in press; Zhang and Liou 1998, Hacker et al. 1997). Although mechanisms for mineral exsolution in VHP phases are unknown, each lamellae-bearing host mineral preserves information on the composition and conditions of formation of the homogenous precursor phase, and the inferred P-T path during decompression. Compositional and structural characterization of lamellae-host mineral pairs in turn will encourage experimentation to delineate the P-T conditions and mechanism for the formation of the homogeneous phase and the later annealing under lower grade conditions.

Exsolution of lamellae in VHP minerals may have taken place during nearly isothermal decompression, in contrast to exsolution with falling temperature in primary igneous minerals. The effect of P drop on the genesis of exsolution lamellae, and the significance of their occurrences need to be systematically investigated. However, before experimental studies on VHP minerals are undertaken, characterization of these exsolution features is necessary. Identified exsolved lamellae and host mineral pairs from VHP eclogitic and ultramafic rocks are described below. Thus far, descriptions of such lamellae are of a reconnaissance nature. We hope that this review will lead mineral physicists and experimental geochemists to pursue systematic studies of compositions and structures of host mineral and lamellae pairs, and stabilities of mantle minerals under deep-seated conditions. Such investigations will bridge the gap between crustal petrogenesis and mantle processes of concern to many Earth scientists.

Ilmenite rods in olivine

Micron-sized ilmenite rods in olivine have been recently identified in garnet peridotites from several VHP terranes. Such occurrences were overlooked prior to the paper by Dobrzhinetskaya et al. (1996); these authors proposed that the Alpe Arami garnet lherzolite in the Central Alps may have been metamorphosed under deep upper mantle conditions, based on the discovery of (1) micron-scale $FeTiO_3$ rods in olivine, (2) a unique lattice preferred orientation of olivine, and (3) reported TiO_2 contents of olivine of 7,000-20,000 ppm. TEM study revealed that the titanate rods are topotactic with host olivine, parallel to its [010] direction, and have crystallized with four structures: ilmenite, and three previously unrecognized phases interpreted as intermediate between ilmenite and the dense perovskite structure. Because all previously reported occurrences of $FeTiO_3$ inclusions in olivine are reported as ilmenite plates, the Alpe Arami rods are distinctive and were hypothesized to have exsolved at 10-15 GPa (300-450 km) as perovskite, followed by variable conversion to ilmenite. Green and Dobrzhinetskaya (1997) further noted that the unique preferred orientation of Alpe Arami olivine may have formed during recrystallization of a wadsleyite or ringwoodite-bearing protolith. Since the electrifying report by Dobrzhinetskaya et al. (1996), at least three other independent groups have begun working on such assemblages. If a garnet peridotite body such as the Alpe Arami complex indeed was derived from the mantle transition zone, its occurrence would have profound significance.

TiO2 content of olivines from Sulu garnet peridotites. Hacker et al. (1997) reported a new occurrence of ilmenite rods in olivine from the Chijiadian garnet lherzolite in the Sulu terrane. This garnet peridotite (Ol + En + Di + Grt) experienced a metamorphic P of at least >3 GPa, as indicated by the presence of coesite in eclogite enclosed within the lherzolite (Zhang et al. 1994). The compositions of the ultramafic phases give no indication of transition-zone P: Al partitioning between enstatite and garnet indicates a P of 4.1 ± 0.6 GPa (Zhang et al. 1994), and the distribution of Ca between olivine and diopside suggests ~4.4 GPa. Fe-Mg partitioning among garnet, enstatite, and diopside indicates a T of ~875 ± 50°C. The same sorts of measurements on the Alpe Arami garnet lherzolite have been interpreted to indicate similar maximum P ~ 4 GPa for this Alpine garnet peridotite (Ernst 1977, 1978).

The olivine grains in the Chijiadian lherzolite are characterized by a fairly homogeneous distribution of rod-like inclusions (Fig. 21A; also see Fig. 2 of Hacker et al. 1997). The rods with composition $Fe_{0.82}Mg_{0.15}Mn_{0.03}TiO_3$ are more magnesian than the Alpe Arami iron titanates, which are $Fe_{0.94}Mg_{0.06}TiO_3$. The rods are nearly square prisms 0.2-0.5 μm across, whereas the third dimension is typically ~20 μm; these prisms are smaller than the Alpe Arami rods. TEM images indicate that inclusions in the Chijiadian olivines also have long axes parallel to [010] of the host olivine crystals. Thus the ilmenite rods at Alpe Arami are not an isolated occurrence.

Hacker et al. (1997) measured the TiO_2 content of Chijiadian and Alpe Arami olivines directly by electron probe microanalysis and secondary-ion mass spectrometry. The bulk TiO_2 content of the Chijiadian olivine is 269 ± 188 ppm as determined by the defocused beam method, and the TiO_2 content of inclusion-free portions of the olivine grains is 100 ± 20 ppm, as determined by secondary-ion mass spectrometry. Equivalent analysis of Alpe Arami olivine revealed bulk olivine TiO_2 contents of ~220 ppm. Hence, the TiO_2 contents of Alpe Arami and Chijiadian olivines are not 7,000-20,000 ppm, but within the range of values reported from worldwide garnet lherzolite xenoliths (Hervig et al. 1986) and cannot be used to argue that Alpe Arami had its origin in the mantle transition zone. Less than about 300 ppm of TiO_2 contents was independently obtained by Nakajima and Ogasawara (1997) for olivines from the Rongcheng garnet peridotites of the Sulu terrane, with estimated P-T conditions at 4-6 GPa and 780-820°C (Hiramatsu et al. 1995).

Electron diffraction patterns of FeTiO3 rods in olivine. The second argument made by Dobrzhinetskaya et al. (1996) for an ultradeep origin is that some iron titanate rods have previously unrecognized crystal structures. Their electron diffraction patterns were subsequently explained by dynamical diffraction between overlapping olivine and ilmenite structures, without requiring novel $FeTiO_3$ structures (Hacker et al. 1997). The topotaxial relations of ilmenite rods in Alpe Arami olivine were also examined by Risold et al. (1997) using selected-area electron diffraction. The unrecognized (ortho-rhombic) structure (β-ilmenite) reported by Dobrzhinetskaya et al. (1996) could be indexed as a trigonal structure. Rislod et al. (1996) measured less than 400 ppm TiO_2 for olivines from both Alpe Arami and Cima di Gagnone in the Central Alps, and concluded that the ilmenite rods formed as exsolution from Ti-rich olivine during cooling and decompression from about 3 GPa, 800°C.

Experimental study of TiO2 solubility in olivine. Three independent experimental studies of the solubility of TiO_2 in olivine as a function of P have been undertaken. Phase relations relevant to titanate, olivine, and enstatite are shown in Figure 22. Dobrzhinetskaya et al. (1996) and Green et al. (1997) used natural olivine + ilmenite as starting materials, and conducted experiments at 1700°K in the α-olivine field at 6 GPa and in the β-phase field at 14 GPa. Both experiments produced abundant melt from which olivine + ilmenite symplectites crystallized. The 6 GPa olivines contain 0.15 to 0.25 wt % TiO_2, whereas olivine polymorphs at 14 GPa contained 0.5 to 0.7 wt % TiO_2. The later

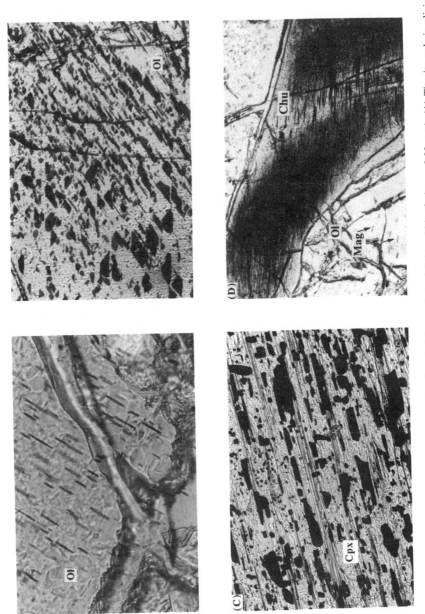

Figure 21. Photomicrographs of exsolution lamellae in VHP minerals (plane polarized light; width of view = 0.32 mm). (A) Titanite rods in olivine from Ronychen peridotite (Zhang and Liou 1998). (B) Magnetite lamellae exsolved from Dabie harzburgitic olivine (Zhang et al. in press). (C) Ilmenite lamellae plates in clinopyroxene from Sulu pyroxenite (Zhang and Liou 1998). (D) Magnetite lamellae exsolved in Ti-clinohumite from Dabie harzburgitic olivine (Zhang et al. in press).

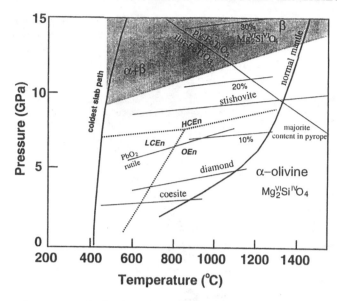

Figure 22. P-T diagram showing phase relations relevant to titanates, olivine, enstatite, rutile for garnet peridotite and eclogite after Akaogi et al. 1989, 1992, 1995), Pacalo and Gasparik (1990) and Mehta et al (1994). Coldest subduction zone and normal mantle geotherms after Hellfrich et al. (1989) and Mercier and Carter 1975). The 10, 20, 30% isopleths of majorite content in pyrope are after Gasparik (1990). The polymorphs of orthoenstatite, (OEn)-high-P, clinoentatite, (CEn)-low-P, clinopyroxene, (LEn), are after Angel and Hugh-Jones (1994).

values approach the lower range of the estimated TiO_2 contents of Alpe Arami olivine suggested by them and support their idea that these olivines are of ultradeep origin.

Ulmer and Trommsdorff (1997) used synthetic magnesio-ilmenite (X_{Mg} = 0.6) and an oxide mix of olivine (X_{Mg} = 0.9) in the proportions 5:95 respectively (+ brucite as the source of H_2O) as starting materials, and ran experiments at 1-10 GPa and 1200-1600°C. These P-T conditions cross the reaction $Mg_2SiO_4 + TiO_2 = MgSiO_3 + MgTiO_3$, thus leading to the paragenesis Ol + Rt + Ilm at 1-6 GPa and Ol + En + Ilm at 9 GPa. Coarse-grained homogeneous olivine crystals >150 μm in size were produced. Olivine contains a constant amount of TiO_2, 0.135 ± 0.01 wt % at 1400°C, and 0.105 ± 0.01 wt % at 1200°C; extrapolating these values to 800°C results in a TiO_2-content of olivine saturated with ilmenite and/or rutile on the order of 0.04-0.06 wt %, corresponding to the amount measured by Risold et al. (1997). These experiments demonstrate the lack of P dependence of TiO_2 in α-olivine; rather, the TiO_2 content of olivine increases with increasing T.

Okamoto et al. (1997) conducted reconnaissance experiments at 6, 8, and 10 GPa at 1200°C and 13 GPa at 1300°C using 90 wt % San Carlos olivine + 10 wt % rutile. Recrystallized olivines contain about 0.25 wt % TiO_2, and Mg-ilmenite formed in 13 GPa products. One experiment at 15 GPa yielded β-olivine containing about 0.15 wt % TiO_2. Although higher Ti solubilities in α-olivine were reported than values determined by Ulmer and Tromsdorff (1997), they are far less than those reported by Green et al. (1997).

Magnetite lamellae in olivine and clinohumite

Occurrence. Magnetite lamellae in olivine have been described from a few mafic-ultramafic rocks and are associated with other silicate phases (Goode 1974, Ambler and Ashley 1977, Moseley 1984, Arif and Jan 1993). Magnetite exsolution lamellae in olivine and Ti-clinohumite from a Dabei harzburgite and a garnet pyroxenite have been recently identified (Zhang et al., in press). Light-brown magnetite lamellae are common in olivine grains, except along their margins. The lamellae vary from a trace to 1.5 vol%, based on backscattered electron images, and are 3 to 75 μm long, 2-30 μm wide, and 0.5-2 μm thick (Fig. 21B). These thin magnetite lamellae consist mainly of Fe with 2-3 wt % MgO, 0.73 wt % NiO and 0.45 wt % SiO_2; minor MgO and SiO_2 may reflect excitation from the host olivine. Ti-clinohumite grains 0.1 to 1.5 μm in size in Dabie harzburgite and ortho-pyroxenite also exhibit exsolved textures (Fig. 21D). Abundant black to deep-brown magnetite lamellae are restricted to the core and mantle of the Ti-clinohumite crystals, and show a distinct preferred orientation. These Ti-clinohumites are not replaced by ilmenite + olivine. The mechanism of magnetite exsolution in Ti-clinohumite is not known.

Crystal structure. Cell parameters and crystallographic orientations of magnetite lamellae and host olivine from the Dabie VHP rocks were determined (Zhang et al., in press). The host olivine has orthorhombic cell parameters: $a = 6.007(5)$ Å, $b = 10.249(8)$ Å and $c = 4.757(2)$ Å, $V = 239.42(18)$ Å^3. The magnetite is cubic: $a = 8.411(2)$ Å, and $V = 594.95(38)$ Å^3. The topotaxial intergrowth between olivine and magnetite is

$$[220]_{Mag} \| [200]_{Ol} \quad [111]_{Mag} \| [3\bar{3}1]_{Ol} \quad [11\bar{1}]_{Mag} \| [331]_{Ol} \quad [242]_{Mag} \| [2\bar{2}0]_{Ol} .$$

Such topotaxial intergrowths and homogeneous distributions of magnetite lamellae in host olivine suggest an exsolution origin.

Discussion. Formation of magnetite or a symplectic intergrowth of magnetite with silica or a silicate mineral in olivine from igneous rocks has been interpreted as due to serpentinization (Arif and Jan 1993), as due to decomposition during cooling of Fe^{3+}-bearing high-T olivine (Moseley 1984), or as due to oxidation (Haggerty and Baker 1967, Champness 1970, Putnis 1979, Rietmeuer 1996). Among these hypotheses, oxidation of olivine has been commonly used to explain such intergrowths in olivine from gabbro and mafic-ultramafic cumulate complexes. Experimental studies of olivine oxidation also have revealed that oriented hematite- and magnetite-like phases + amorphous silica are precipitated in Mg-Fe olivine at 500-800°C, and coarser Fe-oxide + a more ordered silica phase is produced at 1000°C (Champness 1970). In both natural occurrence and experimental results, the magnetite lamellae are invariably associated with silica or a silicate mineral, such as orthopyroxene or, rarely, andradite-rich garnet (Champness 1970, Blanchard and Cunningham 1974, Putnis 1979, Rietmeuer 1996). For an oxidation processes, a topotaxial relationship between magnetite lamellae and the host olivine is not required. The symplectic exsolution of magnetite and augite in olivine from the Rhum pluton, the Cuilin and the Bushveld complexes, and the Skaergaard intrusion has been suggested as due to decomposition of high-T Fe^{3+}-bearing olivine during cooling (Moseley 1984). These olivines have wide compositional ranges (Fo_{86-0}) and contain ferric iron; the reaction of $3 Fe_{4/3}^{3+}SiO_4 + Fe_2SiO_4 + 4 X_2SiO_4 = 2 Fe_3O_4 + 4 X_2Si_2O_6$ (here X = Ca, Mg, Fe) was held responsible for such intergrowths. The silicate phase constitutes twice the magnetite in the reaction product, and is roughly consistent with estimates from TEM micrographs.

In contrast, the host olivine with magnetite exsolution lamellae in the Dabie VHP harzburgite is characterized by: (1) a forsterite-rich (Fo_{93}) composition, (2) a topotaxial relationship with coarse-grained oriented magnetite lamellae, (3) an even distribution of magnetite lamellae, and (4) lack of an accompanying silicate phase. These magnetite lamellae are confined to the cores and mantles of host olivines with homogenous bulk compositions. In addition, they are crosscut by very thin fractures filled with fine-grained serpentine + randomly-oriented magnetite. Such a relation suggests that these lamellae formed earlier than brittle deformation and incipient serpentinization of the olivine. These features cannot be explained by serpentinization inasmuch as serpentinization was initiated along grain boundaries and fractures, and the resulting magnetite dust is randomly distributed throughout the bulk of the olivines.

Zhang et al (in press) suggest exsolution of magnetite from a very HP olivine phase as a spinel solid solution: Fe_3O_4-$(Fe,Mg)_2SiO_4$. Although Fe^{3+} and Cr^{3+} of intermediate ionic radii prefer octahedral sites, Fe^{3+} may have been accommodated in tetrahedral sites by the substitution of $2Fe^{3+}$ for $(Mg,Fe)^{VI}Si^{IV}$ at high T and P. Olivine could contain 1-1.5 vol% magnetite; this implies that the original olivine could have contained 1.0 wt.% Fe_2O_3. Such a high Fe_2O_3 content is difficult to accommodate in the forsterite structure at low P. If the original "olivine" was actually wadsleyite (β phase) with a distorted-spinel structure, it would readily accommodate Fe_3O_4 as spinel solid solution along the binary join Fe_3O_4-$(Fe,Mg)_2SiO_4$. The β phase is stable at P > 13 GPa, according to experimental studies of the olivine-spinel transformation in the system Mg_2SiO_4-Fe_2SiO_4 for forsterite (Fo > 90%) (Ito and Takahashi 1989, Akaogi et al. 1989, Katsura and Ito 1989).

It is possible that transformation of the wadsleyite to olivine occurred during decompression, and the excess Fe_3O_4 exsolved to form lamellae. The observed topotaxial relations between the magnetite lamellae and olivine are consistent with such an exsolution process. Ross et al. (1992) synthesized a spinelloid phase, $Fe_{5.2}Si_{0.8}O_8$ ($Fa_{40}Mt_{60}$) at 1200°C, 7 GPa. More recent experiments in the system Fe_2SiO_4-Fe_3O_4 by Woodland and Angel (1998) yield a new spinelloid phase ($Fe_{2.45}Si_{0.55}O_4$) with the wadsleyite structure, suggesting that significant amounts of Fe^{3+} can be incorporated into a wadsleyite-type structure through substitution of $2 Fe^{3+} = Fe^{2+} + Si^{4+}$. The difference in the stability of Fe^{3+}-substituted wadsleyite (5.0 to 6.0 GPa at 1100-1200°C) and $(Mg,Fe)_2SiO_4$ wadsleyite (i.e. 12 GPa at 1200°C) suggests that the addition of Fe^{3+} could act to stabilize $(Mg,Fe)_2SiO_4$ wadsleyites to lower P compared with the Fe^{3+}-free system; such a relationship could affect the depth of the 410-km seismic discontinuity in the mantle (Woodland and Angel 1998).

Ilmenite rods in clinopyroxene

Two distinct clinopyroxenes occur in some relatively fresh olivine garnet pyroxenites (Di + Ol + Grt + Ilm) from the Sulu VHP terrane. Most matrix-forming diopsides have a uniform grain size of 0.3-0.5 mm, contain inclusions of ilmenite and Ti-clinohumite, and are less than 0.1 wt % in TiO_2 content. However a few coarse-grained diopside porphyroblasts contain exsolved ilmenite and Mg-Al-Cr titanomagnetite rods (Zhang and Liou 1998) (Fig. 21C). Based on the volumes and compositions of the exsolved phase and host, the original diopside may have contained 0.5-1.0 wt % TiO_2; such Ti-rich diopside has been experimentally synthesized (e.g. I. McGregor, pers. comm. 1997).

Abundant coarse, lamellar intergrowths of ilmenite and clinopyroxene have been described from kimberlite and ultramafic nodules from Lesotho and South Africa (e.g. Dawson and Reid 1970, Ringwood and Lovering 1970, Boyd and Nixon 1973). Clinopyroxene in these intergrowths forms single crystals up to 20 cm in diameter and is sieved with exceedingly regular, coarse laminations of ilmenite. These intergrowths may have formed by the breakdown of a VHP Precambrian Cpx (Ringwood and Lovering 1970, Boyd and Nixon 1973) or by eutectoid crystallization from kimberlitic magma (e.g. Wyatt 1977). Figure 21C shows lamellar intergrowths of ilmenite and diopside from a VHP garnet pyroxenite from the Sulu terrane. The ilmenites possess finer lamellae and have higher Fe contents (21-24 mol % $MgTiO_3$) than Mg-ilmenites in kimberlitic intergrowths (35-50 mol % $MgTiO_3$).

K-feldspar lamellae in clinopyroxene

One of the most significant compositional characteristics of some clinopyroxene crystals from kimberlite and diamond-bearing rocks is their substantial K_2O contents (Sobolev et al. 1971, Sobolev and Shatsky 1989, Harlow and Veblen 1991, Shatsky et al. 1995, Zhang et al. 1997). In fact, the K_2O content of Ca-Na pyroxenes has been considered to be a potential barometer for VHP rocks (e.g. Schmidt 1996, Okamoto and Maruyama 1998).

100 μm

Figure 23. K-feldspar lamellae in diopside from diamond-bearing dolomitic marble of the Kokchetav Massif (crossed polarizers; width of view = 0.76 mm) (Zhang et al. 1997).

Occurrence and K_2O content of clinopyroxene. K-feldspar lamellae exsolved from host clinopyroxene in both matrix and as inclusions in garnet have been described by Sobolev and Shatsky (1990) from Kokchetav diamondiferous rocks; the estimated K_2O content of the pre-exsolved Cpx is up to 1.55 wt %. Similar Cpx inclusions have been reported from eclogitic garnets in kimberlites (Harlow and Veblen 1991). Most analyzed Kokchetav clinopyroxene grains contain negligible K_2O; this reflects the bulk-rock composition (e.g. Zhang et al. 1997). Diopsides in dolomitic marble display variable K_2O contents and amounts of exsolved K-feldspar lamellae in a single thin section. Some diopside grains lack "coarse" exsolved K-feldspar lamellae and contain 0.17 wt % K_2O, whereas other grains exhibit "coarse" K-feldspar lamellae (Fig. 23) and range in K_2O content from 0.2 to 0.7 wt %, with an average of 0.47 wt %. The lamellae are too narrow to be analyzed; electron beam analyses on Kfs + Di yield about 4.8 wt % K_2O. Based on back-scatter electron and characteristic wavelength elemental images, the amount of the K-feldspar lamellae was estimated to be about 6.88 % based on 4900 points in an area of 2336 μm^2. Assuming end-member composition for the exsolved K-feldspar lamellae, primary diopsidic pyroxene should contain 1.56 wt % K_2O (Zhang et al. 1997) which is identical to the analysis of 1.55 wt % for clinopyroxene from diamond-bearing pyroxene-carbonate reported by Sobolev and Shatsky (1990) and the estimated values (>1.2 wt %) for Cpx by Shatsky et al. (1995).

Experimental results: In spite of large ionic radius (K^+ = 1.51 Å), K^+ evidently can be accommodated in clinopyroxene from kimberlite and diamond-bearing VHP metamorphic rocks (e.g. Rickard et al. 1989, Sobolev et al. 1991, Zhang et al. 1997). In order to calibrate the K_2O content of clinopyroxene as a function of P and clinopyroxene composition, several experiments have been recently performed (Ryabchikov and Ganeev 1990, Doroshev et al. 1992, Luth 1992, 1995, Harlow 1997, Okamoto and Maruyama

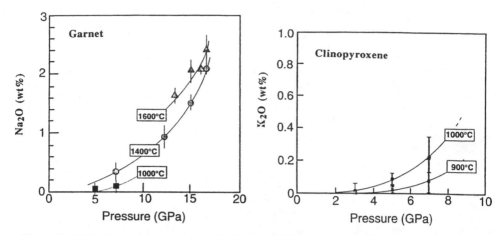

Figure 24. (a) Na_2O contents of garnets synthesized at 1000°C and from Ono and Yasuda at 1400°C and at 1600°C. Vertical bars denote uncertainties. (b) K_2O content in clinopyroxene as a function of pressure and temperature. The vertical bar denotes the concentration variation measured from matrix clinopyroxene (modified after Okamoto and Maruyama 1998).

1998). Clinopyroxene with 3 wt % K_2O has been synthesized in the $KAlSi_2O_6$-$CaMgSi_2O_6$ system (Luth 1995). The maximum K_2O content is 4.7 wt % in Cpx ($Di_{38}KO_{39}K_{Cpx22}En_1$) formed from a 50:50 Di + Ko ($NaCrSi_2O_6$) mixture at 10 GPa, 1400°C. The K uptake is proportional to the P as well as the Cpx composition, but is unrelated to the temperature of formation (Harlow 1997).

Okamoto and Maruyama (1998) calibrated the K_2O content of clinopyroxene as a potential geobarometer using Dabie coesite-bearing eclogite (VHP303), or this eclogite + 1.1 wt % K_2O (UHK303), for re-equilibration experiments at P = 2.6 to 7 GPa and T = 900-1200°C. All run products consisted of garnet, sodic augite, with or without a SiO_2 phase, and phengite. Textural evidence indicates that the solidus lies between 900 and 1000°C at 5 GPa, and about 1000°C at 3 and 7 GPa for VHP 303. The solidus T decreases to 1000°C at 3-5 GPa and below 900°C at 5 GPa in UHK 303. Garnets of all run products have relatively higher Alm + Sps (24-30 mol%) contents and lower grossular contents (12-16 mol%) with constant pyrope contents (58-61%) compared to the starting garnet. The Na content of these garnets formed at 1000°C increases from slightly below the detection limit at 5 GPa to ~0.06 wt % Na_2O at 7 GPa; a pronounced P-dependence of Na_2O content in garnet has been reported for higher T synthetic garnets (1400 and 1600°C) by Ono and Yasuda (1996) as shown in Figure 24A. The stable Ca-Na pyroxenes are sodic augite rather than omphacite at T > 900°C and P = 3-7 GPa. As shown in Figure 24B at T = 900°C, the K_2O content of Cpx increases from 0.07 to 0.23 wt % with increasing P from 3 to 7 GPa. A similar pressure dependence of the K_2O content of clinopyroxene is observed at 1000°C; it increases from below the detection limit at 3 GPa, through 0.13 wt % at 5 GPa to 0.36 wt % at 7 GPa.

Applying Figure 24B as a geobarometer, the maximum 1.2 wt % K_2O content for Cpx inclusions in garnets from the Kokchetav diamond-grade eclogite yields P estimates exceeding 7 GPa, whereas the Dabie Cpx containing less than 0.1 wt % K_2O yields P less than 3 GPa (Okamoto and Maruyama 1998). Tsuruta and Takahashi (1997) also reported K_2O solubility in clinopyroxenes drastically increasing up to 1 wt % at 7 GPa (>1400°C); such higher K_2O concentrations are due to 400°C higher experimental T than runs conducted by Okamoto and Maruyama (1998).

Silica rods in omphacite and clinopyroxene

Strongly oriented quartz needles (~5-20 μm wide and ~10-200 μm long) have been identified in clinopyroxenes of both eclogites and garnet clinopyroxenites of several VHP terranes (e.g. Smith 1984, 1988; Bankun-Czubaow 1992, Gayk et al. 1995, Zhang et al. 1995c, Zhang and Liou 1998, Tsai 1998). Figure 25 illustrates two examples of oriented quartz lamellae in omphacite from Sulu eclogite (Zhang and Liou 1998) and in the Cpx core of a granulite-overprinted eclogite from the northern Dabie terrane (Tsai 1998). The quartz lamellae line up in a single crystallographic orientation, and appear to bear a topotaxial relation with the host Cpx. Based on the shape and orientation of the SiO_2 rods, higher SiO_2 contents of the host Cpx compared to the needle-free matrix, and the lack of free quartz in the matrix, Tsai (1998) interpreted the quartz needles to be exsolution products of a precursor, supersilicic clinopyroxene which contained excess silica at peak metamorphic conditions. Supersilicic Cpx was defined by Smith and Cheeney (1980) to have $(Si + Ti) > (Ca + Mg + Fe^{2+} + Mn + Ni-2Na)$. Since the volume percentage of the SiO_2 needles in the host clinopyroxene varies from grain to grain, and the needles are not completely visible by backscatter electron imaging, it is difficult to quantify the excess SiO_2 in the precursor Cpx.

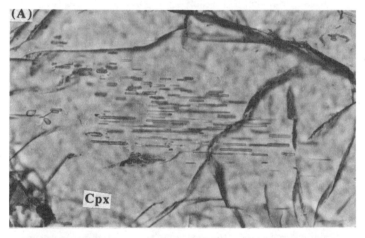

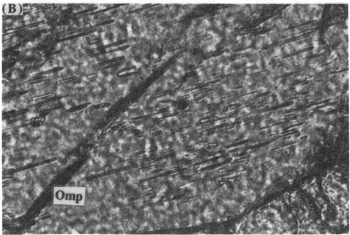

Figure 25. Photomicrographs of SiO_2 lamellae in VHP clinopyroxene (plane polarizer; width of view = 0.32 mm). (A) Quartz needles in Cpx core of retrograded eclogite of northern Dabie terrane (Tsai 1998). (B) Quartz lamellae in omphacite from Sulu eclogite (Zhang and Liou 1998).

Two stages of Cpx crystallization may have occurred in the retrograde eclogite: the cores represents a supersilicic mineral that formed at peak VHP conditions, and the rims reflect a granulite facies overgrowth. From core-to-rim, the Cpx composition shows depletion of Na and Si and enrichment of Fe (Tsai 1998).

Exsolved SiO_2 phases in clinopyroxene also have been observed in quartz-eclogites (e.g. Smith and Cheeney 1980, Bankun-Czubarow 1992) and HP "eclogite granulites" (Gyak et al. 1995) from localities where VHP minerals have been identified. Smith (1988) suggested that the existence of supersilicic Cpx requires high P and high T conditions.

Several experiments have clearly demonstrated that pressure is the most important factor required to stabilize supersilicic Cpx. For example, Mao (1976) documented that (1) jadeite-CaTsch clinopyroxene can contain up to 7.5 wt % excess SiO_2 at P up to 4.0 GPa at high T (1100-1700°C), and (2) the incorporation of excess SiO_2 in Cpx is highly P-sensitive. Similar indications have also been confirmed from diopside-CaTsch ($CaAl_2SiO_6$) pyroxene by Khanukhova et al. (1976) and by Wood and Henderson (1978) at 2.5-3.5 GPa, 1200-1450°C, and from diopside-Ca-eskola ($Ca_{0.5}AlSi_2O_6$) Cpx by Zharikov et al. (1984) at 3.5-7 GPa, 1200°C. In a synthesis experiment for silica-rich Na-Mg pyroxene at 10-15 GPa and 1600°C, Angel et al. (1988) showed that the incorporation of excess silica could stabilize the Na-Mg pyroxene to higher P (possibly up to 20 GPa) and that VHP Na-Mg Cpx contains both tetrahedrally and octahedrally coordinated silicon.

In summary, on the basis of natural occurrence of silica rods in several VHP terranes and experimental results mentioned above, it is concluded that the topotaxial growth of quartz lamellae in eclogitic Ca-Na clinopyroxene are best explained by exsolution from a former supersilicic Cpx stabilized at VHP conditions.

Monazite lamellae in apatite

Apatite occurs as a common minor phase in VHP rocks. In some apatite-bearing garnet peridotites with high REE concentrations from the Dabie and Kokchetav Massifs, monazite was identified as an accessory rare-earth phosphate with formula $REE^{3+}(PO_4)^{3-}$ (e.g. Zhang et al. 1997a, Liou and Zhang 1998). Monazite occurs as a matrix phase, as small inclusions 0.04-0.08 mm long within garnet, ortho- and clinopyroxene and as thin lamellae in apatite in garnet clinopyroxenite (Grt + Di, 1-2 vol% monazite + Rt), Ap-bearing garnet pyroxenites (Grt + Opx + Cpx + 1 vol % Ap + rare monazite, Rt and allanite), and coesite-bearing omphacitite (Omp + Mz + Rt + Coe) of the Dabie terrane. Apatites contain variable amounts of monazite lamellae (Fig. 26); the monazites show variable compositions (Zhang et al. 1997a). These lamellae have relatively lower LREE and higher MREE and HREE than matrix monazites, and appear to have a topotaxial relationship to the host apatite.

For most rocks, small amounts of REE are included in apatites under VHP conditions inasmuch as apatites are major sinks of these elements. Similar to other trace phases described above, monazite lamellae may have exsolved from the apatite host during decompression. The size, abundance, and composition of monazite lamellae depend on the bulk rock REE content. For extremely REE-rich lithologies such as those of the Dabie Shan, discrete monazite grains occur as a matrix mineral as well as inclusions in addition to monazite lamellae in apatite.

OTHER UNUSUAL MINERALS

In the Dora Maira Massif, inclusions of purple, pleochroic ellenbergerite in pyrope megacrysts from a whiteschist has been described (Chopin 1986, Chopin et al. 1986). This phase forms a complete solid-solution series between silicate and phosphate end-members

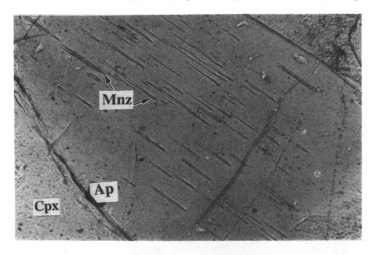

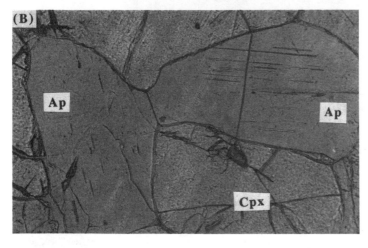

Figure 26. Photomicrographs showing monazite-(Ce) lamellae (Mzn) in apatite (Ap) from Dabie clinopyroxenite (plane polarized light; width of view = 0.32 mm for A, 0.78 mm for B).

as shown in Figure 27. The silicate end-member contains 8 Si, 7 Mg, 6 Al and about one Ti + Zr with 10 hydroxyl p.f.u., whereas the phosphate end-member contains no Si, Al, or Ti and 8 P, 14 Mg and 9 hydroxyl p.f.u. Intermediate compositions have been analyzed; zoning from core to rim has been established (as shown by the arrows). Thus far, ellenbergerite has been reported only in the pyrope-whiteschist from the Dora Maira Massif (e.g. Simon et al. 1997). The rare occurrence may be related to a very restricted P-T stability field for ellenbergerite, as shown in Figure 28. At lower P, ellenbergerite is replaced by Tlc + Chl + Ky + Rt, whereas at higher T, the assemblage Py + Tlc + Ky + Rt is stable. With increasing phosphate substitution, its stability field expands toward lower pressure. In fact P-ellenbergerite occurs as rims of the Ti-variety and as late stage phase in the whiteschist; it has also been reported from Norway. Phase relations of several polymorphs in the MgO-P_2O_5-H_2O system have been investigated (Brunet et al. 1998): results indicate that most of the polymorphs are stable at low pressures within the quartz stability field. However, one wagnerite crystal from the Dora Maira Massif with inclusions of coesite and coesite pseudomorphs (Fig. 6E of Brunet et al. 1998) may be a VHP phase.

Several other minor VHP phases in the Dora Maira Massif have been described; these

Ellenbergerite (Chopin & Sobolev, 1995)

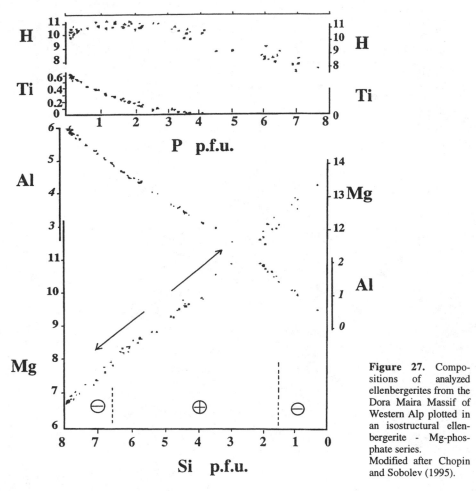

Figure 27. Compositions of analyzed ellenbergerites from the Dora Maira Massif of Western Alp plotted in an isostructural ellenbergerite - Mg-phosphate series. Modified after Chopin and Sobolev (1995).

include bearthite, $[(Ca_2Al(PO_4)_2OH]$, as a matrix phase as well as inclusions in both metapelites and pyrope-phengite quartzite (Chopin et al. 1992), and corundum and Mg-staurolite (X_{Mg} = 0.85-0.95) in metapelitic rocks (Chopin et al. 1991). Near-end-member magnesiochloritoid (up to 97 mol % end-member) occurs along with ellenbergerite, talc, chlorite, and kyanite as prograde inclusions in pyrope megablasts (Simon et al. 1997). Most coarse pyrope crystals in this terrane are homogeneous and are extremely magnesian (Prp 90-Prp98). The compositions of these unusual phases are listed in Table 4.

Corundum as a matrix mineral and Mg-staurolite as a retrograde phase after garnet have also been reported in some Sulu eclogitic rocks from Donghai by Enami and Zang (1988). Calderite garnet, $(Mn_3Fe_2^{3+}Si_3O_{12})$, which is stable only at P > 3.0 GPa (Schreyer and Baller 1981), has been described in VHP metamanganiferous quartzite from the Western Alps by Reinecke et al. (1992). Similarly, the high-P polymorph of $CaSiO_3$, wollastonite II, (or walstromite), is stable at P above 3.5 GPa at 800°C and 3.0 GPa at 1000°C (Essene 1974, Huang and Wyllie 1975b); this phase may occur in some VHP

Table 4. Representative compositions of unusual VHP minerals.

Mineral:	The Dora Maira Massif										The Dabie-Sulu Terrane			
	Pyrope	Ti-Elb	Ti-Elb	P-Elb	P-Elb	Wagnerite	Bearthite	Mg-Staur	Mg-Cht	Talc	Al-Ttn	Al-Ttn	Sr-Ep	Sr-Zo
SiO2	44.37	39.30	39.86	32.51	29.71	0.02	0.30	30.69	27.31	61.87	31.27	31.48	37.60	38.70
P2O5		0.44	0.06	8.27	13.67	43.80	44.32							
TiO2	0.05	4.10	2.06	0.52	0.31	0.15		0.08	0.01	0.04	28.46	28.40		
ZrO2		0.00	2.04	2.10	0.15	0.03								
SrO							3.53						2.66	3.04
Al2O3	25.57	25.24	25.07	20.41	19.22	0.00	15.91	57.56	46.13	0.75	7.36	7.78	28.40	31.30
Fe2O3													5.09	1.91
FeO	0.52	0.20	0.33	0.43	0.49	0.75	0.03	0.73	1.08	0.28	1.15	0.38		
MnO	0.00						0.04	0.04	0.09	0.00				
MgO	30.03	22.48	22.25	25.72	27.32	48.30	0.12	8.80	17.28	30.39	0.28	0.10	0.14	0.04
CaO	0.18					0.12	33.04	0.03	0.00	0.00	28.67	29.73	22.00	22.40
K2O														
Total	100.76	91.76	91.67	89.96	90.87	93.17	97.25	97.93	91.90	93.33	97.19	97.87	95.99	97.47
Si	2.97	7.91	7.99	6.70	5.76	0.00	0.03	7.97	2.02	7.97	1.04	1.04	3.00	3.00
P		0.07	0.01	1.44	2.24	1.01	3.97							
Ti	0.00	0.62	0.31	0.08	0.05	0.00		0.02	0.00	0.00	0.71	0.70		
Zr		0.00	0.20	0.21	0.01									
Sr							0.22			0.03			0.12	0.14
AlIV								0.03						
AlVI	2.02	5.99	5.92	4.96	4.39	0.00	1.98	17.60	4.02	0.09	0.29	0.30	2.67	2.86
Fe+3	0.03									0.03	0.03	0.01	0.31	0.11
Fe+2	0.00	0.03	0.06	0.08	0.08	0.02	0.00	0.16	0.07	0.00				
Mn		0.00						0.01	0.01					
Mg	3.00	6.74	6.65	7.90	7.89	1.95	0.02	3.41	1.91	5.84	0.01		0.02	0.01
Ca	0.01					0.00	3.74	0.01			1.02	1.05	1.88	1.86
Total	8.03	21.36	21.14	21.37	20.42	2.99	9.96	29.20	8.02	13.95	3.10	3.10	7.70	7.88
Ref.	1	2	1	2	1	3	4	1	1	1	5	6	7	7

1. Simon et al. (1997); 2. Chopin et al. (1986); 3. Brunet et al. (1998); 4. Chopin et al. (1993); 5. Zhang et al. (1995c); 6. Cong et al. (1995c); 7. Nagasaki and Enami, 1998)

impure marbles, and should be looked for.

According to X_{Al} (= Al/(Al+Fe+Ti)) values, titanite is classified as high-Al (X_{Al} > 0.25) and low-Al varieties (X_{Al} < 0.25) (Oberti et al. 1985). High-Al titanite occurs in several coesite-bearing eclogites from the Western Gneiss Region (Smith 1988) and the Dabie-Sulu terrane (e.g. Cong et al. 1995, Zhang et al. 1995c, 1996). Titanites in Sulu coesite-bearing eclogites contain up to 7.35 wt % Al_2O_3, and those from the Dabie Mountains reach 12.7 wt %; their X_{Al} ranges from 0.27 to 0.50. High-Al titanites from VHP marbles contain abundant Al_2O_3 (7.2-10.5 wt %; X_{Al} = 0.28-0.40), F (3.1-3.2 wt %) and considerable P (P_2O_5 = 0.3 wt %) and Mg (MgO = 0.3 wt %), whereas secondary low-Al titanites from the same samples contain negligible P and Mg. A positive relationship between (P + Mg) and (Si + Al) is evident; such a relationship suggests that P replaces Si in the tetrahedral site and Mg replaces Al in the octahedral site through the paired substitution, P + Mg = Si + Al, similar to the ellenbergerite solid solution series shown in Figure 27.

It should be pointed out that a host of peculiar minerals and mineral compositions (38 in all) were described in detail by Smith a decade ago in VHP rocks from the Norwegian coesite-eclogite province. Their occurrences, structures, and compositions were listed in Table 1.4 of Smith (1988) and documented in many photomicrographs of this extensive treatment.

CONCLUSIONS

This review summarizes published and unpublished data regarding index, hydrous, carbonate, and lamellae minerals from VHP eclogitic and ultramafic rocks mainly from the Dabie-Sulu terrane of east-central China. Petrogenetic grids for mafic rocks and experimentally determined P-T fields for these phases have been illustrated. We emphasize that some of these laboratory-grown phases and mineral assemblages were merely synthesized from appropriate compositions under VHP conditions, and their P-T field boundaries have not been reversed. Synthesis does not necessarily equate with stability. A reasonable demonstration of chemical equilibrium can only be accomplished in cases where it can be proven that the synthetic run products are homogeneous and that the occurrence is independent of P-T path. Another major problem concerns rates of transformation: degrees to which VHP equilibrium assemblages have been attained are very much a function of reaction kinetics, and as yet this topic has not been fully addressed in laboratory studies.

Experimental investigations of the pelitic system $K_2O-MgO-Al_2O_3-SiO_2-H_2O$ have revealed possible occurrences of several OH-bearing phases in VHP rocks (see Schreyer 1988a, 1995 for summaries). These include MgMgAl-pumpellyite [$Mg_5Al_5Si_6O_{21}(OH)_7$] (Fockenberg 1998), Mg-carpholite [$MgAl_2Si_2O_6(OH)_4$], "piezotite" (or phase Pi) [$Al_3Si_2O_7(OH)$], (3.5-5.5 GPa, T< 650°C) (Wunder and Schreyer 1992), K-cymrite [$KAlSi_3O_8(H_2O)_{1.0-0.9}$] (Massonne 1991), and OH-Topaz [$Al_2SiO_4(OH)_2$] (Wunder et al. 1993); their stability fields are shown in Figure 28 together with Phase A [$Mg_7Si_2O_8(OH)_6$] (Luth 1995, Pawley and Wood 1996) and lawsonite [$CaAl_2Si_2O_8(OH)_2$] (Poli and Schmidt 1995). Many of these OH-bearing phases should occur at 500-750°C (Fig. 28) near 4-10 GPa, but only along extremely cold geotherms considerably less than 5°C/km. These P-T fields lie within a "forbidden zone" inasmuch as the 5°C/km geotherm has been regarded as the lowest one realized on the Earth (Schreyer 1988a,b). New computations regarding the temperature distribution within subduction zones (e.g. Peacock 1990) indicate that rocks might be subducted along subduction-zone gradients as low as 3°C/km. These low geotherms are restricted to the inner parts of rapidly buried continental or oceanic crust in long-lasting subduction zones (e.g. Wunder et al. 1993). Since such conditions are essentially transient and will inevitably be followed by a period of thermal relaxation, all these low-T phases have little chance to survive the

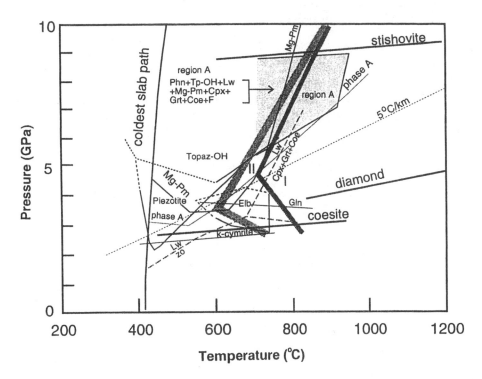

Figure 28. P-T diagram showing phase relations and stabilities of several VHP index, hydrous and unusual phases. The geotherm for the coldest subduction slab is after Hellfrich et al. (1989). The stability curve of diamond-graphite is after Chatterjee (1991), coesite-quartz after Hemingway et al. (1998), coesite-stishovite after Yagi and Akimoto (1976), lawsonite after Poli and Schmidt (1995), MgMgAl-pumpellyite after Fockenberg (1998), Phase A after Wunder (1996), piezotite after Wunder and Schreyer (1992), ellenbergereite after Chopin and Sobolev (1995), and K-Cymrite after Massonne (1991). The P-T field for region A for coexisting pumpellyite (Pm), lawsonite (Lw), OH-topaz in the presence of Cpx + Grt + Coe is after Domanik and Holloway (1996).

temperature increase on return to the surface through erosional and tectonic processes. However, one possibility for their preservation evidently is as minute inclusions in high mechanical-strength, impervious containers like garnet, zircon, or diamond (Chopin and Sobolev 1995).

The wealth of new phases and assemblages described for whiteschists and metapelitic rocks (e.g. Chopin and Sobolev 1995) was predicted by Schreyer (1968) based on reconnaissance VHP experiments. Similar experiments for mafic, carbonate, and quartzofeldspathic compositions—common rock types of VHP terranes—should be performed; the results may lead to additional findings of new very high pressure metamorphic minerals. According to Sobolev et al. (1992), zircon and garnet are the best containers of primary VHP minerals in metamorphic rocks (also see Liou et al. 1997c, Tabata et al. 1998). Systematic characterization of crystalline inclusions employing various analytical instrumentation including the laser micro-Raman spectrometer and micro-synchrotron beam analysis may lead to discovery of based on reconnaissance VHP minerals as relict inclusions in garnet, zircon, and other strong, unreactive minerals (e.g. Schertl and Schreyer 1996).

Olivine, garnet, and pyroxene of garnet peridotites contain minute mineral inclusions as yet unidentified that may turn up to be VHP minerals or dense hydrous, magnesian silicates previously synthesized only in diamond-cell or multi-anvil experiments (e.g. see Bose and Navrotsky 1998). Discovery of natural representatives of these synthetic phases would be a major advance in understanding mantle processes including the role of hydrous phases as storage sites for H_2O. The search for mantle minerals in VHP garnet peridotites should be conducted with the same techniques (micro-Raman spectroscopy, synchrotron X-ray diffraction, and TEM) that were used by the original investigators to identify their experimental run products. Such studies will certainly bridge the gap between mantle petrology and mineral physics.

Exsolution intergrowths are common in minerals of VHP rocks as described above. However, the mechanisms for mineral exsolution in VHP phases are poorly known. Each lamellae-bearing host mineral preserves information concerning the compositions and physical conditions of the homogeneous precursor phase, as well as the inferred P-T path during decompression. Compositional and structural characterization of lamellae-host mineral pairs should provide important new constraints on physical conditions of crystallization/recrystallization. Such studies could encourage subsequent experimentation to delineate the P-T conditions and mechanisms for their formation.

Indeed, VHP eclogitic and ultramafic rocks still hold many secrets remained to be explored. Recent research on regional VHP rocks containing coesite and diamond has produced a dramatic restructuring of our understanding of continental-collision processes. Extremely rare occurrence of intragranular coesite suggests that the lack of fluid during rapid exhumation of VHP rocks may have played a more important role than the pressure-vessel effect. Geologists need to intensify their search for such tiny inclusions in order to uncover more completely the extent and magnitude of profound subduction and exhumation of segments of the continental crust in collisional orogens (e.g. Liou et al. 1997c).

ACKNOWLEDGMENTS

Our study of VHP metamorphism and tectonics has been supported by NSF grants EAR 92-04563, 93-04480, 95-06468, and 97-25347 and by the Stanford China Industrial Affiliates Program. Unpublished photomicrographs of diamond inclusions in Kokchetav zircon from I. Katayama of Tokyo Institute of Technology and quartz rods in Dabie Cpx from Chin-ho Tsai of Stanford were of great help. We thank our coworkers whose data are included in this review: B. Cong, Brad Hacker, Chin-ho Tsai, Yoshi Ogasawara, Soichi Omori, Kazu Okamoto, and Tzen F. Yui. This note was critically reviewed by Bjorn Mysen. We thank the above individuals and institutes for their support and assistance.

REFERENCES

Ahrens TJ (1989) Water storage in the mantle. Nature 342:122-123
Akaogi M, Ito M, Navrotsky A (1989) Olivine-modified spinel-spinel transitions in the system Mg_2SiO_4-Fe_2SiO_4 calorimetric measurements, thermochemical calculation, and geophysical application. J Geophys Res 94:1567-1568
Akaogi M, Kusaba K, Susaki JL, Yagi T, Matsui M, Kikegawa T, Yusa H, Ito E (1992) High-pressure high-temperature stability of α–PbO_2-type TiO_2 and $MgSiO_3$ majorite: calorimetric and in situ X-ray diffraction studies. *In:* Syono Y, Manghnani MH (eds) High Pressure Research: Applications to Earth Planet Sci, p 447-455, Am Geophys Union, Washington, DC
Akaogi M, Yusa H, Shiraishi K, Suzuki, T (1995) Thermodynamic properties of α–quartz, coesite, and stishovite and equilibrium phase relations at high pressure and high temperature. J Geophys Res 100:22337-22347
Ambler EP, Ashley PM (1977) Vermicular orthopyroxene-magnetite symplectites from the Wateranga Layered mafic intrusion, Queensland, Australia. Lithos 10:163-172

Ames L, Tilton, GR, Zhou GZ (1993) Timing of collision of the Sino-Korean and Yangtze cratons. Geology 21:339-342

Ames L, Zhou G, Xiong B (1996) Geochronology and geochemistry of ultrahigh-pressure metamorphism with implications for collision of the Sino-Korean and Yangtze cratons, central China. Tectonics 15:472-489

Angel RJ, Gasparik T, Ross NL, Finger LW, Prewitt CT, Hazen RM (1988) A silica-rich pyroxene phase with six-coordinated silicon. Nature 335:156-158

Angel RJ, Hugh-Jones DA (1994) Equations of state and thermodynamic properties of enstatite pyroxenes. J Geophys Res 99:19777-19783

Aoki K, Fujino K, Akaogi M (1976) Titanochodrodite and Ti-clinohumite derived from the upper mantle in the Bull Park kimberlite, Arizona, USA. Contrib Mineral Petrol 56:243-253

Arai S (1978) Chromian spinel lamellae in olivine from the Iwanai-dake peridotite mass, Hokkaido, Japan. Earth Planet Sci Lett 39:267-273

Arif M, Jan MQ (1993) Chemistry of chromite and associated phases from the Shangla ultramafic body in the Indus suture zone of Pakistan. *In:* Treloar PJ, Searle MP (eds) Himalayan Tectonics. Geol Soc Spec Publ 74:101-112

Ashworth JR (1979) Two kinds of exsolution in chondritic olivine. Mineral Mag 43:535-538

Austrheim H (1998) Influence of fluid and deformation on metamorphism of the deep crust and consequences for the geodynamics of collision zones. *In:* Hacker B, Liou JG (eds) When Continents Collide: Geodynamics and Geochemistry of Ultrahigh-Pressure Rocks. Kluwer Academic & Lippincott Raven

Avigad D (1992) Exhumation of coesite-bearing rocks in the Dora Maira massif (western Alps, Italy). Geology 20:947-950

Baker J, Matthews A, Mattey D, Rowley DB, Xue F (1997) Fluid-rock interaction during ultra-high pressure metamorphism, Dabie Shan, China. Geochim Cosmochim Acta 61:1685-1696

Bakun-Czubarow N (1992) Quartz pseudomorphs after coesite and quartz exsolutions in eclogitic omphacites of the Zlote Mountains in the Sudetes (SW Poland). Archiwum Mineral 48:3-25

Banno S, Wallis SR, Hirajima T (1997) From the depths of continental collision belt: Discovery and significance of UHP metamorphic rocks. Kagaku 67:39-47

Beane R, Liou JG, Coleman RG, Leech ML (1995) Petrology and retrograde P-T path for eclogites of the Maksyutov Complex, Southern Ural Mountains, Russia. Island Arc 4:254-266

Berman RG (1988) Internally-consistent thermodynamic data for minerals in the system Na$_2$O-K$_2$O-CaO-MgO-FeO-Fe$_2$O$_3$-Al$_2$O$_3$-SiO$_2$-H$_2$O-CO$_2$. J Petrol 29:445-522

Biellmann C, Gillet Ph, Guyot F, Peyronneau J, Reynard B (1993) Experimental evidence for carbonate stability in the Earth's lower mantle. Earth Planet Sci Lett 118:31-41

Blanchard MB, Cunningham GC (1974) Artificial meteor ablation studies: Olivine. J Geophys Res 79:3973-90

Blundy JD, Brodholt JP, Wood BJ (1991) Carbon-fluid equilibria and the oxidation state of the upper mantle. Nature 349:321-324

Bohlen SR, Boettcher AL (1982) The quartz-coesite transformation: a pressure determination and the effects of other components. J Geophy Res 87:7073-7078

Bohlen SR, Mosenfelder J (1995) The coesite to quartz transformation: Nature vs. experiment. EOS Trans Am Geophys Union 76:F531

Bohlen SR, Montana A, Kerrick DM (1991) Precise determinations of the equilibria kyanite = sillimanite and kyanite = andalusite and a revised triple point for Al$_2$SiO$_5$ polymorphs. Am Mineral 76:677-680.

Bose K, Ganguly J (1993) Stability of talc at high pressures: Experimental determination, retrieval of thermodynamic properties, and applications to subduction processes. Geol Soc Am Abstr with Program 25:A-213-214

Bose K, Ganguly J (1995) Quartz-coesite transition revisited: Reversed experimental determination at 500-1200°C and retrieved thermodynamic properties. Am Mineral 80:231-238

Bose K, Navrotsky A (1998) Thermochemistry and phase equilibria of hydrous phases in the system MgO-SiO$_2$-H$_2$O: Implications for volatile transport to the mantle. J Geophys Res 103:9713-9719

Boyd FR, Nixon PH (1973) Origin of the ilmenite-silicate nodules in kimberlites from Lesotho and South Africa. in Nixon PH (ed) Lesotho Kimberlites. Lesotho Nat Develop Corp 254-268

Brey G, Brice WR, Ellis DJ, Green DH, Harris KL, Ryabchikov ID (1983) Pyroxene-carbonate reactions in the upper mantle. Earth Planet Sci Lett 62:63-74

Brey GP, Köhler T (1990) Geothermobarometry in four lherzolite II. New thermobarometers, and practical assessment of existing thermobarometers. J Geol 31:1553-1378

Bruckner HK (1998) A sinking intrusion model for the introduction of garnet-bearing peridotites into continent collision orogens. Geology 26:631-634

Brunet F, Chopin C, Seifert F (1998) Phase relations in the MgO-P$_2$O$_5$-H$_2$O system and the stability of phosphoellenbergerite: petrological implications. Contrib Mineral Petrol 131:54-70

Bundy FP (1980) The P, T phase and reaction diagram for elemental carbon. J Geophys Res 85:6930-6936

Caby R (1994) Precambrian coesite from northern Mali: first record and implications for plate tectonics in the trans-Saharan segment of the Pan-African belt. Eur J Mineral 6:235-244

Carlson WD, Rosenfeld JL (1981) Optical determination of topotactic aragonite-calcite growth kinetics: metamorphic implications. J Geol 89:615-638

Carswell DA (1990) Eclogites and the eclogite facies: definitions and classification. In: DA Carswell (ed) Eclogite Facies Rocks. Blackie, New York, p 1-13

Carswell DA, O'Brien PJ, Wilson RN, Zhai M (1997) Thermobarometry of phengite-bearing eclogites in the Dabie Mountains of central China. J Meta Geol 15:239-252

Champness PE (1970) Nucleation and growth of iron oxides in olivine (Mg,Fe)$_2$SiO$_4$. Mineral Mag 37:790-800

Chao ECT, Shoemaker EM, Madsen BM (1960) First natural occurrence of coesite. Science 132:220-222

Chesnokov BV, Popov VA (1965) Increasing of volume of quartz grains in eclogites of the South Urals. Dol Akad Nauk USSR 162:176-178

Chopin C (1981) Talc-phengite: a widespread assemblage in high-grade pelitic blueschists of the Western Alps. J Petrol 22:628-650

Chopin C (1984) Coesite and pure pyrope in high-grade blueschists of the western Alps: a first record and some consequences. Contrib Mineral Petrol 86:107-118

Chopin C (1986) Phase relationships of ellenbergerite, a new high-pressure Mg-Al-Ti-silicate in pyrope-coesite-quartzite from the Western Alps. Geol Soc Am Mem 164:31-42

Chopin C, Sobolev N (1995) Principal mineralogic indicators of UHP in crustal rocks. In: Coleman RG, Wang X (eds) Ultrahigh-Pressure Metamorphism, Cambridge Univ Press, p 96-131

Chopin C (1987) Very-high-pressure metamorphism in the western Alps: implications for subduction of continental crust. Phil Trans Roy Soc London Series A 321:183-197

Chopin C, Henry C, Michard A (1991) Geology and petrology of the coesite-bearing terrain, Dora Maira massif, Western Alps. Eur J Mineral 3:263-291

Chopin C, Klaska R, Medenbach O, Dron D (1986) Ellenbergerite, a new high-pressure Mg-Al(Ti-Zr)-silicate with a novel structure based on face-sharing octahedra. Contrib Mineral Petrol 92:316-321

Chopin C, Brunet F, Gebert W, Medenbach O, Tillmanns E (1993) Bearthite, Ca$_2$Al[PO$_4$]$_2$(OH), a new mineral from high-pressure terranes of the western Alps. Schweiz Mineral Petrogr Mitt 73:1-9

Claoue-Long JC, Sobolev NV, Shatsky VS, Sobolev AV (1991) Zircon response to diamond-pressure metamorphism in the Kokchetav massif, USSR. Geology 19:710-713

Cong BL (1996) Ultrahigh-pressure Metamorphic Rocks in the Dabieshan-Sulu Region of China. Science Press, p 224

Cong BL, Zhai M, Carswell DA, Wilson RH, Wang Q, Zhao Z, Windley BF (1995) Petrogenesis of ultrahigh-pressure rocks and their country rocks in Shuanghe of Dabieshan Mountains, central China. Eur J Mineral 7:119-138

Cuthbert SJ, Harvey MA, Carswell DA (1983) A tectonic model for the metamorphic evolution of the Basal Gneiss Complex, Western South Norway. J Meta Geol 1:63-90

Dachille F, Zeto RJ, Roy R (1963) Coesite and stishovite: Stepwise reversal transformations. Science 140:991-993

Dawson JB, Reid AM (1970) A pyroxene-ilmenite intergrowth from the Monastery mine, South Africa. Contrib Mineral Petrol 26:296-301

Dobretsov NL, Shatsky VS, Sobolev NV (1995a) Comparison of the Kokchetav and Dabie Shan metamorphic complexes: coesite- and diamond-bearing rocks and UHP-HP accretional-collisional events. Int'l Geol Rev 37:636-656

Dobretsov NL, Sobolev NV, Shatsky VS, Coleman RG, Ernst WG (1995b) Geotectonic evolution of diamondiferous paragneisses, Kokchetav complex, northern Kazakhstan-the geologic enigma of ultrahigh-pressure crustal rocks within a Paleozoic foldbelt. Island Arc 4:267-279

Dobrzhinetskaya LF, Eide EA, Larsen RB, Sturt BA, Tronnes RG, Smith DC, Taylor WR, Posukhova TV (1995) Microdiamond in high-grade metamorphic rocks of the western gneiss region, Norway. Geology 23:597-600

Dobrzhinetskaya L, Green HW, Wang S (1996) Alpe Arami: a peridotite massif from depths of more than 300 kilometers. Science 271:1841-1846

Domanik KJ, Holloway JR (1996) The stability and composition of phengitic muscovite and associated phases from 5.5 to 11 GPa: Implications for deeply subducted sediments. Geochim Cosmochim Acta 60:4133-4150

Doroshev AM, Sobolev NV, Brey G (1992) Experimental evidence of high-pressure origin of the potassium-bearing pyroxenes. 29th Int'l Geol Cong 2:602

Dymek RF, Boak JI, Brothers SC (1988) Titanian chondrodite- and titanian clinohumite-bearing metadunite from the 3800 Ma Isua supracrustal belt, West Greenland: Chemistry, Petrology and origin. Am Mineral 73:547-558

Eide E, McWilliams MO, Liou JG (1994) ^{40}Ar/^{39}Ar geochronology and exhumation of high-pressure to ultrahigh-pressure metamorphic rocks in east central China. Geology 22:601-604

Enami M, Zang Q (1988) Magnesian staurolite in garnet-corundum rocks and eclogite from the Donghai District, Jiangsu Province, east China. Am Mineral 73:48-56

Engi M, Lindsley DH (1980) Stability of titanian clinohumite: experiments and thermodynamic analysis. Contrib Mineral Peterol 72:415-424

Ernst WG (1977) Mineralogic study of eclogitic rocks from Alpe Arami, Lepontine Alps, southern Switzerland. J Petrol 18:371-398

Ernst WG (1978) Petrochemical study of lherzolitic rocks from the Western Alps. J Petrol 19:341-392

Ernst WG (1988) Tectonic history of subduction zones inferred from retrograde blueschist P-T paths. Geology 16:1081-1084

Ernst WG, Liou JG (1995) Contrasting plate-tectonic styles of the Quiling-Dabie-Sulu (Alpine-type) and Franciscan (Pacific-type) metamorphic belts. Geology 23:353-356

Ernst WG, Liou JG, Coleman RG (1995) Comparative petrotectonic study of five Eurasian ultrahigh-pressure metamorphic complexes. Int'l Geol Rev 37:191-211

Ernst WG, Peacock, SM (1996) A thermotectonic model for preservation of ultrahigh-pressure mineralogic relics in metamorphosed continental crust. in Subduction: *In:* Bebout GE et al. (eds) Top to Bottom. Am Geophys Union Mono 96:171-178

Ernst WG, Maruyama S, Wallis S (1997) Buoyancy-driven, rapid exhumation of ultrahigh-pressure metamorphosed continental crust: Proc Nat'l Acad Sci 94:9532-9537

Ernst WG, Mosenfelder JL, Leech ML, Liu J (1998) H$_2$O recycling during continental collision: Phase-equilibrium and kinetic considerations. *In:* Hacker B, Liou JG (eds) When Continents Collide: Geodynamics and Geochemistry of Ultrahigh-Pressure Rocks. Kluwer Academic & Lippincott Raven Publishers

Essene E (1974) High-pressure transformations in CaSiO$_3$. Contrib Mineral Petrol 45:247-250

Evans BW, Trommsdorff V (1978) Petrogenesis of garnet lherzolite, Cima di Gagnone, Lepontine Alps. Earth Planet Sci Lett 40:415-424

Fockenberg T (1998) An experimental study of the pressure-temperature stability of MgMgAl-pumpellyite in the system MgO-Al$_2$O$_3$-SiO$_2$-H$_2$O. Am Mineral 83:220-227

Franz G, Spear FS (1983) High pressure metamorphism in siliceous dolomites from the central Tauern Window, Austria. Am J Sci 283A:396-413

Gaspar JC (1992) Titanian clinohumite in the carbonatites of the Jacupiranga complex, Brazil: mineral chemistry and comparison with titanian clinohumite from other environments. Am Mineral 77:168-178

Gasparik T (1990) Phase relations in the transition zone. J Geophys Res 95:15751-15769

Gayk T, Kleinschrodt R, Langosch A, Seidel E (1995) Quartz exsolution in clinopyroxene of high-pressure granulite from the Munchberg Massif. Eur J Mineral 7:1217-1220

Gebauer D, Schertl HP, Brix M, Schreyer W (1997) 35 Ma old ultrahigh-pressure metamorphism and evidence for very rapid exhumation in the Dora Maira Massif, Western Alps. Lithos 41:5-24

Gillet Ph (1993) Stability of magnesite MgCO$_3$ at mantle pressure and temperature conditions: A Raman spectroscopic study. Am Mineral 78:1328-1331

Gillet Ph, Ingrin J, Chopin C (1984) Coesite in subducted continental crust: P-T history deduced from an elastic model. Earth Planet Sci Lett 70:426-436

Goode ADT (1974) Oxidation of natural olivines. Nature 248:500-501

Green HW, Dobrzhinetskaya L (1997) Response (Determine the origin of ultrahigh-pressure lherzolites by Hacker et al 1997) Science 278:704-707

Green HW, Dobrzhinetskaya L, Riggs EM, Jin ZM (1997) Alpe Arami: a peridotite massif from the mantle transition zone? Tectonophysics 279:1-21

Griffin WL, Brueckner HK (1985) REE, Rb-Sr, Sm-Nd studies of Norwegian eclogites. Chem Geol 52:249-271

Hacker BR, Peacock SM (1995) Creation, preservation, and exhumation of UHPM rocks. *In:* Coleman RG, Wang X (eds) Ultrahigh-Pressure Metamorphism, Cambridge Univ Press, p 159-181

Hacker BR, Wang Q (1995) Ar/Ar geochronology of ultrahigh-pressure metamorphism. Tectonics 14:994-1006

Hacker BR, Kirby SH, Bohlen SR (1992) Metamorphic petrology: calcite = aragonite experiments. Science 258:110-112

Hacker BR, Ratschbacher L, Webb L, Ireland T, Walker D, Dong S (in press) Orogen-scale architecture of the ultrahigh-pressure Dabie-Hong'an-Tongbai Shan, China. Earth Planet Sci Lett

Hacker BR, Zhang RY, Liou JG Hervig RL (1997) Determine the origin of ultrahigh-pressure lherzolites. Science 278:702-704

Haggerty SE, Baker I (1967) The alteration of olivine in basaltic and associated lavas. Part I. High-temperature alteration. Contrib Mineral Petrol 16:233-257

Harley SL, Carswell DA (1995) Ultradeep crustal metamorphism: a prospective view. J Geophys Res 100:8367-8380

Harlow GE (1997) K in clinopyroxene at high pressure and temperature: An experimental study. Am Mineral. 82:259-269

Harlow GE, Veblen DR (1991) Potassium in clinopyroxene inclusion from diamonds. Science 251:652-655

Hemingway BS, Bohlen SR, Hankins WB, Westrum Jr EF, Kuskov OL (1998) Heat capacity and thermodynamic properties for coesite and jadeite, reexamination of the quartz-coesite equilibrium boundary. Am Mineral 83:409-418

Hervig RL, Smith JV (1982) Temperature-dependent distribution of Cr between olivine and pyroxenes in lherzolite xenoliths. Contrib Mineral Petrol 81:184-189

Hervig RL, Smith JV, Dawson JB (1986) Lherzolite xenoliths in kimberlites and basalts; petrogenetic and crystallochemical significance of some minor and trace elements in olivine, pyroxenes, garnet and spinel. Trans Royal Soc (Edinburgh) Earth Sci. 77:181-201

Hirajima T, Ishiwatari A, Cong BL, Zhang RY, Banno S, Nozaka T (1990) Identification of coesite in Mengzhong eclogite from Donghai county, northeastern Jiangsu Province, China. Mineral Mag 54:579-584

Hirajima T, Zhang RY, Li JJ, Cong BL (1992) Nyböite from the Donghai area, Jiangsu Province, eastern China. Mineral Mag 56:37-46

Hirajima T, Wallis SR, Zhai M, Ye K (1993) Eclogitized metagranitoid from the Sulu ultrahigh-pressure (UHP) province, eastern China. Proc Japan Acad 69:249-254

Hiramatsu N, Banno S, Hirajima T, Cong B (1995) Ultrahigh-pressure garnet lherzolite from Chijiadian, Rongcheng County, in the Su-Lu region of eastern China. Island Arc 4:324-333

Holland TJB (1978) High water activities in the generation of high pressure kyanite eclogites of the Tuern window, Austria. J Geol 87:1-27

Holland TJB (1979) Experimental determination of the reaction paragonite = jadeite + kyanite + water, internally consistent thermodynamic data for part of the system $Na_2O-Al_2O_3-SiO_2-H_2O$, with applications to eclogites and blueschists. Contrib Mineral Petrol 68:293-301.

Holland TJB (1980) The reaction albite = jadeite + quartz determined experimentally in the range 600-1200°C. Am Mineral 65:129-134

Holland TJB (1988) Preliminary phase relations involving glaucophane and applications to high pressure petrology: new heat capacity and thermodynamic data. Contrib Mineral Petrol 99:134-142

Huang WL, Wyllie PJ (1975a) Melting reactions in the system $NaAlSi_3O_8-KAlSi_3O_8-SiO_2$ to 35 kilobars, dry and with excess water. J Geol 83:737-748

Huang WL, Wyllie PJ (1975b) Melting and subsolidus phase relations for $CaSiO_3$ to 35 Kb pressure. Am Mineral 60:213-217

Huang WL, Wyllie PJ (1984) Carbonation reactions for mantle lherzolite and harzburgite. Proc 27th Int'l Geol Cong Moscow 9:455-473

Ito E, Takahashi E (1989) Postspinel transformations in the system $Mg_2SiO_4-Fe_2SiO_4$ and some geophysical implications. J Geophys Res 94:10637-10646

Jahn BM (1998) Geochemical and isotopic characteristics of UHP eclogites of the Dabie orogen: implications for continental subduction and collisional tectonics. In: Hacker B, Liou JG (eds) When Continents Collide: Geodynamics and Geochemistry of Ultrahigh-Pressure Rocks. Kluwer Academic & Lippincott Raven Publishers.

Jagoutz E, Shatsky VS, Sobolev NV (1990) Sr-Nd-Pb isotopic study of ultrahigh PT rocks from Kokchetav massif. EOS Trans Am Geophys Union 71:1707

Jenkins DM (1987) Synthesis and characterization of tremolite in the system $H_2O-CaO-MgO-SiO_2$. Am Mineral 72:707-715

Jenkins DM, Clare AK (1990) Comparison of the high-temperature and high-pressure stability limits of synthetic and natural tremolite. Am Mineral 75:358-366

Katsura T, Ito E (1989) The system $Mg_2SiO_4-Fe_2SiO_4$ at high pressure and temperatures: precise determination of stabilities of olivine, modified spinel, and spinel. J Geophys Res 94:15663-15670

Katsura T, Ito E (1990) Melting and subsolidus phase relations in the $MgSiO_3-MgCO_3$ system at high pressures: implications to evolution of the Earth's atmosphere. Earth Planet Sci Lett 99:110-117

Katsura T, Tsuchida Y, Ito E, Yagi T, Utsumi W, Akimoto S (1991) Stability of magnesite under the lower mantle conditions. Proc Japan Acad 67:57-60

Khanukhova LT, Zharikov VA, Ishbulatov RA, Litvin YA (1976) Excess silica in solid solutions of high-pressure clinopyroxenes as shown by experimental study of the system $CaMgSi_2O_6$-$CaAl_2SiO_2$ at 35 kilobars and 1200°C. Doklady Earth Sci Sect 229:170-172

Kienast JR, Lombardo B, Biino G, Pinardon JL (1991) Petrology of very-high-pressure eclogitic rocks from the Brossasco-Isasca Complex, Dora-Maira Massif, Italian western Alps. J Meta Geol 9:19-34

Kraft S, Knittle E, Williams Q (1991) Carbonate stability in the Earth's mantle: A vibrational spectroscopic study of aragonite and dolomite at high pressures and temperatures. J Geophys Res 96:17997-18009

Kretz R (1983) Symbols for rock-forming minerals. Am Mineral 68:277-279

Krogh EJ (1988) The garnet-clinopyroxene Fe-Mg geothermometer—a reinterpretation of existing experimental data. Contrib Mineral Petrol 99:44-48

Krogh EJ, Carswell DA (1995) HP and UHP eclogites and garnet peridotites in the Scandinavian Caledonides. in Coleman RG, Wang X (eds) Ultrahigh Pressure Metamorphism, p 244-298

Kushiro I, Satake H, Akimoto S (1975) Carbonate-silicate reactions at high pressures and possible presence of dolomite and magnesite in the upper mantle. Earth Planet Sci Lett 28:116-120

Lappin MA, Smith DC (1978) Mantle-equilibrated orthopyroxene eclogite pods from the Basal gneisses in the Selje district, western Norway. J Petrol 19:530-584

Leech ML, Ernst WG (1998) Graphite pseudomorphs after diamond? A carbon isotope and spectroscopic study of graphite cuboids from the Maksyutov Complex, south Ural Mountains, Russia. Geochem Cosmochim Acta 62:2143-2154

Lennykh VI, Valizer PM, Beane R, Leech M, Ernst WG (1995) Petrotectonic evolution of the Maksyutov Complex, southern Urals, Russia: Implications for ultrahigh-pressure metamorphism. Int'l Geol Rev 37:584-600

Li S, Chen Y, Cong B, Zhang Z, Zhang RY, Liu D, Hart SR, Ge N (1993) Collision of the North China and Yangtze blocks and formation of coesite-bearing eclogite: timing and processes. Chem Geol 109:70-89

Li W, Lu R, Yang H, Prewitt CT, Fei Y (1997) Hydrogen in synthetic coesite crystals. EOS Trans Am Geophys Union 78:F736

Liou JG, Zhang RY (1995) Significance of ultrahigh-P talc bearing eclogitic assemblages. Mineral Mag 59:93-102

Liou JG, Zhang RY (1996) Occurrence of intergranular coesite in Sulu ultrahigh-P rocks from China: Implications for fluid activity during exhumation. Am Mineral 81:1217-1221

Liou JG, Zhang RY (1998) Petrogenesis of ultrahigh-P garnet-bearing ultramafic body from Maowu, the Dabie Mountains, Central China. Island Arc 7:115-134

Liou JG, Zhang RY, Ernst WG (1995) Occurrences of hydrous and carbonate phases in ultrahigh-pressure rocks from east-central China: Implications for the role of volatiles deep in cold subduction zones. Island Arc 4:362-375

Liou JG, Zhang RY, Ernst WG (1997a) Lack of fluid during ultrahigh-P metamorphism in the Dabie-Sulu region, eastern China. Proc 30th Int'l Geol Cong 17:141-156

Liou JG, Zhang RY, Jahn BM (1997b) Petrogenesis of ultrahigh-pressure jadeite quartzite from the Dabie region, East- central China. Lithos 41:59-78

Liou JG, Maruyama S, Ernst WG (1997c) Seeing mountain through a grain of garnet? Sciences 276:48-49

Liou JG, Maruyama S, Cong B (1998a) Introduction to geodynamics for high- and ultrahigh-pressure metamorphism. Island Arc 7:1-5

Liou JG, Zhang R, Ernst WG, Liu J, McLimans R (1998b) Mineral parageneses in the Piampaludo eclogitic body, Gruppo di Voltri, Western Ligurian Alps. Schweiz Mineral Petrogr Mitt 78:317-335

Liou JG, Zhang RY, Eide EA, Wang XM, Ernst WG, Maruyama S (1996) Metamorphism and tectonics of high-P and ultrahigh-P belts in the Dabie-Sulu region, eastern central China. *In:* Yin A, Harrison MT (eds) Ruby Volume VIII: 300-344, Tectonic Development of Asia: Cambridge Univ Press

Liu J, Bohlen SR, Ernst WG (1996) Stability of hydrous phases in subducting oceanic crust. Earth Planet Sci Lett 143:161-171

Luth RW (1992) Potassium in clinopyroxene at high pressure: experimental constraints. EOS Trans Am Geophys Union 73:608

Luth RW (1993) Diamonds, eclogites, and the oxidation state of the Earth's mantle. Science 261:66-68

Luth RW (1995) Potassium in clinopyroxene at high pressure. EOS Trans Am Geophys Union 76:F711

Mao HK (1971) The system jadeite ($NaAlSi_2O_6$)-anorthite ($CaAl_2Si_2O_8$) at high pressures. Carnegie Inst Year Book 69:163-168

Maruyama S, Liou JG (1998) Initiation of UHP metamorphism and its significance on the Proterozoic/Phanerozoic boundary. Island Arc 7:6-35

Maruyama S, Liou JG, Terabayashi M (1996) Blueschists and eclogites of the world and their exhumation. Int'l Geol Rev 38:490-596

Massonne HJ, Schreyer W (1987) Phengite geobarometry based on the limiting assemblage with K-feldspar, phlogopite, and quartz. Contrib Mineral Petrol 96:212-224

Massonne HJ, Schreyer W (1989) Stability field of the high-pressure assemblage talc + phengite and two new phengite barometers. Eur J Mineral 1:391-410

Massonne HJ (1995) Experimental and petrogenetic study of UHPM. *In:* Coleman RG, Wang X (eds) Ultrahigh-pressure metamorphism. Cambridge Univ Press, Cambridge, p 33-95

Massonne HJ (1998) Ultra-high pressure metamorphism of rocks from the gneiss-eclogite unit of the Saxonian Erzgebirge, Germany, Supplement EOS Trans Am Geophys Union 79:W129

McGetchin TR, Besancon JR (1973) Carbonate inclusions in mantle derived pyropes. Earth Planet Sci Lett 18:408-410

Medaris LG, Carswell DA (1990) The petrogenesis of Mg-Cr garnet peridotites in European metamorphic belts. *In:* Carswell DA (ed) Eclogite Facies Rocks, Chapman and Hall, New York, p 260-290

Medaris LG, Beard BL, Johnson CM, Valley JW, Spicuzza MJ, Jelinek E, Misar Z (1995) Garnet pyroxenite and eclogite in the Bohemian Massif: geochemical evidence for Variscan recycling of subducted lithosphere. Geol Rundsch 84:489-505

Mehta AK, Leinenweber K, Navrotsky A, Akaogi M (1994) Calorimetric study of high pressure polymorphism in $FeTiO_3$; stability of perovskite phase. Phys Chem Minerals 21:207-212

Michard A, Henry C, Chopin C (1995) Structures in UHPM rocks: A case study from the Alps. *In:* Coleman RG, Wang X (eds) Ultrahigh-pressure metamorphism. Cambridge Univ Press, Cambridge, p 132-158

Mitchell RH (1979) Manganoa magnesian ilmenite and titanian clinohumite from the Jacupiranga carbonatite, Sao Paulo, Brazil. Am Mineral 63:544-547

Monié PJ, Chopin C (1991) $^{40}Ar/^{39}Ar$ dating in coesite-bearing and associated units of the Dora-Maira massif, Western Alps. Eur J Mineral 3:239-262

Moseley D (1984) Symplectic exsolution in olivine. Am Mineral 69:139-153

Mosenfelder JL, Bohlen SR, Hankins WB (1994) Kinetics of the coesite to quartz transformation. EOS Trans Am Geophys Union 75:702

Mosenfelder JL, Bohlen SR (1995) Kinetics of the quartz to coesite transformation. EOS Trans Am Geophys Union 76:531-532

Mosenfelder JL, Bohlen SR (1997) Kinetics of the coesite to quartz transformation. Earth Planet Sci Lett 153:133-147

Mottana A, Carswell DA, Chopin C, Obserhansli R (1990) Eclogite facies mineral parageneses. *In:* Carswell DA (ed) Eclogite Facies Rocks. Blackie, p 14-52

Nakajima Y, Ogasawara Y (1997) Garnet lherzolite with olivine-Ti mineral intergrowth from Rongcheng in the Sulu UHP terrane, eastern China. EOS Trans Am Geophys Union 78:F787

Nagasaki A, Enami M (1998) Sr-bearing zoisite and epidote in ultrahigh-pressure (UHP) metamorphic rocks from the Sulu province, eastern China: An important Sr reservoir under UHP conditions. Am Mineral 83:240-247

Nickel KG, Green DH (1985) Empirical geothermobarometry for garnet peridotites and implications for the nature of the lithosphare, kimberlites and diamonds. Earth Planet Sci Lett 73:158-170

Obata M, Morten L (1987) Transformation of spinel lherzolite to garnet lherzolite in ultramafic lenses of the Austridic Crystalline Complex, Northern Italy. J Petrol 28:599-623

Oberti R, Ross G, Smith DC (1985) X-ray crystal structure refinement studies of the TiO = Al(OH,F) exchange in high-aluminous-titanites. Terra Cognita 5:428

Ogasawara Y, Liou JG, Zhang RY (1995) Petrogenetic grid for ultrahigh-P metamorphism in the model system $CaO-MgO-SiO_2-CO_2-H_2O$. Island Arc 4:240-253

Ogasawara Y, Liou JG, Zhang RY (1997) Thermochemical calculation of $logf_{O_2}$-T-P stability relations of diamond-bearing assemblages in the model system $CaO-MgO-SiO_2-CO_2-H_2O$. Rus Geol Geophys 38:587-598

Ogasawara Y, Zhang RY, Liou JG (1998) Petrogenesis of dolomitic marbles from Rongcheng in the Sulu UHP metamorphic terrane, eastern China. Island Arc 7:82-97

Okamoto K, Maruyama S (1998) Multi-anvil re-equilibration experiments of a Dabie Shan ultrahigh-pressure eclogite within the diamond-stability fields. Island Arc 7:52-69

Okamoto K, Maruyama S (in press) The high pressure stability limits of lawsonite in the MORB + H_2O system. Am Mineral

Okamoto K, Maruyama S, Liou JG (1997) Experimental study of TiO_2 content in olivine under ultrahigh-pressure conditions. EOS Trans Am Geophys Union 78:F761

Okay AI (1993) Petrology of a diamond and coesite-bearing metamorphic terrane. *In:* Dabie Shan, China. Eur J Mineral 5:659-673

Okay AI (1994) Sapphirine and Ti-clinohumite in ultra-high-pressure garnet-pyroxenite and eclogite from Dabie Shan, China. Contrib Mineral Petrol 116:145-155

Olafsson M, Eggler DH (1983) Phase relations of amphibole, amphibole-carbonate, and phlogopite-carbonate peridotite: petrologic constraints on the asthenosphere. Earth Planet Sci Lett 64:305-315

Omori S, Liou JG, Zhang RY, Ogasawara Y (1998) Petrogenesis of impure dolomitic marble from the Dabie Mountains, central China. Island Arc 7:98-114

Pacalo REG, Gasparik T (1990) Reversals of the orthoenstatite-clinoenstatite transition at high pressures and high temperature. J Geophys Res 95:15853-15858

Parkinson CD, Miyazaki K, Wakita K, Barber AJ, Carswell DA (1998) An overview and tectonic synthesis of the pre-Tertiary very-high-pressure metamorphic and associated rocks of Java, Sulawesi and Kalimantan, Indonesia. Island Arc 7: 184-200

Pan Y, Fleet ME, Macrae ND (1993) Oriented monazite inclusions in apatite porphyroblasts from the Hemlo gold deposit, Ontario, Canada. Mineral Mag 57:697-707

Pawley AR (1992) Experimental study of the compositions and stabilities of synthetic nyböite and nyböite-glaucophane amphiboles. Eur J Mineral 4:171-192

Pawley AR (1994) The pressure and temperature stability limits of lawsonite: Implications for H_2O recycling in subduction zones. Contrib Mineral Petrol 118:99-108

Peacock SM (1990) Fluid processes in subduction zones. Science 248:229-345

Peacock SM (1991) Numerical simulation of subduction zone pressure-temperature-time paths: constraints on fluid production and arc magmatism. Philos Trans Roy Soc London A-335:341-353

Peacock SM (1993) Large-scale hydration of the lithosphere above subducting slabs. *In:* Touret JLR, Thompson AB (eds) Fluid-Rock Interactions in the Deeper Continental Lithosphere. Chem Geol 108:49-59

Pearson DG, Davies GR, Nixon PH (1993) Geochemical constrains on the petrogenesis of diamond facies pyroxenites from the Beni Bousera peridotite massif, North Morocco. J Petrol 34:125-172

Pearson DG, Davies GR, Nixon PH (1995) Orogenic ultramafic rocks of UHP (diamond facies) origin. *In:* Coleman RG, Wang X (eds) Ultrahigh-Pressure Metamorphism, Cambridge Univ Press, Cambridge, p 456-510

Pearson DG, Davies GR, Nixon PH, Milledge HJ (1989) Graphitized diamonds from a peridotite massif in Morocco and implications for anomalous diamond occurrences. Nature 338:60-62

Philippot P (1993) Fluid-melt-rock interaction in mafic eclogites and coesite-bearing metasediments: Constraints on volatile recycling during subduction. Chem Geol 108:93-112

Poli S, Schmidt MW (1995) H_2O transport and release in subduction zones: Experimental constraints on basaltic and andesitic systems. J Geophys Res 100:22299-22314

Ponomarenko AI, Spetsius ZV (1978) Coesite of eclogites from the Udachnaya kimberlite pipe. XI Int'l Mineral Assoc Moskva, Nauka 2:23

Powell R (1985) Regression diagnostics and robust regression in geothermometer/geobarometer calibration: the garnet-clinopyroxene geothermometer revisited. J Meta Geol 3:327-342

Putnis A (1979) Electron petrography of high-temperature oxidation in olivine from the Rhum layered Intrusion. Mineral Mag 43:293-296

Redfern SAT, Wood BJ, Henderson CMB (1993) Static compressibility of magnesite to 20 GPa: implications for $MgCO_3$ in the lower mantle. Geophys Res Lett 20:2099-2102

Reinecke T (1991) Very-high-pressure metamorphism and uplift of coesite-bearing metasediments from the Zermatt-Saas zone, Western Alps. Eur J Mineral 3:7-17

Reinecke T, Van der Klauw S, Stockhert B (1992) Ultrahigh-pressure metamorphism of oceanic crust in the western Alps. 29th Int'l Geol. Cong Kyoto Japan Abstr 2:599

Rickard RS, Harris JW, Gurney JJ, Cardoso P (1989) Mineral inclusions in diamonds from the Koffiefontein Mine. *In:* Ross J et al. (eds) Kimberlite and related rocks, 2: Their mantle/crust setting, diamonds, and diamond exploration, Geol Soc Austra Spec Pub 14:1054-1062

Rietmeuer FJM (1996) Cellular precipitates of iron oxide in olivine in a stratospheric interplanetary dust particle. Mineral Mag 60:877-885

Ringwood AE, Lovering JF (1970) Significance of pyroxene-ilmenite intergrowths among kimberlite xenoliths. Earth Planet Sci Lett 7:371-375

Risold AC, Trommosdorff V (1996) Alpe Arami and Cima di Gagnone garnet peridotites: observations contradicting the hypothesis of ultra deep origin. EOS Trans Am Geophys Union 77:F761

Risold AC, Trommosdorff V, Reusser E, Ulmer P (1997) Genesis of $FeTiO_3$ inclusions in garnet peridotites from the central Alps. TERRA 9:28-29

Ross NL, Reeder RJ (1992) High-pressure structural study of dolomite and ankerite. Am Mineral 77:412-421

Rowley DB, Xue F, Tucker RD, Peng ZX, Baker J, Davis A (1997) Ages of ultrahigh pressure metamorphism and protolith orthogneisses from the eastern Dabie Shan: U/Pb zircon geochronology. Earth Planet Sci Lett 151:21-36

Ross II CR, Armbruster T, Canil D (1992) Crystal structure refinement of a spinelloid in the system Fe_3O_4-Fe_2SiO_4. Am Mineral 77:507-511

Rubie DC (1990) Role of kinetics in the formation and preservation of eclogite. *In:* Carswell DA (ed) Eclogite Facies Rocks. p 111-140, Chapman and Hall, New York

Rumble D (1998) Stable isotope geochemistry of ultrahigh pressure rocks. *In:* Hacker B, Liou JG (eds) When Continents Collide: Geodynamics and Geochemistry of Ultrahigh-Pressure Rocks. Kluwer Academic & Lippincott Raven Publishers

Rumble D, Yui TF (1998) The Qinglongshan oxygen and hydrogen isotope anomaly near Donghai in Jiangsu Province, China. Geochim Cosmochim Acta

Ryabchikov ID, Ganeev II (1990) Isotopic substitution of potassium in clinopyroxene at high pressures. Geokhimija 1:3-12

Schertl H, Schreyer W, Chopin C (1991) The pyrope-coesite rocks and their country rocks at Parigi, Dora Maira massif, western Alps: detailed petrography, mineral chemistry and PT-path. Contrib Mineral Petrol 108:1-21

Schertl H, Okay AI (1994) Coesite inclusion in dolomite of Dabie Shan, China: Petrological and geological significance. Eur J Mineral 6:995-1000

Schertl H, Schreyer W (1996) Mineral inclusions in heavy minerals of the ultrahigh-pressure metamorphic rocks of the Dora-Maira massif and their bearing on the relative timing of the petrological events. Geophyhs Monog 95:331-342

Schmidt MW, Poli S (1994) The stability of lawsonite and zoisite at high pressure: Experimental in CASH to 92 kbar and implications for the presence of hydrous phases in subducted lithosphere, Earth Planet Sci Lett 124:105-118

Schmidt MW (1996) Experimental constraints on recycling of potassium from subducted oceanic crust. Science 272:1927-1930

Schreyer W (1968) A reconnaissance study of the system MgO-Al_2O_3-SiO_2-H_2O at pressures between 10 and 25 kbar. Carnegie Inst. Washington Year Book 66:380-392

Schreyer W (1988a) Experimental studies on metamorphism of crustal rocks under mantle pressures. Mineral Mag 51:1-26

Schreyer W (1988b) Subduction of continental crust to mantle depths: Petrological evidence. Episodes 11:97-104

Schreyer W (1995) Ultradeep metamorphic rocks: The retrospective viewpoint. J Geophys Res 100:8353-66

Schreyer W, Baller TH (1981) Calderite, $Mn_3Fe_2Si_3O_{12}$, a high-pressure garnet. Proc. XII M.A. Meeting, Novosibirk, 1978. Experi Mineral 68-77

Selverstone J, Franz G, Thomas S (1992) Fluid heterogeneities in 2 GPa eclogites: implications for element recycling during subduction. Contrib Mineral Petrol 112:341-357

Sharp ZD, Essene EJ, Hunziker JC (1993) Stable isotope geochemistry and phase equilibria of coesite-bearing whiteschists, Dora Maira Massif, western Alps. Contrib Mineral Petrol 114:1-12

Shatsky VS, Jagoutz E, Kozmenko OA, Blinchik TM, Sobolev NV (1993) Age and genesis of eclogites from the Kokchetav massif (Northern Kazakhstan). Russian Geol Geophys 34:40-50

Shatsky VS, Sobolev NV, Vavilov MA (1995) Diamond-bearing metamorphic rocks of the Kokchetav massif (northern Kazakhstan). *In:* Coleman RG, Wang X (eds) Ultrahigh-Pressure Metamorphism, p 427-455. Cambridge Univ Press, Cambridge

Simon G, Chopin C, Schenk V (1997) Near-end-member magnesiochloritoid in prograde-zoned pyrope, Dora-Maira massif, western Alps. Lithos 41:37-57

Smith DC (1977) Titanochondrodite and Ti-clinohumite derived from the upper mantle in the Buell Park kimberlite, Arizona, USA: a discussion. Contrib Mineral Petrol 61:213-215

Smith DC (1984) Coesite in clinopyroxene in the Caledonides and its implications for geodynamics. Nature 310:641-644

Smith DC (1988) A review of the peculiar mineralogy of the Norwegian coesite-eclogite province, with crystal-chemical, petrological, geochemical and geodynamical notes and an extensive bibliography. *In:* Smith DC (ed) Eclogites and Eclogite-facies Rocks. p 1-206

Smith DC, Cheeney RF (1980) Oriented needles of quartz in clinopyroxene: evidence for exsolution of SiO_2 from a non-stoichiometric supersilicic "clinopyroxene". 26th Int'l Geol Cong Paris France 145

Smyth JR (1977) Quartz pseudomorphs after coesite. Am Mineral 62:828-830

Smyth JR, Hatton CJ (1977) A coesite-sanidine grospydite from the Robert Victor kimberlite. Earth Planet Sci Lett 34:284-290

Smyth JR (1994) A crystallographic model for hydrous wadsleyite (β-Mg_2SiO_4): An ocean in the Earth's interior. Am Mineral 79:1021-1024

Sobolev NV, Shatsky VS (1990) Diamond inclusions in garnets from metamorphic rocks. Nature 343:742-746

Sobolev NV, Shatsky VS, Valilov MA, Goryainov SV (1991) Coesite inclusion in zircon from diamondiferous gneiss of Kokchetav massif-first find of coesite in metamorphic rocks in the USSR territory. Doklady Akademii Nauk SSSR 321:184-188

Sobolev NV, Shatsky VS, Vavilov MA, Goryainov SV (1994) Zircon from ultrahigh-pressure metamorphic rocks of folded regions as an unique container of inclusions of diamond, coesite and coexisting minerals. Doklady Akademii Nauk 334:488-492

Song Y, Frey F (1989) Geochemistry of peridotite xenoliths in basalts from Hannuoba, eastern China: implications for subcontinental mantle heterogeneity. Geochim Cosmochim Acta 53:97-113

Spear FR (1993) Metamorphic phase equilibria and pressure-temperature-time paths. Mineral Soc Am Mono 1:799

Tabata H, Yamauchi K, Maruyama S, Liou JG (1998) Tracing the extent of ultrahigh-pressure metamorphic terrane: A mineral-inclusion study of zircons in gneisses from the Dabie Mountains. *In:* Hacker B, Liou JG (eds) When Continents Collide: Geodynamics and Geochemistry of Ultrahigh-Pressure Rocks. Kluwer Academic & Lippincott Raven Publishers

Thompson AB (1992) Water in the Earth's upper mantle. Nature 358:295-301

Tilton GR, Schreyer W, Schertl HP (1991) Pb-Sr-Nd isotopic behavior of deeply subducted crustal rocks from the Dora-Maira Massif, Western Alps-II: what is the age of the ultrahigh-pressure metamorphism? Contrib Mineral Petrol 108:22-33

Trommsdorff V, Evans BW (1980) Titanian hydroxyl-clinohumite: formation and breakdown in antigorite rocks (Malenco, Italy). Contrib Mineral Petrol 72:229-242

Tsai CH (1998) Petrology and geochemistry of mafic-ultramafic rocks in the north Dabie Complex, central-eastern China. PhD dissertation, Stanford Univ, Stanford, CA

Tsuruta K, Takahashi E (1998) Melting study of an alkali basalt JB-1 up to 10 GPa: Behavior of potassium in the deep mantle. Phys Earth Planet Inter

Ulmer P, Trommsdorff V (1997) Titanium content of mantle olivine: an experimental study to 10 GPa. TERRA 9:39

Van der Molen I, Van Roermund HLM (1986) The pressure path of solid inclusions in minerals: the retention of coesite inclusions during uplift. Lithos 19:317-324

Wang X, Liou JG (1991) Regional ultrahigh-pressure coesite-bearing eclogitic terrane in central China: evidence from country rocks, gneiss, marble, and metapelite. Geology 19:933-936

Wang X, Liou JG (1993) Ultrahigh-pressure metamorphism of carbonate rocks in the Dabie Mountains, central China. J Meta Geol 11:575-588

Wang X, Liou JG, Maruyama S (1992) Coesite-bearing eclogites from the Dabie Mountains, Central China: Petrology and P-T path. J. Geol 100:231-250

Wang X, Zhang RY, Liou JG (1995) Ultrahigh-pressure metamorphic terrane in eastern central China. *In:* Coleman RG, Wang X (eds) Ultrahigh-pressure metamorphism. Cambridge Univ Press, Cambridge, p 356-390

Wang Q, Ishiwatari A, Zhao Z, Hirajima T, Enami M, Zhai M, Li J, Cong BL (1993) Coesite-bearing granulite retrograded from eclogite in Weihai, eastern China: a preliminary study. Eur J Mineral 5:141-152

Welch MD, Pawley AR (1991) Tremolite: New enthalpy and entropy data from a phase equilibrium study of the reaction tremolite + 2 diopside + 1.5 orthoenstatite + ß-quartz + H_2O. Am Mineral 76:1931-1939

Wain A (1997) New evidence for coesite in eclogite and gneisses: Defining an ultrahigh-pressure province in the Western Gneiss region of Norway. Geology 25:927-930

Wood BJ (1974) The solubility of alumina in orthopyroxene coexisting with garnet. Contrib Mineral Petrol 46:1-15

Wood BJ, Banno S (1973) Garnet-orthopyroxene and orthopyroxene-clinopyroxene relationships in simple and complex systems. Contrib Mineral Petrol 42:109-124

Wood BJ, Henderson CMB (1978) Compositions and unit-cell parameters of synthetic non-stoichiometric tschermakitic clinopyroxene. Am Mineral 63:66-72

Woodland AB, Angel RJ (1998) Crystal structure of a new spinelloid with the wadsleyite structure in the system Fe_3O_4-Fe_2SiO_4 and implications for the Earth's mantle. Am Mineral 83:404-408

Wunder B (1996) Gleichgewichtsexperimente im system MgO-SiO_2-H_2O: Volaufige stabilitatsfelder von klinohumit ($Mg_9Si_4O_{16}(OH)_2$), chondrodit ($Mg_5Si_2O_8(OH)_2$) und phase A ($Mg_7Si_2O_8(OH)_6$). Berichte der deutschen Mineralogischen Gesellschaft, Beiheft 1 zum. Eur J Mineral 8:321

Wunder B, Rubie DC, Ross CR, Medenbach O, Seifert F, Schreyer W (1993) Synthesis, stability and properties of $Al_2SiO_4(OH)$: A fully hydrated analog of topaz. Am Mineral 78:285-297

Wyatt BA (1977) The melting and crystallization behavior of a natural clinopyroxene-ilmenite intergrowth. Contrib Mineral Petrol 61:1-9

Wyllie PJ, Huang WL, Otto J, Byrnes AP (1983) Carbonation of peridotites and decarbonation of siliceous dolomites represented in the system $CaO-MgO-SiO_2-CO_2$ to 30 kbar. Tectonophys 100:359-88

Xu ZQ (1997) Scientific significance and site-selected researches of the Chinese first continental scientific drill hole. Int'l workshop on the pre-site selection of scientific drilling in the Dabie-Sulu UHPM region, in Qingdao City, China, 16-20

Xu S, Okay AI, Ji S, Sengor AMC, Su W, Liu Y, Jiang L (1992) Diamond from the Dabie Shan metamorphic rocks and its implication for tectonic setting. Science 256:80-82

Yang J, Godard G, Kienast JR, Lu Y, Sun J (1993) Ultrahigh-pressure (60 kbar) magnesite-bearing garnet peridotites from northeastern Jiangsu, China. J Geol 101:541-554

Ye K, Hirajima T, Ishiwatari A, Guo J, Zhai M (1996) Significance of interstitial coesite in eclogite at Yangkou, Qingdao City, eastern China. Chniese Sci Bull 41:1407-1408 (in Chinese)

Yui TF, Rumble D, Lo CH (1995) Unusually low ^{18}O ultrahigh-pressure metamorphic rocks from the Sulu terrain, eastern China. Geochem Cosmochem Acta 59:2859-2864

Yui TF, Rumble D, Chen CH, Lo CH (1997) Stable isotope characteristics of eclogites from the ultrahigh-pressure metamorphic terrain, east-central China. Chem Geol 137:135-147

Zhang RY, Liou JG (1994a) Coesite-bearing eclogite in Henan Province, central China: detailed petrography, glaucophane stability and PT-path. Eur J Mineral 6:217-233

Zhang RY, Liou JG (1994b) Significance of magnesite paragenesis in ultrahigh-P metamorphic rocks. Am Mineral 79:397-400

Zhang RY, Liou JG (1996) Significance of coesite inclusions in dolomite from eclogite in the southern Dabie mountains, China. Am Mineral 80:181-186

Zhang RY, Liou JG (1997) Partial transformation of gabbro to coesite-bearing eclogites from Yangkou, the Sulu terrane, eastern China, J Meta Geol 15:183-202

Zhang RY, Liou JG (1998) Ultrahigh-pressure metamorphism of the Sulu terrane, eastern China: A prospective view. Continental Dynamics

Zhang RY, Liou JG, Cong BL (1994) Petrogenesis of garnet-bearing ultramafic rocks and associated eclogites in the Su-Lu ultrahigh-pressure metamorphic terrane, China. J Meta Geol 12:169-186

Zhang RY, Liou JG, Cong BL (1995a) Ultrahigh-pressure metamorphosed talc-, magnesite- and Ti-clinohumite-bearing mafic-ultramafic complex, Dabie Mountains, east central China. J Petrol 36:1011-1038

Zhang RY, Liou JG, Ernst WG (1995b) Ultrahigh-pressure metamorphism and decompressional P-T path of eclogites and country rocks from Weihai, eastern China. Island Arc 4:293-309

Zhang RY, Liou JG, Tsai CH (1996a), Petrogeneses of a high-temperature metamorphic belt: a new tectonic interpretation for the north Dabieshan, Central China. J Meta Geol 14:319-333.

Zhang RY, Liou JG, Ye K (1996b) Petrography of UHPM rocks and their country rock gneisses. In: Cong B (ed) Ultrahigh-Pressure Metamorphic Rocks in the Dabieshan-Sulu Region of China. p 49-68

Zhang RY, Liou JG, Ye K (1996c) Mineralogy of UHPM rocks. In: Cong B (ed) Ultrahigh-Pressure Metamorphic Rocks in the Dabieshan-Sulu Region of China. p 106-127.

Zhang RY, Hacker B, Liou JG (1997a) Exsolution in ultrahigh-P minerals from the Dabie-Sulu (China) and Kokchetav (Kazakhstan) terranes. Terra Nova 9:43

Zhang RY, Hirajima T, Banno S, Cong B, Liou JG (1995c) Petrology of ultrahigh-pressure rocks from the southern Sulu region, eastern China. J Meta Geol 13:659-675

Zhang RY, Rumble D, Liou JG, Wang QC (1998) Low delta ^{18}O ultrahigh-P garnet mafic-ultramafic rocks from Dabieshan, China. Chemical Geol.

Zhang RY, Shu JF, Mao HK, Liou JG (in press) Magnetite lamellae in olivine and clinohumite from Dabie UHP ultramafic rocks, east-central China. Am Mineral

Zhang RY, Liou JG, Coleman RG, Ernst WG, Sobolev NV, Shatsky VS (1997b) Metamorphic evolution of diamond-bearing and associated rocks from the Kokchetav Massif, northern Kazakhstan. J Meta Geol 15:479-496.

Zharikov VA, Ishbulatov RA, Chudinovskikh LT (1984) High-pressure clinopyroxene and the eclogite barrier. Sov Geol Geophys 25:53-61

Zheng YF, Fu B, Cong BL, Li SG (1996) Extreme ^{18}O depletion in eclogite from the Su-Lu terrane in east China. Eur J Mineral 8:317-323

Zheng YF, Fu B, Li YL, Xiao Y, Li SG (1998) Oxygen and hydrogen isotope geochemistry of ultrahigh-pressure eclogites from the Dabie Mountains and the Sulu terrane. Earth Planet Sci Lett 155:113-129

Chapter 3

THE UPPER MANTLE
NEAR CONVERGENT PLATE BOUNDARIES

Bjorn O. Mysen

Geophysical Laboratory and Center for High-Pressure Research
Carnegie Institution of Washington
5251 Broad Branch Road, NW
Washington DC 20015

Peter Ulmer

Institute für Mineralogie und Petrologie
ETH-Zentrum
CH-8092 Zürich, Switzerland

Jürgen Konzett

Geophysical Laboratory and Center for High-Pressure Research
Carnegie Institution of Washington
5251 Broad Branch Road, NW
Washington DC 20015

Max W. Schmidt

CNRS–UMR 6524
Magmas et Volcan
5 rue Kessler
63039-Clermont-Ferrand, France

INTRODUCTION

Magmatic, metamorphic, hydrothermal, and dynamic processes at and near convergent plate boundaries offer unique opportunities to examine the physics and chemistry of materials recycling in the earth.

The upper mantle near convergent plate boundaries (Fig. 1) typically consists of the following major tectonic units. There is the subducting slab itself, which at the beginning of descent into the underlying upper mantle, consists of pelagic sediments, oceanic crust of basalt, dolerite and gabbro, hydrated and pristine oceanic upper mantle. This slab undergoes series of metamorphic reactions, and may possibly melt, as it descends into the upper mantle and perhaps even togreater depth.

Above the subducting slab is an upper mantle wedge (Fig. 1). The material in this zone consists of peridotite that may be altered by ingress of fluids or melts from the subducting slab. This metasomatism might affect the chemical composition of the upper mantle wedge itself, both in regard to major and trace element chemistry. This alteration in bulk composition, in turn, results in the formation of minerals not found in peridotite elsewhere in the Earth's upper mantle. Furthermore, melting processes ultimately responsible for formation of overlying crust as well as near-surface volcanic and hydrothermal activity, most likely originate in the mantle wedge immediately above the subducting slab (Fig. 1).

Above the upper mantle wedge, there is crust, often referred to as an island arc although in continent-continent collision environments, the surface manifestation will be

0275-0279/98/0037-0003$05.00

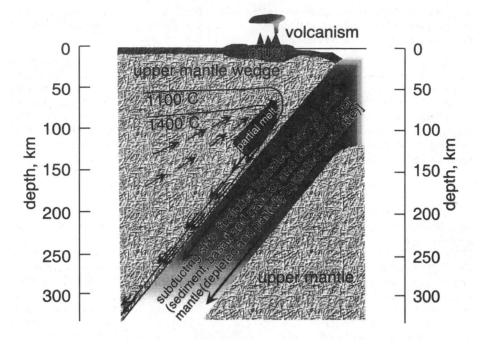

Figure 1. Schematic representation of the principal tectonic units in the upper mantle and crust near convergent plate boundaries. Arrows denote flow directions. Solid lines with temperatures illustrate isotherms.

uplift and mountain chain formation. The crust in island arcs consists of lower and intermediate portions that include igneous intrusives dominated by intermediate granitoids, but does also include more mafic components such as gabbro, in particular at at greater depth. Active volcanism, often, but not always, of an explosive nature, is a typical feature in island arcs. The explosive nature of this volcanism can be traced to H_2O released during dehydration of the subducting slab during its descent into the mantle.

Metamorphic reactions that occur in the descending slab and in the overlying mantle wedge during plate descent require identification and characterization through field and experimental studies. To this end, experimental data on element partitioning behavior and phase relations among relevant minerals during hydration and dehydration and during partial melting have been the focus of much research.

Study of the relevant mineral assemblages provides the key for establishing the mechanisms by which the bulk compositions, temperature and pressure control the mineralogy of both the subducting slab and the overlying mantle wedge. Viable physical and chemical models for upper mantle near convergent plate boundaries can be tested only with experimentally-determined partitioning of trace and major elements between minerals, melts and fluids, by detailed laboratory studies of phase equilibria, volatile solubility in melts and silicate solubility in fluids, and by thermochemical properties of the phases and phase assemblages involved.

In view of the chemical complexity of natural materials, information such as that illustrated above can realistically be obtained by integrated research efforts that include field observations, laboratory experiments, and theoretical modeling. Significant experimental progress has been made in this area. In this review, some of the relevant natural observations will be combined with published results that offer experimental constraints on the mineralogy and physicochemical properties of the upper mantle.

In this chapter, we will review the most important natural, experimental, and theoretical information relevant to the mineralogy of the upper mantle near convergent plate boundaries. Information from natural observations will be summarized first. This summary will be followed by a discussion of experimental data on stability of minerals and mineral assemblages. Some aspects of materials transport from the subducting slab to the overlying mantle, and the consequences of this transport for the chemical composition and phase relations in the overlying mantle will also be examined. The chemical composition of the minerals and mineral groups discussed in this chapter are summarized in Table 1.

SEISMIC DEFINITION OF SUBDUCTION

The upper mantle beneath convergent plate boundaries can be defined from seismic data. Results from recent advances in seismic tomography reveal that compared with the surrounding mantle the descending slab as a seismically faster, and probably cooler tongue descending into the mantle, perhaps including the lower mantle (van der Hilst et al. 1991, Kerr 1997). In other words, the propagation velocity of the sesimic waves typically is greater in the subducting slab than in the surrounding peridotite upper mantle. Seismic tomography is a three dimensional method to image these velocities. The high-velocity zones can be clearly delineated in cross-sections across central Japan to perhaps slightly greater depth than 400 km (Fig. 2) and in Kamchatka to depths exceeding 1000 km (Fig. 1). The exact depth to which a descending slab may be identified is, however, under discussion (e.g, Silver et al. 1986, Cormier 1989, Jeanloz and Williams, this volume).

Often the subduction zone itself is marked by a double seismic zone (e.g. Abers 1996). An upper zone marks the interface between the descending slab and the overlying peridotite wedge (Fig. 3). About 20-40 km into the slab itself, there appears to be a second layer, a layer that most probably reflects the compositional heterogeneity within the subducting slab. The distance between the two zones is insufficient to mark the upper and lower boundary of the subducting slab.

The velocity structure may be converted to mineralogical changes with depth. Attention has been paid in particular to the uppermost portions of the subducting lab. This portion presumably includes metamorphosed sediments and oceanic crust. To this end Bina and Helffrich (1992) employed the phase relations suggested for a hydrated metabasalt system (e.g. Peacock 1993). Whereas the basalt-to-eclogite transformation is commonly invoked to model the seismic data, Bina and Helffrich (1992) found that the seismic data were consistent with mineralogies including hydrous phases such as lawsonite, glaucophane, tremolite and zoisite. Helffrich (1996) was also able to model a low-velocity layer near the top of the subduction zone to depth of about 65 km as a lawsonite blueschist with a transition to eclogite assemblages at greater depth.

NATURAL EXAMPLES OF THE UPPER MANTLE
NEAR CONVERGENT PLATE BOUNDARIES

Natural geologic, mineralogical and petrological information relevant to the upper mantle near convergent plate boundaries can be examined in a number of geologic settings. These include (1) accretionary subduction complexes such as illustrated by the Fransiscan Complex, California, (2), high and very high-grade metamorphic terranes, (3) peridotite massifs that include well known localities such as Ronda, Lherz, Josephine, Horoman, and

Table 1. Compositions of minerals used in text.[†]

10Å-phase	$Mg_3Si_4O_{10}(OH)_2 \cdot 2H_2O$
albite	$NaAlSi_3O_8$
amphibole	$AX_2Y_5Z_9O_{22}(OH,F)_2$ where
	$A = Na,K; \quad X = Na,Ca,Fe^{2+},Mg$
	$Y = Mg,Fe^{2+},Fe^{3+},Al; \quad Z = Si,Al$
anthophyllite	$Mg_7Si_8O_{22}(OH)_2$
antigorite (m = 17)	$Mg_{48}Si_{34}O_{85}(OH)_{62}$
apatite	$Ca_3(PO_4)_2(OH,F,Cl)$
brucite	$Mg(OH)_2$
calcite	$CaCO_3$
chlorite (clinochlor)	$Mg_5Al_2Si_3O_{10}(OH)_8$
chondrodite	$Mg_5Si_2O_8(OH)_2$
clinohumite	$Mg_9Si_4O_{16}(OH)_2$
clinopyroxene	XYZ_2O_6, where
	$X = Na,Ca; \quad Y = Mg,Fe^{2+},Fe^{3+},Al; \quad Z = Si,Al$
cordierite	$Mg_2Al_4Si_5O_{18}$
corundum	Al_2O_3
diamond	C
diaspore	$AlO(OH)$
diopside	$CaMgSi_2O_6$
ellenbergerite	$[Mg,(Ti,Zr)]_2Mg_6(Al,Mg)_6(Si,P)_8O_{29}(OH)_{10}$
enstatite	$MgSiO_3$
epidotite	$X_2Y_3Si_3(O,OH,F)_{13}$
	$X = Ca, REE^{3+},Fe^{2+}; \quad Y = Al, Fe^{3+}, Fe^{2+}. Ti$
forsterite	Mg_2SiO_4
garnet	$(Ca,Fe^{2+},Mg)_2(Al,Si)_3O_{12}$
graphite	C
humite	$Mg_7Si_3O_{12}(OH)_2$
ilmenite	$(Fe,Mg)TiO_3$
jadeite	$NaAlSi_2O_6$
K-richterite	$K_2Ca(Mg,Fe^{2+})_5Si_8O_{22}(OH)_2$
kyanite	Al_2SiO_5
lawsonite	$CaAl_2(Si_2O_7)(OH)_2H_2O$
lizardite	$(Mg,Al)_3(Si,Al)_2O_5(OH)_4$
magnesiowustite	$(Mg,Fe)O$
magnetite	Fe_3O_4
MgMgAl-pumpellyite	$Mg_4(Al,Mg)Al_4Si_6O_{21}(OH)_7$
orthopyroxene	XYZ_2O_6, where
	$X = Na,Ca; \quad Y = Mg,Fe^{2+},Fe^{3+},Al; \quad Z = Si,Al$
norbergite	$Mg_3SiO_4(OH)_2$
OH-topaz	$Al_2SiO_4(OH)_2$
paragonite	$Na_2Al_4Al_2Si_6O_{20}(OH)_4$
phengite	$K_2Al_2(Mg,Fe)_2Si_8O_{20}(OH)_4$
phase A	$Mg_7Si_2O_8(OH)_6$
phase B	$Mg_{12}Si_4O_{19}(OH)_2$
phase C	$Mg_{10}Si_3O_{14}(OH)_4 \ (?)^3$
phase D	$MgSi_{2-x}H_{2+4x}O_6$
phase E	$Mg_{2.3}Si_{1.25}H_{2.4}O_6$
phase F	$MgSi_{2-x}H_{2+4x}O_6^*$
phase G	$MgSi_{2-x}H_{2+4x}O_6^*$
phlogopite	$K_2(Mg,Fe^{2+})_6Si_6Al_2O_{20}()H,F)_4$
quartz	SiO_2
pyrope	$Mg_3Al_2Si_3O_{12}$
rutile	TiO_2
serpentine (chrysotile, lizardite)	$Mg_3Si_2O_5(OH)_4$

Table 1 (continued).

sphene	$CaTiSiO_5$
spinel	$MgAl_2O_4$
staurolite	$(Mg,Fe^{2+})_2(Al,Fe^{3+},_9O_6(SiO_4)_4(O,OH)_2$
stishovite	SiO_2
talc	$Mg_3Si_4O_{10}(OH)_2$
Ti-chondrodite	$Ti_{0.5}Mg_{4.5}Si_2O_9(OH)$
Ti-clinohumite (x = 0.5)	$Ti_{0.5}Mg_{8.5}Si_4O_{17}(OH)$
tremolite	$Ca_2Mg_5Si_8O_{22}(OH)_2$

[†] Mineral names are abbreviated in many of the figures. For each figure, the individual abbreviations are defined in the figure captions.
[#] Phase C is often considered the same phase at superhydrous B. See text for further discussion of this.
[*] Although phases D, F and G in the literature have been suggested to be individual, distinct phases, it is now considered that these may well be the same phase. See text for further discussion.

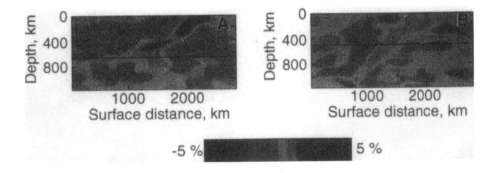

Figure 2. Vertical mantle sections across the Izu Bonin (A) and Kurile Arcs (B). Both sections were developed by inversion of ISC P data. Modified after van der Hilst et al. (1991).

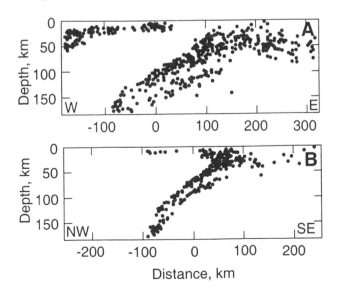

Figure 3. Cross sections of double seismic zones from Honshu, Japan (A) and the Alaska Peninsula (B). Data compiled by Abers (1996).

Lanzo, and (4) upper mantle xenoliths entrained in island arcs volcanics.

Accretionary subduction complexes

The Fransiscan complex can be viewed as a typical example of an accretionary subduction complex in which there is evidence for both prograde and retrograde metamorphism (e.g. Ernst 1971, 1988). In the Fransiscan complex, tectonic blocks, which record metamorphism up to eclogite grade are common, whereas examples of in situ eclogite metamorphism are less frequent. Such eclogite, consisting of garnet, glaucophane, phengite, albite, quartz and rutile and may record peak metamorphism at pressures near 1 GPa and temperatures near 300°C (Oh and Liou 1990). Among other distinctive rock types are blueschists and garnet amphibolite. The blueschsists consist mostly of glaucophane, epidote, phengite, sphene, quartz, garnet, and chlorite. The principal phases in garnet amphibolite are garnet, edenitic hornblende, and epidote with minor quartz, rutile and phengite. A typical pressure-temperature path for the Fransiscan Complex is shown in Figure 4.

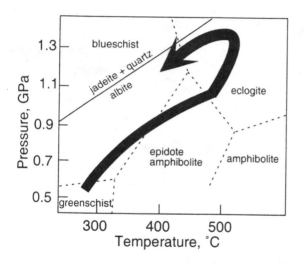

Figure 4. Pressure-temperature path of in situ eclogite facies rocks from Jenner, Fransiscan Complex. Modified after Oh and Liou (1990) with albite breakdown curve from Newton and Smith (1967).

High and very high pressure terranes

High- and very-high pressure metamorphic terranes have been identified world-wide following the discovery of coesite (Chopin 1984) and microdiamonds (Sobolev and Shatsky 1990) in crustal rocks associated with continental collision zones. These rocks have been subject to intense study, and a detailed review of the geological setting and mineral assemblages can be found elsewhere in this volume (Liou et al., this volume). Among the most well-known are the Dora Maira Massif in the Western Alps (Chopin et al. 1991, Kienast et al. 1991), the Sulu-Dabie Complex in East Central China (Schertl and Okay 1994, Zhang and Liou 1996, Liou et al., this volume), the Western Gneiss Region, Norway (Medaris and Carswell 1990), and the Kokchetav Region, Kazakhstan (Ernst et al. 1995, Maruyama et al. 1996). These rocks are unique in that pressures as high as perhaps in excess of 5 GPa are recorded in some of the mineral assemblages.

The terranes include meta-sediments, mafic and minor ultramafic sequences. All rock types show evidence of high to very high-pressure metamorphic conditions. The very high-pressure conditions are indicated by the occurrence of coesite or pseudomorphs after coesite, ellenbergerite, microdiamonds, phengite, zoisite, and pure or nearly pure pyrope garnet.

Metasediments occur either as interlayers with eclogite or as host country rocks of both eclogites and garnet peridotites. The metasediments include quartzofeldspathic gneiss together with quartzite, pelite, marble, garnet-quartz jadeite rock, calc-silicate rocks, and whiteschist. The classic locality of coesite and whiteschist, the Dora Maria Massif of the Western Alps, also includes unusual minerals such ellenbergerite (e.g. Chopin 1986).

Kyanite quartzite consists of quartz, kyanite, zoisite, pyrite, and rutile. Quartzite consists of quartz and kyanite, with smaller amounts of zoisite, phengite and epidote as well as minor garnet and rutile (Zhang et al. 1995). Garnet-jadeite quartz rocks are composed of jadeite, garnet, quartz, minor rutile, and apatite with or without chlinozoisite (Liou et at 1997).

In the Dabie-Sulu complex, China, some eclogite-bearing marbles contain inclusions of quartz pseudomorphs after coesite and calcite pseudomorphs after aragonite in garnet, and coesite inclusions in dolomite (Schertl and Okay 1994, Zhang and Liou 1996). Mg-rich calcite, diopside, forsterite, and minor Ti-clinohumite and phlogopite occur in the matrix.

Eclogite is the ultimate product of very high-pressure metamorphism of gabbros, basaltic flows and tuffs, and diabasic sills. These rocks exhibit the most varied mineral assemblages and provide perhaps the most detailed information of the petrogenetic history among the rocks found in high- and very high-pressure metamorphic terranes. Eclogite contains garnet + omphacite + rutile as essential phases and occurs in all very high-pressure terranes. Eclogite in gneiss has inclusions of coesite. Quartz pseudomorphs after coesite are common in garnet and omphacite, more rarely in kyanite, zoisite, or epidote. Hydrous phases such as phengite, epidote/zoisite, and talc, are also among the very high-pressure minerals.

Most eclogites within ultramafic rocks are bimineralic garnet + omphacite with minor rutile. A few contain minor quartz, kyanite, phengite, or amphibole (Zhang et al. 1994, Zhang and Liou 1998). Corundum-bearing garnetite and eclogite occurs as lenses in the Zhimafang garnet lherzolite in China (Enami and Zang 1988).

Peridotite massifs

Peridotite massifs occur worldwide. The most well known (and also the largest) probably is that of Ronda, Southern Spain (e.g. Obata 1980). Others include Beni Bousera in neighboring Morocco (Pearson et al. 1989), Lherz in the French Pyreenes (Hall and Bennet 1979, Bodinier et al. 1987), Lanzo, Western Italian Alps (e.g. Venturini et al. 1994), the Bohemian massif, Germany (Medaris et al. 1995). These massifs consist of peridotite, harzburgite, and wherlite and minor abundance of more mafic compositions occurring as veins. Hydrous minerals, in particular amphibole, are common in these veins.

The peridotite massifs were most likely emplaced at relatively shallow depth and at comparatively high temperature. Those conditions notwithstanding, the massifs contain lithologies ranging from plagioclase peridotite, through spinel to garnet peridotite (e.g. Obata 1980, Suen and Frey 1986). There is also evidence of high-pressure crystal fractionation (Suen and Frey 1986) perhaps at pressures exceeding 2 GPa.

These phase assemblages are consistent with equilibration in the pressure range between 1 and 3 GPa. Recent descriptions of possible graphitized diamond in garnet pyroxenite from Ronda and Beni Bousera may place an even higher pressure limit (Davies et al. 1993, Pearson et al. 1993 see also Slodkevitch 1982, Zhang et al. 1994, Krogh and

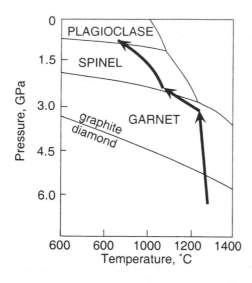

Figure 5. Schematic temperature-pressure diagram of phase relations relevant to Ronda massif. GARNET, SPINEL, and PLAGIOCLASE denote the stability fields of garnet peridotite, spinel peridotite, and plagioclase peridotite. Thick arrows denote the pressure-temperature path of the massif. Modified after Davies et al. (1993).

Carswell 1995). Based on these lithologies together with the evidence for possible diamond a pressure-temperature path such as that shown in Figure 5 has been suggested for the Ronda massif (Davies et al. 1993).

Upper mantle xenoliths

Upper mantle xenoliths entrained in basaltic and related rocks in arc environments are rare compared with the mantle xenolith suites found in alkali basalt and kimberlite erupted in other terranes. Crustal xenoliths are, on the other hand, more common (Aoki and Kuno 1973). Nevertheless, extensive studies on the xenolith suites from Itchinomegata, Japan (e.g. Aoki. 1987), from Alaska and the Kurile chain (e.g. Swanson et al. 1987, DeLong et al. 1975), the Philippines (e.g. Maury et al. 1992, Vidal et al. 1989) and from the Western United States (e.g. Mukhopadhyay and Manton 1994, Riter and Smith 1996) offer a glimpse of the upper mantle wedge above subducting oceanic plates.

The ultramafic xenoliths commonly are spinel peridotites (olivine + orthopyroxene + clinopyroxene + spinel) reflecting equilibration pressures in the 1-2 GPa range and at temperatures near 1000±200°C (Takahashi 1986a, Swanson et al. 1987). Garnet-bearing peridotite xenoliths (olivine + orthopyroxene + clinopyroxene + garnet), which are common in the xenolith suite in kimberlite, are absent in the xenoliths sampled from island arc environments. In some xenoliths, glass inclusions, perhaps reflecting incipient melting, have been observed (e.g. Takahashi 1986a). Notably, the spinel peridotite inclusions commonly contain several percent pargasitic amphibole. This amphibole often show evidence of partial breakdown to diopside + spinel + olivine ± plagioclase ± magnetite. The breakdown assemblage also sometimes contains Al-rich glass (Aoki and Shiba 1973) with SiO_2 contents that may reach 65 wt % (Fig. 6). Phlogopite is among other rare hydrous phases (Ionov and Hofmann 1995, Ertan and Leeman 1996) in certain xenoliths suites from the Baikal region, Germany and the Western United States.

Natural evidence for the role of H_2O in subduction zones

In the oceanic and continental upper mantle most evidence points to low H_2O abundances, perhaps in the range 100-500 ppm H_2O (e.g. Jambon and Zimmermann 1990,

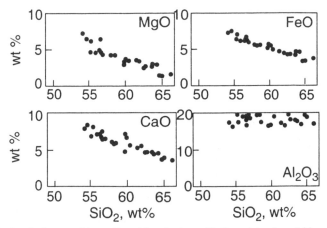

Figure 6. Bulk chemical compositions of partial melts in peridotite nodules from Ichinome Crater, Japan. Data from Takahashi (1986).

Thompson 1992). The upper mantle near convergent plate boundaries, on the other hand, appears to be considerably more H_2O-rich.

The natural evidence for H_2O-rich conditions derive in part from the H_2O-rich mineral parageneses in exhumed metamorphic terranes (see above). Additional evidence exists in metasomatically altered xenoliths, in fluid inclusions in metamorphic minerals, in the H_2O content of arc magmas and of melt inclusions in phnocrysts from arc environments. The melting phase relations of arc magmas appear consistent with high water activity (Kushiro 1972, 1990). Finally, it is commonly concluded that high H_2O activity would also be reflected in higher oxygen fugacities in island arc upper mantle environments as compared with regions of the upper mantle where H_2O plays a less dominant role (e.g. Wood et al. 1990, Arculus 1994).

Metasomatically altered xenoliths

Metasomatic alteration of the mantle wedge overlying a subducting slab through invasion of aqueous fluids or hydrous melts can be seen in mineral assemblages (e.g. Ertan and Leeman 1996, Johnson et al. 1996), and in the trace and isotope geochemistry of arc-derived ultramafic xenoliths (e.g. Maury et al. 1992, Ionov and Hofmann 1995). Recent studies of B/Be systematics have been used to estimate the proportion of hydrous slab components in the source region of arc magmatism (e.g. Bebout et al. 1993).

The most obvious example of influx of H_2O to the mantle wedge is found in the occurrence of sodic amphiboles (pargasite) frequently observed in the xenolith suites (e.g. Aoki 1987, Swanson et al. 1987). Occasionally, phlogopite can also be found (Swanson et al. 1987, Ionov and Hofmann 1995). Overgrowth of olivine by orthopyroxene rims has been taken as evidence of transport by of silica from the subducting slab of silica rich fluids (Ertan and Leeman 1996). Amphibole inclusions within these rims is additional evidence for the presence of aqueous fluids.

Trace element patterns in xenoliths have also been employed to infer a slab signature in the overlying mantle. For example, there is a relative depletion of high field strength elements and in particular Nb and Ta relative to large ion lithophile elements and light to middle Rare Earth elements. relative to primitive mantle composition (Ionov and Hofmann 1995; see also Fig. 7) as well as enrichment of other highly incompatible elements expected

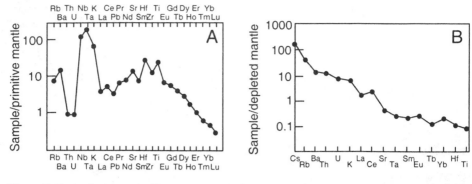

Figure 7. Trace element patters from metasomatized xenoliths from subduction environments. (A) From amphibole in veins in a nodule from Western Baikal Region, Russia (data from Ionov and Hofmann 1995). (B) Average xenolith composition for xenoliths from the Luzon Arc, Philippines. Data from Maury et al. (1992).

to enter a fluid or a hydrous melt formed in the subducting slab. Equally compelling evidence can be seen in the rare earth element patterns in spinel lherzolite xenoliths from the Philippines, suggesting perhaps garnet in the source region of the metasomatizing fluid (Fig. 7).

H_2O content of arc magmas and melt inclusions in arc-derived minerals

The H_2O budget during partial melting in the subducting slab or in the overlying peridotite mantle wedge may be inferred from analysis of H_2O contents of erupted magma (glass) or from H_2O content of glass trapped in phenocrysts (Harris et al. 1984, Schiano et al. 1995, Sobolev and Chaussidon 1996). The former approach suffers from the need to assume that the analyzed water contents do indeed reflect the H_2O content of the melt and was not altered during crystallization processes or by other processes subsequent to formation of the glass. It is not clear, therefore, how precise such estimates might be. Nevertheless, felsic magma associated with arc volcanism and explosive volcanism suggest that these melts may have had on the order of 5 wt % H_2O at least during the later stages of the magmatic evolution (e.g. Rutherford et al. 1985), whereas suggestions in the range 1-3 wt % H_2O have been made for primitive basalt and basaltic andesite (Sakuyama 1979)

The H_2O content in glass inclusions trapped in phenocrysts during magma crystallization may provide snapshots in time of the water content of the magma itself. High water contents were observed in glass inclusions in olivine from arc magmas (Harris et al. 1984; see Fig. 8). The negative correlation between forsterite content of the olivine phenocrysts and the H_2O content of the glass inclusion may suggest, however, that those olivine trapped liquid during fractional crystallization. Interestingly, in a re-analysis of some of those glasses, Sisson and Layne (1993) reported up to 6.2 wt % H_2O. A recent study of hydrous silica-rich glass inclusions in spinel-bearing harzburgite from the Philippines (Schiano et al. 1995) do, however, indicate H_2O contents of the magma near 5 wt %.

Water contents inferred from experimental phase equilibria

Phase relations of arc magmas (or any other magmatic system) reflect bulk composition, temperature, pressure, activities of volatiles and the fugacity of oxygen. Fingerprinting techniques where these variables have been altered in laboratory experiments so as to duplicate bulk compositions and mineral assemblages of arc magmas, may be employed to infer water contents in their melting source regions. In this manner, Eggler

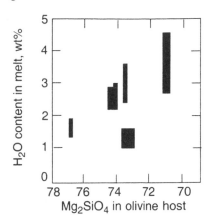

Figure 8. H_2O concentration in glass inclusions in olivine phenocrysts from the October 1974 of the Fuego eruption as a function of Mg_2SiO_4 content of the host olivine. Size of symbol denotes uncertainties. Data from Harris et al. (1984).

(1972) concluded, for example, that Paricutin andesite crystallized from a lava with about 2.2 wt % H_2O. Gaetani et al. (1993) in a conceptually similar experimental approach concluded that as much as 6 wt % H_2O might be dissolved in magmas during crystallization of arc magmas. In a series of partial melting experiments aimed at understanding the melting behavior beneath the Japanese arc, Kushiro (1990) concluded that basaltic melts probably contained on the order of 1 wt % H_2O, whereas magnesian andesite melt formed by direct partial melting of the peridotite beneath the Japanese arc might contain as much as 8 wt % H_2O.

Water dissolved in magmas also affects equilibrium mineral compositions and those can be used to deduce H_2O budgets. For example, Yoder et al. (1957) showed that high water pressure causes plagioclase to begin to crystallize at lower temperatures than in dry systems. The plagioclase melting loop under hydrous conditions would also favor crystallization of more An-rich plagioclase than that crystallizing from water-free plagioclase melt compositions. The plagioclase commonly found in arc basalts (An_{90-95}) is more anorthite-rich than the plagioclase found in Mid-Ocean Ridge Basalt (MORB). This difference in An-content may be attributed, therefore, to a higher water fugacity in arc basalts than in ridge basalts.

From the plagioclase-melt exchange equilibrium

$$K_d = (Ca/Na)^{plagioclase}/(Ca/Na)^{melt} \tag{1}$$

Sisson and Grove (1993) found that for dry melts at pressures below 2.0 GPa, this K_d is always less than 2.5. The K_d increases with increasing water content in the melt, and reaches 5.5 for water-saturated liquids at 0.2 GPa (Sisson and Grove 1993). These results quantify the observation that at a given temperature and bulk composition, plagioclase coexisting with hydrous melts is more An-rich than plagioclase coexisting with dry melts.

Hornblende phenocrysts, together with anorthite-rich plagioclase and wollastonite-rich clinopyroxene, are classic phase equilibrium indicators of high water pressure. Merzbacher and Eggler (1984) showed that 4 to 5 wt % dissolved water is necessary to stabilize hydrous hornblende in basaltic to dacitic melts. Hence, the occurrence of hydrous hornblende phenocrysts in natural rocks requires that at least several weight percent water must have been dissolved in the coexisting melt.

Oxygen fugacity in the mantle near convergent plate boundaries

The oxygen fugacity in upper mantle peridotite can be calculated based on the simple equilibrium:

$$2 \, Fe_3O_4 + 3 \, Fe_2Si_2O_6 = 6 \, Fe_2SiO_4 + O_2 \tag{2}$$

(Wood et al. 1990, Brandon and Draper 1996, Woodland et al. 1992). An example from spinel harzburgite from the Western United States (Fig. 9) reveal that compared with abyssal peridotites as well as xenolith data from cratonic kimberlites, the ultramafic xenoliths in arc lavas are ~2 orders of magnitude higher values of oxygen fugacity. This observation is consistent, therefore, with the H_2O-rich environment associated with continental margins as compared abyssal and continental upper mantle (Wood et al. 1990, Arculus 1994).

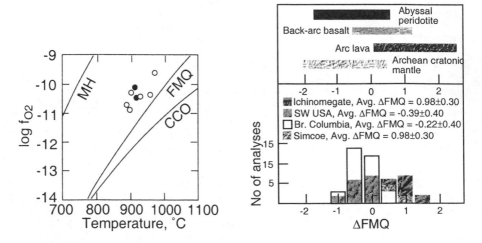

Figure 9. Values of oxygen fugacity, f_{O2}, calculated from spinel crystal chemistry in peridotite from Simcoe, Washington, USA (left panel) at 1.5 GPa pressure. Closed symbols: websterite, open symbols: harzburgite. MH = magnetite-hematite oxygen buffer trajectory, QFM = quartz-fayalite-magnetite oxygen buffer trajectory, and CCO = carbon-carbon monoxide-carbon dioxide oxygen fugacity trajectory. The upper right panel shows the range of oxygen fugacity, calculated relative to that of the quartz-magnetite-fayalite buffer (QFM),calculated for upper mantle and source regions of upper mantle magmas. The lower right panel shows distribution in calculated oxygen fugacities for several different subduction zone environments.

PHASE RELATIONS AMONG METASEDIMENTS, GABBRO, AND ULTRAMAFIC COMPOSITIONS NEAR SUBDUCTION ZONES

The subduction of oceanic lithosphere, composed of peridotitic mantle, plutonic cumulates and gabbros, basaltic volcanics and pelitic and carbonate sediments, is the most efficient material recycling process in Earth's upper mantle and transition zone. Water-rich basaltic compositions represented by gabbroic and volcanic rocks together with pelitic compositions corresponding to ocean floor sediments have been considered the principal H_2O-reservoirs in the subducted oceanic lithosphere. In the basaltic component, pargasitic amphibole is often considered the dominant H_2O-carrier (e.g. Tatsumi 1986, Davies and Stevenson 1992), whereas in the pelitic metasediments, phengitic white mica frequently has been proposed as an additional H_2O repository (e.g. Massone and Schreyer 1989, Schmidt 1996, Domanik and Holloway 1996).

If the thickness of oceanic crust is used as a guide to estimate the fraction of sediment

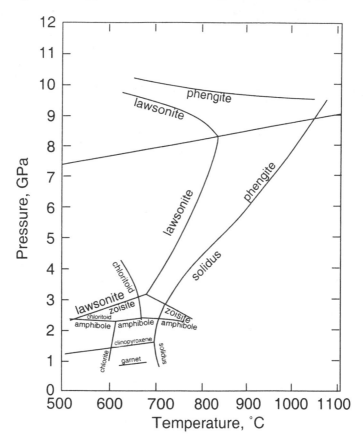

Figure 10. Phase relations for H_2O-saturated mid-ocean ridge basalt. Experimental data compiled by Schmidt and Poli (1998).

and gabbroic components in the subducting slab, perhaps <20% of the slab volume would be made up from oceanic crust and sediments. Nevertheless, before considering the volumetrically more important ultramafic component (subducted oceanic lithosphere), a brief summary of the phase relations and mineral assemblages expected in subducted sediments and gabbroic compositions is warranted.

Phase relations in metabasalt, basalt and metasediments

Phase relationships of mid-ocean ridge basalt at water-saturated conditions as compiled from the experimental data of Schmidt and Poli (1998) (>3.0 GPa), Apted and Liou (1983) (<0.8 GPa), Poli (1993) (0.8-2.4 GPa) and Pawley and Holloway (1993) (2.0-3.0 GPa) are shown in Figure 10. Assemblages from natural blueschists were employed for the lowest-temperature and -pressure conditions.

At blueschist conditions (10-50 km depth), the phase assemblage in metabasalt is lawsonite + glaucophane + chlorite + albite (or jadeite, depending on exact pressure and temperature) ± phengite (Fig. 10). As the temperature is increased lawsonite reacts to produce epidote or zoisite (Fig. 10), whereas chlorite decomposes to form principally garnet. Glaucophane is transformed to barroisitic amphibole. At temperatures above 600°C, an 1.5-2.2 GPa, an amphibole-bearing eclogite results. At pressures above about 2.4 GPa and temperatures below 660°C (corresponding to about 70 km depth), amphibole breaks down to chloritoid and blueschist transforms to lawsonite-eclogites (Fig. 10). Amphibole-bearing eclogite is transformed to zoisite-bearing eclogite. At pressures near 3.0 GPa, zoisite reaches its maximum stability in metabasalt resulting in an almost dry eclogite at conditions above 3.0 GPa 700°C (Fig. 10) corresponding to a depth near 100 km in the subducting metabasaltic slab component.

At lower temperatures, lawsonite persists. Lawsonite reaches its maximum temperature stability at 830°C and 8.4 GPa (Fig. 10). At pressures above the zoisite stability, lawsonite decomposes through a continuous reaction of the type

$$\text{lawsonite} + \text{diopside} + \text{garnet}_1 = \text{garnet}_2 + \text{coesite/stishovite} + H_2O \tag{3}$$

where garnet$_2$ is richer in grossular and pyrope component than garnet$_1$ (Schmidt and Poli 1998). In the coesite stability field, this reaction has a positive ΔV and a positive dP/dT slope which steepens with increasing pressure (Fig. 10). When stishovite replaces coesite (above ~8 GPa), ΔV of Reaction (3) becomes negative, which results in a negative dP/dT slope for the reaction. The maximum pressure stability of lawsonite in subduction zones is, therefore, to 8-9 GPa. In a potassium-bearing metabasalt, phengite forms ubiquitously to pressures of 10-11 GPa (Domanik and Holloway 1996, Schmidt 1996, Ono 1998).

Other minor hydrous phases in metabasalt are paragonite, talc, and staurolite. Paragonite forms at 1.4±0.2 GPa (at 500-650°C) and decomposes near 2.2 GPa and 500-700°C (Fig. 10). The conditions of talc and staurolite occurrence are relatively restricted, although talc might be an important high pressure phase in Mg-gabbros.

Although metasediments contribute only a small portion to the water budget, they are important for hosting certain elements or isotopes characteristic for the sediment signature in arc magmas (K, Rb, Ba, Sr, [10]Be, see, for example, Morris et al. 1990, Plank and Langmuir 1993).

Phase relations of metasediments relevant to the conditions in subduction zones were reported by Nichols et al. (1994), Fockenberg (1995, 1998a,b), and Domanik and Holloway (1996). Chlorite + kyanite and staurolite + quartz are stable only to pressures of about 2.0 GPa (~60 km depth). At higher pressures hydrous assemblages comprise talc + chloritoid + phengite or talc + chloritoid + phengite. Between 4 and 5 GPa talc + kyanite breaks down to and Mg-Al-pumpellyite or OH-topaz. The stability field of lawsonite in these metasediments is similar to that in mafic systems (Domanik and Holloway 1996). The storage of H_2O in sediments to more than 200 km depth is ensured by phengite, lawsonite, MgAl-pumpellyite, and temperatures as high as 900°C.

Phase relations in ultramafic compositions

The ultramafic, crustal cumulates and the harzburgitic to lherzolitic mantle underlying the oceanic crust have only recently been subject of detailed experimental study even though in the subducting oceanic lithosphere such materials may comprise as much as 80 % of the total slab volume.

Several observations have led to this reorientation of experimental effort. Ultramafic rocks occurring in ophiolites reveal that intensive serpentinization during metamorphism in the oceanic environment (see O'Hanley 1996 for review of information). Intensely serpentinized peridotites in a variety of tectonic settings as reported in the vicinity of both fast (e.g. East Pacific Rise, Frueh-Green et al. 1995) and slow (e.g. Southwest Indian Ridge, Muller et al. 1997) spreading ridges, as well as along passive continental margins (e.g. Galicia margin, Boillot et al. 1988). The oceanic Mohorovitchik discontinuity (MOHO) may in fact correspond to the limit between serpentinized and non-serpentinized peridotites, and, therefore, represents a serpentinization front (e.g. Coulton et al. 1995, Muller et al. 1997).

The discovery of deeply serpentinized oceanic mantle several km below the ocean opens the possibility of very cold subduction paths. Evidence for such cold, subducted hydrated oceanic lithosphere is present in several high pressure/low temperature eclogite terranes in the Alps and elsewhere (e.g. Liguria, Scambelluri et al. 1995; Zermatt-Saas Zone, Barnicoat and Fry 1986; Cerro de Almirez, Sierra Nevada, Spain, Trommsdorff et al. 1998) which all include serpentinites associated with basaltic eclogites. Thermo-mechanical modelling (Peacock 1990, Davies and Stevenson 1992, Furukawa 1993) indicates considerably cooler environments in the interior of slabs than at the slab-mantle interface. Temperatures of 500°C at 200 km depth might be realized within subducting slabs. The implication that the top part of the ultramafic rocks of the subducted oceanic lithosphere may not only be hydrated but also cold during subduction requires a reevaluation of the stability of hydrous minerals in such compositions and their potential role as H_2O-carriers to great depth.

Antigorite is the most important hydrous phase at upper greenschist to eclogite or amphibolite facies conditions in ultramafic compositions (e.g. Trommsdorff 1983). The phase relations of antigorite control the H_2O-budget at the initiation of subduction. The breakdown of antigorite and the formation of post-antigorite hydrous phases might control of H_2O-transport at deeper parts of the subducting oceanic lithosphere.

Chrysotile and lizardite stability. Phase equilibrium studies of chrysotile and lizardite are limited to pressures exceeding 0.5 GPa. Most of these studies were synthesis experiments to delimit the stability of brucite + serpentine and that of the serpentine alone.

Several studies exist at low pressures, starting with the work of Bowen and Tuttle (1949) in the pure $MgO-SiO_2-H_2O$ (MSH) system and Yoder (1952) on the stability of aluminous lizardite in the $MgO-Al_2O_3-SiO_2-H_2O$ (MASH) system. Summaries on these low-pressure phase relations are provided by the paper of Chernosky et al. (1988) and O'Hanley (1996).

A summary of the high pressure phase relations for the two serpentine group minerals chrysotile and lizardite is shown in Figure 11. The major difference between the data in this figure and those inferred from field observations (Ulmer and Trommsdorff 1998) is the lack of the direct breakdown of serpentine to the anhydrous assemblage forsterite + enstatite + H_2O, corresponding to the reactions;

$$\text{antigorite} = 14 \text{ forsterite} + 20 \text{ enstatite} + 31 \text{ } H_2O \tag{4}$$

Instead, a transformation reaction such as

$$5 \text{ chrysotile} = 6 \text{ forsterite} + \text{talc} + 9 \text{ } H_2O \tag{5}$$

has been observed for a wide pressure range from 0.05 GPa (e.g. Bowen and Tuttle 1949, Caruso and Chernosky 1979, Chernosky 1982) to pressures as high as 6 GPa (Pistorius 1963, Kitahara et al. 1966, Yamamoto and Akimoto 1977, Khodyrev and Agoshkov 1986).

Below 4-5 GPa the talc + forsterite stability field is limited by the reaction:

$$\text{forsterite} + \text{talc} = 5 \text{ enstatite} + H_2O. \tag{6}$$

The experimental brackets of Kitahara et al. (1966), Yamamoto and Akimoto (1977) and

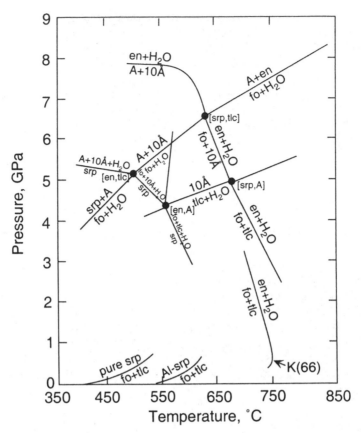

Figure 11. Synthesis of experimental results for chrysotile and lizardite (shown as srp) modified after Ulmer and Trommsdorff (1998). The curve marked K(66) is from Khitihara et al. (1966). The experimental data used in the synthesis are from Khitahara et al. (1966), Akimoto and Yamamoto (1977), Caruso and Chernosky (1979), Chernosky (1982), and Khodyrev and Agoshkov (1986).

Abbreviations: en = enstatite, A = phase A, fo = forsterite, 10Å = 10Å phase, srp = serpentine (lizardite or chrysotile), tlc = talc.

Khodyrev and Agoshkov (1986) are consistent with the exception of Reaction (6), the breakdown of talc + forsterite to enstatite + H_2O, which is about 40°C lower according to Kitahara et al. (1966) compared with Yamamoto and Akimoto (1977).

The results of the lower-pressure studies of Chernosky (1982) and Caruso and Chernosky (1979) have been added to Figure 11 to emphasize the importance of the starting material investigated. Chernosky (1982) used synthetic chrysotile in the MgO-SiO_2-H_2O system. His experimental brackets are consistent with the work of Bowen and Tuttle (1949) and nearly consistent with the high pressure investigation of Kitahara et al. (1966) who employed a similar synthesis technique for chrysotile.

Caruso and Chernosky (1979) investigated the breakdown of synthetic aluminous lizardite, containing 0.5 clinochlore-component corresponding to the chemical formula $Mg_{5.5}Al_{1.0}Si_{3.5}O_{10}(OH)_8$. Aluminium expands the serpentine mineral stability field. The difference between the two curves (pure chrysotile and aluminous lizardite) is in the order of 120°C.

At high pressures, various parageneses are observed along the serpentine breakdown curve and beyond. The phase assemblages involving the 10Å-phase (Bauer and Sclar 1981) and phase A (Ringwood and Major 1967) observed by Yamamoto and Akimoto (1977) and Khodyrev and Agoshkov (1986) are consistent with each other and with the reaction boundaries drawn on Figure 11. The following additional reactions were deduced by Yamamoto and Akimoto (1977) and are shown on this figure:

$$talc + 2 \ H_2O = 10Å\text{-phase} \tag{7}$$

$$5 \ chrysotile = 6 \ forsterite + 10Å\text{-phase} + 7 \ H_2O \tag{8}$$

$$11 \ chrysotile = 3 \ phase \ A + 4 \ 10Å\text{-phase} + H_2O \tag{9}$$

$$3 \ chrysotile + phase \ A = 8 \ forsterite + 9 \ H_2O \tag{10}$$

$$5 \ phase \ A + 3 \ 10Å\text{-phase} = 22 \ forsterite + 24 \ H_2O \tag{11}$$

$$forsterite + 10Å\text{-phase} = 5 \ enstatite + 3 \ H_2O \tag{12}$$

$$phase \ A + 5 \ 10Å\text{-phase} = 22 \ enstatite + 18 \ H_2O \tag{13}$$

$$phase \ A + 3 \ enstatite = 5 \ forsterite + 3 \ H_2O \tag{14}$$

Stability relations of antigorite. Experimental studies on the stability of antigorite and its breakdown products have been difficult in part because successful synthesis of antigorite in the $MgO\text{-}SiO_2\text{-}H_2O$ system is only very rarely achieved. Ishii and Saito (1973) synthesized antigorite from silica-rich alkaline solutions under small excess H_2O conditions in the pure $MgO\text{-}SiO_2\text{-}H_2O$ system. Johannes (1975) grew antigorite from an oxide mix with seeds of (Fe,Al)-bearing natural antigorite. Wunder and Schreyer (1997) used presynthesized brucite and talc as starting material and run durations of 5 days at 5 GPa and 500°C resulting in antigorite plus minor chrysotile and brucite or talc. This mixture was later used in the study of Wunder and Schreyer (1997) on the stability of antigorite.

Phase equilibrium data on the stability of antigorite and its breakdown products are shown in Figure 12. Evans et al. (1976) reported experimental results that delimit the breakdown of antigorite to forsterite + talc + H_2O,

$$Antigorite = 18 \ forsterite + 4 \ talc + 27 \ H_2O \tag{15}$$

to 1.5 GPa. The growth of antigorite at low pressures was very sluggish and Evans et al. (1976) concluded that the breakdown reaction probably occurs to the high temperature side of their bracketing experiments.

Evans et al. (1976) observed a higher temperature stability for antigorite than for either chrysotile or low-Al lizardite (see above, Fig. 11). At 1 GPa the difference is ~90°C. Ulmer and Trommsdorff (1995), Wunder and Schreyer (1997) and Bose and Navrosky (1998) extended the study of antigorite stability to higher pressures. Ulmer and Trommsdorff (1995) used natural, well-crystallized antigorite which contained ~2 wt % of combined Al, Fe^{3+}, and Cr. The breakdown of antigorite up to 8 GPa was investigated with a similar technique as by Evans et al. (1976). Natural antigorite + brucite was mixed with 10% olivine + enstatite. Wunder and Schreyer (1997) used two different starting materials. These were mixtures of synthetic $MgO\text{-}SiO_2\text{-}H_2O$ antigorite, forsterite, enstatite, and 20 wt % H_2O, and mixtures of natural antigorite, containing ~1.7 wt % trivalent cations, forsterite, enstatite, and H_2O. Bose and Navrotsky (1998) used natural antigorite and synthetic forsterite, enstatite and phase A (results not shown on Fig. 12). The experimental data in Figure 12 reveal a considerable, yet unresolved, discrepancy between the two experimental determinations of the high pressure stability of antigorite. Up to 4 GPa the Wunder and Schreyer (1997) curve for the reaction,

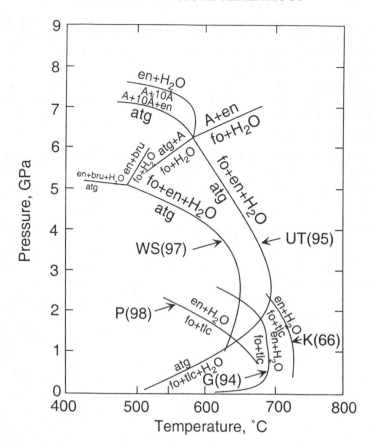

Figure 12. Synthesis of experimental results for antigorite modified after Ulmer and Trommsdorff (1998). The data base used by Ulmer and Trommsdorff (1998) included experimental results from Khitahara et al. (1966), Evans et al. (1976), Chernosky et al. (1985), Guggenheim (1994), Ulmer and Trommsdorff (1995), Wunder and Schreyer (1997). Where differences exist between determination of individual reactions, these are marked. K(66)—Khitahara et al. (1966), WS(97)—Wunder and Schreyer (1997), UT(95)—Ulmer and Trommsdorff (1995), and G(94)—Guggenheim (1994) and P(98)—Pawley (1998).
Abbreviations: en = enstatite, A = phase A, 10Å = 10Å phase,
fo = forsterite, atg = antigorite, bru = brucite, tlc = talc.

Antigorite = 14 forsterite + 20 enstatite + 31 H_2O (16)

is ~40°C below that of Ulmer and Trommsdorff (1995). At higher pressure the curve from Wunder and Schreyer (1997) bends into a strongly negative slope so that the difference becomes 100°C at 5 GPa. As a result, in the case of Wunder and Schreyer (1997), very cold subduction paths of 500°C at 5 GPa (160 km) would be necessary to carry any H_2O in ultramafic rocks deeper than this choke point (Kawamoto et al. 1995)[1] of complete dehydration of antigorite to forsterite plus enstatite. Only chlorite could possibly survive to high temperatures. From the results of Ulmer and Trommsdorff (1995), the choke point is located around 6 GPa and 600°C, where antigorite transforms to other H_2O-bearing

---

[1] The choke point refers to maximum depth along the subduction zone beyond which the temperature and pressure conditions are such that no H_2O-bearing phase would be stable and H_2O could be transported in hydrous phases to greater depth.

assemblages containing phase A and possibly the 10Å phase. The latter results have been confirmed by the experimental study of Bose and Navrosky (1998), who determined a similar antigorite stability at high pressures as Ulmer and Trommsdorff (1995); their choke point is located at ~620°C and 5.9 GPa.

A second important choke point is invariant point [srp.tlc] in Figure 12, which is the termination of the stability field of forsterite plus talc. The experimental data of Kitahara et al. (1966) and Guggenbuehl (1994) for the reaction

$$\text{forsterite} + \text{talc} = \text{enstatite} + \text{H}_2\text{O}, \tag{17}$$

are indicated in Figure 12. Kitahara et al. (1966) used presynthesized talc, forsterite and enstatite, as well as oxide mixtures; Guggenbuehl (1994, unpublished diploma thesis performed at ETH Zurich, as reported by Ulmer and Trommsdorff 1998) used synthetic forsterite, enstatite and talc as well as natural talc. Pawley (1998) also used natural and synthetic talc and synthetic forsterite and enstatite. Akimoto and Yamamoto (1977) also examined Reaction (17), and found it to occur at even higher temperatures than those reported by Kitahara et al. (1966). Akimoto and Yamamoto (1977) observed forsterite + talc at 715°C and 2.9 GPa. Berman et al. (1986) considered the experiments by Kitahara et al. (1966) as metastable due to the short experimental run times and the fine-grained poorly crystallized starting material.

The experiments by Khitahara et al. (1966) are not consistent with the low-pressure brackets obtained by Chernosky et al. (1985) for Reaction (17). This reaction is, however, metastable with respect to Mg-amphibole at low pressures (Ulmer and Trommsdorff 1998) Mg-amphibole is not shown in Figure 12 because its stability field is not relevant for subduction. It is only stable under low pressure–high temperature amphibolite to granulite facies conditions not encountered during subduction zone metamorphic conditions.

The preliminary, reversed experiments of Guggenbuehl (1994) resulted in a lower-temperature stability of forsterite + talc and its metastable extension into the antigorite stability field than Khitahara et al. (1966). The difference between Guggenbuehl (1994) and Kitahara et al. (1966) is ~40°C. The experimental curve derived by Guggenbuehl (1994) is compatible with the low pressure brackets determined by Chernosky et al. (1985). The brackets obtained by Pawley (1998) are considerably lower than all previous studies. This would result in an ever smaller stability field for forsterite + talc, reducing the significance of talc as a potential H_2O-carrier in subduction zones even further. All three studies are reversal experiments using either synthetic starting materials or mixtures of synthetic and natural starting material; the reason for the large discrepancy (~140°C at 2.5 GPa) is not evident to date.

The experiments performed by Evans et al. (1976), Guggenbuehl (1994) and Ulmer and Trommsdorff (1995) and repeat experiments at 690°C and 710°C at 1.8 GPa by Ulmer and Trommsdorff (1998) have been used to draw a consistent phase diagram (Fig. 12). It should be emphasized, however, that perfect agreement between experiments in the MgO-SiO_2-H_2O system (Kitahara et al. 1966, Chernosky et al. 1985, Guggenbuehl 1994, Wunder and Schreyer 1997, Pawley 1998) and the natural system (Evans et al. 1976, Ulmer and Trommsdorff 1995, Bose and Navrostsky. 1998) is neither necessary nor expected. The presence of Fe^{2+} and its fractionation between the various silicate phases can account for as much as 20°C difference in the reaction temperatures alone (see Fig. 3; Ulmer and Trommsdorff 1998). The effect of trivalent cations on the stability of antigorite has not been studied as yet. Antigorite can contain several wt % Fe_2O_3 (O'Hanley 1996) up to 5 wt % Al_2O_3 and up to 0.5 wt % Cr_2O_3 (Trommsdorff and Evans 1974, Trommsdorff et al. 1998).

Phase relations in mantle compositions with additional components

The simplified MgO-SiO_2-H_2O and FeO-MgO-SiO_2-H_2O systems represent ~95-96 wt % of the total oxide components of harzburgitic to lherzolitic mantle and have, therefore, been the target of most experimental and theoretical studies on the stability of H_2O-bearing minerals relevant for the mantle. The remaining 4 to 6 wt % includes Al_2O_3, CaO, Cr_2O_3, Na_2O, NiO and TiO_2. The following sections provides a short discussion of the phase relations in Al-bearing systems and on the stability of Ti- and F-bearing hydrates.

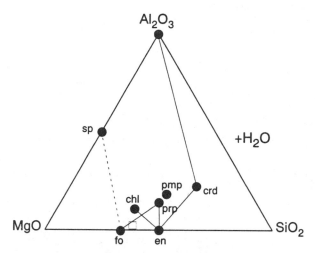

Figure 13. Compatibility relations among phases relevant for hydrous mantle (open square) projected from H_2O onto the plane MgO-Al_2O_3-SiO_2. Modified after Ulmer and Trommsdorff (1998). Abbreviations: sp = spinel, pmp = pumpellyite, crd = cordierite, chl = chlorite, prp = pyrope, fo = forsterite, en = enstatite.

Chlorite stability in Al-bearing systems. Phase relations in the aluminous system MgO-Al_2O_3-SiO_2-H_2O system provide a basis to examine the role of Al to stabilize additional hydrous phases in the mantle compositions during subduction (Fig. 13). Chlorite is the major aluminous hydrated phase in harzburgitic mantle in this system. A number of studies have delimited the stability of chlorite and its breakdown assemblages to pressures exceeding 7 GPa (Fawcett and Yoder 1966, Chernosky 1974, Staudigel and Schreyer 1977, Jenkins and Chernosky 1986, Fockenberg 1995).

The phase equilibrium data on the stability of chlorite are summarized in Figures 14 and 15. Chlorite has a large stability field. Its maximum temperature stability in the pressure range 0.5 to 5 GPa exceeds that of antigorite in the MgO-SiO_2-H_2O system (Fig. 12). At low pressures, cordierite is the breakdown product of clinochlore. This phase will not be stable in a typical mantle composition as long as Ca and Na stabilize plagioclase. At intermediate pressures (0.3 to 2 GPa) spinel + forsterite + enstatite form the breakdown assemblage of clinochlore, and at higher pressures (2 to 5 GPa), forsterite + spinel + pyrope-garnet is the mineral assemblage formed upon dehydration of chlorite.

At pressures above 5 to 5.5 GPa and temperatures below 750°C, a new hydrated phase is observed in the MgO-Al_2O_3-SiO_2-H_2O system (at clinochlore composition). This is a MgAlSi-hydrate with the pumpellyite structure (Schreyer and Maresch 1991, Fockenberg 1995, 1998a,b). The experimental data in Figure 14 are quite consistent among the different experimental studies except for the high pressure data of Staudigel and Schreyer

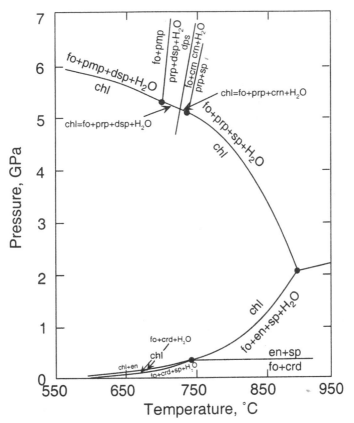

Figure 14. Synthesis of experimental results in the system $MgO\text{-}Al_2O_3\text{-}SiO_2\text{-}H_2O$ emphasizing chlorite stability relations modified after Ulmer and Trommsdorff (1998). The experimental data used in the synthesis are from Fawcett and Yoder (1966), Chernosky (1974), Staudigel and Schreyer (1977), Jenkins and Chernosky (1986), and Fockenberg (1995).

Abbreviations: chl = chlorite, fo = forsterite, sp = spinel, crd = cordierite, en = enstatite, prp = pyrope, pmp = pumpellyite, dsp = diaspore, crn = corundum.

(1977) at 3.5 GPa and those of Fockenberg (1995) at 3.3 GPa. The clinochlore stability determined by Fockenberg (1995) is 20-30°C lower than that of Staudigel and Schreyer (1977). Unlike antigorite, which has a composition near the mantle composition in terms of MgO (+ FeO) and SiO_2, this is not the case for chlorite. Clinochlore, representing the typical chlorite composition found in hydrated harzburgites in high pressure metamorphic terranes, is displaced towards the Al_2O_3 corner of the $MgO\text{-}Al_2O_3\text{-}SiO_2$ compositional triangle (Fig. 13) relative to the hydrated mantle bulk composition. Consequently, the absolute stability of chlorite on its own composition may not represent its stability in a mantle-like composition. The mantle bulk composition is located below the tie-line forsterite-pyrope (fo-prp on Fig. 13). Only if this tie-line is broken, but the crossing tie-line enstatite + chlorite occurs, can chlorite be present in a mantle composition.

The occurrence of chlorite is limited by reactions that stabilize the chlorite + enstatite assemblage. At high pressures and low temperatures, Fockenberg (1995) observed the paragenesis forsterite + diaspore + Mg-pumpellyite + H_2O. Mg-pumpellyite can only occur if pyrope is not stable. Fockenberg (1995) also investigated the stability of pyrope + H_2O,

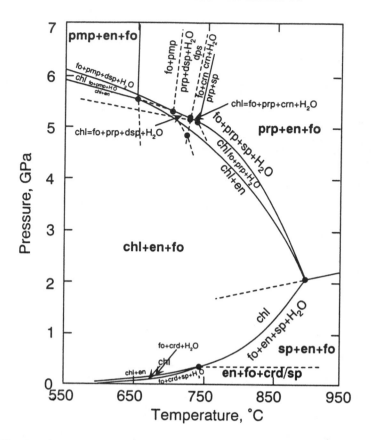

Figure 15. Experimental results in the system $MgO-Al_2O_3-SiO_2-H_2O$ relevant to hydrous mantle composition (see square in Fig. 13) modified after Ulmer and Trommsdorff (1998). The experimental data used in the synthesis are from Fawcett and Yoder (1966), Chernosky (1974), Staudigel and Schreyer (1977), Jenkins and Chernosky (1986), and Fockenberg (1995).
Abbreviations: chl = chlorite, fo = forsterite, sp = spinel, crd = cordierite, en = enstatite,
prp = pyrope, pmp = pumpellyite, dsp = diaspore, crn = corundum.

which lies at ~100°C lower temperature than the maximum stability of Mg-pumpellyite.

Ulmer and Trommsdorff (1998) conducted a topological analysis of a part of the $MgO-Al_2O_3-SiO_2-H_2O$ system (projected from MgMgAl-pumpellyite; not considering quartz/coesite saturated assemblages) in order to establish the stable phase assemblages containing chlorite and MgMgAl-pumpellyite for mantle-like compositions (Fig. 15). The exact locations of the enstatite + chlorite reactions shown on Figure 15 are not known.

Chernosky (1986) determined the reaction

chlorite + enstatite = forsterite cordierite + H_2O (18)

and found only very small differences to the chlorite breakdown

chlorite = forsterite + cordierite + spinel + H_2O.   (19)

Fockenberg (1995) reported an experiment at 5.4 GPa and 700°C containing the assemblage chlorite + forsterite + pumpellyite + diaspore + enstatite near the final breakdown of chlorite and an experiment at 5.5 GPa and 640°C where chlorite + enstatite + pumpellyite was observed.

The stability field for chlorite and Mg-pumpellyite in hydrous mantle compositions is shown in Figure 15. The stability field of each phase is smaller than the maximum stability on its own stoichiometry. This effect is much smaller for chlorite than for Mg-pumpellyite.

Below 5 GPa the temperature difference between the chlorite and antigorite breakdown curves is large (200°C at 2.5 GPa). This implies that in the warmer parts of a descending slab or in general during slow and therefore hotter subduction, some of the H_2O stored in chlorite will survive the antigorite breakdown. Chlorite contains about 20 to 25% of the total H_2O (2 to 2.5 wt % H_2O) stored in a hydrated lherzolite to harzburgite composition in the stability field of forsterite + antigorite + chlorite. The critical conditions of complete dehydration to the anhydrous assemblage pyrope + enstatite + forsterite occurs at 5.6 GPa and 650°C which compares with 6 GPa and 600°C for the MgO-SiO_2-H_2O system.

Ti-clinohumite stability in TiO_2- and F-bearing systems. Titanium- or F-rich clinohumite occurs in chlorite, spinel and garnet-peridotites from the Alps, Liguria, Dabie-Shan and other ultramafic massifs (Trommsdorff and Evans 1980, Scambelluri et al. 1995, Zhang et al. 1995). Ti-clinohumites are always subordinate. The pure hydroxyl-endmembers of the humite group minerals, such as hydroxyl-clinohumite and chondrodite, are only stable in the MgO-rich part of the MgO-SiO_2-H_2O system at pressures of less than 4-5 GPa. At higher pressure the forsterite-H_2O tie-line is not existent, and Mg-rich phases with Mg/Si ratios > 2 coexist with enstatite (Ulmer and Trommsdorff 1998, Luth 1995, Pawley and Wood 1996, Wunder 1998).

In systems that contain TiO_2 or F, humite group minerals can coexist with enstatite over a wide pressure range as evidenced by the field occurrences of OH-F Ti-clinohumite. Engi and Lindsley (1980) have studied the stability and breakdown of a Ti-clinohumite, containing the maximum amount of TiO_2 ($x_{Ti} = 0.5$), to 2.5 GPa. They observed the breakdown of Ti-clinohumite (Ti-chu) to olivine + ilmenite + H_2O. Ti-clinohumite was stable to 600°C at 2.5 GPa. Weiss (1997) extended the range to 8 GPa and studied the influence of variable TiO_2-contents and F-contents in the system.

The stability fields for clinohumite containing TiO_2, FeO and F are shown in Figure 16. Fluorine expands the clinohumite stability field to much higher temperatures (+ 700°C at 2.5 GPa, Fig. 16). Even small amounts of F will have a prominent effect on the stability of clinohumite. Ti-undersaturated OH-clinohumite or F-OH-Ti-bearing clinohumite form divariant reaction fields: Ti-and/or F-undersaturated OH-clinohumites react to olivine + ilmenite + Ti/F-richer clinohumite + H_2O over 60-100°C (Fig. 16). Such divariant reaction fields are also observed in nature (Evans and Trommsdorff 1983).

Titanium-saturated hydroxyl-clinohumite is the most common clinohumite composition found in ultramafic rocks. At 2 GPa the breakdown of Ti-clinohumite occurs above that of antigorite and above 3.5 GPa to chlorite. The quantity of Ti-clinohumite in mantle rocks is low and is limited by the TiO_2 content of peridotitic mantle which does not exceed 0.1 to 0.15 wt %. Even if all this TiO_2 is stored in Ti-clinohumite, its potential as an H_2O-carrier is small, with a maximum amount of H_2O of 1000 to 1500 ppm. The Ti-rich humite phase, however, may significantly control the high field strength element (HFSE) concentrations of fluids emanating from the peridotite during the breakdown of antigorite or chlorite. Like other Ti-rich phases, Ti-clinohumite strongly concentrates HFSE elements (Weiss 1997) and thus contributes to the HFSE-depleted characteristic for island-arc magmas generated above subduction zones.

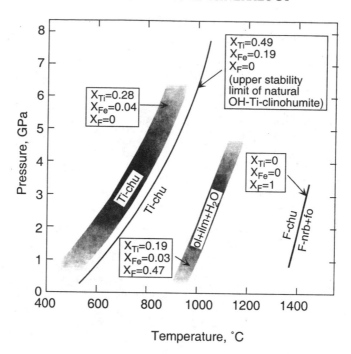

Figure 16. Experimental curves for the breakdown of hydroxy-Ti-clinohumite, fluorine-hydroxy-Ti-clino-humite together with that of synthetic fluorine-clinohumite modified after Weiss (1997). The shaded fields are those where reactants and products coexist.

Abbreviations: chu = clinohumite, ol = olivine, ilm = ilmenite, nrb = norbergite.

H$_2$O in nominally anhydrous minerals

Determinations of the OH-contents of nominally anhydrous minerals, such as olivine, pyroxenes, and garnet (for a review see Rossman 1996) show that considerable amount of H$_2$O (up to several thousand ppm) can be accommodated in their structure. Kohlstedt et al. (1996), using infrared spectroscopy to determine the H$_2$O contents, examined hydrated α-, β-, and γ-(Mg,Fe)$_2$SiO$_4$ [Mg/(Mg+Fe) ~ 0.9, San Carlos olivine]. Experiments were conducted at 1000-1100°C in the pressure range 2.5-19 GPa. They found that in the α-phase the H$_2$O content increased with increasing pressure reaching about 1200 ppm at the α-β phase boundary near 13 GPa. They further concluded that β-(Mg,Fe)$_2$SiO$_4$ might contain as much as 2.4 wt % H$_2$O in the 14-16 GPa range at 1100°C. Even higher H$_2$O contents (up to 2.7 wt % H$_2$O) were found in γ-(Mg,Fe)$_2$SiO$_4$ at 19.5 GPa and 1100°C. Smyth and Kawamoto (1996) identified a new structural type of wadsleyite with high H$_2$O-content that is stable to very high temperatures. Thus H$_2$O contained in nominally anhydrous phases may be transported deeply into the earth upper mantle and beyond.

WATER IN PERIDOTITE

The principal hydrous phases in H$_2$O-saturated peridotite to ~8 GPa are serpentine, phase A, chlorite, talc, and amphibole (Fig. 17). Stabilities of serpentine, phase A, and talc are expected to be almost identical for depleted or fertile peridotite because these phases are nearly Ca and Na-free and contain only minor amounts of Al$_2$O$_3$. Their stabilities have been determined experimentally on natural mineral compositions (Evans et al. 1976, Ulmer and

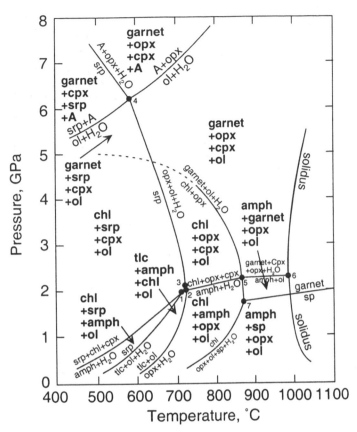

Figure 17. Univariant equilibria relevant to hydrated peridotite. Phase fields are shown in **bold** letters. Experimental data compiled by Schmidt and Poli (1998).
    Abbreviations: amph = amphibole, A = phase A, chl = chlorite, cpx = clinopyroxene, garnet = garnet, ol = olivine, opx = orthopyroxene, srp = serpentine, sp = spinel, tlc = talc

Trommsdorff 1995). In natural peridotites serpentine forms during low grade metamorphic hydration, its maximum temperature stability is 720°C at 2.1 GPa (Fig. 17) Serpentine (antigorite) decomposes to talc + olivine + H_2O as a function of increasing temperature below 1.9 GPa (Evans et al. 1976) to orthopyroxene + olivine + H_2O between 1.9 and 6.2 GPa, and to phase A + orthopyroxene + H_2O at higher pressures (Ulmer and Trommsdorff 1995, Bose and Navrotsky 1998).

In subducted peridotite, phase A (~11.8 wt % H_2O) replaces serpentine at pressures between 6 and 7 GPa through a water-conserving reaction (Fig. 17). Although phase A could form at lower pressures through olivine + H_2O = phase A + serpentine, the free water necessary for this reaction would not be available in peridotite subducted in a descending slab. Above 6.2 GPa, phase A + orthopyroxene decompose with increasing temperature to enstatite + H_2O. This reaction has a moderate positive slope in P-T space (Luth 1995).

Chlorite is compositionally near clinochlore in hydrated peridotite. Synthetic clinochlore decomposes between 0.3 and ~2 GPa to olivine + orthopyroxene + spinel +

H_2O with an upper temperature stability of 870°C (Jenkins and Chernosky 1986). At higher pressure (above 2.1 GPa) the chlorite stability in peridotite is limited by the reaction

$$\text{chlorite + orthopyroxene = olivine + garnet + } H_2O, \tag{20}$$

which has been determined experimentally to 4 GPa (Fockenberg 1995, Pawley 1996). Although considerable uncertainty remains, it may be deduced from experiments on the terminal stability of chlorite most likely the stability of chlorite + orthopyroxene does not exceed about 5 GPa. The stability of chlorite outlined in Figure 17 probably is a maximum.

In average mantle compositions, the talc (4.7 wt % H_2O) + olivine phase assemblage has a limited stability field less than 100°C wide. This assemblage decomposes to enstatite + H_2O at 690°C between 1 and 2 GPa (Ulmer and Trommsdorff 1995).

A compilation of all experimental studies on amphibole in water-saturated peridotite compositions yields the following picture. Near the water saturated solidus (~1000°C, at 2-3 GPa), amphibole is pargasitic and decomposes between 2 and 3 GPa. In harzburgite, calcic amphibole decomposes at 2.2 GPa, in lherzolite at 2.5-2.8 GPa and in pyrolite at 2.8-3.0 GPa (Mysen and Boettcher 1975, Millhollen et al. 1974, Green 1973).

The exact stoichiometry of the amphibole decomposition reactions is complicated by the change of amphibole composition with temperature. From experiments and natural xenoliths at temperatures near the peridotite solidus amphibole is pargasitic. In low temperature-high pressure metamorphic peridotites, amphibole is aluminous tremolite. Figure 17 is valid for amphibole compositions to be expected in natural peridotite, however, reaction topologies are different for endmembers such as tremolite or edenite.

DENSE HYDROUS MAGNESIUM SILICATES (DHMS)

As discussed above and summarized in Figure 17 the serpentine mineral antigorite is a major alteration product of ultramafic regions of the oceanic lithosphere and it has been shown to have a significant high pressure stability. The high-pressure stability limit of antigorite is marked by the formation of the first of the alphabetic phases, phase A at temperatures below about 600°C at a pressure near 6 GPa:

$$\text{antigorite + phase A = enstatite + fluid} \tag{21}$$

At temperatures greater than 600°C antigorite dehydrates to produce olivine, enstatite and fluid. Provided that this choke point can be overcome, at increasing pressure, additional dense hydrous magnesium silicates (DHMS) may be stabilized.

Since the original report on the occurrence of phase A as a possible hydrous high-pressure phase (Ringwood and Major 1967), a number of additional dense and hydrous phases has been reported (phases B, C, D, E, F, and G). There remains considerable doubt, however, that all these phases are thermodynamically stable at high pressure and temperature. It has been suggested, for example, that phases E, F, G, and D may not be distinct phases but may simply be the same phase (Frost 1998). In that same study, it was suggested that phase C and superhydrous B might be the same phase.

Among the other phases (phase A, B, D, and superhydrous B), phase A is the first phase to occur as the pressure is increased above that of the stability of antigorite (see, for example, Figs. 11 and 12). Phase A probably transforms to phase D and possibly phase E, depending on temperature as the pressure is increased further (Yamamoto and Akimoto 1977, Burnley and Navrotsky 1996, Pawley and Wood 1996). At even higher pressure, phase D or phase D together with superhydrous phase B might be stable (Gasparik 1993,

Ohtani et al. 1995).

Experiments at pressures required for the formation of these phases in the system $MgO\text{-}SiO_2\text{-}H_2O$ are conducted at such high pressures that the coexisting aqueous solutions are greatly enriched in magnesium silicate components (Fujii et al. 1996, Stadler et al. 1998). Precipitation of phases during quenching of such fluids remains a major problem. Therefore, until phase relations in this system are conducted in situ at high pressure and temperature, the stability relations, and perhaps even actual thermodynamic stability cannot be established with certainty. Reported stabilities of such phases should be considered with caution until in situ experiments have been conducted and the phase boundaries are reversed.

EXPERIMENTAL STUDIES OF METASOMATIC ALTERATION OF THE MANTLE WEDGE

Materials transport—silicate solubility in aqueous solutions

As discussed in sections above, there is pervasive evidence for transport of major, minor and trace elements from the subducting slab to the overlying mantle wedge. The alteration of the bulk chemistry of the peridotite in the mantle wedge will also result in changes in its mineralogy. This realization has led to several experimental studies aimed to determine the partitioning of trace elements between silicate melts, minerals and aqueous solutions (e.g. Mysen 1981, Schneider and Eggler 1996, Brenan et al. 1995). Alteration of trace element patterns of source regions of partial melts in the mantle wedge beneath subduction zones can be traced in magmas erupted in island arcs (e.g. Morris et al. 1990, Plank and Langmuir 1993). Before discussing these changes, we will review briefly some of the experimental data needed to characterize the materials transfer process from the subducting slab to the overlying upper mantle peridotite wedge.

Aqueous fluids released from a slab subducting into peridotite upper mantle, can dissolve significant quantities of silicate components. For example, Manning (1994) found for the simple dissolution reaction,

$$SiO_2(quartz) \Leftrightarrow SiO_2(aqueous\ fluid) \tag{22}$$

There is profound dependence of the solubility of SiO_2 reaching molalities near 1.5 at 900°C and about 1 GPa (Fig. 18). Manning (1994) found that the solubility is a comparatively simple function of temperature and the density of H_2O:

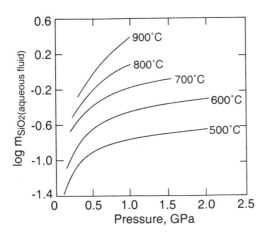

Figure 18. Solubility of SiO_2 in aqueous fluid (log molality) in equilibrium with crystalline quartz as a function of temperature and pressure. Data from Anderson and Burnham (1965), Manning (1994).

$$\text{Log } m_{SiO_2}(\text{fluid}) = 4.262 - (5764.2/T) + (1.7513 \cdot 10^6/T^2) - (2.2869 \cdot 10^8/T^3)$$
$$+ [2.8454 - (1006.9/T) + (3.5689 \cdot 10^5/T^2)] \log \rho_{H_2O} \qquad (23)$$

where $m_{SiO_2}(\text{fluid})$ is the molal SiO_2 content in aqueous fluid, T is temperature (K), and ρ_{H_2O} is the density of H_2O.

Mysen (1998, in prep) extended studies of this nature to include alkalis and alumina in order closer to represent the chemical compositions of fluids expected to emanate from dehydrating phases in subducting oceanic slabs. He found solubilities of K-Al-silicate materials in aqueous solutions in the pressure range 0.8-2.0 GPa at temperatures above 1000°C to range to be between ~1 and ~12 mol % (Fig. 19) with the solubility being a positive function of both pressure and temperature. He also noted that the silicate solubility increased with increasing alkali content and diminished as the system became aluminous (Fig. 20). Significant concentrations of aluminosilicate components remain, however, in the fluid. These data are consistent with the potential for alkali aluminosilicate enrichment in the peridotite upper mantle wedge even at comparatively shallow depth (<30 km). Alkaline earths, including in particular MgO, might dissolve in the fluids as the pressure is increased above perhaps 6 GPa (Stalder et al. 1998).

Silica enrichment and mineralogical alteration during flushing of the mantle above subducting slabs

Natural evidence exists to suggest that simple transformation of olivine to orthopyroxene may be the result of silica metasomatism in a peridotite mantle wedge (e.g. Riter and Smith 1996, Wenrich et al. 1995, McInnes 1996). An experimental examination of the kinetics associated with such processes recently was conducted by Iizuka et al. (1998). They used a simplified mineral couple to examine the behavior in a dehydrating slab and the adjacent mantle. Starting materials consisted of two portions—model slab and mantle minerals are glaucophane and olivine, respectively. The starting glaucophane was separated from blueschist facies rock of the Franciscan metamorphic zone, California, which is an ancient subducted slab (e.g. Ernst 1988). The other side (mantle portion) was olivine (Fo_{90}) from San Carlos, Arizona.

The critical observation in these experiments was that as glaucophane became partially dehydrated as a function of time in the pressure range 1.5-4.0 GPa, and enstatite zone grew at the expense of olivine. The growth rate, k, of this zone was determined and found to be significantly dependent on pressure;

$$k \ (m^2/s) = (0.76 \pm 0.40)\Delta P - (17.79 \pm 1.29), \qquad (24)$$

where pressure is in GPa. The growth rate is insensitive to temperature. By using the pressure-dependence of the reaction constant, (Eqn. 24), a two dimensional migration velocity for an enstatite-reaction-front into the mantle olivine as a function of pressure was estimated. In this calculation, it was assumed that a subducting oceanic stab descends and releases of an aqueous fluid at a constant rate. The model was developed as follows. The sub-arc mantle receives the aqueous fluid with silica in solution from subducting oceanic slab continuously and at a constant rate. Trace element and isotopic data from across-arc in the Quaternary volcanic rocks in the northeastern Japan, the Kurile, and the Izu-Bonin arcs (Shibata and Nakamura 1997, Ishikawa and Tera 1997, Ishikawa and Nakamura 1994) are consistent with this assumption. A temperature gradient in the sub-arc mantle was not considered because the experimental data show the reaction constant, k, depends principally upon the pressure with only a very weak temperature dependence. The temperature was fixed at 850°C for the purpose of the calculation.

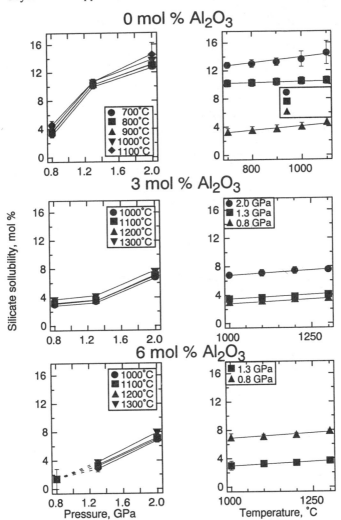

Figure 19. Solubility of aluminosilicate in aqueous fluid (silicate solubility, wt %) as a function of temperature, pressure, and Al_2O_3 concentration in the system. The experiments were conducted with water-saturated aluminosilicate melt in equilibrium with aluminosilicate-saturated aqueous fluid with compositions on the join $K_2Si_4O_9$-$K_2(KAl)_4O_9$ with the individual bulk compositions identified with their mol % Al_2O_3 in the system. Data from Mysen (1998) and Mysen (in preparation).

Figure 21 shows an estimate of silica-metasomatism to form enstatite from olivine in the sub-arc mantle calculated for geologic time scales. For example, the time required to make an orthopyroxene replacement front from olivine at 135 km and 100 km slab depth to 65 km depth in the overlying upper mantle is ~10 and ~8 million years, respectively.

Phase relations among K-rich, hydrous phases -results of metasomatism of peridotite?

Experimental work in the system K_2O-CaO-MgO-Al_2O_3-SiO_2-H_2O (KCMASH) has shown that phlogopite + clinopyroxene ± orthopyroxene assemblages break down to form

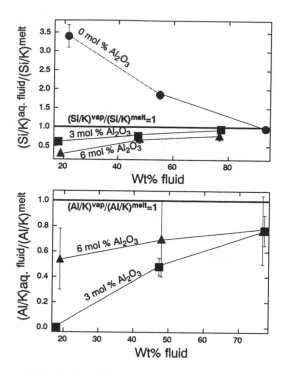

Figure 20. (Si/K) and (Al/K) distribution between coexisting water-saturated aluminosilicate melt and aluminosilicate-saturated aqueous fluid as a function of bulk Al_2O_3 concentration in the system. The experiments were conducted with water-saturated aluminosilicate melt in equilibrium with aluminosilicate-saturated aqueous fluid with compositions on the join join $K_2Si_4O_9$-$K_2(KAl)_4O_9$. The wt % fluid indicates the weight percent aluminosilicate-saturated fluid present in the experiments. The thick lines denote the situation if solution of aluminosilicate components in the fluid was congruent. Data from Mysen (1998) and Mysen (in preparation).

potassium-richterite + garnet + olivine at P 6-7 GPa or ~180-210 km depth, thus making phlogopite-bearing peridotites potential host rocks for potassium amphibole (Sudo and Tatsumi 1990, Luth 1997). This observation leads to the suggestion that in a subalkaline bulk composition [2] K-amphibole can only be stable in mantle regions where $(K_2O + H_2O)$-metasomatism is active or where metasomatized phlogopite-bearing mantle can be transported to a depth exceeding the upper pressure stability limit of phlogopite (e.g. Sato et al. 1997).

At present, no K-richterite-bearing peridotites have been found in which the amphibole is in textural equilibrium with the lherzolitic host assemblage. Nevertheless, the widespread occurrence of K-richterite in equilibrium with a garnet lherzolite assemblage in subduction zone peridotites has been proposed by Tatsumi (1989), Tatsumi et al. (1991) and Tatsumi and Eggins (1995) based on the stability of phlogopite in the peridotitic mantle wedge above subducting slabs and experimental results of Sudo and Tatsumi (1990). Tatsumi and Eggins (1995) proposed that fluids released by the phlogopite-to-potassium amphibole reaction can trigger the mantle melting responsible for arc-volcanism above subduction zones.

---

[2] A subalkaline bulk composition is one where Al > (Na+K).

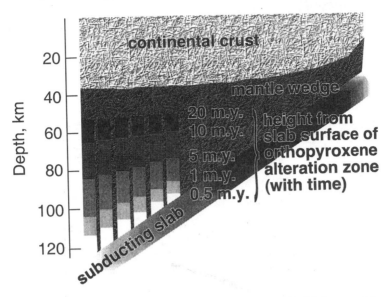

Figure 21. Height of orthopyroxene alteration column from subducting slab into overlying peridotite mantle as a function of time since onset of dehydration and alteration and initial depth of the slab surface. Data and calculations from Iizuka et al. (1998).

Only reconnaissance studies have been available to date on the stability field of K-richterite in subalkaline bulk compositions relevant to normal lherzolitic mantle compositions (Trønnes et al. 1988, Trønnes 1990, Harlow 1995). In order to locate the pressure-temperature conditions of the reaction by which phlogopite breaks down to form amphibole precisely and to determine the phase compositions across the phlogopite-amphibole transition, experiments were conducted by Konzett and Ulmer (1998) using subalkaline bulk compositions representative of peridotitic mantle.

Konzett and Ulmer (1998) reported experimental data on K-richterite and phlogopite stability by using two different starting materials. These were a simplified KNCMASH bulk composition. In addition, a spinel lherzolite from Mont Briançon, French Massif Central was used (Downes 1987) because it represents a moderately fertile bulk composition slightly higher in modal clinopyroxene than average spinel lherzolite (Maaløe and Aoki 1977). This composition is less fertile than KLB-1 (Takahashi 1986b). Thirty wt % olivine (Fo_{91}) was subtracted from the lherzolite starting composition in order to increase the amount of pyroxene relative to olivine. Although an excess of orthopyroxene with respect to phlogopite is present, phlogopite could survive the amphibole formation and coexist with K-richterite as a result of exhaustion of jadeite-component in clinopyroxene. In order to avoid this less realistic scenario, 0.4 wt % Na_2O was added to the lherzolite starting material

Phase relations in K_2O-Na_2O-CaO-MgO-Al_2O_3-SiO_2-H_2O (KNCMASH)

In the subalkaline KNCMASH system K-richterite is stable at pressures > 6.0 GPa (Fig. 22). Below the K-richterite-in curve, the stable assemblage is phlogopite + clinopyroxene + orthopyroxene + garnet + fluid (Fig. 22). Olivine joins the assemblage at pressures near the K-richterite-in curve. The position of the K-richterite-in reaction curve was located between 6.0 and 6.5 GPa at 800°C and between 6.5 and 7.0 GPa at 1100°C.

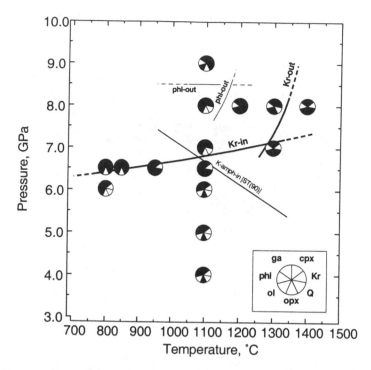

Figure 22. P-T diagram of experimental results in the KNCMASH system. K-amphibole-in-reaction reported by Sudo and Tatsumi (1990) is shown as ST(90). Phases present in the experimental charges are represented by black sectors within the run symbol, phases not detected are denoted by white sectors (inset lower right). (modified from Konzett and Ulmer 1998)

Abbreviations: phl = phlogopite, ga = garnet, cpx = clinopyroxene,
Kr = K-richterite, Q = quartz, opx = orthopyroxene, ol = olivine.

Hence, the reaction has a positive slope with $dP/dT < 3.3 \cdot 10^{-3}$ GPa/°C.

With orthopyroxene as a possible reactant phase, the changes of mineral assemblage associated with the appearance of K-richterite are consistent with a reaction,

$$\text{phlogopite} + \text{clinopyroxene} + \text{orthopyroxene} = \text{mphibole} + \text{garnet} + \text{olivine} + \text{fluid} \quad (25)$$

as originally proposed by Sudo and Tatsumi (1990). The high temperature stability limit of K-richterite is reached between 1300 and 1400°C at 8.0 GPa and at <1300°C at 7.0 GPa, amphibole breakdown produces an anhydrous lherzolite assemblage garnet + olivine + clinopyroxene + orthopyroxene. Due to the surplus of phlogopite with respect to orthopyroxene, phlogopite coexists with amphibole between 7.0 and 8.0 GPa at 1100°C. Phlogopite finally breaks down in the absence of orthopyroxene between 8.0 and 9.0 GPa at 1100°C. This is accompanied by an abrupt increase of potassium concentration of the amphibole. The schematic reaction is

$$\text{phlogopite} + \text{clinopyroxene}_{ss} = \text{K-richterite}_{ss} + \text{garnet}_{ss} + \text{olivine} + \text{fluid} \quad (26)$$

Phlogopite breakdown between 1100° and 1200°C at 8.0 GPa again produces K-richterite component in amphibole and is also responsible for the re-appearance of orthopyroxene in subsolidus experiments (cf. Fig. 22). The upper T-stability of

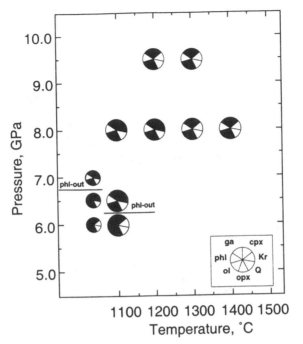

Figure 23. Pressure-temperature diagram summarizing experimental results using a natural lherzolite starting composition. Phases present in the experimental charges are represented by black sectors within the run symbol, phases not detected are denoted by white sectors (inset lower right). (modified from Konzett and Ulmer 1998) Small circles: experiments conducted in single Pt capsules. Large circles: experiments conducted in double capsules with graphite capsule inside Pt capsule. Experiments marked.
Abbreviations: phl = phlogopite, ga = garnet, cpx = clinopyroxene,
Kr = K-richterite, Q = quartz, opx = orthopyroxene, ol = olivine.

K-richterite at 8.0 GPa is reached between 1300° and 1400°C. Beyond these conditions amphibole breaks down to form a lherzolitic assemblage plus small amounts of fine-grained quench products.

Phase relations in the natural lherzolite system

In contrast to the experiments in the system KNCMASH, no free fluid phase was present in the natural lherzolite system below the pressure-temperature conditions of the K-richterite-in curve. Further, the influence of iron has to be considered. The stability of K-richterite in the lherzolite system is shifted slightly towards lower pressures compared with KNCMASH (Figs. 23 and 24). K-richterite first appears and coexists with phlogopite at 6.5 GPa. At 1100°C, the pressure interval of coexisting phlogopite + amphibole is <1.0 GPa (Fig. 23). The shift of the K-richterite-in curve probably reflects partitioning of Fe^{2+} into garnet which is part of the product assemblage of phlogopite breakdown. The persistence of phlogopite to 6.5 GPa is ascribed to a high Fe^{3+}/Fe^{2+} ratio, which stabilizes phlogopite with respect to amphibole. This assumption is supported by unusually high forsterite contents of coexisting olivine and pronounced tetrahedral cation deficiency in phlogopite analyses when normalized to 11 oxygens and stoichiometric OH. To reduce Fe^{3+} in the experimental charges, experiments were re-run at 1100°C using double capsules with an inner graphite liner (Knzett and Ulmer 1998). This led to the disappearance of phlogopite at 6.5 GPa and 1100°C along with a significant change in the mineral

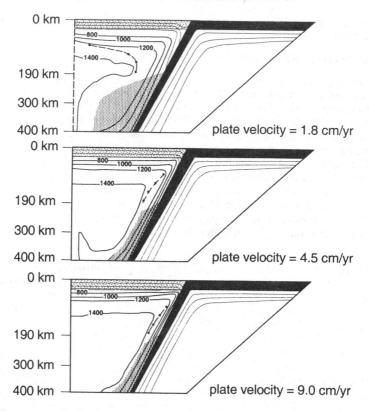

Figure 24. Temperature distribution in the mantle wedge above a subducting slab according to the model of Davis and Stevenson (1992) for a subduction angle of 60° and three different subduction velocities. The shaded area in the wedge represents an approximate potential stability field of K-richterite. The upper temperature-stability limit of K-richterite is <1400°C as determined in the KNCMASH system and was (arbitrarily) placed at 1300°C in the absence of unambiguous evidence for K-richterite melting in the lherzolitic system (see text for discussion). The dashed line connecting black dots is an approximate upper temperature-stability limit for phlogopite in lherzolite extrapolated from experimental data of Wendlandt and Eggler (1980). Lines with numbers refer to isotherms and corresponding temperatures. Modified after Konzett and Ulmer (1998).

compositions. At 8.0 GPa K-richterite disappears between 1200 and 1300°C. The K-richterite breakdown does not produce detectable quench phases nor is there any significant change in the composition of lherzolite phases between 1100 and 1300°C. At 9.5 GPa no crystalline hydrous potassic phase was found at 1200 and 1300°C regardless of whether phlogopite or K-richterite-bearing starting material was used. X-ray mapping revealed that within the experimental charge K is concentrated in clinopyroxene and to a smaller degree dispersed along grain boundaries. The highest K-concentration was found along grain boundaries of graphite grains within an ~20 μm wide zone at the interface between graphite capsule and experimental charge. The complete absence of K-richterite in experiments at 9.5 GPa would imply a negative slope of the reaction that defines the high temperature breakdown of K-richterite. This slope is opposite to that observed in the synthetic KNCMASH system (Figs. 22 and 23). A possible explanation for this discrepancy is, that in the natural system, increasing oxidation of graphite forms CO_2 with increasing pressure and temperature. This CO_2 would reduce the activity of H_2O in the fluid and destabilize K-richterite, and possibly form trace amounts of alkaline CO_2-rich fluids or melts (e.g. Sweeney 1994).

Hydrous potassic phase stability in subduction zone settings

Subducting slabs are heat sinks. The temperatures in the mantle wedge above a descending slab are anomalously cold compared to other mantle regions. Thermal models for subduction zones (e.g. Davis and Stevenson 1992, Furukawa 1993) predict steady state temperatures of 800 to 1000°C at the slab/mantle interface at depths near 180 km, with isotherms parallel to the subducting slab. Pressure-temperature conditions in part of the mantle wedge are thus within the K-richterite stability field below a depth of ~180 km or 6 GPa. The mantle volume in which K-richterite is potentially stable varies with the spacing of isotherms. According to the theoretical model of Davis and Stevenson (1992), increasing subduction velocity leads to a hotter wedge with a tighter spacing of the isotherms. Therefore, fast subduction should reduce the mantle volume in which K-richterite is stable (Fig. 24). Different subduction angles and varying thickness of the slab, on the other hand, have little influence on the T-distribution in the wedge in Davis and Stevenson's (1992) model.

Both phlogopite-breakdown in peridotites dragged down along the slab/mantle interface and interaction between mantle peridotites and K-rich fluids or melts at pressures above the stability limit of phlogopite can produce K-richterite in subduction zones. Mechanisms for the generation of K-enriched fluids from subducting slabs within the phlogopite stability field include: (1) breakdown of Ca-amphibole in the subducted slab (e.g. Tatsumi and Eggins 1995); (2) decomposition of K-feldspar + phlogopite (Massonne 1992); (3) continuous phengite dehydration/K-MORB-melting (Schmidt 1996, Domanik and Holloway 1996); and (4) hybridization between hydrous silicic melts and peridotitic mantle (Sekine and Wyllie 1982, Wyllie and Sekine 1982). Together, these mechanisms can form phlogopite in a large wide pressure-temperature interval within the mantle wedge. The occurrence of alkali- and silica-rich aluminous glass inclusions coexisting with phlogopite or amphibole in mantle minerals in xenoliths from modern and ancient subduction zone settings (Schiano et al. 1995, Szabo et al. 1996) provides evidence for the presence of hydrous melts/fluids migrating through the mantle wedge above subducting slabs. Experiments by Draper and Green (1997) show that these alkali and silica-rich aluminous glasses have phlogopite on their liquidus under water-saturated conditions and thus would precipitate phlogopite during migration through the mantle wedge. The presence of phlogopite in arc magma sources is further supported by phlogopite/calcic amphibole peridotites occurring in arc-related ophiolites (Arai and Takahashi 1990) and by K-rich lamprophyric rocks in some young subduction zones (Esperança and Holloway 1987, Hochstaedter et al. 1996, Luhr 1997). Because phlogopite is stable to higher temperatures than is pargasite (cf. Wendlandt and Eggler 1980, Mengel and Green 1989), phlogopite can survive low-pressure partial melting of hydrated peridotite that extracts the Ca-amphibole. A possible source for K-rich fluids or hydrous melts at pressures exceeding the upper pressure stability of phlogopite is the breakdown of phengite to K-hollandite or melting of K-rich mid-oceanic ridge basalt (Schmidt 1996, Domanik and Holloway 1996, Sorensen et al. 1997).

CONCLUDING REMARKS

In contrast to the oceanic and continental upper mantle, H_2O is a pervasive constituent in the upper mantle near convergent plate boundaries. The physicochemical processes responsible for the formation and evolution of this portion of the Earth's upper mantle is in many ways governed directly by the presence of large quantities of H_2O. These high H_2O

contents are critical to understand the mineralogy and the behavior of the upper mantle near convergent plate boundaries. A combination of field observations, experimental phase equilibrium studies and thermodynamic calculations yield the conclusion, however, that only a limited number of hydrous phases are important H_2O-carriers in subducting hydrous mantle and in an overlying metasomatically altered peridotite mantle wedge. At the initial stage of subduction antigorite is the dominant H_2O-repository, with subordinate chlorite and tremolite. Tremolite, which only constitutes an H_2O-carrier in Ca-rich lherzolitic compositions, reacts to diopside at less than 80 km. Depending on the temperature-pressure path followed by the subducting lithosphere antigorite and chlorite are stable to 150 to 200 km. At 200 km depth and low temperatures of 600-650°C both, antigorite and chlorite react to assemblages containing high pressure hydrate phases. At greater depth H_2O will be stored in phase A, which has a very large temperature stability field at high pressures. Eventually phase E can transport H_2O into the mantle transition zone. If the critical conditions of 6 GPa and 600°C for antigorite reacting to forsterite + enstatite + H_2O are bypassed at the high-temperature side, the assemblage forsterite + enstatite + H_2O is directly formed from antigorite, and H_2O can only be stored in nominally anhydrous phases.

The coexistence of several hydrous minerals in the peridotitic part of the subducting slab may result to a more or less continuos supply of H_2O to the overlying mantle wedge below 200 km. Continuous H_2O-fluxing of the overlying mantle wedge down to at least 200 km (6 GPa) is therefore expected. This flux most likely result in the formation of hydrous potassic minerals such as potassium richterite and phlogopite and possibly an increased modal abundance orthopyroxene and the expense of olivine. This mantle also retains a trace and isotopic signature of the subducting slab. Partial melting then most probably occurs as the result of melting point depression in the peridotitic mantle due to the influx of H_2O into the hottest part of the wedge. This melt, ultimately observed in island arc volcanism, reflects the major, minor, trace element and isotopic signatures of the subducting slab as well as the original composition of the upper mantle wedge as this existed prior to metasomatic alteration.

REFERENCES

Abers GA (1996) Plate structure and the origin of double seismic zones. *In:* G Bebout (ed) Subduction: Top to Bottom, 223-228. Am Geophys Union Washington, DC

Anderson GM, Burnham CW (1965) The solubility of quartz in supercritical water. Am J Sci 263:494-511

Aoki K (1987) Japanese island arcs: xenoliths in alkali basalts, high-alumina basalts, andcalc-alkaline andesites and dacites. *In:* PH Nixon (ed) Mantle Xenolths, p 319-333. John Wiley & Sons

Aoki K, Shiba I (1973) Pyroxenes from lherzolite inclusions of Itinome-gata, Japan. Lithos 6:41-51

Aoki K, Kuno H (1973) Gabbro-Quartz Diorite Inclusions from Izu-Hakone Region, Japan. Bull Volcanologique 36:164-173

Apted MK, Liou JG (1983) Phase relations among greenschist, epidotite-amphibolite, and amphibolite in basaltic systems. Am J Sci 283A:328-354

Apted MJ, Liou JG (1983) Phase relations among greenschist, epidotite-amphibolite, and amphibolite in a basaltic system. Am J Sci 283-A:328-354

Arai S, Takahashi I (1990) Formation and compositional variation of phlogopite in the Horoman peridotite complex, Hokkaido, northern Japan: implications for origin and fractionation of metasomatic fluids in the upper mantle. Contrib Mineral Petrol 101:165-175

Arculus RJ (1994) Aspects of magma genesis in arcs. Lithos 33, 189-208.

Bauer JF, Sclar CB (1981) The "10-Å phase" in the system $MgO-SiO_2-H2O$ Am Mineral 66, 576-585.

Barnicoat A, Fry N (1986) High-pressure metamorphism of the Zermatt-Saas ophiolite zone, Switzerland. Lithos 21:227-236

Bebout G, Ryan JG, Leeman WP (1993) B-Be systematics in subduction-related metamorphic rocks: characterization of the subducted component. Geochim Cosmochim Acta 57, 2227-2237.

Berman RG, Engi M, Greenwood HJ, Brown TH (1986) Derivation of internally-consistent thermodynamic data by the technique of mathematical programming: a review with application to the system MgO-SiO_2-H_2O. J Petrol 27:1331-1364

Bina CR, Helffrich GR (1992) Calculations of elastic properties from thermodynamic equation of state principles. Ann Rev Earth Planet Sci 20:527-552

Bodinier JL, Guiraud M, Fabries J, Dostal J, Dupuy C (1987) Petrogenesis of layered pyroxenites from the Lherz, Freychinede and Prades ultramafic bodies (Ariege, French Pyrenees). Geochim Cosmochim Acta 51:279-290

Boillot G, Girardeau J, Kornprobst J (1988) Rifting of the Galicia margin: Crustal thinning and emplacement of mantle rocks on the seafloor. Proc Ocean Drilling Program, Scientific Results, 103:741-756

Bose K, Navrotsky A (1998) Thermochemistry and phase equilibria of hydrous phases in the system MgO-SiO_2-H_2O: Implications for volatile transport to the mantle. J Geophys Res 103:9713-9719

Bowen NL, Tuttle OF (1949) The system MgO-SiO_2-H_2O. Geol Soc Am Bull 60:439-460

Brandon AD, Draper DS (1996) Constraints on the origin of the oxidationstate of mantle overlying subducion zones: An example from Simcoe, Washington, USA. Geochim Cosmochim Acta 60: 1739-1749

Brenan JM, Shaw HF, Ryerson FJ, Phinney DL (1995) Mineral-aqueous fluid partitioning of trace elements at 900°C and 2.0 GPa: Constraints on the trace element geochemistry of mantle and deep crustal fluids. Geochim Cosmochim Acta 59:3331-3350

Burnley PC, Navrotsky A (1996) Synthesis of high-pressure hydrous magnesium silicates: Observations and analysis. Am Mineral 81:317-326

Caruso LJ, Chernosky JV (1979) The stability of lizardite. Canadian Mineral 17, 757-769.

Chernosky JV (1974) The upper stability of clinochlore at low pressure and the free energy of formation of Mg-cordierite. Am Mineral 59:496-507

Chernosky JV (1982) The stability of clinochrysotile. Canadian Mineral 20:19-27

Chernosky JV, Day HW, Caruso LJ (1985) Equilibria in the system MgO-SiO_2-H_2O: experimental determination of the stability of Mg-anthophyllite. Am Mineral 70:223-236

Chopin C (1986) Phase relationships of ellenbergerite,a new high-pressure Mg-Al-Ti-silicate in pyrope-coesite-quartzite from the western Alps. Geol Soc. Am Memoirs 164:31-42

Chopin C, Henry C, Michard A (1991) Geology and petrology of the coesite-bearing terrain, Dora Maira massif, westernAlps. Eur J Mineral 3:263-291

Cormier VF (1989) Seismic body waveforms as constraints on deep slab structure. EOS Trans Am Geophys Union 70:389

Coulton AJ, Harper DS, O'Hanley, DS (1995) Oceanic versus emplacement age serpentinisation in the Josephine ophiolite: Implication for the nature of the Moho at intermediate and slow spreading ridges. J Geophys Res 100:22,245-22,260

Davis HJ, Stevenson DJ (1992) Physical Model of Source Region of Subduction Zone Volcanics. J Geophys Res 97:2037-2070

Davies GR, Nixon, PH, Pearson DG, Obata M (1993) Tectonic implications of graphitized diamonds from the Ronda peridotite massif, Southern Spain. Geology 21:471-474

DeLong SE, Hodges FN, Arculus RJ (1975) Ultramafic and mafic inclusions, Kanaga Island, Alaska, and the occurrence of alkaline rocks in island arcs. J Geol 83:721-736

Domanik KJ, Holloway JR (1996) The stability and composition of phengitic muscovite and associated phases from 5.5 to 11 GPa: Implications for deeply subducted sediments. Geochim Cosmochim Acta 60:4133-4151

Downes H (1987) Relationship between geochemistry and textural typ in spinel lherzolites, Massif Central and Languedoc, France. *In:* PH Nixon (ed) Mantle Xenoliths, p 125-133. John Wiley & Sons

Draper DS, Green TH (1997) P-T Phase Relations of Silicic, Alkaline, Aluminous Mantle-Xenolith Glasses Under Anhydrous and C-O-H Fluid-saturated Conditions. J Petrol 38:1187-1224

Eggler DH (1972) Water saturated and undersaturated melting relations of a Paricutin andesite and an estimate of water content in natural magma. Contrib Mineral Petrol 34:261-271

Engi M, Lindsley DH (1980) Stability of tianian clinohumite: Experiments and thermodynamic analysis. Contrib Mineral Petrol 72:415-424

Ernst WG, Liou JG, Coleman RG (1995) Contrastingplate-tectonic styles of the Quiling-Dabie-Sulu (apline-type) and Fransiscan (Pacific-type) metamorphic belts. Int"l Geol Rev 37, 191-211.

Ernst WG (1971) Metamorphic zonation in the presumably subducted lithospheric plates from Japan, California and the Alps. Contrib Mineral Petrol 34:43-59

Ernst WG (1988) Tectonic history of subduction zones inferred from retrograde blueschist P-T paths. Geology 16:1081-1084

Ertan IE, Leeman WP (1996) Metasomatism of Cascades subarc mantle: Evidencefrom a rare phlogopite orthopyroxenite xenolith. Geology 24:451-454

Esperança S, Holloway JR (1987) On the origin of some mica-lamprophyres: experimental evidence from a mafic minette. Contrib Mineral Petrol 95:207-216

Evans BE, Tromsdorff V (1983) Fluorine-hydroxyl titanian clinohumite in Alpine recrystallized garnet peridotite: Compositional controls and petrologic significance. Am J Sci 283-A:355-369

Evans BW, Johannes W, Oterdoom H, Trommsdorf V (1976) Stability of chrysotile and antigorite in the serpentine multisystem. Schweiz mineral petrogr Mitt 56:79-93

Fawcett JJ, Yoder HS (1966) Phase relationships of chlorites in the system $MgO-Al_2O_3-SiO_2-H_2O$. Am Mineral 51:353-380

Fei Y, Saxena SK, Navrotsky A (1990) Internally consistent thermodynamic data and equilibrium phase relations for compounds in the system $MgO-SiO_2$ at high pressure and high temperature, J Geophys Res 95:6915-6928

Fockenberg T (1995) New experimental results up 100 kbar in the system $MgO-Al_2O_3-SiO_2-H_2O$ (MASH): Preliminary stability fields of chlorite, chloritoid, staurolite, MgMgAl-pumpellyite, and pyrope. Bochumer geologische und geotechnische Arbeiten 44:39-44.

Fockenberg T (1998a) An experimental study of the pressure-temperature stability of MgMgAl-pumpellyite in the system $MgO-Al_2O_3-SiO_2-H_2O$. Am Mineral 83:220-227.

Fockenberg T (1998b) An experimental study of the P-T stability of Mg-staurolite in the system $MgO-Al_2O_3-SiO_2-H_2O$. Contrib Mineral Petrol 130:187-198

Frost DJ (1998) The stability of dense hydrous magnesium silicate phases in the Earth's transition zoneandupper mantle. In: Y-W Fei, C Bertka, BO Mysen (eds) Mantle Petrology: Field Observations and High-Pressure Experimentation. Geochemical Society (in press)

Fujii T, Mibe K, Masuda A (1996) Composition of fluid coexisting with olivine and pyroxene at high pressure: The role of water on the differentiation of the mantle. Misasa Seminar on Evolutionary Processes of Earth and Planetary Materials, p 37-38. Inst Study Earth's Interior, Okayama University, Misasa, Japan

Furukawa Y (1993) Depth of decoupling plate interface and thermal structure under arcs. J Geophys Res 98:20005-20013

Gaetani GA, Grove TL, Bryan WB (1993) The influence of water on the petrogenesis of subductin-related igneous rocks. Nature 365:332-334

Gasparik T (1993) The role of volatiles in the transition zone, J Geophys Res 98:4287-4299.

Green DH (1973) Contrasted melting relations in a pyrolite upper mantle under mid-ocean ridge, stable crust and island arc environments. Tectonophysics 17:285-297

Hall CA, Bennet, VC (1979) Significance of lherzolite at the Etang de Lherz, Central Pyrenees, Southern France. Earth Planet Sci Lett 45:349-354

Harlow GE (1995) K-amphibole and Mica Stability in K-rich Environments at high P and T. EOS Trans Am Geophys Union 76:298

Harris DM Jr, Anderson AT (1984) Volatiles H_2O, CO_2, and Cl in a subduction related basalt. Contrib Mineral Petrol 87:120-128

Hochstaedter AG, Ryan JG, Luhr JF and Hasenaka T (1996) On B/Be ratios in the Mexican Volcanic Belt. Geochim Cosmochim Acta 60:613-628

Iizuka Y, Mysen BO, Nakamura E (1998) Experimental study on slab-mantle interactions: Implicationsfor the silica-metasomatism in sub-arc mantle and isalnd arc magma genesis. Chemical Geol (in press)

Ionov DA, Hofmann AW (1995) Nb-Ta-rich mantle amphiboles and micas. Implications for subduction-related metasomatic trace element fractionations. Earth Planet Sci Lett 131:341-356

Iishi K, Saito M (1973) Synthesis of antigorite. Am Mineral 58:915-919

Ishikawa T, Nakamura E (1994) Origin of the slab component in arc lavas from across-arc variation of B and Pb isotopes. Nature 362:739-743

Ishikawa T, Terra F (1997) Source, composition and distribution of the fluid in the Kurile mantle wedge: Constraints from across-are variations of B/Nb and B isotopes. Earth Planet Sci Lett 152:123-138

Jambon A, Zimmermann JL (1990) Water in oceanic basalts: Evidence for dehydration of rcycled crust. Earth Planet Sci Lett 101:323-331

Jenkins DM, Chernosky JV (1986) Phase equilibria and crystallochemical properties of Mg-chlorite. Am Mineral 71:924-936.

Johannes W (1975) Zur Synthese und thermischen Stabilität von Antigorit. Fortschr Mineral 53:36

Johnson KE, Davis AM, Bryndzia LT (1996) Contrasting styles of hydrous metasomatism in the upper mantle: an ion probe investigation. Geochim Cosmochim Acta 60:1367-1386

Kawamoto T, Leinenweber K, Hervig RL, Holloway JR (1995) Stability of hydrous minerals in H_2O-saturated KLB-1 peridotite up to 15 GPa. In: Farley K (ed) Volatiles in the Earth and Solar System, p 229-239. American Institute of Physics

Kerr RA (1997) Deep-sinking slabs stir the mantle. Science 275:613-615

Khodyrev O Yu, Agoshkov VM (1986) Phase transitions in the $MgO-SiO_2-H_2O$ system at 40-80 kbar. Geochemistry International 23:47-52

Kienast JR, Lombardo B, Biino G, Pinardon JL (1991) Petrology of very-high-pressure eclogitic rocks from the Brossasco-Isasca Complex, Dora Maira massif, Italian western Alps. J Metam Geol 9:19-34

Kitahara, S, Takenouch, IS, Kennedy, GC (1966) Phase relations in the system $MgO-SiO_2-H_2O$ at high temperatures and pressures. Am J Sci 254:223-233

Kohlstedt DL, Keppler H and Rubie DC (1996) Solubility of water in the alpha , beta and gamma phases of $(Mg,Fe)_2SiO_4$. Contributions to Mineralogy and Petrology 123:345-357.

Konzett J, Ulmer P (1998) The stability of hydrous potassic phases in lherzolitic mantle-an experimental study to 9.5 GPa in simplified and natural bulk compositions. J Petrol, (in press)

Krogh EJ, Carswell DA (1995) HP and UHP eclogitesand garnet peridotites in the scandiavian Caledonides. *In:* RG Coleman, X Wang (eds) Ultrahigh pressure metamorphism, p 244-298

Kushiro I (1972) Effect of water on the composition of magmas formed at high pressures. J Petrol 13:311-334

Kushiro I (1990) Partial melting of mantle wedge and evolution of island arc crust. J Geophys Res 95:15929-15939

Liou JG, Zhang R, Ernst WG, Liu J, McLimans R (1998) Mineral parageneses in the Piampaludo eclgogitic body, Gruppo diVoltri, Western Ligurian Alps. Schweiz mineral petrogr Mitt 78:317-335

Liou JG, Zhang RY, Jahn BM (1997) Petrogenesis of ultrahigh-pressure jadeite quartzite from the Dabie region, East-central China. Proc 30th Int'l Geol Congr 17:141-156

Luhr JF (1997) Extensional Tectonics and the Diverse Primitive Volcanic Rocks in the Western Mexican Volcanic Belt. Canadian Mineral 35:473-500

Luth RW (1995) Is phase A relevant to the Earth's mantle? Geochim Cosmochim Acta 59:679-682

Luth RW (1997) Experimental study of the system phologopite-diopside from 3.5 to 17 GPa. Am Mineral 62:1198-1209

Maaløe S, Aoki K-I (1977) The Major Element Composition of the Upper Mantle Estimated from the Composition of Lherzolites. Contrib Mineral Petrol 63:161-173

Manning, CE (1994) The solubility of quartz in H_2O in the lower crust and upper mantle. Geochim Cosmochim Acta 58:4831-4840

Maruyama S, Liou JG, Terabayashi M (1996) Blueschists and eclogites of the world and their exhumation. Int'l Geol Rev 38:490-596

Massonne H-J (1992) Evidence for low-temperature ultrapotassic siliceous fluids in subduction zone environments in the system $K_2O-MgO-Al_2O_3-SiO_2-H_2O$ (KMASH). Lithos 28:421-435

Massone HJ, Schreyer W (1989) Stability field of the high-pressure assemblage talc + phengite and two new phengite barometers. Eur J Mineral 1:391-410

Maury RC, Defant MJ, Joron J-L (1992) metasomatism of the sub-arc mantle inferred from trace elements in Philippine xenoliths. Nature 360:661-663

McInnes, B (1996) Fluid-peridotite interactions in mantle wedge xenoliths (abstr.). EOS Trans Am Geophys Union 77:282

Medaris KG, Beard BL, Johnson CM, Valley JW, Spicuzza MJ, Jelinek E, Misar Z (1995) Garnet pyroxenite and eclogitefrom the Bohemian massif: geochemical evidence for Variscan recycling of subducted llithosphere. Geol Rundschau 84:489-505

Medaris LG, Carswell DA (1990) The petrogenesis of Mg-Cr garnet peridotites in Eurpean metamorphic belts. *In:* DA Carswell (ed) Eclogite Facies Rocks, p 260-290. Chapman and Hall

Mengel, K, Green, DH (1989) Stability of amphibole and phlogopite in metasomatised peridotite under water-saturated and water-undersaturated conditions. *In:* J Ross (ed) 4th Int'l Kimberlite Conf, Australian J Earth Sci Spec Publ 14:571-581

Merzbacher C, Eggler DH (1984) A magmatic geohygrometer: application to Mount St. Helens and other dacitic magmas. Geology 12:587-590

Millhollen GL, Irving AJ, Wyllie PJ (1974) Melting interval of peridotite with 5.7 percent water to 30 kilobars. J Geol 82:575-587

Morris JD, Leeman WP, Tera F (1990) The subducted component in island arc lavas: constraints from Be isotopes and B-Be systematics. Nature 344:31-36

Mukhopadhyay B, Manton WI (1994) Upper-mantle fragments from beneath Sierra Nevada batholith: Partial fusion, fractionalcrystallization, and metasomatism in a subduction-related ancient lithosphere. J Petrol 35:1417-1450

Muller MR, Robinson CJ, Minshull TA, White RS, Bickle MJ (1997) Thin crust beneath ocean drilling program borehole 735B at the Southwest Indian Ridge. Earth Planet Sci Lett 148:93-107

Mysen BO (1981) Rare earth element partitioning between minerals and (H_2O + CO2) vapor as a function of pressure, temperature and vapor composition. Carnegie Inst Washington Year Book 80:347-350

Mysen BO (1998) Interaction between aqueous fluid and silicate melt in the pressure and temperature regime of the Earth's crust and upper mantle. Neues Jahrb Mineral 172:227-244

Mysen, BO, Boettcher, AL (1975) Melting of a hydrous mantle. II Geochemistry of crystals and liquids formed by anatexis of mantle peridotite at high pressures and high temperatures as a function of controlled activities of water, hydrogen and carbon dioxide. J Petrol 16:549-590

Newton RC, Smith JV (1967) Investigations concerning the breakdown of albite at depth in the earth. J Geol 75:268-286

O'Hanley DS (1996) Serpentinites—Records of tectonic and petrological history. Oxford University Press

Obata M (1980) The Ronda peridotite: garnet-, spinel-, and plagioclase-lherzolitefacies and the P-T trajectories of a high-temperature mantle intrusion. J Petrol 21:533-572

Oh C-W, Liou JG (1990) Metamorhic evolution of two different eclogites inthe Fransiscan complex, California, USA. Lithos 25:41-53

Ohtani E, Shibata T, Kubo T, Kato T (1995) Stability of hydrous phases in the transition zone and the uppermost part of the lower mantle. Geophys Res Lett 22:2553-2556

Ono S (1998) Stability limits of hydrous minerals in sediment and mid-ocean ridge basalt compositions: Implications for water transport in subduction zones J Geophys Res 103:18,253-18,267

Pawley AR, Wood BJ (1996) The low-pressure stability of phase A, $Mg_7Si_2O_8(OH)_6$. Contrib Mineral Petrol 124:90-97

Pawley AR, Wood BJ (1995) The high-pressure stability of talc and 10Å phase: Potential storage sites of H_2O in subduction zones. Am Mineral 80:998-1003

Pawley AR, Holloway JR (1993) Water sources for subduction zone magmatism: new experimental constraints. Science 260:664-667

Pawley AR (1998) The reaction talc + forsteriste = enstatite + H_2O: New experimental results and petrological implications. Am Mineral 83:51-57

Pawley AR (1996) High pressure stability of chlorite: a source of H_2O for subduction zone magmatism. Terra 8:50

Peacock SM (1990) Numerical simulations of metamorphic pressure-temperature-time paths and fluid production in subducting slabs. Tectonics 9:1197-1211

Peacock SM (1993) The importance of blueschist to eclogite dehydration reactions in subducting oceanic crust. Geol Soc Am Bull 105:684-694

Pearson DG, Davies GR, Nixon PH, Milledge MJ (1989) Graphitized diamonds from a peridotite massif in Morocco and implications for anomalous diamond occurrences. Nature 338:60-62

Pearson DG, Davies, GR, Nixon, PH (1993). Geochemical constraints on the petrogenesis of diamond facies pyroxenites from the Beni Bousera massif, North Morocco. J Petrol 34:125-17

Pistorius CWFT (1963) Some phase relations in the system $MgO-SiO_2-H_2O$ to high pressures and temperatures. Neues Jahrb Mineral Monatsh 11:283-293

Planck T, Langmuir CH (1993) Tracing trace elements from sediment input to volcanic output at subduction zones. Nature 362:739-743

Poli S (1993) The amhibolite-eclogite transformation: an experimental study on basalt. Am J Sci 293:1061-1107

Ringwood AE, Major A (1967) High-pressure reconnaissance investigation in the system $Mg_2SiO_4-MgO-H_2O$ Earth Planet Sci Lett 2:130-133

Riter, JCA, Smith D (1996) Xenolith constraints on the thermal history of the mantle below the Colorado Plateau. Geol 24:267-270

Rossman, GR (1996) Studies of OH in nominally anhydrous minerals. Phys Chem Minerals 23:299-304

Rutherford MJ, Sigurdsson H, Carey S, Davis A (1985) The May 18, 1980, Eruption of Mount St. Helens—I Melt composition and Experimental Phase Equilibria. J Geophys Res 90:2929-2947

Sakuyama M (1979) Lateral variations of H_2O contents in quarternary magmas of Northeastern Japan. Earth Planet Sci Lett 43:103-111

Sato K, Katsura T, Ito E (1997) Phase relations of natural phlogopite with and without enstatite up to 8 GPa: implication for mantle metasomatism. Earth Planet Sci Lett 146, 511-526

Scambelluri M, Muentener O, Hermann J, Piccardo GB, Trommsdorff V (1995) Subduction of water into the mantle: History of an alpine peridotite. Geology 23:459-462

Schertl H, Okay AI (1994) Coesite inclusion in dolomite of Dabie Shan, China: Petrological and geological significance. Eur J Mineral 6:995-1000

Schiano P, Clocchiatti N, Shimizu N, Maury RC, Jochum KP, Hofmann AW (1995) Hydrous, silica-rich melts in the sub-arc mantle and their relationship with erupted arc lavas. Nature 377:595-600

Schmidt MW, Poli S (1998) What causes the position of the volcanic front? Experimentally based water budget for dehydrating slabs and consequences for arc magma generation. Earth Planet Sci Lett, (in press)

Schmidt MW (1996) Experimental constraints on recycling of potassium from subducted oceanic crust. Science 272:1927-1930

Schneider ME, Eggler DH (1986) Fluids in equilibrium with peridotite minerals: Implications for mantle metasomatism. Geochim Cosmochim Acta 50:711-724

Schreyer W, Maresch WV, Baller T (1991) A new hydrous, high-pressure phase with a pumpellyite structure in the system MgO-Al$_2$O$_3$-SiO$_2$-H$_2$O *In:* Perchuk LL (ed) Progress in Experimental Petrology, p 47-64. Cambridge University Press

Sekine T, Wyllie PJ (1982) Phase Relationship in the System KAlSiO$_4$-Mg$_2$SiO$_4$-SiO$_2$-H$_2$O as a model for Hybridization between Hydrous Siliceous Melts and Peridotite. Contrib Mineral Petrol 79:368-374

Shibata T, Nakamura E (1997) Across-arc variations of isotope and trace element compositions from Quaternary basaltic volcanic rocks in northeastern Japan: Implications for interaction between subducted oceanic slab and mantle wedge. J Geophys Res 102:8051-8064

Silver PG, Beck SL, Wallace TC, Meade C, Myers S, James E, Kuehnel R (1995) Rupture characteristics of the deep Bolivian earthquake of June 9, 1994 and the mechanism of deep-focus earthquakes. Science 268:69-73

Sisson TW, Grove TL (1993) Temperatures and H$_2$O contents of low-MgO high-alumina basalts. Contrib Mineral Petrol 113:167-184

Sisson TW, Layne GD (1993) H$_2$O in basalt and basaltic andesite glass inclusions from four subduction-related volcanoes. Earth Planet Sci Lett 117:619-637

Slodkevitch VV (1982) Geochemical constraints of diamond into graphite. Mineral, Soc. USSR, 1:13-33.

Sobolev A, Chaussidon M (1996) H$_2$O concentrations in primary melts from supra-subduction zones and mid-ocean ridges: implications for H$_2$O storage and recycling in the mantle. Earth Planet Sci Lett 137:45-55

Sorenson SS, Grossman JN, Perfit MR (1997) Phengite-hosted LILE enrichment in eclogite and related rocks: implications for fluid-mediated mass transfer in subduction zones and arc magma genesis. J Petrol 38:3-34

Stalder R, Ulmer P, Thompson AB, Gunther D (1998) Experimentaldetermination of second critical endpoints in fluid/melt systems. Mineral Mag 62A:1441-1442

Staudigel H, Schreyer W (1977) The upper thermal stability of clinochlore, Mg$_5$Al[AlSi$_3$]O$_{10}$(OH)$_8$, at 10-35 kb P$_{H_2O}$. Contrib Mineral Petrol 61:187-198

Sudo A, Tatsumi Y (1990) Phlogopite and K-amphibole in the upper mantle: Implications for magma genesis in subduction zones. Geophys Res Lett 17:29-32

Suen CJ, Frey FA (1987) Origins of the mafic and ultramafic rocks ofthe Ronda peridotite. Earth Planet Sci Lett 85:183-205

Swanson SE, Kay SM, Brearley M, Scarfe CM (1987) Arc and back-arcxenolithsin Kurile-Kamchatka and western Alaska. *In:* PH Nixon (ed) Mantle Xenoliths, p 303-318. John Wiley & Sons

Sweeney RJ (1994) Carbonatite melt compositions in the Earth's mantle. Earth Planet Sci Lett 128:259-270

Szabo C, Bodnar RJ, Sobolev AV (1996) Metasomatism associated with subduction-related, volatile-rich silicate melt in the upper mantle beneath the Nógrád-Gömör Volcanic Field, Northern Hungary/Southern Slovakia: Evidence from silicate melt inclusions. Eur J Mineral 8:881-899

Takahashi E (1986a) Genesis of calc-alkali andesite magma in a hydrous mantle-crust boundary: petrology of lherzolite xenoliths from the itchinomegata crater, Oga peninsula, Northeast Japan, Part II. J Volcan Geothermal Res 29:355-395

Takahashi E (1986b) Melting of a dry peridotite KLB-1 up to 14 GPa: Implications on the origin of peridotitic upper mantle. J Geophys Res 91:9367-9382

Tatsumi Y (1986) Origin of subduction zone magmas based on experimental petrology. *In:* L Perchuk, I Kushiro (eds) Physics and Chemistry of Magmas, p 268-301. Springer-Verlag

Tatsumi Y (1989) Migration of fluid phases and genesis of basalt magmas in subduction zones. J Geophys Res 94:4697-4707

Tatsumi Y, Eggins S (1995) Subduction Zone Magmatism. Blackwell Scientific Publications

Tatsumi Y, Murasaki M, Arsadi EM, Nohda S (1991) Geochemistry of Quaternary lavas from NE Sulawesi: transfer of subduction components into the mantle wedge. Contrib Mineral Petrol 107:137-140

Thompson AB (1992) Water in the Earth's upper mantle. Nature 358:295-302

Trommsdorff V (1983) Metamorphose magnesiumreicher Gesteine: Kritischer Vergleich von Natur, Experiment und thermodynamischer Datenbasis. Fortschr Mineral 61:283-308

Trommsdorff V, Evans BW (1974) Alpine metamorphism of peridotitic rocks. Schweiz mineral petrograph Mitt 54:333-352

Trommsdorff V, Evans BW (1980) Titanium hydroxyl clinohumite: Formation and breakdown in antigorite rocks (Malenco, Italy). Contrib Mineral Petrol 72:229-242

Trommsdorff, V, Lopez Sanchez-Vizcaino, V, Gomez-Pugnaire, MT, Müntener, O (1998) High pressure breakdown of antigorite to spinifex-textured olivine and orthopyroxene, SE Spain. Contrib Mineral Petrol 132:139-148

Trønnes RG, Takahashi, E, Scarfe CM (1988) Stability of K-richterite and phlogopite to 14 GPa. EOS Trans Am Geophys Union 69:1510-1511

Trønnes RG (1990) Low-Al, high-K amphiboles in subducted lithosphere from 200-400 km depth: Experimental Evidence. EOS Trans Am Geophys Union 71:1587

Ulmer, P, Trommsdorff, V (1995) Serpentine stability to mantle depths and subduction-related magmatism. Science 268:858-861

Ulmer P, Trommsdorff V (1998) Phase relations of hydrous mantle subducting to 300 km. *In:* Y-W Fei, C Bertka, BO Mysen (eds) Mantle Petrology: Field Observations and High-Pressure Experimentation, Geochemical Society (in press)

van der Hist R, Engdahl R, Spakman W, Nolet G (1991) Tomographic imaging of subducted lithosphere below northwest Pacific islandarcs. Nature 353:37-43

Venturini G, Martinotti G, Armando G, Barbero M, Hunziker JK (1994) The Central Sesia Lanzo Zone (western Italian Alps): new field observations and lithostratigraphic subdivisions. Schweiz mineral petrograph Mitt 74:115-125

Vidal P, Dupuy C, Maury R, Richard M (1989) Mantle metasomatism above subduction zones: Traceelement and radiogenic isotope characteristics of peridotitexenoliths from Batan Ilsand (Philippines). Geology 17:1115-1118

Wang X, Liou JG (1991) Regional ultrahigh-pressurecoesite-bearing eclogitic terrane in central China: evidence from country rocks, gneiss, marble, and metapelite. Geology 19:933-936

Weiss M (1997) Clinohumites: A field and experimental study. Swiss Federal Institute of Technology (ETH), Zürich

Wendlandt RF, Mysen BO (1980) Melting phase relations of natural peridotite + CO_2 as a function of degree of partial melting at 15 and 30 kbar. Am Mineral 65:37-44

Wenrich, KJ, Billingsley, GH and Blackerby, BA (1995) Spatial migration and compositional change of Miocene-Quaternary magmatism in the western Grand Canyon. J Geophys Res 100:10417-10440

Wood BJ, Bryndzia LT, Johnson KE (1990) Mantle oxidation and its relationship to tectonic environment and fluid speciation. Nature 249:337-345

Woodland AB, Kornprobst J, Wood BJ (1992) Oxygen thermobarometry oforogenic lherolite massifs. J Petrol 33, 203-230

Wunder B (1998) Equilibrium experiments in the system $MgO-SiO_2-H_2O$ (MSH): stability fields of clinohumite-OH [$Mg_9Si_4O_{16}(OH)_2$], chondrodite-OH [$Mg_5Si_2O_8(OH)_2$) and phase A [$Mg_7Si_2O_8(OH)_6$]. Contrib Mineral Petrol 132:111-120

Wunder B, Schreyer W (1997) Antigorite: High-pressure stability in the system $MgO-SiO_2-H_2O$(MSH). Lithos 41:213-227

Wyllie PJ, Sekine T (1982) The Formation of Mantle Phlogopite in Subduction Zone Hybridization. Contrib Mineral Petrol 79:375-380

Yamamoto K, Akimoto S-I (1977) The system $MgO-SiO_2-H_2O$ at high pressures and temperatures—stability field for hydroxyl-chondrodite, hydroxyl-clinohumite and 10-Å phase. Am J Sci 277:288-312

Yang H, Prewitt CT, Frost DJ (1997) Crystal structure of the dense hydrous magnesium silicate, phase D. Am Mineral 80:998-1003

Yoder HS (1952) The $MgO-Al_2O_3-SiO_2-H_2O$ system and the related metamorphic facies. Am J Sci (Bowen Vol), p 569-627

Yoder HS, Stewart DB, Smith JR (1957) Ternary feldspars. Carnegie Inst Washington Year Book 56: 206-214

Zhang RY, Liou JG (1994) Petrogenesis of garnet-bearing ultramafic rocks and associated eclogites in the Sulu ultrahigh pressure metamorphic terrane, China. J Metamorphic Geol 12:169-186

Zhang RY, Liou JG (1998) Ultrahigh-pressure metamorphism in the Sulu terrane, easten China: A prospective view. Continental Dynamics, (in press)

Zhang RY, Liou JG, Coney BL (1995) Talc-, magnesite-, and Ti-clinohumite-bearing ultrahigh-pressure meta-mafic and ultramafic complex in the Dabie Mountains, China. J Petrol 36:1011-1037

Chapter 4

MINERALOGY AND COMPOSITION
OF THE UPPER MANTLE

William F. McDonough and Roberta L. Rudnick

Department of Earth and Planetary Sciences
Harvard University, 20 Oxford Street
Cambridge, Massachusetts 02138

INTRODUCTION

The mineralogy and composition of the mantle is principally constrained from the geology and petrology of ultramafic rocks, as well as from seismological and geodynamical observations on the Earth's interior and meteoritical analogies. Over the latter half of this century our understanding of the uppermost mantle has increased greatly through a vast number of petrological, geochemical and isotopic studies on upper mantle rocks and from numerous experimental laboratory investigations that have attempted to reproduce the physical conditions of the upper mantle for a range of rock compositions.

There exist compositional distinctions between continental and oceanic crust and so by analogy there are likely to be compositional differences between the lithospheric mantle that underlies the continents and that beneath the oceans. This is borne out by studies of peridotites from these different settings. There may also be compositional differences between Archean and post-Archean continental crust (Taylor and McLennan 1985) and so one might expect compositional differences in the underlying lithospheric mantle. With these ideas in mind we will examine the mantle samples that are available from these different regions.

Surface heat flow and global seismic tomography for oceanic and continental domains show that the oceans have thinner and hotter lithosphere than the continents. Thus, our present day view of the upper mantle is one of considerable lateral heterogeneity in its chemical and physical properties that can be directly related to its geological history. The present review will concentrate on the findings from studies of upper mantle rocks derived from a wide variety of tectonic environments and their relevance to a range of geological, geophysical, and cosmochemical observations and issues.

ROCKS OF THE UPPER MANTLE

The upper mantle is a peridotitic metamorphic complex dominated by olivine, with decreasingly lesser amounts of orthopyroxene, clinopyroxene and an aluminous phase (plagioclase, spinel or garnet depending on pressure, Fig. 1). In addition, there are lesser quantities of mafic rocks (i.e. eclogites and pyroxenites) that derive from (1) the recycling of oceanic crust into the mantle or (2) the precipitation of pyroxene $\pm$ spinel $\pm$ garnet during the ascent of magmas to the surface. There are three modes of upper mantle sampling that have given us differing insights into the spectrum of upper mantle lithologies: (1) exposures of tectonically emplaced sections of the mantle (i.e. ophiolites and orogenic peridotites), (2) xenoliths in magmas (i.e. kimberlite- and basalt-hosted samples) and (3) fragments of serpentinized peridotite dredged and/or drilled along mid-ocean ridges. Another interesting window into the mantle is that provided by mineral inclusions within diamonds. These minerals carry with them a wealth of information about a greater spectrum of processes and lithologies in the mantle than seen in other mantle samples. The formation

0275-0279/98/0037-0004$05.00

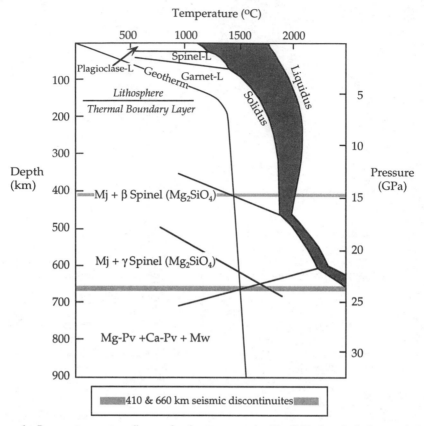

Figure 1. Pressure-temperature diagram for the upper mantle. The fields for plagioclase-, spinel- and garnet-facies lherzolite are shown at the top of the mantle (e.g. "Plagioclase-L"). Abbreviations: Mj = majorite, Mg-Pv = magnesium perovskite, Ca-Pv = calcium perovskite, Mw = magnesiowüstite. The major phase changes in upper mantle mineralogy are coincident with the 410 and 660 km seismic discontinuities (shaded horizonatal lines) given the mantle geotherm. Solidus and liquidus are shown for an anhydrous peridotite. The lithosphere thermal boundary layer marks the transition from an adiabatic thermal regime in the convecting mantle to a conductive thermal regime in the lithosphere. This thermal boundary layer is shallow beneath the mid-ocean ridges, leading to melting by adiabatic upwelling. In contrast, the lithospheric geotherm is steeper beneath ancient continents, leading to a deep lithospheric thermal boundary layer (see Rudnick et al. 1998 and references therein). Modified from Philpotts (1990).

of diamond in the mantle begins at about 150 km depth and continues to considerable depth (the deepest limits are not well constrained). Mineral inclusions in diamonds are diverse and indicate a very broad depth range of mineral entrapment, including some mineral inclusions that may have lower mantle origins. We will briefly review the spectrum of upper mantle lithologies, their mineralogies and geochemical characteristics. This review is not exhaustive but we attempt to provide sufficient referencing so that the interested reader can locate references appropriate to different aspects of upper mantle petrology and geochemistry.

Peridotites

Some of the earliest ideas about "rocks of the Earth's interior" came from studies that integrated petrology, geophysics and cosmochemistry. Meteorite studies revealed a spectrum of rock types ranging from stones to metals (iron meteorites) to mixtures of the two (chondrites) and it was recognized that these diverse rock types might be representative of the Earth. Also, various lines of physics indicated that rocks of the Earth's interior have high densities, due to their high pressure and temperature origins in contrast to typical crustal rocks. Characteristically, upper mantle rocks also tend to have fewer phases than crustal rocks, due to the restricted range in compositions found for peridotites and their higher temperatures of equilibration.

Initial concepts of the mineralogy and composition of the upper mantle were developed through integrating petrological, geophysical and cosmochemical observations. For example, a classic paper by Henry Washington (1925) on the chemical composition of the Earth characterizes the Earth's interior as being made up of a series of shells extending from the core to the crust based on (1) mineralogy and petrology of meteorites, (2) a density profile for the planet from the recent findings of Adams and Williamson (1923) and Williamson and Adams (1923) and (3) the mineralogy and petrology of rocks found at the Earth's surface. As Washington viewed it, there was a dense core of Fe-Ni, similar to iron meteorites, overlain by a mantle composed of three shells of increasing density made up mostly of olivine and other phases, which was overlain by a thin granitic and basaltic crust. Washington and others thus had a pretty fair estimate of the situation for the turn of the century. Over the latter portion of this century we have been able to refine this view and place better quantitative constraints on the phases and their proportions.

Geophysical constraints. Global seismological models for the mantle (e.g. PREM of Dziewonski and Anderson 1981) constrain to a first order the chemical and mineralogical properties of this region. The average density of the mantle immediately below the Mohorovicic seismic discontinuity (i.e. the base of the crust) is ~3.3 g cm^{-3}, which is consistent with a mineralogy dominated by olivine, pyroxene, and a minor aluminous phase. In addition, seismic studies show that the mantle has a significant degree of anisotropy, suggesting that olivine makes up a considerable fraction of the mantle (the *c*-axis of olivine transmits sound waves faster than its other axes).

For some time there were some who supported a view that the upper mantle is not composed of peridotite, but instead may be composed of eclogite, a bi-mineralic assemblage of garnet and jadeite-rich (Na-Al-rich) pyroxene with a density of about 3.3-3.8 g cm^{-3}. Eclogite is a high pressure metamorphic equivalent of basalt, which is a common rock type at the Earth's surface. Importantly however, few eclogites have modal olivine and in those that do it is only a minor constituent. This latter observation is significant and is consistent with the observation that eclogites, even highly deformed ones, lack significant seismic anisotropy (Fountain et al. 1994), a fundamental feature of the upper mantle. In addition, it is generally accepted that eclogitic source regions are incapable of producing basalts, because to generate such melt compositions would require complete melting of the source region. There are, however, some eclogites that can be partially melted to produce some basaltic compositions, but these melts are not typical of basalts. Given this, it is difficult to support the view that the upper mantle is eclogitic. Thus, even though eclogites, pyroxenites and other lithologies are found in association with peridotites, the upper mantle is dominantly composed of peridotite, a rock with a considerable amount of olivine.

Petrology. Peridotites are typically composed of olivine, orthopyroxene, clinopyroxene and an aluminous phase that is either plagioclase, spinel or garnet, depending on the pressure and temperature of equilibration (Fig. 1). The stability fields of plagioclase, spinel and garnet peridotites are dependent on bulk composition. The fields shown in Figure 1 are representative of a fertile bulk composition; with increasing bulk rock Cr content (i.e. as the peridotite becomes more refractory) the higher pressure stability field of spinel is expanded (O'Neill 1981). A typical reaction for the transition between the plagioclase and spinel peridotite facies is:

$$\text{olivine} + \text{plagioclase} + \text{Cr-spinel} = \text{orthopyroxene} + \text{clinopyroxene} + \text{Al-spinel} \quad (1)$$

(Green and Hibberson 1970, Kushiro and Yoder 1966) and for the transition from spinel to garnet peridotite facies is:

$$\text{spinel} + \text{orthopyroxene} = \text{olivine} + \text{garnet} \quad (2)$$

(Green and Ringwood 1967, MacGregor 1974).

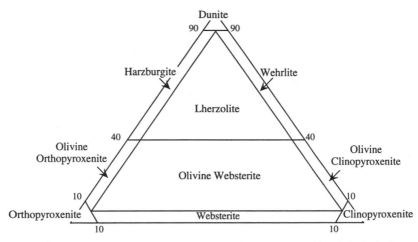

Figure 2. The classification ternary for ultramafic and mafic rocks with <5% plagioclase, spinel, amphibole or garnet (Streckeisen 1976). Peridotites have >40% olivine. The top ternary shows the field for most peridotites along with the melting trend from lherzolite (L) to harzburgite (H) to dunite (D).

A mineralogical classification diagram for peridotites is given in Figure 2 (opposite). This diagram is constructed by projecting from the aluminous phase onto a ternary constituting the three most abundant phases in peridotites: olivine, orthopyroxene and clinopyroxene. The shaded region in the top ternary shows the field for most samples of upper mantle peridotite. The trend marked by arrows from lherzolite (L) to harzburgite (H) to dunite (D) is representative of a typical mantle melting trend. As the degree of melting increases, the compositions of residual peridotites and their minerals change along this trend. Most notable is the marked depletion of clinopyroxene with melting.

Representative major element compositions of peridotites are given in Table 1 along with a model composition for the initial mantle, Primitive Mantle (or Pyrolite). Figure 3 shows the modal abundances of these various peridotite compositions in both the spinel- and garnet-facies. The compositional spectrum from lherzolite to dunite shows marked decreases in CaO, Al_2O_3, Na_2O, SiO_2 and TiO_2 and increases in MgO and NiO (Fig. 4). These variations in peridotites are due to the extraction of basalts with high CaO, Al_2O_3, and NaO and low MgO and NiO contents; this relationship between basalt and peridotite is shown in Figure 4 using the compositions given in Table 1 and nominal compositions for 2 basalts (MORB and alkali basalt from an ocean island). The compositions of residual peridotites and basalts are complementary to one another, relative to the Primitive Mantle composition.

Mineralogy. Here we make brief observations on the chemistry of the dominant minerals of the upper mantle. The mineral compositions given in Tables 2-8 were selected to cover the range of bulk compositions from fertile lherzolite to depleted harzburgite from a variety of occurrences including (1) abyssal peridotite, (2) ophiolite, (3) the Zarbargad ultramafic massif in the Red Sea, a recently uplifted section of the mantle, (4) spinel lherzolite and harzburgite xenoliths brought up by alkali basalts, (5) a garnet lherzolite xenolith brought up by an alkali basalt and (6) a garnet harzburgite xenolith brought up in kimberlite.

Olivines. Representative olivine analyses are given in Table 2. Of all the minerals in peridotites these have the simplest chemistry with fosterite (Mg_2SiO_4) content generally between Fo_{88} and Fo_{92}, ranging as high as Fo_{94}. Olivines are typically slightly less magnesian than the bulk rock Mg# ($\equiv$ atomic proportions of Mg/(Mg+Fe)). Peridotitic olivines typically have several hundred to a thousand ppm Ca (which can be expressed as a larnite component, Ca_2SiO_4) (Simkin and Smith 1970). Ca exchange between clinopyroxene and olivine has been utilized as a geobarometer for spinel peridotites because these rocks do not have any other recognized pressure sensitive reaction (see (Köhler and Brey 1990) and references therein). However, most attempts to use Ca as a geobarometer have met with failure or widespread skepticism because this reaction is strongly temperature sensitive, the bane of any barometer. In addition, as pointed out by Köhler and Brey (1990), it is essential that one takes care to obtain high quality Ca data to estimate pressure.

Orthopyroxenes. Representative orthopyroxene analyses are given in Table 3. The alumina contents of orthopyroxenes in peridotites from the spinel and plagioclase facies vary as a function of the bulk rock Al_2O_3 and thus with the degree of refractoriness of the peridotite (Fig. 5). For peridotites from the garnet facies, alumina content of ortho-pyroxenes are lower and vary as a function of pressure. In general, the CaO contents of orthopyroxenes vary as a function of temperature, while their Na_2O and TiO_2 contents reflect the bulk composition of the peridotite.

Clinopyroxenes. Representative clinopyroxene analyses are given in Table 4; these clinopyroxenes are Cr-rich diopsides, which are typical of peridotites, whereas those that

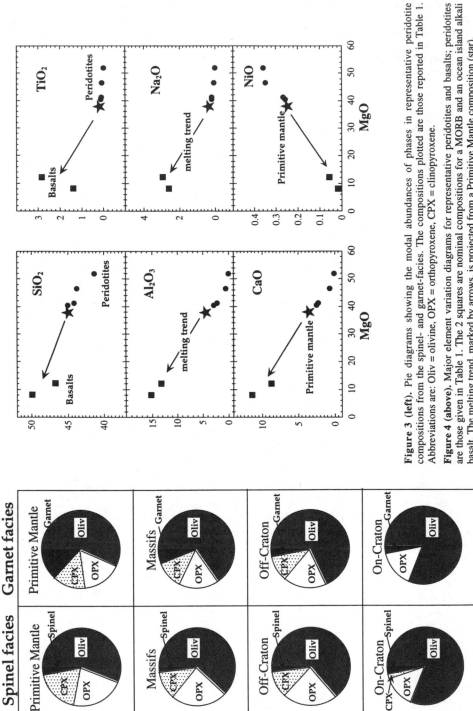

Figure 3 (left). Pie diagrams showing the modal abundances of phases in representative peridotite compositions from the spinel- and garnet-facies. The compositions plotted are those reported in Table 1. Abbreviations are: Oliv = olivine, OPX = orthopyroxene, CPX = clinopyroxene.

Figure 4 (above). Major element variation diagrams for representative peridotites and basalts; peridotites are those given in Table 1. The 2 squares are nominal compositions for a MORB and an ocean island alkali basalt. The melting trend, marked by arrows, is projected from a Primitive Mantle composition (star).

Table 1. Compositions of Primitive Mantle and peridotites from different settings

	Primitive Mantle	Massifs lherzolite	Off-Craton lherzolite	On-Craton* harzburgite	89-774 dunite
SiO_2	45.00	45.05	44.16	43.8	41.30
TiO_2	0.20	0.13	0.09	0.08	0.00
Al_2O_3	4.45	2.89	2.25	0.56	0.10
Cr_2O_3	0.38	0.37	0.39	0.38	0.28
FeO	8.05	8.05	8.14	7.71	6.26
MnO	0.135	0.13	0.14	0.12	0.08
MgO	37.80	40.25	41.05	46.3	51.88
NiO	0.25	0.26	0.27	0.35	0.36
CaO	3.55	2.51	2.27	0.75	0.16
Na_2O	0.36	0.2	0.21	0.07	0.032
K_2O	0.029	0.010	0.020	0.030	0.02
P_2O_5	0.021	0.02	0.03	0.01	0.05
Mg #	89.3	89.9	90	91.4	93.7

Spinel Peridotite Facies Mineralogy

olivine	56	62	62	81	98
orthopyroxene	22	24	22	14	--
clinopyroxene	19	12	11	2	1
spinel	3	2	2	3	1

Garnet Peridotite Facies Mineralogy

olivine	57	66	68	83	98
orthopyroxene	16	17	18	15	--
clinopyroxene	14	12	11	--	1
garnet	13	6	3	2	1[#]

Data sources: Primitive Mantle (McDonough and Sun 1995), Massifs, Off-Craton and On-Craton (McDonough 1994, Rudnick et al. 1998) and dunite, 89-774 (Rudnick et al. 1993).

*Table 9 presents an alternative model composition for Cratonic mantle.

[#] 1% spinel, not garnet; such a refractory peridotite is unlikely to have garnet.

Table 2. Representative analyses of olivine

	Abyssal 8-6/2-8	Ophiolite TM 1245	Zarbargad Z-37	Sp. Lherz. Ib/8	Sp. Harz. D50	Gt. Lherz. 313-1	Gt. Harz. 674
SiO_2	40.32	40.69	39.59	40.32	40.63	41.05	40.08
TiO_2	0.01	n.d.	0.00	0.011	0.004	n.d.	n.d.
Cr_2O_3	0.00	0.00	0.03	0.03	0.05	n.d.	0.04
FeO	8.59	9.09	10.72	9.80	8.50	9.64	8.62
MnO	0.16	n.d.	0.17	0.16	0.10	n.d.	0.10
NiO	0.40	0.41	0.33	0.335	0.336	0.29	0.33
MgO	50.58	49.83	49.14	47.96	49.89	48.97	50.10
CaO	0.05	0.04	0.13	0.14	0.11	0.06	0.05
Na_2O	n.d.	n.d.	n.d.	0.014	0.007	n.d.	n.d.
Mg#	91.2	90.6	89.0	89.6	91.2	89.9	91.1

Data sources: Abyssal (Shibata and Thompson 1986), Ophiolite (Suhr and Robinson 1994), Zarbargad (Bonatti et al. 1986), Sp. Lherz (Spinel Lhezrolite) and Sp. Harz. (Spinel Harzburgite) (McDonough et al. 1992), Gt. Lherz. (Garnet Lherzolite) (Ionov et al. 1993), and Gt. Harz. (Garnet Harzburgite) (Rudnick et al. 1994).

Table 3. Representative analyses of orthopyroxene

	Abyssal	Ophiolite	Zarbargad	Sp. Lherz.	Sp. Harz.	Gt. Lherz.	Gt. Harz.
	8-6/6	TM 1245	Z-37	Ib/8	D50	313-1	674
SiO_2	56.37	57.58	54.03	53.33	55.58	55.41	57.10
TiO_2	0.03	0.00	0.15	0.178	0.077	0.19	0.14
Al_2O_3	2.03	0.92	3.97	6.58	4.20	3.85	1.42
Cr_2O_3	0.72	0.52	0.26	0.47	0.86	0.55	0.50
FeO	5.42	5.29	7.11	6.59	5.50	5.88	4.94
MnO	0.16	0.11	0.18	0.15	0.13	n.d.	0.09
NiO	0.10	0.04	0.02	0.108	0.104	0.09	0.10
MgO	33.5	34.93	33.13	31.16	31.61	32.94	35.00
CaO	2.11	1.03	0.67	1.34	1.33	0.75	0.51
Na_2O	n.d.	0.04	0.01	0.193	0.089	0.14	0.24
Mg#	91.6	92.1	89.1	89.3	91.0	90.8	92.6

Data sources and references as in Table 2.

Table 4. Representative analyses of clinopyroxene

	Abyssal	Ophiolite	Zarbargad	Sp. Lherz.	Sp. Harz.	Gt. Lherz.	Gt. Harz.
	8-6 10	TM 1245	Z-37	Ib/8	D50	313-1	674
SiO_2	52.87	55.85	50.87	51.01	52.43	52.82	54.5
TiO_2	0.03	0.02	0.61	0.492	0.177	0.5	0.50
Al_2O_3	2.67	1.14	6.47	8.34	4.87	5.67	4.65
Cr_2O_3	1.10	0.97	0.68	0.72	1.32	1.29	2.70
FeO	2.34	1.77	2.62	4.04	3.12	2.78	3.05
MnO	0.10	0.09	0.08	0.14	0.08	n.d.	0.08
NiO	0.07	0.04	0.01	0.045	0.055	n.d.	0.05
MgO	17.83	17.94	15.13	16.31	16.69	15.41	15.30
CaO	22.45	22.58	21.79	17.07	20.10	19.04	16.10
Na_2O	n.d.	0.50	1.23	1.5	0.788	1.92	3.18
Mg#	93.0	94.7	91.0	87.6	90.4	90.7	89.8

Data sources and references as in Table 2.

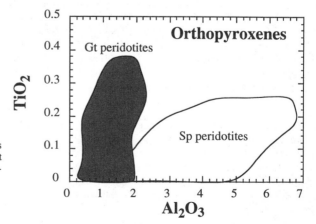

Figure 5. TiO_2-Al_2O_3 variations seen in orthopyroxenes from garnet (shaded) and spinel (white) peridotites.

are precipitated from ascending basaltic magmas are Ti-rich augites. For spinel and plagioclase facies peridotites, the alumina, Na_2O and TiO_2 contents of clinopyroxene reflect the fertility of the peridotite (Fig. 6). Again, for garnet peridotites the aluminum contents of clinopyroxenes are lower and vary as a function of pressure. The Cr_2O_3 contents of peridotitic clinopyroxenes are generally high and, along with NiO contents, increase as the bulk peridotite becomes more refractory. For plagioclase, spinel and garnet facies peridotites, the Mg-Fe exchange between the pyroxenes is a reliable geothermometer (Brey and Köhler 1990, Wells 1977).

Spinel. Representative spinel analyses are given in Table 5. Spinel is both a mineral and a mineral group name; peridotitic spinels typically vary in composition between 4 end member components: spinel ($MgAl_2O_4$), hercynite ($FeAl_2O_4$), Magnesio-chromite ($MgCr_2O_4$) and chromite ($FeCr_2O_4$). In general, the variation in Mg# and Cr# (atomic proportions of Cr/Cr+Al) seen in spinels and chromites from peridotites (Fig. 7) reflects the degree of melt depletion experienced by a peridotite. Typically, lherzolites have spinels with high Al and low Cr contents (e.g. samples Z-37 and Ib/8 in Table 5), whereas harzburgites have chromites with low Al and high Cr contents. Interestingly, chromites found as inclusions in diamonds tend to be very refractory, as reflected by their high Cr#, but they can show a wide range of Mg#, unlike that seen in peridotites (Fig. 7).

One way to trace the melting path followed by peridotites is to compare the Fo contents of olivines and the Cr# of spinels (Fig. 8). Preß et al. (1986), Arai (1987) and others have shown that there is generally a rapid increase in the Fo contents of olivines with little change in the Cr# of spinels during the initial stages of melting. This situation changes rapidly once olivines evolve to about Fo_{90} and above. At this point there is limited subsequent change in the Fo content of olivine for a considerable increase in the Cr# of the spinel.

Finally, one can occasionally find peridotites with either coexisting spinel and garnet or coexisting plagioclase and spinel. When present, these phases are commonly found touching or in proximity of one another, often with a texture suggesting a reaction relationship that has not yet gone to completion. For example, plagioclase mantling spinel is a manifestation of the type of reaction given in equation (1), and such reactions will also have a marked effect on the spinel's Cr#.

Garnet. Representative garnet analyses are given in Table 6. These analyses include garnets from xenoliths brought up by alkali basalts and kimberlites. The 313-3 garnet is from a fertile peridotite brought up by a recent alkali basalt from the Vitim plateau, Russia (Ionov et al. 1993). This suite of xenoliths is dominated by fairly fertile lherzolites. In contrast, most of the other garnets in Table 6 are from more depleted peridotites from older portions of the continents, with F9 being a Ca-poor, Cr-rich garnet with a striking lilac color referred to as a "G10 garnet" (Stebbins and Dawson 1977). G10 garnets (Fig. 9) are used as diamond indicator minerals (Sobolev 1977). An analysis of a garnet from a sheared garnet peridotite (Nixon and Boyd 1973) is included here (PHN 1595), although it is recognized that these garnets can show considerable compositional zoning from core to rim.

For garnet peridotites the exchange of Al between orthopyroxene and garnet is a pressure-dependent reaction that is only mildly sensitive to temperature (MacGregor 1974, Nickel and Green 1985, Brey and Köhler 1990). Together with an estimate of temperature (e.g. the 2 pyroxene geothermometer) one is able to determine the pressure and temperature from which the peridotite was extracted, allowing construction of a pressure-temperature array for suites of garnet bearing peridotites.

Amphiboles and phologopites. Representative amphibole and phlogopite analyses are given in Table 7. There is quite a range in compositions for hydrous minerals in peridotites. It is beyond the scope of this chapter to review the diversity and origins of micas and amphiboles in peridotites; the interested reader is referred to two review volumes on peridotites and their minerals (Menzies and Hawkesworth 1987, Nixon 1987).

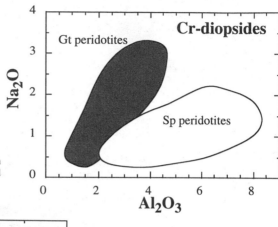

Figure 6. Na_2O-Al_2O_3 variations seen in clinopyroxenes from garnet- (shaded) and spinel- (white) peridotites.

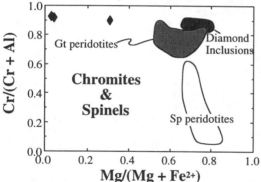

Figure 7. Mg#-Cr# variations seen in spinels and chromites from garnet- and spinel-peridotites and for chromite inclusions in diamonds.

Table 5. Representative analyses of spinel

	Abyssal	Ophiolite	Zarbargad	Sp. Lherz.	Sp. Harz.	Gt./Sp. Lh.	Gt. Harz.
	8-6/4-1	TM 1245	Z-37	Ib/8	D50	313-3*	674
TiO_2	0.07	0.36	0.09	0.26	0.273	0.45	1.19
Al_2O_3	26.83	30.62	56.98	57.19	38.24	44.79	14.80
Cr_2O_3	43.35	33.29	9.17	10.09	28.30	23.05	51.10
FeO	14.94	16.12	12.92	11.79	14.63	13.60	11.70
MnO	0.26	0.21	0.11	0.14	0.11	n.d.	0.17
NiO	0.13	0.19	0.42	0.35	0.22	0.18	0.10
MgO	14.69	13.54	19.85	21.03	18.17	18.92	15.30
Mg#	63.3	59.6	73.0	75.8	68.6	71.0	69.7
Cr#	52.0	42.2	9.7	10.6	33.2	25.7	69.8

Data sources and references as in Table 2, except 313-3 (Ionov et al. 1993)

*there are no spinels in 313-1, this example is from a similarly fertile lherzolite

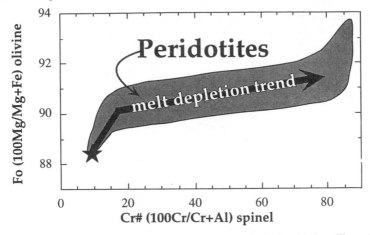

Figure 8. Cr# of spinels vs. Mg# in olivines seen from garnet and spinel peridotites. The melt depletion trend, marked by arrows, is projected from a Primitive Mantle composition (star).

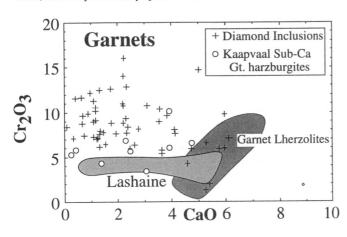

Figure 9. CaO-Cr_2O_3 variations seen in garnets from cratonic peridotites (including sub-calcic garnets from harzburgite xenoliths from the Kaapvaal and Tanzanian (Lashaine) cratons) and "G10" garnet inclusions in diamonds. See Rudnick et al. (1994) for data sources.

Table 6. Representative analyses of garnets

	Gt. Lherz.	Gt. Harz.	Sh. Gt. Lherz.	Gt. Harz.	"G10"
	313 -3	674	PHN 1595[#]	PHN 1596[#]	F9*
SiO_2	42.53	41.70	41.93	42.77	41.40
TiO_2	0.12	0.26	0.04	0.70	n.d.
Al_2O_3	23.62	21.10	18.81	20.63	15.10
Cr_2O_3	1.18	4.27	6.80	2.64	11.60
FeO	7.21	6.59	6.69	6.43	5.87
MnO	0.24	0.26	0.46	0.25	0.28
MgO	20.77	22.00	19.03	22.05	24.10
CaO	4.82	3.95	7.18	4.26	0.65
Na_2O	n.d.	0.05	0.01	0.08	n.d.
Mg#	83.5	85.4	83.3	85.8	87.8

Data sources and references as in Table 2, except [#]Nixon and Boyd (1973) and *Gurney et al. (1980)

Incongruent melting of hydrous phases occurs when peridotite xenoliths are transported to the surface. Frey and Green (1974) documented this phenomena petrographically and chemically for several peridotite xenoliths brought up by alkali basalts. The rapid heating and decompression of the peridotite typically leads to the breakdown of the hydrous phases to glass and daughter crystals that usually include olivine, spinel, clinopyroxene ± plagioclase (Yaxley et al. 1997).

Table 7. Representative analyses of amphibole and phlogopite

Amphibole				Phlogopite					
	Z-37	PKP	2642	674	314-56	GPP	PKP	2640	2925
SiO_2	41.81	55.5	45.19	40.00	38.83	40.91	42.84	38.10	40.30
TiO_2	4.49	0.41	1.28	2.42	5.4	0.56	0.68	3.50	0.40
Al_2O_3	12.46	0.97	15.43	11.70	15.95	13.78	10.75	17.50	15.00
Cr_2O_3	1.63	0.47	0.97	2.65	0.47	0.94	0.33	1.40	1.20
FeO	4.51	2.34	2.04	5.27	4.53	3.08	3.87	4.00	4.40
MnO	0.14	n.d.	<0.03	0.05	n.d.	n.d.	n.d.	n.d.	n.d.
NiO	0.09	n.d.	n.d.	n.d.	0.28	0.13	n.d.	n.d.	n.d.
MgO	16.65	22.74	17.98	24.50	20.68	24.87	26.2	21.30	24.60
CaO	11.8	6.64	10.11	0.42	n.d.	n.d.	n.d.	n.d.	n.d.
Na_2O	3.16	3.86	3.47	0.66	0.49	0.47	0.18	0.50	1.30
K_2O	0.9	4.68	0.04	7.22	9.36	9.38	10.01	9.80	9.20
Mg#	86.6	94.5	93.9	89.1	88.9	93.4	92.2	90.3	90.8

Data sources: Z-37 (Bonatti et al. 1986), PKP and GPP (Erlank et al. 1987), 2642, 2640 and 2925 (Frey and Green 1974), 674 (Rudnick et al. 1994), 314-56 (Ionov et al. 1993).

Table 8. Representative analyses of plagioclase

	Z-37	2925	Dibi -3	Z-45	Z--94
SiO_2	52.4	54.10	53.88	46.48	46.30
Al_2O_3	30.59	28.00	29.71	33.40	34.10
FeO	0.23	0.70	0.22	0.11	0.15
CaO	11.28	10.30	11.75	15.99	17.50
Na_2O	5.09	5.30	4.86	2.18	1.70
K_2O	0.05	0.80	0.06	0.04	0.04
An	54.9	50.2	57.2	78.6	85.6

Data sources: Z-37, Z-45 and Z-94 (Bonatti et al. 1986), 2925 (Frey and Green 1974), and Dibi-3 (Nixon 1987).

Plagioclase. Representative plagioclase analyses are given in Table 8. Plagioclase compositions in peridotites typically range between An_{50} to An_{80}, where An is the ratio of anorthite ($CaAl_2Si_2O_8$) to albite ($NaAlSi_3O_8$) component. For plagioclases in peridotites from the Zarbargad ultramafic massif in the Red Sea the composition of plagioclase is correlated with certain textural relationships (Bonatti et al. 1986). In contrast, however, others report that there is no significant correlation between the compositions of plagioclases and their textural setting (Rampone et al. 1993). Plagioclase found as secondary crystals in glass patches in some peridotite xenoliths can have sodic compositions (e.g. sample 2925 and Dibi 3 in Table 8), which can be related to the relatively high Na content of former amphibole (e.g. Frey and Green 1974).

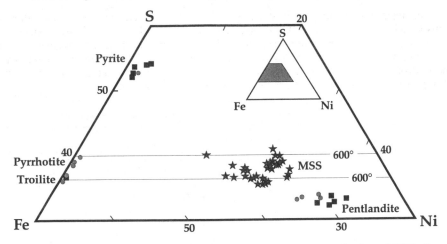

Figure 10. The Fe-Ni-S ternary diagram for sulfides from spinel peridotites (after Lorand 1989). Squares and circles represent measured compositions which formed by exsolution from monosulfide solid solution (MSS), here shown as re-integrated compositions (stars).

Sulfides. Sulfides of Cu-Fe-Ni solid solutions are common accessory phases in upper mantle peridotites from both xenolithic and massif occurrences and are important hosts of siderophile and chalcophile elements in the upper mantle. They occur as inclusions within silicates, as interstitial phases between silicates and as round to oval inclusions decorating healed fractures in olivine (Andersen et al. 1987, Dromgoole and Pasteris 1987, Lorand 1989, MacRae 1979, Szabo and Bodnar 1995). The latter are often associated with a brown glass or CO_2-rich fluid inclusions (Andersen et al. 1987, Dromgoole and Pasteris 1987, Lorand 1989, MacRae 1979). Their modal abundance correlates with fertility of the peridotite and with proximity to mafic veins (Dromgoole and Pasteris 1987, Lorand 1989), implying depletion during partial melting and re-introduction by the passage of mafic magmas through peridotite host rock.

A large range of sulfide compositions have been reported from upper mantle peridotites (e.g. phyrrotite, pentlandite, chalcopyrite, troilite pyrite, millerite, cubanite) (Fig. 10). In general, these phases reflect exsolution from primary mono-sulfide solid solution (MSS), forming phyrrotite, pentlandite, pyrite, cubanite and chalcopyrite. In addition, weathering or low temperature alteration produces additional phases (pyrite, millerite, mackinawite). Determining the sulfur fugacity of the upper mantle from the sulfide abundances in peridotite samples is fraught with problems associated alteration and weathering in the case of massif peridotites and oxidation and loss during decompression for xenolithic peridotites (Lorand 1987, 1990, 1993). The good anti-correlations seen between sulfur and MgO in massif peridotites provide probably the best estimate for upper mantle sulfur abundances (Lorand 1993).

Mafic rocks

In any occurrence of peridotite samples one will also find mafic lithologies (i.e. rocks made up of mostly mafic minerals with <40% olivine), demonstrating that the upper mantle is not solely peridotitic. The difficult part, however, is establishing the proportion of peridotitic to non-peridotitic material in the mantle. These mafic components reveal some of the diverse magmatic and tectonic processes of the upper mantle (e.g. see Irving (1980), Suen and Frey (1987), Bodinier et al. (1987), and Kelemen et al. (1992) and references therein). In the case of ophiolites and massif peridotites (see Section 3 for description of these bodies) mafic layers occur as concordant to discordant bodies interspersed within peridotites. Generally, the mafic component of a massif peridotite body can be reasonably estimated from exposure, constituting on the order of 5% by volume, but this number can vary considerably. In contrast, any attempt to estimate the in-situ proportions of mafic to ultramafic material in mantle xenolith suites brought up by kimberlites or alkali basalts faces considerable (if not insurmountable) problems. Such problems include uncertainties in (1) xenolith settling characteristics during their entrainment, (2) xenolith preservation characteristics during transport and subsequent weathering at the surface, (3) the sampling processes of the entraining magma—was it biased toward ascent paths where previous magmas have traversed?, and (4) what proportion of the mafic xenoliths are from the lower crust versus from the subjacent mantle? Finally, is the proportion of mafic material seen in these suites of upper mantle samples representative of the entire upper mantle (or whole mantle for that matter), or might these sample suites be representative of only the uppermost reaches of the mantle? At a global scale this question is comparable to that which asks if there is gross scale compositional heterogeneities in the mantle with depth. For example, the Transition Zone (400-660 km depth) is considered by some to be a region of the mantle that has a considerably higher proportion of mafic or basaltic component relative to the peridotitic upper mantle (Anderson and Bass 1986, Ringwood 1994).

The sizes of mafic layers range considerably from a few centimeters to meters thick. Many have sharp contacts with surrounding peridotite. Their compositions vary considerably and include: pyroxenites, websterites, gabbros, eclogites, amphibolites, MARIDs (Mica-Amphibole-Rutile-Ilmenite-Diopside xenoliths), glimmerites, and various other exotic lithologies. Various forms of pyroxenites, websterites, gabbros, and amphibolites are typically found in massifs and ophiolites, along with other layers including dunites and wehrlites. MARID and glimmerites are generally restricted to xenolith suites, particularly those from kimberlites.

In many instances, mafic rocks from the upper mantle do not have the compositions of typical basalts seen at the Earth's surface, and many studies conclude that these mafic rocks are products of crystal accumulation and peridotite-melt interactions in the mantle (Frey and Prinz 1978, Irving 1980, Kelemen et al. 1992, Suen and Frey 1987, Waters 1987, Wilshire and Nielson-Pike 1975). There are, however, many other examples of pyroxenites and eclogites that, because of their diverse oxygen isotopic compositions, are either samples of basalts that were once at the Earth's surface and have been recycled back into the mantle or are derived from such basalts. For example, the Robert's Victor kimberlite pipe contains a high proportion of eclogites with oxygen isotopic compositions indicating that these eclogites have been subjected to surface or near surface alteration processes (MacGregor and Manton 1986). In another example, Pearson et al. (1991) showed that pyroxenites containing graphite pseudomorphs after diamond from the Beni Bousera Massif have considerable variations in their oxygen isotopic compositions, and suggested that they are crystallized products of melts derived from altered recycled oceanic crust.

Inclusions in diamonds

The minerals inside diamonds provide clues regarding the conditions (P, T, f_{O_2} and f_{S_2}) under which diamonds crystallized and may be the only samples we have of the lower mantle. The negative diamond shape adopted by most of these inclusions is taken as evidence for their syngenetic origin (Bulanova 1995). Sulfides are by far the dominate type of mineral inclusion, followed by olivine, garnet, chromite, clinopyroxene and orthopyroxene (Bulanova 1995, Harris and Gurney 1979, Meyer 1987). The chemical composition of most diamond inclusions can be used to assign a "paragenesis" to the diamond. For example, diamonds containing olivine (typically Fo_{93} and above), enstatite, Cr-diopside, subcalcic, Cr-rich (G10) garnets (Fig. 9), chromite (Cr#>75, see Fig. 7), pentlandite and Ni-rich monosulfide solid solution (mss) are interpreted to have crystallized in peridotitic rocks. Omphacite, low Cr pyrope, low Ni mss (<15 wt.%), kyanite and coesite inclusions are interpreted to have formed in an eclogitic host rock.

Rare but potentially significant inclusion types include staurolite (Daniels et al. 1996), interpreted as a metastable recycled crustal phase; native Fe, wüstite, hematite and magnesite (Bulanova 1995, Daniels and Gurney 1991, Stachel et al. 1998, Wang et al. 1996), which indicate a range of f_{O_2} conditions during diamond growth. High-pressure phases recovered from diamond includes majorite (Moore and Gurney 1985) and closely associated ferripericlase and enstatite (interpreted as former Mg-perovskite), which may be the first samples of the Earth's lower mantle yet recovered (Harte and Harris 1994; Harris et al. 1997).

THE SAMPLES

There are three principle sources of upper mantle rocks: (1) orogenic massifs, packages of ultramafic rocks meters to kilometers in scale that were tectonically obducted onto the continental crust during continental convergence, (2) peridotitic xenoliths, centimeter scale samples carried in rapidly erupted host volcanics such as alkali basalts and kimberlites and (3) oceanic, or abyssal peridotites, dredged from fracture zones that penetrate below the oceanic Moho or uplifted in ophiolite sequences associated with marginal basins. The dominant petrologic, geochemical and geochronological features of each of these occurrences is summarized briefly below.

Orogenic massifs

These consist of large, several to hundreds of square kilometer bodies composed dominantly of peridotite with layers and lenses of pyroxenites [see Menzies et al. (1991) *Journal of Petrology*, Special Orogenic Lherzolite issue, for a recent review]. The best studied of these are the European and north African massifs, lying in the Betic-Rif chain around the Mediterranean (Ronda and Beni Bousera) or in the Pyrennes (Lherz, Ariege, Prades) and Alps (Lanzo, Balmuccia, Liguria). Other well known occurrences include the Horoman massif (Japan), Tinaquillo massif (Venezuela), the Lizard massif (southern England), and the Zarbargad massif (Red Sea). In general, these peridotites were emplaced into the continental crust along Phanerozoic zones of convergence and have experienced variable degrees of deformation and retrograde metamorphism (including serpentinization) associated with their uplift (van der Wal and Vissers 1993). Hence, the original derivation pressures of these massifs can be hard to delineate. Relict garnet-facies mineralogy and graphite pseudomorphs after diamond in the Betic-Rif massifs attest to an ultra high pressure derivation for at least some portions of these bodies (Davies et al. 1993, Pearson et al. 1989).

Dominant rock types in orogenic massifs are lherzolites, with less abundant harzburgite, and pyroxenite. The abundance of lherzolites reflects the relatively fertile bulk compositions of these bodies, which have average Mg# only slightly higher than primitive mantle (Table 1), but show depletions in Al_2O_3 and CaO, indicative of 6-10% melt extraction from a model primitive mantle composition (Frey 1984, McDonough and Frey 1989). Discordant dunite layers occur locally, but generally do not have high Mg# and are interpreted as reaction zones around feeder dikes of basaltic magmas (Kelemen et al. (1992) and references therein). With few exceptions, massif peridotites are depleted in incompatible trace elements, with chondrite normalized REE patterns that show light REE depletions (see McDonough and Frey (1989), and references therein), features expected of residues of partial melting, and similar to that expected of MORB residues. The unradiogenic Os isotopic compositions of some of these peridotite bodies (Reisberg and Lorand 1995, Reisberg et al. 1991), however, points to ancient melt extraction, indicating that they resided for long time periods within the subcontinental lithosphere before their tectonic emplacement into the crust.

Mantle xenoliths

Xenoliths of upper mantle origin occur in a variety of host volcanics, of which alkali basalts and ultrapotassic magmas (kimberlites, lamproites and minettes) dominate (Nixon 1987). Ultrapotassic hosts erupt preferentially through stable cratons, whereas alkali basalts are generally found in younger, tectonically active regions of the continents (e.g. western USA, eastern Australia, eastern China). The xenoliths carried by these two types of hosts show differences in both mineralogy and chemical composition which are summarized briefly below. Various reviews of mantle xenolith occurrences can be found in the literature, the interested reader is referred to Nixon (1987), Boyd (1989), McDonough (1990), Griffin et al. (1998), as well as the proceedings of the International Kimberlite Conferences (Ahrens et al. 1975, Boyd 1979, Gurney and Richardson 1998, Kornprobst 1984, Meyer et al. 1994, Ross et al. 1986, Sobolev 1977) for a wealth of information related to mantle xenoliths.

Phanerozoic alkali basalt xenolith suites. Mantle xenoliths carried in alkali basalts are dominated by spinel-facies peridotite, chiefly lherzolite with lesser amounts of harzburgite, wehrlite and, rarely, dunite. Pyroxenites also occur but are subordinate to peridotites in overall abundance (see Wilshire et al. (1988), for a statistical treatment of xenoliths from the southwestern USA). Pyroxenites have received special attention, however, due to the fact that they commonly contain garnet and allow estimates of regional geotherms, which are typically quite hot (e.g. O'Reilly and Griffin 1984). Plagioclase-bearing peridotite xenoliths are rare, occurring only in regions of anomalously thin crust (<30 km thick) (e.g. Zipfel and Wörner 1992). Interestingly, garnet-bearing peridotites are also uncommon in alkali basalts -- a few garnet peridotite-rich localities have been described (Vitim plateau—Ionov et al. 1993, Pali Aike—Stern et al. 1999), but garnet peridotites occur only rarely in most Phanerozoic xenolith-rich regions such as eastern Australia and eastern China. It is not clear whether their scarcity reflects thin lithosphere in these regions or the failure of the basalts to sample the deeper portions of the lithosphere (>50-60 km deep, Fig. 1).

Like massif peridotites, basalt-hosted peridotite xenoliths are relatively fertile, with an average composition slightly more refractory than that of the average massif peridotite (Table 1). However, in contrast to massif peridotites, many spinel lherzolites from alkali basalts are enriched in incompatible trace elements, reflecting re-enrichment of these elements after partial melting (Frey and Green (1974), and see McDonough and Frey (1989) and references therein). Not surprisingly, isotopic systems based on these incompatible elements (Rb-Sr, Sm-Nd, U-Pb) show a large range in values (Hawkesworth et al. 1990), which may reflect complex combinations of the original isotopic composition of the lithosphere and the source of the metasomatic melts. For these reasons, it has often proven difficult to make robust age determinations from these isotopic systematics. Recent Re-Os studies have demonstrated that mantle lithosphere formation is roughly synchronous with crustal age (McBride et al. 1996, Handler et al. 1997).

Cratonic xenolith suites. Mantle xenoliths from continental cratons are mainly sampled in kimberlite or lamproite diatremes that erupt through Precambrian crust. These same pipes carry diamonds to the Earth's surface and have thus been the subject of

Table 9. Compositions of Primitive Mantle and peridotites from Archean Cratons

	Primitive Mantle	Deformed, High T[†]	PHN 1611[#]	Kaapvaal Coarse-Granular*	Average Cratonic Peridotite
SiO_2	45.00	44.4	44.3	46.6	43.8
TiO_2	0.20	0.14	0.25	0.04	0.08
Al_2O_3	4.45	1.46	2.79	1.46	0.56
Cr_2O_3	0.38	0.35	0.61	0.40	0.38
FeO	8.05	8.00	10.2	6.24	7.71
MnO	0.135	0.12	0.13	0.12	0.12
MgO	37.80	44.0	37.8	44.1	46.3
NiO	0.25	0.26	0.26	0.27	0.35
CaO	3.55	1.07	3.31	0.79	0.75
Na_2O	0.36	0.05	0.34	0.09	0.07
K_2O	0.029	0.05	0.14	n.d.	0.030
P_2O_5	0.021	0.01	0.00	n.d.	0.01
Mg #	89.3	90.7	86.8	92.6	91.4

Data sources: [†]Median composition of 110 high-temperature, deformed peridotites from South Africa; [#]Boyd and Nixon (1973), *Boyd (1989) and Cratonic (McDonough 1994, Rudnick et al. 1998)

numerous scientific inquiries, in proportions opposite to their volumetric significance. Like the alkali basalt mantle xenolith suites, cratonic xenoliths are dominated by peridotites, with less common eclogite (garnet + omphacite), pyroxenite and rare metasomatic lithologies (glimmerites, MARID xenoliths, alkremites, etc.).

For decades, studies of the mantle xenoliths cast onto the diamond mine tailings in the Kaapvaal craton provided a paradigm of cratonic mantle (Ahrens et al. 1975, Boyd 1989, Menzies and Hawkesworth 1987, Nixon 1987, Nixon and Boyd 1973). These studies demonstrated that:

(1) the geothermal gradient in the Kaapvaal craton is low, consistent with a surface heat flow of 40-45 mW/m^2 (the average surface heat flow observed in Archean cratons, Nyblade and Pollack 1993). These geotherms, if representative of present-day thermal conditions, also restrict the thickness of the cratonic keels to depths of 250 kms (Rudnick and Nyblade 1998).

(2) two distinctive suites of peridotites occur in the Kaapvaal craton: refractory, low temperature peridotites (equilibrated at 150 km depth in both spinel and garnet facies) with coarse-granular textures and more fertile, high temperature garnet peridotites (equilibrated at 150 km depth) with sheared or porphyroclastic textures.

(3) the low temperature cratonic peridotites are more refractory than mantle xenoliths from Phanerozoic regions; they are depleted in Al_2O_3 and CaO and have significantly higher Mg# compared to estimates of the Primitive Mantle composition (Table 1). Most importantly, Kaapvaal peridotites are severely depleted in FeO (Table 9), leading to very low intrinsic density (Boyd 1989, Hawkesworth et al. 1990).

(4) cratonic mantle is ubiquitously metasomatized, as evidenced by either discrete metasomatic minerals (e.g. phlogopite, K-richterite, secondary clinopyroxene, ilmenite and rutile; (Dawson 1987, Erlank et al. 1987, Harte et al. 1987)), or by "cryptic" metasomatism where light REEs and other Large Ion lithophile Elements (LILEs) are enriched without the presence of metasomatic phases.

(5) cratonic peridotites have high modal orthopyroxene content when compared to off-

craton or oceanic peridotites, or to the residues of high degree partial melting at low pressures (Boyd 1989). This is reflected in the high SiO_2 contents and low Mg/Si of Kaapvaal peridotites. Despite much attention and the advance of many creative hypotheses (e.g. Boyd 1989, Kelemen et al. 1992, Herzberg 1993, 1998; Rudnick et al. 1994), the origin of this silica enrichment has remained enigmatic. Residues of high pressure (5-7 GPa) melting are more Si-rich than those of lower pressure melting (Kinzler and Grove 1998, Walter 1998a,b), but the most silica-rich residues still do not match the degree of silica enrichment observed in the most orthopyroxene-rich peridotites from Kaapvaal. As shown below, however, this feature may not be pervaisive in Archean cratons.

(6) cratonic mantle was stabilized by melt extraction during the Archean (Carlson et al. 1998b, Pearson et al. 1995a, Pearson et al. 1995b, Walker et al. 1989).

The petrological and geochemical features of cratonic peridotites summarized above, coupled with geophysical evidence of low surface heat flow (Nyblade and Pollack 1993) and seismically fast uppermost mantle (Jordan 1975, 1979, 1988) has led to the concept of Archean cratons being stabilized by a thick (up to 250 km), cold but buoyant lithospheric keel (Boyd 1989, Hoffman 1990, Jordan 1988). Recent studies of peridotites from other Archean cratons (i.e. Siberia—Boyd 1984, Boyd et al. 1997, Griffin et al. 1996, Sobolev 1977, Tanzania—Dawson et al. 1997, Lee and Rudnick 1998, Reid et al. 1975, Rudnick et al. 1994, Wyoming—Carlson et al. 1998a, Hearn and McGee 1984, Slave—Kopylova et al. 1998, Superior—Meyer et al. 1994, Schulze 1996, Zimbabwe—Stiefenhofer et al. 1997, and North Atlantic—Bernstein et al. 1998) have largely substantiated this picture.

However, certain features of the Kaapvaal paradigm do not appear to hold up elsewhere. For example, certain styles of metasomatism have not been observed elsewhere (i.e. K-richterite bearing xenoliths). More importantly, the silica enrichment seen in the Kaapvaal coarse-granular peridotites is observed in the Siberian craton, but not to the same extent in other cratons (Fig. 11). The moderate amounts of silica enrichment seen in Tanzanian (Lee and Rudnick 1998, Rudnick et al. 1994) peridotites are consistent with high pressure melt extraction (Walter 1998a,b). Peridotites from other cratons apparently lack Si enrichment altogether (Bernstein et al. 1998, Schmidberger and Francis 1999).

A review of cratonic peridotites would not be complete without a discussion of the significance of the sheared, high temperature peridotites. Samples with these same remarkable textures and compositions have been recovered from every craton where significant sampling has been carried out. Their more fertile compositions and sheared textures led Nixon and Boyd (1973) to suggest that they may be pieces of subcontinental asthenosphere. However, this hypothesis has not held up in light of more recent investigations. Firstly, most high temperature peridotites, although Fe-rich relative to the coarse-granular peridotites, are still quite depleted in CaO and Al_2O_3 (Table 9). This holds also for the famous sample, PHN1611, which has been the subject of several experimental investigations aimed at understanding melting in the Primitive Mantle (e.g. Scarfe and Takahashi 1986). Moreover, the garnets in these samples show strong enrichments of certain trace elements on their rims (Ti, Zr, Hf, ect.), indicative of metasomatic overprinting by basaltic or picritic magmas shortly before their entrainment in the host kimberlite (Smith and Boyd 1987, Griffin et al. 1989). Finally, Os isotopic investigations of these samples show them to be as equally unradiogenic as the low temperature, coarse-granular peridotites, demonstrating that these samples have been isolated from the convective mantle for billions of years (Pearson et al. 1995). In summary, rather than samples of fertile, sublithospheric mantle, the combined evidence cited above shows these samples to be ancient, refractory peridotite that was metasomatically overprinted near the base of the lithosphere shortly before the entrainment of the xenolith in the kimberlite.

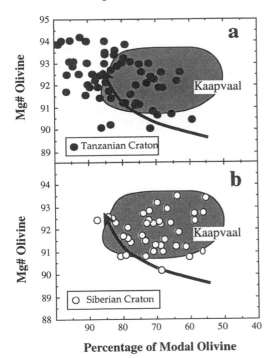

Figure 11. Modal olivine versus Mg# of olivine, with the trend defined by low pressure melt residues—"oceanic trend," after Boyd (1989). Shaded field encompasses data for spinel-peridotites and both low and high temperature garnet-peridotites from the Kaapvaal craton. (a) Data for garnet and spinel peridotites from the Tanzanian craton (Rudnick et al. 1994, Lee and Rudnick 1998). (b) Data for garnet-and spinel-peridotites from the Siberian craton (Boyd et al. 1997). A significant number of the Kaapvaal and Siberian peridotites have low modal olivine, due to the high abundance of orthopyroxene.

Oceanic peridotites

Partially or completely serpentinized peridotites have been dredged, or drilled, from the mid-ocean-ridges, particularly along escarpments of oceanic fracture zones (Bonatti et al. 1971, Miyashiro et al. 1969). Studies of these peridotites show (1) that peridotite compositions are comparable between oceans, (2) that mineralogical and chemical variations seen in peridotites are explicable in terms of varying degrees of melt extraction and (3) that there is a good correlation between the regional compositions of peridotites and the major element compositions of spatially associated dredged basalts (Dick and Fisher 1984, Dick et al. 1984, Michael and Bonatti 1985). The average chemical composition of abyssal peridotites is difficult to establish because these samples are so significantly altered by low temperature processes at the ridges. Detailed petrographic analyses, however, has provided modal mineralogical abundances for these samples and from this one can model the major element compositions of abyssal peridotites. Dick et al. (1984) reported the mineralogy of a typical abyssal peridotite from the spinel facies as having about 75% olivine, 21% orthopyroxene, 3.5% clinopyroxene and 0.5% spinel. Abyssal peridotites are harzburgites and are more refractory than average massif or off-craton peridotites (Table 1), but not as depleted as the average Cratonic peridotite (Table 1). Dick et al. (1984) and Michael and Bonatti (1985) noted that many of the abyssal peridotites from the North Atlantic are more depleted than the average abyssal peridotite, particularly those from regions near known hotspots (ocean island magmatism). They used this correlation to support the view that there exists regional variations in upper mantle thermal structure and composition.

Johnson et al. (1990) used an ion microprobe to analyze the trace element compositions of clinopyroxenes from abyssal peridotites. They showed that most clinopyroxenes have significant depletions in the light REE, much more so than what is observed for massif peridotites. With this and other trace element data Johnson et al. (1990) showed that it was not possible to explain the chemical characteristics seen in abyssal peridotites by batch melting processes and concluded that fractional melting was key in the production of MORBs. This and subsequent studies on minerals from abyssal peridotites have played a significant role in shaping our understanding of the melting process in the mantle beneath ridges.

Finally, there are other localities where one can find peridotites in the oceanic environment. A number of ocean islands have xenolith-bearing alkali basalts (e.g. Hawaii, Samoa). These peridotites are nearly all from the spinel facies and have mineralogical and chemical compositions that are comparable to those found in off-cratonic peridotite suites. In other instances there are exposed ultramafic complexes on oceanic islands (e.g. St. Paul's rocks in the equatorial Atlantic). St. Paul's rocks is not far from the mid-Atlantic ridge and is a spinel peridotite massif with considerable amounts of primary amphibole (i.e. kaersutitic hornblende). This peridotite complex is made up of lherzolites and harzburgites that show extensive chemical and isotopic evidence for metasomatism (Roden et al. 1984). The complex has been likened to as a source for alkali basalts (Roden et al. 1984).

A MODEL COMPOSITION FOR THE UPPER MANTLE

Samples of the upper mantle have played a significant role in the development of model compositions for the Earth's Primitive Mantle. The Primitive Mantle, also known as the Bulk Silicate Earth or Pyrolite, is a model composition for the whole of the silicate portion of the Earth, including the crust and mantle. For most of this century studies have integrated observations from petrology, geophysics and cosmochemistry to develop compositional models for Earth reservoirs. As it became recognized that ultramafic bodies, such as orogenic peridotites, are large slices of the mantle that were tectonically thrust up onto the crust, these rocks and the mantle xenoliths brought up by basalts and kimberlites were used to develop models for the mineralogy and composition of the upper mantle. In 1962, Ted Ringwood developed a model for the composition for the upper mantle and proposed that this composition is similar to a mixture of 1 part basalt to approximately 4 parts dunite (i.e. refractory peridotite) (Ringwood 1962). This ratio of melt to residual mantle was constrained from the relative proportions of MgO, Al_2O_3 and CaO in chondrites, primitive meteoritic materials from which planets were assembled. At the same time this compositional model was also required to be consistent with seismological and heat flow data for the mantle.

From the early 1960s onward there have been refinements of the Pyrolite model, along with an increasing number of studies on peridotites and on the generation of basalts. From all of this, our understanding of the composition of the Primitive Mantle has improved greatly. A notable development occurred with the work of Jagoutz et al. (1979) which reported the major, minor and trace element compositions of a selected suite of unweathered peridotite samples. These samples were chosen because of their freshness and because they had relatively fertile compositions, reflecting the loss of little to no basalt. By examining the chemical systematics of these lherzolites, they were able to propose a more complete compositional model for the Primitive Mantle, including an estimate for many trace elements. The major element model for the Primitive Mantle given by Jagoutz et al. (1979) matched that of Ringwood's Pyrolite model (1975). A second notable development occurred at about this same time with the studies of Sun and Nesbitt who reported the compositions of a spectrum of komatiites, MgO-rich melts commonly found in Archean terrains (Sun and Nesbitt 1977, Sun 1982 Sun and Nesbitt 1977). These lavas are produced by high degrees of melting and thus provide windows into the composition of the Archean mantle. These authors estimated a composition for the Primitive Mantle that was essentially equivalent to that reported by Jagoutz et al. (1979) and Ringwood (1975). These studies thus provide multiple lines of support for our estimate of the mineralogical and chemical composition of the upper mantle.

A major issue in geochemistry is how well the upper mantle composition reflects that of the lower mantle. It is generally accepted that the seismic discontinuities at 410 km and 660 km depth are the result of phase changes and are not compositionally induced (see also Fig. 1). We believe the upper and lower mantle have similar bulk compositions based on a number of different geophysical observations (see reviews in later chapters of this volume, including those by Bina and Masters). In addition, a compositionally stratified mantle is difficult to maintain in light of the evidence for whole mantle convection (Hager and Richards 1989, Davies and Richards 1992, Grand 1994, van der Hilst et al. 1997), which implies significant mass exchange across the 660 km seismic discontinuity.

ACKNOWLEDGMENTS

We thank Rus Hemley for the invitation to contribute to this volume, albeit on short notice. Review comments from Cin-Ty Lee and Matthias Barth are appreciated. We thank Alayne Moody and Jamie Gonzales for their very capable assistance in the preparation of this manuscript. The authors acknowledge grants from the Earth Science Division of the National Science Foundation in support of our studies of upper mantle rocks.

REFERENCES

Adams, LH and Williamson, ED (1923) The compressibility of minerals and rocks at high pressures. J Franklin Inst 195:475-487

Ahrens, LH, Dawson JB, Duncan AR, Erlank, AJ (1975) Proc 1st Int'l Kimberlite Conf, 940 p Pergamon Press, Oxford

Andersen T, Griffin WL, O'Reilly, SY (1987) Primary sulphide melt inclusions in mantle-derived megacrysts and pyroxenites. Lithos 20:279-295

Anderson DL, Bass JD (1986) Transition region of the Earth's upper mantle. Nature 320:321-328

Arai S (1987) An estimation of the least deplted spinel peridotite on the basis of olivine-spinel mantle array. Neues Jahrb Mineral Monatsh 8:347-354

Bernstein S, Kelemen PB, Brooks CK (1998) Depleted spinel harzburgite xenoliths in Tertiary dykes from East Greenland: Restites from high degree melting. Earth Planet Sci Lett 154:221-235

Bodinier JL, Guiraud M, Fabriès J, Dostal J, Dupuy C (1987) Petrogenesis of layered pyroxenites from the Lherz, Freychinède and Prades ultramafic bodies (Ariège, French Pyrénées). Geochim Cosmochim Acta 51:279-290

Bonatti E, Honnorez J, Ferrara G (1971) Peridotite-gabbro-basalt complex from the equatorial Mid-Atlantic Ridge. Phil Trans R Soc Lond A268:385-402

Bonatti E, Ottonello G, Hamlyn PR (1986) Peridotites from the Island of Zabargad (St. John), Red Sea: Petrology and Geochemistry. J Geophys Res 91:599-631

Boyd FR (1979) Garnet lherzolite xenoliths from the kimberlites of East Griqualand, South Africa. Carnegie Inst Wash Yrbk 79:296-302

Boyd FR (1984) Siberian geotherm based on lherzolite xenoliths from the Udachnaya kimberlite, USSR Geol 12:528-530

Boyd FR (1989) Compositional distinction between oceanic and cratonic lithosphere. Earth Planet Sci Lett 96:15-26

Boyd FR, Pokhilenko NP, Pearson, DG, Mertzman SA, Sobolev NV, Finger LW (1997) Composition of the Siberian cratonic mantle: evidence from Udachnaya peridotite xenoliths. Contrib Mineral Petrol 128:228-246

Brey GP, Köhler, T (1990) Geothermobarometry in four-phase lherzolites II. New thermobarometers, and practical assessment of existing thermobarometers. J Petrol 31:1353-1378

Bulanova GP (1995) The formation of diamond. J Geochem Explor 53:1-23

Carlson, RW, Irving AJ, Hearn BCJ (1998a) Peridotite Xenoliths from the Williams Kimberlite, Montana: Implications for Delamination of the Wyoming Craton Lithosphere. Abstracts 7th Int'l Kimberlite Conf, 132-134

Carlson RW, Pearson DG, Boyd FR, Shirey SB, Irvine G, Menzies AH, Gurney JJ (1998b) Re-Os

Systematics of Lithosphere Peridotites: Implications for Lithosphere Formation and Preservation. *In:* Gurney, JJ, Richardson, SR (eds), Proc 7th Int'l Kimberlite Conf (in press) Red Barn, Cape Town

Daniels LRM, Gurney JJ (1991) Oxygen fugacity constraints on the southern African lithosphere. Contrib Mineral Petrol 108:154-161

Daniels LRM, Gurney JJ, Harte B (1996) A crustal mineral in a mantle diamond. Nature 379:153-156

Davies G, Richards M (1992) Mantle convection. J Geol 100:151-206

Davies GR, Nixon PH, Pearson DG, Obata M (1993) Tectonic implications of graphitized diamond in the Ronda peridotite, S. Spain. Geol 21:471-474

Dawson JB (1987) Metasomatized harzburgites in kimberlite and alkaline magmas: enriched restites and "flushed" lherzolites. *In:* Menzies MA, Hawkesworth CJ, (eds) Mantle Metasomatism, p 125-144. Academic Press, London

Dawson, JB, James D, Paslick C, Halliday AM (1997) Ultrabasic potassic low-volume magmatism and continental rifting in north-central Tanzania: association with enhanced heat flow. Proc 6th Int'l Kimberlite Conf, Russian Geol Geophys 38:69-81

Dick HJB, Fisher RL (1984) Mineralogical studies of the residues of mantle melting: Abyssal and alpine-type peridotites. Kimberlites II: The Mantle and Crust- Mantle relationships, 295-308

Dick HJB, Fisher RL, Bryan WB (1984) Mineralogic variability of the uppermost mantle along mid-ocean ridges. Earth Planet Sci Lett 69:88-106

Dromgoole EL, Pasteris, JD (1987) Interpretation of the sulfide assemblages in a suite of xenoliths from Kilbourne Hole, New Mexico. Geol Soc Am Spec Paper 215:25-46

Dziewonski AM, Anderson, DL (1981) Preliminary Reference Earth Model. Phys Earth Planet Int 25:297-356

Erlank AJ, Waters FG, Hawkesworth CJ, Haggerty SE, Allsopp HL, Richard RS, Menzies MA (1987) Evidence for mantle metasomatism in peridotite nodules from the Kimberley pipes, South Africa. *In:* Menzies MA Hawkesworth CJ (eds), Mantle Metasomatism, p 221-312 Academic Press, London

Fountain DM, Boundy TM, Austrheim H, Rey P (1994) Eclogite-facies shear zones—deep crustal reflectors? Tectonophys 232:411-424

Frey FA (1984) Rare earth element abundances in upper mantle rocks. *In:* Henderson P (ed) Rare Earth Element Geochemistry, p 153-203 Elsevier, Amsterdam

Frey FA, Green DH (1974) The mineralogy, geochemistry and origin of lherzolite inclusions in Victorian basanites. Geochim Cosmochim Acta 38:1023-1059

Frey FA, Prinz M (1978) Ultramafic inclusions from San Carlos, Arizona: petrologic and geochemical data bearing on their petrogenesis. Earth Planet Sci Lett 38:129-176

Grand SP (1994) Mantle shear structure beneath the Americas and surrounding oceans. J Geophys Res 99:11591-11621

Green DH, Hibberson W (1970) The instability of plagioclase in peridotite at high pressure. Lithos 3:209-221

Green DH, Ringwood, AE (1967) An experimental investigation of the gabbro to eclogite transformation and its petrological applications. Geochim Cosmochi Acta 31:767-833

Griffin WL, Kaminsky FV, Ryan CG, O'Reilly SY, Win TT, Ilupin IP (1996) Thermal state and composition of the lithospheric mantle beneath the Daldyn kimberlite field, Yakutia. Tectonophys 262:19-33

Griffin WL, O'Reilly SY, Ryan CG (1998) The composition and origin of sub-continental lithospheric mantle. In Fei Y, Bertka CB, Mysen, BO (eds) Mantle Petrology: Field Observations and High-Pressure Experimentation, Special Publication in honor of F.R. Boyd, (in press) The Geochemical Society, San Antonio

Griffin WL, Smith D, Boyd FR, Cousens DR, Ryan CG, Sie SH, Sutter GF (1989) Trace-element zoning in garnets from sheared mantle xenoliths. Geochim Cosmochim Acta 53:561-567

Gurney JJ, Harte B (1980) Chemical variations in upper mantle nodules from South African kimberlites. Phil Trans R Soc London 297:273-293

Gurney JJ, Richardson SR (1998) Proc 7th Int'l Kimberlite Conf Red Roof Design, Cape Town, (in press)

Hager B, Richards M (1989) Long-wavelength variations in the Earth's geoid: physical models and dynamical implications. Phil Trans R Soc Lond A328:309-327

Handler MR, Bennett VC, Esat TM (1997) The persistence of off-cratonic lithospheric mantle: Os isotopic systematics of variably metasomatised southeast Australian xenoliths. Earth Planet Sci Lett 151:61-75

Harris J, Hutchinson MT, Hursthouse M, Light M, Harte B (1997) A new tetragonal silicate mineral occurring as inclusions in lower-mantle diamonds. Nature 387:486-488

Harris JW, Gurney, JJ (1979) Inclusions in diamond. *In:* Field JE (ed) Properties of Diamond p 555-591 Academic Press, New York

Harte B, Harris JW (1994) Lower mantle mineral associations preserved in diamonds. Mineral Mag 58:384-385

Harte B, Winterburn PA, Gurney JJ (1987) Metasomatic and Enrichment Phenomena in Garnet Peridotite Facies Mantle Xenoliths from the Matsoku Kimberlite Pipe, Lesotho. *In:* Menzies MA, Hawkesworth CJ (eds) Mantle Metasomatism, p 145-220 Academic Press, London

Hawkesworth CJ, Kempton PD, Rogers NW, Ellam RM, van Calsteren PW (1990) Continental mantle lithosphere, and shallow level enrichment processes in the Earth's mantle. Earth Planet Sci Lett 96:256-268

Hearn BC, McGee ES (1984) Garnet Peridotites from Williams Kimberlite, North-Central Montana, U.S.A. *In:* Kornprobst J (ed) Kimberlites II: The mantle and crust-mantle relationships, p 57-70 Elsevier, Amsterdam

Herzberg C (1998) Formation of Cratonic Mantle as Plume Residues and Cumulates. *In:* Fei Y, Bertka CB, Mysen BO (eds) Mantle Petrology: Field Observations and High-Pressure Experimentation, Special Publication in honor of F.R. Boyd, (in press) The Geochemical Society, San Antonio

Herzberg CT (1993) Lithosphere peridotites of the Kaapvaal craton. Earth Planet Sci Lett 120:13-22

Hoffman PF (1990) Geological constraints on the origin of the mantle root beneath the Canadian shield. Phil Trans R Soc Lond A331:523-532

Ionov DA, Ashchepkov IV, Stosch H-G, Witt-Eickschen G, Seck, HA (1993) Garnet Peridotite Xenoliths from the Vitim Volcanic Field, Baikal Region: the Nature of the Garnet-Spinel Peridotite Transition Zone in the Continental Mantle. J Petrol 34:1141-1175

Irving AJ (1980) Petrology and geochemistry of composite ultramafic xenoliths in alkalic basalts and implications for magmatic processes within the mantle. Am J Sci 280A:389-426

Jagoutz E, Palme H, Baddenhausen H, Blum K, Cendales M, Dreibus G, Spettel B, Lorenz V, Wänke H (1979) The abundances of major, minor and trace elements in the earth's mantle as derived from primitive ultramafic nodules. Proc Lunar Planet Sci Conf 10:2031-2050

Johnson KTM, Dick HJB, Shimizu N (1990) Melting in the oceanic mantle: an ion microprobe study of diopsides in abyssal peridotites. J Geophys Res 95:2661-2678

Jordan TH (1975) The continental tectosphere. Rev Gephys Space Phys 13:1-12

Jordan TH (1979) Mineralogies, densities and seismic velocities of garnet lherzolites and their geophysical implications. *In:* Boyd FR, Meyer, HOA (eds) The mantle sample: Inclusions in kimberlites and other volcanics, p 1-14 Am Geophys Union, Washington, DC

Jordan TH (1988) Structure and formation of the continental lithosphere. J Petrol Special Lithosphere Issue:11-37

Kelemen PB, Dick HJB, Quick JE (1992) Formation of harzburgite by pervasive melt/rock reaction in the upper mantle. Nature 358:635-639

Kinzler RJ, Grove TL (1998) Origin of Depleted Cratonic Harzburgite by Deep Fractional Melt Extraction and Shallow Olivine Cumulate Infusion *In:* Gurney JJ, Richardson SR (eds) Proc 7th Int'l Kimberlite Conf (in press) Red Roof Design, Cape Town

Köhler TP, Brey GP (1990) Calcium exchange between olivine and clinopyroxene calibrated as a geothermobarometer for natural peridotites from 2 to 60 kb with applications. Geochim Cosmochim Acta 54: 2375-2388

Kopylova MG, Russell JK, Cookenboo H (1998) Upper-mantle stratigraphy of the Slave craton, Canada: insights into a new kimberlite province. Geol 26:315-319

Kornprobst J (1984) Kimberlites II: The mantle and crust-mantle relationships Elsevier, Amsterdam, 393 p

Kushiro I, Yoder HS (1966) Anorthite-fosterite and anorthite-enstatite reactions and their bearing on the basalt-eclogite transformation. J Petrol 7:337-362

Lee C-T, Rudnick, RL (1998) Compositionally stratified cratonic lithosphere: petrology and geochemistry of peridotite xenoliths from the Labait tuff cone, Tanzania. *In:* Gurney JJ, Richardson SR (eds) Proc 7th Int'l Kimberlite Conf (in press) Red Roof Design, Cape Town

Lorand JP (1987) Caractères minéralogiques et chimiques généraux des microphases du système Cu-Fe-Ni-S dans les roches du manteau supérieur: exemples d'hétérogeneites en domaine subcontinental. Bull Soc Geol Fr 8:643-657

Lorand, JP (1989) The Cu-Fe-Ni sulfide component of the amphibole-rich veins from the Lherz and Freychinede spinel peridotite massifs (Northeastern Pyrenees, France): A comparison with mantle-derived megacrysts from alkali basalts. Lithos 23:281-298

Lorand JP (1990) Are spinel lherzolite xenoliths representative of the abundance of sulfur in the upper mantle? Geochim Cosmochim Acta, 54:1487-1492

Lorand JP (1993) Comment on "Content and isotopic composition of sulphur in ultramafic xenoliths from central Asia" by D.A. Ionov, J. Hoefs, K.H. Wedepohl and U. Weichert. Earth Planet Sci Lett 119:627-634

MacGregor ID (1974) The system $MgO-Al_2O_3-SiO_2$: Solubility of Al_2O_3 in enstatite for spinel and garnet peridotite. Am Mineral 59:110-119

MacGregor ID, Manton, WI (1986) Roberts Victor eclogites: ancient oceanic crust? J Geophys Res

91:14063-14079

MacRae ND (1979) Silicate glass and sulfides in ultramafic xenoliths, Newer basalts, Victoria, Australia. Contrib Mineral Petrol 68:275-280

McBride JS, Lambert DD, Greig A, Nicholls IA (1996) Multistage evolution of Australian subcontinental mantle: Re-Os isotopic constraints from Victorian mantle xenoliths. Geol 24:631-634

McDonough WF (1990) Constraints on the composition of the continental lithospheric mantle. Earth Planet Sci Lett 101:1-18

McDonough WF (1994) Chemical and isotopic systematics of continental lithospheric mantle. *In:* Meyer HOA, Leonardos OH (eds) Proc 5th Int'l Kimberlite Conf p 478-485 CPRM, Brasilia

McDonough WF, Frey FA (1989) REE in upper mantle rocks. *In:* Lipin B, McKay GR (eds) Geochemistry and Mineralogy of Rare Earth Elements, p 99-145 Mincralogical Society of America, Chelsea, Michigan

McDonough WF, Stosch H-G, Ware N (1992) Distribution of titanium and the rare earth elements between peridotitic minerals. Contrib Mineral Petrol 110:321-328

McDonough WF, Sun S-s (1995) The composition of the Earth. Chem Geol 120:223-253

Menzies MA, Dupuy C, Nicolas A (1991) Orogenic lherzolites and mantle processes. J Petrol Spec Issue, 306 p

Menzies, MA, Hawkesworth CJ (1987) Mantle Metasomatism. Academic Press, London, 472 p

Meyer, HOA (1987) Inclusions in diamond. *In:* Nixon PH (ed) Mantle Xenoliths, p 501-522 John Wiley & Sons, Chichester, England

Meyer HOA, Waldman MA, Garwood BL (1994) Mantle xenoliths from kimberlite near Kirkland Lake, Ontario. Can Mineral 32:295-306

Michael PJ, Bonatti E (1985) Peridotite composition from the North Atlantic: regional and tectonic variations amd implications for partial melting. Earth Planet Sci Lett, 73:91-104

Miyashiro A, Shido F, Ewing M (1969) Composition and origin of serpentinites from the Mid-Atlantic Ridge near 24° and 30° north latitude. Contrib Min Petrol 23:117-127

Moore RO, Gurney, JJ (1985) Pyroxene solid solution in garnets included in diamond. Nature 335:784-789

Nickel KG, Green DH (1985) Empirical geothermobarometry for garnet peridotites and implications for the nature of the lithosphere, kimberlites and diamonds. Earth Planet Sci Lett 73:158-170

Nixon PH (ed) (1987) Mantle Xenoliths, J Wiley & Sons, New York, 844 p

Nixon PH, Boyd FR (1973) Petrogenesis of the granular and sheared ultrabasic nodule suite in kimberlites. *In:* Nixon PH (ed) Lesotho Kimberlites, p 48-56 Lesotho National Development Corporation

Nyblade AA, Pollack HN (1993) A global analysis of heat flow from Precambrian terrains: implications for the thermal structure of Archean and Proterozoic lithosphere. J Geophys Res 98:12207-12218

O'Neill HSC (1981) The transition between spinel lherzolite and garnet lherzolite, and its use as a geobarometer. Contrib Min Petrol, 77:185-194

O'Reilly SY, Griffin WL (1984) A xenolith-derived geotherm for southeastern Australia and its geophysical implications. Tectonophys 111:41-63

Pearson, DG, Carlson RW, Shirey SB, Boyd FR, Nixon PH (1995a) The stabilisation of Archaean lithospheric mantle: A Re-Os isotope study of peridotite xenoliths from the Kaapvaal craton. Earth Planet Sci Lett 134: 341-357

Pearson DG, Davies GR, Nixon PH, Greenwood PB, Mattey DP (1991) Oxygen isotope evidence for the origin of pyroxenites in the Beni Bousera peridotite massif, North Morocco: derivation from subducted oceanic lithosphere. Earth Planet Sci Lett 102:289-301

Pearson DG, Davies GR, Nixon PH, Middedge, HJ (1989) Graphitized diamonds from a peridotite massif in Morocco and implications for anomalous diamond occurrences. Nature 338:60-62

Pearson DG, Shirey SB, Carlson RW, Boyd FR, Pokhilenko NP, Shimizu N (1995b) Re-Os, Sm-Nd and Rb-Sr isotope evidence for thick Archaean lithospheric mantle beneath the Siberian craton modified by multi-stage metasomatism. Geochim Cosmochim Acta, 59:959-977

Philpotts AR (1990) Principles of igneous and metamorphic petrology Prentice Hall, Englewood Cliffs, 498 p

Preß S, Witt G, Seck HA, Eonov D, Kovalenko VI (1986) Spinel peridotite xenoliths from the Tariat Depression, Mongolia. I: Major element chemistry and mineralogy of a primitive mantle xenolith suite. Geochim Cosmochim Acta 50:2587-2599

Rampone E, Piccardo GB, Vannucci R, Bottazzi P, Ottolini L (1993) Subsolidus reactions monitored by trace element partitioning: the spinel- to plagioclase-facies transition in mantle peridotites. Contrib Mineral Petrol, 115:1-17

Reid AM, Donaldson CH, Brown RW, Ridley WI, Dawson JB (1975) Mineral chemistry of peridotite xenoliths from the Lashaine volcano, Tanzania. *In:* Ahrens LH, Dawson JB, Duncan AR, Erlank AJ (eds) Physics and Chemistry of the Earth, p 525-543 Pergamon Press, New York

Reisberg L, Lorand, J-P (1995) Longevity of sub-continental mantle lithosphere form osmium isotope systematics in orogenic peridotite massifs. Nature 376:159-162

Reisberg LC, Allegre CJ, Luck JM (1991) The Re-Os systematics of the Ronda Ultramafic Complex of southern Spain. Earth Planet Sci Lett 105:196-213

Ringwood AE (1962) A model for the Upper Mantle. J Geophys Res 67:857-867

Ringwood AE (1975) Composition and petrology of the earth's mantle, McGraw-Hill, New York, 618 p

Ringwood AE (1994) Role of the transition zone and 660 km discontinuity in mantle dynamics. Phys Earth Planet Inter 86:5-24

Roden MK, Hart SR, Frey FA, Melson WG (1984) Sr, Nd and Pb isotopic and REE geochemistry of St. Paul's Rocks: the metamorphic and metasomatic development of an alkali basalt mantle source. Contrib Mineral Petrol 85:376-390

Ross J, Jaques AL, Ferguson J, Green DH, O'Reilly SY, Danchin RV, Janse AJA (1986) Kimberlites and Related Rocks. Volume 2. Their mantle/crust setting, diamonds and diamond exploration. Blackwell Scientific, London, 1271 p

Rudnick RL, McDonough WF, Chappell BW (1993) Carbonatite metasomatism in the northern Tanzanian mantle: petrographic and geochemical characteristics. Earth Planet Sci Lett 114:463-475

Rudnick RL, McDonough WF, O'Connell RJ (1998) Thermal structure, thickness and composition of continental lithosphere. Chem Geol 145:399-415

Rudnick, RL, McDonough WF, Orpin A (1994) Northern Tanzanian peridotite xenoliths: a comparison with Kaapvaal peridotites and evidence for carbonatite interaction with ultra-refractory residues. *In:* Meyer HOA, Leonardos OH (eds) Proc 5th Int'l Kimberlite Conf p 336-353 CPRM, Brasilia

Rudnick RL, Nyblade A (1998) The thickness and heat production of Archean lithosphere: Constraints from xenolith thermobarometry and surface heat flow. *In:* Fei Y, Bertka CB, Mysen BO (eds) Mantle Petrology: Field Observations and High-Pressure Experimentation, Special Publication in honor of F.R. Boyd, *(in press)* The Geochemical Society, San Antonio

Scarfe CM, Takahashi E (1986) Melting of garnet peridotite to 13 GPa and the early history of the upper mantle. Nature 322:354-356

Schmidberger SS, Francis D (1999) Nature of the mantle roots beneath the North American Craton: Mantle xenolith evidence from Somerset Island kimberlites. Lithos (in press)

Schulze DJ (1996) Kimberlites in the vicinity of Kirkland lake and Lake Timiskaming, Ontario and Québec. *In:* LeCheminant AN, Richardson DG, DiLabio RNW, Richardson KA (eds) Searching for Diamonds in Canada, p 73-78 Geol Surv Canada, Open File Rept, Ottawa

Shibata T, Thompson G (1986) Peridotites from the Mid-Atlantic Ridge at 43° N and their petrogenetic relation to abyssal tholeiites. Contrib Mineral Petrol 93:144-159

Simkin T, Smith, JV (1970) Minor-element distribution in olivine. J Geol 78:304-325

Smith D, Boyd FR (1987) Compositional heterogeneities in a high-temperature lherzolite nodule and implications for mantle processes. *In:* Nixon PH (ed) Mantle Xenoliths, p 551-561 John Wiley & Sons, Chichester

Sobolev NV (1977) Deep-seated inclusions in kimberlites and the problem of the composition of the upper mantle American Geophysical Union, Washington, DC, 279 p

Stachel T, Harris JW, Brey GP (1998) Rare and unusual mineral inclusions in diamonds from Mwadui, Tanzania. Contrib Mineral Petrol 132:34-47

Stebbins WE, Dawson JB (1977) Statistical comparison between pyroxenes from kimberlites and their associated xenoilths. J Geol 85:433-449

Stern CR, Kilian R, Olker B, Hauri EC, Kyser TK (1999) Evidence from mantle xenoliths for thin (<100 km), essentially rootless continental lithopshere below the Phanerozoic crust of southernmost South America. Lithos, *(in press)*

Stiefenhofer J, Viljoen KS, Marsh JS (1997) Petrology and geochemistry of peridotite xenoliths from the Letlhakane kimberlites, Botswana. Contrib Mineral Petrol 127:147-158

Streckeisen A (1976) To each plutonic rock its proper name. Earth Sci Rev 12:1-33

Suen CJ, Frey FA (1987) Origins of the mafic and ultramafic rocks in the Ronda peridotite. Earth Planet Sci Lett 85:183-202

Suhr G, Robinson PT (1994) Origin of mineral chemical stratification in the mantle section of the Table Mountain massif (Bay of Islands Ophiolite, Newfoundland, Canada). Lithos 31:81-102

Sun S-s (1982) Chemical composition and origin of the earth's primitive mantle. Geochim Cosmochim Acta 46: 179-192

Sun S-s, Nesbitt RW (1977) Chemical heterogeneity of the Archaean mantle, composition of the Earth and mantle evolution. Earth Planet Sci Lett 35:429-448

Szabo CS, Bodnar RJ (1995) Chemistry and origin of mantle sulfides in spinel peridotite xenoliths from alkaline basaltic lavas, Nograd-Gomor Volcanic Field, northern Hungary and Southern Slovakia. Geochim Cosmochim Acta 59:3917-3927

Taylor SR, McLennan SM (1985) The Continental Crust: its Composition and Evolution Blackwell,

Oxford, 312 p

van der Hilst RD, Widiyantoro S, and Engdahl, ER (1997) Evidence for deep mantle circulation from global tomography. Nature 386:578-584

van der Wal, D, Vissers RLM (1993) Uplift and emplacement of upper mantle rocks in the western Mediterranean. Geol 21:1119-1122

Walker RJ, Carlson, RW, Shirey SB, Boyd, FR (1989) Os, Sr, Nd and Pb isotope systematics of southern African peridotite xenoliths: implications for the chemical evolution of subcontinental mantle. Geochim Cosmochim Acta 53:1583-1595

Walter MJ (1998a) Melting of Garnet Peridotite and the Origin of Komatiite and Depleted Lithosphere. J Petrol 39:29-60

Walter MJ (1998b) Melting Residues of Garnet Peridotite and the Origin of Cratonic Lithosphere. In: Fei Y, Bertka CB, Mysen BO (eds) Mantle Petrology: Field Observations and High-Pressure Experimentation, Special Publication in honor of F.R. Boyd (in press) The Geochemical Society, San Antonio

Wang A, Pasteris JD, Meyer HOA, Dele-Duboi ML (1996) Magnesite-bearing inclusion assemblage in natural diamond. Earth Planet Sci Lett 141:293-306

Washington HS (1925) The chemical composition of the Earth. Am J Sci 9:351-378

Waters FG (1987) A suggested origin of MARID xenoliths in kimberlites by high pressure crystallization of an ultrapotassic rock such as lamproite. Contrib Mineral Petrol 95:523-533

Wells PRA (1977) Pyroxene thermometry in simple and complex systems. Contrib Mineral Petrol 62:129-139

Williamson ED, Adams LH (1923) Density distribution in the Earth. J Wash Acad Sci 13:413-428

Wilshire HG, Meyer CE, Nakata JK, Calk LC, Shervais JW, Nielson JE, Schwarzman EC (1988) Mafic and ultramafic xenoliths from the western United States United States Government Printing Office, Washington, DC, 179 p

Wilshire HG, Nielson-Pike JE (1975) Upper-mantle diapirism: evidence from analogous features in alpine peridotite and ultramafic inclusions in basalts. Geol 3:467-470

Yaxley GM, Kamenetsky V, Green DH, Falloon TJ (1997) Glasses in mantle xenoliths from western Victoria, Australia, and their relevance to mantle processes. Earth Planet Sci Lett 148:433-446

Zipfel J, Wörner G (1992) Four- and five-phase peridotites from a continental rift system: evidence for upper mantle uplift and cooling at the Ross Sea margin (Antarctica). Contrib Mineral Petrol 111:24-36

Chapter 5

PHASE TRANSFORMATIONS AND SEISMIC STRUCTURE IN THE UPPER MANTLE AND TRANSITION ZONE

Carl B. Agee

NASA Johnson Space Center
Houston, Texas 77058

INTRODUCTION

The Earth's mantle is divided globally into layers having distinct seismic properties. The transition zone (hereafter denoted "TZ") is located in the deeper portion of the upper mantle between approximately 400 and 670 km depth. It is characterized by rapidly increasing seismic wave velocities with depth (Fig. 1; Dziewonski and Anderson 1981). The wave velocity increases are of two sorts, discontinuous jumps over comparatively narrow distances (perhaps tens of km or less) at the bounds of the transition zone at 400 km and 670 km and a more or less continuous, relatively steep, gradient within the transition zone (Walck 1984, Walck 1985, Grand 1984). The values of 400 km and 670 km for the discontinuities are now commonly referred to as the *410* and the *660*. In most seismic models the *410* falls within range 380-420 km depth, while the *660* can span the interval 640-680 km. The mineralogic nature of the discontinuities and high velocity gradients have been the focus of lively, ongoing debate. The debate has often been portrayed as a choice between two simple models of the mantle. One model proposes that the transition zone seismic velocities are a result of first order phase transitions in an isochemical medium similar in composition to peridotite xenoliths (e.g. Ringwood 1979, Bina and Wood 1987; see McDonough and Rudnick, this volume). The alternative view has been that the transition zone is characterized by abrupt radial changes in mantle chemistry (e.g. Anderson and Bass 1986). These choices have implications not only for understanding the present state of the mantle transition zone but for the whole Earth and its formation.

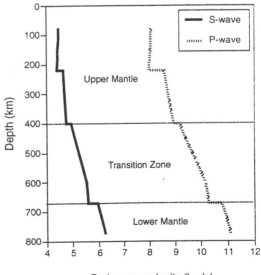

Figure 1. Body wave velocity versus depth diagram defining the bounds of the upper mantle transition zone in the Preliminary Reference Earth Model (PREM) (after Dziewonski and Anderson 1981). Numerous subsequent studies indicate that the 400 and 670 km disconitinuities are, on average, closer to 410 and 660 km.

0275-0279/98/0037-0005$05.00

Our knowledge of the upper mantle transition zone comes from seismological observations and a number of indirect lines of evidence from mineral physics, phase equilibria, and cosmochemical constraints. Strictly taken, the mineralogy and petrology of the transition zone is unknown. The transition zone occupies a deep, inaccessible part of the Earth. Geologists have never held a proven, unadulterated piece of the transition zone in their hands. There is no evidence that pristine samples from below 410 km are ever exposed at the Earth's surface. Geobarometry shows that chemical equilibration of most upper mantle xenoliths occurred at depths less than 200 km (see, for example, Mercier and Carter 1975, Hervig et al. 1986). A few studies (e.g. Haggerty and Sautter 1990) have argued that some xenoliths retain traces of their formation in the transition zone or even the lower mantle, however none of these samples contains the original, supposed, high-pressure mineralogy. The closest we come to tangible specimens from the transition zone are mineral inclusions in diamond. Moore and Gurney (1991) have discovered majorite garnet, a probable transition zone mineral in diamond. No study to date has offered more than a fragmentary glimpse of possible rock types residing in the transition zone.

After taking this strong dose of reality, one might ask: "How can a chapter be written on the mineralogy of the transition zone if no rock samples from this region exist?" The answer is, of course, that we can only attempt to describe what the mineralogy of transition zone might be. The goal of this work is to produce a sketch of the transition zone based on the information that is currently available. Because the information on this subject is incomplete there are many unresolved questions, conflicting hypotheses, and gaps in understanding. It goes without saying that this reference is not intended to be the final word on the subject. No doubt some aspects of the transition zone will remain enigmatic far into the future. On the other hand, the optimistic view is that there is still much we can learn about certain features of the transition zone without access to tangible samples. The challenge is to piece the fragmentary evidence together in a sensible fashion.

The first line of evidence that is explored here comes from phase equilibria and physical properties of olivine and the Mg_2SiO_4-Fe_2SiO_4 high-pressure polymorphs. Regardless of whether the transition zone is enriched or depleted in olivine-component, the connection between the high-pressure properties of "olivine" and seismic discontinuities and velocities is key in understanding the upper mantle transition zone.

OLIVINE AND THE Mg_2SiO_4-Fe_2SiO_4
HIGH-PRESSURE PHASE TRANSITIONS

"...the thicknesses of the transitions are more likely to be found from petrology and physical chemistry than from seismology..."   Jeffreys (1936)

The properties of the Earth's upper mantle are often equated with the properties of olivine, one of its common and perhaps most abundant minerals. This comes from the assumption that the average mantle composition is an olivine-rich peridotite. Ringwood (1975) proposed that the bulk mantle is a peridotite composition with 60-65% olivine, which he named "pyrolite". If the mantle is olivine-rich, then it follows that the main features of the mantle can be described by the high pressure and high temperature behavior of olivine. Olivines form a continuous solid solution between pure phases forsterite Mg_2SiO_4 and fayalite Fe_2SiO_4. They belong to the orthorhombic crystal system, however at high pressures olivines transform into denser, spinel structured polymorphs. These transformations may be directly related to seismic discontinuities in the mantle. Therefore, high pressure polymorphism and phase equilibria of "olivine" is highly relevant to understanding the nature of the transition zone. The idea that olivine transforms to denser structured polymorphs at high pressure has been considered as far back as the 1920s.

Many references attribute the original proposal to Bernal (1936). However the first experimental evidence for high pressure olivine polymorphism was provided by Ringwood who observed the transformation in Fe, Ni, and Co olivines at relatively low pressures (Ringwood 1958, 1962, 1963). Further experimental studies (Ringwood 1966) confirmed that the high-pressure polymorphic transformations of Mg_2SiO_4-rich olivine proceeds with pressure first to the closer packed "β"-phase (modified spinel structure, mineral name: wadsleyite) and then to "γ"-spinel (mineral name: ringwoodite).

High pressure polymorphism in Mg_2SiO_4

The olivine mantle model has, in its simplest form, the composition of Mg_2SiO_4 (forsterite, Fo_{100}). Mg_2SiO_4 is a convenient compositional shorthand for expressing high pressure phase transitions and the behavior of an Mg-rich, peridotite mantle. The simple Mg_2SiO_4 version of the mantle is attractive in that it allows phase equilibria to be presented in a single component system; complex behavior such as solid solution and non-ideal mixing can be neglected. In the case of Mg_2SiO_4 the same phase transitions, olivine to β-phase to γ-spinel, occur as in natural, multicomponent systems such as peridotite. We will first consider some recent experimental phase equilibria on Mg_2SiO_4 as they apply to the 410, 520, and 660 km discontinuities in the transition zone.

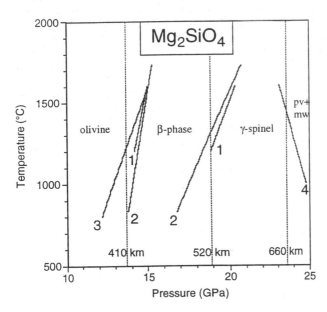

Figure 2. Pressure-temperature diagram showing phase boundaries for Mg_2SiO_4. Phase boundaries are from [1] Katsura and Ito (1989), [2] Akaogi et al.(1989), [3] Morshima et al. (1994), and Ito and Takahashi (1989). Depths 410, 520, and 660 km are from pressure versus depth tabulation in the Preliminary Reference Earth Model (Dziewonski and Anderson 1981) pv = perovskite, mw = magnesiowüstite.

The lowest pressure polymorphic phase transition in Mg_2SiO_4 is olivine to β-phase. Figure 2 shows the approximate location of this transition from three different methods of determination. Ito and Katsura (1989) performed quenching experiments in an octahedral multi-anvil device, Akaogi et al. (1989) calculated the phase boundaries from calorimetric measurements, and Morishima et al. (1994) made in situ synchrotron x-ray diffraction measurements in a cubic anvil apparatus. The three proposed boundaries of the olivine to β-phase transition have markedly different slopes and at 1000°C the location of the boundary ranges from about 12.8 to 14 GPa. At higher temperatures (~1600°C) there is very good agreement between the three different studies that place the transition at approximately 15 GPa. Morishima et al. argue that their in situ study is superior to the multi-anvil quenching method. They were able to measure pressure by an internal marker and directly observed the kinetic behavior of the polymorphic phase transformation at P and T.

What connection does the Mg_2SiO_4 phase diagram have to the seismic velocity discontinuity observed at a depth of ~410 km? If one assumes that the olivine to β-phase transition is largely responsible for the jump in seismic velocity, then it can offer useful information about the composition and temperature of the mantle at this depth. Let us assume that a depth of 410 km in the mantle is equivalent to 13.7 GPa as given by PREM (Dziewonski and Anderson 1981). Then we consider a mantle of pure Mg_2SiO_4 and ask: at what temperatures do the three proposed olivine to β-phase boundaries intersect a pressure of 13.7 GPa? The shallowest sloping phase boundary (Morishima et al.) fixes the 410 km (13.7 GPa) PREM depth at a temperature of 1230°C. The data of Akaogi et al., which has the steepest sloping phase boundary, indicates a temperature of ~800°C. The Katsura and Ito boundary, with an intermediate slope, gives a temperature of ~980°C. The latter two temperatures are probably too low to be consistent with estimates of mantle geotherms based on petrologic constraints such as barometry/thermometry of peridotite xenoliths and mantle adiabats for MORB genesis (Mercier and Carter 1975, Herzberg 1992, Nisbet et al. 1993). The Morishima et al. boundary temperature, though closer to mantle estimates from other methods, is probably about 70-170° too low.

Multi-anvil quenching experiments and calorimetry measurements have also provided data on the location and slope of the β-phase to γ-spinel transition in Mg_2SiO_4 (Fig. 2). These methods are in good agreement; in situ measurements such as those of Morishima et al. do not exist at present. The polymorphic β-phase to γ-spinel transition cannot be attributed to the 410 km discontinuity because the temperature of the transition at 13.7 GPa is at very low or even negative values. A more reasonable manifestation of the β-phase to γ-spinel transition is the proposed 520 km discontinuity present in some seismic models (see, for example, Shearer 1990, Gaherty and Jordan 1996), however other petrologic factors may also be responsible for this discontinuity

At high enough pressures the spinel-structured polymorph of Mg_2SiO_4 becomes thermodynamically unstable. This so-called "post-spinel breakdown" is described by the univariant reaction:

Mg_2SiO_4 (γ-spinel) to $MgSiO_3$ (perovskite) + MgO (periclase)

The reaction was first discovered experimentally by Liu (1976) using a diamond anvil cell. Shortly thereafter the reaction was confirmed with the multi-anvil device (Ito 1977, Ito and Matsui 1978).

In order to determine pressures and temperatures of this phase transformation more precisely, Ito and Takahashi (1989) performed multi-anvil quenching experiments on Mg_2SiO_4 up to 25 GPa. Their data show that this phase transformation boundary has a negative slope described as P(GPa) = 27.6 -0.0028 T(°C) (Fig. 2). The reason for the negative slope can be understood from the Clausius-Clapeyron equation that defines the slope (dP/dT) of the univariant phase boundary at equilibrium as

$$\frac{dP}{dT} = \frac{S_{pv+mw} - S_\gamma}{S_{pv+mw} - V_\gamma}$$

In this case $V_{pv+mw} - V_\gamma$ is the molar volume change and $S_{pv+mw} - S_\gamma$ is the entropy change in the reaction from γ-spinel to perovskite + periclase. P-V-T data (compiled by Anderson 1989) show that the combination of perovskite+periclase has a smaller molar volume than does γ-spinel giving a positive value for $V_{pv+mw} - V_\gamma$. What does this fact and the negative value for dP/dT indicate? It implies that entropy of a mole of perovskite + periclase is greater than the entropy of a mole of γ-spinel and that $S_{pv+mw} - S_\gamma$ is positive. This result has also been confirmed by calorimetric measurements. Ito et al. (1990) found that the value for $S_{pv+mw} - S_\gamma$ is indeed positive and falls within the range 7-23 J K^{-1} mol^{-1}, giving dP/dT = -0.004 ± 0.002 GPa K^{-1}.

Based on the Ito and Takahashi (1989) and the Ito et al. (1990) boundaries, we can ask the same type of hypothetical question as for the *410*. What is the temperature at the 660 km discontinuity in a pure Mg$_2$SiO$_4$ mantle? If we assume that a PREM depth of 660 km is equal to a pressure of 23.6 GPa, then the calculated temperature is 1425°C. This value is higher than any of the temperatures calculated for the 410 km and is therefore consistent with the idea that the geotherm has a positive dT/dP slope in the transition zone. The depth assigned to the *660* is somewhat arbitrary. The PREM model as well as the widely cited seismic models of Grand and Helmberger (1984) and Walck (1984, 1985) position the discontinuity at 670 km, however the resolution of body wave data probably allow the boundary to be anywhere in the range between 640 and 700 km. This variation is not only due to lack of precision, seismic reflection studies indicate that *660* may have significant topography, indicating that some depth variability may be real. Because the γ-spinel to perovskite + periclase phase boundary has a negative slope, the temperature of the transition decreases with increasing depth.

There appears to be a broad correlation between the phase transitions in Mg$_2$SiO$_4$ and the major seismic discontinuities that define the transition zone. Though this composition can be a convenient shorthand description, it is by no means a plausible rock type for the transition zone. The likelihood that the mantle consists of pure Mg$_2$SiO$_4$ or any other monomineralic aggregate is extremely remote. Below we discuss the effect of more complex chemical compositions on phase transitions. The first compositional effect on phase transitions we consider is that of ferrous iron in olivine.

Phase equilibria of the system Mg$_2$SiO$_4$-Fe$_2$SiO$_4$

Studies of peridotite xenoliths indicate that the upper mantle molar FeO/(FeO+MgO) is approximately equal to a value of 0.90 (Jagoutz et al. 1979, Maaløe and Aoki 1977). Accordingly, average olivine compositions in these rocks are a solid solution of about 90 mol% Mg$_2$SiO$_4$ and 10 mol% Fe$_2$SiO$_4$ (Fo$_{90}$). A number of workers have investigated the system Mg$_2$SiO$_4$-Fe$_2$SiO$_4$ at transition zone conditions. Their data confirm the importance of a ferrous iron effect on the pressures and temperatures of olivine polymorphism. The most comprehensive early experimental study on this system was carried out by Akimoto and Kawada (see Akimoto 1987) and Kawada 1977. Later, additional measurements were produced by Katsura and Ito (1989) and Akaogi et al. (1989). Ito and Takahashi (1989) have also investigated Fo$_{90}$ and other Mg$_2$SiO$_4$-Fe$_2$SiO$_4$ mixtures at post-spinel P-T conditions.

Figure 3 summarizes the data from these studies in the form of a pressure-composition diagrams at 1200, 1600, and 2000°C. The phase transitions in the system Mg$_2$SiO$_4$-Fe$_2$SiO$_4$ at high pressures and temperatures are governed by two intervals of eutectic solid solution. The lower pressure eutectic solid solution involves the transformations between olivine, β-phase and γ-spinel. The second, higher pressure, eutectic solid solution consists of transformations between γ-spinel, perovskite, magnesiowüstite, and stishovite.

There are aspects of the eutectic solid solutions in Figure 3 that have important implications for the petrology of the transition zone. Except for the pure end-members Mg$_2$SiO$_4$ and Fe$_2$SiO$_4$, all other compositions undergo phase transitions that have some finite pressure-temperature interval. Within these intervals, called "binary phase loops,"

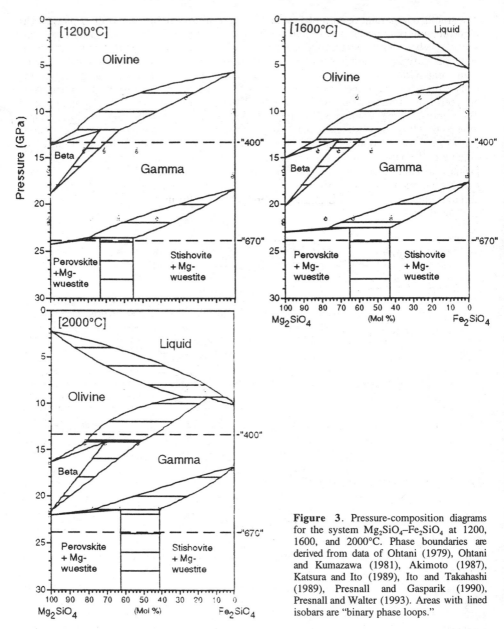

Figure 3. Pressure-composition diagrams for the system Mg_2SiO_4–Fe_2SiO_4 at 1200, 1600, and 2000°C. Phase boundaries are derived from data of Ohtani (1979), Ohtani and Kumazawa (1981), Akimoto (1987), Katsura and Ito (1989), Ito and Takahashi (1989), Presnall and Gasparik (1990), Presnall and Walter (1993). Areas with lined isobars are "binary phase loops."

two or more phases coexist at the equilibrium pressures and temperatures during the transformation. If one assumes that Mg_2SiO_4-Fe_2SiO_4 phase transitions are responsible for mantle discontinuities, then the presence of binary phase loops in the phase diagram place constraints on the structural details of the discontinuities. The position and size of the binary phase loops in Figure 3 should control not only the location of seismic discontinuities, but also their radial width.

Another important feature of Figure 3 is the limited stability range of β-phase. This high-pressure polymorph is only stable in Mg_2SiO_4-rich compositions. Consider a path that starts in the olivine stability field and moves to greater depth. Compositions of approximately Fo_{85} and greater convert to β-phase and then convert to γ-spinel. Olivines with compositions less than Fo_{85} first convert to γ-spinel. This first conversion to γ-spinel does not always mark the end of the spineloid transformations. Those with compositions between approximately Fo_{85} and Fo_{55} have a more complicated P-T transformation path. The original olivine volume will partially convert to γ-spinel as the binary phase loop is traversed. Then, depending on the bulk composition, either olivine or γ-spinel will be consumed in a reaction to form β-phase at the eutectic. From there, with pressure, the volume of β-phase increases for a short interval. Ultimately, with increasing pressure, the β-phase reacts to form all γ-spinel. For compositions of Fo_{55} and less, the polymorphic transformation is straightforward; olivine converts completely to γ-spinel over an uninterrupted P-T interval; β-phase is thermodynamically unstable in such Fe_2SiO_4-rich compositions.

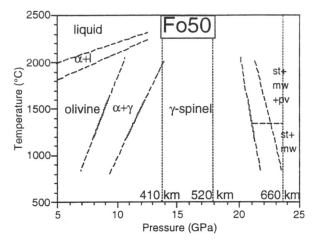

Figure 4. Pressure-temperature diagram for the composition Fo_{50} (50 mol% Mg_2SiO_4–50 mol% Fe_2SiO_4) Phase boundaries are estimated from Katsura and Ito (1989) and Ito and Takahashi (1989). α = olivine, β = β-phase, γ = γ-spinel, pv = perovskite, mw = magnesiowüstite, st = stishovite. Depths are from PREM (Dziewonski and Anderson 1981).

Fe_2SiO_4-rich compositions.

The broad negative gradient of the olivine to γ-spinel binary phase loop in Figure 3 indicates that the pressure of the phase transition decreases with increasing Fe_2SiO_4 content. This fact is a strong limitation on the amount Fe_2SiO_4 that can be in the olivine solid solution at a 410 km depth in the upper mantle. For example at 1200°C pure Fe_2SiO_4-olivine converts to γ-spinel at ~5.8 GPa. This pressure corresponds to a PREM depth of ~180 km, obviously far too shallow for the 410 discontinuity. For an olivine composition of Fo_{50} the transition at this temperature begins at about 7.5 GPa or ~230 km; also too shallow. If the temperature is raised to 1600 or 2000°C it causes the transition in Fo_{50} to occur at 9 GPa (275 km) or 10.5 GPa (320 km) respectively. The temperature required to achieve ~13.7 GPa in this composition is not well constrained by experimental data. Figure 4 suggests that the 13.7 GPa olivine to γ-spinel transition in Fo_{50} occurs at temperatures far in excess of 2000°C. Temperatures of 2000°C or more are much higher than estimates for the mantle at 410 km depth. Furthermore, it is likely that Fo_{50} would be partially or completely molten at such temperatures, which is inconsistent with seismic velocity observations.

The kilometer width of the olivine to γ-spinel phase loop is an even more compelling argument against Fe_2SiO_4-rich olivines being compositions responsible for the 410 km discontinuity. For Fo_{50} the pressure interval of the binary phase loop is 3-4 GPa. This would require a discontinuity with radial width of 100-125 km which is inconsistent with seismic reflection data. Clearly, from phase equilibria constraints, Fe_2SiO_4-rich compositions are not attractive candidates for causing the *410*.

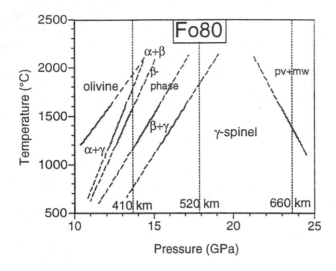

Figure 5. Pressure versus temperature diagram for the composition Fo_{80} (80 mol % Mg_2SiO_4–20 mol % Fe_2SiO_4). Phase boundaries are from Katsura and Ito (1989) and Ito and Takahashi (1989). α = olivine, β = β-phase, γ = γ-spinel, pv = perovskite, mw = magnesiowüstite. Depths are from PREM (Dziewonski and Anderson 1981).

Even some compositions richer in Mg_2SiO_4 such as Fo_{80} are inconsistent with the location of seismic discontinuities and thermal estimates of the transition zone. Figure 5 illustrates that the *410* would have to be at a temperature of 1900°C to coincide with the olivine to γ-spinel transition in this composition. But the temperature of the breakdown of γ-spinel to perovskite + magnesiowüstite at *660* would be at 1375°C. The result is a steeply negative temperature gradient within the transition zone. This would preclude convection and heat flow upward in the mantle. A further obstacle in having a Fo_{80} transition zone is that it would produce a broad and complex discontinuity structure near 410 km. Figure 5 indicates that a narrow field of β-phase stability exists between olivine and γ-spinel. This could translate into a series of seismic velocity jumps or perhaps a steep velocity gradient over a depth of perhaps 150 km.

Fo_{90}

The olivine composition of Fo_{90} (defined as solid solution of 90 mol% Mg_2SiO_4 and 10 mol% Fe_2SiO_4) is probably most relevant to the transition zone, and this holds for at least two important reasons. First, as mentioned above, Fo_{90} is the approximate average of olivines in peridotite xenoliths. Second, at "reasonable" mantle temperatures, olivines of ~Fo_{90} can undergo conversion to β-phase at pressures equivalent to the depth of the 410 km discontinuity (~13.7 GPa). Figures 3 and 6 show that the olivine to β-phase transition in this composition occurs at 410 km over a fairly narrow pressure range. The β-phase to γ-spinel transformation occurs in the mid-transition zone over slightly broader pressure range and could be correlated with the 520 km discontinuity. According to Ito and Takahashi (1989) the breakdown of γ-spinel to perovskite + magnesiowüstite has a pressure range so narrow that it is indistinguishable from zero. In Fo_{90} it takes place at 660 km (23.6 GPa) and ~1325°C. In an isochemical mantle the pressure interval widths of these three phase transitions must be consistent with the width of transition zone seismic discontinuities. If there are large discrepancies between the phase boundary and the seismic discontinuity thicknesses, then alternative, layered mantle compositions may be required. Another important consideration, as alluded to above, is the temperature of the coincident phase transition/seismic discontinuity.

The first of the three phase transitions to consider is olivine to β-phase. The pressure interval width of the olivine to β-phase "phase loop" for composition Fo_{90} may place an important constraint on whether the phase transition actually causes the *410*. The most widely cited data set for this transition is that of Katsura and Ito (1989); some of the features of Figure 3, for instance, are derived from their results. Though their results are crucial to understanding the phase equilibria at transition zone discontinuities, the actual data coverage is fairly sparse. They constructed pressure-Mg_2SiO_4-Fe_2SiO_4 binary phase diagrams for 1200° and 1600°C from a total of 54 experiments. However, only a three experiments at 1600°C actually bracket the pressure-width of the olivine to β-phase loop relevant to ~ Fo_{90} at the *410*. This bracket is 14.6-14.1 GPa or an equivalent of a ~15 km thickness. The pressure-widths of the olivine to β-phase at 1200°C and the β-phase to γ-spinel at 1200° and 1600°C are inferred from the general shape of the adopted phase boundaries but are not rigorously constrained by bracketing experiments. The phase boundaries in Katsura and Ito (their Fig. 6) should be considered preliminary. Unfortunately these phase diagrams tend to be viewed by the casual reader as firmly established facts rather than works in progress. From their phase diagram, Katsura and Ito concluded that *400*, if caused by the olivine to β-phase transition, should have a thickness of 11-19 km. They offer a range of kilometer-widths because the data is interpreted as showing a narrowing of the phase loop with increasing temperature. In other words, the discontinuity should be thinner at high temperatures and thicker at lower temperatures. Figure 5 shows the kilometer-width of the olivine to β-phase transition taken directly from the phase diagrams of Katsura and Ito. To be precise, the width of the Katsura and Ito olivine to β-phase transition is 22 km at 1200°C and 13 km at 1600°C. If we assume that the middle of the olivine + β-phase field occurs exactly at 400 km depth, then the temperature of the mantle must be 1320°C there with a width of 20 km. If the discontinuity is assigned to a deeper level, for instance 410 km, then the mantle temperature should be correspondingly higher at 1425°C, while a shallower discontinuity requires colder temperatures. If the mantle near *410* is more depleted in Fe_2SiO_4 than Fo_{90}, then the transformation will be narrower still and at a slightly deeper level. On the other hand if that region contains more Fe_2SiO_4 than Fo_{90}, then the transformation will be broader (>>20 km) and at a shallower level (as discussed above).

The location and width of the β-phase to γ-spinel transition can also be compared with the 520 km discontinuity (PREM pressure of 17.9 GPa). If we assume that the *520* is the result of, and occurs in the middle of the β-phase + γ-spinel stability field, then from the Katsura and Ito data, the mantle temperature there is 1475°C, with a width of 24 km. The relatively broad interval for the β-phase + γ-spinel transition in the Fo_{90} phase diagram is consistent with a diffuse *520* rather than a sharp reflector. It is interesting to note that the *520* can be difficult to observe in some seismic data sets. Perhaps this can be related to the broad depth interval of the β-phase + γ-spinel transition, though other possibilities exist.

Transition zone geotherms can be calculated if phase transitions do indeed induce discontinuities and the P-T locations of the phase transitions are known. For example, Figure 6 shows, that between the *410* and the *520*, the mantle temperature should increase by 50 degrees, if the mantle olivine composition is Fo_{90} (curve 'a' in Fig. 6). This translates to a geotherm with slope 0.45°/km. If we consider the deeper parts of the transition zone, then the transformation boundary of γ-spinel to perovskite+ magnesio-

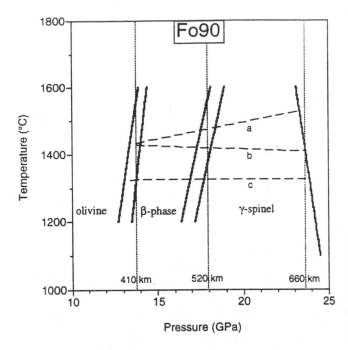

Figure 6. Pressure-temperature phase diagram for Fo_{90} after data from Katsura and Ito (1989) and Ito and Takahashi (1989). Dashed lines are hypothetical transition zone geotherms for discontinuity depths of (a) 410, 520, 655 km, (b) 410, 512, 660 km and (c) 400, 505, 670 km.

wüstite is encountered. As noted previously, this phase transformation boundary is attributed to the *660* and it has a negative P-T slope. Figure 6 shows that a shallow *660* will be at higher temperatures than a deeper *660*. If we assume that the *660* is actually at a depth of 660 km or a PREM pressure of 23.6 GPa, then the mantle temperature there should be approximately 1410°C. This result would require the *410* to be hotter than the *660*, implying a negative geotherm (curve 'b' in Fig. 6). Alternatively, the transition zone would be isothermal if the phase boundaries occur at 410, 505, and 670 km (curve 'c'). A negative geotherm or an isothermal geotherm would indicate that the transition zone is not adiabatic, implying stable stratification and absence of convection. On the other hand, the *660* may, on average, be actually located at a shallower level and hence the mantle temperature there would be higher. For example, if we simply extend the slope of 0.45°/km from *520*, the γ-spinel to perovskite + magnesiowüstite is intersected at 1530°C and 23.2 GPa or a PREM depth of 654 km. Another possibility is that the P-T position of the γ-spinel to perovskite + magnesiowüstite boundary from Ito and Takahashi is incorrect. If the boundary is actually shifted to higher pressures, then the dT/dP slope of a geotherm with discontinuities at *410* and *660* will be positive.

The slope and position of the hypothetical geotherms in Figure 6 do not exactly match estimates for the average transition zone geotherm obtained from other methods. One method of estimating the temperature profile in the mantle is from geobarometry and geothermometry of minerals in mantle xenoliths. Figure 7 shows one set of such estimates from the work of Mercier and Carter (1975) on pyroxenes. Their estimates show a very steep geotherm for both suboceanic and subcontinental mantle. Extrapolation of their data to deeper levels along any reasonable positive sloping trajectory is in poor agreement with the P-T locations of the "olivine" phase transitions. Such extrapolations will most likely intersect the peridotite solidus before the *410* is reached.

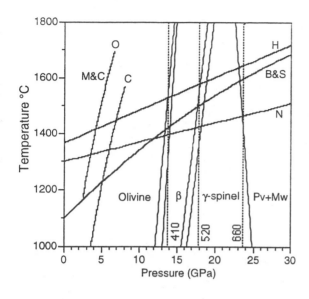

Figure 7. Temperature-pressure diagram for Fo_{90} and mantle geotherms from different authors. M&C is from Mercier and Carter (1975) and is based on geothermometry and geobarometry of mantle xenoliths. The "O" is for the suboceanic mantle and the "C" is for the subcontinental mantle. H and N are from Herzberg (1992) and Nisbet et al. (1994) and are based on estimates of MORB adiabats and the solidus of peridotite at high pressure. B&S (Brown and Shankland 1981) is based on estimates of thermodynamic parameters in the Earth from seismic profiles. The depths of 410, 520, and 660 are the kilometer equivalents of pressure from PREM. Actual geotherms would show perturbations in slope at phase boundaries due to the change in enthalpy of the phase transition, but are neglected here for simplicity.

Another method of estimating mantle temperatures is obtained from mid-ocean ridge basalts (MORB). Here it is assumed that MORB is produced from adiabatic upwelling and melting of fertile mantle peridotite. Given average MORB eruption temperature, the coordinates of the mantle melting interval at high pressure, and properties such as the heat capacity (C_P), thermal expansion (α), and density of solids and melts (ρ), an adiabatic P-T trajectory can be calculated from the relation below.

$$\left(dT/dP\right)_S = \frac{\alpha T}{\rho C_P}$$

Geotherms "H" (Herzberg 1992) and "N" (Nisbet et al. 1993) in Figure 7 correspond to MORB adiabats of approximately 1370°C and 1300°C respectively. These adiabats intersect the middle of the olivine to β-phase transition at 1530° and 14.0 GPa (417 km) and 1395°C and 13.7 GPa (409 km). Both favor the location of the discontinuity to be deeper than 400 km. The slopes of the H and N geotherms are comparatively shallow and resemble the positive410 geotherm in Figure 6 (curve 'a'). Projection of geotherms H and N through the transition zone places the 520 at 533 and 518 km and the 660 at 648 and 659 km, respectively.

Brown and Shankland (1981) calculated a geotherm based on seismological models and estimates of thermodynamic properties of mantle minerals which is shown as curve B&S in Figure 7. This geotherm has a steeper slope than that of H and N and has phase transformation depths of 410, 527, and 651 km. It is interesting to note that geotherms H, N, and B&S all give transition zone thicknesses less than the original PREM value of 270 km, all have the 410 at 409 or greater and the 660 at 659 km or less.

Geotherms H, N, and B&S of Figure 7 argue strongly for phase transitions in Fo_{90} as being the cause of the seismic discontinuities in the transition zone. However, the smooth curves of these geotherms lack some degree of realism because they do not take into consideration the thermodynamic effects of phase transitions given by the relationship

$$T_{equilibrium} = \frac{\Delta H_{transition}}{\Delta S_{transition}}$$

where ΔH is the change in enthalpy and ΔS is the change in entropy of the equilibrium phase transition. If ΔS is positive, then ΔH must also be positive, the reaction is endothermic and heat is absorbed. If ΔS is negative, then ΔH must also be negative, the reaction is exothermic and heat is released. In the case of the olivine to β-phase transition the reaction is exothermic and a temperature increase is expected with depth across the *410*. Conversely, the γ-spinel to perovskite+magnesiowüstite phase transition is an endothermic reaction and a temperature decrease should occur with depth across the *660*. It is inaccurate to label the phase boundaries at *410* and the *660* as exothermic and endothermic respectively. The phase boundaries represent equilibrium reactions that can run either direction when perturbed by a change in pressure or temperature. When the phase boundary is crossed by going "down" into the Earth (higher pressure) changes in enthalpy and entropy have opposite signs than when crossing the boundary by going "up" in the Earth (to lower pressures). Hence the β-phase to olivine reaction is endothermic and the transformation of perovskite + magnesiowüstite to γ-spinel is exothermic (Fig. 8).

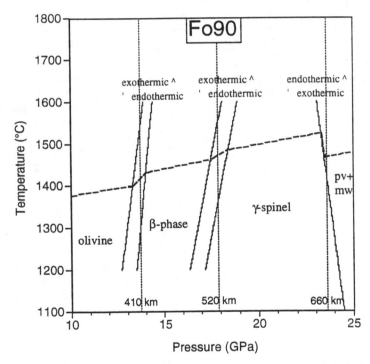

Figure 8. Temperature-pressure diagram for phase transitions in Fo$_{90}$ with a hypothetical adiabatic mantle geotherm. The arrows show the directions in which the transitions are either exothermic or endothermic. For example, the phase transformation γ-spinel to perovskite + magnesiowüstite is endothermic if the reaction proceeds from left to right, whereas it is exothermic is the reaction proceeds from right to left. pv = perovskite, mw = magnesiowüstite.

Figure 8 gives a schematic P-T trajectory of an adiabatic geotherm in the transition zone that includes the enthalpic and entropic effects of the Fo$_{90}$ phase transitions. The overall slope of the geotherm is similar to those of N and H in Figure 7, but it increases in temperature at *410* and *520* and decreases in temperature at *660* corresponding to differences in sign of the change in enthalpy and entropy for the transitions olivine to β-phase, β-phase to γ-spinel, and γ-spinel to perovskite+magnesiowüstite. The pertur-bations in temperature in the geotherm of Figure 8 should have an effect on the seismic velocities on either side of the phase transition. For example, the abruptly higher temperatures just below the *410* should reduce the magnitude of the jump in seismic velocity across the transition boundary. Conversely, the large temperature decrease below *660* should magnify the seismic velocities there. It must be emphasized that the geotherm in Figure 8 is an adiabatic path and assumes convective flow across the transition zone discontinuities. Should convective flow be impeded or should there be layering in the transition zone, then a different geotherm can be expected. For example, if flow is impeded at the *660*, then the thermal structure could be the opposite as in Figure 8. Cold downwellings associated with subduction would cause lower than average temperatures to be located just above the *660*, while hot upwellings in the lower mantle would cause a temperature increase in some regions just below the *660*.

Physical properties in the system Mg_2SiO_4-Fe_2SiO_4

The physical properties in the system Mg_2SiO_4-Fe_2SiO_4 should be considered "a work in progress". In some cases values have been firmly established by multiple studies using a variety of measurement techniques. The most reliable data are generally density and bulk modulus values of minerals of olivine. Other values especially for β-phase and γ-spinel are no better than "educated guesses" and are likely to change as new measurements become available in the future. In particular data on the shear modulus and its temperature and pressure derivatives are probably the poorest constrained. Where no direct experimental data exist the values are adopted from Duffy and Anderson (1989). It should be noted that both isentropic and isothermal bulk moduli values are encountered in the literature and are distinguished by

$$K_S = -V \left(\frac{\partial V}{\partial P} \right)_S$$

where K_S is the isentropic bulk modulus at the reference pressure (usually 1 bar), V is the molar volume and $(\partial V / \partial P)_S$ is the change in molar volume with pressure at constant entropy. Conversion from K_s to the isothermal bulk modulus (K_T) can be calculated using the following relationship

$$K_T = \frac{K_S}{(1 + \alpha \gamma T)}$$

where α is the coefficient of thermal expansion and is defined as

$$\alpha = \frac{1}{V} \left(\frac{\partial V}{\partial T} \right)_P$$

with V the molar volume and T the temperature at constant pressure. Gamma (γ) is the thermal Grüneisen parameter and is defined as

$$\gamma = -\frac{\partial \ln \theta}{\partial \ln V} = \frac{\alpha K_S}{\rho C_P}$$

with θ the Debye temperature, and C_P the specific heat at constant pressure.

Olivine. Olivine is based on the orthorhombic lattice and belongs to the 2/m 2/m 2/m or rhombic di-pyramidal class and the Pbnm space group. The structure of olivine can be described as a hexagonal-close-packed arrangement of O^{-2}, with octahedrally coordinated

Table 1. Physical properties in the system Mg_2SiO_4-Fe_2SiO_4
At 1 bar and 25°C (where applicable).

	Olivine	β-phase	γ-spinel
Density ρ (g/cc)	$3.222+(1.182X_{FA})$	$3.47+(1.24X_{FA})$	$3.55+(1.30X_{FA})$
Molar volume V (cc/g)	$43.68+(2.59X_{FA})$	$40.53+(2.72X_{FA})$	$39.66+(2.38X_{FA})$
Thermal expansion α (K^{-1})	2.68×10^{-5}	2.15×10^{-5}	1.93×10^{-5}
Bulk modulus K_s (GPa)	129	174	184
Shear modulus G (GPa)_82-(31X_{FA})_114-(41X_{FA})_119-(41X_{FA})__dK$_s$/dP_4.2[a]_4.8[d]_5.0__d G/dP_1.4[a]_1.7[d]_1.7[f]__-dK$_s$/dT (GPa/°)_0.016[bc]_0.018_0.017[f]__-dG/dT (GPa/°)_0.014[bc]_0.014_0.014__Gr ueneisen parameter γ_1.14-(0.06X_{FA})_1.32_1.21+(0.31X_{FA})__D ebye temperature θ (K)_924-(236X_{FA})		$974-(203X_{FA})$	$1017-(212X_{FA})$

Values are from Duffy and Anderson (1989) and Anderson (1989) and from the following references:
a. Duffy et al. (1995)
b. Sumino et al. (1989)
c. Isaak et al. (1989)
d. Gwanmesia et al. (1990)
f. Rigden et al. (1991)
X_{FA}=Mole fraction Fe_2SiO_4

Mg^{+2} and Fe^{+2}, and tetrahedrally coordinated Si^{+4}. There exists a complete solid solution between pure-phase endmembers forsterite (Mg_2SiO_4) and fayalite (Fe_2SiO_4).

Olivine density is strongly dependent on composition and increases linearly with the fayalite/forsterite ratio. In contrast, the bulk modulus of olivine is not particularly sensitive to variations in fayalite-forsterite proportions. Regardless of composition all olivines are expected to have reference pressure bulk moduli (K_s) of approximately 129 GPa (Table 1). The shear modulus (G) on the other hand is quite sensitive to composition and decreases markedly with fayalite content and is described as a linear relationship in Table 1. This behavior underscores the importance of shear wave data for elucidating compositional variations in the mantle. The relevance of shear velocities for minerals and mantle seismic models is discussed in a subsequent section.

One of the earliest studies on elastic properties of olivine can be attributed to Kumazawa and Anderson (1969). They measured ultrasonic velocities in single crystals of forsterite and olivine Fo_{93} at room temperature and 1 bar of pressure. More recently investigators have performed elastic property measurements at elevated pressures and temperatures. Pressure derivative of the bulk modulus (dK/dP) for a single crystal of natural mantle olivine (San Carlos, $Fo_{90.5}$) was measured by Webb (1989) up to 3 GPa. Her value of 4.6 is somewhat smaller than dK/dP = 5.1 from Kumazawa and Anderson (1969). An even lower value of dK/dP = 4.2 for pure forsterite was obtained at compressions up to 6 GPa (Yoneda and Morioka 1992). Zaug et al. (1993) used impulsive stimulated scattering in a diamond anvil cell (DAC) up to 12.5 GPa and derived a value of dK/dP = 4.56, essentially the same as that of Webb (1989). Duffy et al. (1995) were the first to measure elastic properties of olivine at transition zone pressures. They employed Brillouin scattering in a diamond cell up to 16 GPa on crystal of pure forsterite and found dK/dP = 4.2±0.2.

Data on the shear modulus (G) and its pressure derivative (dG/dP) for olivine are more variable. Zaug et al.(1993) give G = 77.6 GPa and dG/dP = 1.71. Duffy et al. (1995), on the other hand, report G = 81.6 GP and dG/dP = 1.4. At first glance, the differences between the various values for elastic moduli and their pressure derivatives may seem small or insignificant. Small differences between values of dK/dP or dG/dP are important because the pressure range of olivine stability is extensive. Hence, the change in the bulk modulus from the 1-bar reference value to the corresponding pressure of the transition zone and the *410* discontinuity can be comparatively large. The difference between a dK/dP of 4.2 and 5.1 translates into a ~6% spread in K_s at 13.8 GPa. Calculation of P-wave velocity of olivine at that pressure using dK/dP = 4.2 yields a value that is ~3% lower than if dK/dP = 5.1. As demonstrated below, a 3% difference in P-wave velocity at the *410* is equivalent to a variation of 20-25% olivine content in the mantle.

The temperature effect on the bulk and shear moduli (dK_s/dT, dG/dT) of forsterite has been measured at 1-bar by Sumino et al. (1977) and Isaak et al. (1989) using the rectangular parallelepiped resonance method. The temperature range covered by Sumino et al. (190°-400°C) for pure forsterite, though smaller than that of Isaak et al. (300-1700K), gives values (dK_s/dT = -0.016 Gpa K^{-1}; dG/dT = -0.014 GPa K^{-1}) almost identical to the more recent study. Isaak (1992) extended this work to iron bearing olivines (Fo_{90-92}) over a temperature range 300-1500K and found only a small effect of Fe_2SiO_4 component on the temperature derivatives.

The variation of Mg_2SiO_4-Fe_2SiO_4 in olivine does have a significant effect on the shear modulus. Combining the forsterite-rich data described above with measurements of fayalite by Sumino (1979) a strong decrease in G with fayalite content is observed. At 300K the shear moduli of forsterite and fayalite are 82 and 51 GPa respectively. In contrast, the effect of Mg_2SiO_4-Fe_2SiO_4 on the bulk modulus is apparently quite weak. The data suggest that the difference between forsterite and fayalite may be about 5%, though it is unclear if the effect of Fe_2SiO_4 actually increases or decreases the bulk modulus. The fact that Fe_2SiO_4 content of olivine affects the shear modulus much more than the bulk modulus provides a potentially useful constraint on the composition of the mantle. The ratio V_P/V_S, where V_P is the P-wave velocity given by

$$V_P = \sqrt{\frac{K_s + \frac{4}{3}G}{\rho}}$$

and V_S is the S-wave velocity given by

$$V_S = \sqrt{\frac{G}{\rho}}$$

can be used as an indicator of Fe_2SiO_4 variability. Since V_P depends on the bulk modulus, the shear modulus, and density, while V_S depends only on the shear modulus and density, it follows that an increase in fayalite content in the mantle, which decreases G but not K_s, will cause both V_P and V_S to decrease, but will cause an increase in V_P/V_S. For example, V_P/V_S is ~15% higher if olivine is pure Fe_2SiO_4 rather than pure Mg_2SiO_4. On the other hand, it seems improbable that the fayalite endmember is a realistic olivine composition for the mantle. Hence the actual V_P/V_S variations due to ferrous iron variability would probably be confined to a narrower range than that theoretically possible.

Increasing temperature also decreases the shear and bulk moduli of olivine. The effect of temperature on V_p/V_s, however, is weak compared to the effect fayalite content. V_p/V_s increases only about 2% between 1000° and 2000°C. This is roughly equivalent to the increase in V_p/V_s caused by increasing olivine fayalite content from Fo_{90} to Fo_{70}. Temperature induced increases in V_p/V_s differ from those caused by fayalite induced increases because the former is accompanied by a decrease in density and the latter an increase in density. Also the absolute values of V_p and V_s decrease with fayalite content much more dramatically than with temperature. Therefore temperature variations in the mantle, in principle, could be distinguished from variations in fayalite content.

All mantle minerals are to some degree anisotropic and show directional properties. Olivine, because of its orthorhombic symmetry, is markedly anisotropic in terms of properties such as optics, thermal conductivity, and elasticity. Those familiar with optical mineralogy will recall that orthorhombic crystals have three principal rays or refractive indices. In contrast, elastic properties belong to a fourth ranked tensor, which results in nine independent acoustic moduli for orthorhombic olivine. This in turn gives a 25 and 22% variation about the mean velocity of P- and S-waves, depending on crystallographic orientation of olivine (Anderson 1989). In fact, it is widely believed that parts of the shallow upper mantle (<200 km depth) are strongly anisotropic due to flow alignment of olivine and pyroxene crystals in peridotite. When present as a rock fabric, aligned anisotropic crystals can cause seismic velocity directional variations that are comparable to temperature or composition effects. As an example, natural dunites with aligned olivines can have acoustic velocity directional variations of ~5% (Anderson 1989).

Anisotropy is not limited to crystals of lower symmetry, even minerals of the isometric crystal system, which are optically isotropic, can display anisotropic acoustic behavior. A notable example is γ-spinel which is distinctly anisotropic for S-waves and moderately so for P-waves. Therefore, mineral anisotropy could potentially, have a significant effect on seismic velocity in the transition zone and elsewhere in the mantle. At present, there is no tangible evidence of significant seismic anisotropy in the transition zone. Of course this does not prove the absence of anisotropy; it may exist, especially in regions of upwelling, downwelling, and at shear-boundaries. For instance, subducted slabs in the transition zone may preserve preferred direction fabrics that were formed in the shallow mantle. Also, if the transition zone is convectively layered, then boundary regions between the layers could have flow-aligned anisotropy. Until such features are confirmed, it is assumed, for this presentation, that the bulk transition zone is, on average, an aggregate of randomly oriented grains, which cancels the effects of elastic anisotropy. Accordingly, seismic velocity with depth profiles presented in subsequent sections feature only acoustically isotropic behavior.

β-phase (wadsleyite). β-phase (mineralogic name: wadsleyite), like olivine, also belongs to the orthorhombic crystal system (Imma space group), with octahedrally coordinated Mg^{+2} and Fe^{+2}, and tetrahedrally coordinated Si^{+4}, however the O^{-2} are in nearly cubic-close-packed arrangement. β-phase is restricted to the Mg_2SiO_4-rich part of the Mg_2SiO_4-Fe_2SiO_4 binary system (Fig. 3) and is the high pressure polymorph that transforms from forsteritic olivines. The maximum Fe_2SiO_4 content of β-phase is ~25 mole percent. The polymorphic transformation of olivine to β-phase is widely accepted as the dominant mechanism in producing the seismic velocity jump at the 410 km discontinuity. Accordingly, accurate knowledge of the physical properties of β-phase can provide some of the most crucial constraints on the petrology of the transition zone.

Unfortunately, experimental determination of the properties of β-phase is much more challenging than for olivine. One of the main experimental hurdles is simply synthesizing sufficient quantities of β-phase to measure. In order to stabilize β-phase one must subject a sample of $(Mg,Fe)_2SiO_4$ to pressures of ~15 GPa at ~1000°C. Octahedral multi-anvil devices are commonly employed for this task. The synthesis experiment must allow enough time for the sample to react completely and this usually requires at least several hours at high pressure and temperature. The synthesized sample can then be extracted from the multi-anvil device and prepared for measurement. Perhaps the most challenging aspect of measuring elastic properties in β-phase is accomplishing this feat at high temperature. The temperature derivatives of bulk and shear moduli of β-phase may be difficult to determine at 1-bar, because the phase is metastable at low pressures. The structure of β-phase can be quenched from high-pressure synthesis experiments if this done at ~25°C. However, should the sample of β-phase be heated at 1-bar to say ~1000°C or higher, then it should readily convert back to olivine. Hence, studies to measure dK_S/dT and dG/dT of β-phase by the rectangular parallelepiped resonance method at very high temperatures (as in the olivine studies described above) are not feasible. Even acoustic measurements at 1-bar and room temperature could be plagued by metastable breakdown if the sample is disturbed sufficiently. A better strategy is to maintain the β-phase sample under confining pressure while the measurement is performed. This requirement restricts the experimental design to be accommodated within a piston-cylinder, multi-anvil, or a diamond-anvil device.

β-Mg_2SiO_4 has a density of 3.47 g cm^{-3} at 1-bar and 25°C, which is approximately 7.8% greater than the value for Mg_2SiO_4 olivine. Beta-phase density data were obtained from early x-ray diffraction studies on synthetic samples by Mizukami et al. (1975), Suzuki et al. (1980), and Horiuchi and Sawamoto (1981). Pioneering efforts by Sawamoto et al. (1984) provided some of the first measurements of β-phase elastic properties. Their data showed K_S = 174 GPa, a value 35% greater than that of olivine. More recently, Gwanmesia et al. (1990) determined dK/dP = 4.8 and dG/dP = 1.7 using ultrasonic pulse interferometry on hot-pressed polycrystals of β-phase in a piston-cylinder device. These experiments were performed at room temperature, so dK/dT and dG/dT were not determined. Interestingly, their experimentally determined values were very similar to the estimates from systematics of condensed silicates and oxides by Duffy and Anderson (1989). As new data are reported, it appears that the Duffy and Anderson estimates have been remarkably good for elastic moduli pressure and temperature derivatives. Fei et al. (1992) carried out x-ray diffraction measurements in a diamond-anvil cell on β-phase with composition Fo_{84} at simultaneous high pressure and high temperature (up to 800K and 26 GPa) and reported dK_T/dT = -0.027 GPa/°. Meng et al. (1993) performed x-ray diffraction measurements on Mg_2SiO_4 β-phase at simultaneous high P and T (872 K and 7.6 GPa) in a cubic anvil device and also determined dK_T/dT = -0.027 GPa/°. They recalculated this value to an isentropic dK_S/dT = -0.018 GPa/° using β-phase thermal expansion data of Suzuki et al. (1980). Again, this value is identical to that estimated by Duffy and Anderson (1989). At this time there have been no direct measurements of the temperature derivative of the β-phase shear modulus (dG/dT). In lieu of future work, the value dG/dT = -0.014 Gpa K^{-1}, from Duffy and Anderson (1989) is adopted.

Compositional effects on the β-phase elastic moduli are poorly known. Duffy and Anderson (1989) estimate that variations in the Fe/Mg have similar effects on β-phase shear modulus as is the case for olivine (Table 1). Of course, Fe_2SiO_4-rich β-phase is not stable, therefore large decreases in shear modulus due to composition should not be expected. It is interesting to note that when it coexists with olivine, in the two-phase field near the *410* (Fig. 3), β-phase has a higher Fe_2SiO_4 content. However, this compositional effect on seismic velocity in the mixed phase region is very small compared to the non-linear effect of the binary phase loop asymmetry (see below). Experiments have also shown that β-phase can accommodate trace amounts of the hydroxyl (OH⁻) anion. At present there are no data on the bulk and shear moduli of hydroxyl-bearing β-phase.

γ-spinel (ringwoodite). Gamma spinel is the highest pressure Mg_2SiO_4-Fe_2SiO_4 polymorph. A mineralogic form of γ-spinel (ringwoodite) was first discovered in the impact shocked Tenham meteorite. Gamma spinel is isostructural with $MgAl_2O_4$ spinel having isometric symmetry and belongs to the *Fd3m* space group. The olivine to spinel phase transition can be described as a rearrangement of oxygen atoms from hexagonal-close-packing to cubic-close-packing; no coordination change of Si, Mg, and Fe cations occurs. Figure 3 suggests that at temperatures below 1200°C, there exist complete solid solutions between Mg_2SiO_4 and Fe_2SiO_4 spinels within a narrow range of isobars near 20 GPa. At temperatures of 1200°C and greater, the solid solution is interrupted by Mg_2SiO_4-rich β-phase, olivine (lower pressures), perovskite + magnesiowüstite (higher pressures) and in the Fe_2SiO_4-rich part of the binary diagram by magnesiowüstite + stishovite. The transformation of β-phase to γ-spinel has important implications for the petrology of the transition zone because its presence should contribute to a rate of change increase of seismic velocities in the depth range 500-550 km.

Not surprisingly, the elastic properties of γ-spinel are less well known than for β-phase and olivine. Measurements become increasingly more difficult to perform with pressure. Following the earlier x-ray diffraction work of Mizukami et al. (1975) and Sasaki et al. (1982), Weidner et al. (1984) first used Brillouin scattering to determine the elastic moduli of a single crystal of γ-spinel that had been synthesized at 22 GPa and 2200°C at Nagoya University. Their measurements, performed at 1-bar and room temperature on a sample approximately 80 microns in diameter, gave $K_S = 184$ GPa and G = 119 GPa. The bulk modulus was significantly smaller than that of Mizukami et al. (1975), leading Weidner et al. (1984) to conclude that the β-phase to γ-spinel transition was not sufficient to cause the jump in seismic velocity associated with the 520 km discontinuity. They noted that the bulk and shear moduli of γ-spinel were only about 5% greater than β-phase and when combined with the density difference, amounted to an aggregate P and S-wave velocity difference of approximately 1%. Rigden et al. (1991) performed ultrasonic interferometry on synthetic polycrystalline aggregates of γ-spinel in a piston-cylinder device as a function of pressure up to 3 GPa at room temperature. From their measurements $dK_S/dP = 5$ and $dG/dP = 1.7$ were derived. Rigden et al. were impressed by the fact that the pressure derivatives of γ-spinel were very similar those of β-phase and olivine. This might lead one to assume that the temperature derivatives of the Mg_2SiO_4-Fe_2SiO_4 polymorphs are also similar. Ridgen et al. noted that if this assumption is valid, then the olivine content of the mantle at *410* would be lower than that in pyrolite. Furthermore, they calculated that a discrepancy exists in required olivine content between P- and S-wave data. Rigden et al. concluded that dG/dT for β-phase and γ-spinel should be a much larger negative value than for olivine in order to remedy the P- and S-wave discrepancy. They also preferred a larger dK_S/dT for β-phase and γ-spinel to be consistent with the olivine content of pyrolite (60-65%). As discussed above, however, more recent work has confirmed that β-phase dK_S/dT is similar to that of olivine (Fei et al. 1992, Meng et al. 1993). This demonstrates the weakness of the circular reasoning associated with requiring mineral elastic properties to conform to preconceived notions of the composition and structure of the mantle. Rigden et al. also contended that the if the dK_S/dT and dG/dT of γ-spinel were larger than that of olivine, then the *520* discontinuity could be explained as the β-phase to γ-spinel transition. Though this speculation cannot be ruled out, we prefer the systematic estimates of Duffy and Anderson (1989) in lieu of future work on dK_S/dT and dG/dT of γ-spinel. Below we explore the possibility that the ephemeral nature of *520* may be attributed to a small (~1%) P- and S-wave velocity difference between β-phase and γ-spinel.

THE "410" KILOMETER DISCONTINUITY

"...sudden increase in gradient at about 413 km depth, followed by a continual diminution in gradient within this transition layer until the fairly steady gradient of layer D is reached. The curves [P and S wave velocity distributions] suggest the fairly sudden occurrence of a proportion of new material immediately below layer B, the proportion increasing fairly rapidly in the upper part of layer C and the new material becoming fairly rapidly predominant."   Bullen (1940)

The 410 kilometer discontinuity is a relatively narrow layer at the top of the transition zone that is associated with an abrupt increase in seismic velocities. The *410* along with the *660* yield the firmest information on the mineralogy of the transition zone. If one assumes that the *410* is caused by the olivine to β-phase transformation, then seismic determinations can be translated into chemical composition and temperature. Specifically, the following four relationships exist:

- The size of the velocity jumps gives the olivine content at the *410*.
- The depth of the discontinuity (topography) gives the temperature at the *410*.
- Thickness of the discontinuity gives temperature and iron-content of olivine at the *410*.
- Absolute values of the seismic velocities on either side of the discontinuity give a range of solutions for the proportions of olivine (β-phase), pyroxene, and garnet at the *410*.

If the *410* is not caused by olivine to β-phase, then the first three of the above relationships do not hold. In that case the *410* could be rationalized by a change in rock type. This explanation has been proposed by Anderson and Bass (1986). Though a global compositional change at the *410* cannot be ruled out, the olivine to β-phase is favored by most workers because of its simplicity.

The size of the velocity jump at *410*

The sizes of P- and S-wave velocity jumps at the *410* are probably the best indicators of olivine content in that part of the transition zone (Duffy and Anderson 1989, Duffy et al. 1995). In forsteritic olivines, such as those present in peridotite xenoliths, the increase in velocities of olivine to β-phase is approximately 10% for compressional waves and 12% shear waves (see Fig. 9). Therefore, as examples, a *410* with 70% olivine should show an

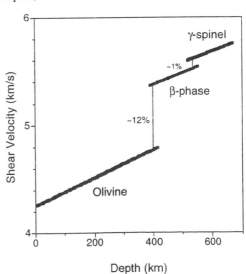

Figure 9. Shear wave velocity versus depth diagram illustrating the S-wave velocity jumps associated with the olivine to β-phase and the β-phase to γ-spinel. The velocities were calculated along a 1400°C adiabat after Duffy and Anderson (1989).

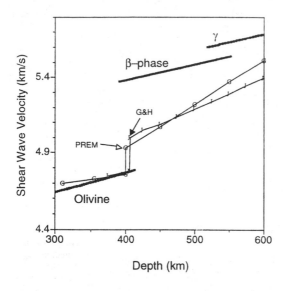

Figure 10. Shear wave velocity versus depth diagram illustrating the differences between S-wave velocity jump associated with the olivine to β-phase and the velocity jumps at the *410* in seismic models PREM (□) and Grand and Helmberger (●).

S-wave velocity jump of 8.4%, and a *410* with only 35% olivine should show a 4.2% change. This assumes that the other minerals present, such as pyroxenes and garnet, do not have abrupt changes in elastic properties nor do they change abruptly in modal abundance at the *410*.

Most 1-D global seismic velocity models have ΔV_P and ΔV_S at *410* that are too low for an olivine-rich transition zone. For instance, Walck (1984, 1985) models P-wave velocity increases at 390 and 410 km as 4.9% and 3.9% respectively. Grand and Helmberger (1984) assign an S-wave velocity increase to 405 km of 4.6% (Fig. 10). PREM (Dziewonski and Anderson 1981) has modest velocity jumps of 2.6% for P-waves and 3.6% for S-waves. These differences with the mineralogic $\Delta V_{(P,S)}$ have been used as evidence that the mantle in the vicinity of the *410* is depleted in olivine component (see, for example, Duffy and Anderson 1989). At first glance the $\Delta V_{(P,S)}$ discrepancy seems to require 50% or less olivine component at the *410*—39-49% olivine to match the P-wave model of Walck and 38% to match the S-wave model of Grand and Helmberger. The match with PREM requires only 26-30% olivine. This is significantly lower than the olivine content of pyrolite (~60%) or the average olivine content of upper mantle peridotite xenoliths and poses a serious challenge to homogeneous mantle models.

There are some possible explanations that would allow olivine contents >50% at the *410*. One way that this could be accomplished is by assigning β-phase a larger negative value for the temperature derivative of the shear modulus than the value of dG/dT = -0.14 for olivine. For example, if dG/dT = -0.20 for β-phase, then the S-wave velocity jump at 410 may be accounted for by an olivine component concentration greater than 50%. As mentioned above, dG/dT of β-phase has not been measured, therefore this possibility remains to be tested.

Another possibility is that the seismic models cited underestimate the size of the velocity jump at *410*. This is conceivable because these models partition the entire velocity increase through the transition zone between the *410* jump, the *660* jump, and the slope (dV$_{s,p}$/dz) between the *410* and *660*. If, for instance, the *660* jump were smaller in the seismic models or dV$_{s,p}$/dz smaller, then *410* jump could be adjusted to a larger value and,

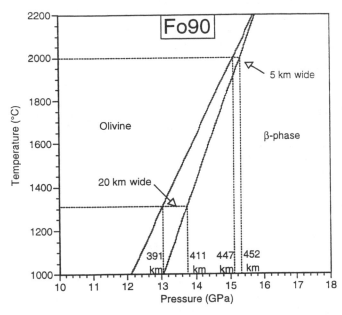

Figure 11. Temperature-pressure diagram showing the variable width of the olivine to β-phase transition based on experiments by Katsura and Ito (1989). At 1320°C the transition is approximately 20 km wide with a median depth of 401 km. At 2000°C the transition is approximately 5 km wide at a median depth of 450 km.

in turn, yield a higher olivine content.

More recently, examination of travel times and amplitudes of underside reflections from upper mantle discontinuities (SS precursors) using more than 13,000 high quality broad-band and long-period transverse component seismograms (Gu et al. 1998) shows that the jump at the *410* may be greater than previously thought. The new data show the S-wave velocity jump ratio of *410/660* (amplitude ratio) to be between 0.75 and 0.90 rather than the value of 0.54 in PREM or the Grand and Helmberger value of 0.62. One way of converting this ratio to a new revised *410* velocity jump is to assume that old model values for the *660* are correct and adjust only the *410* according to the higher ratio. Hence the revised S-wave velocity jump may be as much as 6.7% (the Grand and Helmberger corrected value), giving an olivine content of 56% at the *410*.

The thickness of the *410*

If the *410* exists largely as the result of olivine to β-phase, then the depth over which it occurs should correspond directly to the olivine to β-phase binary phase loop in the system Mg_2SiO_4-Fe_2SiO_4. In most seismic models the discontinuity is expressed as an abrupt jump in velocity over a relatively narrow width. Workers have undertaken short-period seismic reflection studies to image the sharpness of the velocity change at *410* (see, for example,: Petersen et al. 1993, Benz and Vidale 1993, Yamazaki and Hirahara 1994, Vidale et al. 1995, Neele 1996). These investigations have generally concluded that the *410* is quite narrow, perhaps only ~5 km thick. Surprisingly, this is too narrow to be in agreement with a discontinuity caused by olivine to β-phase in a mantle composition with Fo_{90}. Earlier we noted that this mineralogic phase transition covers a pressure interval equivalent to a thickness of 20 km at 400 km depth and 1320°C. However, Figure 11 shows that the width of the olivine to β-phase transition narrows with pressure (depth) and temperature. Hence, greater depths and higher temperatures are consistent with a thinner discontinuity. It is difficult to invoke a phase transition as narrow as 5 km for two important reasons. First, the temperature at which the olivine to β-phase transition is 5 km thick is approximately 2000°C. This temperature is near or above the solidus in Fo_{90} and in average peridotite

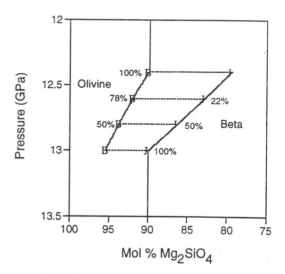

Figure 12. Pressure-composition diagram showing some isobaric tie-lines (dotted) for coexisting olivine/β-phase equilibrium pairs at 1200°C. At pressures less than 12.4 GPa only olivine is stable, at pressures greater than 13 GPa only β-phase is stable. At the intermediate pressures 12.6 and 12.8 GPa the phase mixes are 78% olivine + 22% β-phase, and 50% olivine + 50% β-phase respectively. The asymmetry of the two-phase loop should cause a non-linear increase in seismic velocities across the *410*. Phase boundaries are from Figure 3, bulk composition is Fo$_{90}$.

composition. Second, because of the positive P-T slope of the olivine to β-phase transition, a high temperature, narrow, transition would occur at around 450 km. This is approximately 30-50 km deeper than most models for the average depth of the *400*.

The thickness mismatch is enigmatic, however it has not yet caused most workers to abandon the olivine to β-phase transition as explanation for the *410*. The idea of a phase change at *410* is simple, elegant, and hence widely accepted. Proponents argue that there may be other effects which make the *410* appear thin. For example, the relationship between pressure and transformation volume is nonlinear in the olivine plus β-phase stability field (Fig. 12). This is caused by the asymmetric shape of olivine to β-phase phase loop and as a result, at 1200°C, 50% of the transformation actually occurs in the pressure interval 12.8 to 13.0 GPa or a width of 6 km. Whether this effect is sufficient to cause a sharp seismic discontinuity remains to be determined.

An possible way of reconciling the sharpness of the *410* with the Mg$_2$SiO$_4$-Fe$_2$SiO$_4$ phase diagram is to adjust the Fe/Mg of that part of the mantle to lower values. For pure forsterite (Mg$_2$SiO$_4$) the transformation occurs at a single pressure, creating a infinitesimally thin discontinuity. If the composition in the vicinity of the *410* were (Mg$_{0.95}$Fe$_{0.05}$)$_2$SiO$_4$-olivine (Fo$_{95}$), then a narrow discontinuity would be consistent with "expected" temperatures and depths (Fig. 13). For Fo$_{95}$ at 1350°C the transformation is spread over ~0.25 GPa, equivalent to a *410* of about 7 km thick. Such a requirement has far reaching implications because it suggests a heterogeneous or layered upper mantle—recall that on average the shallow (<200 km) peridotite mantle has ~Fo$_{90}$ olivines. This layering would be gravitationally unstable relative to the Fo$_{90}$ shallow mantle unless the iron depleted *410* also was enriched in a dense phase such as garnet.

A third explanation has been proposed by Stixrude (1997), where theory suggests that the shape of the olivine to β-phase loop is sensitive to presence of coexisting pyroxene and garnet. Stixrude argues that in such multicomponent system the width of the phase loop will be approximately half that in the simple system Mg$_2$SiO$_4$-Fe$_2$SiO$_4$. Future experimental work will hopefully test this promising hypothesis.

A fourth explanation for the relatively sharp *410* is that of compositional layering. In this case phase diagrams in the system Mg$_2$SiO$_4$-Fe$_2$SiO$_4$ have little relevance and

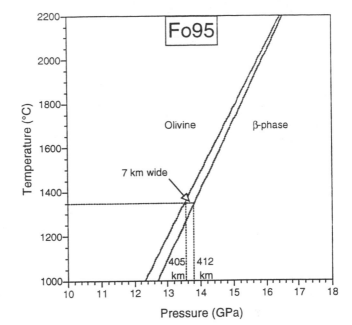

Figure 13. Temperature-pressure diagram showing the variable width of the olivine to β-phase transition based on experiments by Katsura and Ito (1989). At 1350°C the transition is approximately 7 km wide with a median depth of 408 km.

discontinuity sharpness is governed by abruptness of the change in rock type. The most commonly proposed layering at the *410* is eclogite or a garnet rich transition zone overlain by peridotite or olivine rich upper mantle (see, for example, Anderson and Bass 1986). In the very broadest of terms this is a reasonable explanation for seismic velocity profiles above and below the *410*, because olivine is good match for the upper mantle and garnet a good match the transition zone. This is also a petrologically appealing stratification because it could result from either primordial (magma ocean solidification) or current (subducted ocean crust) mantle differentiation. The main problem with a compositional boundary at *410* may be maintaining its sharpness and modest global topography in an actively convecting mantle. It is unclear how likely it is to have a uniformly sharp compositional boundary that would exist everywhere in the mantle at the same depth. This configuration may be a reasonable explanation for the *660*, but probably only if the upper and the lower mantles are in separate and isolated convection patterns. There is strong evidence that the *410* is not a barrier to convection (at least not in subduction zones). Hence, a sharp compositional boundary at the *410* may not exist. On the other hand, this does not rule out smaller amplitude radial and lateral compositional variations in the upper mantle and transition zone. Interestingly, a recent study of P-wave reflections in the transition zone beneath southern Africa (Xu et al. 1998) suggests that here *410* is wider than previously thought and indicating that *410* thickness may also depend on tectonic setting. If so, then one could argue that the thicker *410* beneath southern Africa could be caused by lower temperatures or higher iron content than the global average.

Absolute velocities near the *410*

Seismic models give absolute values for P- and S-wave velocities at transition zone discontinuities. At face value, this might seem the strongest constraint on mantle mineralogy because it requires a match for the combined elastic moduli and the elastic moduli pressure and temperature derivatives of all phases present. Unfortunately, these derivative quantities, especially those related to temperature, as mentioned above, are

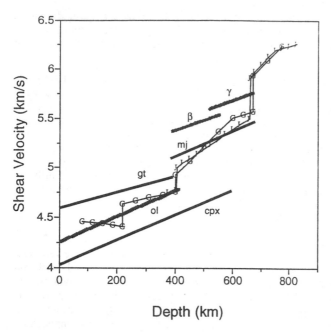

Figure 14. Shear velocity versus depth diagram comparing the velocities of upper mantle minerals with seismic velocity models PREM (□) and Grand and Helmberger (●).

ol = forsteritic olivine,
gt = pyropic garnet,
cpx = clinopyroxene,
mj = majorite garnet,
β = beta phase,
γ = gamma spinel.
Mineral curves are calculated along a 1400°C adiabat after Duffy and Anderson (1989). All mineral elastic properties are from Duffy and Anderson (1989) except majorite garnet, which is from Rigden et al. (1991).

poorly known. Furthermore, the mantle geotherm must be known in order to match seismic velocities with mineral velocities. Even with complete knowledge of mineral seismic velocities, and the temperature in the mantle, a unique solution for mantle mineralogy may not exist without some additional information or assumptions. This comes from the probability that three or more mineral phases with similar elastic properties may coexist near the *410*. In such cases it could be impossible to distinguish between various mixtures of the mineralogic components. As an example, consider the components olivine, pyroxene, and garnet. Figure 14 shows that above *410* a pure olivine mantle is quite a good match for seismic velocity models. On the other hand, because the seismic models lie between pyroxene and garnet a perfect match for this region is obtained by a particular mix of these two minerals with no olivine. But, with all three mineral phases present many combinations are possible and no unique solution is available. A similar situation exists below *410*. There the transition zone seismic velocity is well matched by a mantle made of pure majorite garnet. However, because the required velocity also lies between β-phase and pyroxene, a mantle of with a particular mix of these two minerals, and no majorite garnet, gives a perfect match. In short, absolute seismic velocities cannot be used to distinguish between a compositionally homogeneous upper mantle and an upper mantle with an abrupt compositional change at *410*—additional lines of evidence are needed.

Peridotite (pyrolite) at the *410*?

Peridotite xenoliths brought to the surface in volcanic eruptions or in tectonic uplift have been widely recognized as samples of the Earth's upper mantle. Peridotites span a relatively large compositional range that undoubtedly reflects the diverse petrologic histories of these rocks. They are usually a mineralogic mixture of the three major phases olivine, orthopyroxene, and clinopyroxene. Minor phases in peridotites can be plagioclase, spinel, garnet, hornblende, and phlogopite mica. Particular types of peridotite with approximately 60% olivine, 20% clinopyroxene, and 20% orthopyroxene, namely the lherzolites, have been proposed as examples of so-called "fertile" mantle. The term fertile implying that these peridotites, when partially melted, will produce a basaltic magma equivalent to that erupted at mid-ocean ridges. In order to supply large volumes of mid ocean ridge basalt (MORB) over the history of the Earth it has been hypothesized that much of the Earth's mantle must consist of fertile peridotite.

Ringwood took this concept a step further (see, for example, Ringwood 1975). He realized that even the most fertile peridotite samples had probably experienced some amount of igneous processing during their residence in the mantle. He therefore proposed a hypothetical precursor or primitive mantle composition called pyrolite. Pyrolite can be imagined as the original or primordial mantle composition before any igneous processing had taken place. The composition of pyrolite is nearly the same as the silicate material (or more precisely the refractory "lithophile" elements minus "siderophile" or iron loving elements now in the core) that accreted from the solar nebula to form the Earth. In Ringwood's model, pyrolite is not an igneous rock, rather, it is actually a metamorphosed solar nebula sediment. Furthermore, the composition of pyrolite was designed to account for MORB genesis. It is a mixture of 80% harzburgite (olivine + orthopyroxene peridotite) and 20% basalt.

Because the fertile peridotite and pyrolite models for the mantle have received wide acceptance in the geochemical and geophysical communities much experimental work has been carried out on these compositions. An important issue in the geochemical evolution of the Earth concerns the question of how much peridotite or pyrolite exists in the mantle. Is most of the mantle really equivalent to pyrolite composition or it simply representative of the shallow lithospheric mantle? The answer to this question has far reaching implications for understanding Earth formation and evolution.

Phase equilibria studies reveal the inventory of stable mineral assemblages in peridotite at high pressures and temperatures (see, for example, Akaogi et al. 1979, Takahashi 1986, Ito and Takahashi 1987, Takahashi and Ito 1987, Zhang and Herzberg 1994, Ohtani et al. 1995). These can be used to compare the seismic velocity of peridotite in the transition zone with seismic models.

Much of the phase equilibria data on peridotite is at temperatures near to or above the solidus. The main reason for this is that chemical equilibrium is more readily achieved if silicate melt is present in the sample. The melt acts as a flux for species transport during element exchange reactions. Therefore unzoned, near equilibrium crystals can be grown in a much shorter time interval when silicate melt is present. Such experiments can be as short as a few minutes. A secondary effect that enhances reaction kinetics are the higher temperatures of supersolidus conditions. Subsolidus experiments can require many hours, if not days, to reach chemical equilibrium. Unfortunately, success rate of longer experimental runs is much lower because of thermocouple contamination or mechanical failure of the tungsten carbide multi-anvil experiments. Another problem with performing subsolidus experiments on peridotite is that there are usually multiple phases which must equilibrate at a given pressure and temperature. Some phases will have slower diffusion, reaction, and crystal growth rates than others. Therefore it can be expected that larger errors may be present in subsolidus peridotite data than in simpler systems such Mg_2SiO_4-Fe_2SiO_4. The data from Zhang and Herzberg's KLB-1 peridotite experiments bracket the olivine to β-phase transition near the solidus. In contrast to some other peridotite studies, their data is in very good agreement with a linear extrapolation of the Katsura and Ito olivine to β-phase boundaries. In lieu of future work on subsolidus phase relations we adopt the Zhang and Herzberg results as a reasonable representation of likely transition zone mineralogies for peridotite composition (Fig. 15).

Figure 15 shows that olivine + clinopyroxene + garnet coexist above *410* (~13.4 GPa)

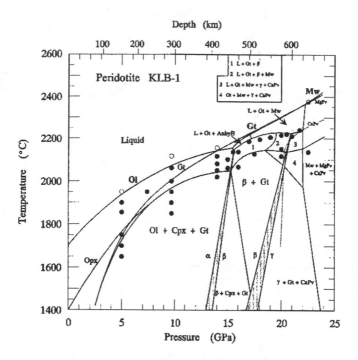

Figure 15. Peridotite P-T phase diagram from Zhang and Herzberg (1994).

in a peridotite mantle. The same phase assemblage is present below *410* except for the fact that β-phase is stable instead of olivine. If the composition of the mantle at *410* is peridotite, then these phases combined in their modal proportions should account for the seismic velocities and the discontinuity. As an example, Figure 16 gives the inventory of the phase proportions of an idealized pyrolite mantle. For simplicity, both olivine and β-phase are assumed to make up 60% of their respective stability regions. In reality, there maybe a small deviation from 60% for either phase due primarily to small differences in Fe-Mg distribution coefficients with majorite garnet and pyroxene. Near *410* the second most abundant phase is majorite garnet at ~25%, followed lastly by clinopyroxene at ~15%. In the regions above and below *410* the proportions of majorite garnet and clinopyroxene change significantly. At shallower levels clinopyroxene reaches a value of ~28% and garnet decreases to ~12%. Majorite garnet reaches its maximum of ~40% at ~500 km depth. Below 500 km depth majorite garnet decreases as $CaSiO_3$-perovskite becomes stable.

Figures 17 and 18 show S- and P-wave velocities in the vicinity of *410* calculated by combining the elastic properties of the phases given in Figure 16. The elastic properties and the method of calculation (along a 1400°C adiabat) is adopted from Duffy and Anderson (1989). The diagrams show, as mentioned above, that a composition with 60% olivine has a velocity jump that is too large to match widely cited seismic models. On the other hand, the 1400°C adiabat calculation gives an excellent match for the absolute values for model S- and P-wave velocities on the shallow side of *410*. This, however requires that the deep side of *410* to be faster than seismic models if it is composed of pyrolite. The choice of a 1400°C adiabat is fairly arbitrary as models based on MORB eruption temperatures and phase equilibria experiments may vary by several hundred degrees. If we select a lower

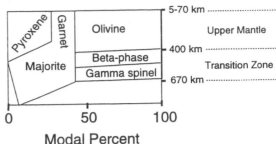

Pyrolite Mantle

Figure 16. Mineralogic make-up of an upper mantle based on the pyrolite model along a 1400°C adiabat. At 410 the phase distribution is 60% olivine/β-phase, 25% majorite garnet, and 15% pyroxene.

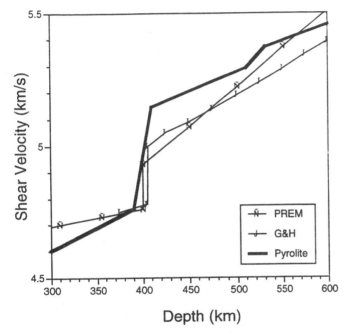

Figure 17. Shear velocity versus depth diagram comparing pyrolite velocity with seismic velocity models PREM (Ñ) and Grand and Helmberger (1984) (J). The bold line represents the velocity of pyrolite based on the mineralogic proportions in Figure 16. The method of velocity calculation is after Duffy and Anderson (1989).

temperature, 1300°C adiabat (see curve of Nisbet et al. 1994 in Fig. 7), which gives a temperature of ~1350°C at *410*, then both the shallow and deep sides of a pyrolite *410* would be too fast for seismic models. If an adiabat hotter than 1400°C is chosen, then the shallow side of *410* would be too slow and the deep side would, at a particular value be a good match for the seismic models.

Eclogite at *410*?

Peridotites are not the only rock type that come from the mantle. Eclogites, pyroxene-garnet rich rocks, make up about ten percent of mantle xenoliths. It is well established that basalt transforms to eclogite at high pressure, hence some or all of the eclogite xenoliths may be remnants of subducted basaltic crust that has resurfaced. Alternatively, some eclogite may be primitive in the sense that it formed during the early differentiation of the mantle. Could the dominance of peridotite xenoliths be due to a shallow level sampling bias? Could eclogite be more common in the transition zone? One way to test these possibilities is to determine whether the sharp discontinuity at *410* can be caused by phase

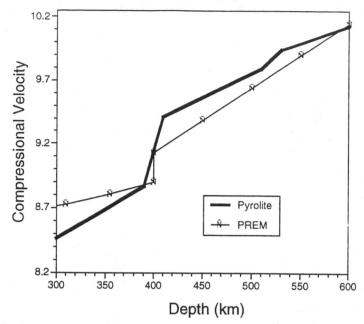

Figure 18. Compressional velocity (P-wave) versus depth diagram comparing pyrolite velocity with seismic velocity model PREM (open squares). The bold line represents the velocity of pyrolite based on the mineralogic proportions in Figure 16. The method of velocity calculation is after Duffy and Anderson (1989).

transformations in eclogite.

Phase equilibria experiments by Irifune (1987) and others show that there is strong pressure effect on the solid solution between orthopyroxene (low-calcium) and clinopyroxene (high-calcium) and between clinopyroxene and garnet. With increasing pressure a gradual increase in the solubility of orthopyroxene in clinopyroxene occurs. At a pressure of 8-9 GPa (275 km depth) orthopyroxene is no longer stable and the only pyroxene that exists is clinopyroxene. In the pressure range 10-17 GPa (~300 to 500 km depth) clinopyroxene undergoes a continuous solid solution reaction with pyropic garnet to form majorite garnet; clinopyroxene is no longer stable at ~500 km. It seems that neither of these transformations occurs over the narrow interval that defines the discontinuity at *410*. The orthopyroxene-clinopyroxene transformation can also be ruled out simply because it occurs at depth that is far too shallow for *410* and because there is a relatively small density and velocity difference between the two phases. Clinopyroxene is significantly less dense and "slower" than majorite garnet but the transformation is spread over ~200 km, while *410* is probably narrower than 10 km. It is concluded that a homogeneous eclogite mantle does not exist on both sides of *410*.

Compositional layering at *410*?

It is mentioned above that the simplest mineralogic match for seismic velocities in the upper mantle is olivine above *410* and garnet below *410*. This simple mineralogy is also a very good match for the velocity jump at *410* (Fig. 19). Naturally, the mantle is unlikely to be this simple, however the simple comparison is still useful because it illustrates that radial compositional heterogeneities are compatible with seismic velocity profiles. The major drawback in compositional layering at *410* is that it requires long-lived, globally isolated

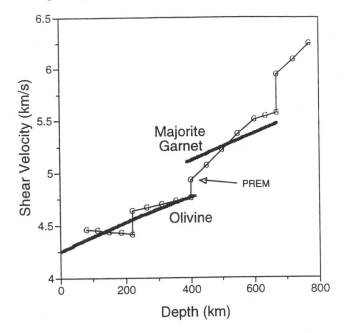

Figure 19. Shear wave velocities of olivine and majorite garnet compared to PREM. The mineral curves are calculated along a 1400°C adiabat after Duffy and Anderson (1989). Majorite garnet elastic properties are from Rigden et al. (1991).

convection patterns. There is strong evidence that subducted slabs pass relatively unhindered through *410*, hence such isolation seems unlikely. If there are compositionally distinct regions in the upper mantle, their interface probably is not the cause of *410*. This conclusion does not mean that the simple concept of shallow-olivine and deep-garnet upper mantle is false. It simply means that a sharp compositional boundary probably does not exist in the upper mantle. Ohtani (1985) and Agee and Walker (1988) argued that a garnet-rich transition zone and an olivine-rich shallow mantle are reasonable outcomes of primordial differentiation. If such compositionally distinct regions still exist then, they may grade into one another.

THE "520" KILOMETER DISCONTINUITY AND THE MID-TRANSITION ZONE

"There is no serious basis for supposing that the whole interior has the composition of olivine; about half the silicates of stony meteorites are pyroxenes."   Birch (1951)

The 520 kilometer discontinuity (hereafter *520*) is a relatively modest seismic discontinuity when compared to *410* and *660*. In fact, until recently *520* was not included in seismic velocity models (e.g. PREM). The pioneering seismic reflection work of Peter Shearer (1990) has led to a general acceptance that *520* does indeed exist, though there remains some controversy on its global extent.

The 520 kilometer discontinuity is probably a relatively narrow layer in the mid-transition zone. It is associated with a small increase in seismic velocity, however the width and depth of *520* are still not that well known. It may be similar or slightly wider than *410*.

Its depth may range from 500 to 540 km. In similar fashion to *410*, if one assumes that the *520* is caused by the β-phase to γ-spinel transformation, then seismic determinations can be translated into chemical composition and temperature. Because of its weak and ephemeral nature, *520* may actually be a very sensitive gauge of "olivine" content in that part of the mantle.

If *520* is not caused by β-phase to γ-spinel, then its presence, absence or size would give less direct information about the olivine content of the mid-transition zone. Other possibilities for *520* are (1) the phase transformation clinopyroxene + majorite garnet to majorite garnet + calcium silicate perovskite or (2) a change in rock type (i.e. compositional layering).

In addition to *520*, the mid-transition zone is characterized by seismic velocity profiles with particular slopes and trajectories. In seismic models these velocity slopes are largely governed by the size of the velocity jumps assigned to *410* and *660*. However, if these jumps were known very precisely, then the mid-transition velocity profiles would provide important constraints on composition and dynamics of that region. Below we discuss the significance of the apparent steep velocity slope suggested by seismic models of the transition zone.

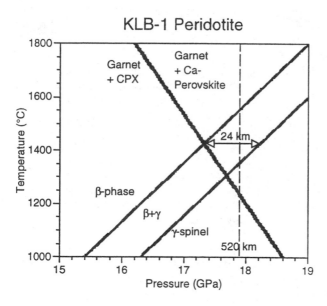

Figure 20. Portion of the KLB-1 peridotite P-T phase diagram after Zhang and Herzberg (1994) showing the phase boundaries in the vicinity of *520*. The width of the β-phase to γ-spinel is adopted from Katsura and Ito (1989).

Which phase transformation could cause *520*?

First let us assume that *520* is not a compositional layer, but is caused by a phase transformation in a locally homogeneous mantle. We refer to the peridotite P-T phase diagram to elucidate which phase boundaries could be responsible for this mild seismic discontinuity. Figure 20 illustrates the two likeliest possibilities: (1) β-phase to γ-spinel and (2) clinopyroxene + majorite garnet to majorite garnet + calcium silicate perovskite.

β-phase to γ-spinel has been discussed above and is probably the most widely accepted cause of *520*. The ratios G/ρ and K_S/ρ for γ-spinel are slightly larger than for β-phase (Table 1) giving a net increase in shear and compressional velocity with pressure across the phase boundary. Accordingly, if the mantle at this depth is rich in olivine component, then *520* should show a change in seismic velocity that is slightly less than a 1%. Figure 20 indicates that in peridotite the phase boundary is approximately 24 km wide and is at a temperature of ~1450°C at an equivalent depth of 520 km (17.9 GPa). These coordinates are in very good agreement with estimates of mantle adiabats discussed earlier. The very small velocity change associated with β-phase to γ-spinel has led several workers to doubt that this phase transition could be responsible for *520*. Initial results (Shearer 1990) on the size and the sharpness of the jump at *520* suggested that it could be as much as a 3% velocity change over a very narrow interval (only a few kilometers). This of course is in poor agreement with the β-phase to γ-spinel transition. Recent studies have still not placed firm constraints on the precise dimensions of *520*, however it appears that *520* is not universally detectable in underside reflection data of the transition zone (Gossler and Kind 1996, Gu et al. 1998). This suggests that the velocity change at *520* may indeed be as small as 1% or less.

Figure 20 shows that the phase boundary clinopyroxene + majorite garnet to majorite garnet + calcium silicate perovskite (Cpx+Gt to Gt+CaPv) occurs at a pressure equivalent to *520* if the mantle temperature there is ~1250°C. At first glance it may seem that Cpx+Gt to Gt+CaPv is a viable candidate to explain *520*. The K_S and G of CaPv are much higher than that of Cpx (Table 2). Therefore, at constant T-P-X, the right hand side of the transition reaction will have larger P- and S-wave velocities. Furthermore, the phase boundary occurs as a single line on the P-T phase diagram, hence the transition does not appear to be "smeared out" over some interval as is the case for β-phase γ-spinel. One might expect a sharp or abrupt discontinuity from this sort of phase boundary—interestingly, this is probably not the case.

Whether a discontinuity at *520* is produced or not will depend on two additional factors. First, the composition of the mantle in the vicinity of *520* must contain enough calcium to stabilize sufficient amounts of both Cpx and CaPv. As an extreme example, if *520* were very rich in $CaSiO_3$ and nearly all of the mineralogy were either Cpx or CaPv, then the seismic velocity jump at the phase boundary might be as much as 30%. Of course, the mantle is probably not particularly rich in $CaSiO_3$; CaO content of peridotite is approximately ~3 wt%. Phase equilibria studies show that CaPv makes up only about 5-10% of the total mineralogy in such compositions at the pressure appropriate to *520*. One would not expect much more of a seismic velocity jump than roughly 1-3%—similar to that of the β-phase to γ-spinel transition.

The second factor involves the partitioning of calcium in majorite garnet. Phase equilibria studies have also shown that Cpx becomes increasingly soluble in majorite garnet with pressure and as the phase boundary Cpx+Gt to Gt+CaPv is approached Cpx chemical components are increasingly accommodated in the majorite garnet structure (Fig. 21). Once the phase boundary is traversed, the $CaSiO_3$ in majorite gradually exolves to form CaPv. The exolution does not produce the maximum amount of CaPv right at the phase boundary—some additional pressure, possibly reaching 30 GPa or an equivalent depth of ~800 km is required for a CaPv maximum. Hence there is a significant pressure interval on either side of the phase boundary where majorite garnet makes up most of the mid-transition zone "non-olivine" mineralogy. The net effect of the gradual disappearance of Cpx and appearance of CaPv is to "smear out" the seismic velocity change associated with the Cpx+Gt to Gt+CaPv phase transition. One might expect a region that has a more rapidly changing seismic velocity with depth than "normal" mantle, but a distinct seismic discontinuity or reflector seems less likely.

There remains the possibility that both above mentioned phase transitions combine to cause *520*. Figure 21 shows that in such a case *520* should occur at a rather shallow depth

Table 2. Physical properties of some mid-transition zone phases
at 1 bar and 298 K (where applicable).

	Garnet $Mg_3Al_2Si_3O_{12}$ $Fe_3Al_2Si_3O_{12}$ $Ca_3Al_2Si_3O_{12}$	*Majorite* $MgSiO_3$ $FeSiO_3$	*Clinopyroxene* $CaMgSi_2O_6$ $CaFeSi_2O_6$	*Ca-perovskite* $CaSiO_3$
Density (g/cm³)	$3.561(X_{Py})$ + $4.318(X_{Al})$ + $3.617(X_{Gr})^a$	3.522^b + $0.973(X_{FeSiO3})$	3.277 + $1.73(X_{CaFeSi2O6})^d$	4.23^e
Thermal expansion $(K^{-1}) \times 10^{-5}$	2.38	2.38	2.55	2.5
Bulk modulus, K_S (GPa)	$175(X_{Py})$ + $176(X_{Al})$ + $169(X_{Gr})$	160^b	113	286^e
Shear modulus, G (GPa)	$90(X_{Py})$ + $98(X_{Al})$ + $104(X_{Gr})$	90^b	67	159^f
dK_S/dP	4.9	5.8^c	4.5	4.0
dG/dP	1.4	2.9^c	1.7	1.9
$-dK_S/dT$ (GPa/°)	0.021	0.021	0.013	0.027
$-dG/dT$ (GPa/°)	0.010	0.010	0.010	0.023
Gruneisen parameter, γ	$1.24(X_{Py})$ + $1.06(X_{Al})$ + $1.05(X_{Gr})$	1.24	$1.06-0.11(X_{CaFeSi2O6})$	1.96
Debye temperature, θ (K)	$981(X_{Py})$ + $909(X_{Al})$ + $904(X_{Gr})$	$949-129(X_{FeSiO3})$	$941-96(X_{CaFeSi2O6})$	917

Values are from Duffy and Anderson (1989) and Anderson (1989) and from the following references:
a. Skinner (1956), Akimoto and Akaogi (1977)   b. Pacalo and Weidner (1997)
c. Rigden et al. (1994)   d. Cameron et al. (1973)
e. Mao et al. (1989)   f. From Mg-perovskite
g. Assuming K/G = 1.8

between 504-528 km (17.3-18.2 GPa and 1450°C). With a width 24 km, this the narrowest possible discontinuity that could be caused by the combined phase transitions. If we assume a lower temperature of 1250°C, then the discontinuity would be shallower still and wider, spreading from 485-520 km (16.5-17.9 GPa).

Finally, it must be noted that a compositional change or layer could cause *520*. We discussed above the problem of maintaining a global stratification in the upper mantle. In the case of *410*, this is an unlikely scenario because of the continuous and relatively uniform nature of that discontinuity and the fact that subducted slabs penetrate it. As discussed in the next section, *520* is not always present or detectable in seismic reflection studies. The intermittent nature of *520* could indicate that it represents isolated lenses or layers of "fast" material rich in garnet and/or CaPv.

Is an intermittent *520* symptomatic of variable "olivine" content?

It has been recently argued that the velocity jump at *410* may be greater beneath mid-

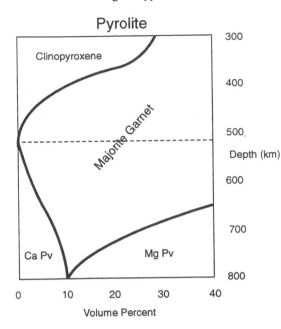

Figure 21. Portion of a diagram showing the mineralogic make-up of a pyrolite mantle. The diagram excludes the "olivine" component of pyrolite. The horizontal dashed line denotes *520*. The temperature is assumed to be abiabatic and have a value of 1250°C at *520*.

age ocean crust than beneath continental shields (Gu et al. 1998). One way of explaining these observations is that there are lateral variations in the olivine-component content of the upper mantle. The picture that is consistent with this idea is a suboceanic mantle at *410* rich in olivine relative to the subshield *410*. The data of Gu et al. also give information about the variable nature of *520*. Underside S-wave reflections (SdS) show the existence of a weak but detectable *520* in regions beneath mid-age ocean crust, but no reflectors are present (or detectable) at this depth beneath continental shields. The continental shield data of Gu et al. are dominated by bounce points beneath the North American craton and one could argue locally anomalous behavior. Interestingly, underside P-wave reflections beneath southern Africa (Xu et al. 1998) also fail to detect a *520*, whereas work by Shearer (1993) sampling the transition zone beneath the Indian Ocean, show a undisputed *520*. Though more work is needed to establish this ocean-shield dichotomy, it seems that surface tectonics and composition may influence or extend the to great depths in the mantle.

An intermittent *520* is compatible with the variable olivine content hypothesis used to explain variations in *410*. The seismic data suggest different amounts of olivine, but they do not give absolute values for olivine content. What might be a realistic range for olivine variation in the transition zone? As an example, if the suboceanic *520* is assumed to have the composition of pyrolite, then the olivine content of the mantle there would be ~60%, and the corresponding S-wave velocity increase for of β-phase to γ-spinel should be about 0.6%. Such an increase may just be resolvable with SdS reflection data and is represented in the feature at -3.25 minutes that Gu et al. show in their slowness profile of the suboceanic mantle. If the olivine content at *520* beneath continental shields is below the pyrolite level—perhaps ~30% of the mantle mineralogy, then it is conceivable that seismic imaging techniques would fail to detect the corresponding 0.3% increase in S-wave velocity. Hence the intermittent appearance of *520* may be a sensitive measure of compositional variations in the mid-transition zone.

The seismic reflection data of Gu et al. suggest distinct lateral variations in olivine content in the transition zone, but it is also interesting to consider the extent to which olivine varies radially in the upper mantle. It is possible that, for a given vertical column, the olivine content of *410* and *520* are equal and that radial variations, if they exist, are only in the shallow (<410 km) mantle. On the other hand, a subshield upper mantle that possesses an olivine-content gradient is also consistent with petrologic and seismic observations. Thus one could envisage a subshield mantle root changes in composition with depth. In other words, the top of the mantle, constrained by xenolith data is olivine- and pyroxene-rich (harzburgite). The harzburgite zone may extend to depths of 200 km or more, but by *410* is olivine-depleted. The depletion may increase further with depth and reach below *520*.

The $dV_{S,P}/dZ$ in the mid-transition zone

The seismic velocity gradient in the transition zone, should in principle, match the velocity combined velocity gradients of the minerals present is that region. At present, this is not the case—for example the S-wave gradient calculated for model peridotite between 410 and 660 km is much shallower than the gradients from the seismic models of PREM (Anderson and Dziewonski 1980, Grand and Helmberger (1984). Figure 22 illustrates this further by showing the gradients calculated for the mineralogic components in pyrolite. It is clearly shown that all of the components have shallower slopes than the seismic models.

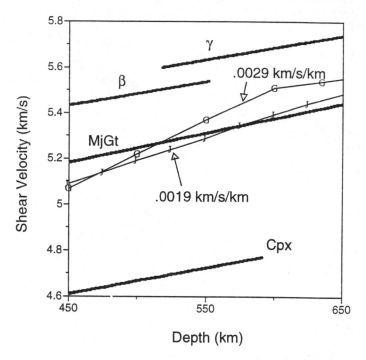

Figure 22. Shear velocity versus depth diagram showing the difference in slope between some transition zone seismic models and mantle minerals.

The increasing amount of high velocity material with depth associated with the conversion of cpx and garnet to the majorite structure had, at one time, been considered a possible candidate for to improving the gradient match. Experimental studies by Irifune et al. (1986) and Gasparik (1990) showed that this reaction is completed at depths shallower than 500 km and hence could not be called upon to produce the observed gradient throughout the transition zone. Also ruled out for explaining the gradient are the velocity increases associated with the β-phase to γ-spinel transformation which are now known to be confined within the range of ~505 to ~545 km (Katsura and Ito 1989). Duffy and Anderson (1989) speculated that the observed velocity increases with depth could be caused by a chemical gradient in the transition zone. Rigden et al. (1991) suggested that anelastic relaxation in the shallow transition zone might produce the observed steeper S-wave gradient, though they noted that the compositional heterogeneity proposal of Duffy and Anderson (1989) could also explain the steep gradient.

THE "660" KILOMETER DISCONTINUITY

The 660 km discontinuity forms the boundary between the upper mantle and lower mantle. Here, at this boundary, we conclude the discussion on the mineralogy of the upper mantle transition zone. Bina (this volume) discusses the hypotheses and geophysical constraints on the deeper part of the mantle. Before concluding however, some brief discussion of the *660* is in order.

The *660* is characterized by a sharp jump in seismic velocities, occurring over a narrow range, perhaps less than 5 km (Lees et al. 1983). The experiments of Ito and Takahashi (1989) showed that the abrupt discontinuity may be explained by an isochemical phase transformation of γ-spinel to $(Mg,Fe)SiO_3$-perovskite+magnesiowüstite. However, because the same transformation can occur in peridotite, chondrite, komatiite, and a number of other hypothetical mantle compositions, a chemical boundary or abrupt change in rock type at *660* may be difficult to distinguish from an isochemical *660*. The proportions of participating phases would be different for each composition, thus very accurate pressure and temperature derivatives of elastic moduli for $(Mg,Fe)SiO_3$-perovskite and magnesio-wüstite, if they were available, combined with accurate data on the size and width of *660*, might allow a unique compositional determination.

The size of the velocity jump at *660*, according to most seismic models, is roughly twice as large as that of *410*. For example, PREM gives ΔV_P and ΔV_S of 4.8 and 6.6% respectively for *660* and only 2.6 and 3.6% for *410*. In some seismic models the velocity discontinuity is followed by a relatively high velocity gradient between 650 and 750 km (Grand and Helmberger 1984). This gradient has been attributed to the phase transformation majorite garnet to $(Mg,Fe)SiO_3$-perovskite in presence of aluminum, which occurs over a larger pressure interval than of γ-spinel to $(Mg,Fe)SiO_3$-perovskite + magnesiowüstite.

If there is a transition zone barrier to convection, then it is more likely to be found at *660*, rather than at *410* or *520*. It is still unclear however, to what extent the 660 is a barrier to mantle convection or if it is a location of mantle compositional layering. Arguments for and against a *660* barrier come from thermodynamics, physical properties of mantle materials, geochemical models of the Earth, and seismic tomography. Compelling arguments favoring a convection barrier at 660 is the negative dT/dP slope of the γ-spinel to $(Mg,Fe)SiO_3$-perovskite+magnesiowüstite transformation (Fig. 8) and the effect of suducted slabs on the *660* topography.

Subducted slabs at *660*

Features of the transition zone that we know exist, but have been omitted from the discussion up to now, are of course, subducted slabs. A central question concerning slabs is their fate in the transition zone. Are they deposited near the *660* and contribute to the gross structure of that region or are they merely transient features that ultimately pass into the lower mantle?

Conditions in slabs, both thermal and chemical, are expected to be different than in surrounding mantle, and these factors will determine the relative buoyancy. For example the existence of positive and negative dT/dP phase boundaries can translate into a shallower and deeper phase transitions in a cold slab either enhancing or inhibiting descent relative to surrounding mantle. The chemical nature of the mantle below the 660 km will also affect the relative buoyancy of slabs. Jeanloz (1991) has pointed out that a small, but in terms of density, significant increase in ferrous iron content at 660 km would be difficult if not impossible to detect from seismic velocity data. Such a chemically induced density increase combined with the negative dT/dP associated with γ-spinel to perovskite+magnesiowüstite transformation might be sufficient to prohibit slab penetration into the lower mantle.

Ringwood and Irifune (1988) proposed an idealized slab with a 6.5 km top layer of basalt (cpx-rich) underlain by thicker layers of residual harzburgite (82% olivine, 18% opx) and pyrolite (62% olivine, 38% opx, cpx, and garnet). During descent the *pyrolite* layer is resorbed into the surrounding upper mantle by convective circulation and the slab enters the transition zone as a 30-40 km thick package of 20% basalt and 80% harzburgite. The model proposes that young, thin oceanic plates undergo rapid thermal equilibration and can be gravitationally trapped between 600 and 700 km. Thick, cold, mature slabs may pass through the *660*, however if they first encounter a layer of ancient trapped oceanic lithosphere between 600 and 700 km they will experience buoyant resistance. The cold slab buckles and thickens eventually forming a *megalith* that protrudes into and becomes entrained in a lower mantle, downward moving, convection current.

The Ringwood and Irifune hypothesis is supported by some tomographic imaging of subduction zones. van der Hilst et al. (1991) have shown that slabs beneath Japan and Izu Bonin arcs seem to be deflected at the 660 km discontinuity while the slabs beneath Kurile and Mariana descend into the lower mantle. Fukao et al. (1992) found that the subducting slab beneath Southern Kurile to Bonin arcs becomes subhorizontal near *660*, and extends in this orientation for more than 1000 km, where it connects to a high velocity blob at ~800 km depth. Global topography near *660* has been reported by Shearer and Masters (1992) and more recently by Gu et al. (1998), that is consistent with the *megalith* of Ringwood and Irifune. They propose that the broad regional depressions observed in the *660* are correlated with the locations of subduction zones, though whether deeper penetration in to the lower mantle occurs or not, remains equivocal. Woodward et al. (1992) interpret the *660* as a boundary that may, under some circumstances, be breached by upwelling or downwelling convection currents, and under other circumstances, be a impenetrable barrier that sustains layering. For example, Grand (1994) has argued that such a subduction breach of the *660* is found to extend well into the lower mantle beneath South America. It is easy to imagine that a transition zone with layered structure as envisaged in some mantle models may be preserved in particular regions of the Earth's interior, while in other locations, the 660 may have been breached by slabs. If any slabs are gravitationally trapped near the *660*, then chemical layering would, by definition, exist in the transition zone.

REFERENCES

Agee CB, Walker D (1988) Mass balance and phase density constraints on early differentiation of chondritic mantle. Earth Planet Sci Lett 90:144-156

Akaogi M, Akimoto S (1979) High-pressure phase equilibria in a garnet lherzolite, with special reference to Mg^{2+}-Fe^{2+} partitioning among constituent minerals. Phys Earth Planet Inter 15: 31-51

Akaogi M, Ito E, Navrotsky A (1989) Olivine-modified spinel-spinel transitions in the system Mg_2SiO_4-Fe_2SiO_4: calorimetric measurements, thermochemical calculation, and geophysical implication. J Geophys Res 94: 15671-15685

Akimoto S (1987) High-pressure research in geophysics: past, present, and future. In: Manghnani MH, Syono Y (eds) High-Pressure Research in Mineral Physics, p 1-13, Terra, Tokyo

Anderson DL (1989) Theory of the Earth. 366 p, Backwell, Boston

Anderson DL, Bass JD (1986) Transition region of the Earth's upper mantle. Nature 320:321-328

Bernal JD (1936) Geophysical discussion. Observatory 59:268

Benz HM, Vidale JE (1993) The sharpness of upper mantle discontinuities determined from high-frequency reflections. Nature 365:147-150.

Bina CR, Wood BJ (1987) Olivine-spinel transitions: experimental and thermodynamic constraints and implications for the nature of the 400-km discontinuity J Geophys Res 92:4853-4866

Birch F (1951) Elasticity and constitution of the Earth's interior. Trans New York Acad Sci Ser II, 14: 72-76

Brown JM, Shankland TJ (1981) Thermodynamic parameters in the Earth as determined from seismic profiles. Geophys J R Astron Soc 66:579-596

Bullen KE (1940) The problem of the Earth's density variation. Bull Seismo Soc Am 30:235-250

Cameron ME, Sueno S, Prewitt CT, Papike JJ (1973) High-temperature crystal chemistry of acmite, diopside, hedenbergite, jadeite, spodumene, and ureyite. Am Mineral 58:594-618

Dziewonski AM, Anderson DL (1981) Preliminary Reference Earth Model. Phys Earth Planet Inter 25:297-356

Duffy TS, Zha CS, Downs RT, Mao HK, and Hemley, RJ (1995) Elastic constants of forsterite Mg_2SiO_4 to 16 GPa. Nature 378: 170-173

Duffy TS, Anderson DL (1989) Seismic velocities in mantle minerals and the mineralogy of the upper mantle. J Geophys Res 94:1895-1912

Fei, Y., Mao, HK, Shu JF, Parthasarathy G, Bassett, WA (1992) Simultaneous high P-T x-ray diffraction study of β-$(Mg,Fe)_2SiO_4$ to 26 GPa and 900K. J Geophys Res 97:4489-4495

Fukao Y, Obayashi M, Inoue H, Nenbai N (1992) Subducting slabs stagnant in the mantle transition zone. J Geophys Res 97:4809-4822

Gaherty JB, Jordan TH, Gee LS (1996) Seismic structure of the upper mantle in a central Pacific corridor. J Geophys Res 101:22291-22310

Gasparik T (1990) Phase relations in the transition zone. J Geophys Res 95:15751-15769

Gossler J, Kind R (1996) Seismic evidence for very deep roots of continents. Earth Planet Sci Lett 138: 1-13

Grand SP (1994) Mantle shear structure beneath the Americas and surrounding oceans. J Geophys Res 99:11591-11621

Grand SP, Helmberger DV (1984) Upper mantle shear structure of North America. Geophys J R Astron Soc 76:399-438

Gu Y, Dziewonski AM, Agee CB (1998) Global topography of transition zone discontinuities: deep shadows of the continents? Earth Planet Sci Lett 157:57-68

Gwanmesia GD, Rigden S, Jackson I, Liebermann RC (1990) Pressure dependence of elastic wave velocity for β- Mg_2SiO_4 and the composition of the Earth's mantle Science 250:794-79

Haggerty SE, Sautter V (1990) Ultradeep (greater than 300 kilometers), ultramafic upper mantle xenoliths. Science 248:993-996

Hervig RL, Smith JV, Dawson, JB (1986) Lherzolite xenoliths in kimberlites and basalts: petrogenetic and crystallochemical significance of some minor and trace elements in olivine, pyroxenes, garnets, and spinel. Trans Royal Soc Edinburgh: Earth Sciences 77:181-201

Herzberg C (1992) Depth and degree of melting of komatiites. J Geophys Res 97:4521-4540

Horiuchi H, Sawamoto H (1981) β-Mg_2SiO_4: single-crystal x-ray diffraction study. Am Mineral 66:568-575

Irifune T (1987) An experimental investigation of pyroxene-garnet transformation in a pyrolite composition and its bearing on the constitution of the mantle. Phys Earth Planet Inter 45:324-336

Ito E (1977) The absence of oxide mixture in high-pressure phases of Mg-silicates. Geophys Res Lett 4:72-74

Ito E, Matsui Y (1978) Synthesis and crystal-chemical characterization of $MgSiO_3$ perovskite. Earth Planet Sci Lett 38:443-450

Ito E, Takahashi E (1989) Postspinel transformations in the system Mg_2SiO_4-Fe_2SiO_4 and some geophysical implications. J Geophys Res 94:637-646

Ito E, Akaogi M, Topor L, and Navrotsky A (1990) Science 249:1275-1278.

Isaak D (1992)

Isaak D, Anderson OL, Goto T (1989) Elasticity of single crystal forsterite measured to 1700 K. J Geophys Res 94:5895-5906

Jagoutz E, and eight co-authors (1979) The abundances of major, minor, and trace elements in the Earth's mantle as derived from primitive ultramafic nodules. Proc Lunar Planet Sci Conf X: 2031-2050

Jeanloz R (1991) Effects of phase transitions and possible compositional changes on the seismological structure near 650km depth. Geophys Res Lett 18:1743-1746

Jeffreys H (1936) On the materials and density of the Earth's crust. Proc. Nat. Acad. Sci. 15:50-61

Katsura T, Ito, E (1989) The system Mg_2SiO_4-Fe_2SiO_4 at high pressures and temperatures: precise determination of stabilities of olivine, modified spinel, and spinel. J Geophys Res 94:15663-15670

Kawada K (1977) The system Mg_2SiO_4-Fe_2SiO_4 at high pressures and temperatures in the earth's interior. PhD thesis, University of Tokyo

Kumazawa M, Anderson OL (1969) Elastic moduli, pressure derivatives of single crystal olivine and single crystal forsterite. J Geophys Res 57:227-286

Lees AC, Bukowinski MST, Jeanloz R (1983) Reflection properties of phase transition and compositional change models of the 670 km discontinuity. J Geophys Res 88:8145-8159

Liu L (1976) The post-spinel phases of forsterite. Nature 262:770-772

Maaloe S, Aoki K (1977) The major element composition of the upper mantle estimated from the composition of lherzolites. Contrib Mineral Petrol 63:167-173

Mao HK, Chen LC, Hemley RJ, Jephcoat AP, Wu Y, Bassett WA (1989) Stability and equation of state of $CaSiO_3$-perovskite to 134 GPa. J Geophys Res 94:17890-17894

Meng Y, Weidner DJ, Gwanmesia GD, Liebermann RC, Vaughan MT, Wang Y, Leinenweber K, Pacalo RE, Yeganeh-Haeri A, Zhao, Y (1993) In situ high P-T X-ray diffraction studies on three polymorphs (α,β,γ) of Mg_2SiO_4 J Geophys Res 98:22199-22207

Mercier JC, Carter NL (1975) Pyroxene geotherms. J Geophys Res 80:3349-3362

Mizukami S, Ohtani A, Kawai N, Ito E (1975) High-pressure x-ray diffraction studies of β- and γ-Mg_2SiO_4. Phys Earth Planet Inter 10:177-182

Morishima H, Kato T, Suto M, Ohtani E, Urakawa S, Utsumi W, Shimomura O, T. Kikegawa, T (1994) The phase boundary between α- and β-Mg_2SiO_4 determined by in situ x-ray observation. Science 265:1202-1203

Neele F (1996) Sharp 400-km discontinuity from short-period P reflections. Geophys Res Lett 23:419-422

Nisbet EG, Cheadle MJ, Arndt NT, Bickle MJ (1993) Constraining the potential temperature of the Archaean mantle: a review of the evidence from komatiites. Lithos 30:291-307

Ohtani E (1979) Melting relation of Fe_2SiO_4 up to about 200 kbar. J Phys Earth 27:189-208

Ohtani E (1985) The primordial terrestrial magma ocean and its implication for stratification of the mantle. Phys Earth Planet Inter 38:70-80

Ohtani E, Nagata Y, Kato T (1995) Melting relations of peridotite and density cross over in planetary mantles. Chem Geol 120:207-222

Ohtani E, Kumazawa M (1981) Melting of forsterite Mg_2SiO_4 up to 15 GPa. Phys Earth Planet Inter 27:32-38

Pacalo REG, Weidner DJ (1997) Elasticity of majorite, $MgSiO_3$ tetragonal garnet. Phys. Earth Planet Inter 99:145-154

Petersen N, Vinnik L, Kosarev G, Kind R, Oreshin S, Stammler K (1993) Sharpness of the mantle discontinuities, Geophys Res Lett 20: 859-862

Presnall DC, Gasparik T (1990) Melting of enstatite ($MgSiO_3$) from 10 to 16.5 GPa and the forsterite (Mg_2SiO_4)- majorite ($MgSiO_3$) eutectic at 16.5 GPa: implications for the origins of the mantle. J Geophys Res 95:15771-15777

Presnall DC, Walter MJ (1993) Melting of forsterite, Mg_2SiO_4, from 9.7 to 16.5 GPa. J Geophys Res 98:19777-19783

Rigden SM, Gwanmesia GD, Fitz Gerald JD, Jackson I, Liebermann RC (1991) Spinel elasticity and seismic structure of the transition zone of the mantle. Nature 354:143-145

Ringwood AE (1958) The constitution of the mantle II. Further data on the olivine-spinel transition. Geochim Cosmochim Acta 26:18-29

Ringwood AE (1962) Prediction and confirmation of the olivine to spinel transition in Ni_2SiO_4 Geochim Cosmochim Acta 26:457-469

Ringwood AE (1963) Olivine-spinel transformation in cobalt orthosilicate. Nature 198:79-80

Ringwood AE (1975) Composition and petrology of the Earth's mantle. p 618 McGraw-Hill, New York

Ringwood AE (1979) Origin of the Earth and Moon. p 295 Springer Verlag, New York

Ringwood AE, Irifune T (1988) Nature of the 650-km seismic discontinuity: implications for mantle dynamics. Nature 331:131-136

Ringwood AE, Major A (1966) Synthesis of Mg_2SiO_4-Fe_2SiO_4 spinel solid solutions. Earth Planet Sci Lett 1:241-245

Sasaki C, Prewitt CT, Sato Y, Ito E (1982) Single-crystal x-ray study of γ-Mg_2SiO_4. J Geophys Res 87:7829-7832

Sawamoto H, Weidner DJ, Sasaki S, Kumazawa M (1984) Single crystal elastic properties of the modified (beta) phase of magnesium orthosilicate. Science 224:749-751

Shearer PM (1990) Seismic imaging of upper-mantle structure with new evidence for a 520-km discontinuity. Nature 344:121-126

Shearer PM (1993) Global mapping of upper mantle reflectors from long-period SS precursors. Geophys J Int'l 115:878-904

Shearer PM, Masters GT (1992) Global mapping of topography on the 660-km discontinuity. Nature 355:791-796

Skinner BJ (1956) Physical properties of end-members of the garnet group. Am Mineral 14:428-436

Stixrude L (1997) Structure and sharpness of phase transitions and mantle discontinuities. J Geophys Res 102:14835-14852

Sumino Y (1979) The elastic constants of Mn_2SiO_4, Fe_2SiO_4, and Co_2SiO_4 and the elastic properties of olivine group minerals at high temperatures. J Phys Earth 27:209-238

Sumino Y, Nishizawa O, Goto T, Ohno I, Ozima M (1977) Temperature variation of elastic constants of single-crystal forsterite, between $-190°C$ and $900°C$. J Phys Earth 25:377-392

Suzuki, I, E. Ohtani E, Kumazawa M (1980) Thermal expansion of modified spinel β-Mg_2SiO_4. J Phys Earth 28:273-280

Takahashi E (1986) Melting of a dry peridotite KLB-1 up to 14 GPa: implications on the origin of peridotitic upper mantle. J Geophys Res 91:9367-9782

Takahashi E, Ito E (1987) Mineralogy of mantle peridotite along a model geotherm up to 700 km depth. In: Manghnani MH, Syono Y (eds) High-Pressure Research in Mineral Physics, p 427-437, Terra, Tokyo

van der Hilst R, Engdahl R, Spakman W, Nolet G (1991) Tomographic imaging of subducted lithosphere below northwest Pacific island arcs. Nature 353:37-43

Vidale JE, Ding XY, Grand SP (1995) The 410-km-depth discontinuity: a sharpness estimate from near-critical reflections, Geophys Res Lett 22:2557-2560

Walck MC (1984) The P-wave upper mantle structure beneath and active spreading center the Gulf of California. Geophys J R Astron Soc 76:697-723

Walck MC (1985) The upper mantle beneath the north-east Pacific rim: A comparison with the Gulf of California Geophys J R Astron Soc 81:243-276

Webb SL (1989) The elasticity of upper mantle orthosilicates olivine and garnet to 3 GPa. Phys Chem Mineral 16:684-692

Weidner DJ, Sawamoto H, Sasaki S, Kumazawa M (1984) Single-crystal elastic properties of the spinel phase of Mg_2SiO_4. J Geophys Res 89:7852-7860

Woodward RL, Forte AM, Dziewonski AM, Peltier WR (1992) Scrutinizing mantle tomographic images. Eos Trans Am Geophys Union 73:317

Xu F, Vidale JE, Earle PS, Benz HM (1998) Mantle discontinuities under southern Africa from precursors to P' P'$_{df}$. Geophys Res Lett 25:571-574

Yamazaki A, Hirahara K (1994) The thickness of upper mantle discontinuities, as inferred from short-period J-Array data. Geophys Res Lett 21:1811-1814

Yoneda A, Morioka M (1992) Pressure derivatives of elastic constants of single crystal forsterite. In: Syono Y, Manghnani MH (eds) High-Pressure Research: Application to Earth and Planetary Science, p 207-214 Terra Scientific/Am Geophys Union, Washington, DC

Zaug JM, Abramson EH, Brown JM, Slutsky LJ (1993) Sound velocities in olivine at Earth mantle pressures. Science 260:1487-1489

Zhang J, Herzberg CT (1994) Melting experiments on anhydrous peridotite KLB-1 from 5.0 to 22.5 GPa. J Geophys Res 99:17729-17742

Chapter 6
LOWER MANTLE MINERALOGY AND
THE GEOPHYSICAL PERSPECTIVE

Craig R. Bina

Department of Geological Sciences
Northwestern University
Evanston, Illinois 60208

INTRODUCTION

A variety of observations (Jeanloz 1995, Helffrich and Wood 1996, Irifune and Isshiki 1998, Agee this volume) suggest that the upper mantle may be largely peridotitic in bulk composition, perhaps approaching the composition of the model pyrolite (Ringwood 1989). What can be said about the composition and mineralogy of the lower mantle? Certainly we expect the mineralogy of the lower mantle to differ from that of shallower regions, if only due to high-pressure phase transformations, but why might the bulk composition of the lower mantle differ from that of the overlying material? Early partial melting of the mantle (Herzberg and O'Hara 1985) may have resulted in large-scale differentiation between upper and lower mantle, although elemental partitioning data appear not to support such a model (Kato et al. 1988). Diffusive (Garlick 1969, Bina and Kumazawa 1993) or convective (Weinstein 1992) processes acting across regions of phase transition may have generated chemical separation between upper and lower mantle, but such processes would occur over inordinately long time scales in the absence of fluid phases (Mao 1988). Perhaps more fundamentally, if chondritic meteorites (Anders and Grevesse 1989) are taken as representative of the cosmochemistry of the solar nebula from which the planet condensed, then a bulk earth whose whole mantle is of pyrolite composition (Ringwood 1989) is deficient in silicon. Unless the excess silicon was taken up in the core or volatilized during planetary accretion (Ringwood 1975), or unless the chondritic model is not an appropriate model for the bulk earth (McDonough and Sun 1995), then the lower mantle should be enriched in silica relative to a pyrolitic upper mantle. Early models of solar cosmochemistry indicated an iron deficit, suggesting that the lower mantle might be enriched in iron relative to the upper mantle (Anderson 1989a), but subsequent calibration of photospheric spectra (Holweger et al. 1990, Biémont et al. 1991) brought solar abundances into agreement with those of chondrites, thus removing the cosmochemical argument for iron enrichment.

Aside from a few diamond inclusions that appear to represent low-pressure, back-transformation products of lower mantle mineral assemblages (Kesson and Fitz Gerald 1991, Harte and Harris 1994, Harte et al. 1994), the lower mantle is not amenable to direct sampling. Thus, the mineralogy and composition of this region must be inferred via geophysical remote sensing, from (for example) observations of seismic waves and electric fields at the surface. The processing of raw geophysical data lies beyond the scope of this review, but an outline of how geophysical models are constructed (and exposure to their associated terminology) can give valuable insight into potential uncertainties. Furthermore, the models (e.g. three-dimensional seismic velocity structures) which result from inversion of these data can be used to construct models of lower mantle composition and mineralogy. Ideally, the resulting lower mantle mineralogical models could then be tested by using them to predict the primary geophysical observations, in the sort of forward modeling approach already applied in the transition zone (Helffrich and Bina 1994). Many uncertainties remain in our understanding of the geophysical properties of minerals under lower mantle conditions, so that conclusions tentatively drawn

0275-0279/98/0037-0006$05.00

herein may be subject to change in the wake of future measurements. However, while we may not yet be "in possession of the talismans which are to open to thee the mineral kingdoms and the centre of the earth itself" (Beckford 1786), the topics reviewed herein should provide a framework for addressing unanswered questions as more data become available.

RADIAL DENSITY AND VELOCITY PROFILES

Geophysical background

Deep earth structure is investigated seismologically by studying the propagation of waves excited by earthquakes (or nuclear explosions). These can be thought of as body waves (generally with periods of order 1 s), traveling along specific paths from source to receiver, or as superpositions of normal modes of free oscillation (generally with periods of order 100 s) of the entire planet. (They can also be thought of as surface waves, but these are seldom employed for lower mantle study.)

1-D velocities. The speeds with which body waves travel through the earth's interior are simply related to the density and elastic moduli of the material:

$$V_P^2 = (K_S + \tfrac{4}{3}\mu) / \rho \ , \quad V_S^2 = \mu / \rho$$

where V_P is compressional wave (P-wave) velocity, V_S is shear wave (S-wave) velocity, ρ is density, μ is shear modulus, and K_S is adiabatic bulk modulus (used instead of the isothermal modulus K_T because the characteristic time of the passage of seismic waves is far shorter than that of thermal diffusion). P-waves, therefore, travel faster than S-waves in the same medium. When comparing the results of seismological studies to laboratory analyses of minerals, it is sometimes desirable to extract K_S in isolation from μ, because the former is often better constrained experimentally under deep mantle conditions than the latter. To this end, the "bulk sound velocity" V_ϕ is defined:

$$V_\phi^2 \equiv V_P^2 - \tfrac{4}{3} V_S^2 = K_S / \rho$$

Although no wave actually travels at this speed in the solid earth (Bina and Silver 1997), V_ϕ can be thought of as the speed of a sound wave in an equivalent liquid (because μ therein is always zero), and it can be measured in the laboratory from the slope of the (adiabatic) static compression curve. As mineralogical equations of state for $\mu(P,T)$ improve in quality, use of V_ϕ in earth models should decline in favor of the more physically realizable V_P and V_S.

An initial goal of body-wave studies is to determine functions $V_P(z)$ and $V_S(z)$ which describe the variation in seismic wavespeed with depth z. Because velocities generally increase with depth, seismic waves refract and bend toward the surface, with more the steeply descending waves traveling a greater distance from the source before reaching the receiver (Fig. 1a). Since the bending of the wave paths depends upon the velocity gradients, the dependence of the travel-times of particular seismic waves (e.g. direct S-waves, core-reflected P-waves, etc.) between source and receiver upon the distance between source and receiver (sometimes plotted as a "τ-p" rather than a time-distance relation) can be inverted for models of velocities as functions of depth (Lay and Wallace 1995). Like most inversions, the resulting velocity models are non-unique. Because they exhibit correlated uncertainties, one cannot draw simple errors bars on the velocity profiles; a change in velocity at one depth must be compensated by a different change at another depth. Furthermore, such inversions can exhibit strong dependence upon the starting model, so that a discontinuity in the initial profile, for example, may not vanish in the final profile even if its presence is not required by the data. Thus, uncertainties are more often illustrated by plotting several different velocity profiles that all fit the data equally well, sometimes beginning from different starting models (Kennett et al. 1995, Montagner and Kennett 1996). In addition, the sensitivity of seismic waves to structure is frequency-dependent, so that shorter wavelengths are required to resolve finer features of the velocity profiles.

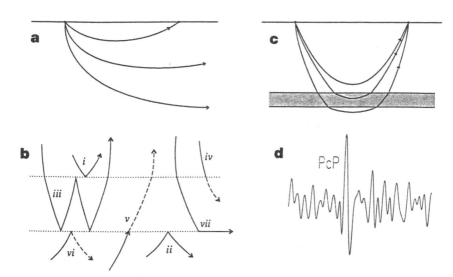

Figure 1. Seismic waves: Ray paths refract due to increase of velocity with depth (a). For sharp velocity gradients, deeper paths undergo greater refraction, resulting in a triplication (b). Boundary interactions phases (c) may reflect (i, ii), reverberate (iii), convert between P and S (iv, v), both reflect and convert (vi), or diffract (vii) at velocity discontinuities. Complete waveforms (d) contain more information than simply the travel times of identifiable waves (e.g. *PcP*).

Fine structure. Fine features such as rapid velocity changes can be resolved by study of additional properties of travel-time vs. distance behavior. At a given source-receiver distance, for example, a single direct *P*-wave will generally arrive at the receiver, and more steeply descending waves will arrive at more distant receivers (Fig. 1a). In a zone of rapid increase in velocity, however, more steeply descending waves will have their paths bent much more strongly than the more shallowly descending waves, so that two or three distinct pulses, separated in time, may arrive successively at the receiver from the gradient zone (Fig. 1b), in a phenomenon known as a "triplication." It was by observing such triplications that the major seismic velocity discontinuities in the mantle were first detected (Byerley 1926, Jeffreys 1936). A region of decrease in velocity, on the other hand, bends the wave paths oppositely, resulting in a "shadow zone" over which no pulse at all arrives from the gradient region (Lay and Wallace 1995).

Triplications and shadow zones arise by refraction of seismic waves as they pass through gradient zones. More detailed investigation of rapid changes in velocities with depth can be performed through study of "boundary interaction phases" (Clarke 1993): waves which are reflected, converted, or diffracted at gradient zones (Fig. 1c). Thus, waves may reflect off the topside (i) or underside (ii) of a boundary (Shearer 1991a), with underside reflections often appearing as precursors to waves which penetrated the boundary only later to reflect off of the underside of some shallower boundary (such as the earth's surface). Waves which reflect multiple times between boundaries (iii) are known as "reverberations," and because they spend so much time in the region between boundaries they can serve as sensitive probes of the properties of such zones (Revenaugh and Jordan 1991a,b). When a *P*- or *S*-wave is incident upon a boundary, the individual wave may give rise to both transmitted *P*- and *S*-waves (with the restriction that *S*-waves cannot propagate in liquids) in a process known as "mode conversion," and such converted waves may arise on either downgoing (iv) or upgoing (v) paths. Such conversions may also generate reflected *P*- and *S*-waves, so that a boundary interaction phase may be both converted and reflected (vi). Finally, a wave incident upon a boundary may be "diffracted" so that it travels horizontally along the boundary itself (vii) before rising once again to a receiver (Wysession et al. 1992, Silver and Bina 1993, Garnero et al. 1993). Because such waves undergo fundamental changes in character at the boundary, their properties are sensitive to those of the boundary itself rather than only to average properties across a broader region, thus making them valuable probes of mantle fine structure.

For some time, seismologists focused upon the travel times of individual, identifiable waves, such as the distinct "*PcP*" reflection of a *P*-wave off of the core-mantle boundary (Fig. 1d). While study of such travel times reveals much about deep earth structure, much more information is contained in the complete waveforms, information such as the relative amplitudes of waves and the existence of interaction phases from previously unsuspected boundaries. Originally, synthetic waveforms were calculated only in a forward modeling approach, for relatively simple comparison with observed waveforms, but advances in numerical modeling now permit actual iterative "waveform inversion" for earth structure (Nolet 1990, Zielhuis and Nolet 1994, van der Lee and Nolet 1997).

All of these aspects of body-wave analysis can be brought to bear upon the study of rapid changes in seismic velocity or "discontinuities." For such features, important questions concern the depth, polarity (or sign), sharpness (or width), and magnitude (or brightness) of the associated velocity contrast. The depth at which a velocity contrast occurs can be determined from travel time vs. distance observations, but such estimates depend upon the models assumed for shallow velocity structure (Walck 1984). Better depth measurements can be obtained from boundary interaction phases by comparison to reference waves which undergo no boundary interaction, by comparison to reference waves with free surface interactions, or by measuring the differential times between arrivals of different boundary interaction phases. The polarity of a velocity contrast, whether it is a velocity increase or decrease, can be determined by looking for triplications or shadow zones in time-distance curves, or by comparing the upswing or downswing of boundary interaction phases to those of reference waves with either no boundary interaction or a free surface interaction. The sharpness of a velocity contrast, in terms of the depth extent over which it is largely complete, can be estimated from the frequency range over which observable boundary interaction phases are detectable: broader boundaries will not reflect well at shorter wavelengths. Sharpness can also be estimated from the spatial range over which a triplication is observable (Melbourne and Helmberger 1998), although this requires dense spatial coverage such as that afforded by an array. While relative amplitudes of boundary interaction phases can also provide sharpness information, their interpretation is generally model-dependent. Finally, the magnitude of a velocity contrast can be estimated from the change in slope of time-distance curves at triplications, but it is better determined from the amplitudes of boundary interaction phases relative to reference waves. Such amplitudes yield impedance contrasts, so that assumptions about densities are required to convert them to velocities, but the dependence of reflection coefficients upon density is relatively weak for waves with grazing paths (Lay and Wallace 1995). Both discontinuity sharpness and magnitude are generally more difficult to resolve than depth and polarity. Furthermore, the apparent magnitude, sharpness and depth measured by these methods will depend upon the frequencies of the seismic waves employed (Helffrich and Bina 1994).

Global averages. Radial velocity profiles can be constructed for specific regions (Grand and Helmberger 1984), or an attempt can be made to construct a globally-averaged profile (which may not correspond to real structure at any particular point on the earth). Global profiles generally suffer from a bias toward the structure beneath continents, because most receivers are located on land. Global profiles are also designed for differing purposes. While some (e.g. SP6 of Morelli and Dziewonski 1993) may represent attempts to model the actual globally-averaged structure of the earth, many (e.g. IASP91 of Kennett and Engdahl 1991, AK135 of Kennett et al. 1995) are designed primarily as "machines" for generating travel times for waves useful in locating earthquakes. Indeed, these latter often come with warnings attached, advertising their "relatively weak constraints on the details of the velocity distribution," noting that a "variety of styles of model give a similar level of fit to the data," and urging caution in interpreting them as any sort of "simple average of the earth" (Kennett and Engdahl 1991). The properties of discontinuities, in particular, are usually poorly resolved by such models. A picture of the global extent and properties of discontinuities can, however, be obtained by global stacking of a large number of seismograms in such a manner that signals representing laterally coherent structure are mutually reinforced while incoherent noise is damped through destructive interference (Shearer 1991b, Astiz et al. 1996).

Information on deep earth structure which is complementary to that obtained from body wave studies can also be gleaned by analysis of the earth's normal modes of oscillation as recorded on seismograms. Normal modes are equivalent to superpositions of many different body waves, and both the frequencies and amplitudes of these waves determine the resultant normal modes. As noted above in the context of body waves, the amplitudes (controlled by various reflection and transmission coefficients) depend upon impedance contrasts rather than simply velocity contrasts. Thus, measurements of the eigenfrequencies of such free oscillations allow construction of radial profiles of density (Masters 1979) as well as of P- and S-wave velocities (and also provide constraints upon anisotropy and anelasticity, as discussed later). Furthermore, modes of varying degrees (or spatial frequency) possess different profiles (or "kernels") of depth-sensitivity (Lay and Wallace 1995). Uncertainties in lower mantle ρ of the order of 0.10 g/cm^3 may arise due to choice of starting model and treatment of anisotropy and attenuation (Montagner and Kennett 1996). While such analyses yield good V_S resolution (1066A and 1066B of Gilbert and Dziewonski 1975), the low frequencies of most normal mode studies generally fail to resolve uncertainties in the fine structure of velocity and density profiles (Montagner and Kennett 1996).

Body wave data may be added to normal mode data in joint inversions for radial structure, thus providing improved V_P resolution as well (PREM of Dziewonski and Anderson 1981). However, differences in geographic biases, frequency bands, and sensitivity to anelasticity and anisotropy generally lead to systematic misfits with one data set or another. Recently, methods of reconciling the body wave and normal mode data sets have been explored, by first using body waves to constrain radial structure and discontinuities and then using the eigenperiods of normal modes to extract density, anelasticity, and anisotropy structure (Montagner and Kennett 1996).

Mineralogical interpretation

Given models of the radial variations of ρ, V_P, and V_S in the lower mantle, our goal is to calculate velocity and density profiles for a variety of candidate compositions, so as to ascertain the compositions whose predicted elastic properties best fit the seismological models. This may be done either by "adiabatic decompression" of lower mantle properties for comparison to aggregate mineral properties at zero pressure or by extrapolation of mineral properties to lower mantle conditions. We adopt the latter approach here, so as to minimize the problems of parameter covariance associated with the former (Bukowinski and Wolf 1990), but recent work has shown how consistent results may be obtained from both approaches (Jackson 1998a). In accordance with a large body of experimental data, we shall begin by assuming that the lower mantle consists largely of assemblages of (Mg, Fe)SiO$_3$ silicate perovskite and (Mg,Fe)O magnesiowüstite, so that we can characterize a candidate composition by the two molar ratios:

$$X_{Mg} = Mg / (Mg + Fe) \quad , \quad X_{Pv} = Si / (Mg + Fe)$$

Equations of state. For each candidate composition, we wish to calculate densities and seismic wave velocities at a variety of lower mantle pressures and temperatures, so we require thermoelastic equations of state for the relevant minerals (Bina and Helffrich 1992, Duffy and Wang this volume, Liebermann this volume). Fundamentally, the required equations of state (EoS) must provide volume V as an explicit or implicit function of pressure P and temperature T. From such a $V(P,T)$ function, we can calculate both density ρ, from $\rho \equiv M / V$ where M is the molar mass, and bulk modulus K_T, from $K_T \equiv -V(\partial P / \partial V)_T$. A variety of EoS parameterizations may be employed, but it is common to write an EoS for the volume implicitly as $P(V,T)$, thus allowing the following differential expansion:

$$P(V,T) - P(V_0, T_0) = \int_{V_0}^{V} \left(\frac{\partial P}{\partial V}\right)_{T_0} d\hat{V} + \int_{T_0}^{T} \left(\frac{\partial P}{\partial T}\right)_V d\hat{T}$$

The first term on the right-hand side is often called the "cold" pressure, and its integrand can be written as $-K_T / \hat{V}$, evaluated at the initial temperature T_0, where K_T is the isothermal bulk modulus. The integral is usually evaluated via a finite strain formalism, such as the third-order Birch-Murnaghan equation:

$$P_{cold} = 3K_{T0}f(1 + 2f)^{5/2}(1 - \xi f)$$

$$f \equiv \tfrac{1}{2}[(V_0 / V)^{2/3} - 1] \quad , \quad \xi \equiv -\tfrac{3}{4}(K'_{T0} - 4)$$

where the reference isothermal bulk modulus K_{T0} and its pressure derivative K'_{T0} are evaluated at T_0 and V_0. Higher-order versions of the Birch-Murnaghan equation may also be employed. Similarly, a logarithmic strain formalism (Poirier and Tarantola 1998) could be used instead of this Eulerian finite strain formalism.

The second term on the right-hand side is generally called the "thermal" pressure, and its integrand can be written as αK_T or, equivalently, as $\gamma C_V / V$, evaluated at the final volume V, where α is the volume coefficient of thermal expansion, γ the Grüneisen parameter, and C_V the isochoric heat capacity. The integral can be evaluated using either notation, but the former generally leads to a class of EoS involving integrals of the Anderson-Grüneisen parameter,

$$\delta_T \equiv \frac{-1}{\alpha K_T}\left(\frac{\partial K_T}{\partial T}\right)_P = \frac{-K_T}{\alpha}\left(\frac{\partial \alpha}{\partial P}\right)_T$$

and some empirical form of polynomial expansion of $\alpha(T)$, such as:

$$\alpha(T) \approx \alpha_0 + \alpha_1 T + \alpha_2 T^{-2}$$

Arbitrary polynomial expansions can exhibit pathological behavior (e.g. negative temperature derivatives) at high temperatures and thus may not be optimal for extrapolating beyond the range of experimental data, although the example given above works well for $\alpha_0 > 0$, $\alpha_1 > 0$, and $\alpha_2 < 0$ because inflection points then lie in the domain of negative temperatures.

The latter notation yields a different class of EoS, perhaps the simplest of which is the Mie-Grüneisen-Debye type:

$$P_{thermal} \approx \frac{\gamma_D}{V}[E_{th}(V,T) - E_{th}(V,T_0)]$$

$$E_{th}(V,T) \equiv 9nRT\left(\frac{\theta_D}{T}\right)^{-3} \cdot \int_0^{\theta_D/T} \frac{t^3}{e^t - 1}\,dt$$

$$\theta_D \equiv \theta_{D0}(V_0/V)^{\gamma_D}, \quad \gamma_D \equiv \gamma_{D0}(V_0/V)^{-q}$$

where θ_{D0} and γ_{D0} are the reference Debye temperature and Grüneisen parameter, respectively, and q gives the volume dependence of the latter. This lattice dynamical approach to the EoS need not be restricted to the form of a Debye model, however. The thermal pressure can also be calculated from an explicit density of states, obtained from spectroscopic data, for example (Navrotsky 1994). Because such formalisms attempt to model the lattice dynamical behavior of crystals at high temperatures, they can prove useful for extrapolating beyond the range of experimental data for Debye-like (Anderson 1998) or more complex solids. Alternatively, a thermal pressure EoS can be obtained from simulation techniques such as molecular dynamics.

Derivation of such mineral equations of state involves fitting values to unknown coefficients by inverting from $V(P,T)$ measurements. Since these coefficients represent various derivatives of V, care must be take to ensure that coverage of the P-T space is adequate to constrain the values in question. Thus, while a given range of $V(P)$ data may be sufficient to constrain volume (and therefore density) over a certain range of pressure, the same data may not be sufficient to constrain bulk modulus (and therefore seismic velocity) or other P- and T-derivatives over that range. Such EoS uncertainties can often be fruitfully examined by determining the trade-offs among the various unknown coefficients to be fit to the data (Bell et al. 1987, Bina 1995).

Ideally, we would also like to have an additional equation of state for the shear modulus, $\mu(P,T)$, to allow us to calculate independent P- and S-wave velocities, V_P and V_S, at a variety of lower mantle pressures and temperatures. Such relations for shear moduli, however, currently remain relatively poorly constrained under lower mantle conditions. For this reason, the bulk sound velocity V_ϕ is often employed instead of the P- and S-wave velocities, because V_ϕ is independent of the shear modulus, as noted above. (The adiabatic and isothermal bulk moduli, employed in seismic velocities and equations of state, respectively, are related by $K_S = K_T(1 + T\alpha\gamma)$). Because shear moduli in many cases appear to be significantly more sensitive to variations in temperature and chemistry than bulk moduli, however, important advances in our understanding of the lower mantle can be expected to accompany improved constraints upon shear EoS in the future.

P, T, and K profiles. Having obtained suitable equations of state, for $V(P,T)$, $K_T(P,T)$, and possibly $\mu(P,T)$, we also require a profile of pressure $P(z)$ as a function of depth, which we can calculate by radial integration of the equation of hydrostatic equilibrium, $dP/dr = -\rho(r)g(r)$, where $\rho(r)$ is obtained from a seismological reference model such as AK135 (Kennett et al. 1995). Using the same $\rho(r)$ profile, $g(r)$ is calculated from Newton's law of gravitation, $g(r) = GM(r)/r^2$, where G is the gravitational constant and $M(r)$ the mass of that portion of the earth contained within a sphere of radius r.

Next, we require a profile of temperature $T(z)$ as a function of depth. It is common to begin by assuming such temperature profiles to be approximately adiabatic:

$$(\partial T / \partial P)_S = \alpha VT / C_P = \gamma T / K_S$$

Thus, given a starting temperature at the top of the lower mantle, we need merely perform a depth integration of:

$$dT / dz = g\alpha T / C_P = g\gamma T / V_\phi^2$$

where any slightly superadiabatic gradient can be accommodated simply by introducing a superadiabaticity factor:

$$\Delta T = \Delta P(\partial T / \partial P)_S(1 + f_{sup})$$

The olivine phase diagram indicates that, for the breakdown of γ silicate spinel (ringwoodite) to coincide with the 660-km seismic discontinuity, the temperature at 655 km should be approximately 1900 K (Ito and Katsura 1989). We can allow for uncertainty in this estimate by examining three different cases, where the starting temperature T_{LM} at 660 km is either 1800, 2000, or 2200 K.

Finally, because both $(Mg, Fe)SiO_3$ silicate perovskite (pv) and $(Mg,Fe)O$ magnesiowüstite (mw) are solid solutions, we must constrain the elemental partitioning (Fei this volume) of Mg and Fe between these two coexisting phases. While the relevant molar partitioning coefficient,

$$K_{Mg-Fe}^{pv-mw} \equiv \frac{X_{Mg}^{mw} / X_{Fe}^{mw}}{X_{Mg}^{pv} / X_{Fe}^{pv}} = \frac{(1 - X_{Fe}^{mw})X_{Fe}^{pv}}{(1 - X_{Fe}^{pv})X_{Fe}^{mw}}$$

is commonly assumed to remain constant, we allow $K_{Mg-Fe}^{pv-mw}(P,T,X_{Mg})$ to vary by performing a trilinear fit to recent experimental partitioning data (Mao et al. 1997) that exhibit a dependence on P, T, and X_{Mg} (Fig. 2).

A worked example. Our fitting procedure is now straightforward. First, we assume a value of T_{LM}. Then we choose a candidate composition, characterized by values of X_{Mg} and X_{Pv}, to test. At each depth z we determine P and T. We then calculate K_{Mg-Fe}^{pv-mw} and thereby obtain X_{Fe}^{mw} and X_{Fe}^{pv}. Using these mineral proportions, along with the (Debye-like) EoS functions for each of the component minerals, we compute ρ for the mineral assemblage and the aggregate V_ϕ (or, ideally, V_P and V_S). We compare the latter to $\rho(z)$ and $V_\phi(z)$ from the reference seismological model and calculate the root-mean-square (RMS) misfit. We repeat this procedure for a set of (X_{Mg}, X_{Pv}) compositions and perhaps for several different values of T_{LM}.

There are two simple ways of examining the resulting set of misfit values to determine best-fitting lower mantle compositions. First, we can ask which uniform composition yields the minimum misfit summed over the full depth range of the lower mantle. This method provides, for each value of T_{LM}, two sets of ellipses in X_{Mg}-X_{Pv} space within which compositions yield RMS misfits to the seismological models for $\rho(z)$ and $V_\phi(z)$, respectively, that fall below some threshold misfits (Fig. 3). The orientations of the major axes of the misfit ellipses reveal the relative sensitivities of the two parameters: misfit to ρ is primarily sensitive to changes in X_{Mg} but not X_{Pv}, while misfit to V_ϕ is sensitive to changes in either. This graphically illustrates the fact that attempts to fit mantle density alone cannot constrain the silica content of the lower mantle; it is the intersection of these two ellipses which provides tight bounds on allowable mantle compositions. Indeed, the $\rho(z)$ and $V_\phi(z)$ misfit ellipses can be combined in an RMS sense (with, for simplicity, equal weighting) to give a single ellipse within which all compositions yield a joint RMS misfit to the seismological model that falls below some chosen threshold (Fig. 4). The variation in best-fit composition with changes in T_{LM} shows that, in general, a hotter lower mantle requires a greater silica content in order to match globally averaged seismological models. Changes in assumed mineral EoS functions will, of course, also affect the resulting best-fit compositions, as will changes in reference seismic models (Birch 1952, Jackson 1983, Knittle et al. 1986, Jeanloz and Knittle 1989, Bina and Silver 1990, Stixrude et al. 1992, Hemley et al. 1992, Zhao and Anderson 1994, Jackson and Rigden 1998, Bina and Silver 1997, Jackson 1998a).

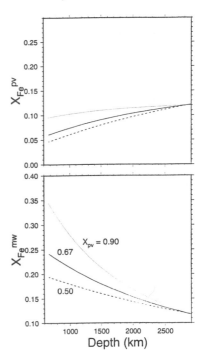

Figure 2 (right). Depth-dependent Fe-Mg partitioning between silicate perovskite (pv) and magnesiowüstite (mw). Mole fractions of Fe in coexisting pv and mw, for three bulk compositions of varying silica content, are derived from the data of Mao et al. (1997).

Figure 3 (below, left). Contours of RMS misfit between seismologically derived and mineralogically calculated profiles of $\rho(z)$ (dashed) and $V_\phi(z)$ (solid), constraining allowable lower mantle compositions in terms of iron and silica content. Contour interval is 0.5%. Triangle denotes pyrolite composition. T_{LM} is 2000 K. Perovskite thermal EoS from line 2 of Table 1 of Bina (1995) with cold EoS from Fabrichnaya (1995).

Figure 4 (below, right). Contours of joint RMS misfit between seismologically derived and mineralogically calculated profiles of $\rho(z)$ and $V_\phi(z)$, combined in an equally-weighted RMS sense, constraining allowable lower mantle compositions in terms of iron and silica content. Contour is 1.0%. Triangle denotes pyrolite composition. T_{LM} is 1800, 2000, and 2200 K (left to right). Perovskite thermal EoS from line 2 of Table 1 of Bina (1995) with cold EoS from Fabrichnaya (1995).

RMS% misfit to ρ & V$_\phi$

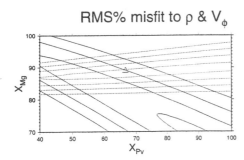

Joint RMS% misfit to ρ & V$_\phi$

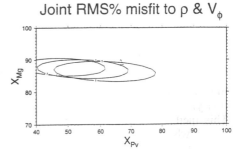

Second, we can ask which composition yields the minimum misfit at each individual depth, thus revealing how both the apparent best-fit lower mantle composition and the overall quality of the fit vary with depth (Fig. 5a). For T_{LM} of 2000 K, the effective value of $K_{Mg-Fe}^{pv-\bar{m}w}$ rises steadily from around 0.25 to near 1. While the best-fit value of X_{Mg} remains near that of an upper mantle pyrolite composition, the best-fit value of X_{Pv} begins near that of pyrolite and then falls gradually to lower values. The quality of the fit remains good throughout. The gradual drift in best-fit X_{Pv} value with increasing depth probably reflects, at least in part, the failure of our chosen EoS to faithfully describe the elastic behavior of lower mantle minerals as we extrapolate to great depths beyond the range of measured data. For a markedly different (non-Debye-like) EoS for silicate perovskite (Fig. 5b), the best-fit value of X_{Mg} remains near that of an upper mantle pyro-lite composition, with somewhat greater scatter. The best-fit value of X_{Pv}, on the other hand, begins significantly enriched in silica relative to pyrolite and then falls steeply to lower values. The quality of the fit gradually improves to about 1700 km depth and remains good thereafter. The steep decline in best-fit X_{Pv} value with increasing depth probably also reflects faults in the extrapolatory behavior of our EoS, but the problems with this EoS appear to be more severe than in the previous case. Such defects are also suggested by the surprisingly poor quality of fit for the initially high values of X_{Pv}, even at relatively shallow depths.

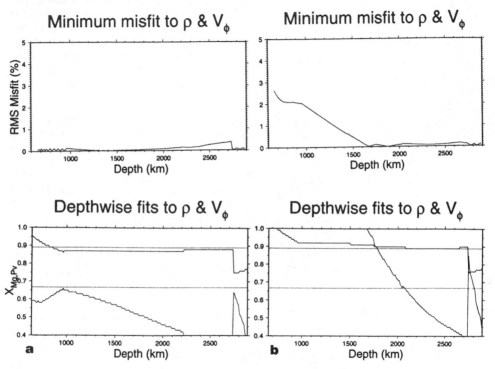

Figure 5. Joint RMS misfit between seismologically derived and mineralogically calculated ρ and $V_\emptyset$, combined in an equally-weighted RMS sense, and best-fit compositions, in terms of iron and silica content, all as functions of depth in the lower mantle, for T_{LM} of 2000 K. Fitting performed for Debye-like (a) and non-Debye-like (b) perovskite equations of state (thermal EoS from line 2 and line 1, respectively, of Table 1 of Bina 1995, with cold EoS from Fabrichnaya 1995). Dotted lines denote pyrolite composition.

It is clear from these analyses that the best-fit compositions deviate markedly at the very top and bottom of the lower mantle. This is not surprising, because thermal, miner-alogical, and chemical phenomena in these regions give rise to significant uncertainties and lateral variations in the velocity gradients and discontinuities of seismological mod-els. Complex structure in the topmost lower mantle may arise from several sources, including temperature-induced topography on the phase boundary marking the top of the lower mantle (Bina and Helffrich 1994), possible stability of silicate ilmenite in the top-most lower mantle in cold regions of subduction (Reynard et al. 1996, Reynard and Rubie 1996, Irifune et al. 1996), and extension of the garnet stability field into the topmost lower mantle due to Al-Fe^{3+} coupling between ringwoodite, silicate perovskite, and gar-net phases (Wood and Rubie 1996). The bottommost lower mantle (Jeanloz and

Williams this volume) is believed to be a region of chemical changes, arising from interactions with the core or accumulation of subducted material, and may also feature additional high-pressure phase changes. Hence, these two regions are generally omitted from the calculations when using the first method detailed above to determine which uniform composition yields the minimum misfit summed over the lower mantle.

Further work. Many secondary effects have been neglected in the above analyses. For example, consideration of oxygen fugacity and defect chemistry may introduce Fe^{3+} substitution for Fe^{2+} or non-stoichiometry in magnesiowüstite, and charge-coupled substitutions may allow uptake of Al and Fe^{3+} and possibly Na into silicate perovskite. The effects of such solid solution phenomena on elastic properties remain poorly constrained at present. The accessory cubic $CaSiO_3$ perovskite phase has also been ignored here, but recent studies (Wang et al. 1996) suggest that its effects upon the aggregate elastic properties of lower mantle assemblages should be minimal. Given suitable measurements of elastic properties for such phases, these effects could be accounted for in future analyses by incorporation of additional molar ratios, such as X_{Ca} or X_{Al}, into the compositional parameterization.

Other potential phase changes have also been neglected in these analyses. Both $(Mg, Fe)SiO_3$ and $CaSiO_3$ perovskites may undergo structural distortions under deep mantle conditions, and high-pressure phase changes observed in FeO may also occur in lower mantle $(Mg,Fe)O$. Stabilization of higher pressure phases of SiO_2 or Al_2O_3 or pressure-induced changes in the maximum solubility of $FeSiO_3$ may induce disproportionation of perovskites into mixed oxides under certain conditions. Order-disorder phenomena may play a role in elemental partitioning. Additional experimental information on the relevant stability fields and equations of state will be required to fully treat such effects, but both the apparent absence of globally coherent seismic velocity discontinuities within the bulk of the lower mantle and the absence of evidence for major transformations in recent experiments under deep lower mantle conditions (Kesson et al. 1998) suggest that such effects may be more important in understanding local or lateral variations in temperature, composition, or seismic velocity than in explaining globally averaged models.

LATERAL SEISMIC VELOCITY HETEROGENEITY

Geophysical background

Proceeding from a model of how velocities and densities vary radially in a particular region or in a globally averaged earth, we now ask how these properties vary laterally in the earth as well. To our list of questions about discontinuities (depth, polarity, sharpness, magnitude), we now add lateral variability of all of these properties, including the question of over what spatial extent the velocity contrast is detectable at all. One obvious approach is to look for differences between radial profiles constructed for different regions. However, this is complicated by the dependence of the radial profiles upon their various starting models and by the fact that variations in mantle structure trade off against assumed shallow structure (Walck 1984). Given a large enough geographic distribution of boundary interaction points, however, such a method can be applied to the study of discontinuities, and constraints upon depth variations or "topography" of discontinuities can be further enhanced by evidence of focusing or defocusing in the amplitudes of boundary interaction phases (van der Lee et al. 1994).

Tomographic imaging. An increasingly popular method, however, is the direct inversion of travel times or other data for two- or three-dimensional structure, in a method known as "tomography." Seismic tomography employs the delay times or waveforms of identifiable seismic waves to map the spatial distribution of velocity perturbations relative to a reference model (Su et al. 1994, Li and Romanowicz 1996, Grand et al. 1997, van der Hilst et al. 1997). When interpreted conservatively, the 3-D maps or "images" that result from tomographic modeling can powerfully illuminate both radial and lateral heterogeneity.

Tomography is subject to a variety of resolution limits (Spakman and Nolet 1988, Spakman et al. 1989) and other distortions (van der Hilst et al. 1993, Liu et al. 1998). In particular, the amplitudes of velocity perturbations may not be particularly well constrained, leading to possible underestimation of velocity anomalies, due to a variety of effects. The coarse parameterization of small-scale structure, for example, means that an anomaly smaller than the "cell size" of the model volume elements will be distributed as an average velocity perturbation across the entire cell: a 10-km-wide velocity gradient cannot be resolved in a model whose cells have 100-km edge lengths. Thus, models are sometimes constructed so that anticipated discontinuities (e.g. at 660 km) coincide with the boundaries, rather than penetrate the bodies, of model volume elements (or "voxels").

Another issue concerns the density and orientation of rays sampling any given cell (Fig. 6a). Cells with higher "hit counts" — i.e. those which are sampled by a greater number of rays (which are actually more like tubes in that they sample finite cross-sectional areas or "Fresnel zones" which depend upon frequency) — will have a greater influence upon the inversion, so that velocity perturbations contained in such cells may be better resolved. However, if a given ray samples a velocity anomaly (Fig. 6a), producing a delayed arrival at a receiver, the anomaly could be located anywhere along that ray path, resulting in "streaking" or "smearing" along rays which sample structure in a preferred direction. Most inversions seek to minimize such streaking by incorporating rays which intersect at various angles, so that the placement of the anomaly will be constrained by whether or not the various intersecting rays are delayed as well. Recent developments also allow the use of irregularly sized cells whose geometry can evolve during the inversion (Sambridge and Gudmundsson 1998).

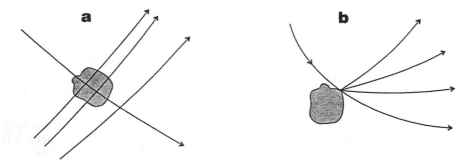

Figure 6. Seismic imaging: Tomographic resolution is limited by (a) the density of rays which sample a velocity anomaly and the fraction of those which intersect each other. Seismic waves may scatter energy (b) off of heterogeneities.

Subtler issues involve the choice of reference model (van der Hilst et al. 1989, Fukao et al. 1992). Ray paths computed from the reference model are used to locate velocity anomalies, but these paths should be recomputed for the perturbed model (VanDecar et al. 1995, Wolfe et al. 1997). Neglect of such corrections may skew results from a "checkerboard test," commonly used to estimate spatial resolution, in which a preselected pattern of velocity perturbations is numerically sampled by the same ray paths associated with the real data, and the resulting synthetic data are inverted to discover how well the original pattern is recovered. The uniform distribution of rays arising from a smooth reference model may recover the input pattern well, but use of the perturbed model may reveal areas of less uniform resolution. Furthermore, tomographic images constructed through linear perturbation to homogeneous (essentially 1-D) reference models may yield inhomogeneities larger than are strictly suitable for linear perturbation theory. Recent developments allow inversions to be performed relative to heterogeneous (i.e. 3-D) reference models (Geller and Hara 1993) and without restriction to weak perturbations (Cummins et al. 1997).

Thus far we have presupposed complete knowledge of the locations of sources and receivers. The uncertainties attending earthquake locations, however, require joint inversion for source mislocation along with velocity structure, resulting in reduced amplitudes for perturbations. Furthermore, the inherent non-uniqueness of inversions means that a variety of velocity structures will fit the data equally well, so one must adopt some sort of measure to penalize certain types of structures in favor of others. To account for the variety of resolution issues discussed above, some sort of damped solution is generally sought. While "norm" damping favors models which differ minimally from the reference model, biasing models toward small perturbation amplitudes, "gradient" damping favors smooth models with minimal variations between adjacent cells. Combinations of these methods can be employed (van der Hilst et al. 1997), with some subjective choice of relative weighting between the two.

Despite these various resolution issues, most of which can be treated by suitable weighting and damping (or "regularization"), the results of tomographic inversions show a remarkable consensus in structural features, such as slab penetration into the lower mantle. Other images of lower mantle heterogeneity suggest sheet-like downwellings (Grand et al. 1997, van der Hilst et al. 1997) and regional deep mantle discontinuities with significant topography (Kawakatsu and Niu 1994, le Stunff et al. 1995, Niu and Kawakatsu 1997). Indeed, efforts at mineralogical interpretation of 3-D structure may most profitably focus upon such robust observations, demonstrated by agreement between independent inversions involving different data sets, damping choices, or other assumptions. The largest remaining uncertainties in tomographic imaging probably concern the amplitudes of velocity perturbations and the apparent scale lengths of heterogeneity. Additional errors may arise, however, from approximations in the theory, such as the common assumption of isotropic media. Given that neglect of velocity anisotropy (discussed below) can have significant effects upon inversion for 1-D reference models (Montagner and Kennett 1996), it is reasonable to expect similar effects in inversion for 3-D models.

Other observations. While a classical tomographic inversion is based solely upon travel times (Zhou and Clayton 1990, Grand 1994), additional constraints upon 3-D velocity structure can be obtained by expanding the variety of constraints to include both local and teleseismic data (Zhao et al. 1994), a priori constraints derived from boundary interaction phases (Zhao and Hasegawa 1993), actual waveforms, or free oscillation data. Indeed, by incorporating such a variety of data, one can extract constraints on the 3-D structure not only of velocities but also of density, attenuation, and anisotropy. However, expanding the number of free parameters can lead to instability in the inversion, especially if factors such as the differing frequency ranges or polarizations of various components of the data set are not properly taken into account.

It can be useful to check the regional structures illuminated by tomography by using another technique. For example, waveform modeling of triplicated seismic waves in the northwest Pacific (Tajima et al. 1998) yields smaller volumes for velocity anomalies than obtained from tomography in the same region. Another method, complementary to tomography, employs scattering theory, using data collected across a seismographic array to locate a volume that consistently scatters energy (Fig. 6b) from several sources (Kaneshima and Helffrich 1998). Scattering observations yield information on the spectrum of spatial heterogeneities, and joint likelihood methods may then be employed to constrain the locations of individual scatterers. Such scattering studies can detect energy from smaller features than are presently resolvable by tomography, and comparisons between results from the two methods can further constrain the size of mantle velocity anomalies (Castle 1998).

Global seismological observations suggest that the lower mantle is more laterally homogeneous than the upper mantle (Gudmundsson et al. 1990) and that the characteristic spatial wavelengths of heterogeneity increase near the bottom of the lower mantle (Su et al. 1994). While evidence for strong (>2%) V_S heterogeneity at large (~200 km) scales in the lower mantle appears to be sparse (Nolet and Moser 1993), weaker (~2%) lower mantle V_P heterogeneity at small (<8 km) scales may be common (Hedlin et al. 1997), and strong heterogeneities may be present at different length scales.

Mineralogical interpretation

For any given seismic velocity anomaly, it is important to determine the polarity, magnitude, and sharpness of velocity contrasts at the boundary of the region, as well as (if possible) the spatial extent of the anomalous region. We can gain an idea of what sort of phenomena might give rise to such anomalies by investigating the partial derivatives of mineral elastic properties under lower mantle conditions.

P-, T-, and X-dependence. The temperature dependence of velocity, $(\partial V_\phi / \partial T)_{P,X}$, is about -2×10^{-4} km/s/K at 1000 km, falling to -1×10^{-4} km/s/K at 2000 km (varying slightly with composition). Thus, a ΔT of 100 K yields a ΔV_ϕ of -0.01 to -0.02 km/s, about the same as that arising from ΔP due to decreasing z by 10 km, because the pressure dependence of velocity, $(\partial V_\phi / \partial P)_{T,X}$, is ~0.03 km/s/GPa. Dynamical implications should also be considered when attempting to interpret velocities in terms of purely thermal anomalies. Given that $(\partial \rho / \partial T)_{P,X}$ is about -9×10^{-5} g/cm^3/K at 1000 km, rising to -7×10^{-5} g/cm^3/K at 2000 km (varying slightly with composition), a ΔT of 100 K would also yield a $\Delta \rho$ of -0.007 to -0.009 g/cm^3, so that inference of large thermal anomalies might also require the presence of significant density (i.e. buoyancy) anomalies which could prove dynamically unstable. Time scales should also be considered when interpreting velocity anomalies of thermal origin, especially if the putative thermal anomalies are believed to have arisen from the subduction process. For example, a slab thermal anomaly of ~1000 K would yield a lower mantle V_ϕ anomaly of a few ~0.1 km/s. If the velocity change were localized over ~10 km, the temperature anomaly would drop to nearly 1/3 of its initial value in about 10^6 years: $(time) \approx (length)^2 / (diffusivity)$. For slab material to reach a depth of 1000 km at a subduction rate of 10 cm/yr, however, would require 10^7 years. Thus, a ΔV_ϕ anomaly of a few ~0.1 km/s over 10 km, located below a depth of 1000 km, is unlikely to be due solely to a thermal anomaly associated with subducted material.

Time scales are not so relevant to anomalies of chemical origin, given the extremely long time scales of solid-state diffusion (Zindler and Hart 1986, Hart and Zindler 1989). The Mg-Fe composition dependence of velocity, $(\partial V_\phi / \partial X_{Mg})_{P,T}$, is about 1.1 km/s at 1000 km, rising to 1.4 km/s at 2000 km, so that a 0.01 increase in X_{Mg} yields ΔV_ϕ of ~0.01 km/s (varying slightly with Si content). Thus, small changes in local Mg/Fe ratio can produce velocity anomalies of the same general magnitude as those due to 100-K temperature perturbations. Similarly, the Si composition dependence of velocity, $(\partial V_\phi / \partial X_{Pv})_{P,T}$ is about 0.9 km/s at 1000 km, falling to 0.7 km/s at 2000 km (varying slightly with Mg content), but this latter systematicity may appear deceptively simple.

With increasing z, Si sensitivity for the Mg component falls while that for the Fe component rises, so that the bulk Si sensitivity (which falls for Mg_{90}) might actually rise with increasing z for a more Mg-poor composition. Nonetheless, to a first approximation, a 0.01 increase in X_{Pv} yields ΔV_ϕ of ≤ 0.01 km/s. So that, again, small changes in local Si content can produce velocity anomalies comparable to those arising from 100-K temperature anomalies. To the extent that lateral velocity variations may be due to compositional heterogeneity, such variations in major element chemistry need not necessarily correspond to the observed minor or trace element geochemical heterogeneity in the mantle (Hoffmann 1997). Nonetheless, either type of distributed chemical heterogeneity may persist in relatively unmixed form over geological time scales even in an actively convecting mantle (Davies 1990, Kellogg 1992, 1993; Metcalfe et al. 1995).

The dependence of velocity upon change in phase, $(\partial V_\phi / \partial X^{phase})_{P,T}$, may also contribute to velocity anomalies in important ways. A local thermal anomaly, for example, may shift a volume of material into a different stability field via a phase transformation, and these two effects can conspire to double the effective velocity anomaly (Bina 1998a). For this reason, phase diagrams should be checked for proximity of local P,T conditions to a reaction boundary when evaluating the potential effects of thermal anomalies. An attempt to interpret such a velocity anomaly in terms of temperature alone would yield a ΔT which exceeded the actual value by 100%. Furthermore, chemical anomalies may have similarly complex effects. A local increase in Fe content, for example, if it exceeded the maximum solubility of $FeSiO_3$ in perovskite under the prevailing P,T conditions (Fei et al. 1996, Wang et al. 1997, Mao et al. 1997), might induce breakdown to oxides, with consequent dramatic changes in V_ϕ.

V_P / V_S *relations.* Ideally, we would interpret lateral heterogeneity in terms of actual V_P and V_S instead of our proxy V_ϕ. However, such interpretations remain difficult at present, because the various derivatives of the shear moduli, especially under lower mantle P,T conditions, remain poorly known. This is particularly unfortunate, because V_S and the V_P / V_S ratio, sometimes expressed in terms of Poisson's ratio σ:

$$\sigma \equiv \frac{3K_S - 2\mu}{2(3K_S + \mu)} = \frac{1}{2} \frac{(V_P / V_S)^2 - 2}{(V_P / V_S)^2 - 1}$$

appear to be more sensitive to temperature and compositional changes than either V_P or V_ϕ. Global seismological studies indicate that $d \ln V_S / d \ln V_P$, a measure of the relative lateral heterogeneity in V_S and V_P, rises from about 1.7 to 2.6 with increasing depth in the lower mantle (Robertson and Woodhouse 1996), increasing to more than 3 in the mantle's bottommost few hundred kilometers (Bolton and Masters 1996). If we assume that all such lateral heterogeneity is due to thermal anomalies, then from ratios of temperature derivatives at constant pressure and composition we obtain:

$$\left(\frac{\partial \ln V_S}{\partial \ln V_P}\right)_{P,X} = \frac{\dfrac{K_S}{\mu} + \dfrac{4}{3}}{\dfrac{K_S}{\mu}\left(\dfrac{1 - \delta_S}{1 - \zeta}\right) + \dfrac{4}{3}}$$

where

$$\delta_S \equiv \frac{-1}{\alpha K_S}\left(\frac{\partial K_S}{\partial T}\right)_P , \quad \zeta \equiv \frac{-1}{\alpha \mu}\left(\frac{\partial \mu}{\partial T}\right)_P$$

For reasonable mineralogical values of these parameters at the top of the lower mantle (Anderson 1989b), this ratio should adopt values of 1 to 2. The seismic observations of larger values thus suggest that the temperature-dependence of μ becomes progressively stronger than that of K_S with increasing depth. However, the similar magnitudes of the various partial derivatives noted above allow ambiguity to remain regarding the relative degrees of thermal or compositional origin for lower mantle velocity heterogeneity, resolution of which awaits better constraints upon δ_S and ζ for perovskite-magnesiowüstite aggregates under deep lower mantle conditions (Stacey 1998).

SEISMIC VELOCITY ANISOTROPY

Geophysical background

Seismic heterogeneity, comprising lateral variations in P- and S-wave velocities from place to place, must further be distinguished from seismic anisotropy, in which velocities of P- and S-waves in any given location may vary with direction of propagation. Indeed, velocity anomalies due to anisotropy can be larger in magnitude than those arising from variations in temperature or composition and can exhibit significant lateral variation (Anderson 1989b). As macroscopic consequences of the fact that many rock-forming minerals exhibit directional variations in both V_P and V_S, seismic constraints upon elastic anisotropy in earth's interior are obtained by observing such phenomena as S-wave splitting (Vinnik et al. 1984, Ando 1984, Fukao 1984, Silver and Chan 1988, 1991; Kaneshima and Silver 1995, Fischer and Wiens 1996, Fouch and Fischer 1996), diffracted waves (Vinnik et al. 1995), converted waves (Vinnik and Montagner 1996), and free oscillations (Montagner and Kennett 1996). In attempting to differentiate the seismological signature of anisotropy from that of heterogeneity, it is important to distinguish the manner in which the seismic anisotropy is observed. In particular, it is useful to distinguish between observation of propagation and polarization anisotropy (Silver 1996).

Propagation and polarization. Propagation anisotropy is perhaps the simplest manifestation of the fact that V_P and V_S vary with direction of propagation. By observing the travel times of seismic waves that have sampled the same region but have traveled through it in different directions along different paths, the fast and slow directions for each of V_P and V_S may be deduced. However, because these waves have traveled along different paths, each may also have sampled different material outside the region of overlap. Thus, such observations admit a significant trade-off between anisotropy and heterogeneity (lateral variation in isotropic structure).

Polarization anisotropy, on the other hand, is a manifestation not only of anisotropy but also of the fact that particle motions for S-waves may be plane polarized. Hence, even along a single path in an anisotropic medium, S-waves will travel with different velocities V_S depending upon their polarization directions, described by the orientations of their particle motions within a plane quasi-orthogonal to that path. One consequent effect is that an initially arbitrarily polarized S-wave passing through a homogeneous anisotropic medium will be split into waves traveling at different speeds with different polarizations (Fig. 7a), a phenomenon known as "shear-wave splitting." Thus, because polarization anisotropy can be characterized by utilizing a single path, such analyses exhibit no trade-off between anisotropy and heterogeneity for the splitting of S-waves (although they do experience some trade-off for surface waves in the upper mantle).

Symmetry frameworks. Observations of seismic anisotropy are interpreted within the framework of simplified symmetry models (Crampin 1984, Babuska and Cara 1991, Silver 1996), which usually assume hexagonal (i.e. cylindrical) symmetry of the elastic properties of aggregate mantle material. Whereas an isotropic medium possesses 2 independent elastic moduli, one with hexagonal symmetry has 5 (compared to 9 for orthorhombic, 13 for monoclinic, or 21 for triclinic). One such model is that of transverse isotropy (also called radial anisotropy), in which the bulk material is assumed to possess hexagonal symmetry about a vertical axis (Fig. 7a), such as might arise from a simply layered fabric. For such a symmetry, horizontally traveling waves will consist of a horizontally polarized *P*-wave (*PH*) and two *S*-waves (horizontally polarized *SH* and vertically polarized *SV*) of differing velocities. Vertically traveling waves, on the other hand will comprise a vertically polarized *P*-wave (*PV*) and two *S*-waves, the latter of which are orthogonally polarized in the horizontal plane and possess identical velocities equal to that of the horizontally traveling *SV*-wave. If the anisotropy is due to planar layering of isotropic media, the horizontally polarized *S*- and (horizontally propagating) *P*-waves will have faster velocities than their vertically polarized counterparts, but this need not be true for for anisotropy arising from inherently anisotropic crystals. Such anisotropy can be characterized using a variety of parameterizations. One commonly used to represent earth models (Montagner and Kennett 1996) consists of specifying the two velocities V_{PH} and V_{SV} along with three anisotropic parameters: ξ, given by the ratio $(V_{SH} / V_{SV})^2$ and analogous to mineralogical c_{66} / c_{44}; ϕ, given by $(V_{PV} / V_{PH})^2$ and analogous to c_{33} / c_{11}; and η, which describes velocity variations in off-axis propagation directions and is analogous to $c_{31} / (c_{11} - 2c_{44})$; where ξ, ϕ, and η are all unity for an isotropic medium.

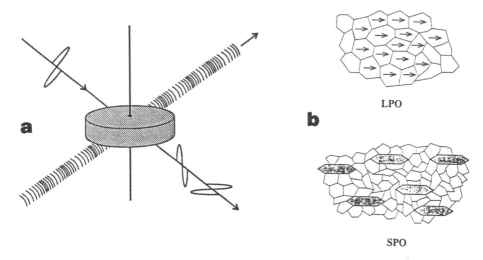

Figure 7. Elastic anisotropy: Cylindrically symmetric seismic anisotropy (a), showing a propagating *P*-wave and a split *S*-wave. Elastic anisotropy in mineral aggregates may arise (b) from lattice preferred orientation (LPO) of anisotropic crystals or from shape preferred orientation (SPO) of aligned inclusions with distinct elastic properties.

Another symmetry model, often employed in lithospheric studies, is that of azimuthal anisotropy. The most common form of this model is fundamentally a model of transverse isotropy rotated through 90°, in which the bulk material is assumed to possess hexagonal symmetry about a horizontal axis, such as might arise as oceanic lithosphere spreads away from a mid-ocean ridge forming a preferred orientation lineation (Christensen and Crosson 1968). In this case, horizontally traveling waves propagating along the symmetry axis will consist of a *P*-wave and two *S*-waves of equal velocities, but those propagating in other directions, whether horizontally or vertically, will comprise a *P*-wave and two *S*-waves of differing velocities. It is in this framework that observations of *S*-wave splitting from nearly vertically propagating shear waves are interpreted. For example, a particular *S*-wave, *SKS*, becomes plane polarized at the core-mantle boundary and travels nearly vertically upwards through the mantle. If it passes through anisotropic material, it becomes split into two orthogonally polarized *S*-waves traveling at different speeds. From observations of such waves as they arrive at the surface, both the polarization direction (in the horizontal plane) of the fast S-wave and the delay time between the fast and slow arrivals can be determined, where the latter is a function of both the magnitude of velocity anisotropy and the physical extent of anisotropic material (Silver 1996).

While transverse isotropy is usually assumed in modeling the whole earth, shear-wave splitting provides evidence of azimuthal anisotropy in the upper mantle. Clearly, these simplified models cannot both be correct for the same region, although transverse isotropy can be viewed as a spatial average of azimuthal anisotropy if the symmetry axis for the latter is randomly distributed in the horizontal plane. Furthermore, individual mantle minerals usually do not possess a symmetry as simple as hexagonal (e.g. orthorhombic olivine with 9 moduli), although they may exhibit a simpler effective symmetry in aggregates. Surface waves have been used to image upper mantle anisotropy in a more complex framework involving both radial and azimuthal components, with results that suggest that S-wave azimuthal anisotropy becomes insignificant below 300 km (Montagner and Tanimoto 1991). Wave interaction with non-planar interfaces can further complicate the signature of anisotropy, as can non-uniformity of symmetry axis orientation. Indeed, when S-wave splitting occurs in regions of varying anisotropy, the orthogonal polarization planes can rotate about the propagation direction, complicating interpretation of the apparent fast polarization direction (Babuska and Cara 1991). Nonetheless, interpretation of seismic data within these simple uniformly hexagonal frameworks generally works well (e.g. effectively eliminating SKS energy from the transverse components of seismograms and reducing elliptical to linear particle motions) and provides a useful integral constraint on overall anisotropy. Complications can arise when the apparent splitting parameters vary with the direction of the incoming waves, which may signal deviation from a simple hexagonal model or violation of the assumption of a strictly vertical symmetry axis, or when they vary with the initial polarization of the incoming waves from a single direction, which may indicate the presence of multiple anisotropic layers (Silver and Savage 1994). Of course, vertically propagating S-waves yield poor resolution of the depth of any anisotropic material, so special geometries involving combinations of waves traveling along different paths must be used to determine the depths of anisotropic regions.

Lower mantle isotropy. The bulk of the lower mantle, however, appears to be largely elastically isotropic (Kaneshima and Silver 1995, Fischer and Wiens 1996, Fouch and Fischer 1996, Montagner and Kennett 1996). Any significant level of azimuthal anisotropy should be evident when integrated over the large path lengths of seismic waves in the lower mantle, yet none is observed (Meade et al. 1995). Measurable anisotropy is not evident except in the lowermost few hundred kilometers of the mantle, where there is evidence of transversely isotropic (radially anisotropic) material, in which $V_{SH} > V_{SV}$, with significant lateral variations in the magnitude of the anisotropy (Kendall and Silver 1996, Matzel et al. 1996, Garnero and Lay 1997). The possibility of some weak radial anisotropy in the uppermost lower mantle (660-1000 km), in which $V_{SV} > V_{SH}$, has also been suggested (Montagner and Kennett 1996) from modeling of normal modes.

Mineralogical interpretation

The apparent isotropy of the bulk of the lower mantle is somewhat puzzling, given that magnesiowüstite and silicate perovskite, the minerals believed to comprise the bulk of the lower mantle, both exhibit elastic anisotropy (Hemley and Cohen 1996, Karato 1997) and, given what is currently known about the pressure- and temperature-dependence of the relevant elastic moduli (Karato 1998a), should do so through most of the lower mantle. Macroscopic seismic expression of such anisotropy, however, would require preferred orientation of anisotropic crystals (Fig. 7b). Thus, the absence of seismic anisotropy in the lower mantle suggests that the minerals assemblages therein do not develop lattice preferred orientation (LPO). While at least portions of the upper mantle appear to deform via a dislocation creep mechanism (Karato and Wu 1993), the conditions of stress and grain size in the lower mantle may fall within the regime governed by the diffusion creep mechanism (Karato 1998a), in which LPO fabrics do not form. Any local or regional deviations from this isotropy might then arise from corresponding regional variations in stress or grain size, reflecting lateral variations in composition, mineralogy, temperature, or deformation history. On the other hand, lower mantle minerals may fail to develop LPO even when deformed within the dislocation creep regime, a hypothesis suggested by deformation experiments on magnesium silicate perovskite (Meade et al. 1995)

Mantle flow. While lithospheric anisotropy is commonly interpreted in terms of preferred orientation of olivine fabrics reflecting spreading or mantle flow (Ribe 1989, Russo et al. 1996), mineralogical interpretation of any limited anisotropy in the deep mantle is still in its infancy. Anderson (1989b) suggested that LPO of silicate ilmenite in subducting slabs could contribute some anisotropy to areas of the topmost lower mantle. Karato (1998a, 1998d) recently used a numerical method (Montagner and Nataf 1986) to calculate the effective elastic moduli arising from lattice preferred orientation in lower mantle mineral assemblages, incorporating current estimates of the significant pressure- and temperature-dependence of mineral anisotropy (Isaak et al. 1990, Karki et al. 1997a, 1997b). Invoking the relative magnitudes of V_{SH} and V_{SV} as observed along certain source-receiver paths, he proposed interpretation of narrow anisotropic layers at the top and bottom of the lower mantle in terms of LPO induced by horizontal flow. In addition to LPO of anisotropic crystals, however, radial anisotropy can also arise from shape preferred orientation (SPO) of aligned inclusions of distinct elastic properties (Fig. 7b). Because observation of $V_{SH} > V_{SV}$ at the core-mantle boundary is consistent with effective horizontal layering, Karato (1998a) interpreted this as due to horizontally aligned SPO. Given the large differences in shear moduli between silicate perovskite and magnesiowüstite, either inclusions of one in the other or inclusions of melt in an aggregate could satisfy the requirements of such SPO.

ANELASTIC ATTENUATION AND VISCOSITY

Geophysical background

Thus far we have considered the elastic properties of the mantle, but the mantle is not perfectly elastic (Weidner this volume). Furthermore, neglect of anelastic effects when inverting seismic data can significantly affect velocity gradients and fine structure in the resulting models (Kennett 1975, Karato 1993).

Attenuation. Anelasticity at the frequency of seismic waves is often described by a "quality factor" Q, where the "attenuation" Q^{-1} describes the fractional energy loss per oscillation. Attenuation can be measured by examining the amplitude decay of body waves, such as S- and P-waves reflected one or more times from the core. However, even in the absence of anelastic attenuation, the amplitudes of body waves decay due to simple geometrical spreading of the wavefronts, so this effect first must be accounted for in order to estimate any anelastic attenuation. Since geometrical spreading will be significantly affected by the presence of lateral velocity variations, there is an inherent trade-off between resolving heterogeneity and attenuation (Bhattacharyya et al. 1996, Bhattacharyya 1998). Normal modes of free oscillation (Fig. 8a), in which the earth undergoes spheroidal and toroidal deformations, can also be used to measure anelastic attenuation. The effects of attenuation are evident both in the quality factors for the amplitudes of individual modes (e.g. CORE11 of Widmer et al. 1993) and in the velocity dispersion (frequency-dependence of velocities, Fig. 8b) which must attend any finite Q (Montagner and Kennett 1996), both of which can be determined from the spreading of spectral peaks (Durek and Ekström 1997). While fundamental modes possess low resolution in the lower mantle, the use of higher-order overtone modes (Okal and Jo 1990) can markedly improve models of this region. However, spectral peak spreading also accompanies velocity heterogeneity, so again there is an inherent trade-off between resolution of lateral heterogeneity and anelasticity (Romanowicz 1987).

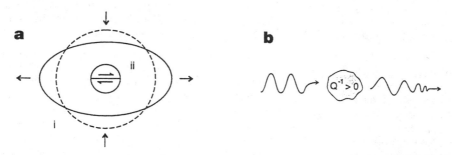

Figure 8. Anelasticity: Normal modes of free oscillation (a) may be spheroidal (i) or toroidal (ii). Attenuation of seismic waves (b) results in both a decrease in amplitude and a dispersion of high- and low-frequency components.

Attenuation can occur either in bulk (isotropic volume changes) or in shear (isochoric deformation) strain, measured by Q_κ and Q_μ, respectively. Because there are few data that constrain the former, and because most attenuation mechanisms operate largely in shear, Q_κ is often assumed to be infinite (Q_κ^{-1} is zero) while normal mode data are inverted for radial models of Q_μ (Montagner and Kennett 1996). Bulk and shear attenuation can be mapped directly into quality factors for V_P and V_S, given by Q_P and Q_S, respectively:

$$Q_S^{-1} = Q_\mu^{-1} \ , \quad Q_P^{-1} = LQ_\mu^{-1} + (1 - L)Q_\kappa^{-1} \ , \quad L \equiv \tfrac{4}{3}(V_S / V_P)^2$$

In many radial models of attenuation, Q increases with depth. However, the magnitude and even the sign of the lower mantle Q_μ gradient exhibits significant dependence upon choice of starting model (Montagner and Kennett 1996). What does seem clear is that Q_μ in the lower mantle (> 400) exceeds that in the upper mantle (< 200).

While attenuation observed in the laboratory usually varies with frequency, most seismological models assume that Q is independent of frequency, and the fact that such models can satisfy available data attests to the poor resolving power of current models of anelasticity (Anderson 1989b). Evidence of frequency dependence of Q can be found in the observation that, over the range of frequencies characteristic of body waves, energy at higher frequencies is attenuated more strongly than at lower frequencies. Furthermore, Q at the lower frequencies of normal modes cannot be simply extrapolated to even lower tidal frequencies. Moreover, the frequency dependence of Q varies with depth. In the lower mantle, body waves are attenuated less strongly than are longer-period (i.e. lower-frequency) normal modes. In the upper mantle and in the bottommost lower mantle, on the other hand, strong attenuation extends to higher frequencies. Such observations suggest the presence of a high-frequency cut-off (Sipkin and Jordan 1979), leading to a model of an "absorption band" for Q whose frequency limits shift with pressure and temperature and hence with depth (Anderson and Given 1982).

Above and beyond such changes in Q with depth, lateral variations in Q can amount to 50-100%. Thus, 1-D models of attenuation are fundamentally unsatisfactory, especially given the dependence of inversions for velocity structure upon accurate representation of Q structure (Montagner and Kennett 1996). While 3-D models of Q are now beginning to emerge (Romanowicz 1994, 1998), they are still generally based on the assumption of frequency-independence. Complicating such efforts are the effects of lateral heterogeneity, because scattering of energy off of heterogeneities can also result in significant apparent attenuation of seismic energy. Attempts at distinguishing intrinsic attenuation from such scattering are best undertaken at low frequencies, so that small scatterers have minimal effect, but this limits potential spatial resolution.

Viscosity. While anelasticity at seismic frequencies appears as attenuation, anelastic behavior at much longer periods manifests itself in the form of viscosity v. Radial viscosity models are generally constructed from one or both of two data sets. The first data set consists of records of sea-level changes associated with post-glacial rebound of the lithosphere, and these data constrain the absolute value of mantle viscosity (Haskell 1935, 1936; Fang and Hager 1996). The second method is somewhat more complicated and constrains only relative viscosity changes (King 1995, Forte and Mitrovica 1996). Viscosity models are constructed so that the gravity-driven flow induced by a specified 2- or 3-D distribution of density heterogeneities in the mantle will generate dynamic topography that reproduces observed long-wavelength geoid anomalies (Hager et al. 1985, Forte and Woodward 1997). Further constraints are sometimes imposed based upon observed plate velocities (Lithgow-Bertelloni and Richards 1998). In addition to geoid observations, the second method requires as data a model of mantle density heterogeneity. This is usually constructed by taking a tomographic image of mantle seismic velocities and mapping it into an image of densities using empirical depth-dependent scaling functions between density and velocity (Hager et al. 1985, Kido and Čadek 1997), thereby assuming that all lateral velocity variations arise solely from lateral variations in temperature rather than chemistry. In terms of elastic moduli, this amounts to assuming values for ratios of temperature derivatives at constant pressure and composition:

$$\left(\frac{\partial \ln \rho}{\partial \ln V_\phi}\right)_{P,X} = \frac{-2}{(1 - \delta_S)} \ , \quad \left(\frac{\partial \ln \rho}{\partial \ln V_S}\right)_{P,X} = \frac{-2}{(1 - \zeta)}$$

where δ_S and ζ were defined above.

The results of such inversions are non-unique, possess poorer resolution in the lower than in the upper mantle, and exhibit dependence upon choice of starting model (King and Masters 1992, King 1995, Mitrovica 1996). However, they generally indicate a viscosity increase of about an order of magnitude in the upper part of the lower mantle, and they suggest the possibility of a narrow zone of low viscosity in the topmost lower mantle (Hager and Richards 1989, King 1995, Mitrovica 1996, Kido and Čadek 1997). While the resolving power of post-glacial rebound studies drops off rapidly with depth in the lower mantle, such data have been reconciled with long-wavelength geoid analyses. Both data sets can be fit with models which exhibit a viscosity jump at 1000 km (Forte and Mitrovica 1996). This viscosity transition however, could also be a smooth gradient, because the limited resolution of the inversion and of the tomographic models does not allow fine structure to be reliably distinguished (Kido and Čadek 1997). The largest uncertainties in such analyses may arise from their implicit assumption that there are no lateral viscosity variations. Just as in the case of attenuation, such lateral variations in viscosity may in fact be extremely large, changing by orders of magnitude in response to temperature anomalies of hundreds of degrees. Such changes could significantly impact the simple flow models derived under the assumption of lateral homogeneity, requiring changes in the inverted viscosity structure in order to fit the geoid observations. Recent studies of the impact of lateral viscosity variations upon geoid calculations, however, have thus far revealed only second-order effects (Ravine and Phipps Morgan 1993, Zhang and Christensen 1993, Zhang and Yuen 1995).

Mineralogical interpretation

Karato (1998b) has proposed a theory to distinguish mechanistically between short-term and long-term rheology, in which the elastic deformation of seismic wave attenuation arises from "micro-glide" migration of geometrical kinks while the transient and steady-state creep associated with viscosity arises from "macro-glide" continuing nucleation of kinks. He has also suggested (Karato 1998c) that the shorter time scales of post-glacial rebound may result in smaller effective viscosities relative to processes of mantle convection, due to a transition from inter-granular creep to steady-state creep at the high strains of mantle convection.

Attenuation. Seismic attenuation, Q^{-1}, in minerals can be a complicated function of temperature, pressure, stress, frequency, grain size, and water content (Anderson 1989b, Karato 1998b, Karato and Jung 1998, Jackson 1998b). In the lower mantle, low values of Q_S may indicate that the rheology thereof is controlled by the magnesiowüstite phase, which is weaker than the presumably more abundant silicate perovskite phase (Getting et al. 1997). Very low values of Q_S, on the other hand, would be suggestive of local melting (Williams and Garnero 1996). However, few inferences regarding chemical composition of the lower mantle can currently be drawn from Q models with confidence. Perhaps the greatest utility of imaging radial and lateral variations in mantle attenuation lies in its potential for resolving the ambiguities in the thermal and compositional origin of velocity anomalies as imaged by seismic tomography (Romanowicz 1997). Since attenuation is so much more sensitive to temperature than is velocity, any tomographic low-velocity anomalies that are not accompanied by high local attenuation may be suspected of having compositional components instead of being purely thermal in nature.

Viscosity. As for the ratios of the partial derivatives of density and velocity with respect to temperature, used in extracting viscosity models from the long-wavelength geoid as described above, for most minerals δ_S is 5 ± 2 and ζ is 6 ± 2 (Anderson 1989b) near the top of the lower mantle, in reasonable agreement with the magnitudes of the density-velocity scaling functions commonly employed (Kido and Čadek 1997). However, if δ_S actually falls slightly with depth (Anderson 1998) then the ratio of partial derivatives should increase with depth, rather than decreasing or remaining constant as is sometimes assumed (Kido and Čadek 1997).

The issue of lower mantle viscosity also raises the question of the effective viscosity of mineral aggregates. In particular, as noted above, magnesiowüstite appears to be significantly weaker than silicate perovskite under lower mantle conditions. Although perovskite should be the more abundant phase, it is not clear which phase will control the aggregate lower mantle viscosity, because this depends upon the unknown geometry adopted by the weaker phase during deformation (Karato 1997).

Just as with attenuation, the strong temperature dependence of viscosity may prove useful in resolving temperature-composition trade-offs in images of density or velocity anomalies. Viscosity variations may also provide evidence of effects of phase transformations on rheology. Proximity to phase changes which involve softening of lattice vibrational modes can result in enhanced creep and thus lower viscosity (Poirier 1981, Rubie and Brearley 1994, Karato 1997). In slab and plume material, latent heat release from material undergoing exothermic transitions can result in locally low viscosities (Karato 1997, Bina 1998b), as can weakening due to grain size reduction which may accompany phase transitions at low temperature (Karato and Li 1992, Karato 1997).

Indeed, the issue of grain size provide a potential link between mantle viscosity and anisotropy. A local reduction in grain size that leads to diffusion creep and an absence of LPO anisotropy might also be expected to induce rheological weakening and consequent locally low viscosities.

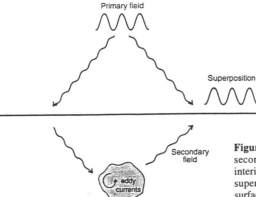

Figure 9. Changing external magnetic fields induce secondary eddy currents in the earth's conductive interior, generating secondary magnetic fields that are superposed upon the external inducing field at the surface.

ELECTRICAL CONDUCTIVITY

Geophysical background

Electrical conductivity in the earth is measured by studying the frequency-dependent electromagnetic response (Roberts 1986) represented by temporal variations in the geomagnetic field. Such temporal variations in external magnetic fields induce secondary eddy currents within conductive pathways in the earth's interior (Fig. 9), and the superposition upon the external inducing field of the secondary magnetic fields generated by these eddy currents can be observed at the surface (Schultz and Larsen 1987). Variations with periods of less than a year arise from solar-atmospheric interactions, inducing consequent telluric currents in the earth, and constrain average conductivity to depths of about 1500 km. Below this depth in the lower mantle, longer period variations in the geomagnetic field, such as the spectrum of secular variation presumably arising from core-mantle interactions or the variations caused by the sunspot cycle, must be used to constrain conductivity (Bott 1982). Inversion of such data suffers from the usual non-uniqueness of the resulting models, but one robust result appears to be that electrical conductivity increases by 1 or 2 orders of magnitude from the upper to the lower mantle. Furthermore, electrical conductivity increases with depth in the lower mantle by about an order of magnitude (from ~1 to ~10 S/m) from top to bottom (Shankland et al. 1993, Petersons and Constable 1996, Honkura et al. 1996). Finally, additional constraints upon electrical conductivity in the lowermost mantle can be inferred from such rotational effects of electromagnetic stresses across the core-mantle boundary as length-of-day variations (Stewart et al. 1995) and nutation of the earth's polar axis (Buffett et al. 1997). The magnitude of the increase in electrical conductivity near the core-mantle boundary remains a matter of debate, because inversions directly constrain conductance rather than conductivity. Because thickness of the highly conductive layer is poorly constrained, the length-of-day constraints on the electromagnetic torque between core and lowermost mantle can be satisfied if the bottom few hundred kilometers of the mantle have a conductivity of 100-1000 S/m (Stewart et al. 1995), but a thinner layer of even higher conductivity is also possible.

Radial profiles of electrical conductivity have been studied for some time (Lahiri and Price 1939). Nonetheless, it is clear that electrical conductivity in the earth exhibits significant lateral heterogeneity (Schultz 1990, Schultz and Larsen 1990). Recently constructed 3-D inverse models (Schultz and Pritchard 1995, 1998) suggest, however, that lateral heterogeneity in the lower mantle may be less extreme than the order of magnitude variations seen in the upper mantle. As in the case of seismic tomography, electromagnetic data are most sensitive to bulk electrical properties rather than to sharp interfaces (Everett and Schultz 1996). Evidence of persistent, long-wavelength, lateral heterogeneity in the geomagnetic field may be attributable either to thermal anomalies or to local regions of high electrical conductivity in the lower mantle (Gubbins 1988, 1994; Johnson and Constable 1998). Thus, mantle compositional heterogeneity may be mapped through electrical as well as seismic methods (Johnson and Constable 1997, Forsyth et al. 1998), especially if thermal anomalies can be independently identified through anelastic effects.

Mineralogical interpretation

While some have suggested bulk iron-enrichment of the lower mantle to account for conductivity increases (Li and Jeanloz 1991a), others have argued that they are due to intrinsic effects of high pressures and temperatures on lower mantle silicates (Peyronneau and Poirier 1989, Shankland et al. 1993). In general, such changes in electrical conductivity are probably controlled by interplay between oxygen fugacity, volatiles (e.g. CO_2, H_2O), and compositional (e.g. ferric iron) variations (Li et al. 1993). If, for example, Fe^{3+} can be accommodated at low levels within the most abundant phases, then the activity of the ferric component will remain low and so will the fugacity of oxygen (f_{O_2}). On the other hand, if Fe^{3+} can be accommodated only within minor secondary phases, the concentration and activity of the ferric components will be high, driving up the f_{O_2}.

In the upper mantle, between 200 and 300 km depth, oxygen fugacity falls to the point where carbonates are reduced to diamond, and the moderate solubility of ferric iron in transition zone minerals suggests that f_{O_2} should continue to fall at greater depths (Wood et al. 1996). Indeed, recent work (Woodland and Angel 1998) demonstrates significant stability of ferric iron in the wadsleyite (β-phase) structure, implying a low oxygen fugacity in the transition zone. Recent observations of extensive ferric iron solubility in aluminous silicate perovskite (Wood and Rubie 1996, McCammon 1997) support the continuation of such low oxygen fugacities and diamond stability into the lower mantle, thus allowing the electrical conductivity in this region to be controlled by the dominant perovskite phase. This is in contrast to earlier suggestions (Wood et al. 1996), based upon a presumed low iron solubility in perovskite, that concentration of iron in the magnesiowüstite phase should lead to increased oxygen fugacity and a return to carbonate stability. This latter scenario would have awkwardly required that the electrical conductivity of the lower mantle be controlled by the secondary magnesiowüstite phase, but the revelation of significant Fe^{3+} solubility in silicate perovskite supports the simpler scenario in which the primary perovskite phase controls the bulk electrical conductivity. Moreover, additional recent work (Mao et al. 1997) suggests that iron-magnesium partitioning between these two lower mantle phases is dependent upon pressure, temperature, and composition, implying that the iron content of silicate perovskite should increase substantially with depth, presumably leading to concomitant electrical conductivity increases. Other studies, however, have found no such systematic depth dependence of Fe-Mg partitioning (Kesson et al. 1998).

Water may also play a role in controlling electrical conductivity (Karato 1990, Li and Jeanloz 1991b), with any H_2O entering the lower mantle from the transition zone in the form of hydrous phases residing in subducting lithosphere (Bose and Navrotsky 1998, Navrotsky this volume). The Fe-Al coupling noted above, however, remains important.

The solubility of H in stishovite rises with increasing Al substitution (Pawley et al. 1993), and the solubility of H in ferromagnesian silicate perovskite may also increase with increasing Fe and Al substitution (Meade et al. 1994). However, the apparent absence of broadened seismic discontinuities in the transition zone (Wood 1995) and the absence of OH in olivine and orthopyroxene inclusions from diamonds (McMillan et al. 1996) suggest that the H_2O content of the transition zone and lower mantle may be generally quite low.

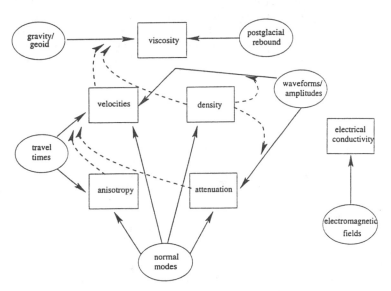

Figure 10. Schematic flowchart illustrating geophysical observations (ovals) and the deep mantle properties which they constrain (rectangles). Solid arrows indicate directions of constraints; dotted arrows show secondary feedback constraints.

CONCLUDING REMARKS

In constructing models of properties of the deep mantle, there is considerable interplay between various geophysical constraints (Fig. 10). The travel times of seismic waves, of course, serve to constrain velocity structure and velocity anisotropy. Normal modes of oscillation, while they also constrain velocities and anisotropy, further serve to constrain the structure of density and anelastic attenuation. The amplitudes of waveforms also constrain velocities and attenuation. On the other hand, both gravity anomalies expressed in the geoid and sea level changes associated with post-glacial rebound combine to constrain viscosity structure. Electromagnetic field variations constrain deep electrical conductivity. Several of the constraints provided by these observations, however, involve significant feedback. Information on the structure of attenuation and anisotropy can significantly affect the determination of velocity structure, for example. Furthermore, models of density are required in order to extract information on velocities and attenuation from the amplitudes of waveforms, and both density and laterally heterogeneous velocity models are necessary in order to constrain viscosity structure from the long-wavelength geoid.

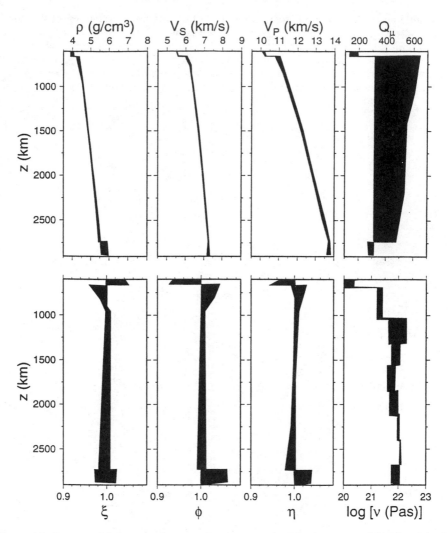

Figure 11. Ranges of lower mantle properties as constrained by inverse models of geophysical observations: density (ρ), seismic wave velocities (S-wave V_S, P-wave V_P) shear quality factor (Q_μ, inverse of shear anelastic attenuation), radial anisotropy (P-wave ϕ, S-wave ξ, off-axis η), logarithm of viscosity (ν), assembled from results of Montagner and Kennett (1996) and Forte and Mitrovica (1996).

Such geophysical models (Fig. 11) provide a variety of constraints upon lower mantle physical conditions. Seismic velocities, density, and electrical conductivity all depend upon temperature as well as upon composition and mineralogy, with electrical conductivity being particularly sensitive to oxygen fugacity through its effects upon ferric iron content. Elastic anisotropy also depends upon temperature and composition, but this property is also sensitive to strain and to grain size. The ambiguity between thermal and compositional origins for lower mantle geophysical anomalies can be investigated through study of anelastic properties, in the form of viscosity and seismic wave attenuation. Both of these properties, while also sensitive to stress, grain size, and composition (especially volatile content), are strongly dependent upon temperature, so that a thermal origin for deep mantle anomalies should have a strong anelastic signature. Resolution of this thermal-compositional ambiguity will also be facilitated by improved equations of state for shear moduli at high temperatures and pressures, because this will allow more thorough analysis of whether coupled variations in *P*- and *S*-wave velocities (i.e. Poisson's ratio) are consistent with isochemical thermal perturbations. Thermal anomalies also will generate buoyancy anomalies which should be consistent with the dynamics of mantle flow.

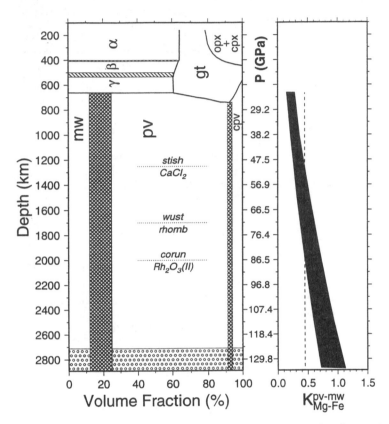

Figure 12. Schematic diagram (after the style of Ringwood 1989) of relative mineral proportions in the mantle (left). Phases are olivine (α), wadsleyite (β), ringwoodite (γ), ortho- and clinopyroxene (opx+cpx), garnet-majorite (gt), magnesiowüstite (mw), ferromagnesian silicate perovskite (pv), and calcium silicate perovskite (cpv). Hatched regions indicate divariant phase transitions in olivine polymorphs. Crosshatched zones denote uncertainty in phase proportions of mw and cpv. Dotted lines indicate approximate pressures of oxide transitions. Also shown (right) is range (shaded region) of partitioning coefficient (K_{Mg-Fe}^{pv-mw}) of Mao et al. (1997) for best-fit lower mantle compositions computed herein and constant value (dashed line) adopted by Kesson et al. (1998).

For a lower mantle composed primarily of silicate perovskite and magnesiowüstite (Fig. 12), current data do not appear to require significant enrichment in iron or silica relative to a pyrolitic upper mantle. However, the dependence of elastic properties upon calcium, aluminum, and ferric iron content has not yet been fully incorporated into such analyses, and both the uppermost and lowermost portions of the lower mantle deviate significantly from this simple model. Coupled solubility of aluminum and ferric iron in silicate perovskite may permit coexistence of garnet and silicate perovskite throughout the uppermost ~100 km of the lower mantle (Wood and Rubie 1996), while γ silicate spinel (ringwoodite) disproportionates to silicate perovskite and magnesiowüstite over a much narrower depth interval (Akaogi et al. 1998).

Beneath cold subduction zones, the disappearance of ringwoodite may be depressed by several tens of kilometers (Bina and Helffrich 1994), and at very low temperatures (below 1000 K) ringwoodite may briefly transform to magnesiowüstite and stishovite before subsequently forming silicate perovskite and magnesiowüstite (Akaogi et al. 1998). As for the transition between garnet and silicate perovskite in the uppermost lower mantle, this may be interrupted by a broad region of stability for silicate ilmenite (Anderson 1989b, Reynard et al. 1996, Reynard and Rubie 1996, Irifune et al. 1996, Vacher et al. 1998) before eventual formation of silicate perovskite.

Despite the apparent absence of major seismic velocity discontinuities of global extent in the lower mantle, additional phase changes may play a role in lower mantle properties (Fig. 12). Lateral variations in temperature or Fe/Mg ratio, for example, may cause local breakdown of silicate perovskite to mixed oxides (Wang et al. 1997). With regard to the oxides, FeO undergoes transformation to a rhombohedral structure at pressures corresponding to ~1700 km (Mao et al. 1996), perhaps shifting to shallower depths in (Mg,Fe)O magnesiowüstite. Furthermore, SiO_2 transforms from stishovite to a $CaCl_2$ structure at pressures corresponding to those at ~1250 km (Tsuchida and Yagi 1989, Kingma et al. 1995). Finally, Al_2O_3 transforms from corundum to a $Rh_2O_3(II)$ structure at pressures corresponding to ~2000 km (Marton and Cohen 1994, Funamori and Jeanloz 1997), with potential consequences for stability, elasticity, elemental partitioning, and buffering of oxygen fugacity in phases such as ferromagnesian silicate perovskite in which alumina exhibits significant solubility.

ACKNOWLEDGMENTS

I thank Michael Hedlin, George Helffrich, Rus Hemley, Catherine Johnson, Scott King, Fred Marton, Emile Okal, Paul Silver, Fumiko Tajima, Rob van der Hilst, and Doug Wiens for helpful comments. I acknowledge the support of NSF grant INT-9603234. Figures 1, 6, 7a, 8, and 9 were drafted by Eric Yates and Figures 7b and 10 by Fred Marton.

REFERENCES

Akaogi M, Kojitani H, Matsuzaka K, Suzuki T (1998) Postspinel transformations in the system Mg_2SiO_4-Fe_2SiO_4: Element partitioning, calorimetry, and thermodynamic calculation. In: MH Manghnani, T Yagi (eds) Properties of Earth and Planetary Materials at High Pressure and Temperature p 373-384 American Geophysical Union, Washington, DC
Anders E, Grevesse N (1989) Abundance of the elements: Meteoritic and solar. Geochim Cosmochim Acta 53:197-214
Anderson DL, Given JW (1982) Absorption band Q model for the Earth. J Geophys Res 87:3893-3904
Anderson DL (1989a) Composition of the Earth. Science 243:367-370
Anderson DL (1989b) Theory of the Earth. Blackwell, Boston
Anderson OL (1998) Thermoelastic properties of $MgSiO_3$ perovskite using the Debye approach. Am Mineral 83:23-35
Ando M (1984) ScS polarization anisotropy around the Pacific ocean. J Phys Earth 32:179-196
Astiz L, Earle PS, Shearer PM (1996) Global stacking of broadband seismograms. Seismo Res Lett 67:8-18
Babuska V, Cara M (1991) Seismic Anisotropy in the Earth. Kluwer, Dordrecht
Beckford W (1786) Vathek. Reprinted by Ballantine, New York, 1971
Bell PM, Mao HK, Xu JA (1987) Error analysis of parameter-fitting in equations of state for mantle minerals. In: MH Manghnani, Y Syono (eds) High-Pressure Research in Mineral Physics, p 447-454 Terra, Tokyo
Bhattacharyya J (1998) Investigation of biasing factors in the measurement of body wave attenuation in the mantle using long-period waveforms. Pure Appl Geoph (in press)
Bhattacharyya J, Masters G, Shearer P (1996) Global variations of shear wave attenuation in the upper mantle. J Geophys Res 101:22273-22289

Biémont E, Baudoux M, Kurucz RL, Ansbacher W, Pinnington EH (1991) The solar abundance of iron: A "final" word! Astron Astrophys 249:539-544

Bina CR (1995) Confidence limits for silicate perovskite equations of state. Phys Chem Min 22:375-382

Bina CR (1998a) Free energy minimization by simulated annealing with applications to lithospheric slabs and mantle plumes. Pure Appl Geophys 151:605-618

Bina CR (1998b) A note on metastable latent heat release in deep seismogenesis. Earth Planets Space (submitted)

Bina CR, Helffrich GR (1992) Calculation of elastic properties from thermodynamic equation of state principles. Ann Rev Earth Planet Sci 20:527-552

Bina CR, Helffrich G (1994) Phase transition Clapeyron slopes and transition zone seismic discontinuity topography. J Geophys Res 99:15853-15860

Bina CR, Kumazawa M (1993) Thermodynamic coupling of phase and chemical boundaries in planetary interiors. Phys Earth Planet Inter 76:329-341

Bina CR, Silver PG (1997) Bulk sound travel times and implications for mantle composition and outer core heterogeneity. Geophys Res Lett 24:499-502

Birch F (1952) Elasticity and constitution of the Earth's interior. J Geophys Res 57:227-286

Bolton H, Masters G (1996) A region of anomalous d ln V_S/d ln V_P in the deep mantle. Eos Trans Am Geophys Union 77:F697

Bose K, Navrotsky A (1998) Thermochemistry and phase equilibria of hydrous phases in the system MgO-SiO_2-H_2O: Implications for volatile transport to the mantle. J Geophys Res 103:9713-9719

Bott MHP (1982) The Interior of the Earth. Elsevier, New York

Buffett BA, Herring TA, Mathews PM (1997) New constraints on the structure of the core-mantle boundary from observations of the Earth's nutations. Eos Trans Am Geophys Union 78:F2

Bukowinski MST, Wolf GH (1990) Thermodynamically consistent decompression: Implications for lower mantle composition. J Geophys Res 95:12583-12593

Byerley P (1926) The Montana earthquake of June 28, 1925. Bull Seis Soc Am 16:209-265

Castle JC (1998) Imaging mid-mantle discontinuities: Implications for mantle chemistry, dynamics, rheology, and deep earthquakes. PhD dissertation, University of Washington, Seattle

Christensen NI, Crosson RS (1968) Seismic anisotropy in the upper mantle. Tectonophys 6:93-107

Clarke, TJ (1993) The complete ordered ray expansion 2: Multi-phase body wave tomography. Geophys J Int'l 115:435-444

Crampin S (1984) An introduction to wave propagation in anisotropic media. Geophys J R Astr Soc 76:17-28

Cummins PR, Takeuchi N, Geller RJ (1997) Computation of complete synthetic seismograms from laterally heterogeneous models using the Direct Solution Method. Geophys J Int'l 130:1-16

Davies GF (1990) Mantle plumes, mantle stirring, and hotspot chemistry. Earth Planet Sci Lett 99:94-109

Durek JJ, Ekström G (1997) Investigating discrepancies among measurements of traveling and standing wave attenuation. J Geophys Res 102:24529-24544

Dziewonski AM, Anderson DL (1981) Preliminary reference Earth model. Phys Earth Planet Inter 25:297-356

Everett ME, Schultz A (1996) Geomagnetic induction in a heterogeneous sphere: Azimuthally symmetric test computations and the response of an undulating 660-km discontinuity. J Geophys Res 101:2765-2783

Fabrichnaya OB (1995) Thermodynamic data for phases in the FeO-MgO-SiO_2 system and phase relations in the mantle transition zone. Phys Chem Min 22:323-332

Fang M, Hager BH (1996) The sensitivity of post-glacial sea level to viscosity structure and ice-load history for realistically parameterized viscosity profiles. Geophys Res Lett 23:3787-3790

Fei Y, Wang Y, Finger LW (1996) Maximum solubility of FeO in $(Mg, Fe)SiO_3$-perovskite as a function of temperature at 26 GPa: Implication for FeO content in the lower mantle. J Geophys Res 101:11525-11530

Fischer KM, Wiens DA (1996) The depth distribution of mantle anisotropy beneath the Tonga subduction zone. Earth Planet Sci Lett 142:253-260

Forsyth DW, Scheirer DS, Webb SC, Dorman LM, Orcutt JA, Harding AJ, Blackman DK, Phipps Morgan J, Detrick RS, Shen Y, Wolfe CJ, Canales JP, Toomey DR, Sheehan AF, Solomon SC, Wilcock WSD (1998) Imaging the deep seismic structure beneath a mid-ocean ridge: The MELT experiment. Science 280:1215-1218

Forte AM, Mitrovica JX (1996) New inferences of mantle viscosity from joint inversion of long-wavelength mantle convection and post-glacial rebound data. Geophys Res Lett 23:1147-1150

Forte AM, Woodward RL (1997) Seismic-geodynamic constraints on three-dimensional structure, vertical flow, and heat transfer in the mantle. J Geophys Res 102:17981-17994

Fouch MJ, Fischer KM (1996) Mantle anisotropy beneath northwest Pacific subduction zones. J Geophys Res 101:15987-16002

Fukao Y (1984) Evidence from core-reflected shear waves for anisotropy in the Earth's mantle. Nature 309:695-698

Fukao Y, Obayashi M, Inoue H, Nenbai M (1992) Subducting slabs stagnant in the mantle transition zone. J Geophys Res 97:4809-4822

Funamori N, Jeanloz R (1997) High-pressure transformation of Al_2O_3. Science 278:1109-1111

Garlick GD (1969) Consequences of chemical equilibrium across phase changes in the mantle. Lithos 2:325-331

Garnero EJ, Grand SP, Helmberger DV (1993) Low P-wave velocity at the base of the mantle. Geophys Res Lett 17:1843-1846

Garnero EJ, Lay T (1997) Lateral variation in lowermost mantle shear wave anisotropy beneath the north Pacific and Alaska, J Geophys Res 102:8121-8136

Geller RJ, Hara T (1993) Two efficient algorithms for iterative linearized inversion of seismic waveform data. Geophys J Int'l 115:699-710

Getting IC, Dutton SJ, Burnley PC, Karato S, Spetzler HA (1997) Shear attenuation and dispersion in MgO. Phys Earth Planet Inter 99:249-257

Gilbert JF, Dziewonski AM (1975) An application of normal mode theory to the retrieval of structure parameters and source mechanism from seismic spectra. Phil Trans R Soc Lond A278:187-269

Grand SP (1994) Mantle shear structure beneath the Americas and surrounding oceans. J Geophys Res 99:11591-11621

Grand SP, Helmberger DV (1984) Upper mantle shear structure of North America. Geophys J R Astr Soc 76:399-438

Grand SP, van der Hilst RD, Widiyantoro S (1997) Global seismic tomography: Snapshot of convection in the Earth. Geol Soc Am Today 7(4):1-7

Gubbins D (1988) Thermal core-mantle interactions and the time-averaged palaeomagnetic field. J Geophys Res 93:3413-3420

Gubbins D (1994) Geomagnetic polarity reversals: A connection with secular variation and core-mantle interaction? Rev Geophys 32:61-83

Gudmundsson O, Davis JH, Clayton RW (1990) Stochastic analysis of global traveltime data: Mantle heterogeneity and random errors in the ISC data. Geophys J Int'l 102:25-43

Hager BH, Clayton RW, Richards MA, Comer RP, Dziewonski AM (1985) Lower mantle heterogeneity, dynamic topography and the geoid. Nature 313:541-545

Hager BH, Richards MA (1989) Long-wavelength variations in Earth's geoid: physical models and dynamical implications. Phil Trans R Soc Lond A328:309-327

Hart S, Zindler A (1989) Constraints on the nature and development of chemical heterogeneities in the mantle. *In:* WR Peltier (ed) Mantle Convection: Plate Tectonics and Global Dynamics p 261-387 Gordon and Breach, New York

Harte B, Harris JW (1994) Lower mantle associations preserved in diamonds. Mineral Mag 58A:384-385

Harte B, Hutchison MT, Harris JW (1994) Trace element characteristics of the lower mantle: An ion probe study of inclusions in diamonds from São Luiz, Brazil. Mineral Mag 58A:386-387

Haskell NA (1935) The motion of a fluid under a surface load, I. Physics 6:265-269

Haskell NA (1936) The motion of a fluid under a surface load, II. Physics 7:56-61

Hedlin MAH, Shearer PM, Earle PS (1997) Seismic evidence for small-scale heterogeneity throughout the Earth's mantle. Nature 387:145-150

Helffrich G, Bina CR (1994) Frequency dependence of the visibility and depths of mantle seismic discontinuities. Geophys Res Lett 21:2613-2616

Helffrich GR, Wood BJ (1996) 410 km discontinuity sharpness and the form of the olivine alpha-beta phase diagram: Resolution of apparent seismic contradictions. Geophys J Int'l 126:F7-F12

Hemley RJ, Cohen RE (1996) Structure and bonding in the deep mantle and core. Phil Trans R Soc Lond A354:1461-1479

Herzberg CT, O'Hara MJ (1985) Origin of mantle peridotite and komatiite by partial melting. Geophys Res Lett 12:541-544

Hoffmann AW (1997) Mantle geochemistry: The message from oceanic volcanism. Nature 385:219-229

Holweger H, Heise C, Kock M (1990) The abundance of iron in the Sun derived from photospheric Fe II lines. Astron Astrophys. 232:510-515

Honkura Y, Abe T, Matsushima M (1996) Global electrical conductivity distribution in the mantle based on the P_1^0 approximation. Eos Trans Am Geophys Union 77:F167

Irifune T, Isshiki M (1998) Iron partitioning in a pyrolite mantle and the nature of the 410km seismic discontinuity. Nature 392:702-705

Irifune T, Koizumi T, Ando J (1996) An experimental study of the garnet-perovskite transformation in the system $MgSiO_3$-$Mg_3Al_2Si_3O_{12}$. Phys Earth Planet Inter 96:147-157

Isaak DG, Cohen RE, Mehl MJ (1990) Calculated elastic and thermal properties of MgO at high pressure and temperatures. J Geophys Res 95:7055-7067

Ito E, Katsura T (1989) A temperature profile of the mantle transition zone. Geophys Res Lett 16:425-428

Jackson I (1983) Some geophysical constraints on the chemical composition of the Earth's lower mantle. Earth Planet Sci Lett 62:91-103

Jackson I (1998a) Elasticity, composition and temperature of the Earth's lower mantle: A reappraisal.

Geophys J Int'l 134:291-311

Jackson I (1998b) Laboratory measurements of seismic wave dispersion and attenuation: Recent progress. *In:* S Karato (ed) Mineral Physics and Seismic Tomography (submitted) American Geophysical Union, Washington, DC

Jackson I, Rigden SM (1998) Composition and temperature of the Earth's mantle: Seismological models interpreted through experimental studies of Earth materials. *In:* I Jackson (ed) The Earth's Mantle: Composition, Structure and Evolution p 405-460 Cambridge University, Cambridge

Jeanloz R (1995) Earth dons a different mantle. Nature 378:130-131

Jeanloz R, Knittle E (1989) Density and composition of the lower mantle. Phil Trans R Soc Lond A328:377-389

Jeffreys, H (1936) Geophysical discussion. Observatory 59:267-268

Johnson CL, Constable CG (1997) The time-averaged geomagnetic field: Global and regional biases for 0-5 Ma. Geophys J Int'l 131:643-666

Johnson CL, Constable CG (1998) Persistently anomalous Pacific geomagnetic fields. Geophys Res Lett 25:1011-1014

Kaneshima S, Helffrich G (1998) Detection of lower mantle scatterers northeast of the Mariana subduction zone using short-period array data. J Geophys Res 103: 4825-4838

Kaneshima S, Silver PG (1995) Anisotropic loci in the mantle beneath central Peru. Phys Earth Planet Inter 88:257-272

Karato S (1990) The role of hydrogen in the electrical conductivity of the upper mantle. Nature 347:272-273

Karato S (1993) Importance of anelasticity in the interpretation of seismic tomography. Geophys Res Lett 20:1623-1626

Karato S (1997) Phase transformations and rheological properties of mantle minerals. *In:* DJ Crossley (ed) Earth's Deep Interior, The Doornbos Memorial Volume p 223-272 Gordon and Breach, Amsterdam

Karato S (1998a) Seismic anisotropy in the deep mantle, boundary layers and the geometry of mantle convection. Pure Appl Geophys 151:565-587

Karato S (1998b) A dislocation model of seismic wave attenuation and micro-creep in the Earth: Harold Jeffreys and the rheology of the solid Earth. Pure Appl Geophys (in press)

Karato S-I (1998c) Micro-physics of post glacial rebound. GeoResearch Forum 3/4:351-364

Karato S (1998d) Some remarks on the origin of seismic anisotropy in the D" layer. Earth Planets Space (in press)

Karato S, Jung H (1998) Water, partial melting and the origin of the seismic low velocity and high attenuation zone in the upper mantle. Earth Planet Sci Lett 157:193-207

Karato S, Li P (1992) Diffusion creep in perovskite: Implications for the rheology of the lower mantle. Science 255:1238-1240

Karato S, Wu P (1993) Rheology of the upper mantle: A synthesis. Science 260:771-778

Karki BB, Stixrude L, Clark SJ, Warren MC, Ackland GJ, Crain J (1997a) Structure and elasticity of MgO at high pressure. Am Mineral 82:51-60

Karki BB, Stixrude L, Clark SJ, Warren MC, Ackland GJ, Crain J (1997b) Elastic properties of $MgSiO_3$ perovskite at lower mantle pressures. Am Mineral 82:635-638

Kato T, Ringwood AE, Irifune T (1988) Experimental determination of element partitioning between silicate perovskites, garnets and liquids: Constraints on early differentiation of the mantle. Earth Planet Sci Lett 89:123-145

Kawakatsu H, Niu F (1994) Seismic evidence for a 920-km discontinuity in the mantle. Nature 371:301-305

Kellogg LH (1992) Mixing in the mantle. Ann Rev Earth Planet Sci 20:365-388

Kellogg LH (1993) Chaotic mixing in the earth's mantle. Advan Geophys 34:1-33

Kendall JM, Silver PG (1996) Constraints from seismic anisotropy on the nature of the lowermost mantle. Nature 381:409-412

Kennett BLN (1975) The effects of attenuation on seismograms. Bull Seis Soc Am 65:1643-1651

Kennett BLN, Engdahl ER (1991) Traveltimes for global earthquake location and phase identifications. Geophys J Int'l 105:429-465

Kennett BLN, Engdahl ER, Buland R (1995) Constraints on seismic velocities in the Earth from travel times. Geophys J Int'l 122:108-124

Kesson SE, Fitz Gerald JD (1991) Partitioning of MgO, FeO, NiO, MnO and Cr_2O_3 between magnesian silicate perovskite and magnesiowüstite: Implications for the origin of inclusions in diamond and the composition of the lower mantle. Earth Planet Sci Lett 111:229-240

Kesson SE, Fitzgerald JD, Shelley JMG (1998) Mineralogy and dynamics of a pyrolite lower mantle. Nature 393:252-255

Kido M, Čadek O (1997) Inferences of viscosity from the oceanic geoid: Indication of a low viscosity zone below the 660-km discontinuity. Earth Planet Sci Lett 151:125-137

King SD (1995) Models of mantle viscosity. *In:* TJ Ahrens (ed) Mineral Physics and Crystallography: A Handbook of Physical Constants p 227-236 American Geophysical Union, Washington, DC

King SD, Masters G (1992) An inversion for radial viscosity structure using seismic tomography. Geophys Res Lett 19:1551-1554

Kingma KJ, Cohen RE, Hemley RJ, Mao HK (1995) Transformation of stishovite to a denser phase at lower mantle pressures. Nature 374:243-245

Knittle et al. (1986) Knittle E, Jeanloz R, Smith GL (1986) Thermal expansion of silicate perovskite and stratification of the Earth's mantle. Nature 319:214-216

Lahiri B, Price A (1939) Electromagnetic induction in non-uniform conductors, and the determination of the conductivity of the Earth from terrestrial magnetic variations. Phil Trans R Soc Lond 237:509-540

Lay T, Wallace TC (1995) Modern Global Seismology. Academic, San Diego, California

le Stunff Y, Wicks CW Jr, Romanowicz B (1995) P'P' precursors under Africa: Evidence for mid-mantle reflectors. Science 270:74-77

Li X, Jeanloz R (1991a) Effect of iron content on the electrical conductivity of perovskite and magnesiowüstite assemblages at lower mantle conditions. J Geophys Res 96:6113-6120

Li X, Jeanloz R (1991b) Phases and electrical conductivity of a hydrous silicate assemblage at lower-mantle conditions. Nature 350:332-334

Li X, Ming L-C, Manghnani MH, Wang Y, Jeanloz R (1993) Pressure dependence of the electrical conductivity of $(Mg_{0.9}Fe_{0.1})SiO_3$ perovskite. J Geophys Res 98:501-508

Li XD, Romanowicz B (1996) Global mantle shear velocity model developed using nonlinear asymptotic coupling theory. J Geophys Res 101:22245-22272

Lithgow-Bertelloni C, Richards MA (1998) The dynamics of Cenozoic and Mesozoic plate motions. Rev Geophys 36:27-78

Liu X-F, Tromp J, Dziewonski AM (1998) Is there a first-order discontinuity in the lowermost mantle? Earth Planet Sci Lett (in press)

Mao HK (1988) The 670-km discontinuity in the mantle: A bulk chemical composition boundary driven by phase transformation. Eos Trans Am Geophys Union 69:1420

Mao HK, Shen G, Hemley RJ (1997) Multivariable dependence of Fe/Mg partitioning in the Earth's lower mantle. Science 278:2098-2100

Mao, HK, Shu J, Fei Y, Hu J, Hemley RJ (1996) The wüstite enigma. Phys Earth Planet Inter 96:135-145

Marton FC, Cohen RE (1994) Prediction of a high-pressure phase transition in Al_2O_3. Am Mineral 79:789-792

Masters G (1979) Observational constraints on the chemical and thermal structure of the Earth's deep interior. Geophys J R Astr Soc 57:507-534

Matzel E, Sen MK, Grand SP (1996) Evidence for anisotropy in the deep mantle beneath Alaska. Geophys Res Lett 23:2417-2420

McCammon CA (1997) Perovskite as a possible sink for ferric iron in the lower mantle. Nature 387: 694-696

McDonough WF, Sun S (1995) The composition of the Earth. Chem Geol 120:223-253

McMillan PF, Hemley RJ, Gillet P (1996) Vibrational spectroscopy of mantle minerals. In: MD Dyar, C McCammon, MW Schaefer (eds) Mineral Spectroscopy: A Tribute to Roger G. Burns p 175-213 Geochemical Society

Meade C, Reffner JA, Ito E (1994) Synchrotron infrared absorbance measurements of hydrogen in $MgSiO_3$ perovskite. Science 264:1558-1560

Meade C, Silver PG, Kaneshima S (1995) Laboratory and seismological observations of lower mantle isotropy. Geophys Res Lett 22:1293-1296

Melbourne T, Helmberger D (1998) Fine structure of the 410-km discontinuity. J Geophys Res 103:10091-10102

Metcalfe G, Bina CR, Ottino JM (1995) Kinematic considerations for mantle mixing. Geophys Res Lett 22:743-746

Mitrovica JX (1996) Haskell (1935) revisited. J Geophys Res 101:555-569

Montagner J-P, Kennett BLN (1996) How to reconcile body-wave and normal-mode reference earth models. Geophys J Int'l 125:229-248

Montagner J-P, Tanimoto T (1991) Global upper mantle tomography of seismic velocities and anisotropies. J Geophys Res 96:20337-20351

Morelli A, Dziewonski AM (1993) Body wave traveltimes and a spherically symmetric P- and S-wave velocity model. Geophys J Int'l 112:178-194

Navrotsky A (1994) Physics and Chemistry of Earth Materials. Cambridge University, Cambridge

Niu F, Kawakatsu H (1997) Depth variation of the mid-mantle seismic discontinuity. Geophys Res Lett 24:429-432

Nolet G (1990) Partitioned waveform inversion and two-dimensional structure under the network of autonomously recording seismographs. J Geophys Res 95:8499-8512

Nolet G, Moser TJ (1993) Teleseismic delay times in a 3-D Earth and a new look at the S-discrepancy. Geophys J Int'l 114:185-195

Okal EA, Jo B-G (1990) Q measurements for phase X overtones. Pure Appl Geophys 132:331-362

Pawley, AR, McMillan PF, Holloway JR (1993) Hydrogen in stishovite, with implications for mantle water content. Science 261:1024-1026

Petersons HF, Constable SC (1996) Global mapping of the electrically conductive lower mantle. Geophys Res Lett 23:1461-1464

Peyronneau J, Poirier J-P (1989) Electrical conductivity of the Earth's lower mantle. Nature 342:537-539

Poirier J-P, Tarantola A (1998) A logarithmic equation of state. Phys Earth Planet Inter (in press)

Poirier J-P (1981) Martensitic olivine-spinel transformation and plasticity of the mantle transition zone. *In:* FD Stacey, MS Paterson, A Nicolas (eds) Anelasticity in the Earth p 113-117 American Geophysical Union, Washington, DC

Ravine MA, Phipps Morgan J (1993) Geoid effects of lateral viscosity variation near the top of the mantle: A 2-D model. Earth Planet Sci Lett 119:617-625

Revenaugh J, Jordan TH (1991a) Mantle layering from ScS reverberations: 1. Waveform inversion of zeroth-order reverberations. J Geophys Res 96:19749-19762

Revenaugh J, Jordan TH (1991b) Mantle layering from ScS reverberations: 4. The lower mantle and core-mantle boundary. J Geophys Res 96:19811-19824

Reynard B, Fiquet G, Itié J-P, Rubie DC (1996) High-pressure X-ray diffraction study and equation of state of $MgSiO_3$-ilmenite. Am Mineral 81:45-50

Reynard B, Rubie DC (1996) High-pressure high-temperature Raman spectroscopic study of ilmenite-type $MgSiO_3$. Am Mineral 81:1092-1096

Ribe, NM (1989) Seismic anisotropy and mantle flow. J Geophys Res 94:4213-4223

Ringwood, AE (1975) Composition and Petrology of the Earth's Mantle. McGraw-Hill, New York

Ringwood, AE (1989) Constitution and evolution of the mantle. Spec Pub Geol Soc Australia 14:457-485

Roberts RG (1986) The deep electrical structure of the Earth. Geophys J R Astr Soc 85:583-600

Robertson GS, Woodhouse JH (1996) Constraints on lower mantle physical properties from seismology and mineral physics. Earth Planet Sci Lett 143:197-205

Romanowicz B (1987) Multiplet-multiplet coupling due to lateral heterogeneity: Asymptotic effects on the amplitude and frequency of the Earth's normal modes. Geophys J R Astron Soc 90:75-100

Romanowicz B (1994) Anelastic tomography: A new perspective on the upper mantle thermal structure. Earth Planet Sci Lett 128:113-121

Romanowicz B (1997) 3D models of elastic and anelastic structure in the mantle. Eos Trans Am Geophys Union 78:F466

Romanowicz B (1998) Attenuation tomography of the Earth's mantle: A review of current status. Pure Appl Geophys (in press)

Rubie DC, Brearley AJ (1994) Phase transformations between β and γ $(Mg, Fe)_2SiO_4$ in the Earth's mantle: Mechanisms and rheological implications. Science 264:1445-1448

Russo RM, Silver PG, Franke M, Ambeh WB, James DE (1996) Shear-wave splitting in northeast Venezuela, Trinidad, and the eastern Caribbean. Phys Earth Planet Inter 95:251-275

Sambridge M, Gudmundsson O (1998) Tomography with irregular cells. J Geophys Res 103:773-781

Schultz A (1990) On the vertical gradient and lateral heterogeneity in mid-mantle electrical conductivity. Phys Earth Planet Inter 64:68-86

Schultz A, Larsen JC (1987) On the electrical conductivity of the mid-mantle: 1—Calculation of equivalent scalar magnetotelluric response functions. Geophys J R Astr Soc 88:733-761

Schultz A, Larsen JC (1990) On the electrical conductivity of the mid-mantle: 2—Delineation of heterogeneity by application of extremal inverse solutions. Geophys J R Astr Soc 101:565-580

Schultz A, Pritchard G (1995) Inversion for the three-dimensional structure of the Earth's mantle: Rapid convergence through spectral methods, stiff solvers and interpolation and integration on a convex hull. *In:* Three-Dimensional Electromagnetics, Proceedings of an International Symposium in Honor of Jerry Hohmann p 429-452 Schlumberger-Doll Research, Ridgefield, Connecticut

Schultz A, Pritchard G (1998) A three-dimensional inversion for large-scale structure in a spherical domain. *In:* M Oristaglio, B Spiess (eds) Three-Dimensional Electromagnetics (in press) Society of Exploration Geophysics, Tulsa, Oklahoma

Shankland TJ, Peyronneau J, Poirier J-P (1993) Electrical conductivity of the Earth's lower mantle. Nature 366:453-455

Shearer PM (1991a) Constraints on upper mantle discontinuities from observations of long-period reflected and converted phases. J Geophys Res 96:18147-18182

Shearer PM (1991b) Imaging global body wave phases by stacking long-period seismograms. J Geophys Res 96:20353-20364

Silver PG (1996) Seismic anisotropy beneath the continents: Probing the depths of geology. Ann Rev Earth Planet Sci 24:385-432

Silver PG, Bina CR (1993) An anomaly in the amplitude ratio of SKKS/SKS in the range 100-108° from portable teleseismic data. Geophys Res Lett 20:1135-1138

Silver PG, Chan WW (1988) Implications for continental structure and evolution from seismic anisotropy. Nature 335:34-39

Silver PG, Chan WW (1991) Shear-wave splitting and subcontinental mantle deformation. J Geophys Res 96:16429-16454

Silver PG, Savage M (1994) The interpretation of shear-wave splitting parameters in the presence of two anisotropic layers. Geophys J Int'l 119:949-963

Sipkin S, Jordan TH (1979) Frequency dependence of Q_{ScS}. Bull Seism Soc Am 69:1055-1079

Spakman W, Nolet G (1988) Imaging algorithms, accuracy and resolution in time delay tomography. In: NJ Vlaar, G Nolet, MJR Wortel, SAPL Cloetingh (eds) Mathematical Geophysics p 155-187 Reidel, Dordrecht

Spakman W, Stein S, van der Hilst R, Wortel R (1989) Resolution experiments for NW Pacific subduction zone tomography. Geophys Res Lett 16:1097-1100

Stacey FD (1998) Thermoelasticity of a mineral composite and a reconsideration of lower mantle properties. Phys Earth Planet Inter 106:219-236

Stewart DN, Busse F H, Whaler KA, Gubbins D (1995) Geomagnetism, Earth rotation and the electrical conductivity of the lower mantle. Phys Earth Planet Inter 92:199-214

Su W, Woodward RL, Dziewonski AM (1994) Degree 12 model of shear velocity heterogeneity in the mantle. J Geophys Res 99:6945-6980

Tajima F, Fukao Y, Obayashi M, Sakurai T (1998) Evaluation of slab images in the northwestern Pacific. Earth Planet Space (in press)

Tsuchida Y, Yagi T (1989) A new, post-stishovite high-pressure polymorph of silica. Nature 340:217-220

Vacher P, Mocquet A, Sotin C (1998) Computation of seismic profiles from mineral physics: The importance of the non-olivine components for explaining the 660 km depth discontinuity. Phys Earth Planet Inter 106:275-298

VanDecar JC, James DE, Assumpçaõ M (1995) Seismic evidence for a fossil mantle plume beneath South America and implications for plate driving forces. Nature 378:25-31

van der Hilst RD, Engdahl ER, Spakman W (1989) Importance of the reference model in linearized tomography and image of subduction below the Caribbean plate. Geophys Res Lett 16:1093-1096

van der Hilst RD, Engdahl ER, Spakman W (1993) Tomographic inversion of P and pP data for aspheric mantle structure below the northwest Pacific region. Geophys J Int'l 115:264-302

van der Hilst RD, Widiyantoro S, Engdahl ER (1997) Evidence for deep mantle circulation from global tomography. Nature 386:578-589

van der Lee S, Nolet G (1997) Seismic image of the subducted trailing fragments of the Farallon plate. Nature 386:266-269

van der Lee S, Paulssen H, Nolet G (1994) Variability of P660s phases as a consequence of topography of the 660 km discontinuity. Phys Earth Planet Inter 86:147-164

Vinnik LP, Kosarev GL, Makeyeva LI (1984) Anizotropiya litosfery po nablyudeniyam voln SKS and SKKS. Dokl Akad Nauk USSR 278:1335-1339

Vinnik L, Montagner J-P (1996) Shear wave splitting in the mantle Ps phases. Geophys Res Lett 23:2449-2452

Vinnik LP, Romanowicz B, le Stunff Y, Makeyeva L (1995) Seismic anisotropy in the D" layer. Geophys Res Lett 22:1657-1660

Walck MC (1984) The P-wave upper mantle structure beneath an active spreading center: The Gulf of California. Geophys J R Astron Soc 76:697-723

Wang Y, Martinez I, Guyot F, Liebermann RC (1997) The breakdown of olivine to perovskite and magnesiowüstite. Science 275:510-513

Wang Y, Weidner DJ, Guyot F (1996) Thermal equation of state of $CaSiO_3$ perovskite. J Geophys Res 101:661-672

Weinstein SA (1992) Induced compositional layering in a convecting fluid layer by an endothermic phase transition. Earth Planet Sci Lett 113:23-39

Widmer R, Masters G, Gilbert F (1993) Spherically symmetric attenuation within the Earth from normal mode data. Geophys J Int'l 104:541-553

Williams Q, Garnero EJ (1996) Seismic evidence for partial melt at the base of Earth's mantle. Science 273:1528-1530

Wolfe CJ, Bjarnason ITh, VanDecar JC, Solomon SC (1997) Seismic structure of the Iceland mantle plume. Nature 385:245-247

Wood BJ (1995) The effect of H_2O on the 410 km seismic discontinuity. Science 268:74-78

Wood BJ, Pawley A, Frost DR (1996) Water and carbon in the Earth's mantle. Phil Trans R Soc Lond A354:1495-1511

Wood BJ, Rubie DC (1996) The effect of alumina on phase transformations at the 660-kilometer discontinuity from Fe-Mg partitioning experiments. Science 273:1522-1524

Woodland AB, Angel RJ (1998) Crystal structure of a new spinelloid with the wadsleyite structure in the system Fe_2SiO_4-Fe_3O_4 and implications for the Earth's mantle. Am Mineral 83:404-408

Wysession ME, Okal EA, Bina CR (1992) The structure of the core-mantle boundary from diffracted waves. J Geophys Res 97:8749-8764

Zhao D, Hasegawa A (1993.) P wave tomographic imaging of the crust and upper mantle beneath the Japan Islands. J Geophys Res 98:4333-4353

Zhao D, Hasegawa A, Kanamori H (1994) Deep structure of Japan subduction zone as derived from local, regional and teleseismic events. J Geophys Res 99:22313-22329

Zhao Y, Anderson DL (1994) Mineral physics constraints on the chemical composition of the Earth's lower mantle. Phys Earth Planet Inter 85:273-292

Zhou H-W, Clayton RW (1990) P and S wave travel time inversion for subducting slab under the island arcs of the northwest Pacific. J Geophys Res 95:6829-6851

Zhang S, Christensen UR (1993) Some effects of lateral viscosity variations on geoid and surface velocities induced by density anomalies in the mantle. Geophys J Int'l 114:531-547

Zhang S, Yuen DA (1995) The influence of lower mantle viscosity stratification on 3D spherical-shell mantle convection. Earth Planet Sci Lett 132:157-166

Zielhuis A, Nolet G (1994) Shear-wave velocity variations in the upper mantle beneath central Europe. Geophys J Int'l 117:695-715

Zindler A, Hart S (1986) Chemical geodynamics. Ann Rev Earth Planet Sci 14:493-571

Chapter 7

THE CORE–MANTLE BOUNDARY REGION

Raymond Jeanloz

Department of Geology and Geophysics
University of California
Berkeley, California 94720

Quentin Williams

Department of Earth Sciences
University of California
Santa Cruz, California 95064

INTRODUCTION

The boundary between the Earth's mantle and core draws attention because of the large contrast in properties across this region. The seismologically observed changes in density and wave-velocities, for example, are significantly greater than across the air-rock (or air-seawater) interface at the Earth's surface (Table 1). Moreover, the difference in materials across the boundary, with predominantly crystalline rock above and liquid iron alloy below, is among the most profound in the Earth. In this sense, the core-mantle boundary can be considered the primary "surface" of the planet, and it is simply because of remoteness that it has attracted less study than the top of the Earth's crust .

In addition to the contrast in horizontally averaged properties (e.g. average density as a function of depth), the region near the core-mantle boundary is also notable for its strong heterogeneity. Figure 1, summarizing the results of global seismic-tomography studies, clearly shows that the degree of lateral heterogeneity is greatest at the bottom as well as the top of the mantle. If one considers heterogeneous properties to be the result of complex geological processes, the top and bottom of the mantle can be loosely viewed as being among the most dynamic regions of the planet. This view is certainly compatible with plate

Table 1. Contrast in properties across Earth's primary structural boundaries.

Property[a]	Earth's surface[b]	Core–mantle boundary	Inner–outer core boundary
Density, $\Delta\rho$ (Mg/m^3)	1.0-2.6	4.3	0.6
Seismic wave velocities (km/s)			
Compressional, V_P	0.7-2.3	5.7	0.6
Shear, ΔV_S	0.0-3.2	7.3	3.4
Electrical conductivity, $\Delta\sigma$ (S/m)	>0-10^2	~10^3-10^4	~0

(a) Global averages based on the Preliminary Reference Earth Model (Dziewonski and Anderson 1981) and a combination of geomagnetic observations and laboratory measurements of electrical conductivity (e.g. Merrill et al. 1996).

(b) Ranges include the differences between seawater and rock being considered as defining the Earth's surface.

0275-0279/98/0037-0007$05.00

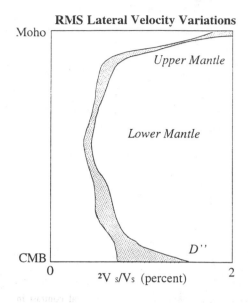

RMS Lateral Velocity Variations

Moho

Upper Mantle

Lower Mantle

D''

CMB

0                    $^2V_s/V_s$ (percent)   2

Figure 1. Seismologically-observed heterogeneity, quantified as the root-mean-square variation in shear-wave velocity at each given depth between the Moho and the core–mantle boundary (CMB), exhibits an increase in the lowermost ~300 km of the mantle which can be associated with the seismologists' D" region (after Li and Romanowicz 1996). The values shown correspond to a "1-σ" variation if the velocities follow a gaussian distribution (i.e. ~3% variation at the ~96% probability 2-σ level). The extreme range of velocity variations at any depth is larger than shown because (1) the figure gives a global average of the variations; (2) the models do not have high resolution (i.e. there is spatial aliasing of small-scale velocity variations); and (3) it is likely that the true distribution of velocities is non-gaussian. Studies of wave-front perturbations (observed using seismic arrays) provide independent evidence of enhanced scattering, and therefore enhanced heterogeneity of seismic-wave velocities, in the lowermost mantle (Bataille et al. 1990; see also Bataille and Lund 1996, Vidale and Hedlin 1998, Shearer et al. 1998).

tectonics and associated geological processes, such as volcanism, seismicity and ore formation observed at the top of the mantle; by analogy, one can imagine that the base of the mantle may be just as geologically active.

Why major structural boundaries of the Earth, such as the top and bottom of the mantle, are both likely to be especially heterogeneous and dynamically active regions is easily understood. A large vertical contrast in density, whether between rock (or seawater) and air at the surface or between iron alloy and silicate rock at depth, ensures that there is virtually no flow across the relevant boundary (Fig. 2). As a result, the pattern of flow must be essentially horizontal just above and below the core-mantle boundary, whereas vertical flow (upwelling and downwelling) dominates in regions away from the boundary. Vertical heat transfer, the process allowing the deep interior to cool, thus depends on conduction across the boundary layers, the regions of predominantly horizontal flow at the top of the core and the base of the mantle (similarly at the Earth's surface, the lithosphere can be associated with the boundary layer at the top of the mantle). The regions away from the boundary, in contrast, are characterized by heat being vertically advected along with the large-scale convection of material; the flow is thus adiabatic throughout the bulk of the mantle and core, away from the boundary layers (e.g. Jeanloz and Morris 1986).

Because the vertical transfer of mass (chemical components), like momentum (stress) and energy (heat), is by the relatively slow process of diffusion across the boundary layers, it is possible for significant lateral heterogeneities to build up just above and below the major boundaries of the Earth (Table 1). Away from the boundary layers, the three-dimensional nature of flow (vs. the flow restricted to two dimensions within the boundary layers) can more readily circulate, mix and disperse heterogeneities. Furthermore, lateral variations in properties (density, in particular) are expected to induce large-scale dynamical processes. For example, buoyant upwelling of hot rock and sinking of cold (therefore dense) subducted lithosphere can be viewed as the processes driving mantle convection and the resultant plate-tectonic motions observed at the surface. Thus, lateral heterogeneities concentrated at a major boundary can play an important role in help-ing to drive or at least strongly modify the large-scale flow patterns of the mantle and core.

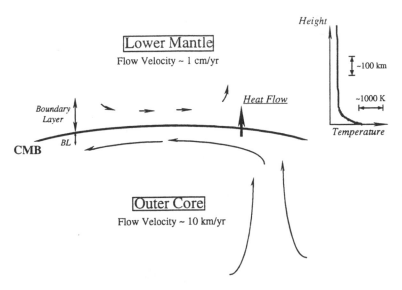

Figure 2. Schematic cross section of a major structural boundary involving a large vertical contrast in density in or on the planet, illustrated here by the core–mantle boundary (CMB), along with the corresponding horizontally averaged temperature profile (geotherm shown in inset). The flow field must become approximately horizontal in thin regions, the boundary layers (BL), on either side of the boundary. Tectonic flow of the mantle is approximately 10^6 slower than that of the outer core, based on geodetic observations at the surface and on modeling the time-varying geomagnetic field (Merrill et al. 1996). This difference in velocities is probably due to the large difference in viscosity between the outer core and lower mantle, estimated to be more than a factor of 10^{20}, which also leads to the mantle-bottom and core-top boundary layers having very different thicknesses (as schematically indicated). In addition to the kinematic boundary layers illustrated here (based on the predominantly horizontal nature of the flow near the boundary), it is also possible to define thermal, compositional ("chemical") and other boundary layers, all of which generally have different thicknesses. The thermal boundary-layer thickness, for example, is given by the diffusion length $\delta_{TBL} \sim (\kappa L/u)^{1/2}$, where L and u are the dimension and velocity of the flow and κ is the thermal diffusivity of the material. Thus, although the thermal diffusivity of the core is expected to be about an order of magnitude larger than that of the mantle, the large difference in flow velocities implies that the thermal boundary layer is much thinner for the top of the core (<1 km) than the bottom of the mantle ($\sim 10^2$ km).

It is the prospect of investigating a geologically active region of the planet that motivates special interest in the core-mantle boundary region. In addition, the past 15 years have witnessed dramatic improvements in geophysical observations, as well as both experimental and theoretical simulations of this region deep inside the Earth (e.g. Lay et al. 1998a, Gurnis et al. 1998). Detailed seismological imaging, along with improved geodetic and geomagnetic observations reveal traces of what would seem to be the results of significant geological activity at the base of the mantle. Correspondingly, the laboratory experiments suggest that intense chemical and physical interactions are essentially inevitable at the core-mantle boundary.

MINERALOGY OF THE CORE-MANTLE BOUNDARY REGION

Lower-mantle mineralogy and temperature

Without detailed elaboration, the classical arguments constraining the bulk composition of the mantle can be summarized as follows. Direct samples of the top-most mantle are observed to be peridotitic (approximately 50% olivine, 40% pyroxene, plus aluminous and other minor phases). Furthermore, peridotite exhibits acoustic properties (V_P, V_S and density) matching those observed seismologically through the mantle, once one takes into account the increase of pressure and temperature with depth, including the mineralogical transformations induced by pressure (Ringwood 1976, Liu and Bassett 1986). That peridotite is roughly similar to the composition of chondritic meteorites, thought to be the remnants of materials from which terrestrial planets accumulated, and can produce basalt upon partial melting at low pressures (i.e. in accord with the extensive eruption of basalts observed at mid-ocean ridges) serves to bolster the conclusion that the bulk composition of the mantle is approximately peridotitic. The detailed composition, such as the exact values for the Mg/Si and Mg/Fe ratios of the mantle, is uncertain but does not matter for the purposes of the present discussion (e.g. Knittle et al. 1986, Hemley et al. 1992, Jackson 1998).

Table 2. Potential mineral constituents of the core–mantle boundary region.

Material	Structure	Maximum observed stability range	Technique
$(Mg,Fe)SiO_3$	Orthorhombic perovskite	127-135 GPa at >2000 K	Laser-heated diamond cell[a]
$CaSiO_3$	Cubic perovskite	135 GPa T unknown	Laser-heated diamond cell[b]
SiO_2	$CaCl_2$-structured	124 GPa at >1273 K	Laser-heated diamond cell[c]
MgO	NaCl-structured	to 227 GPa at 300 K	Diamond cell[d]
$Mg_{0.6}Fe_{0.4}O$	NaCl-structured	to 200 GPa at ~3700 K	Shock-wave[e]
FeO	NiAs-structured	100 GPa at 900 K	Externally heated diamond cell[f]
$FeSi$	B20-structured	49 GPa at 1500 K	Laser-heated diamond cell[g]

(a) Knittle and Jeanloz (1987), Kesson et al. (1998); (b) Kesson et al. (1998), Mao et al. (1989), (c) Tsuchida and Yagi (1989); (d) Duffy et al. (1995); (e) based on no observations of volumetric discontinuities on shock compression: Vassiliou and Ahrens (1982), temperature estimate from Svendsen and Ahrens (1987); (f) Fei and Mao (1994); (g) Knittle and Williams (1995).

The solid-state phase equilibria of peridotitic assemblages at lower mantle conditions are fairly well-constrained to pressures of about 60-85 GPa, corresponding to depths of at least ~1500-2000 km depth (Table 2). Experiments have been carried out both on synthetic and natural samples, starting as component oxides and/or glass, single minerals or even rocks, that have been examined to pressures sometimes exceeding those of the core-mantle boundary. As a result of these studies, the likely crystalline phases present in material of peridotitic bulk composition have been determined to be $(Mg,Fe)SiO_3$ orthorhombic perovskite, $CaSiO_3$ cubic perovskite and $(Mg,Fe)O$ magnesiowüstite. In particular, the stability of silicate perovskite has been verified under lower-mantle conditions through studies utilizing a variety of different starting materials (Knittle and Jeanloz 1987, O'Neill and Jeanloz 1990, O'Neill, et al. 1993, Kesson et al. 1998, Serghiou et al. 1998). Basalt is likewise known to break-down to a two-perovskite assemblage, with about 10-20% stishovite under deep mantle conditions (Kesson et al. 1994, Faust and Knittle 1996,

Hirose et al. 1998). Although minor in amount for peridotite-like bulk compositions, the presence of stishovite (or its $CaCl_2$-structured high pressure phase: Kingma, et al. 1995) is potentially significant due to its relatively high seismic-wave velocity.

Regarding high-temperature phase equilibria, there are limited constraints on the probable liquidus phase of olivine- and of model-mantle assemblages to pressures of 50 GPa. Diamond-cell experiments to 55 GPa (Williams 1990) are compatible with multi-anvil experiments to 27 GPa (Ito and Katsura 1992) in showing that (Mg,Fe)O is probably the liquidus phase of perovskite-magnesiowüstite assemblages to depths of about 1500 km. Although a report that perovskite is the liquidus phase at pressures near 35 GPa does exist (Campbell et al. 1992), we tentatively conclude that magnesiowüstite is likely to be the predominant liquidus phase throughout the bulk of the lower mantle. Studies designed to constrain the temperature at which lower-mantle assemblages melt are similarly limited to olivine compositions. Zerr, et al. (1998) measured an upper bound on the eutectic temperature of an olivine sample to 70 GPa, and extrapolated this result to 135 GPa to obtain a solidus temperature of ~4000 K at CMB conditions. Similarly, Holland and Ahrens (1997) measured the temperature at which Mg_2SiO_4 apparently melts under shock loading as 4300 (±300) K at 130 GPa. This was inferred to represent eutectic melting of the periclase-perovskite system, but neither the amount of melting nor the identity of coexisting solid phases (if any) is known for these experiments.

Many other detailed but significant aspects of deep-mantle phase equilibria remain uncertain. For example, there are reports that $(Mg,Fe)SiO_3$-perovskite dissociates to (Mg,Fe)O and SiO_2 at pressures above 60-80 GPa (Meade et al. 1995b, Saxena et al. 1996), but it is unclear the degree to which partial melting or metastability have influenced either these experiments or the studies indicating that silicate perovskite appears to be stable to the base of the mantle. There are too few experiments at pressures above 60 GPa to separate the effects of sample composition (both bulk composition and the possible presence of impurities), temperature and kinetic hindrances on the reported results. Similarly, the degree to which iron partitions between coexisting $(Mg,Fe)SiO_3$-perovskite and (Mg,Fe)O-magnesiowüstite has only been examined to pressures of ~50 GPa (Kesson and Fitzgerald 1991, Mao et al. 1997).

The average temperature throughout the lower mantle is determined by starting with the petrologically determined temperature of ~1850 K at the 410 km discontinuity, which extrapolates to about 2000 K at the top of the lower mantle (Jeanloz and Richter 1979, Akaogi et al. 1989). A range of fluid dynamic simulations have indicated that material exchange between the upper and lower mantle may be time-dependent due to the effects of phase transitions, and that superadiabatic gradients of several hundred degrees could therefore be present within the transition zone (Honda et al. 1993, Tackley et al. 1994, Peltier 1996). As a result, the temperature at the top of the lower mantle can be loosely bounded between 2000 and 3000 K. Adding on the adiabatic gradient through the lower mantle,

$$(\partial T/\partial P)_S = \gamma T/K_S \qquad (1)$$

where K_S is the adiabatic bulk modulus and γ is the Grüneisen parameter (= $\alpha K_S V/C_P$, with α, V and C_P being the thermal expansion, volume and specific heat, respectively) results in an estimated temperature of about 2500-4000 K at the base of the mantle (Fig. 3). This wide range of values includes an uncertainty of at least 200 K due to the various estimates available for the Grüneisen parameter of silicate perovskite, from ~1.3 (Anderson et al. 1996) to 1.8-2.0 (Hemley et al. 1992). Overall, these values are compatible with the solidus temperatures given above and the seismologically-based conclusion that the bulk of

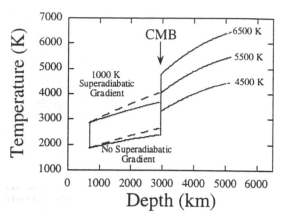

Figure 3. Estimated geotherms within the Earth's core and mantle, with dotted and solid lines within the mantle being for Grüneisen parameters of 1.8 and 1.3, respectively (after Williams 1998). The pressure at the top of the lower mantle, core–mantle boundary, inner-core boundary and center of the Earth is successively 24, 136, 329, 364 Gpa.

the lower mantle is not molten, but there is a suggestion that average temperatures may be close to the melting point in the deep mantle.

Outer-core fluid and temperature at the core-mantle boundary

As with the lower mantle, the general arguments constraining the bulk composition of the core are readily summarized (e.g. Jeanloz 1990): (1) the outer core has both density and elastic properties similar to those of Fe and nearby elements in the Periodic Table; (2) among these candidate elements, Fe is by far the most abundant in the universe (the underlying reason for this is due to the special stability of the ^{56}Fe nucleus); and (3) the generation of the Earth's magnetic field requires that there is a large volume of electrically conducting fluid at great depth (the outer core, being known to be fluid based on seismological observations, is the natural candidate). It thus seems inescapable that the outer core consists predominantly of an iron-rich alloy, perhaps similar in composition to the iron-nickel alloys observed in meteorites.

Although the properties of the core are close to those of Fe, they are not identical: the outer core is about 10% less dense than Fe at comparable pressures and temperatures. Therefore, it is certain that the outer core is not pure iron but an alloy, with sulfur, silicon, oxygen and hydrogen being among the alloying elements that have been proposed. Models favoring one or another of these alloying constituents differ in the assumptions made about when the core acquired its present composition (e.g. prior to and during the formation of the Earth, or as an ongoing contamination process throughout the geological history of the planet). Unfortunately, the wide range of possible alloying elements results in significant uncertainties in the estimated temperature of the core-mantle boundary region.

The temperature throughout the outer core is estimated by extrapolating backward from the temperature at the boundary between the inner core and outer core. Experimentally based estimates of the melting temperature of iron at the inner-core boundary range from 4850 (±200) K to 7600 (±500) K (Brown and McQueen 1982, Williams et al. 1987, Yoo et al. 1993, Saxena et al. 1994, Boehler 1996). Despite this wide range of values, the consequences for estimating the temperature at the core-mantle boundary are comparable to or perhaps even less than the uncertainties due to the effects of alloying.

The issues involving alloying components are two-fold: first, whether these actually shift the melting temperature of core material relative to pure iron; and second, to the degree that they cause melting-point depression of the iron-rich liquid, the amount of melting-point depression is governed by the mole fraction of the alloying components. There is general experimental agreement that oxygen does not produce significant melting-point depression of iron at high pressures (Knittle and Jeanloz 1991b, Boehler 1996), while the available data indicate that sulfur does produce melting-point depression to 90 GPa (Williams and Jeanloz 1990, Boehler 1996). The effect of hydrogen on the melting temperature of iron at high pressures is unknown, but the mole fraction of hydrogen needed to satisfy the density deficit of the core relative to pure iron is ~0.4 (Badding et al. 1992). Assuming a model of ideal melting-point depression, hydrogen could reduce the melting temperature of the outer core by up to ~2500 K relative to pure Fe; for comparison, ideal melting-point depression for a sulfur-bearing core is only 700-1100 K (Williams 1998).

Taking all of these uncertainties into account, the temperature at the inner core-outer core boundary is likely between 4500 and 6500 K (Fig. 3). An adiabatic extrapolation to lower pressures results in a temperature at the top of the core of ~3500-4700 K. A comparison with the adiabatic temperature distribution through the lower-mantle suggests that there needs to be a non-adiabatic temperature increase of about 1000-2000 K across the core-mantle boundary region (e.g. Williams 1998). This is a large temperature offset to accommodate across a set of thermal boundary layers within the Earth, assuming that the mantle above and core below are each homogeneous (Jeanloz and Richter 1979). Therefore, although the estimates of average temperature as a function of depth are mutually compatible (Fig. 3), they do suggest that the core-mantle boundary consists of more than a set of simple thermal boundary layers (e.g. Fig. 2).

Chemical reactions at the core-mantle boundary

What happens when the primary constituents of the outer core and lower mantle, liquid iron alloy and a crystalline assemblage of silicate perovskite plus oxide, are placed in contact with each other at the conditions of the core-mantle boundary? Despite uncertainties in the detailed compositions, phase assemblages and temperatures involved, experiments at ultrahigh pressures and temperatures can directly address this question. What is found is that the liquid metal and crystalline oxides react chemically, at least for pressures above ~20-30 GPa (Knittle and Jeanloz 1989, 1991a).

The occurrence of chemical reactions between core and mantle materials has been demonstrated in several experiments at ultrahigh pressures, including in studies of dried samples (Knittle and Jeanloz 1989, 1991a; Goarant et al. 1992, Poirier et al. 1998). A typical reaction can be written as

$$(Mg_xFe_{1-x})SiO_3 + 3[(1-x)-s]Fe =$$
$$xMgSiO_3 + sSiO_2 + [(1-x)-s]FeSi + [3(1-x)-2s]FeO \qquad (2)$$

where $1-x$ is the amount of iron in the reacting mantle material and s is the amount of silica generated by the reaction (Knittle and Jeanloz 1991a). That such a reaction is expected to proceed to the right at lowermost-mantle pressures and temperatures has also been shown by way of thermochemical analyses (Song and Ahrens 1994, Knittle and Williams 1995). Whether one considers this to be an oxidation-reduction (redox) reaction or a dissolution (alloying) of oxygen into the metal phase is somewhat a matter of semantics at present, because the concepts of valence and charge transfer are currently ill defined at the conditions of the core-mantle boundary. Interestingly, the reaction evolves iron-depleted silicates plus iron alloys as products, which are expected to exhibit distinctly fast and slow seismic velocities, respectively.

Experience at low pressures is quite different, with silicates and liquid iron exhibiting relatively little tendency to react chemically. For example, the steel industry utilizes the relative immiscibility of oxides ("slag") and liquid iron alloy, which is due to the very different bonding character of these two constituents. The elevated pressure of the core-mantle boundary has a profound effect on thermodynamic equilibria, however. This can be understood in general terms by noting that the pressure-volume work done on compressing condensed matter to the 100 GPa (10^6 atm) range is of the order of eV ($\sim 10^5$ J/mole of atoms), comparable to chemical bonding energies (Jeanloz 1989).

More specifically, it has been shown that FeO transforms from a non-metallic compound to a metal alloy at pressures of the deep mantle (Knittle and Jeanloz 1991b), thus documenting that any FeO produced at the core-mantle boundary via (2) is expected to be metallic. One can interpret this as an indication that at high pressures oxygen becomes a metal-alloying component or, more crudely, that it takes on the chalcogenide character of the next elements in its period, S and Se. The point is perhaps most forcefully made by the observation that oxygen itself becomes a metal (even a superconductor) at pressures above 100 GPa (Shimizu, et al. 1998). Using the classical terminology of geochemistry, oxygen apparently transforms from being the archetypal lithophile element at low pressures to becoming siderophile at the pressures and temperatures of the core-mantle boundary. With this in mind, one can imagine that even if the core had initially been pure Fe (or iron-nickel alloy) early in Earth's history, it may inevitably have become contaminated over geological time as the mantle rock slowly dissolves into the liquid metal of the core.

If chemical reactions do occur at the core-mantle boundary, it is also quite plausible that partial melting takes place in this region. In addition to the high temperature of core fluid initiating melting, there is potentially a significant melting-point depression caused by the metal and silicate systems (or their reaction products) being brought into contact. If partial melting does take place, this is likely to enhance the tendency of chemical reaction between (or mutual dissolution of) mantle and core materials. That silicate melts readily undergo chemical reactions with iron-rich material has been experimentally demonstrated to pressures of 24 GPa (Ito et al. 1995), and increased pressure is expected to enhance the thermochemical driving force for such reactions (Knittle and Jeanloz 1991a). Therefore, it is plausible that chemical reactions and partial melting may be intimately connected in the lowermost mantle.

DETAILED OBSERVATIONS OF THE
CORE-MANTLE BOUNDARY REGION

Velocity gradients and the D" region

The traditional seismic definition of the core-mantle boundary layer originated with Bullen (1949), who designated a zone of decreased depth-gradients of seismic velocities at the base of the mantle as D" (the lower mantle, or D layer in travel-time analyses of seismic phases, being split into a D' layer exhibiting typical increases in seismic-wave velocities with depth, overlying the anomalous D" layer). In modern one-dimensional velocity models of the Earth's interior, the decrease in gradient defining the D" layer initiates approximately 150 km above the core-mantle boundary (Dziewonski and Anderson 1981). An average thickness of ~200 kilometers is also compatible with the seismic velocity variations observed in the deep mantle (Fig. 1), and with classical evidence of enhanced seismic-wave scattering near the bottom of the mantle (Cleary and Haddon 1972).

Two improvements are required in order to refine this picture of the lowermost mantle. First, spatial variability must be taken into account, either through tomographic methods or by studying a particular patch of the core-mantle boundary. Second, waveform modeling of the seismograms provides far more detailed information about the velocity structure than does the classical approach of only studying travel times. It is important to recognize that as the lower-mantle structure comes into better focus, the definition of D" as the anomalous layer at the base of the mantle must inevitably vary. For example, a zone of anomalous heterogeneity that is 200 km thick on average, may be 500-800 km thick in some regions and absent (zero thickness) in others. Therefore, we apply a loose definition of D" as being the zone of anomalous seismic properties at the base of the mantle; it need not have a uniform thickness, and it may be discontinuous from one location to another.

Seismic tomography

Tomography depends on having numerous sources and receivers of seismic energy, so that data are collected for multiple sets of intersecting seismic-wave paths. It is then possible to unravel the three-dimensional structure at depth by determining the sign and magnitude of the velocity variation required to match the data for all paths intersecting at a given point (within a given volume) in the mantle. Doing this point-by-point throughout the mantle results in a tomographic model of the structure. The approach is crudely similar to inferring the location and shape of an object by observing the shadows it casts under many different angles of illumination.

The results of global tomography have already been summarized (Fig. 1), but more detailed imaging of the deep mantle has notably improved over the past decade, if for no other reason than that the quality and quantity of seismic recordings have improved. At least one sheet of high-velocity is now convincingly resolved within the mid-lower mantle (van der Hilst et al. 1997), and this slab-like structure has been associated with the subducted Farallon plate.

At depths between 1800 and 2300 km, however, slab-like features are no longer resolved, and the form of mantle heterogeneities appears to be fundamentally altered in comparison with shallower depths (van der Hilst et al. 1998). Indeed, the amplitudes of velocity variations associated with slab-like features in the lower mantle are typically far less than those observed within D": maximum velocity variations of ±0.5% are tomographically resolved in the mid-mantle, while inferred variations of 2-3% or more are commonly observed in D" (van der Hilst et al. 1997, 1998) (Fig. 1). The present state of tomographic imaging thus implies either that slabs generally do not penetrate into the lowermost mantle, or that they are no longer seismologically distinguishable near the core-mantle boundary. Therefore, both the amplitude and depth distribution of slab-related velocity anomalies show that D" structure cannot be attributed to the simple presence or accumulation of subducted slabs, but that it is due to other processes.

Seismic waveform modeling

Whereas tomography is used to reveal the two- and three-dimensional character of structures at depth, a separate constraint on mantle structure is obtained through analyses of the detailed amplitudes of seismic waves. The special advantage to this approach is that seismic waveforms depend fundamentally on the spatial gradient of velocity. In contrast, travel times depend on the integrated velocity along the path of the seismic wave, which conveys far less detailed information about the structures. Because the analysis is so involved, however, it is usually the case that waveform modeling is carried out for a set of records pertaining to a specific region of the mantle. Hence, waveform analysis generally produces a model for a local structure, for example on or near the core-mantle boundary.

Detailed waveform studies of precursors to core-reflected S and P waves demonstrate the presence of considerable variability in the wave-velocities in the region several hundred kilometers above the core-mantle boundary (Lay and Helmberger 1983, Wright et al. 1985, Wysession et al. 1998, Valenzuela and Wysession 1998). Anomalies of ~0.5-5% in shear-

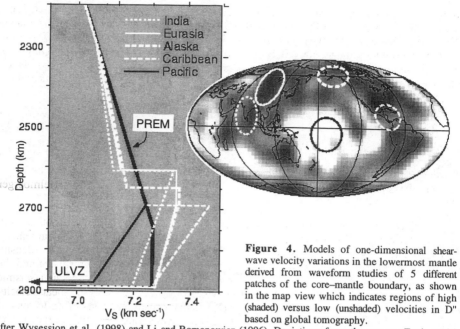

Figure 4. Models of one-dimensional shear-wave velocity variations in the lowermost mantle derived from waveform studies of 5 different patches of the core–mantle boundary, as shown in the map view which indicates regions of high (shaded) versus low (unshaded) velocities in D" based on global tomography. After Wysession et al. (1998) and Li and Romanowicz (1996). Deviations from the average-Earth model PREM (Dziewon-ski and Anderson 1981) begin ~200-600 km above the core–mantle boundary.

and compressional-wave velocities are sporadically observed, with the deviation from one-dimensional average-Earth models (e.g. PREM) starting as much as 500-600 km above the core-mantle boundary (Fig. 4). The observation of negative shear velocity gradients existing in some regions (particularly below the Pacific) for as much as 150 kilometers above the top of the core is also indicated by studies of shear waves diffracted near the core-mantle boundary (Maupin 1994, Ritsema et al. 1997). It is important to recognize that there are uncertainties in the inferred structures both because of intrinsic trade-offs (e.g. between the depth at which a discontinuity is inferred to exist and the overlying velocity gradient: Wysession et al. 1998) and because analyses have relied on 1-dimensional modeling, which is suspect for a medium as heterogeneous as the lowermost mantle.

Seismic constraints on ultra-low velocity zones

Over the past few years, three types of seismic data have indicated the presence of 5-50 km thick zones at the base of the mantle characterized by decreases in compressional- and shear-wave velocities of up to tens of percent. Such large velocity anomalies are essentially unknown in other regions of the mantle, and so have been dubbed ultra-low velocity zones (ULVZ). The first inference of such anomalous patches existing intermittently at the core-mantle boundary came from seismic observations incorporating diffracted compressional waves propagating along the base of the mantle (Garnero et al. 1993, Garnero and Helmberger 1996, Wen and Helmberger 1998a, Garnero et al. 1998, Williams et al. 1998). Also, wave energy is diffracted along the core-mantle boundary because of the large contrast in velocities there and, like surface waves at the top of the mantle, provides a sensitive probe of the boundary region. Finally, core-reflected phases (e.g. PcP and ScP) and their precursors have revealed a substantial decrease in shear velocity in one sub-Pacific spot, with the onset of this anomalous layer being modeled over a depth range of less than 5 km (Mori and Helmberger 1995, Revenaugh and Meyer 1997).

Moreover, the coda of core-traversing compressional waves has been modeled in terms of a large-amplitude and heterogeneous decrease in seismic velocity at the base of the mantle (Vidale and Hedlin 1998), in accord with arrival-time data that are consistent with a ~20 km thick layer exhibiting a compressional wave velocity reduced by 10% or more (Sylvander et al. 1997). A key condition for extracting velocity anomalies over a relatively thin region from such arrival-time data involves accounting for the variations in wave velocities along the path traversed by the seismic ray outside of the region of interest.

The intrinsic heterogeneity of the lowermost mantle is highlighted by observations of reflectivity from the top of ultra-low velocity patches in some places (Mori and Helmberger 1995, Revenaugh and Meyer 1997), and observations of a sharp core-mantle boundary in others (Vidale and Benz 1992). Indeed, a few studies suggest that the ULVZ may be distributed very heterogeneously, with roughly dome-shaped low-velocity regions of 10's to 100's of kilometers lateral extent (Vidale and Hedlin 1998, Wen and Helmberger 1998b).

Trade-offs naturally exist between the inferred thickness, wave velocities and density of any ULVZ (e.g. Garnero and Helmberger 1998). The waveforms are fundamentally dependent on velocity variations (causing refractions), contrasts in impedance (density times velocity, which influences how much energy is reflected within a waveguide) and the product of velocity times thickness which controls the arrival time of individual seismic phases. It has been found that the waveforms typically require contrasts in shear-velocity exceeding those in compressional-wave velocity, and that density increases of up to tens of percent are compatible with the data (Wen and Helmberger 1998a, Garnero and Helmberger 1998). Both the large magnitude of the velocity decreases (tens of percent) and the larger anomaly in shear versus compressional velocity are readily explained by partial melting at the base of the mantle (Williams and Garnero 1996). Altogether, these findings are in remarkable accord with the general picture developed from laboratory studies, namely that a combination of melting and reaction-contamination of the lowermost mantle by the outer-core fluid can readily explain the reduced velocities and increased densities observed at the core-mantle boundary.

Other intriguing results lend support to the idea that partial melting is actually being observed in thin patches at the base of the mantle. For example, Garnero and Helmberger (1998) find tentative evidence for a zone of enhanced absorption of seismic waves in the ULVZ, just as would be expected for propagation through partially molten rock (e.g. Schmeling 1985). In addition, six regions of the mantle have been shown to be underlain by a ULVZ (Fig. 5), amounting to 12% out of the 44% of the core-mantle boundary surface area that has been probed to date, and it has been proposed that there is a strong correlation of the ULVZ patches with the distribution of hot spots at the surface (Williams et al. 1998, Garnero et al. 1998, Lay et al. 1998a). If so, it would appear that the partially-molten patches observed as ultra-low velocity zones at the base of the mantle may play a key role in the genesis of plumes (Williams et al. 1998).

The prospect that seismologists are now imaging phenomena at the core-mantle boundary that are so directly related to igneous and tectonic processes at the Earth's surface is truly revolutionary. The presence of melt in the lowermost mantle would suggest that this zone has a markedly low viscosity, perhaps leading to complex and rapid flow within or throughout the boundary layer. Accordingly, the rate at which mantle material is cycled into proximity with (and away from) the outer core is likely to be rapid, enhancing any chemical interactions between core and mantle.

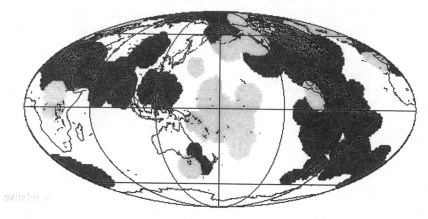

Figure 5. Distribution of regions on the core–mantle boundary that do (light shading) and do not (dark shading) show evidence of an ultra-low velocity zone (ULVZ) (after Garnero et al. 1998). Regions with no shading have not yet been adequately sampled by seismological observations to either reveal or rule out the presence of a ULVZ.

Seismic anisotropy: A signature of deep mantle flow?

Although the splitting of vertically- and horizontally-polarized shear waves deep in the mantle was recognized nearly a quarter century ago (Mitchell and Helmberger 1973), most studies of seismic-wave anisotropy near the core-mantle boundary have been conducted in the last half-decade. This anisotropy is the acoustic analog of birefringence in mineral optics, so it is the both the orientation and magnitude of the anisotropy in shear-wave velocity that must be characterized. For the most part, horizontally polarized shear waves are observed to propagate ~0-3% faster at the base of the mantle than longitudinally polarized waves (Maupin 1994, Kendall and Silver 1996, Lay et al. 1998b), though laterally variable anisotropy (including zones in which SV may be faster than SH) occur beneath the Pacific (Lay et al. 1998a, 1998b). The transverse isotropy can be created by horizontal (but azimuthally variable) alignments of anisotropic minerals and or of heterogeneous inclusions (e.g. patches of melt or contaminants). Unlike these regions near the core-mantle boundary, however, the bulk of the lower mantle shows no evidence of anisotropy (Kaneshima and Silver 1992, Meade et al. 1995a). This suggests that anisotropy may be confined to the thermal boundary layer at the base of the mantle, and reflects the deformation of the lowermost mantle. Indeed, it now seems plausible that the discontinuity typically found ~200-300 km above the core in waveform models based on one-dimensional velocity structures (Fig. 4) is actually due to a change in anisotropy and heterogeneity as one descends through the lowermost mantle (Lay et al. 1998a).

The observation of anisotropy within D" has motivated a number of experimental and theoretical studies in mineral and rock anisotropy at deep-mantle conditions (Duffy et al. 1995, Meade et al. 1995a, Karki et al. 1997, Stixrude 1998). The primary conclusion is that the anisotropy of deep mantle phases is likely to be comparable to or larger than that present at ambient pressures: typical values are ~10-55% for the magnitudes of single-crystal anisotropy of $MgSiO_3$ perovskite, MgO and SiO_2 (e.g. Stixrude 1998). Theoretical calculations indicate that oriented polycrystalline aggregates of these minerals may have anisotropies of 6-8%, with SV > SH for perovskite and periclase, a result with an opposite sign from most seismic studies of D" (Stixrude 1998, Lay et al. 1998b). However, the sign and magnitude of anisotropy depends critically on inferences about the slip systems of deep-mantle minerals, which are only constrained at low pressures for MgO or from studies of analog materials in the case of perovskite and SiO_2 (e.g. Karato et al. 1995). Alternatively, structural anisotropy induced by intercalations of lamellae with differing elastic properties could also produce anisotropy of the observed magnitude. Aligned melt inclusions are particularly effective at generating shear-wave anisotropy (Kendall and Silver 1996, 1998), and any aligned SiO_2-bearing regions could also generate significant anisotropy if the transition from stishovite to $CaCl_2$-structured SiO_2 occurs under lowermost-mantle conditions (e.g. Cohen 1992).

Perhaps the most basic question, however, is: Why does anisotropy occur within D" and apparently not in the overlying mantle (Meade et al. 1995a)? If anisotropy is structural in origin, then chemical heterogeneities may be more abundant within D" or the flow field differs markedly from that within the overlying mantle. The onset of mineralogical anisotropy within D" could, in fact, be directly caused by the presence of a thermal boundary layer at the base of the mantle. For example, the increased temperature within the thermal boundary layer could well change the mineral-deformation mechanisms relative to those of the overlying mantle, thus affecting the flow field or the minerals' response (e.g. preferred alignment) to the flow field (Lay et al. 1998a,b).

IGNEOUS AND METAMORPHIC PROCESSES
OF THE CORE-MANTLE BOUNDARY

The heterogeneity of the lowermost mantle is presumably related to the geological activity of the core-mantle boundary region. Seismological observations reveal large variations in velocities, interpreted as due to partial melt being present at or near the base of the mantle; laboratory experiments suggest that chemical reactions take place between the constituents of the mantle and core. To outline possible scenarios for what may be occurring near the core-mantle boundary, we use analogies with what has been documented geologically in the Earth's upper boundary layer, the lithosphere. Of course, the danger in this approach is that the analogies may be faulty. This is particularly the case as large variations in density, thermal conductivity and viscosity near the core-mantle boundary are likely to strongly influence the dynamics of the lowermost mantle in ways that may not be simply visualized from analogs in the shallower Earth (Manga and Jeanloz 1996, Williams and Garnero 1996, Kellogg 1997, Montague et al. 1998, Williams et al. 1998).

The outer core is liquid, so we can think of the core-mantle boundary in terms of a zone of contact metamorphism and metasomatism. A picture of core liquid infiltrating crystalline mantle rock may be adequate; alternatively, this may be too simplistic, as the recent seismological observations would suggest the presence of a density-stratified mush consisting of partially molten silicate (ULVZ material) and (partially crystallizing?) liquid alloy of the outermost core, reminiscent of anatexis in the crust. The liquid iron alloy of the outermost core can thus infiltrate into the lowermost mantle either through solid-state or through liquid-state processes, just as metasomatic fluids can permeate wall rock surrounding an intrusion or basaltic blebs can be entrained in granitic magma.

The large density contrast between mantle and core must be taken into account, however (Poirier and LeMouël 1992). If there is topography on the core-mantle boundary, the high density of the metal relative to the silicate can be used to help drive infiltration and reaction (Fig. 6) (Jeanloz 1993). For no partial melting of the lowermost mantle, this is analogous to the movement of water in an aquifer. However, the infiltration in this case depends on chemical reactions taking place between the liquid metal and crystalline silicates, because there is no hope of the lowermost mantle having any porosity whatsoever: due to the high pressures and temperatures of the deep mantle, any porosity that might appear is rapidly squeezed shut. If the mantle rock is partially molten, combined

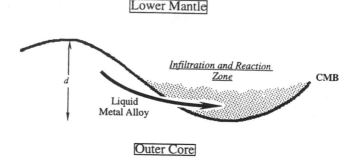

Figure 6. Schematic illustration of how outer-core liquid may infiltrate into the lowermost mantle, assuming that topography of amplitude d is present on the core–mantle boundary. If the metal and silicate react chemically, infiltration-reaction allows the dense metal alloy to permeate "downhill" into the mantle rock from regions of elevated topography to regions where the core–mantle boundary is low. Geodetic and seismological observations are compatible with topography of magnitude $\sim 10^2$-10^3 m (e.g. Buffett 1992, 1996), which would then be the maximum width of the zone of chemical reaction at the base of the mantle (Jeanloz 1990, 1993).

reaction-infiltration including physical mixing can proceed readily. Overall, the rates of such reaction, infiltration and mixing processes are unclear, but they are probably accelerated by the presence of partial melting in the lowermost mantle.

Once a patch of the core-mantle boundary has undergone infiltration and chemical reaction, it is likely that further chemical reaction is initially inhibited because the reactant fluid would now have to traverse through a finite thickness of rock that is already iron-enriched. The core-mantle boundary therefore becomes armored. This type of local armoring is only temporary on a geological timescale, however, because the reacted rock is expected to be entrained within the large-scale convective flow of the entire mantle (e.g. Davies and Gurnis 1986, Sleep 1988, Hansen and Yuen 1989, Olson and Kincaid 1991, Kellogg and King 1993, Kellogg 1997) or local flows associated with the core-mantle boundary layer.

The effects of chemical reactions between mantle and core could therefore be expressed over two distinct lengthscales (Jeanloz 1990, 1993). First, the actual zone of reaction and infiltration is likely to be relatively thin ($\sim 10^2$-10^3 m), and would be expected to develop over geologically short timescales. This zone could have density increases of as much as several tens of percent. Second, over the longer timescales characteristic of large-scale mantle convection, tens to hundreds of millions of years, regions containing infiltrated and reacted mantle material could be swept upward as much as hundreds of kilometers off the core-mantle boundary (the average density of the combined reacted and unreacted mantle material being swept up may be only a few percent different from that of pristine, unreacted mantle). The image one might have is of iron-rich "drifts" of reacted mantle rock being "blown about" by the background wind of mantle convection (analogous to wind-blown snow drifts). Two distinct length- and time-scales are characteristic of analogous geological phenomena, such as the infiltration of seawater to depths of $\sim 10^2$-10^3 m into the oceanic lithosphere over short times, followed by the long-term subduction of hydrated and weathered lithosphere to hundreds of kilometers into the mantle (note that downward infiltration of buoyant water in denser rock corresponds to upward infiltration and sweeping of dense core components in the overlying mantle).

In short, chemical reactions and partial melting are likely to be highly synergistic in the core-mantle boundary region. Infiltration of mantle material by core liquid is likely to trigger or increase the degree of melting in D'', yet even modest degrees of iron contamination of a partially molten zone could stabilize large portions of the ULVZ against buoyant rise (e.g. Knittle 1998). As the percentage of the core-mantle boundary occupied by the ULVZ is comparable to that occupied by continental crust at the surface of the planet (Fig. 5), the stabilization of most of the ULVZ (and perhaps major portions of the thermal boundary layer, as well) through negative buoyancy may be critically important in reducing the volume upwelling out of the lowermost mantle.

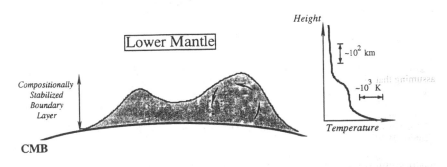

Figure 7. Schematic cross section of a compositionally-stabilized, internally convecting region at the base of the mantle (shaded), along with the corresponding temperature profile (inset). The mantle above and core below the compositionally-distinct zone both convect, as in Figure 2 (not shown here). The multiple thermal boundary layers associated with the internally convecting region can lead to a much larger temperature jump from the deep mantle to the outermost core, in contrast with the simple thermal boundary layer shown at the base of the mantle in Figure 2. Consequently, partial melting can readily occur in the high-temperature regions of the lowermost mantle.

One consequence of dense material being infiltrated and then (partially) swept upward into the mantle is that the thermal boundary layer above the core-mantle boundary may be stabilized (Jeanloz and Richter 1979). That is, because of the presence of a dense component, it becomes much more difficult to sweep lowermost-mantle material up high above the core-mantle boundary. Such chemically-induced stabilization can significantly influence the thermal and dynamic regime of the deep Earth (Fig. 7), resulting in a temperature distribution that is compatible with the estimates of mantle and core temperatures (Fig. 3) which suggest that there is a large (~1000-2000 K) jump in temperature in the core-mantle boundary region (Williams 1998). If the lowermost mantle is chemically homogeneous, a temperature jump is likely to be limited to about 1000 K due to small-scale convection (Loper and Eltayeb 1986); however, if there is contamination of the lowermost mantle by the core, then even temperature increases of 2000 K may be stably maintained across the core-mantle boundary region (Sleep 1988, Farnetani 1997). Possible geotherms may thus include multiple boundary layers (Fig. 7), however it is important to recognize that there is likely to be considerable variability in the temperature at a given depth near the base of the mantle, much as there is near the surface of the planet.

ACKNOWLEDGMENTS

We thank E.J. Garnero and R.J. Hemley for helpful comments and discussions. This work was supported by the National Science Foundation..

REFERENCES

Akaogi M, Ito E, Navrotsky A (1989) Olivine-modified spinel-spinel transitions in the system Mg_2SiO_4-Fe_2SiO_4—Calorimetric measurements, thermochemical calculation, and geophysical application. J Geophys Res 94:15671-15685

Anderson OL, Masuda K, Isaak DG (1996) Limits on the value of δ_T and γ for $MgSiO_3$ perovskite. Phys Earth Planet Inter 98:31-46

Badding JV, Mao HK, Hemley RJ (1992) High-pressure crystal structure and equation of state of iron hydride: Implications for the Earth's core. *In:* Y Syono, MH Manghnani (eds) High-Pressure Research: Application to Earth and Planetary Sciences, p 363-371 American Geophysical Union, Washington, DC

Bataille K, Lund F (1996) Strong scattering of short-period seismic waves by the core-mantle boundary and the P-diffracted wave. Geophys Res Lett 23:2413-2416

Bataille K, Wu RS, Flatté S (1990) Inhomogeneities near the core-mantle boundary evidenced from scattered waves: A review. Pageoph 132:151-173

Boehler R (1996) Melting of mantle and core materials at very high pressures. Phil Trans R Soc Lond A 354: 1265-1278

Brown JM, McQueen RG (1982) The equation of state for iron and the Earth's core. *In:* S Akimoto, MH Manghnani (eds) High-Pressure Research in Geophysics, p 611-623 Center for Academic Publications, Tokyo

Buffett BA (1992) Constraints on magnetic energy and mantle conductivity from the forced nutations of the Earth. J Geophys Res 97:19581-19587

Buffett BA (1996) Effects of a heterogeneous mantle on the velocity and magnetic fields at the top of the core. Geophys J Int 125:303-317

Bullen KE (1949) Compressibility-pressure hypothesis and the Earth's interior. Mon Not R Astr Soc, Geophys Suppl. 5:355-368

Campbell AJ, Heinz DL, Davis AM (1992) Material transport in laser-heated diamond anvil cell melting experiments. Geophys Res Lett 19:1061-1064

Cleary JR, Haddon RAW (1972) Seismic wave scattering near the core-mantle boundary: A new interpretation of precursors to PKP. Nature 240:549-551

Cohen RE (1992) First-principles predictions of elasticity and phase transitions in high pressure SiO_2 and geophysical implications. *In:* Y Syono, MH Manghnani (eds) High-Pressure Research: Application to Earth and Planetary Sciences, p 425-431 American Geophysical Union, Washington, DC

Davies GF, Gurnis M (1986) Interaction of mantle dregs with convection: Lateral heterogeneity at the core-mantle boundary. Geophys Res Lett 13:1517-1520

Duffy TS, Hemley RJ, Mao HK (1995) Equation of state and shear strength at multimegabar pressures: MgO to 227 GPa. Phys Rev Lett 74:1371-1374

Dziewonski AM, Anderson DL (1981) Preliminary reference Earth model. Phys Earth Planet Inter 25:297-356

Farnetani CG (1997) Excess temperature of mantle plumes: The role of chemical stratification across D". Geophys Res Lett 24:1583-1586

Faust J, Knittle E (1996) The stability and equation of state of majoritic garnet synthesized from natural basalt at mantle conditions. Geophys Res Lett 23:3377-3380

Fei Y, Mao HK (1994) In situ determination of the NiAs phase of FeO at high pressure and temperature. Science 266:1678-1680

Garnero EJ, Grand SP, Helmberger DV (1993) Low P wave velocity at the base of the mantle. Geophys Res Lett 20:1843-1846

Garnero EJ, Helmberger DV (1996) Seismic detection of a thin laterally varying boundary layer at the base of the mantle beneath the central-Pacific. Geophys Res Lett 23:977-980

Garnero EJ, Helmberger DV (1998) Further structural constraints and uncertainties of a thin laterally varying ultralow-velocity layer at the base of the mantle. J Geophys Res 103:12495-12509

Garnero EJ, Revenaugh J, Williams Q, Lay T, Kellogg LH (1998) Ultralow velocity zone at the core-mantle boundary. *In:* M Gurnis, ME Wysession, E Knittle, B Buffett (eds) The Core-Mantle Boundary Region, p 319-334 American Geophysical Union, Washington, DC

Goarant F, Guyot F, Peyronneau J, Poirier JP (1992) High-pressure and high-temperature reactions between silicates and liquid iron alloys, in the diamond anvil cell, studied by analytical electron microscopy. J Geophys Res 97:4477-4487

Gurnis M, Wysession ME, Knittle E, Buffett BA, eds (1998) The Core-Mantle Boundary Region. American Geophysical Union, Washington, DC

Hansen U, Yuen DA (1989) Dynamical influences from thermal-chemical instabilities at the core-mantle boundary. Geophys Res Lett 16:629-632

Hemley RJ, Stixrude L, Fei Y, Mao HK (1992) Constraints on lower mantle composition from P-V-T measurements of (Fe,Mg)SiO$_3$-perovskite and (Fe,Mg)O. *In:* Y Syono, MH Manghnani (eds) High-Pressure Research: Application to Earth and Planetary Sciences, p 183-189 American Geophysical Union, Washington, DC

Hirose K, Fei Y, Ma Y, Mao HK (1998) Fate of subducted basaltic crust in the lower mantle. Nature (in press)

Holland KG, Ahrens TJ (1997) Melting of (Mg,Fe)$_2$SiO$_4$ at the core-mantle boundary of the Earth. Science 275:1623-1625

Honda S, Yuen DA, Balachandar S, Reuteler D (1993) Three-dimensional instabilities of mantle convection with multiple phase transitions. Science 259:1308-1311

Ito E, Katsura T (1992) Melting of ferromagnesian silicates under the lower mantle conditions. *In:* Y Syono, MH Manghnani (eds) High-Pressure Research: Application to Earth and Planetary Sciences, p 315-322 American Geophysical Union, Washington, DC

Ito E, Morooka K, Ujike O, Katsura T (1995) Reactions between molten iron and silicate melts at high pressure: Implications for the chemical evolution of Earth's core. J Geophys Res 100:5901-5910

Jackson, I (1998) Elasticity, composition and temperature of the Earth's lower mantle: A reappraisal. Geophys J Inter 134:291-311

Jeanloz, R (1989) Physical chemistry at ultrahigh pressures and temperatures. Ann Rev Physical Chemistry 40:237-259

Jeanloz R (1990) The nature of the Earth's core. Ann Rev Earth Planet Sci 18:357-386

Jeanloz R (1993) Chemical reactions at the Earth's core-mantle boundary: Summary of evidence and geomagnetic implications. *In:* K Aki, R Dmowska (eds) Relating Geophysical Structures and Processes, The Jeffreys Volume, p 121-127 American Geophysical Union, Washington, DC

Jeanloz R, Morris S, (1986) Temperature distribution in the crust and mantle. Ann Rev Earth Planet Sci 14:377-415

Jeanloz R, Richter FM, (1979) Convection, composition and the thermal state of the lower mantle. J Geophys Res 84:5497-5504

Kaneshima S, Silver PG (1992) A search for source-side anisotropy. Geophys Res Lett 19:1049-1052

Karato S-I, Zhang S, Wenk HR (1995) Superplasticity in Earth's lower mantle: Evidence from seismic anisotropy and rock physics. Science 270:458-461

Karki BB, Stixrude L, Clark SJ, Warren MC, Ackland GJ, Crain J (1997) Elastic properties of orthorhombic MgSiO$_3$ perovskite at lower mantle pressures. Am Mineral 82:635-638

Kellogg LH (1997) Growing the Earth's D" layer: Effect of density variations at the core-mantle boundary. Geophys Res Lett 24:2749-2752

Kellogg LH, King SD (1993) Effect of mantle plumes on the growth of D" by reaction between the core and mantle. Geophys Res Lett 20:379-382

Kendall J-M, Silver PG (1996) Constraints from seismic anisotropy on the nature of the lowermost mantle. Nature 381:409-412

Kendall J-M, Silver PG (1998) Investigating causes of D" anisotropy. *In:* M Gurnis, ME Wysession, E Knittle, B Buffett (eds) The Core-Mantle Boundary Region, p 97-118 American Geophysical Union, Washington, DC

Kesson SE, Fitzgerald JD (1991) Partitioning of MgO, FeO, NiO, MnO and Cr$_2$O$_3$ between magnesian silicate perovskite and magnesiowustite: implications for the origin of inclusions in diamond and the composition of the lower mantle. Earth Planet Sci Lett 111:229-240

Kesson SE, Fitzgerald JD, Shelley JMG (1994) Mineral chemistry and density of subducted basaltic crust at lower-mantle pressures. Nature 372:767-769

Kesson SE, Fitzgerald JD, Shelley JMG (1998) Mineralogy and dynamics of a pyrolite lower mantle. Nature 393:252-255

Kingma KJ, Cohen RE, Hemley RJ, Mao HK (1995) Transformation of stishovite to a denser phase at lower mantle pressures. Nature 374:243-245

Knittle E (1998) The solid/liquid partitioning of major and radiogenic elements at lower mantle pressures: Implications for the core-mantle boundary region. *In:* M Gurnis, ME Wysession, E Knittle, B Buffett (eds) The Core-Mantle Boundary Region, p 119-130 American Geophysical Union, Washington, DC

Knittle E, Jeanloz R (1987) Synthesis and equation of state of (Mg,Fe)SiO$_3$ perovskite to over 100 GPa. Science 235:668-670

Knittle E, Jeanloz R (1989) Simulating the core-mantle boundary: An experimental study of high-pressure relations between silicates and liquid iron. Geophys Res Lett 16:609-612

Knittle E, Jeanloz R (1991a) The Earth's core-mantle boundary: Results of experiments at high pressures and temperatures. Science 251:1438-1443

Knittle E, Jeanloz R (1991b) The high presssure phase diagram of Fe$_{0.94}$O, a possible constituent of the Earth's outer core. J Geophys Res 96:16169-16180

Knittle E, Jeanloz R, Smith GS (1986) Thermal expansion of silicate perovskite and stratification of the Earth's mantle. Nature 319:214-216

Knittle E, Williams Q (1995) Static compression of ε-FeSi: An evaluation of reduced Si as a deep Earth constituent. Geophys Res Lett 22:445-448

Lay T, Helmberger DV (1983) A lower mantle S-wave triplication and the velocity structure of D" Geophys J R Astron Soc 75:799-837.

Lay T, Williams Q, Garnero EJ (1998a) The core-mantle boundary layer and deep Earth dynamics. Nature 392:461-468

Lay T, Williams Q, Garnero EJ, Kellogg L, Wysession ME (1998b) Seismic wave anisotropy in the D" region and its implications. In: M Gurnis, ME Wysession, E Knittle, B Buffett (eds) The Core-Mantle Boundary Region, p 299-318 American Geophysical Union, Washington, DC

Li XD, Romanowicz B (1996) Global mantle shear velocity model using nonlinear asymptotic coupling theory. J Geophys Res 101:22245-22272

Liu LG, Bassett WA (1986) Elements, Oxides, and Silicates: High-Pressure Phases with Implications for the Earth's Interior. Oxford University Press, New York

Loper DE, Eltayeb IA (1986) On the stability of the D" layer. Geophys Astrophys Fluid Dyn 36:229-255

Manga M, Jeanloz R (1996) Implications of a metal-bearing chemical boundary layer in D" for mantle dynamics. Geophys Res Lett 23:3091-3094

Mao HK, Chen C, Hemley RJ, Jephcoat AP, Wu Y, Bassett WA (1989) Stability and equation of state of $CaSiO_3$-perovskites to 134 GPa. J Geophys Res 94:17889-17894

Mao HK, Shen G, Hemley RJ (1997) Multivariable dependence of Fe-Mg partitioning in the lower mantle. Science 278:2098-2100

Maupin V (1994) On the possibility of anisotropy in the D" layer as inferred from the polarization of diffracted S waves. Phys Earth Planet Inter 87:1-32

Meade C, Mao HK, Hu JZ (1995b) High-temperature phase transition and dissociation of $(Mg,Fe)SiO_3$ perovskite at lower mantle pressures. Science 268:1743-1745

Meade C, Silver PG, Kaneshima S (1995a) Laboratory and seismological observations of lower mantle isotropy. Geophys Res Lett 22:1293-1296

Merrill RT, McElhinny MW, McFadden PL (1996) The Magnetic Field of the Earth. Academic Press, New York

Mitchell BJ, Helmberger DV (1973) Shear velocities at the base of the mantle from observations of S and ScS. J Geophys Res 78:6009-6020

Montague NL, Kellogg LH, Manga M (1998) High Rayleigh number thermo-chemical models of a dense boundary layer in D". Geophys Res Lett 25:2345-2348

Mori J, Helmberger DV (1995) Localized boundary layer below the mid-Pacific velocity anomaly identified from a PcP precursor. J Geophys Res 100:20359-20365

O'Neill B, Jeanloz R (1990) Experimental petrology of the lower mantle: A natural peridotite taken to 54 GPa. Geophys Res Lett 17:1477-1480

O'Neill B, Nguyen JH, Jeanloz R (1993) Rapid computer analysis of x-ray diffraction films. Am Mineral, 78:1332-1335

Olson P, Kincaid C (1991) Experiments on the interaction of thermal convection and compositional layering at the base of the mantle. J Geophys Res 96:4347-4354

Peltier WR (1996) Phase-transition modulated mixing in the mantle of the Earth. Phil Trans R Soc Lond A 354:1425-1447

Poirier JP, LeMouël JL (1992) Does infiltration of core material in the lower mantle affect the observed geomagnetic field? Phys Earth Planet Int 73:29-37

Poirier JP, Malavergne V, LeMouël JL (1998) Is there a thin electrically conducting layer at the base of the mantle? In: The Core-Mantle Boundary Region. In: M Gurnis, ME Wysession, E Knittle, B Buffett (eds) The Core-Mantle Boundary Region, p 131-137 American Geophysical Union, Washington, DC

Revenaugh JS, Meyer R (1997) Seismic evidence of partial melt within a possibly ubiquitous low velocity layer at the base of the mantle. Science 277:670-673

Ringwood AE (1975) Composition and Petrology of the Earth's Mantle. McGraw-Hill, New York

Ritsema J, Garnero E, Lay T (1997) A strongly negative shear velocity gradient and lateral variability in the lowermost mantle beneath the Pacific. J Geophys Res 102:20395-20411

Saxena SK, Shen G, Lazor P (1994) Temperature in Earth's core based on melting and phase transformation experiments on iron. Science 264:405-407

Saxena SK, Dubrovinsky LS, Lazor P, Cerenius Y, Haggkvist P, Hanfland M, Hu JZ (1996) Stability of perovskite ($MgSiO_3$) in the Earth's mantle. Science 274:1357-1359

Schmeling H (1985) Numerical models on the influence of partial melt on elastic, anelastic and electric properties of rocks. Part I: elasticity and anelasticity. Phys Earth Planet Inter 41:34-57

Serghiou G, Zerr A, Boehler R (1998) (Mg,Fe)SiO$_3$-perovskite stability under lower mantle conditions. Science 280:2093-2095

Shearer PM, Hedlin MAH, Earle PS (1998) PKP and PKKP precursor observations: Implications for the small-scale structure of the deep mantle and core. *In:* M Gurnis, ME Wysession, E Knittle, B Buffett (eds) The Core-Mantle Boundary Region, p 37-55 American Geophysical Union, Washington, DC

Shimizu K, Suhara K, Ikumo M, Eremets MI, Amaya K (1998) Superconductivity in oxygen. Nature 393:767-769

Sleep NH (1988) Gradual entrainment of a chemical layer at the base of the mantle by overlying convection. Geophys J 95:437-447

Song Y, Ahrens TJ (1994) Pressure-temperature range of reactions between liquid iron in the outer core and mantle silicates. Geophys Res Lett 21:153-156

Stixrude L (1998) Elastic constants and anisotropy of MgSiO$_3$ perovskite, periclase and SiO$_2$ at high pressure. *In:* M Gurnis, ME Wysession, E Knittle, B Buffett (eds) The Core-Mantle Boundary Region, p 83-96 American Geophysical Union, Washington, DC

Sylvander M, Ponce B, Souriau A (1997) Seismic velocities at the core-mantle boundary inferred from P waves diffracted around the core. Phys. Earth Planet Inter 101:189-202

Tackley PJ, Stevenson DJ, Glatzmaier GA, Schubert G (1994) Effects of multiple phase transitions in a 3-dimensional spherical model of convection in Earth's mantle. J Geophys Res 99: 15877-15901

Tsuchida Y, Yagi T (1989) A new, post-stishovite high-pressure polymorph of silica. Nature 340:217-220

Valenzuela RW, Wysession ME (1998) Illuminating the base of the mantle with diffracted waves. *In:* M Gurnis, ME Wysession, E Knittle, B Buffett (eds) The Core-Mantle Boundary Region, p 57-71 American Geophysical Union, Washington, DC

van der Hilst RD, Widiyantoro S, Creager KC, McSweeney TJ (1998) Deep subduction and aspherical variations in P-wavespeed at the base of Earth's mantle. *In:* M Gurnis, ME Wysession, E Knittle, B Buffett (eds) The Core-Mantle Boundary Region, p 5-20 American Geophysical Union, Washington, DC

van der Hilst RD, Widiyantoro S, Engdahl ER (1997) Evidence for deep mantle circulation from global tomography. Nature 386:578-584

Vassiliou MS, Ahrens TJ (1982) The equation of state of Mg$_{0.6}$Fe$_{0.4}$O to 200 GPa. Geophys Res Lett 9:127-130

Vidale JE, Benz HM (1992) A sharp and flat section of the core-mantle boundary. Nature 359:627-629

Vidale JE, Hedlin MAH (1998) Intense scattering at the core-mantle boundary north of Tonga: Evidence for partial melt. Nature 391:682-685

Wen L, Helmberger DV (1998a) A two-dimensional P-SV hybrid method and its application to modeling localized structures near the core-mantle boundary. J Geophys Res 103:17901-17918

Wen L, Helmberger DV (1998b) Ultra low velocity zones near the core-mantle boundary from broadband PKP precursors. Science 279:1701-1703

Williams Q (1990) Molten (Mg$_{0.88}$Fe$_{0.12}$)$_2$SiO$_4$ at lower mantle conditions: Melting products and structure of quenched glasses. Geophys Res Lett 17:635-638.

Williams Q (1998) The temperature contrast across D". *In:* M Gurnis, ME Wysession, E Knittle, B Buffett (eds) The Core-Mantle Boundary Region, p 73-81 American Geophysical Union, Washington, DC

Williams Q, Garnero E (1996) Seismic evidence for partial melt at the base of Earth's mantle. Science 273:1528-1530

Williams Q, Jeanloz R (1990) Melting relations in the iron-sulfur system at ultra-high pressures: Implications for the thermal state of the Earth. J Geophys Res 95:19299-19310

Williams Q, Jeanloz R, Bass J, Svendsen B, Ahrens TJ (1987) The melting curve of iron to 250 GPa: A constraint on the temperature at Earth's center. Science 236:181-182

Williams Q, Revenaugh J, Garnero E (1998) A correlation between ultra-low basal velocities in the mantle and hot spots. Science 281:546-549

Wright C, Muirhead KJ, Dixon AE (1985) The P wave velocity structure near the base of the mantle. J Geophys Res 90:623-634

Wysession ME, Lay T, Revenaugh J, Williams Q, Garnero EJ, Jeanloz R, Kellogg LH (1998) The D" discontinuity and its implications. *In:* M Gurnis, ME Wysession, E Knittle, B Buffett (eds) The Core-Mantle Boundary Region, p 273-297 American Geophysical Union, Washington, DC

Yoo CS, Holmes NC, Ross M, Webb DJ, Pike C (1993) Shock temperatures and melting of iron at Earth core conditions. Phys Rev Lett 70:3931-3934

Zerr A, Diegeler A, Boehler R (1998) Solidus of Earth's deep mantle. Science 281:243-246

Chapter 8

THE EARTH'S CORE

Lars Stixrude

Department of Geological Sciences
425 East University Avenue
University of Michigan
Ann Arbor, Michigan 48109

J. Michael Brown

Geophysics Program, AK-50
University of Washington
Box 351650
Seattle, Washington 98195

INTRODUCTION

From the mineralogical point of view, study of the Earth's core presents unique challenges. The most serious is the lack of tangible sample material; we have no samples from the core, nor do we have expectations of ever having any. For now at least, our knowledge of the mineralogy of the core must come from indirect observation and inference. This approach, based on results from seismology, geomagnetism and paleomagnetism, geochemistry, cosmochemistry, and geodynamics, has taught us a great deal about this most remote region of the planet, although many aspects of our picture of the core, for example, its minor element composition, are still uncertain. One remarkable feature of the core is its ability to surprise us despite our limited observational ability—discoveries such as the anisotropy and super-rotation of the inner core overturn long-held assumptions, and force us to re-evaluate our conception of the core's structure and dynamics. For many aspects of core behavior, we still lack even a single model that can demonstrably account for all the relevant observations.

Our knowledge of the core comes from the essential interplay and combination of results from a wide variety of fields. This is illustrated by the arguments that lead to one of the central hypotheses, that the core is primarily composed of iron. This is an inference, although a reasonably secure and certainly unchallenged one, that relies on three observations (Jeanloz 1990). First, iron is remarkably abundant in the universe, in stars, and in meteorites, much more so than in those portions of the Earth that we can sample directly. This abundance is readily understood, in part because iron is the most stable nucleus in the periodic table (Clayton 1983). Unless the Earth's non-volatile bulk composition differs radically from that of the sun and the presumptive remnants of solar system formation, it must contain a large reservoir of iron, which is quantitatively consistent, within broad uncertainties, with the size of the core. Second, the existence of a long-lived, dynamic magnetic field on Earth demands the existence of a dynamo and therefore a large reservoir of a fluid conductor (Merrill et al. 1996). The third set of observations comes from seismology which constrains the density and bulk modulus of the outer core to within a few percent. Comparison with measured equations of state show that only elements with atomic numbers similar to iron can satisfy the observations (Birch 1964). It is worth emphasizing that all three arguments are required to uniquely implicate iron as the primary constituent of the core: the first by itself cannot rule out exceptional accretional histories or vast reservoirs of iron outside the core; to satisfy the second, any highly conductive material would suffice: dynamos in the sun and the outer planets are thought to be contained within layers of metallic hydrogen or other low atomic number fluids; the third is satisfied reasonably well by vanadium.

To the degree that we are secure in our conclusion that the core is primarily composed of iron, we are equally certain that lighter elements must also be present, although in minor amounts. Pure iron does not satisfy the seismological observations of the fluid outer core. Iron is too dense by approximately 10% to satisfy both the observed density and bulk modulus along any geotherm. The identity and amount of the light element or elements is still uncertain. Primarily on cosmochemical grounds, one-third of the elements lighter than argon have been proposed as the most abundant or at least significantly abundant light elements in the core, including H, C, O, Mg, Si, and S (Poirier 1994). It is widely recognized that the way to reduce uncertainties regarding the minor element composition of the core is to investigate the phase stability and physical properties of iron light element solutions at extreme pressures and temperatures. Investigations to date have revealed surprising behavior including the stabilization of iron hydrides and an insulator to metal transition in FeO at high pressure (see Stixrude et al., this volume). There is some evidence that the inner core, as well as the outer core, may also contain light elements, although in smaller proportions (Brown and McQueen 1986).

Before the last decade, most research in geophysics and mineral physics focused on the outer core, by far the largest fraction, and that part of the core where the magnetic field is produced. However, the past ten years have seen a revolution in our understanding and appreciation of the inner core. This small spherical body, only 2400 km in diameter and comparable in mass to our moon, was long seen as essentially inert and featureless. The discovery of anisotropy in the inner core overturned this picture and has led to subsequent developments such as the discovery of super-rotation of the inner core. These results are exciting not only because they were unexpected, but also because they promise to tell us a great deal about the origin and evolution of the core. With these discoveries come new questions relating to the origin of the anisotropy, which is still unkown, the dynamics of the inner core, its growth, and its interaction with the magnetic field.

In this review, we begin with geophysical background concerning observational constraints and current ideas regarding the structure and state of the core from seismology, geomagnetism, and geodynamical considerations. We then provide a brief overview of experimental and theoretical mineralogical approaches to the study of the core. Finally, we organize a review of recent research around three central questions: the composition and temperature of the core, the crystalline structure of the inner core, and the origin of inner core anisotropy.

GEOPHYSICAL BACKGROUND

Structure

Seismology gives us our most complete and detailed picture of the structure of the core. The existence of the inner core is based solely on seismic evidence. Seismology constrains those properties of the core to which elastic wave propagation is sensitive and how these depend on depth and geographic location. It is through these properties that mineralogy makes contact with the core most directly. Our efforts to understand the core are analogous to the classic mineralogical problem of identifying an unknown sample. Just as the unknown is revealed by comparing its measured properties to that of standards, so the composition, temperature, and other properties of the core can be discovered by comparing the seismologically observed properties of the core with those of candidate materials measured in the laboratory. The primary challenges are two-fold. First, because the core is remote, our ability to characterize it through observation is inherently limited. Second, the characterization of candidate core materials is limited in practice by our ability to reproduce in the laboratory the extreme conditions of pressure and temperature that are relevant.

Our knowledge of the structure of the core is based primarily on two types of observations: (1) the travel times and wave forms of body waves (longitudinal, P and shear, S) that traverse the core and (2) the frequencies of free oscillation modes of the Earth that are sensitive to core structure. The theory of body wave propagation and free oscillation is well understood; it is known that travel times and free oscillation frequencies depend on the density and the elastic constants of the Earth's interior. For example, in an isotropic medium, a reasonable first order approximation for much of the Earth, P- and S-wave velocities are

$$V_P = \sqrt{\frac{K_s + 4/3\mu}{\rho}} \qquad (1)$$

$$V_s = \sqrt{\frac{\mu}{\rho}} \qquad (2)$$

where K_s and μ are the bulk and shear moduli, respectively, and ρ is the density. As a result, it is possible to pose an inverse problem in which one constructs a model of the core, consisting of the variation of density and elastic constants with position, in such a way as to satisfy the observations (Fig. 1). There are some fundamental limitations to this inverse problem. For example, because the relevant seismic wavelengths are substantial, there is a limit to which the spatial variation of properties can be resolved. Moreover, the resulting model is non-unique because the observations contain errors and are finite in number. In the case of remote regions, such as the core, this non-uniqueness is

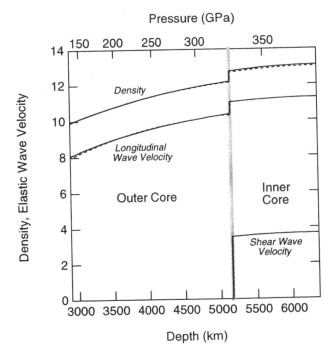

Figure 1. Density and elastic wave velocity for the spherically symmetric structure of the core as determined by seismology. Two seismic models are shown: solid—PREM (Dziewonski and Anderson 1981), dashed—PEM (Dziewonski et al. 1975).

compounded by the need to disentangle the effects on the observations of the overlying mantle from those of the core. Nevertheless, formal resolution analyses demonstrate that some properties of the core are known to within a few percent (Masters and Shearer 1990, Stark et al. 1986).

Outer core. Throughout most of the outer core, the essential structure was established in the 1930s; recent models agree with these early determinations of the *P*-wave velocity structure to within a few percent. The shear wave velocity of the outer core is not resolvably different from zero, indicating that this region is fluid (Masters and Shearer 1990). In this case, the *P*-wave velocity is identical to the bulk sound velocity. More recent studies have also been able to constrain the variation of density in the outer core from observations of normal mode frequencies (Dziewonski and Anderson 1981, Dziewonski et al. 1975). Comparison of density and bulk sound velocity structure allows one to address aspects of the dynamics of the outer core. In a homogeneous, convecting system, these two quantities are related by adiabatic self-compression. Deviations from this state are characterized by the Bullen (1963) inhomogeneity parameter

$$\eta = -\frac{\Phi}{\rho g}\frac{\partial \rho}{\partial r} \tag{3}$$

where $\Phi = K_S/\rho$ is the square of the bulk sound velocity, g is the gravitational acceleration and r is radius. Modern determinations fix this parameter at 1 ± 0.05 (Masters 1979). This is consistent with a convecting outer core but does not uniquely require it. In particular, stable stratification of the outer core cannot be ruled out. The search for stably stratified layers within the outer core has focused primarily on its two boundaries. There are persistent reports of anomalous velocities or velocity gradients just below the core-mantle boundary (Lay and Young 1990) and just above the inner core boundary (Song and Helmberger 1995) which may be associated with stratification. Observation of such layers would be important because they might provide information concerning possible chemical interaction with the mantle or the thermodynamics and fluid dynamics of the crystallization of the inner core (Lister and Buffett 1998).

Inner core. The inner core was discovered by Lehmann (1936) who first recognized that weak arrivals within the *P*-wave shadow zone were of sufficient amplitude to merit explanation as reflections from a new boundary within the Earth's interior. The *P*-wave velocity of the inner core was soon established to within a few percent of modern values. Bullen first argued that the inner core must be solid on the basis of the magnitude of the contrast in *P*-wave velocity at the inner core boundary. This proposal has not seriously been challenged, although other hypotheses for the origin of the velocity contrast have been advanced. For example, Elsasser and Isenberg (1949) suggested that the inner core boundary represents a pressure-induced electronic transition in iron, a hypothesis firmly rejected by subsequent first-principles calculations (Bukowinski 1976b). However, there is still no definitive evidence that the inner core is solid. The best evidence comes from observations of the frequencies of the Earth's free oscillations (Dziewonski and Gilbert 1971). Models that include an inner core with a finite shear wave velocity provide a significantly better fit to the observed frequencies than models with a fluid inner core. Definitive evidence of the inner core's solidity would come from observations of the seismic phase PKJKP which traverses the inner core as a shear wave. There have been a number of claims over the years that this phase had been observed (e.g. Julian et al. 1972), but these have been controversial. Very recently, there has been a revival of interest in the observation of this phase and tentative indications that it may have been observed.

Because of limited resolution, most seismological models of the inner core prior to the last decade were homogeneous, isotropic, and spherically symmetric. In this case, the seismic problem is specified completely by three properties, density, and the velocities of longitudinal (P) and shear (S) wave velocities. In principle seismic observations also depend on the variation with depth of these properties, but the effect is so small as to be poorly constrained; in particular, the observations are consistent with an inner core in a state of adiabatic self-compression (Dziewonski et al. 1975).

Perhaps the most exciting observational discovery in the core in recent years is that the inner core is anisotropic (Creager 1992, Morelli et al. 1986, Tromp 1993, Woodhouse et al. 1986). The most sensitive detection of the anisotropy involves two branches of PKP, one (BC) which traverses the bottom of the outer core, and one that samples the top of the inner core (DF). The travel time difference between these two phases is a sensitive measure of inner core structure because they have very similar propagation paths everywhere except near the inner core boundary. The observation is that the travel time difference between these two phases depends on the direction of propagation with respect to the rotation axis, even after the Earth's ellipticity has been accounted for (Creager 1992). The observations can be explained to first order by assuming that the inner core has an anisotropy of hexagonal symmetry for which *P*-waves travel 3% faster along the spin axis than they do in the equatorial plane. Other observations including direct DF travel times and AB-DF differential travel times (Vinnick et al. 1994), which are sensitive to structure deeper in the inner core, and free oscillation frequencies, are consistent with this simple pattern of anisotropy to first order.

As our ability to observe inner core structure improves, there are indications that further complexity may be superimposed on this first order pattern. Some of these more recent results are not mutually consistent pointing towards the need for further study (see Song 1997 for a recent review). Some observations indicate that the anisotropy may vanish in the uppermost 150 km of the inner core (Song and Hemlberger 1995) or that it may otherwise vary with depth (Su and Dziewonski 1995). A dependence of anisotropy on geographic position has also been proposed (Su and Dziewonski 1995). A recent study finds evidence for hemispherical asymmetry in the pattern of anisotropy: the BC-DF travel time residuals that signal the anisotropy are significantly reduced in the quasi-eastern hemisphere (40°E-180°E) as compared with the other hemisphere (Tanaka and Hamaguchi 1997). There is also evidence for very small-scale heterogeneity (Creager 1997).

Renewed interest in the structure of the inner core has led to new discoveries. It has been found that the structure of the inner core changes on human time scales. By comparing modern observations with older seismic records, the magnitude of the BC-DF travel time difference along nearly identical earthquake-receiver paths is found to vary significantly over a few decades. The origin of this change is not understood. It can be explained by a super-rotation of the inner core; that is, the inner core is proposed to rotate faster than the mantle (Song and Richards 1996). The mechanisms by which the super-rotation is produced are not known, although two hypotheses have been advanced involving electromagnetic (Glatzmaier and Roberts 1996) and viscous (Su et al. 1996) coupling to the overlying outer core.

Energetics

As important as seismology is for our study of the core, it gives us little direct guidance as to the thermal state of the core, its dynamics, the generation of the magnetic field, or its chemical and thermal evolution. Like our knowledge of its major element composition, our picture of the energetics of the core relies on observational and theoretical considerations from a diverse array of fields including studies of planetary accretion, cosmochemistry, geomagnetism, seismology, and mineral physics. Although there is broad consensus on those aspects of the energetics of the core outlined here, it is worth remembering that this picture is considerably less certain than our knowledge of the structure of the core and that new discoveries in any of the fields mentioned above could conceivably overturn some basic tenets.

Consideration of the behavior of a magnetic field in a stationary conductor via the magnetic induction equation shows that the geomagnetic field would decay on a time scale of ~10,000 years in an immobile core. Paleomagnetic observations of a much longer lived field provide direct evidence that the field is continuously regenerated, pointing towards the existence of a magnetic dynamo in the Earth's core. Observations of the present day magnetic field place a lower bound on the energy that must be supplied within the core to drive the dynamo. By integrating the energy contained in the downward continued field and dividing by the free decay time, we find that energy must be delivered at a rate of at least 3×10^{10} W (Merrill et al. 1996). This estimate is a lower bound for at least two reasons: (1) The downward continuation of the field into the conducting core is non-unique. Moreover, a large part of the magnetic energy in the core may be contained in a toroidal field that is unobservable at the surface. Dimensional analysis of the magnetic induction equation indicates that the toroidal field may be larger than the poloidal by a factor $R_m \sim 20$, the magnetic Reynolds number. This estimate may be off by an order of magnitude but does show that the toroidal field in the core may be substantial. (2) Because the dynamo process cannot be perfectly efficient, the energy required to sustain it must be greater than the energy contained in the field. Taking a typical estimate of 15% for the efficiency, a more realistic estimate of the energy required to sustain the geodynamo may be on the order of 4×10^{12} W, of which 85% is released as heat to the overlying mantle. This estimate of the heat flow into the mantle from below is approximately 10% of the observed heat flow at the surface.

We can approach the energetics of the core in another way, by examining the heat that is conducted along the geotherm and that therefore does not contribute to the dynamo process. If we take the simplest possible picture of outer core dynamics, that is one in which this region convects vigorously from top to bottom, we expect the temperature gradient to be nearly adiabatic

$$\frac{\partial T}{\partial r} = -\frac{\gamma g}{\Phi} T \tag{4}$$

where γ is the Grüneisen parameter, a material property that has been measured in liquid iron at core conditions to be 1.5 (Brown and McQueen 1986). The assumption of an adiabatic gradient is consistent with seismological observations, although not uniquely required by them as discussed above. Taking seismologically determined values of at the top of the core, and $T \sim 3500$ K as a reasonable estimate of the temperature at the bottom of the mantle, this implies a gradient of 0.9 K/km and a temperature increase of ~2000 K over the depth of the outer core. The heat conducted along the adiabat is

$$Q = 4\pi R^2 k \frac{\partial T}{\partial r} \tag{5}$$

or 8×10^{12} W, for a typical value of the thermal conductivity $k \sim 60$ W m^{-1} K^{-1}. This is consistent, within uncertainty, with our estimate of the amount of wasted heat on the basis of geomagnetic observations.

The nature of the energy sources that drive the geodynamo have been the source of considerable debate. Possible sources of energy have been summarized in a number of studies (Gubbins 1977, Verhoogen 1980). The most important ones are believed to involve processes primarily occurring at the top and the bottom of the outer core. At the bottom, the inner core grows through incongruent freezing of the overlying liquid. This process provides two distinct energy sources, the latent heat of freezing, and the release of gravitational potential energy as the light element is expelled by the growing inner core. At the top of the outer core, the fluid is being cooled from above; the sinking of this cooler fluid also releases gravitational energy. The relative importance of these three sources of energy is not known. Some analyses indicate that the sources associated with the growth of the inner core are dominant. This has the interesting consequence that terrestrial planets may require a growing inner core in order to have a magnetic field (Stevenson et al. 1983). A more recent analysis of the problem indicates that the negative buoyancy generated at the top of the core is the most important source and that the field can easily be maintained even in the absence of an inner core (Buffett et al. 1996). Part of this uncertainty arises from our relative ignorance of the relevant material properties, and could be reduced by experimental measurements or theoretical predictions. For example, the amount of negative buoyancy generated depends on the thermal expansivity of the outer core fluid. The energy sources associated with the inner core depend on the enthalpy of freezing of the inner core, and the nature of the light element in the core.

Of the fundamental uncertainties associated with this picture, one that is of considerable mineralogical interest is the possibility that some of the common radioactive heat producing elements in the Earth may be partially sequestered in the core. This possibility is suggested by first-principles theory which predicts that potassium becomes a transition metal and a siderophile element at high pressure (Bukowinski 1976a). If this prediction were confirmed, it would violate one of the basic assumptions of our models of core energetics, namely that radioactive decay is not an important energy source because K, Th, and U are lithophile, at least at low pressure.

METHODS

Experimental

There are two primary experimental techniques for generating the extreme pressures and temperatures that are found in the Earth's core. Static experimental techniques, including the laser-heated diamond anvil cell are reviewed in Shen and Heinz (this volume). Here we focus on an alternative method, that of dynamic compression by shock waves.

Shock-wave experiments. In static high pressure experiments quantities are typically measured along either isotherms or isobars. In contrast shock wave (dynamic) data define a unique thermodynamic path, the Hugoniot. The principal Hugoniot is a locus of points achieved by single shocks, i.e. discontinuous steps from an initial to final state. Conserved quantities are mass, momentum and energy, which lead to the Rankine-Hugoniot relationships for compression, pressure, and internal energy:

$$\rho = \rho_0 (1 - u_p / u_s)^{-1} \tag{6}$$

$$P = P_0 + \rho_0 u_s u_p \tag{7}$$

$$E = E_0 + \frac{P_0 u_p}{\rho_0 \mu_s} + \frac{1}{2} \mu_p^2 \tag{8}$$

where subscript 0 refers to the initial state, u_s is the shock wave propagation velocity, u_p is the velocity of particle motion behind the shock front, P is pressure, E is energy.

The physical width of the shock front has received both theoretical and experimental attention. The most highly resolved results suggest that the discontinuity is no more than a few tens of lattice spacings (Franken et al. 1997). Most shock wave experiments have short duration, typically lasting a fraction of a microsecond. Under such conditions, it has been difficult to imagine that thermodynamic equilibrium is achieved or even that phase transitions can occur. Indeed, an early analysis of the thermodynamics of shock waves by Hans Bethe specifically rejected the idea that high-pressure phase transitions could be studied. However, much work has proven this supposition entirely too pessimistic. Phase transitions in many different materials have been investigated by shock-wave techniques. Particularly in the case of elements and simple compounds (e.g. NaCl) the agreement in transition pressures is remarkable. Even in cases where kinetic barriers complicate phase transitions (e.g. graphite-diamond or silicates), shock determined properties of the high pressure phases are, for the most part, in accord with static data.

Thermal equilibrium requires a Boltzman distribution of phonon (vibrational) states. Shock energy is delivered to a material as a longitudinal excitation. In a purely quasi-harmonic material this energy would not be re-partitioned into all appropriate vibrational states. However, at the shock front, where strain gradients are large, anharmonic behavior is enhanced. Molecular dynamic simulations support the idea that vibrational relaxation on a time scale of a few picoseconds can occur in the shock front. Indeed, large strain gradients and initially non-Boltzman vibrational behavior aids in the rapid reorganization of the lattice under conditions appropriate for phase transitions (Holian 1988, 1995; Holian and Lomdahl 1998).

Concerns about differences (as a result of possible unrecognized systematic errors) between static and shockwave data have, to a large extent, been ameliorated by comparisons in cases where data overlap. For example, static compression data at room temperature for ε-iron to 300 GPa (Mao et al. 1990) and an isotherm based on Hugoniot data for ε-iron measured to 200 GPa (Brown and McQueen 1982) are plotted in Figure 2. The static high-pressure data rely on a pressure calibration of the ruby scale which is based on shock-wave Hugoniot data for metals (other than iron). Thus, the agreement shown in the figure could be viewed as resulting from a partially circular argument. However, a more positive interpretation is that both accurate static compression measurements and accurate thermal corrections can be made.

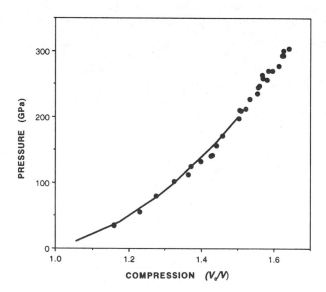

Figure 2. Static compression data at room temperature for ε-iron to 300 GPa (Mao et al. 1990) and (line) an isotherm based on Hugoniot data for ε-iron measured to 200 GPa (Brown and McQueen 1982).

The thermodynamic state in shock experiments, as specified by the Hugoniot relations, is incomplete and ancillary data are required. A difference equation for temperature on the Hugoniot can be derived through combination of basic thermodynamics and the Hugoniot equations:

$$dT = -T\gamma dV/V + [dP(V_0 - V) + dV(P - P_0)]/2C_V \tag{9}$$

This equation (giving temperatures based on thermodynamic integration) requires, in addition to Hugoniot data, constraints for the Grüneisen parameter γ, and the constant volume specific heat C_v. Typically γ and C_v vary slowly with P and T. The Grüneisen parameter is normally found to lie between 1 and 2 and measurements (Brown and McQueen 1986, Chen and Ahrens 1996) give evidence that the product of density and γ for liquid iron varies little with pressure and temperature. The lattice contribution to specific heat at high temperature should be well approximated by $3R$ where R is the gas constant. Electronic contributions are estimated on the basis of quantum-mechanical calculation (Boness et al. 1986). In a core regime of pressure, uncertainties in Hugoniot temperature for iron (incorporating the uncertainties in C_v and γ) are about 10% (Brown and McQueen 1986).

In addition to the traditional measurements of shock and particle velocities, a number of shock techniques provide additional data. Sound velocities in shocked iron can be measured using overtaking rarefaction techniques (Brown and McQueen 1986). If the rarefaction velocity is the bulk sound wave, γ is determined via a finite difference expression. Efforts to more directly measure iron Hugoniot temperatures have required the analysis of thermal emission spectra from an interface between the iron and a transparent anvil. As discussed later, these measurements have been problematic. A particularly useful new direction of research is represented by recent work (Ahrens et al. 1998). By preheating iron to 1573 K, Hugoniot data and release velocities along a new thermodynamic path were obtained.

Theoretical

Theoretical methods as they have been applied to studies of the Earth's interior are reviewed by Stixrude et al. (Chapter 19, this volume). Briefly, these methods fall into three classes that differ greatly in the number and character of the assumptions upon which they are based, the physics that they incorporate, and thus in the quality and security of their predictions. These three classes of theoretical approaches are now outlined.

First-principles methods. These seek to solve the Schrödinger equation in the solid state with a bare minimum of approximations. Most theoretical studies of the Earth's core have been based on density functional theory. Here, the single essential approximation is to the exchange-correlation potential, which represents many-body interactions among the electrons that cannot yet be treated exactly. Two types of approximations to this term are commonly made. First is the local density approximation (LDA) which has been widely studied in geophysics and other fields and has been found to successfully describe the properties of nearly all types of materials including insulators, semi-conductors, and metals, for bulk, surface and defect properties. Although it is widely successful, LDA is known to fail in important ways for some transition metals including iron. For example, it predicts a close-packed ground state of iron where bcc is observed. For this reason, other approximations to the exchange-correlation potential have been developed. One of the most successful and widely studied is the generalized gradient approximation (GGA). The agreement in the case of iron, and most other materials is found to be at least as good, and often better than in the case of LDA. In particular, GGA correctly recovers bcc as the ground state of iron.

Because solution of the quantum-mechanical equations is difficult and costly, the choice of appropriate computational methods is an important issue. The current state-of-the-art is the Linearized Augmented Plane Wave (LAPW) method, which makes no further approximations beyond that to the exchange-correlation potential. First Principles Linearized Muffin Tin Orbitals (FP-LMTO) calculations are similarly precise. A somewhat more approximate strategy is the pseudopotential method. This makes a limited number of additional physically motivated assumptions which permit much more rapid computation. These are the frozen-core approximation—the structure and energies of the electronic core states are assumed to be independent of crystal structure, pressure, or other perturbations to the lattice—and the pseudopotential approximation in which the potential due to the nucleus and core electrons is replaced by a simple smoother object, the pseudopotential, that has the same scattering properties.

Ab initio models. These construct an approximate model of some aspects of the relevant physics, such as the charge density or of the interactions between orbitals. The cost of additional approximation is often outweighed by the increase in computational simplicity and efficiency. For example, ab initio models have been widely used to explore transport properties or the properties of liquids, which are very difficult (costly) to examine with fully first-principles approaches. Moreover, these models often yield insight that is sometimes difficult to extract from more complex and elaborate first-principles calculations. One example is the tight-binding method, which has been successfully applied to the study of iron at high pressure. In this method, the charge density and the wavefunctions do not appear explicitly. Instead, the elements of the Hamiltonian and overlap matrices are approximated by simple parametric functions that are constructed so that the results of first-principles calculations are reproduced.

Semi-empirical methods. These generally do not treat the electrons explicitly, but instead view condensed matter as being composed of atoms. Typically, a simple functional form is adopted to describe the forces acting between pairs or triples of atoms. These models can be parameterized in such a way that subsets of experimental data are reproduced more or less accurately. They have been particularly successful in the context of ionic materials where most of the energy is a sum of pair-wise (Coulombic) interactions, and in simple metals. Transition metals are more problematic however. While the equation of state is a property that can often be represented successfully in terms of atomistic models, other essential aspects of the physics are often lost. Moreover, because they do not account for the electrons, atomistic methods cannot treat many important effects that are essential for our understanding of the core. Among these are magnetism in iron and its variation with pressure, and changes in the electronic structure that lead to effects such as the elastic instability of bcc iron at high pressure.

SOME ISSUES OF CURRENT INTEREST

Composition and temperature

The general framework. A fundamental premise in this discussion is that the thermodynamic state at the inner core boundary (ICB) lies on an equilibrium liquidus. Thus, growth (or shrinkage) of the inner core is linked with changes in temperature (most likely secular cooling, although the reverse can not be entirely discounted). Since recent models implicate the inner core in the stabilization of the geodynamo (Glatzmaier and Roberts 1996) and since some old crustal rocks appear to have primary remnant magnetizations, the notion that Earth has had an inner core for over a billion years would appear to make substantial ICB disequilibria unlikely.

Pressures within Earth are determined as the integral of ρg. Since the total mass of Earth is well determined and since the internal mass distribution is reasonably well constrained, pressure at the inner core boundary is known with reasonable accuracy. Thus, at the ICB pressure of 330 GPa, the melting behavior in iron-dominated systems provides a fixed-point temperature. Inasmuch as iron is typically thought to be more refractory than iron compounds, the melting temperature of pure iron would give an upper boundary.

The fractional temperature difference between the CMB and the ICB, through a well mixed, convecting (adiabatic) outer core is, in the case of constant Grüneisen parameter, given by:

$$\frac{\Delta T}{T} = \gamma \frac{\Delta \rho}{\rho} \tag{10}$$

Using densities from PREM (Dziewonski and Anderson 1981) and $\gamma = 1.5$ (Brown and McQueen 1986, Chen and Ahrens 1996) the core-mantle boundary (CMB) is expected to be 25% to 30% cooler than the ICB; for a nominal CMB temperature of 4000 K the associated ICB temperature is 5500 K.

Based on the preceding discussion, the compositional state and the thermal state of the core are interrelated through equations of state and melting relations for iron and iron-dominated compounds. Any proposed composition gives a predicted melting temperature at the ICB pressure. At this specified *P-T* state, the associated equation-of state must match the seismically determined density and sound velocity at the ICB. An example of this analysis was given by Brown and McQueen (1982).

Shaner (1988) have shown that both metallic and non-metallic fluids obey velocity-density-mean-atomic-weight systematics over a wide range of compression and atomic weight. If correct in detail, different Fe-X mixtures with common mean-atomic weight would have the same velocity-density behavior. Thus, the analyses of composition are intrinsically non-unique and additional constraints must be imposed.

Geochemical arguments. Earth's core likely formed in an early core forming event associated with a magma ocean or during a catastrophic bolloid impact. In either case separation of a metallic core from the mantle could occur under conditions of relatively low pressures. Thus, elemental partitioning experiments to a few tens of GPa are pertinent to the chemistry of the core. Comprehensive overviews have been given by (Poirier 1994) and (Allegre et al. 1995). Although all naturally occurring elements are likely present in the core, carbon, sulfur, oxygen, silicon, and hydrogen remain the principal light elements under consideration. Recent experimental studies show enhanced partioning of hydrogen (Okuchi 1997), sulfur (Jie and Agee 1996), and oxygen (Ito et al. 1995) under modest pressure and temperature. Thermodynamic models have been recently used to support enhanced core partitioning of carbon and hydrogen (Kuramoto and Matsui 1996), and silicon (Xi and Ahrens 1994). In aggregate, these results allow for an uncertain mixture of light elements in sufficient concentration to appropriately alter the density of iron.

Physical arguments. The Fermi surface of iron lies within a broad peak of d-like electronic states. Significant electronic contributions to the thermodynamics of iron result from this topology. Boness et al (1986) showed that this remains generally valid for all hypothetical iron structures at elevated pressure and temperature and that the electronic contributions to the thermodynamics must be taken into account in any analysis of the core. The light elements oxygen and sulfur provide contrasting high pressure behavior that impacts elemental partitioning during core formation, melting temperatures, and equations-of-state. Oxygen does not have a spherical closed-shell electronic configuration. Unfilled

(d-like) electronic states lie well beyond the Fermi energy and are thus unavailable for metallic hybridization in iron–oxygen compounds. That the metallization of FeO appears to be Mott-like supports this idea (see Stixrude et al., this volume). In contrast, sulfur has low-energy unfilled d-states which can participate in the metallization of iron–sulfur compounds. Using LMTO calculations, Boness and Brown (1990) documented an increasing transition metal behavior for sulfur at high pressure and argued on the basis of atomic size and electronic structure that sulfur could form a solid solution with iron under core conditions. First-principles density functional calculations in the GGA approximation (Sherman 1995) further contrast the behavior of iron compounds containing oxygen and sulfur. Fei et al. (1996) have reported experimental evidence for the high pressure metallization of an iron–sulfur compound.

The principal consequences of arguments concerning iron-mixtures lie in the expected compositional effect on melting temperature and on the partitioning of light elements between the inner and outer core. In the most simplistic analysis, elements either do or do not form solid solution with iron. In the absence of significant solid solution, the melting point is expected to decrease with increasing light element abundance. For small concentrations of the light element and under an assumption of ideality, the melting depression for a eutectic system is given by

$$T_m = T_{Fe}/(1 - \ln X) \tag{11}$$

where T_{Fe} is the melting temperature of pure iron and X is the mole fraction of the light element. Perfect eutectic behavior would produce an inner core of pure iron and an estimated melting point depression of about 1000K. Solid-solution behavior, on the other hand, would allow more light element abundance in the inner core. In contrast to the melting of iron-dominated eutectic systems, model melting temperatures in solid-solution systems require constraints for all end-members. Non-ideality introduces additional complexity (Stixrude et al. 1997).

Iron: equations of state and high pressure melting. Both shock wave and static high-pressure data constrain the properties of iron under the pressure conditions of the core. Since shocked iron is melted at pressures above 243 GPa and since Hugoniot temperatures overlap core temperatures in this regime, direct comparisons between Earth and shock wave data require small thermal corrections. However, under inner core conditions of pressure, shocked iron is liquid whereas Earth is solid. In contrast, the highest-pressure static data determine solid iron behavior. Such measurements require substantial corrections for temperature and are not a direct constraint on the liquid equation of state. As previously noted, the Hugoniot data are in agreement with static data extending to 300 GPa. Analyses based on measured (Brown and McQueen 1986) and theoretical (Stixrude et al. 1997) properties of iron indicate that while the inner core is not pure iron, it has a substantially smaller light element component than the outer core.

Melting on the iron Hugoniot at 243 GPa was reported by (Brown and McQueen 1980). A discontinuity in sound velocity gave evidence of melting. A second discontinuty at 200 GPa was interpreted as resulting from the solid-solid ε–γ transition. Brown and McQueen (1982, 1986) estimated an ICB temperature of 5800 ($\pm$500) K on the basis of Hugoniot temperatures determined by thermodynamic integration. The CMB temperature was given as 4000($\pm$500).

The pioneering diamond anvil melt experiments and interface shock temperature measurements, first reported by Williams et al. (1987), suggested substantially higher iron melting temperatures. However, neither these static nor dynamic data have been supported in more recent work. As summarized by Shen et al (1998), three separate laboratories, all using improved laser-heated diamond anvil techniques, find iron melting temperatures at 100 GPa near 3000K. This is approximately 1000 K below the result given by Williams et al. (1987). Subsequent work has also highlighted a number of technical issues associated with the interface temperature measurements. The argument can be advanced that direct shock temperature measurement in iron, as well as thermodynamic integration, also requires the use of theory.

A recent analysis (Ahrens et al. 1998) incorporated recent shock and static data for the melting of iron. This re-examination of both new and existing data gave an estimated melting temperature for iron at the ICB of 5500 (±500) K. This is significantly lower than the previous estimates based on shock interface measurements of 7600 K (Williams et al. 1987) or 6830 K (Yoo et al. 1993). Ahrens et al (1998) also gave a new estimate for the core-side temperature at the CMB of 3930 (±630) K.

Iron compounds: equations of state and high pressure melting. The nearly infinite space of compositions for Fe-X mixtures, where X stands for all light elements in the core has been sparsely sampled by experimental data. Brown et al (1984) explored the equation of state and melting of iron-sulfur systems. They found the outer core well-matched using 10 wt % sulfur. Svendsen et al (1989) modeled both Fe-O and Fe-S data to suggest that the outer core could lie on the FeO side of a eutectic composition. On the basis of first-principal calculations, Sherman (1995) argued that the inner core is more likely to contain sulfur than oxygen. Boehler (1996) found that the melting point depression in the Fe-FeS system decreases at high pressure, supporting enhanced solid-solution behavior at high pressure. Boehler (1992) has also reported melting temperatures in the Fe-O-S system. He suggests that such a system has a melt temperature at the CMB near 3300 K.

Crystalline structure of the inner core

To the extent that the inner core consists of pure, or nearly pure iron, its crystalline structure is determined by the iron phase diagram. While there has been considerable progress in experimental determination of the phase diagram at pressures approaching those of the inner core, the stable phase of iron at inner core conditions cannot yet be uniquely identified on the basis of phase equilibrium measurements. Nor is reasonable extrapolation of measured solid-solid coexistence curves a secure procedure. Aside from the fact that the magnitude of the extrapolation is still large, and cannot be assumed to be linear, the more important reason is that there is not yet universal agreement on the list of possible phases; while some experiments find no evidence for phases other than those long known from low pressure work, others find evidence for new phases that are unique to the high pressure environment, considerably complicating the picture.

This is an important issue from the geophysical and geochemical point of view for a number of reasons. First, it is central to our understanding of the elastic anisotropy of the inner core. The three known phases of iron exhibit markedly different magnitudes and symmetries of elastic anisotropy. One expects that their deformation mechanisms will also differ qualitatively. In a poly-crystalline inner core, each would then produce radically different bulk anisotropies for a given texturing mechanism. Second, the enthalpy of the inner core, and thereby the latent heat of freezing released as the inner core grows may depend strongly on its crystalline structure. This release of latent heat is thought to be a major source of energy that drives the production of the geomagnetic field (see above). Third, the thermodynamics of interaction of a dominantly iron inner core with the light element may be strongly affected. For example, the ability of the inner core to incorporate light elements or other (e.g. radioactive) impurities may depend on its crystalline structure.

There are three phases of iron that have been unambiguously identified. The body-centered cubic (bcc, or α) is a ferromagnetic phase stable at ambient conditions and over a limited range of pressure and temperature (up to 10 GPa, and 1200 K). The bcc phase reappears above its Curie temperature in a narrow stability field (labeled δ) just below melting. The face-centered cubic (fcc, or γ) phase exists at higher temperatures than α (up to melting) and to higher pressures. There is some evidence that this phase also possesses local magnetic moments at low pressure although these are not ordered within its stability field (Macedo and Keune 1988). The hexagonal close-packed (hcp or ε) phase is the high pressure phase, stable to at least 300 GPa at room temperature (Mao et al. 1990). Local magnetic moments are not observed experimentally in this phase (Taylor et al. 1991) although they are predicted at low pressure on the basis of first-principles calculations (Steinle-Neumann et al. 1998). The relative stability of these phases at low pressure can be understood in detail as a competition between magnetic and non-magnetic contributions to the internal energy, differences in vibrational and magnetic entropy, and differences in volume (Moroni et al. 1996). In particular, it is known that the bcc phase owes its stability at low pressure to its large magnetic moment.

There are three distinct lines of evidence pointing to the possible existence of other stable phases of iron that are unique to high pressure. First, a solid-solid phase transition is inferred to occur along the Hugoniot at a pressure of 200 GPa (Brown and McQueen 1986). This phase transition was originally identified with the high pressure continuation of the fcc-hcp boundary on the basis of the topology of the phase diagram as it was then known. However, subsequent measurement of the lower pressure portion of the iron phase diagram rules out this possibility: the fcc-hcp boundary ends at a triple point with melt at much lower pressures (Shen et al. 1998). If a solid-solid phase transition exists at 200 GPa, it must involve at least one as yet undiscovered phase or phase stability field. The evidence for this solid-solid phase transition is indirect: the determination of crystal structure (e.g. by x-ray diffraction) during the brief passage of the shock wave through the sample is not yet possible. Instead the evidence comes from measurements of the longitudinal wave velocity along the Hugoniot. The observation is that the longitudinal wave velocity increases smoothly with compression until a pressure of 200 GPa where it decreases by several percent over a narrow pressure range (<5 GPa) before increasing smoothly again upon further compression. The interpretation is that the Hugoniot crosses a solid-solid phase transition at this pressure for which the high temperature phase has a lower longitudinal wave velocity than the low temperature phase. The anomaly is not attributed to melting; melting is identified with a second decrease in longitudinal wave velocity, nearly identical in magnitude, that is observed at higher pressure (243 GPa).

The second line of evidence that suggests the occurrence of new phases of iron comes from static high pressure experiments. X-ray diffraction patterns measured in the laser-heated diamond anvil cell in situ at high pressure and temperature have been argued to be incompatible with any of the known phases of iron (Andrault et al. 1997, Dubrovinsky et al. 1997, Saxena et al. 1995). The evidence consists of new peaks unattributable to fcc, hcp, or bcc, anomalous splittings of peaks, and the absence of peaks expected from the known phases. The anomalous signal is subtle and the structures proposed on the basis of these observations are very similar to the hcp phase. There are important questions associated with these observations however. In this context, it is worth pointing out that these are pioneering experiments; in no other system have x-ray diffraction experiments been attempted under such extreme conditions. In evaluating the results of these experiments there are several important issues: (1) have temperature and stress gradients within the diamond cell been adequately controlled and characterized (2) are the observations of a stable phase, or possibly a metastable transformation product (3) is the averaging over crystal orientations within the narrow x-ray beam adequate?. Aside from technical experimental details, there are questions concerning the results themselves. Within overlapping regions of pressure and temperature, two mutually incompatible new phases have been proposed by two different laboratories, a double-hexagonal close-packed (dhcp) structure (Dubrovinsky et al. 1997), and an orthorhombically distorted hcp structure (Andrault et al. 1997). Shen et al. (1998) find additional x-ray peaks similar to those found in both of these studies in the vicinity of the the transformation between ε-γ transition, but these disappear when the transition is incomplete: diffraction patterns of the well-crystallized single-phase sample indicate either hcp or fcc structures, depending on the pressure and temperature (see also Shen and Heinz, this volume).

The third line of evidence comes from semi-empirical theoretical calculations. Calculations based on pair potentials indicate that the bcc phase may re-appear as a stable phase at very high temperatures and pressures (Matsui and Anderson 1997, Ross et al. 1990). This result is appealing because the predicted phase transition boundary from hcp to bcc at high pressure is compatible with the solid-solid phase transition inferred from shock wave experiments. However, as discussed above, the atomistic models upon which these calculations are based are highly approximate, and cannot include aspects of the physics which may be crucial. Indeed, these simple models cannot reproduce the known phase diagram at low pressure. Moreover, they lack the physics which drives the mechanical instability of bcc iron at high pressure as revealed by first-principles calculations: a splitting at high pressure of the peak in the electronic density of states at the Fermi level (Stixrude et al. 1998). It is not surprising then that semi-empirical methods incorrectly predict a stable bcc phase.

Although first-principles calculations have not yet identified the stable phase of iron at inner core conditions, they have been able to place some of the first constraints on the problem by showing that the bcc phase is unstable at high pressure. By limiting the list of possible candidate phases, the first-principles calculations allow future work to focus on the two other observed phases of iron, and the search for possible new phases. The simplest elastic distortion of the cubic phases of iron—a stretching along one crystallographic axis—reveals the elastic instability (Stixrude and Cohen 1995a). At low pressure, the total energy as a function of c/a ratio displays two local minima, one corresponding to the bcc structure and one to the fcc structure (Fig. 3). This is in agreement with experiment, which finds both structures elastically stable at low pressure. At high pressure, a completely different picture is revealed by the theoretical results. While the fcc phase is still elastically stable, the bcc phase is elastically unstable. As the bcc lattice is strained by small amounts, its energy is lowered. This means that there is no restoring force that preserves the bcc structure at high pressure. In the presence of infinitesimal thermal fluctuations, the bcc lattice will undergo a spontaneous distortion to the fcc structure. The mechanical instability of the bcc lattice has been confirmed by Söderlind et al. (1996) using a different first-principles technique.

There is no evidence for a stable tetragonal structure in iron at any pressure. Such a structure was proposed by Söderlind et al. (1996) on the basis of the apparent local minimum in total energy for $c/a < 1$ (Fig. 3). It is important to recognize that this minimum is merely a saddle point and does not correspond to a mechanically stable structure (Stixrude et al. 1998). This becomes clear when one considers the orthorhombic strain energy surface. The fcc structure can be derived from the bcc structure in one of two ways: (1) by increasing the c/a ratio or (2) by decreasing c/a and b/a. By decreasing c/a below unity, a structure is generated which is intermediate between bcc and fcc. Ab initio calculations at a series of volumes that span inner core conditions confirm that the apparent local minimum is a saddle point and does not represent a mechanically stable structure.

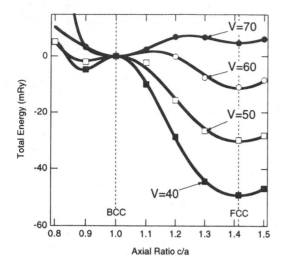

Figure 3. Total energy as a function of uniaxial strain in iron at four different atomic volumes. Results from LAPW calculations in the GGA approximation. V = 70 Bohr³ is similar to the zero pressure volume and V = 48 Bohr³ corresponds to the mean density of the inner core. The lines are polynomial fits to guide the eye. At high pressure ($V < 60$ Bohr³), the energy of bcc is lowered by small changes in the c/a ratio; i.e., the structure is elastically unstable and will spontaneously distort. From Stixrude and Cohen (1995a).

Origin of inner core anisotropy

Elastic anisotropy has been known to exist in the Earth's crust and mantle for some time. In the upper mantle, the anisotropy originates in the preferred alignment (texturing) of an aggregate of intrinsically anisotropic minerals, dominantly olivine and pyroxene. Anisotropy may also originate in heterogeneity on scale lengths much smaller than the seismic wave. This form of anisotropy is common in the crust. Examples of heterogeneous structures that are known to lead to seismic anisotropy include laminated structures that are composed of rocks of contrasting seismic velocity, or rock that is permeated by oriented cracks. In these examples, the components of the heterogeneous structure need not be anisotropic themselves. In the crust and mantle the tectonic processes responsible for the anisotropy can often be understood in terms of brittle or fluid deformation of the medium. Indeed, because of our knowledge of the elastic constants of the constituent minerals and their deformation mechanisms, it is possible to relate the sense of anisotropy observed in the upper mantle to the direction of mantle flow in some detail (Christensen and Salisbury 1979).

In the case of the anisotropy of the inner core, neither the structures responsible for the anisotropy nor the dynamic processes that produce them are known. Whether the anisotropy originates in small-scale heterogeneity or in the preferred alignment of intrinsically anisotropic crystals is unclear. Most attention has focused on explanations in terms of crystalline anisotropy because these have been able to explain at least some of the seismic observations. We will focus on the crystalline anisotropy of iron here and its comparison to that of the inner core. However, it is important to remember that while no convincing explanations in terms of small-scale heterogeneity have yet been developed, that this is no proof against such structures. It is also unknown what dynamic processes might operate in the inner core that are capable of producing either anisotropy or heterogeneity, although several mechanisms have been proposed (Jeanloz and Wenk 1987, Karato 1993, Yoshida et al. 1996).

Elastic anisotropy of iron and the inner core. Explanations of inner core anisotropy in terms of the intrinsic properties of its most abundant constituent are appealing because the known phases of iron are all observed to be elastically anisotropic. The details of the anisotropy, in particular its symmetry and magnitude, are important. The magnitude of the intrinsic anisotropy must be at least as great as that of the inner core if it is to explain the observations – the anisotropy of a homogeneous polycrystalline aggregate can only be less than or equal to that of a single crystal. If the single crystal shows much larger anisotropy, then this indicates the degree of preferred alignment in the inner core. The symmetry of the anisotropy in the inner core is a combination of the symmetry of the constituent single crystals and that of the fabric of the aggregate. Therefore, the fact that the symmetry of inner core anisotropy is that of a hexagonal single crystal to a first approximation does not uniquely implicate a hexagonal phase of iron.

The anisotropy of a single crystal is determined by its elastic constant tensor. For cubic phases, there are three independent elastic constants, for hexagonal, five. The elastic constants are related to the elastic wave (seismic) velocities by the Cristoffel equation (Nye 1985)

$$\rho V^2 = C_{ijkl} n_i w_j n_k w_l \tag{12}$$

where V is the velocity, ρ is the density, **n** and **w** are the propagation and polarization vectors, respectively and **C** is the elastic constant tensor. The elastic anisotropy is determined by calculating the velocity of each elastic wave (one longitudinal or P, two shear or S) for all propagation directions. To determine the elastic anisotropy of a polycrystalline aggregate, one must know the elastic constant tensor of the individual crystals and their texture as specified by the orientation distribution function.

One of the major sources of uncertainty in our studies of inner core anisotropy is our ignorance of the elastic constants of iron at the relevant conditions. Experimentally the relevant pressures and temperatures can be produced by shock wave or static techniques. The difficulty lies in measuring the elastic constant tensor in situ. Theoretically, the primary stumbling block is the tremendous cost of first-principles, high-temperature calculations, especially calculations of the elastic constants which would demand high precision. Despite these difficulties, considerable progress has been made in both theory and experiment. The initial focus has been on constraining the elasticity of iron at high pressure but low temperature. Aside from practical limitations, the reason is that while temperature is expected to be important for understanding the absolute magnitude of the elastic constants, it is not expected to alter the elastic anisotropy qualitatively on the basis of the behavior of other transition metals. Moreover, the largest effect of temperature - thermal expansion - can be removed by comparing the properties of iron with the inner core at constant density, rather than at constant pressure. Nevertheless, studies of high temperature elasticity will be essential for a complete understanding.

The elastic constants of iron have been predicted by ab initio and first-principles theory over a range of pressures (densities) that span those of the inner core (Steinle-Neumann et al. 1998, Stixrude and Cohen 1995b). These calculations focused on the fcc and hcp phases; the bcc phase is not expected to be stable at inner core conditions (see above). The calculations predict that both phases are elastically anisotropic. The P-wave anisotropy of hcp is found to be nearly identical in symmetry and magnitude to that of the bulk inner core: P-wave anisotropy is 3% at the density of the inner core and the fast propagation direction is along the symmetry axis. The anisotropy of fcc is found to be three times greater with a fast direction of P-wave propagation along the cubic diagonal. The anisotropy of S-waves is much greater than for P-waves in both phases, by approximately a factor of three. The relative magnitudes of P- and S-wave anisotropy in hcp and fcc and the directions of fast and slow propagation can be understood on the basis of a nearest-neighbor central force model (Born and Huang 1954).

To apply these single crystal results, one requires a model of polycrystalline texture in the inner core. The simplest possible forms that are consistent with the observed cylindrical symmetry of the inner core are: (1) For hcp: An aggregate with all c-axes aligned with the Earth's spin axis. The effective elastic constants of an inner core composed of such an aggregate are identical to those of the hcp single crystal. (2) For fcc: An aggregate with all [111] axes aligned with the Earth's spin axis, but otherwise randomly oriented. The effective elastic constants of this cylindrically averaged fcc aggregate are readily calculated from the single crystal fcc elastic constants.

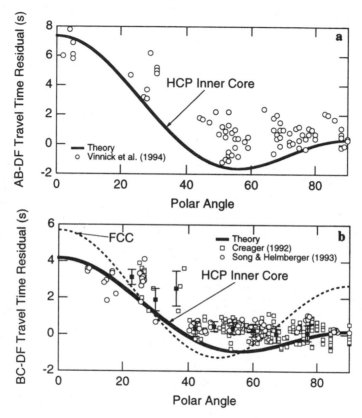

Figure 4. Travel time anomalies as a function of the polar angle ξ between the inner core ray segment and the Earth's spin axis. The theoretically-based models of an inner core consisting of an hcp aggregate (bold line) and a cylindrically averaged fcc aggregate (bold dashed line) are compared with seismological observations of (b) BC-DF anomalies (Creager 1992, Song and Helmberger 1993) and (a) AB-DF anomalies (Vinnik et al. 1995). Calculations were performed for distances of (BC-DF) $\Delta = 150°$ and (AB-DF) $\Delta = 175°$, near the center of the distance ranges represented by the respective observations.

These simple models allow us to compare directly first-principles theory with seismological observation. In particular, we may calculate the P-wave velocity of our inner core model in any direction and from this the BC-DF travel time anomalies that arise from inner core anisotropy (Stixrude and Cohen 1995b). The results show that the hcp model of inner core anisotropy, but not the fcc model, is in good agreement with the observations, particularly those near the pole and near the equator (Fig. 4). The magnitude of the polar (+4 s) and equatorial (<+1 s) travel-time anomalies are well reproduced. There are some discrepancies, such as the magnitude of the off-axis minimum in travel time anomaly centered at $\xi = 56°$. For such a simple model, this is not surprising, as our assumed texture is not likely to be correct in detail and could be further optimized (Sayers 1989). Moreover, our theoretical predictions are not exact and do not include the effect of temperature in any case. The level of agreement is encouraging and shows that (1) inner core anisotropy can be explained to first order by the intrinsic elastic anisotropy of iron, (2) the inner core may be composed of a hexagonal phase, either hcp or a similar structure (e.g. dhcp or slight orthorhombic distortions of hcp), (3) if the inner core is composed of hcp, then the texturing must be strong, with nearly complete alignment of the c-axes.

Recently experimental constraints on the elastic constants of hcp iron have been reported for the first time (Mao et al. 1998). These experiments do not measure the elastic constants directly, but nevertheless represent a major advance because the hcp phase is stable only at high pressure and cannot be measured by standard techniques at ambient conditions. The experimental determinations do not agree with the theoretical predictions. In particular, the pattern of P-wave anisotropy and the magnitude of the S-wave anisotropy differ radically from the theoretical predictions. The primary differences can be traced to the shear elastic anisotropy expressed by the ratio C_{44}/C_{66} (Steinle-Neumann et al. 1998). While this ratio is near unity according to experimental measurements of all other hcp transition metals and for the theoretical predictions of iron, the experiments on hcp iron find a much larger ratio, near 2.5.

The reason for the discrepancy between theory and experiment in this case is still unclear. It cannot be ruled out at present that there are unanticipated physics that are unique to iron among the hcp transition metals and that are not captured by the theory. Another possible explanation is that the indirect experimental measurement is biased in some way, as recognized by Mao et al. (1998). The experiments do not measure the elastic constants but rather the strain induced by a non-hydrostatic stress field in a polycrystalline sample. By making assumptions about the nature of the stress and strain fields in the constituent grains, elastic constants can be deduced. However, the measurement is sensitive to the mechanisms of inelastic as well as elastic strain. Indeed, strong induced textures are observed in these experiments. Slip along basal planes, a common deformation mechanism in hcp metals, can be shown to lead to high apparent values of the ratio C_{44}/C_{66}. Indeed, the only comparison to date between this technique and standard measurements at ambient conditions in an hcp transition metal (Re) shows that the lattice strain technique overestimates this ratio (Duffy et al. 1998). If it could be shown that the new experimental technique is primarily sensitive to deformation mechanisms rather than elastic constants, this would be a valuable contribution to our knowledge of iron with potentially important implications for the dynamics and texturing mechanisms of the inner core.

CONCLUSIONS AND OUTLOOK

Considerable uncertainty still shrouds the outer core and inner core from our complete understanding. Much of this uncertainty is associated with our ignorance of the physics and chemistry of iron and iron alloys at the extreme conditions that are relevant. The last decade has seen dramatic progress in our understanding of the iron phase diagram, the behavior of liquid and solid iron alloys, and the elasticity of iron. The continued development of new theoretical and experimental techniques promise major advances in the next few years. These include new ways of measuring the elasticity and deformation mechanisms of metals at high pressure, advances in situ characterization and measurement in the diamond anvil cell and shock wave environment, and theoretical methods for studying the properties of metals at high temperature.

REFERENCES

Ahrens TJ, Holland KG, Chen GQ (1998) Phase diagram of iron, revised core temperatures. Geophys Res Lett (submitted)

Allegre CJ, Poirier JP, Humler E, Hofmann AW (1995) The chemical composition of the Earth. Earth Planet Sci Lett 134:515-526

Andrault D, Fiquet G, Haydock R (1997) The orthorhombic structure of iron: An in situ study at high-temperature and high-pressure. Science 278:831-834

Birch F (1964) Density and composition of the mantle and core. J Geophys Res 69:4377-4388

Boehler R (1992) Melting of the Fe-FeO and the Fe-FeS systems at high pressure: constraints on core temperatures. Earth Planet Sci Lett 111:217-227

Boehler R (1996) Fe-FeS eutectic temperatures to 620 kbar. Phys Earth Planet Inter 96:181-186

Boness DA, Brown JM (1990) The electronic band structures of iron, sulfur, and oxygen at high pressures and the Earth's core. J Geophys Res 95:21721-21730

Boness DA, Brown JM, McMahan AK (1986) The electronic thermodynamics of iron under Earth core conditions. Phys Earth Planet Inter 42:227-240

Born M, Huang K (1954) Dynamical Theory of Crystal Lattices. Clarendon Press, Oxford

Brown JM, Ahrens TJ, Shampine DL (1984) Hugoniot data for pyrrhotite and the Earth's core. J Geophys Res 89:6041-6048

Brown JM, McQueen RG (1980) Melting of iron under core conditions. Geophys Res Lett 7:533-536

Brown JM, McQueen RG (1982) The equation of state of iron and the Earth's core. The equation of state of iron and the Earth's core. *In:* S Akimoto, MH Manghnani (eds) High Pressure Research in Geophysics. p 611-624 Center for Academic Publications, Tokyo

Brown JM, McQueen RG (1986) Phase transitions, Grüneisen parameter, and elasticity for shocked iron between 77 GPa and 400 GPa. J Geophys Res 91:7485-7494

Buffett BA, Huppert HE, Lister JR, Woods AW (1996) On the evolution of the Earth's core. J Geophys Res 101:7989-8006

Bukowinski MST (1976a) The effect of pressure on the physics and chemistry of potassium. Geophys Res Lett 3:491-494

Bukowinski MST (1976b) On the electronic structure of iron at core pressures. Phys Earth Planet Inter 13:57-66

Bullen KE (1963) An index of degree of chemical inhomogeneity in the Earth. Geophys J Roy Astron Soc 7:584-592

Chen GQ, Ahrens TJ (1996) High-pressure melting of iron: new experiments and calculations. Phil Trans Roy Soc London A 354:1251-1263

Christensen NI, Salisbury MH (1979) Seismic anisotropy in the upper mantle: evidence from the Bay of Islands ophiolite complex. J Geophys Res 84:4601-4610

Clayton DD (1983) Principles of stellar evolution and nucleosynthesis. University of Chicago, Chicago

Creager KC (1992) Aninsotropy of the inner core from differential travel times of the phases PKP and PKIKP. Nature 356:309-314

Creager KC (1997) Inner core rotation rate from small-scale heterogeneity and time-varying travel times. Science 278:1284-1288

Dubrovinsky LS, Saxena SK, Lazor P (1997) X-ray study of iron with in-situ heating at ultra high pressures. Geophys Res Lett 24:1835-1838

Duffy TS, Shen G, Heinz DL, Ma Y, Hemley RJ, Mao HK, Singh AK (1998) Lattice strains in gold and rhenium under non-hydrostatic compression. *In:* RM Wentzcovitch, RJ Hemley, WB Nellis, P Yu (eds) High Pressure Materials Research. in press Materials Research Society, Warrendale, PA

Dziewonski AM, Anderson DL (1981) Preliminary reference Earth model. Phys Earth Planetary Inter 25:297-356

Dziewonski AM, Gilbert JF (1971) Solidity of the inner core of the Earth inferred from normal mode observations. Nature 234:465-466

Dziewonski AM, Hales AL, Lapwood ER (1975) Parametrically simple Earth models consistent with geophysical data. Physics of the Earth and Planetary Interiors 10:12-48

Elsasser WM, Isenberg I (1949) Electronic phase transition in iron at extreme pressure. Phys Rev 76:469A

Fei Y, Prewitt CT (1996) High-pressure behavior of iron sulfide. *In:* AF Hepp, PN Kumta, JJ Sullivan, GS Fischman, AE Kaloyeros (eds), Materials Research Society Symposium Vol 410. p 223-228 Materials Research Society, Pittsburgh

Franken J, Hambir SA, Hare DE, Dlott DD (1997) Shock waves in molecular solids: ultrafast vibrational spectroscopy of the first nanosecond. Shock Waves 7:135-145

Glatzmaier GA, Roberts PH (1996) Rotation and magnetism of Earth's inner-core. Science 274:1887-1891

Gubbins D (1977) Energetics of the Earth's core. J Geophys 43:453-464

Holian BL (1988) Modeling shockwave deformation via molecular dynamics. *In:* SC Schmidt, NC Holmes (eds) American Physical Society Topical Conference on Shock Waves, p 185-190 North Holland, Amsterdam

Holian BL (1995) Atomistic computer simulations of shock waves. Shock Waves 5:149-157

Holian BL, Lomdahl PS (1998) Plasticity induced by shock waves in nonequilibrium molecular-dynamics simulations. Science 280:2085-2088

Ito E, Morooka K, Ujike O, Katsura T (1995) Reactions between molten iron and silicate melts at high pressure: implications for the chemical evolution of Earth's core. J Geophys Res 100:5901-5910

Jeanloz R (1990) The nature of the Earth's core. Ann Rev Earth Planet Sci 18:357-386

Jeanloz R, Wenk H-R (1987) Convection and anisotropy of the inner core. Geophys Res Lett 15:72-75

Jie L, Agee CB (1996) Geochemistry of mantle-core differentiation at high pressure. Nature 381:686-689

Julian BR, Davies D, Sheppard RM (1972) PKJKP. Nature 235:317-318

Karato S (1993) Magnetic-field-induced preferred orientation of iron. Science 262:1708-1711

Kuramoto K, Matsui T (1996) Partitioning of H and C between the mantle and core during the core formation in the Earth: its implications for the atmospheric evolution and redox state of early mantle. J Geophys Res 101:14909-14932

Lay T, Young CJ (1990) The stably-stratified outermost core revisited. Geophys Res Lett 17:1001-1004

Lehmann I (1936) P Bur Centr Seism Internat A 14:3-31

Lister JR, Buffett BA (1998) Stratification of the outer core at the core-mantle boundary. Phys Earth Planet Inter 105:5-19

Macedo WAA, Keune W (1988) Magnetism of epitaxial fcc-Fe(100) films on Cu(100) investigated in situ by conversion-electron Mössbauer-spectroscopy in ultrahigh-vacuum. Phys Rev Lett 61:475-478

Mao HK, Shu J, Shen G, Hemley RJ, Li B, Singh AK (1998) Elasticity and rheology of iron above 200 GPa and the nature of the Earth's inner core. Nature (in press)

Mao HK, Wu Y, Chen LC, Shu JF, Jephcoat AP (1990) Static compression of iron to 300 GPa and $Fe_{0.8}Ni_{0.2}$ alloy to 260 GPa: Implications for composition of the core. J Geophys Res 95:21737-21742

Masters TG (1979) Observational constraints on the chemical and thermal structure of the Earth's deep interior. Geophys J Royal Astron Soc 57:507-534

Masters TG, Shearer PM (1990) Summary of seismological constraints on the structure of Earth's core. J Geophys Res 95:21691-21695

Matsui M, Anderson OL (1997) The case for a body-centered cubic phase (α') for iron at inner core conditions. Phys Earth Planet Inter 103:55-62

Merrill RT, McElhinny MW, McFadden PL (1996) The magnetic field of the Earth: paleomagnetism, the core, and the deep mantle. Academic Press, San Diego

Morelli A, Dziewonski AM, Woodhouse JH (1986) Anisotropy of the inner core inferred from PKIKP travel times. Geophys Res Lett 13:1545-1548

Moroni EG, Grimvall G, Jarlborg T (1996) Free energy contributions to the hcp-bcc transformation in transition metals. Phys Rev Lett 76:2758-2761

Nye JF (1985) Physical Properties of Crystals: Their Representation by Tensors and Matrices. Oxford Univ Press, Oxford, UK

Okuchi T (1997) Hydrogen partitioning into molten iron at high pressure: implications for Earth's core. Science 278:1781-1784

Poirier J-P (1994) Light elements in the Earth's outer core: a critical review. Phys Earth Planetary Inter 85:319-337

Ross M, Young DA, Grover R (1990) Theory of the iron phase diagram at Earth core conditions. J Geophys Res 95:21713-21716

Saxena SK, Dubrovinsky LS, Haggkvist P, Cerenius Y, Shen G, Mao HK (1995) Synchrotron x-ray study of iron at high-pressure and temperature. Science 269:1703-1704

Sayers CM (1989) Seismic anisotropy of the inner core. Geophys Res Lett 16:267-270

Shaner JW, Hixson RS, Winkler MA, Boness DA, Brown JM (1988) Birch's law for fluid metals. *In:* S.C. Schmidt and N.C. Holmes (eds) Shock Waves in Condensed Matter - 1987, p 135-138 Elsevier Science Publishers, Amsterdam

Shen GY, Mao HK, Hemley RJ, Duffy TS, Rivers ML (1998) Melting and crystal structure of iron at high pressures and temperatures. Geophys Res Lett 25:373-376

Sherman DM (1995) Stability of possible Fe-FeS and Fe-FeO alloy phases at high pressure and the composition of the Earth's core. Earth Planet Sci Lett 132:87-98

Söderlind P, Moriarty JA, Willis JM (1996) First-principles theory of iron up to Earth-core pressures: structural, vibrational, and elastic properties. Phys Rev B 53:14063-14072

Song X (1997) Anisotropy of the Earth's inner core. Rev Geophys 35:297-313

Song X, Helmberger DV (1993) Anisotropy of Earth's inner core. Geophys Res Lett 20:2591-2594

Song X, Helmberger DV (1995) A P wave velocity model of the Earth's core. J Geophys Res 100:9817-9830

Song X, Hemlberger DV (1995) Depth dependence of anisotropy of Earth's inner core. J Geophys Res 100:9805-9816

Song X, Richards PG (1996) Seismological evidence for differential rotation of the Earth's inner core. Nature 382:221

Stark PB, Parker RL, Masters G, Orcutt JA (1986) Strict bounds on seismic velocity in the spherical Earth. J Geophys Res 91:13892-13902

Steinle-Neumann G, Stixrude L, Cohen RE (1998) First-principles elastic constants for the hcp transition metals Fe, Co, and Re at high pressure. Phys Rev B (submitted)

Stevenson DJ, Spohn T, Schubert G (1983) Magnetism and thermal evolution of the terrestrial planets. Icarus 54:466-489

Stixrude L, Cohen RE (1995a) Constraints on the crystalline structure of the inner core: Mechanical instability of BCC iron at high pressure. Geophys Res Lett 22:125-128

Stixrude L, Cohen RE (1995b) High pressure elasticity of iron and anisotropy of Earth's inner core. Science 267:1972-1975

Stixrude L, Wasserman E, Cohen RE (1997) Composition and temperature of Earth's inner core. J Geophys Res 102:24729-24739

Stixrude L, Wasserman E, Cohen RE (1998) First-principles investigations of solid iron at high pressure and implications for the Earth's inner core. In: M.H. Manghnani and T. Yagi (eds) Properties of Earth and Planetary Materials at High Pressure and Temperature, p 159-171 American Geophysical Union, Washington, DC

Su W, Dziewonski AM (1995) Inner core anisotropy in three dimensions. J Geophys Res 100:9831-9852

Su W, Dziewonski AM, Jeanloz R (1996) Planet within a planet: rotation of the inner core of Earth. 274:1883-1887

Svendsen B, Anderson WW, Ahrens TJ, Bass JD (1989) Ideal Fe-FeS, Fe-FeO phase relations and Earth's core. Phys Earth Planet Inter 55:154-186

Tanaka S, Hamaguchi H (1997) Degree one heterogeneity and hemispherical variation of anisotropy in the inner core from PKP(BC)-PKP(DF) times. J Geophys Res 102:2925-2938

Taylor RD, Pasternak MP, Jeanloz R (1991) Hysteresis in the high-pressure transformation of bcc-iron to hcp-iron. J App Phys 69:6126-6128

Tromp J (1993) Support for anisotropy of the Earth's core from free oscillations. Nature 366:678-681

Verhoogen J (1980) Energetics of the Earth. National Academy of Sciences, Washington, DC

Vinnick L, Romanowicz B, Breger L (1994) Anisotropy in the center of the inner core. Geophys Res Lett 21:1671-1674

Williams Q, Jeanloz R, Bass J, Svendsen B, Ahrens TJ (1987) The melting curve of iron to 250 GPa: a constraint on the temperature at Earth's center. Science 236:181-182

Woodhouse JH, Giardini D, Li XD (1986) Evidence for inner core anisotropy from free oscillations. Geophys Res Lett 13:1549-1552

Xi S, Ahrens TJ (1994) Pressure-temperature range of reactions between liquid iron in the outer core and mantle silicates. Geophys Res Lett 21:153-156

Yoo CS, Holmes NC, Ross M, Webb DJ, Pike C (1993) Shock temperatures and melting of iron at Earth core conditions. Phys Rev Lett 70:3931-3934

Yoshida S, Sumita I, Kumazawa M (1996) Growth model of the inner core coupled with the outer core dynamics and the resulting elastic anisotropy. J Geophys Res 101:28085-28103

Chapter 9

HIGH-PRESSURE CRYSTAL CHEMISTRY

Charles T. Prewitt

Geophysical Laboratory and Center for High Pressure Research
Carnegie Institution of Washington
5251 Broad Branch Road, NW
Washington, DC 20015

Robert T. Downs

Department of Geosciences
University of Arizona
Tucson, Arizona 85721

INTRODUCTION

The response of earth materials to increasing pressures and temperatures is an area of research that is of interest to many investigators in geochemistry and geophysics. There have been many scientific pioneers including Bernal, Goldschmidt, Bridgman, Birch, and Ringwood, who made essential contributions to understanding how earth materials combine, disassociate, and transform as environmental conditions change. However, it is only in the past 10-20 years that we have had access to the experimental and theoretical tools that allow us to confirm or dispute the ideas of the pioneers and to make a priori predictions of what will happen when a particular mineral composition is subjected to specific conditions of temperature, pressure, or stress. These developments include new instrumentation for x-ray diffraction and spectroscopy, particularly synchrotron sources, but also new laboratory-based systems with improved x-ray optics and detectors. Neutron scattering sources are built and maintained by national governments, but made available to a wide community of scientists, including earth scientists. Raman, infrared, Brillouin, and Mössbauer spectroscopic techniques have improved substantially, techniques for quantitative chemical and isotopic analysis are accurate and reliable, and powerful digital computers with sophisticated software are available almost everywhere. Thus, we have an astounding array of scientific facilities available to us for, we hope, making astounding discoveries.

The most abundant elements in the Earth are O, Si, Fe, and Mg, and consequently, the most abundant minerals in the Earth contain these elements as major components. As a result, much current research at high pressure involves these elements, the phases they form, and the transitions that take place as we simulate Earth's interior in laboratory apparatus or in computers. The phases of magnesium-iron silicates serve as model systems for studies of high-pressure structures, phase transitions, vibrational dynamics, and chemical bonding. In view of the wide-ranging importance of the high-pressure behavior of oxides, silicates, and sulfides, the literature on this subject is extensive and growing, but far from complete. Recent discoveries include new phases, electronic and magnetic transitions, contrasting results from hydrostatic or differential stress, insight to the role of hydrogen, and how specific phases respond to changing conditions. Thus, there is much new interest in the high-pressure behavior of minerals, with important implications for geology, planetary science, materials science, and fundamental physics. The goal of this chapter is to explore the crystal chemical constraints imposed by the high temperatures and high pressures within the Earth and to provide an overview of the dominant phases

0275-0279/98/0037-0009$05.00

resulting from mixtures of the major elements under these conditions. We will also discuss the methods used to study crystal structures of minerals under extreme conditions.

An Appendix is included at the end of this chapter that provides the cell parameters, space groups, and atom coordinates for the principal mineral phases of the mantle and core. Although these are parameters derived at ambient conditions, it is felt that they will be useful to those who want to explore the structures more thoroughly using one of the excellent programs now available for displaying and manipulating crystal structures on personal computers.

HIGH-PRESSURE EFFECTS ON BONDING AND COORDINATION NUMBER

Most physical properties of crystalline materials can be understood by examining their crystal structures and the nature of the bonding, i.e. its crystal chemistry. In order to simplify the chemistry and physics involved in high-pressure crystal chemistry we have assembled a set of high-pressure crystal chemical rules of thumb to guide our understanding. Perhaps the starting point for such a set are the rules put forward by Linus Pauling, known as Pauling's Rules. It is assumed that the reader is familiar with these because they are summarized and discussed in most introductory mineralogy texts (see, for example, *Crystallography and Crystal Chemistry* —Bloss 1994).

Rules of thumb

1 . *A structure usually compresses by displaying the greatest distortion between atoms separated by the weakest bonds.* Imagine that a crystal structure is composed of spheres separated by springs, each with a given strength, or force constant. In general it is sufficient to assume that the springs exist between nearest neighbors and next-nearest neighbors. For example, within SiO_4 groups in quartz, the nearest neighbor springs are between Si and O and next-nearest neighbor springs are between the six pairs of O atoms. These next nearest neighbor interactions can often be thought of in terms of bond bending. So the O-O springs in SiO_4 tetrahedra are associated with O-Si-O angle bending. A structure usually compresses by displaying the greatest distortion between atoms separated by the softest force constants. In quartz, the force constants can be ranked as stiffest for the SiO bond, then the O-Si-O angle, and weakest for the Si-O-Si angle. Therefore the greatest change in the structure with application of pressure will be the Si-O-Si angles.

2 . *Short bonds are the strongest, and long bonds are the weakest.* The force constants between bonded pairs of atoms often can be quickly estimated by examining their separations. Hill et al. (1994) demonstrated that the magnitudes of the bonding force constants in molecules and crystals of nitrides, oxides and sulfides vary in a systematic fashion, with short bonds displaying the largest force constants and long bonds displaying the smallest. The equation for the force constant F_{MX} between a pair of atoms, M and X, is given as:

$$F_{MX} = 7500 \cdot R(MX)^{-5.4} \text{ N/m,}$$

where the bond length, R(MX), is expressed in Ångströms.

3 . *As a given bond compresses it becomes more covalent.* This results from the observation that if the distance between a given pair of atoms decreases then the electron density between the atoms must increase in order to keep the bond stable. Gibbs et al. (1994) have shown that such an increase in the electron density between a bonded pair of atoms comes laterally from the region normal to the bond, and not from along the bond. This makes sense because it is largely the valence electrons that provide the bonding.

4. *Increasing pressure increases coordination number.* Above some point the changes in pressure eventually result in changes in the coordination numbers of the atoms. For example, SiO_4 groups transform to SiO_6 groups with sufficient increase in pressure. An increase in coordination number is usually accompanied by a lengthening of the bonds. For example, $R(Si^{IV}O) = 1.62$ Å in quartz while $R(Si^{VI}O) = 1.78$ Å in stishovite. However, the O—O separation usually decreases: it is 2.63 Å in quartz and 2.51 Å in stishovite. In addition, the change in coordination number is usually accompanied by an increase in the ionic character of the bond.

5. *The oxygen atom is more compressible than the cations.* Total electron density calculations for coesite and stishovite show that the bonded radius of the oxygen atom decreases by 0.20 Å while the radius of the Si atoms decreases by 0.02 Å with the change in coordination from Si^{VI} to Si^{IV} (Nicoll et al. 1994). This is consistent with Pauling's radius ratio rule. For the coordination number of Si to increase from 4 to 6 requires that the ratio of $r(Si)/r(O)$ increase. This can occur only if $r(O)$ decreases relative to $r(Si)$. The reason that oxygen is more compressible than the cations is not related to the size of the atoms but rather to the slope of the electron density in the bonding region. The electron density of oxygen falls off rather rapidly compared to the electron densities around cations, which fall off slowly. Therefore, as a bond is compressed and shortened it appears that the size of the oxygen shrinks relative to the cation.

6. *Angle bending is dependent upon coordination.* Little is known about angle bending force constant systematics. For example, molecular orbital calculations show that Si-O-Si angles are stiffer than Si-O-Al but weaker than Si-O-B (Nicholas et al. 1992). This observation may be related to the distance between cations as discussed earlier in rule #2. However, we also know that if a bridging oxygen atom is coordinated to yet a third atom then the force constant of the angle increases dramatically (Geisinger et al. 1985). Such effects are observed while compressing framework structures such as the feldspar minerals and may also be related to garnet compressibility systematics.

7. *O-O packing interactions are important.* Inter-tetrahedral O-O interactions are known to be very important in understanding the compression of silica framework structures. For example, work by Lasaga and Gibbs (1987) and Boisen and Gibbs (1993) using force field calculations could not replicate experimental compression behavior for silica polymorphs unless strong O-O repulsion terms were included in the energy calculations. All first principles and ionic model calculations include the O-O interactions implicitly, but little systematic work has been done to quantify the magnitude of its importance. For instance, olivines are closest-packed structures and yet the compressibility has been studied in terms of the bonding in the octahedral and tetrahedral sites. It may be that its compressibility is determined principally by O-O interactions.

8. *High-pressure structures tend to be composed of closest-packed arrays of atoms.* Application of pressure on a crystal structure forces the atoms to occupy a smaller volume. Closest-packed structures are the densest arrangement of atoms and therefore high-pressure structures tend to be composed of closest-packed arrays of atoms. For oxide minerals this means that the structures found deep in the Earth are generally closed-packed arrays of oxide atoms, with short-ranged metal-oxygen bonding perturbing the arrangement. Pressure forces the oxygen atoms into more regularly close-packed arrangements, but the cation-oxygen bonds influence how they get there. For example, in their study of kyanite, Yang et al. (1997) found that the arrangement of oxygen atoms became more closest-packed under pressure. They computed a best fitting ideal closest-packed array of oxygen atoms to the observed structure as a function of pressure by minimizing the distance between the observed atomic positions and those in the ideal array by varying the radii and orientation of the ideal array. The result produced a value for the radius of oxygen and a isotropic root-mean square displacement parameter that describe the deviation of the structure from ideal. They found that the oxygen atoms became more closest-packed, with $U_{iso} = 0.0658$ Å at room pressure and $U_{iso} = 0.0607$ Å at 4.5 GPa. Furthermore the radius of the oxygen atom decreases from 1.372 Å to 1.362 Å. The compression was anisotropic, with the most compressible direction oriented parallel to the direction of the closest-packed planes that were furthest apart. Another example is quartz, SiO_2. Quartz traditionally represents the prototype framework structure with its corner-linked SiO_4 groups and is never called a closest-packed structure. Yet as pressure is applied the O atoms tend to arrange themselves in closest packed layers, oriented parallel to (1̄ 20) (Hazen et al. 1989).

9 . *Elements behave at high pressures like the elements below them in the periodic table at lower pressures.* For example, at high pressure Si behaves like Ge does at low pressure. Because of this observation, many studies of germanate compounds have been conducted as high-pressure silicate analogues in days past because sufficiently high-pressure experimental conditions were unattainable. The germanates tend to display structures that are similar to silicates and undergo similar phase transformations at more modest pressures. This can also be understood in terms of softer anions with Pauling's radius ratio rule. Upon application of pressure to, say a SiO_4 group we see that the r(O) decrease at a faster rate than the r(Si) and so r(Si)/r(O) increases. However, at ambient conditions r(Ge) > r(Si) so the ratio of r(Ge)/r(O) is already greater than for Si-O. This rule of thumb may also apply to oxygen and sulfur inasmuch as sulfides at room pressure appear to behave like oxides at high-pressure. For example, it is known that S-S bonding in sulfides is prevalent, e.g. pyrite. Recent work has shown that at high pressures O-O bonding can be found in low quartz (Gibbs et al., submitted).

As with many rules of thumb, ours are not laws of nature because the behavior of crystalline materials can often be quite complicated. These rules are provided as a guide of what first to think about when pressure is applied. Departures from the rules often can indicate the presence of multiple interactions, which are, of course, the most exciting to investigate. For every one of these rules of thumb we could probably find an exception. For example, GeO_2 does not crystallize with the coesite structure. Even though closest packing is the most efficient way to pack atoms, many elemental phases (such as Pb) convert to a body-centered structure at high-pressure. Ideal $MgSiO_3$ perovskite is cubic and can be regarded as composed of a closest-packed array of O and Mg with Si in the octahedral voids. However, it actually is found to be orthorhombic, with distorted closest packing, which deviates even more with pressure. The reasons are interesting, and can be related to the differences in the force constants between O-O, Si-O and Mg-O.

COMPARATIVE COMPRESSIBILITIES

When analyzing the behavior of a crystal structure with changes in pressure, it is useful to understand how the relative sizes of atoms (ions) change. Measurements of the change of a unit cell as pressure increases can be obtained rather easily. However, it is more difficult to determine the change in individual ions, not only because of experimental difficulties, but also because assumptions have to be made about whether the changes take place in cations or anions, or both. There has been extensive discussion in the scientific literature about how one defines the radius of a cation or anion, ranging from empirical assignments of a specific radius to oxygen ions (Shannon and Prewitt 1968), to relatively sophisticated molecular orbital calculations (Nicoll et al. 1994). Except for structures with no variable atom coordinates (e.g. NaCl), information on compression of interatomic distances has to be obtained from accurate crystal structure determinations as a function of pressure (and/or temperature).

In general, compressibility of a cation or an anion is proportional to the size of the ion and the coordination number, and inversely proportional to the charge on the ion. Thus, eight-coordinated Na^+ is much more compressible than tetrahedrally-coordinated Si^{4+}. The systematics of ionic compressibility are described in detail by Hazen and Finger (1979, 1982). They give the relation (converted from Mbar to GPa)

$$\frac{K_p d^3}{z_c} = 750 \pm 20 \text{ GPa-Å}^3$$

where K_p is the polyhedral bulk modulus in GPa d is the mean cation-anion distance in Å, and z_c is the integral formal charge on the cation. The polyhedral bulk modulus is obtained by calculating the volume of two coordination polyhedra in a structure at different pressures and using

$$K_p = -\frac{(V_1 + V_2)}{2}\left(\frac{\Delta P}{\Delta V}\right)$$

where $(V_1 + V_2)/2$ is the average volume of the two polyhedra being compared. This relation can provide semiquantitative insight into how different cation polyhedra will behave under compression, but it should be noted that the assumption here is that the concept of macroscopic ("continuous") moduli can be transferred to the atomic scale.

MECHANISMS OF PHASE TRANSITIONS

One of the more interesting aspects of crystal chemistry at high pressures and temperatures involves phase transitions that are accompanied by changes in physical and chemical properties of the phases involved. Phase transitions are very important in studies of Earth's interior because they involve changes in elastic properties and densities of minerals that can be detected via the analysis of seismic waves created by earthquakes and thus provide information about the inner structure of the Earth. One goal of mineral physics investigations is to study relevant phase changes in the laboratory and to attempt to understand why they take place under specific conditions of pressure, temperature, and varying composition.

From a crystal structural point of view, phase transitions are generally described as *reconstructive* or *displacive*, and from a thermodynamic perspective as *first-order* or *second-order* (Buerger 1961). The fundamental definition of a reconstructive transition involves breaking of bonds and can range from a relatively subtle change in a structure to a very drastic one. In contrast, a displacive transition is one that does not involve breaking of existing bonds and creation of new ones, but only involves a shifting of the atomic positions, possibly with a symmetry change. It is incorrect to equate reconstructive with first-order and displacive with second-order, because there are examples where differences are subtle and it is not possible to make such comparisons. There are, for example, displacive transitions that are accompanied by discontinuities in lattice parameters, which indicates that they are also first-order transitions. Other descriptions of phase transitions include *martensitic* (one involving a tilt or distortion of the structure and mostly applied to metals and alloys) and *soft mode* (where one or more modes in, say, a Raman spectrum disappear at the transition). A more complete discussion of phase transition theory is not warranted in this chapter, but the reader should be aware that these categories do exist and are discussed extensively in the literature.

A major aspect of high-pressure phase transitions is the increase in cation and anion coordination as a function of increasing pressure. Emphasis is usually placed on changes in cation coordination, but anion coordination must also increase, as described by the formula for magnesium silicate perovskite:

$$^8Mg\ ^6Si\ ^4O\ ^5O_2$$

Thus, the total coordination of the cations is $8 + 6 = 14$, and for oxygen is $4*1 + 5*2 = 14$. This relation holds for all phase transitions and it should be noted that in some structures crystallographically-different oxygens or other anions may have different coordination numbers. Shannon and Prewitt (1968) showed that as coordination of an ion increases, the average interatomic distances to that ion also increase, an apparent paradox for a transition where the density of the high-pressure phase is always higher. It works out satisfactorily, however, because the ions in the high-pressure phase must always be packed together more efficiently. Several other examples for $MgSiO_3$ phases are

Enstatite	$^6Mg\ ^4Si\ ^3O_2\ ^4O$	$6*1 + 4*1 = 3*2 + 4*1$
Majorite	$^8Mg_3\ ^6(MgSi)\ ^4Si_3\ ^4O_{12}$	$8*3 + 6*2 + 4*3 = 4*12$
Ilmenite	$^6Mg\ ^6Si\ ^4O_3$	$6*1 + 6*1 = 4*3$

One other feature related to coordination number is the geometrical relation of the coordinating ions to each other. In high-pressure phases the four- and six-coordinating groups almost always form tetrahedra or octahedra. Five-coordinated cations are relatively unusual although one Al site in andalusite can be described as a trigonal bipyramid and Angel et al. (1996) and Kudoh et al. (1998) reported that a phase transition in $CaSi_2O_5$ at about 0.2 GPa results in five-coordinated Si. In $MgSiO_3$ perovskite, oxygen is coordinated by three Mg and two Si in a rectangular pyramid with two Mg and two Si forming the base and the other Mg at the apex. The Mg coordination in this structure is sometimes referred to as dodecahedral (Poirier 1991), but it is perhaps better described as a bicapped trigonal prism with the six closest oxygens forming a trigonal prism and the next two closest oxygens positioned outside two of the prism faces. In contrast to the one in perovskite, the polyhedron around Mg in majorite is a slightly distorted triangular dodecahedron with eight vertices and twelve triangular faces. There are, of course, other geometrical arrangements in high-pressure phases, but these are the ones most common in mantle phases.

ANALYTICAL TECHNIQUES

X-ray diffraction is the most important analytical tool used to gain information on the crystallographic properties of high-pressure phases. The approach used to learn about the crystal chemistry of high-pressure, high-temperature phases depends on the objectives of the investigation and upon the tools available. For example, much information can be derived from examination of quenched high-pressure phases if the material in question retains the high-pressure structure upon quench. However, there are limitations to this approach, including the fact that the unit cell and interatomic distances in a quenched phase are different from the ones in the material before the quench. Furthermore, it is always possible that the symmetry and/or crystal structure undergo significant changes upon quench even though the material remains crystalline—in many examples single crystals remain single through one or more phase transitions as external conditions are varied.

For in situ single-crystal diffraction experiments at high pressures and/or high temperatures, the most common device used to apply pressure is the Merrill-Bassett (Merrill and Bassett 1974) diamond-anvil cell. This cell employs a simple clamping mechanism to push together two brilliant-cut diamonds that are positioned on either side of a hole in a metal gasket that contains the crystal being studied. The crystal is suspended in a pressure medium that is usually a mixture of ethanol and methanol, although other organic liquids, water, or cryogenic liquids of noble elements such as neon, argon or helium are sometimes used when loading a diamond cell. In order to provide as large an opening for incident and diffracted x-rays as possible a beryllium disk that is relatively transparent to x-rays backs each diamond. One disadvantage of this is that the resulting polycrystalline diffraction pattern and associated diffuse scattering from beryllium must be accounted for in the data analysis. Using this approach pressures up to 10 GPa can be obtained to record good-quality single-crystal intensity data on a routine basis; pressures as high as 33 GPa have been reported (Zhang et al. 1998).

Applying simultaneously elevated pressures and temperatures for a single-crystal experiment is more difficult than for either pressure or temperature alone. Heat is generally applied with resistance wire wrapped around the diamonds and temperatures to about 600°C can be reached. Because beryllium and other metals weaken at high temperatures and because beryllium oxide vapor is poisonous, investigators have generally used other materials for the backing disk. One such material is B_4C, which is a ceramic that transmits x-rays, but is brittle and will crack and break at pressures higher than about 2-4 GPa.

Some of the above limitations may be overcome with the growing use of synchrotron radiation for high-pressure work. There have been relatively few high-pressure single-crystal experiments conducted at synchrotron x-ray sources [examples are by Mao and Hemley (1996) on stishovite at 65 GPa and by Loubeyre et al. (1996) on H_2 crystals at 119 GPa], and even fewer that involved recording of accurate diffraction intensities for structure determinations or refinements. However, the production of shorter x-ray wavelengths at synchrotrons, improvements in beam stability, and increasing availability of area x-ray detectors capable of recording high-energy radiation make such investigations appear feasible. The shorter wavelengths make it possible to collect sufficient diffraction data in a more restricted angular range of incident and diffracted x-ray beams. In the future this may allow the construction of diamond cells that do not require beryllium backing disks. Diamonds can be mounted directly on tungsten carbide plates having conical apertures for the x-rays to enter and leave the cell. Not only will this decrease the background x-ray scattering, it should also permit inclusion of resistance heaters to provide temperatures up to at least 800°C.

Powder x-ray diffraction

In contrast to single-crystal investigations at high pressure, there have been relatively few powder structure determinations at high-pressure using conventional laboratory apparatus. For many years, high-pressure powder diffraction experiments were confined to recording changes in cell parameters and detecting the occurrence of phase transitions because the only available diffraction tools were sealed-tube x-ray generators and film or scintillation detectors that resulted in inefficient recording of diffraction patterns. In the past 10 years however, new developments in x-ray optics, x-ray generators, synchrotron x-ray sources, detectors, computer control, and analytical software have changed the situation drastically. Improved diamond cells and the ability to reduce the amount of sample required (~10 μm) because high-intensity micro x-ray beams can be focused to the same size as the sample result in the ability to obtain good diffraction patterns on polycrystalline samples at pressures in the 1-3 megabar range. Although most of these kinds of experiments are now being done at synchrotron sources, Hasegawa and Badding (1997) showed that it is possible to collect satisfactory data with a laboratory rotating-anode generator, an x-ray monochromator, and x-ray film as a detector. With further developments of focusing optics, x-ray generation, and either imaging plates or CCD (charge-coupled device) detectors, it is likely that more investigators will make serious attempts to improve their "conventional" laboratory facilities.

Synchrotron radiation sources

Since the "second-generation" synchrotron sources became available to mineral physicists in the early 1980s, there has been strong interest in using these machines for investigations at high pressure. Fundamentally, there are two distinctly different kinds of experiments that are of interest, one uses polychromatic or "white" x-radiation and the other monochromatic x-radiation. There are advantages and disadvantages to each. White radiation experiments take advantage of the continuous x-ray spectrum produced by a synchrotron such as the one at the National Synchrotron Light Source, Brookhaven National Laboratory. In this kind of experiment, a powder sample is held in a diamond cell that has relatively small angular ports for incident and diffracted x-rays and the detector is placed to record the diffracted pattern. One kind of detector is a "point" detector that measures x-rays at a given angle from the direct beam. A common point detector is a solid-state Ge detector that can record a wide energy range and thus records the x-ray spectrum of intensity as a function of energy. Neither the sample nor the detector has to move during an experiment, but changes in cell parameters and the occurrence of phase transitions are measured easily. A major disadvantage is that it is difficult to impossible to make quantitative use of the diffraction intensities because there are so many poorly defined variables in such experiments. Another problem with the second-generation synchrotron sources is that their x-ray intensities decrease substantially as energy increases toward the values most useful for transmission through the diamond cell. A solution is to use a "wiggler" port at the synchrotron that produces a higher energy x-ray spectrum. Such ports are available at NSLS, the Photon Factory, and SSRL.

Three "third-generation" synchrotron sources were built in the 1990s and are now available for use by high-pressure scientists. These are the European Synchrotron Radiation Facility (ESRF) in Grenoble, France, the Advanced Photon Source (APS) at Argonne National Laboratory in Illinois, and SPring-8 near Aoki, Japan. The advantage of all these machines is that they produce high-energy x-rays from their standard bending magnets and are further enhanced by the use of insertion devices, i.e. undulators and wigglers, that boost the x-ray energies to even higher values. A variety of high-pressure experiments are being pursued at all these facilities, including diamond-cell energy dispersive experiments with white radiation, monochromatic diamond-cell experiments, and experiments with large-volume, multi-anvil apparatus.

Neutron powder diffraction

A few papers have been published on the use of powder neutron diffraction to obtain crystal-chemical data on mineral structures at high pressure. Most of the experiments described in these papers utilized neutrons rather than x-rays because the locations of hydrogen or deuterium atoms in hydrous phases can be found more easily. Negative aspects of neutron diffraction are that it requires more sample than does x-ray diffraction and most of the high-pressure cells designed thus far are limited to pressures less than about 10 GPa, but there are attempts to increase this limit to at least 20 GPa. The pressure cell most widely used for neutron is the Paris-Edinburgh cell designed by Besson (Besson et al. 1992). The primary neutron sources used for experiments with this cell are the Los Alamos Neutron Science Center (LANSCE) at Los Alamos National Laboratory in New Mexico and the pulsed neutron source (ISIS) at the Rutherford-Appleton Laboratory in the United Kingdom.

In addition to locating and refining hydrogen/deuterium atoms with neutrons, another goal has been to investigate how the H-O distances change with pressure. Hydrogen bonds in oxides including silicates are generally in the form O-H...O where the O-H distance is ~0.95 Å, the H...O distance is about 2.3 Å, and the O-H...O angle is in the 160°-170° range. The O-D...O distances and angles are slightly different when D substitutes for H in the same crystal structure. In their study of brucite, $Mg(OD)_2$, to 9.3 GPa, Parise et al. (1994) found that the O-D distance did not change with pressure, but that D became disordered into three equivalent positions around the threefold axis in the trigonal space group and the D...O distance decreases from 2.291 Å at 1 atm. to 1.95 Å at 9.3 GPa. There was no observable change in the overall symmetry.

In another application of neutron diffraction, Lager and Von Dreele (1996) used ISIS to collect data on the hydrogarnet katoite $[Ca_3Al_2(O_4D_4)_3]$ at several pressures up to 9.0 GPa. In this example, the O-D distance decreased from 0.906 Å at one atm. to 0.75 at 9 GPa, while the O...D distance decreased from 2.54 Å to 2.48 Å. At the same time, the O-D...O angle increased from about 137° to 141°. Clearly, it is important to obtain more data on different structures so that it will be possible to develop a coherent picture of hydrogen bonding in oxide minerals at high pressure.

HIGH-PRESSURE PHASES

This section provides information about the major and some of the minor phases that are important in determining the character of Earth's inner core, the lower mantle, the transition zone, and the upper mantle. It is a discussion of crystal-chemical concepts and current research directions that investigators believe are important in directing high-pressure experiments and interpreting the results. We have made no attempt to review every detail for each phase involved, but instead try to give the reader an overview of high-pressure mineralogy and hope that this will encourage further research on a wide range of topics. In contrast to the usual approach of starting with Earth's crust and working downward, we first look at the possible mineralogy of the inner core and proceeding upward from there. This approach is used because the high-pressure phases are generally more simple than low-pressure phases, in that they always adopt some sort of closest packing scheme.

In order to set the stage for describing each major mineral phase in the lower and upper mantles, Figure 1 is a phase diagram from Gasparik (1993) that shows both hydrous and anhydrous magnesium silicates that occur in the mantle. When reading through the various mineral descriptions below, the reader can refer to this diagram to see the pressure-temperature regions where the particular phase is stable.

Iron

At ambient conditions Fe adopts a body-centered cubic (bcc) structure (α–Fe). However, at high temperatures it transforms to γ–Fe, a face-centered cubic, closest-packed structure (fcc), and at high pressures it becomes hexagonal closest-packed (hcp), ϵ–Fe. There is a major controversy about yet another possible phase (Boehler 1990, Saxena et al. 1993) that is designated as β–Fe and appears above about 1000 K and 100 GPa. A group at Uppsala University (Saxena et al. 1995, Dubrovinsky et al. 1997) maintains that this phase exists and has a dhcp structure, meaning that the closest-packed monolayers are arranged as *abacaba,* thereby effectively doubling the c cell edge. Other investigators from the Carnegie Geophysical Laboratory (Shen et al. 1998) did not observe evidence for phases other than ϵ–Fe or γ–Fe in situ, but state that the diffraction patterns of temperature-quenched products at high pressure could be fit to other structures such as dhcp. The reason that this is important is that the proposed pressure-temperature range proposed for the possible β-Fe includes the conditions present in Earth's inner core. Thus, definitive

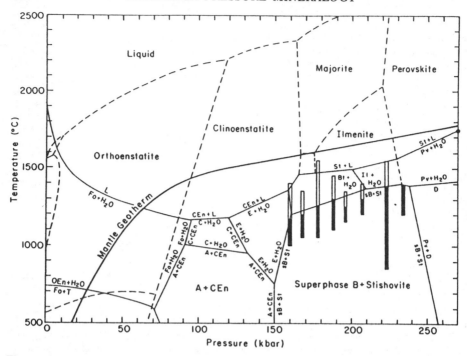

Figure 1. Temperature-pressure phase diagram from the system MgO-SiO$_2$-H$_2$O (solid lines). Dashed lines are phase boundaries in the anhydrous MgSiO$_3$ system. This diagram demonstrates the effect of H$_2$O on phase relations. [Used by permission of the editor of *Journal of Geophysical Research*, from Gasparik (1993), Fig. 5, p. 4294.]

knowledge of the crystal structure of iron in the inner core could be essential for understanding its physical properties such as the anisotropy of transmission of seismic waves through the core. Anderson (1997) gives an interesting discussion of the various points of view on this controversy.

Magnesiowüstite

At ambient conditions both FeO (wüstite) and MgO (periclase) adopt the cubic closest packed structure of rock salt (Fig. 2). This structure can be envisioned as a stacking of closest packed monolayers along the [111] direction. However, at high pressures FeO undergoes several phase transitions whereas MgO does not. The phase believed to occur in

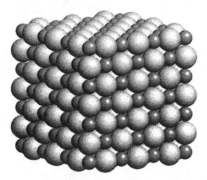

Figure 2. Structure of magnesiowüstite illustrating cubic closest packing of anion and cation layers. The sizes of the atoms are chosen only as a guide to aid in visualization.

the lower mantle is called magnesiowüstite by many investigators, but it actually should be called ferropericlase because all models of the lower mantle assume only 10-20% Fe. However, magnesiowüstite seems to be the dominant terminology for intermediate compositions and we will continue that usage here. In addition, some of the most interesting crystal chemistry results from experiments on FeO at high pressure where it transforms to a NiAs-type structure. However, some high-pressure experiments have been performed on magnesiowüstite such as those by Shu et al. (1998) where intermediate compositions up to $(Fe_{0.60}Mg_{0.40})O$ underwent phase transitions between 25 and 40 GPa at room temperature, but no transition was observed for $(Fe_{0.40}Mg_{0.60})O$ up to 48 GPa..Thus, it is not clear whether or what transitions can be expected for magnesiowüstite compositions of the lower mantle.

Using shock wave and diamond-cell techniques, Jeanloz and Ahrens (1980) and Knittle and Jeanloz (1986) reported a phase transition in FeO at about 70 GPa and temperatures exceeding 1000 K. Knittle and Jeanloz interpreted the shock wave results as evidence for a change in ferrous iron character at high pressures, resulting in a metallic phase. However, they found no such evidence in resistivity measurements in the diamond cell. Zou et al. (1980) and Yagi et al. (1985) found that at low temperature and at pressures above 20 GPa, FeO undergoes a rhombohedral distortion (rhombohedral angle <60°), which increases with pressure. The distortion results in a shortened Fe-Fe distance. The results of first-principles LAPW calculations (Isaak et al. 1993) show that the distortion originates with the onset of Fe-Fe bonding, which has some covalent character and results in a decrease in the interplanar Fe-Fe distances. The magnitude of the Fe-Fe interactions increases with pressure, causing an increase in the rhombohedral distortion (Mazin and Anisimov 1997a). With application of even more pressure a phase transition into a hexagonal structure was observed (Fei and Mao 1994) at high temperatures (600 K at 96 GPa) which was interpreted to be a NiAs-type (B8) structure, the hexagonal analogue of the rocksalt structure. In other words, the transition investigated by Fei and Mao represents a change of symmetry due to a different stacking sequence of the close-packed planes, with the nearest-neighbor Fe-O distances being essentially the same in both structures. In a subsequent development, it was discovered that diffraction patterns of the structure regarded as B8 actually has diffraction intensities consistent with a structure containing both B8 and anti-B8 domains. In the anti-B8 structure, Fe and O are exchanged between non-equivalent crystallographic sites, where Fe has trigonal prismatic and O has octahedral coordination (Mazin et al. 1998).

Perovskite, ilmenite

Perovskite is a mineral with the composition $CaTiO_3$ and originally was thought to be cubic with Ca coordinated by 12 oxygens in a cubo-octahedral geometry and Ti in an octahedron. Further work showed that it is actually orthorhombic and that Ca is coordinated by eight oxygens. It is difficult to know who was the first scientist to realize that $MgSiO_3$ enstatite or Mg_2SiO_4 olivine might transform to the perovskite structure at high pressure, but it was mentioned as a possibility by Ringwood (1962), and Ringwood and Major (1967a) synthesized germanates with the orthorhombic perovskite structure, which was possible with an existing high pressure apparatus. In the first successful experiment on a silicate, Liu (1974) obtained silicate perovskite by starting with pyrope $(Mg_3Al_2Si_3O_{12})$ and laser-heating it in a diamond-anvil cell at 27-32 GPa to produce $MgSiO_3$ perovskite plus corundum. Investigators soon realized that silicate perovskite could be the dominant phase in the lower mantle and, if so, the most abundant mineral in the Earth. The structure of orthorhombic silicate perovskite is shown in Figure 3.

Today, silicate perovskite is synthesized easily in diamond cells and in large-volume,

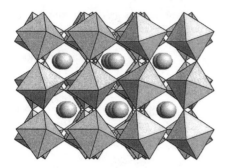

Figure 3. Structure of orthorhombic perovskite viewed down [110]. The SiO_6 groups are illustrated as octahedra, with Mg illustrated as a sphere.

multi-anvil apparatus at pressures of 22 GPa and above. Single crystals as well as powder samples have been made and the crystal structure and elastic properties determined by several different investigators. There have, however, been arguments about the range of stability of perovskite at high temperatures and whether it might break down into MgO and SiO_2 under conditions existing in the lower mantle (Saxena et al. 1996). This was disputed by Mao et al. (1997) and Serghiou et al. (1998), and the exact conditions required for perovskite stability is still an open question. Fei et al. (1996) presented a diagram showing that there is a maximum amount of Fe^{2+} that $(Mg,Fe)SiO_3$ perovskite can accommodate at any given pressure and temperature (see Fig. 6 in Fei, this volume). Mao et al. (1997) and McCammon (1998) found that $(Mg,Fe)SiO_3$ perovskite synthesized in both diamond cells and multi-anvil presses contain a significant amount of Fe^{3+} that can stabilize the structure. This also implies that Al^{3+} can have the same effect and that the amount of Fe^{3+} and Al^{3+} in the lower mantle will have a strong influence on the range of stability of perovskite. It is clear that research on this general subject will continue for the foreseeable future.

No natural samples of silicate perovskite from the lower mantle have been reported although some workers have predicted that it might be found as an inclusion in diamond. However, there are two recent reports of $(Mg,Fe)SiO_3$ ilmenite and more limited evidence for the presence of perovskite in shocked meteorites (Sharp et al. 1997, Tomioka and Fujino 1997). It remains to be seen whether the evidence is strong enough for either of these phases to be approved for official mineral names by the International Mineralogical Association. The ilmenite structure is a derivative of the corundum structure except that it consists of two crystallographically-distinct octahedra, each occupied by Mg and Si, respectively, and is shown in Figure 4.

Stishovite, coesite, quartz

An extensive discussion of silica minerals at high pressure is available in Hemley et al. (1994). Therefore, only a brief summary of the two high-pressure silica phases, stishovite and coesite, is given here.

Stishovite (Fig. 5) is the highest-pressure form of SiO_2 that has been found as a mineral, i.e. in shocked siliceous rocks resulting from meteorite impacts. Stishovite has the rutile structure with Si in octahedral coordination and the octahedra forming edge-shared chains along the c axis that are each connected to four other parallel chains. The crystal structure as a function of pressure has been determined by Sugiyama et al. (1987) to 6 GPa and by Ross et al. (1990) to 16 GPa. It was observed that the compression is anisotropic with the a axis almost twice as compressible as the c axis. The structural response to compression can be considered as mainly polyhedral tilting along with some compression of the SiO bonds, but without appreciable distortion of the octahedra. It came as a surprise that the shared edge O-O separations were not the least compressible. In contrast with earlier workers (Megaw 1973), we find that stishovite can effectively be considered a

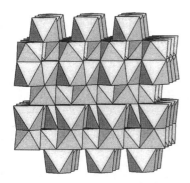

Figure 4. Structure of ilmenite viewed along the *ab* plane showing alternating planes of non-equivalent octahedral layers.

Figure 5. Structure of stishovite viewed down the *c* axis. Note the distorted closest-packed layers of O atoms that align with the edges of the octahedra.

distorted hexagonal closest-packed array of O atoms with Si in octahedral voids. The shared edges of the octahedra ensure distortion. With increasing pressure the O atoms become more closely packed, accounting for the distribution in O-O compression systematics.

A post-stishovite phase with the $CaCl_2$ structure was confirmed experimentally in high-pressure x-ray (Tsuchida and Yagi 1989) and Raman (Kingma et al. 1995) experiments. The latter study also examined the temperature dependence of the transition using calculations based on the potential induced breathing model (PIB[++]) and concluded that any free silica in the lower mantle would have the $CaCl_2$ structure above about 60 GPa. This structure involves a slight tilting of the SiO_6 octahedra, no breaking of bonds, and a symmetry change from the tetragonal space group $P4_2/mnm$ to orthorhombic $Pnnm$.. The polyhedral tilting observed in the transformation to the $CaCl_2$ structure demonstrates that the closest-packing layers should be considered stacked along the *b*-axis direction. Theoretical electron density maps for stishovite at pressure constructed by Gibbs et al. (1998) suggest that the $CaCl_2$ transformation is coincident with the onset of O-O bonding.

Several other post-stishovite phases have been proposed and Teter et al. (1998) describe the various possibilities based on crystal chemical reasoning and first-principles total-energy calculations. Using a laser-heated diamond cell at pressures up to 85 GPa, Dubrovinsky et al. (1997) synthesized a silica phase identified as intermediate between the *a* PbO_2 and ZrO_2 (baddelyite) structures, and El Goresy et al. (1998) found a phase in the shocked SNC meteorite Shergotty whose diffraction pattern is consistent with that of a silica phase with the ZrO_2 structure. The ultimate high-pressure phase of SiO_2, the cubic phase with the $Pa3$ space group, similar to the pyrite structure, has not yet been reported.

Coesite represents the highest-pressure stable polymorph of the tetrahedrally coordinated silica phases (Fig. 6). It forms a structure that in some ways is similar to that of the feldspars. Four membered rings of tetrahedra form chains that run parallel to the *c*-axis. These chains lie on layers that are perpendicular to *b* with a channel separating each chain. Each layer is shifted over the adjoining layers in such a way that chains are always over channels. This is the fundamental way that the framework structure of coesite differs from feldspar. An important feature of the crystal structure that has received much attention is an apparent linear Si-O-Si angle. The nature of this angle has been a subject of considerable debate because the equilibrium Si-O-Si angle is around 144°.

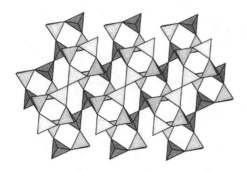

Figure 6. Structure of coesite viewed down the *b* axis. Two layers of the chains of 4-membered tetrahedral rings are displayed. The chains run parallel to the *c* axis.

Levien and Prewitt (1981) determined the crystal structure of coesite as a function of pressure. They demonstrated that the compressibility of coesite is quite anisotropic, with the stiffest direction parallel to the chains and the softest direction coincident with narrowing the widths of the channels. All Si-O-Si angles decrease except for Si1-O1-Si1, which is constrained to be 180°. However, all temperature factors decrease with application of pressure except for O1, which increases significantly. Such an increase is consistent with a bent Si-O-Si angle in which the O1 atom is librating under toroidal motion. Increasing pressure would bend the angle further, thus effectively increasing the temperature factor.

Other metastable high-pressure silica phases have been observed in experiments using coesite (Hemley et al. 1988, Williams et al. 1993), cristobalite (Tsuchida and Yagi 1990, Downs and Palmer 1994, Palmer et al. 1994), and quartz (Kingma et al. 1996) as starting materials. As these are not likely to be significant mantle phases, they are not discussed further here. Those interested should consult Hemley et al. (1994).

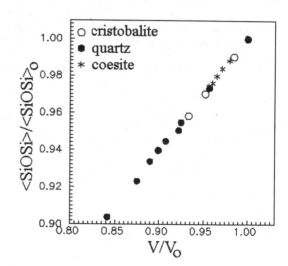

Figure 7. Normalized unit cell volume, V/V_O, vs the normalized average Si-O-Si angle, $\langle\text{Si-O-Si}\rangle/\langle\text{Si-O-Si}\rangle_O$ for quartz, coesite, and cristobalite.

The compression of the tetrahedral silica polymorphs is controlled by the compression of the Si-O-Si angle. Figure 7 shows a plot of the normalized unit cell volume, V/V_O, versus the normalized average Si-O-Si angle, $\langle\text{Si-O-Si}\rangle/\langle\text{Si-O-Si}\rangle_O$, for quartz, coesite and cristobalite (Glinnemann et al. 1992, Levien et al. 1980, Levien and Prewitt 1981,

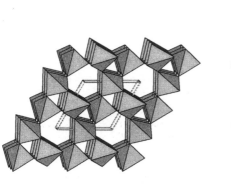

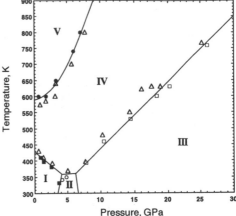

Figure 8. Quartz structure viewed down the *c* axis.

Figure 9 (right). Experimentally determined phase diagram of FeS. FeS I, NiAs- type structure with a (√3*a*,2*c*) unit cell; FeS II, MnP-type structure; FeS III, monoclinic; FeS IV, NiAs-type structure with a (2*a*,*c*) unit cell; FeS V, NiAs-type structure with a (*a*,*c*) unit cell. [Used by permission of the editor of *Science*, from Fei et al. (1995), Fig. 2, p. 1893.]

Downs and Palmer 1994, Hazen et al. 1989). The figure demonstrates that the compression of each of the three structures is controlled in the same way by the angle. The structure of quartz is shown in Figure 8.

FeS

Iron and sulfur are important components of the interiors of the terrestrial planets and it is possible that a high-pressure phase of FeS is present in Earth's lower mantle and the inner core. Under ambient conditions, FeS has the troilite structure (FeS I), but with increasing pressure and/or temperature it undergoes phase transitions to either more symmetric (with temperature) or more distorted (with pressure) structures. Figure 9 is the phase diagram of FeS from Fei et al. (1995) and shows that FeS I transforms to FeS II with the MnP structure at room temperature and 3.5 GPa and to FeS III at about 6.5 GPa. The structure of FeS III was reported by Nelmes et al. (1998) and refinements of FeS IV and V by Fei et al. (1998). These structures are all similar to each other and based on distortions of the hexagonal NiAs structure, except for FeS V, which has the NiAs structure. This structure is based on layers of edge-shared FeS_6 octahedra in which each octahedron shares faces with octahedra in parallel layers above and below. This results in rather short Fe-Fe distances and is important in determining the electronic and magnetic properties of these phases. The Fe atoms form a trigonal prism around each sulfur atom. Figure 10 shows the structures of FeS II (A) and FeS III (B). Note that in FeS II the octahedra are relatively symmetric; by contrast the octahedra in FeS III are highly distorted.

Garnet

Garnet, primarily pyrope and majorite, is an important component in the upper regions of the Earth, accounting for up to 10% of the volume of the upper mantle, and up to 50% of the transition zone (Ita and Stixrude 1992). The chemical formula of garnet can be written $A_3B_2(SiO_4)_3$, with pyrope as $Mg_3Al_2(SiO_4)_3$ and majorite as $Mg_3(Mg,Si)_2(SiO_4)_3 = MgSiO_3$. The structure (Fig. 11) can be viewed as a rather rigid framework of corner sharing octahedra (*B*-site) and tetrahedra, with the *A*-site atom located in interstices that

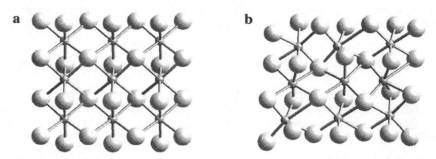

Figure 10. Structures of FeS II (a) and FeS III (b). Both are distorted versions of the NiAs structure.

have a dodecahedral shape. Most garnets display cubic symmetry, however majorite can be tetragonal because of ordering of the two octahedral cations.

Studies of garnet crystal structures at high-pressures by single crystal diffraction have been published on pyrope, grossular, and andradite (Hazen and Finger 1978, 1989; Levien et al. 1979). These studies conclude that the compression of garnets is controlled by the compressibility of the dodecahedral site. For example, the bulk modulus of andradite is 159 GPa and its dodecahedral cation, Ca, has a bulk modulus of 160 GPa. Thus the collapse of the framework, or other words, the tilting of the tetrahedra and octahedra into the dodecahedral site, accomplishes compression.

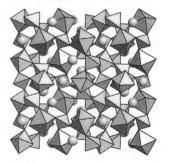

Figure 11. Structure of majorite viewed down the c axis displaying a depth of 1/2 of the unit cell. The small octahedra represent SiO_6 groups, while the larger represent MgO_6 groups. A sphere represents the dodecahedral atoms.

However, in a study of the comparative compressibilities of garnets Hazen et al. (1994) demonstrated that this model does not hold for the majorite type garnets. A garnet with Na in the dodecahedral site was found to be the least compressible garnet ever measured and yet the compressibility of Na is quite high. It was suggested in this study that the compressibility of garnet might, in fact, be controlled by octahedral-O-tetrahedral angle bending, which in turn is modified by bonding of the dodecahedral cation. This is quite analogous to the compressibility mechanism observed for the feldspars. We must conclude that the details about garnet compressibilities need further work.

Spinel

The formula for the spinel type structure is B_2AO_4. Examples are spinel, Al_2MgO_4, magnetite, Fe_3O_4, and ringwoodite, Mg_2SiO_4. The oxygen anions are in a cubic closest-packed arrangement with the B cation in octahedral sites and the A cation in tetrahedral. The octahedra form edge-sharing chains that are linked together by the isolated tetrahedra. All atoms are in special positions, and the only variables for the structure are the cubic cell edge and the x position of the oxygen atom. The cell edge can be determined from the octahedral, R_O, and tetrahedral, R_T, bond lengths as

$$a = \frac{40\,R_T + 8\,\sqrt{33R_O^2 - 8\,R_T^2}}{11\,\sqrt{3}}.$$

This means that compression of the structure "depends" only on the compressibility of the tetrahedral and octahedral bonds. The compression behavior can be separated into 3 categories based on bond compressibilities: (1) the compressibilities of the two bonds are equal, (2) R_T is softer than R_O, (3) R_O is softer than R_T (Finger et al. 1986).

Magnetite offers a good example of a structure with bonds that are roughly equal in strength. This structure compresses primarily by scaling until it transforms to the marokite, $CaMn_2O_4$, structure with Fe^{3+} in octahedral coordination and Fe^{2+} in 8-fold coordination (Fei et al. 1998). The transformation at a pressure near 25 GPa was first reported by Mao et al. (1974). Huang and Bassett (1986) determined a slope of -68°C/GPa for the phase boundary. The crystal structure of magnetite to 4.5 GPa has been determined by Finger et al. (1986) and recently to 30 GPa by Haavik et al. (1998). There has been a wide range of values (190-220 GPa) determined for the bulk modulus, with 220 GPa the latest value reported by Haavik et al. (1998).

The mineral spinel is an example of a structure with soft tetrahedral bonds (Mg-O) and stronger octahedral bonds (Al-O). Its structure has been determined by Finger et al. (1986) to 4.0 GPa and more recently by Pavese et al. (1998) to 3.8 GPa in their study of cation partitioning as a function of P. The result of a softer tetrahedral bond is that the tetrahedra compress more rapidly than the octahedra. Consequently the octahedral site undergoes distortion that increases the length of the shared octahedral edge. In terms of Pauling's rules this leads to a destabilization of the structure. Liu (1975, 1980) demonstrated that Al_2MgO_4 spinel decomposes to a mixture of MgO and Al_2O_3 at pressures above 15 GPa. However, with temperatures above 1000°C and P above 25 GPa, Irifune et al. (1991) observed that Al_2MgO_4 adopts the $CaFe_2O_4$ calcium ferrite structure. This structure is a slight modification of the marokite structure adopted by Fe_3O_4 at high pressure. The bulk modulus was determined as 190-194 GPa with $K'_O = 4$ by diffraction techniques, and 198 GPa by Brillouin spectroscopy (Askarpour et al. 1993). In more recent work using a combination of synchrotron x-ray diffraction with a diamond cell up to pressures of 70 GPa, Funamori et al. (1998) found that $MgAl_2O_4$ spinel transforms first to Al_2O_3 plus MgO, then to the $CaFe_2O_4$ phase, and finally to a new phase having the $CaTi_2O_4$ structure above ~40 GPa.

As an example of a spinel structure with tetrahedral bonds that are stronger than the octahedral ones, we can look at Ni_2SiO_4. The structure was determined by Finger et al. (1979) to 3.8 GPa. They found that the Si-O bond did not vary but the Ni-O bond decreased, as we would predict. Over a given volume decrease the distortion of the octahedral site is much less with this type of spinel than for one with weak tetrahedral bonds. Therefore this structure should be more stable than the one with weak tetrahedral bonds.

The most important spinels as far as the Earth's interior is concerned are the $(Mg,Fe)_2SiO_4$ spinels. The only known natural occurrence of a $(Mg,Fe)_2SiO_4$ spinel is in a meteorite (ringwoodite, Fig. 12), but it is thought to be a major phase in the transition zone. Hazen (1993) collected compressibility data for five different chemistries along the Fe-Mg join. He found that the Mg_2SiO_4 spinel had a $K_O = 184$ GPa, while Fe_2SiO_4 spinel had a $K_O = 207$ GPa. This was taken as a demonstration that although Fe and Mg appear to

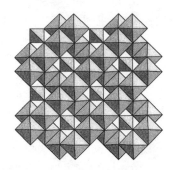

Figure 12. Structure of ringwoodite viewed down the c axis displaying a depth of 1/4 of the unit cell. Note the diagonal crisscrossing chains of MgO_6 octahedra linked together by isolated SiO_4 tetrahedra.

be similar at room conditions, they behave differently at higher pressures. Zerr et al. (1993) studied the compressibility of $(Mg_{0.6}Fe_{0.4})_2SiO_4$ spinel to 50 GPa and obtained $K_O = 183.0$ GPa, K' = 5.4. Irifune et al. (1998) determined the phase boundary for Mg_2SiO_4 spinel in the temperature range of 1400-1800°C and found a negative dT/dP with a transformation pressure of 21 GPa at 1600°C. This spinel transforms to MgO + $MgSiO_3$ perovskite.

Wadsleyite

The mineral known as wadsleyite was first synthesized by Ringwood and Major (1966) who were exploring the Mg_2SiO_4-Fe_2SiO_4 spinel solid solutions at pressures up to 17.5 GPa and temperatures of 1500°C. They were successful in making cubic spinel crystals in compositions containing about 20% Fe and higher, but the product for the Mg_2SiO_4 end member was birefringent and thus of a lower symmetry than cubic. Their conclusion was that this was a new phase and they thought it might be a quench product, i.e., it had the spinel structure under the synthesis conditions but distorted to a lower symmetry upon quenching. It turns out, of course, that they made a new structure that is intermediate in its pressure range between forsterite and ringwoodite. Until this phase was reported as occurring in the Peace River meteorite by Price et al. (1983), it was called by a variety of names including "modified spinel," "beta phase," "beta spinel," and, more rarely, "β-Mg_2SiO_4." All of these terms except the latter are examples of very poor usage because the phase is really different from spinel and is not just a simple distortion of the spinel structure. Price et al. (1983) followed the suggestion of Ringwood that it be named for David Wadsley, a well-known Australian mineralogist and crystallographer. The structure of orthorhombic wadsleyite is given in Figure 13. The structure in this orientation is made of two kinds of layers, one that just contains MgO_6 octahedra plus vacant sites and another that contains MgO_6 octahedra and SiO_4 tetrahedra plus vacant sites. In some respects it is like the hydrous magnesium silicates discussed in the next section and this may be related to the fact that wadsleyite synthesized in the presence of H_2O can

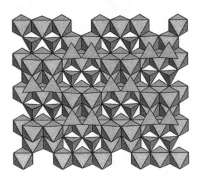

Figure 13. Structure of wadsleyite viewed down [201] showing 2 layers of polyhedra. The lower layer is composed only of octahedra and the upper layer is a mix of tetrahedra and octahedra.

accommodate up to about 3.3% H_2O (Smyth 1994). Smyth and others propose that wadsleyite is a candidate for being a major reservoir for H_2O in Earth's transition zone. In addition to the orthorhombic phase mentioned above, Smyth et al. (1997) described a related monoclinic structure that apparently results from a particular kind of cation ordering.

Dense hydrous magnesium silicates

Water (hydrogen) is stored in the upper mantle in hydrous silicate minerals such as amphibole, phlogopite, and serpentine. High-pressure and high-temperature experiments have shown that such minerals are stable to depths of ~200 km, but at greater depths they decompose and the water contained in them is released. Experimental studies of hydrous magnesium silicates over the past 25 years have demonstrated the existence of a number of hydrous phases with stabilities corresponding to depths much greater than 200 km. If present in the mantle, such materials would have a very significant effect on properties because virtually all mantle models are based on anhydrous minerals such as olivine, spinel, garnet, and silicate perovskite. There is, however, considerable confusion over the nomenclature and identification of these materials because many experimental runs result in multi-phase and/or very fine-grained products.

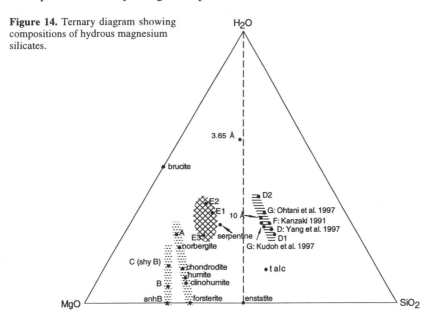

Figure 14. Ternary diagram showing compositions of hydrous magnesium silicates.

HYDROUS MAGNESIUM SILICATES

Figure 14 is a ternary diagram that shows the compositions of various hydrous magnesium silicates that either occur as minerals or have been synthesized in the laboratory. The ones containing octahedral Si, such as phases B, superhydrous B, and phase D are often called dense hydrous magnesium silicates (DHMS). The nomenclature of these high-pressure phases is based on either a description of a prominent line in the x-ray pattern or by an alphabetic naming scheme. The phases described thus far include the 10 Å phase (Sclar et al. 1965a,b), the 3.65 Å phase (Sclar et al. 1967), phases A, B, and C (Ringwood and Major 1967b), D (Yamamoto and Akimoto 1974), D' (Liu 1987), E (Kanzaki 1989 1991), F (Kanzaki 1991), anhydrous B (Herzberg and Gasparik 1989), and superhydrous B (Gasparik 1990). Crystal structure investigations of phases B and anhB

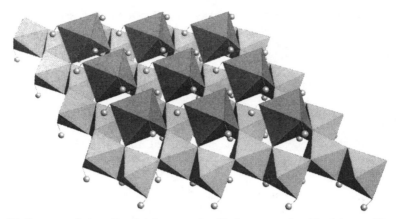

Figure 15. Structure of phase D. One layer contains Mg ions occupying 1/3 of the possible octahedral sites and the other layer contains Si ions occupying 2/3 of the possible octahedral sites.

(Finger et al. 1989 1991) illustrate how hydrogen is incorporated into phase B and how reliable crystal structure information is essential for interpretation of complex high-pressure phase chemistry. No crystal structures have been reported for the 10 Å phase or the 3.65 Å phase, and no authors other than Sclar et al. (1965a) have reported the existence of the latter. The 10 Å phase has been synthesized in polycrystalline form by several investigators and it is thought that its structure is similar to that of talc.

Phase C is probably identical to superhydrous B, an identity that was obscured for many years because the product used for the diffraction pattern of Ringwood and Major (1967b) was not a single phase and the pattern itself was not of high quality. Yamamoto and Akimoto's (1974) phase D turned out to be chondrodite, one of the humite series, but another synthetic product named phase D (Liu 1987) is potentially the most important of these materials because it has the highest pressure stability of any of the DHMS family. The crystal structure of phase D was described by Yang et al.(1997) and is shown in Figure 15. It has the same crystal structure as that of phase G reported by Kudoh et al. (1997) and its high pressure-temperature stability is discussed in several papers (Ohtani et al. 1997, Frost and Fei 1998, Shieh et al. 1998).

Kanzaki (1991) and Kudoh et al. (1995) described the synthesis and characterization of phase F, but it appears that the material reported by Kanzaki (1991) as phase F is actually phase D, and Kudoh et al.'s (1995) phase F is misindexed phase C, i.e. misindexed superhydrous phase B. Kudoh et al. (1997) recognized this error and withdrew their 1995 paper on phase F.

Phase E has a very unusual crystal structure that apparently is characterized by long-range disorder (Kudoh et al. 1989 1993), but is unusual in that its single-crystal diffraction pattern shows sharp spots with no evidence of disorder. Liu et al. (1997) characterize phase E as "the hydrous form of forsterite," but additional work is needed to determine whether this observation is valid or whether the quenched phase E has the same structure as under the original synthesis conditions.

Although several hydrous magnesium silicates such as anthophyllite, serpentine, talc, and the humites are well-known minerals, the synthetic phases are of considerable interest because at least some of them appear to be stable at substantially higher pressures and temperatures than the minerals. Whether or not any of these phases occur naturally is still an open question, but their study does provide a framework in which investigators can develop models that incorporate H_2O, $(OH)^+$, or H^+ as essential constituents of the phase chemistry. Furthermore, development of the crystal chemistry provides important information on the transformation of Si coordination from four to six with increasing pressure and on structural features that permit hydrogen to be retained under extreme conditions. Ahrens (1989) discusses how water was introduced into the Earth and how it might be stored in the mantle over geological time. He points out that it is unlikely that substantial amounts of water are re-introduced to the mantle through subduction processes, but current ideas suggest that a considerable amount of water is present as a result of the early stages of the Earth's accretion. This water has been released through time to form and replenish the oceans, lakes, rivers, and other repositories of water in the crust. An alternate view of how hydrogen might be incorporated into phases stable in the mantle is presented in papers, for example, by Rossman and Smyth (1990), Skogby et al. (1990), McMillan et al. (1991), and Smyth (1987). These authors have identified minute, but significant, amounts of structural hydrogen in nominally anhydrous minerals such as pyroxene and wadsleyite. Their idea is that if the amount of hydrogen found in these phases is representative of the mantle, then a volume of water equivalent to several times that of the present oceans could be generated and thus there is no reason to invoke the presence of hydrous magnesium silicates stable only at high pressures. Currently, we have no evidence to prove or disprove either model, but information being provided in both areas can be used to support further investigations.

Pyroxenes

Pyroxenes are significant phases in the upper mantle and occur in several space groups and compositional ranges. Pyroxene structures are comprised of slabs of octahedra extending along the c axis and connected top, bottom, left, and right to other octahedral slabs by single tetrahedral chains. Orthopyroxene, $(Mg,Fe)SiO_3$, shown in Figure 16 usually contains a small amount of Ca and crystallizes in space group *Pbca*. There are three clinopyroxene structures stable at relatively moderate pressures, clinopyroxene, $(Mg,Fe)SiO_3$, with space group $P2_1/c$ (Fig. 17), several different compositions with space group *C2/c*, diopside-hedenbergite $[Ca(Mg,Fe)Si_2O_6]$, augite $Ca(Mg,Fe,Al)Si_2O_6$, and jadeite ($NaAlSi_2O_6$—Fig. 18), and omphacite, whose ideal composition is 50% jadeite and 50% diopside, but with a different space group, *P2/n*. Recently, however, a high-pressure transition from $P2_1/c$ clinoenstatite to a *C2/c* clinoenstatite was predicted by Pacalo and Gasparik (1990) and confirmed by Angel et al. (1992). The implication of this is that a phase transition from low-P to high-P clinoenstatite could be responsible for the "X-discontinuity" observed in some seismic profiles at about 300 km depth (Woodland 1998).

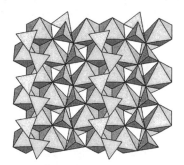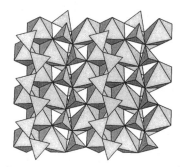

Figure 16 (left). Structure of orthoenstatite looking down the *a* axis at a depth of 0.0–0.3. The *c* axis parallels the direction of the chains. The *M2* site is located in the more distorted octahedra.

Figure 17 (right). Structure of clinoenstatite with the same orientation and depth as for orthoenstatite.

Figure 18. Structure of jadeite with the same orientation and depth as for orthoenstatite.

A similar result was obtained by Yang et al. (1998), who pressurized a single crystal of protopyroxene with composition $(Mg_{1.54}Li_{0.23}Sc_{0.23})Si_2O_6$ and space group *Pbcn* in a diamond cell and observed a room-temperature transition between 2.03 and 2.50 GPa to a structure with space group $P2_1cn$. Even though this is a reconstructive, first-order transition, it takes place at room temperature and is reversible with a ΔV for the transition of 2.6%. Interestingly, no indication of a reconstructive phase transition has been observed in orthopyroxene, although Hugh-Jones and Angel (1994) did note a change in slope of O3-O3-O3 angles at about 4 GPa, indicating that some structural adjustment was taking place as the pressure increased. As can be seen in the phase diagram of Figure 1, further increases in pressure result in $MgSiO_3$ pyroxenes transforming to garnet or ilmenite, depending on the temperature and pressure involved.

Olivine

Olivine is one of the most important minerals in the upper mantle, along with pyroxenes and garnets. The geologically relevant end-member phases are forsterite (Fo), Mg_2SiO_4, and fayalite (Fa), Fe_2SiO_4 with the natural composition being somewhere near $Fo_{90}Fa_{10}$. The transformation of olivine to wadsleyite is widely accepted as the source of the discontinuity that is observed in seismic velocities at 410-km depth (Bernal 1936).

The structure of forsterite (Fig. 19) is based on a distorted hexagonal closest-packed array of oxygen atoms with 1 tetrahedral site (T) and 2 non-equivalent octahedral sites (M_1, M_2). The M_1 site is more distorted than the M_2 site largely because the M_1 octahedron shares 6 of its edges with neighboring polyhedra (2 with M_1, 2 with M_2 and 2 with T) while the M_2 octahedron only shares 3 edges (2 with M_1, and 1 with T). According to Pauling's rules, shared edges are shorter than unshared edges in order to screen cation-cation repulsion. Much research has been directed to understanding olivine structural systematics and the reader is directed to "Orthosilicates," *Reviews in Mineralogy*, Volume 5 for an in-depth discussion.

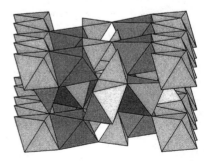

Figure 19. Structure of forsterite with the *b* axis horizontal and the *c* axis parallel to the chains of octahedra, coming out of the page. The illustration include the portion of the structure within the volume defined by $[(-0.2 - 1.2)x, (-0.2-1.2)y, (-0.5-1.5)z]$. The *M2* octahedra are the distorted ones.

Because of the importance of olivine at depth there have been quite a few studies of its compressibility. For example, Olinger, (1977), Kudoh and Takéuchi, (1985), Will et al. (1986), Andrault et al. (1995), Downs et al. (1996) all studied the compressibility of various olivines by determining the cell parameters as a function of pressure, obtaining a value for the bulk modulus around 120-130 GPa. All have found that the compression is anisotropic with the compressibilities of the cell edges ($\beta_a = (1/a)(da/dP)$, etc.) of a, b, and c to be 1.35, 2.70, and 2.10×10^{-3} GPa^{-1}, respectively (Downs et al. 1996). It follows that the relative axial ratios are 1:1.99:1.55 for $a{:}b{:}c$ Kinks in the compressibility curves of the cell edges were reported by Kudoh and Takéuchi (1985), and confirmed by Will et al., (1986) as well as in Raman and infrared spectroscopic studies (Chopelas 1990, Wang et al. 1993, Durben et al. 1993). These studies interpreted the change as an indication of a second-order phase transition or a change in the compression mechanism of forsterite at 10 GPa. However, Downs et al. (1996) suggested that errors may have been introduced in the data sets for these studies because they all used 4:1 methanol:ethanol as a pressure medium and it freezes at pressures near 10 GPa. No deviation from a smooth trend was observed in the Downs et al. (1996) study, which used He as a pressure medium. The smooth trends have also been observed in the state-of-the-art Brillouin experiments on both forsterite and San Carlos olivine to 16 and 32 GPa, respectively (Zha et al. 1996, 1998).

Crystal structures of olivines as a function of pressure have been determined for forsterite (Hazen 1976, Hazen and Finger 1979, Kudoh and Takéuchi 1985), fayalite (Hazen 1977, Kudoh and Takeda 1986), monticellite MgCaSiO$_4$ (Sharp et al. 1987), chrysoberyl Al$_2$BeO$_4$ (Hazen 1987) and LiScSiO$_4$ (Hazen et al. 1996). The anisotropy in the compression of olivine was shown to be chemistry dependent (Hazen et al. 1996) with the a-axis compression controlled by the stacking of M1 and T columns, and b and c controlled by the compressibility of the M2 octahedron. Therefore, a structure such as LiScSiO$_4$ ends up quite isotropic because soft Li are in M1, while stiff Sc are in the M2 site.

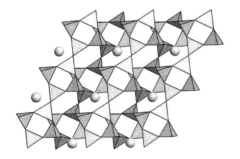

Figure 20. Structure of albite viewed down the b axis. The c axis parallels the chains of four-membered rings. The Na atoms are indicated by spheres. The image includes the volume bounded by $[(-1.0 - 2.0)x, (0.0 - 0.5)y, (-1.0 - 2.0)z]$.

Feldspars

The feldspar group of minerals is the most important of the crustal phases because it makes up about 50-60% of its total volume. The major phases are albite, NaAlSi$_3$O$_8$, K-spar (microcline, orthoclase, sanidine) KAlSi$_3$O$_8$, and anorthite CaAl$_2$Si$_2$O$_8$. The crystal structures of all the feldspars show similar topology, effectively shown in Figure 20 viewed along [010]. The structure can be considered as made of four-membered tetrahedral rings that are linked together to form chains that run parallel to the c-axis. These chains form layers that are separated by channels. The layers are stacked with chains over chains to form pairs, and each pair is shifted to be over the channels of the adjoining pairs. The Na, K, and Ca atoms occupy the channels situated between layers within the pairs.

Single crystal high-pressure structural studies have been completed for low albite (Downs et al. 1994), microcline (Allan and Angel 1997), anorthite (Angel 1988) and reedmergnerite, $NaBSi_3O_8$ (Downs et al. 1998) with bulk moduli of 55, 60, 80, and 69 GPa, respectively. All feldspars display anisotropic compression, with maximum compressibility coinciding with the narrowing of the channels. Each structure compresses by bending of the T-O-T angles, with some getting smaller and some getting larger, leaving the strongly bonded tetrahedra unchanged. On one hand, albite and reedmergnerite compress by bending the O_C0 angle which narrows the channel and causes the chains in different layers to slide over each other. On the other hand, microcline accomplishes the channel narrowing by shearing the four-membered rings. Downs et al. (1998) suggest that the typical framework T-O-T angle bending mechanism is being perturbed by the bonding of the Ca, K, and Na atoms. Consequently, even though each T-O-T angle in anorthite is Al-O-Si, which is softer than Si-O-Si, each of these bridging O atoms is also bonded to one or two Ca atoms. The Ca-O distance (2.48 Å) is shorter than R(K-O) (2.89 Å) and the same as R(Na-O) (2.49 Å), but every O atom is bonded to Ca, so it stiffens the Al-O-Si angles to the extent that anorthite is the least compressible of the feldspars. Albite, reedmergnerite and microcline have the same Al,Si distribution but electron density calculations show that the smaller Na atom is bonded to only 5 oxygens, while K is bonded to seven (Downs et al. 1996). The O_C0 atom in low-albite is not bonded to Na and so the Al-O_C0-Si angle is unconstrained and compresses with the little force. Therefore albite has the smallest bulk modulus. Reedmergnerite compresses similarly to albite because of similar Na-O bonding , but with a larger bulk modulus because B-O-Si angles are stiffer than Al-O-Si. The O_C0 atom in microcline is bonded to K, and this stiffens the structure. This explains why the bulk modulus of microcline is larger than that of albite, in spite of the fact that the unit cell of microcline is significantly larger than that of albite.

Hollandite

Hollandite is a structure-type named after the mineral $BaMn_8O_{16}$ whose structure is indicated in Figure 21. It is constructed of edge-sharing octahedra that form a four-sided eight-membered channel that is capable of containing a low-valence large-radius cation. Together the O atoms and the large cation form a hexagonal closest packed structure, a sort of analogue to the cubic closest packed perovskite. The similarity of the structure to stishovite should remind one of the similarities of coesite and the feldspars. The structure is of considerable interest to the nuclear waste and ion diffusion communities.

Ringwood et al. (1967) transformed sanidine at 900°C and 12 GPa into the hollandite structure with Al and Si randomly occupying the octahedral sites. This was the first oxide structure identified with both Al and Si displaying sixfold coordination and only the

Figure 21. Structure of hollandite viewed down the *c* axis, with ± 1/2 unit cell in the *a-b* plane.

second, after stishovite, with ^6Si. They postulated that because feldspars are the most abundant minerals in the Earth's crust, it is possible that the hollandite structure with ^6Al and ^6Si, may be a common phase within the transition zone of the mantle. Subsequently, Reid and Ringwood (1969) systematically examined the feldspar structure type materials to identify others that would transform to the hollandite structure at high pressures. They were able to synthesize $Sr_xAl_{2x}Si_{4-2x}O_8$ and $Ba_xAl_{2x}Si_{4-2x}O_8$, $x \sim 0.75$ with the hollandite structure, but failed with the Na, Ca, and Rb rich feldspars. Yamada et al. (1984) determined the structure of the $KAlSi_3O_8$ phase by powder diffraction methods and Zhang et al. (1993) reported the structure as a function of pressure.

Madon et al. (1989) reported that their synthesis of $(Ca_{0.5}Mg_{0.5})Al_2Si_2O_8$ at 50 GPa and laser-induced high temperatures exhibited an x-ray powder diffraction pattern consistent with hollandite. Liu (1978) reported the synthesis of $NaAlSi_3O_8$ hollandite at 21-24 GPa from jadeite plus stishovite. Above 24 GPa the $NaAlSi_3O_8$ hollandite transforms to the calcium ferrite structure. Gasparik (1989) synthesized an unknown purple phase in an anvil press from a starting mixture of jadeite, orthoenstatite, and orthopyroxene $En_{44}Py_{56}$ at 1450°C and 16.5 GPa with a Pb flux. Downs et al. (1995) determined this to be a hollandite structure of composition $Pb_{0.8}Al_{1.6}Si_{2.4}O_8$. It thus seems that alumino-silicate hollandite is capable of containing many of the large low valence cations at elevated temperatures and pressures.

REFERENCES

Ahrens TJ (1989) Water storage in the mantle. Nature 342:122-123

Allan DR, Angel RJ (1997) A high-pressure structural study of microcline ($KAlSi_3O_8$) to 7 GPa. Eur J Mineral 9:263-275

Anderson OL (1997) Iron: beta phase frays. Science 278:821-822

Andrault D, Bouhifd MA, Itié JP, Richet P (1995) Compression and amorphization of $(Mg,Fe)_2SiO_4$ olivines: an x-ray diffraction study up to 70 GPa. Phys Chem Minerals 22:99-107

Angel RJ (1988) High-pressure structure of anorthite. Am Min 73:1114-1119

Angel RJ, Allan DR, Miletich R, Finger LW (1997) The use of quartz as an internal pressure standard in high-pressure crystallography. J Appl Cryst 30:461-466

Angel RJ, Chopelas A, Ross NL (1992) Stability of high-density clinoenstatite at upper-mantle pressures. Nature 358:322-324

Angel RJ, Ross NL, Seifert F, Fliervoet TF (1996) Structural characterization of pentacoordinate silicon in a calcium silicate. Nature 384:441-444

Askarpour V, Manghnani MH, Fassbender S, Yoneda A (1993) Elasticity of single-crystal $MgAl_2O_4$ spinel up to 1273 K by Brillouin spectroscopy. Phys Chem Minerals 19:511-519

Bernal JD (1936) The Observatory 59:268

Besson JM, Nelmes RJ, Hamel G, Loveday JS, Weill G, Hull S (1992) Neutron powder diffraction above 10-GPa. Physica B 180:907-910

Bloss FD (1994) Crystallography and Crystal Chemistry. Mineralogical Society of America, Washington, DC, 545 p

Boehler R, von BN, Chopelas A (1990) Melting, thermal expansion, and phase transitions of iron at high pressures. J Geophys Res 95:21731-21736

Boisen MB Jr, Gibbs GV (1993) A modeling of the structure and compressibility of quartz with a molecular potential and its transferability to cristobalite and coesite. Phys Chem Minerals 20:123-135

Buerger MJ (1961) Polymorphism and phase transformations. Fortschr Mineral 39:9-24

Chopelas A (1990) Thermal properties of forsterite at mantle pressures derived from vibrational spectroscopy. Phys Chem Minerals 17:149-156

Downs RT, Andalman A, Hudacsko M (1996) The coordination numbers of Na and K atoms in low albite and microcline as determined from a procrystal electron-density distribution. Am Mineral 81:1344-1349

Downs RT, Hazen RM, Finger LW (1994) The high-pressure crystal chemistry of low albite and the origin of the pressure dependency of Al-Si ordering. Am Mineral 79:1042-1052

Downs RT, Hazen RM, Finger LW, Gasparik T (1995) Crystal chemistry of lead aluminosilicate hollandite: a new high-pressure synthetic phase with octahedral Si. Am Mineral 80:937-940

Downs RT, Palmer DC (1994) The pressure behavior of α-cristobalite. Am Mineral 79:9-14

Downs RT, Yang H, Hazen RM, Finger LW, Prewitt CT (1998) Compressibility mechanisms of alkali feldspars: new data from reedmergnerite. Am Mineral 83 (in press)

Downs RT, Zha C-S, Duffy TS, Finger LW (1996) The equation of state of forsterite to 17.2 GPa and effects of pressure media. Am Mineral 81:51-55

Dubrovinsky LS, Saxena SK, Lazor P (1997) X-ray study of iron with in-situ heating at ultra high pressures. Geophys Res Lett 24:1835-1838

Dubrovinsky LS, Saxena SK, Lazor P, Ahuja R, Eriksson O, Wills JM, Johansson B (1997) Experimental and theoretical identification of a new high-pressure phase of silica. Nature 388:362-365

Durben DJ, McMillan PF, Wolf GH (1993) Raman study of the high-pressure behavior of forsterite (Mg_2SiO_4) crystal and glass. Am Mineral 78:1143-1148

ElGoresy A, Dubrovinsky L, Saxena S, Sharp TG (1998) A new post-stishovite silicon dioxide polymorph with the baddelyite structure (zirconium oxide) in the snc meteorite Shergotty: evidence for extreme shock pressure. Meteoritics Planet Sci 33:A45

Fei Y, Frost DJ, Mao HK, Prewitt CT, Häusermann D (1998) In situ determination of the high-pressure phase of Fe_3O_4 at high pressure and temperature. Am Mineral (in press)

Fei Y, Mao HK (1994) In situ determination of the NiAs phase of FeO at high pressure and temperature. Science 266:1668-1680

Fei Y, Prewitt CT, Frost DJ, Parise JB, Brister K (1998) Structures of FeS polymorphs at high pressure and temperature. In: M Nakahara (ed) Review of High Pressure Science and Technology 7:55-58 Japan Society of High-Pressure Science and Technology, Kyoto

Fei Y, Prewitt CT, Mao HK, Bertka CM (1995) Structure and density of FeS at high pressure and high temperature and the internal structure of Mars. Science 268:1892-1894

Fei Y, Wang Y, Finger LW (1996) Maximum solubility of FeO in $(Mg,Fe)SiO_3$-perovskite as a function of temperature at 26 GPa: implication for FeO content in the lower mantle. J Geophys Res 101:11525-11530

Finger LW, Hazen RM, Hofmeister AM (1986) High-pressure crystal chemistry of spinel ($MgAl_2O_4$) and magnetite (Fe_3O_4): comparison with silicate spinels. Phys Chem Minerals 13:215-220

Finger LW, Hazen RM, Prewitt CT (1991) Crystal structures of $Mg_{12}Si_4O_{19}(OH)_2$(phase B) and $Mg_{14}Si_5O_{24}$ (phase AnhB). Am Mineral 76:1-7

Finger LW, Hazen RM, Yagi T (1986) Crystal structures and electron densities of nickel and iron silicate spinels at elevated temperature or pressure. Am Mineral 64:1002-1009

Finger LW, Ko J, Hazen RM, Gasparik T, Hemley RJ, Prewitt CT, Weidner DJ (1989) Crystal chemistry of phase B and an anhydrous analogue: implications for water storage in the upper mantle. Nature 341:140-142

Frost DJ, Fei Y (1998) Stability of phase D at high pressure and high temperature. J Geophys Res 103:7463-7474

Funamori N, Jeanloz R, Nguyen JH, Kavner A, Caldwell WA, Fujino K, Miyajima N, Shinmei T, Tomioka N (1998) High-pressure transformations in $MgAl_2O_4$. J Geophys Res 103:20813-20818

Gasparik T (1989) Transformation of enstatite-diopside-jadeite pyroxenes to garnet. Contrib Mineral Petrol 102:389-405

Gasparik T (1990) Phase relations in the transition zone. J Geophys Res 95:15751-15769

Gasparik T (1993) The role of volatiles in the transition zone. J Geophys Res 98:4287-4299

Geisinger KL, Gibbs GV, Navrotsky A (1985) A molecular orbital study of bond length and angle variations in framework structures. Phys Chem Minerals 11:266-283

Gibbs GV, Downs JW, Boisen MB Jr (1994) The elusive SiO bond. In: Silica: Physical Behavior, Geochemistry, and Materials Applications. Heaney PJ, Prewiit CT, Gibbs GV (eds) Rev Mineral 29:331-368

Gibbs GV, Rosso KM, Teter DM, Boisen MB, Jr., Bukowinski M (1998) A modeling of the structure and the topology of the electron density distribution of low quartz at pressure: a study of the SiO bond. (submitted to the Larry Bartell issue of J Molecular Struc)

Glinnemann J, King HE Jr, Schulz H, Hahn Th, La Placa SJ, Dacol F (1992) Crystal structures of the low-temperature quartz-type phases of SiO_2 and GeO_2 at elevated pressure. Zeit Krist 198:177-212

Haavik C, Stølen S, Fjellvåg H, Hanfland M, Häusermann D (1998) Equation of state of magnetite and its high-pressure modification. Thermodynamics of the Fe-O system at high pressure. Am Mineral (in press)

Hasegawa M, Badding JV (1997) Rietveld analysis using a laboratory-based high pressure x-ray diffraction system and film-based detection. Rev Sci Instrum 68:2298-2300

Hazen RM (1976) Effects of temperature and pressure on the crystal structure of forsterite. Am Mineral 61:1280-1293

Hazen RM (1977) Effects of temperature and pressure on the crystal structure of ferromagnesian olivine. Am Mineral 62:286-295

Hazen RM (1987) High-pressure crystal chemistry of chrysoberyl, Al_2BeO_4: insights on the origin of olivine elastic anisotropy. Phys Chem Minerals 14:13-20

Hazen RM (1993) Comparative compressibilities of silicate spinels: anomalous behavior of $(Mg,Fe)_2SiO_4$. Science 259:149-280

Hazen RM, Downs RT, Conrad PG, Finger LW and Gasparik T (1994) Comparative compressibilities of majorite-type garnets. Phys Chem Minerals 21:344-349

Hazen RM, Downs RT, Finger LW (1996) High-pressure crystal chemistry of $LiScSiO_4$: an olivine with nearly isotropic compression. Am Mineral 81:327-334

Hazen RM, Finger LW (1978) Crystal structures and compressibilities of pyrope and grossular to 60 kbar. Am Mineral 63:297-303

Hazen RM, Finger LW (1979) Bulk modulus-volume relationship for cation-anion polyhedra. J Geophys Res 84:6723-6728

Hazen RM, Finger LW (1979) Crystal structure of forsterite at 40 kbar. Carnegie Inst Wash Yearb 79:364-367

Hazen RM, Finger LW (1982) Comparative Crystal Chemistry. John Wiley & Sons, New York

Hazen RM, Finger LW (1989) High-pressure crystal chemistry of andradite and pyrope: revised procedures for high-pressure diffraction experiments. Am Mineral 74:352-359

Hazen RM, Finger LW, Hemley RJ, Mao HK (1989) High-pressure crystal chemistry and amorphization of α-quartz. Solid State Comm 72:507-511

Hemley RJ, Prewitt CT, Kingma KJ (1994) High-pressure behavior of silica. *In:* Silica: Physical Behavior, Geochemistry, and Materials Applications. Heaney PJ, Prewiit CT, Gibbs GV (eds) Rev Mineral 29:41-81

Herzberg CT, Gasparik T (1989) Melting experiments on chondrite at high pressures: stability of anhydrous phase B. Eos Trans Am Geophys Union 70:484

Hill FC, Gibbs GV, Boisen, MB Jr (1994) Bond stretching force constants and compressibilities of nitride, oxide, and sulfide coordination polyhedra in molecules and crystals. Struct Chem 6:349-355

Huang E, Bassett WA (1986) Rapid determination of Fe_3O_4 phase diagram by synchrotron radiation. J Geophys Res 91:4697-4703

Hugh-Jones DA, Angel RJ (1994) A compressional study of $MgSiO_3$ orthoenstatite up to 8.5 GPa. Am Mineral 79:405-410

Irifune T, Fujino K, Ohtani E (1991) A new high-pressure form of $MgAl_2O_4$. Nature 349:409-411

Irifune T, Nishiyama N, Kuroda K, Inoue T. Isshiki M, Utsumi W, Funakoshi K, Urakawa S, Uchida T, Katsura T, Ohtaka O (1998) The postspinel phase boundary in $MgSiO_4$ determined by in situ x-ray diffraction. Science 279:1698-1700

Isaak DG, Cohen RE, Mehl MJ, Singh DJ (1993) Phase stability of wüstite at high pressure from first-principles LAPW calculations. Phys Rev B 47:7720-7720

Ita J, Stixrude L (1992) Petrology, elasticity, and composition of the mantle transition zone. J Geophys Res 97:6849-6866

Jeanloz R, Ahrens TJ (1980) Equations of state of FeO and CaO. Geophys.J.R.Astron.Soc., 62:505-528

Kanzaki M (1989) High pressure phase relations in the system $MgO-SiO_2-H_2O$. Eos Trans AGU 508:70

Kanzaki M (1991) Stability of hydrous magnesium silicates in the mantle transition zone. Phys Earth Planet Inter 66:307-312

Kingma KJ, Cohen RE, Hemley RJ, Mao HK (1995) Transformation of stishovite to a denser phase at lower-mantle pressures. Nature 374:243-245

Kingma KJ, Mao HK, Hemley RJ (1996) Synchrotron x-ray diffraction of SiO_2 to multimegabar pressures. High Pressure Res 14:363-374

Knittle E, Jeanloz R (1986) High-pressure metallization of FeO and implications for the Earth's core. Geophys Res Lett 13:1541-1544

Kudoh Y, Finger LW, Hazen RM, Prewitt CT, Kanzaki M, Veblen DR (1993) Phase E: a high-pressure hydrous silicate with unique crystal chemistry. Phys Chem Minerals 19:357-360

Kudoh Y, Kanzaki M (1998) Crystal chemical characteristics of α-$CaSi_2O_5$, a new high pressure calcium silicate with five-coordinated silicon synthesized at 1500°C and 10 GPa. Phys Chem Minerals 25:429-433

Kudoh Y, Nagase T, Mizohata H, Ohtani E, Sasaki S, Tanaka M (1997) Structure and crystal chemistry of phase G, a new hydrous magnesium silicate synthesized at 22 GPa and 1050°C. Geophys Res Lett 24:1051-1054

Kudoh Y, Nagase T, Sasaki S, Tanaka M, Kanzaki M (1995) Phase F, a new hydrous magnesium silicate synthesized at 1000°C and 17 GPa: crystal structure and estimated bulk modulus. Phys Chem Minerals 22:295-299

Kudoh Y, Nagase T, Sasaki S, Tanaka M, Kanzaki M (1997) Phase F, a new hydrous magnesium silicate synthesized at 1000°C and 17 GPa: crystal structure and estimated bulk modulus. Phys Chem Minerals 24:601

Kudoh Y, Takeda H (1986) Single crystal x-ray diffraction study on the bond compressibility of fayalite, Fe_2SiO_4 and rutile, TiO_2 under high pressure. Physica 139&140B:333-336

Kudoh Y, Takéuchi Y (1985) The crystal structure of forsterite Mg_2SiO_4 under high pressure up to 149 kb. Zeit Krist 171:291-302

Lager GA, Von Dreele RB (1996) Neutron powder diffraction study of hydrogarnet to 9.0 GPa. Am Mineral 81:1097-1104

Lasaga AC, Gibbs GV (1987) Applications of quantum mechanical potential surfaces to mineral physics calculations. Phys Chem Minerals 14:107-117

Levien L, Prewitt CT (1981) High-pressure crystal structure and compressibility of coesite. Am Mineral 66:324-333.

Levien L, Prewitt CT, Weidner DJ (1979) Compression of pyrope. Am Mineral 64:805-808

Levien L, Prewitt CT, Weidner DJ (1980) Structure and elastic properties of quartz at pressure. Am Mineral 65:920-930

Liu LG (1974) Silicate perovskite from phase transformations of pyrope-garnet at high pressure and temperature. Geophys Res Lett 1:277-280

Liu LG (1975) Disproportionation of $MgAl_2SiO_4$ spinel at high pressure and temperatures. Geophys Res Lett 2:9-11

Liu LG (1978) High-pressure phase transformations of albite, jadeite and nepheline. Earth Planet Sci Lett 37:438-444

Liu LG (1980)The equilibrium boundary of spinel – corundum + periclase: a calibration curve for pressures above 100 kbar. High-Temp-High Press 12:217-220

Liu LG (1987) Effects of H_2O on the phase behavior of the forsterite-enstatite system at high pressures and temperatures and implications for the Earth. Phys Earth Planet Inter 49:142-167

Loubeyre P, LeToullec R, Hausermann D, Hanfland M, Hemley RJ, Mao HK, Finger LW (1996) X-ray diffraction and equation of state of hydrogen at megabar pressures. Nature 383:702-704

Madon M, Castex J, Peyronneau J (1989) A new aluminocalcic high-pressure phase as a possible host of calcium and aluminum in the lower mantle. Nature 342:422-425

Mao HK, Hemley RJ (1996) Experimental studies of the Earth's deep interior: accuracy and versatility of diamond cells. Phil Trans R Soc London A 354:1315-1333

Mao HK, Shen G, Hemley RJ (1997) Multivariant dependence of Fe-Mg partitioning in the lower mantle. Science 278:2098-2100

Mao HK, Takahashi T, Bassett WA, Kinsland GL, Merrill L (1974) Isothermal compression of magnetite to 320 kbar and pressure-induced phase transformation. J Geophys Res 79:1165-1170

Mazin I, Fei Y, Downs RT, Cohen R (1998) Possible polytypism in FeO at high pressure. Am Mineral 83:451-457.

Mazin II, Anisimov VI (1997) Insulating gap in FeO: correlations and covalency. Phys Rev B 55:12822-12825

McCammon CA (1998) The crystal chemistry of ferric iron in $Fe_{0.05}Mg_{0.95}SiO_3$ perovskite as determined by Mössbauer spectroscopy in the temperature range 80-293 K. Phys Chem Minerals 25:292-300

McMillan PF, Akaogi M, Sato RK, Poe B, Foley J (1991) Hydroxyl groups in β-Mg_2SiO_4. Am Mineral 76:354-360

Megaw HD (1973) Crystal Structures: A Working Approach. W.B. Saunders, Philadelphia

Merrill L, Bassett WA (1974) Miniature diamond anvil pressure cell for single crystal x-ray diffraction studies. Rev Sci Instrum 45:290-294

Nelmes RJ, McMahon MI, Belmonte SA, Allan DR, Gibbs MR, Parise JB (1998) High pressure structures of iron sulphide. In: M Nakahara (ed) Review High-Pressure Science Technology 7:202-204. Japan Society of High Pressure Science and Technology, Kyoto

Nicholas JB, Winans RE, Harrison RJ, Iton LE, Curtiss LA, Hopfinger AJ (1992) *Ab initio* molecular orbital study of the effects of basis set size on the calculated structure and acidity of hydroxyl groups in framework molecular sieves. J Phys Chem 96:10247-10257

Nicoll JS, Gibbs GV, Boisen MB Jr, Downs RT, Bartelmehs, KL (1994) Bond length and radii variations in fluoride and oxide molecules and crystals. Phys Chem Minerals 20:617-624

Ohtani E, Mizobata H, Kudoh Y, Nagase T, Arashi H, Yurimoto H, Miyagi I (1997) A new hydrous silicate, a water reservoir, in the upper part of the lower mantle. Geophys Res Lett 24:1047-1050

Olinger B (1977) Compression studies of forsterite (Mg_2SiO_4) and enstatite ($MgSiO_3$). In: MH Manghnani S Akimoto (eds) High-Pressure Research: Applications in Geophysics. p 325-334 Academic Press, New York

Pacalo REG, Gasparik T (1990) Reversals of the orthoenstatite-clinoenstatite transition at high pressures and high temperatures. J Geophys Res 95:15853-15858

Palmer DC, Hemley RJ, Prewitt CT (1994) Raman spectroscopic study of high-pressure phase transitions in cristobalite. Phys Chem Minerals 21:481-488

Parise JB, Leinenweber K, Weidner DJ, Tan K, Von Dreele RB (1994) Pressure-induced H bonding: neutron diffraction study of brucite, $Mg(OD)_2$, to 9.3 GPa. Am Mineral 79:193-196

Pavese A, Artioli G, Hull S (1998) Cation partitioning versus pressure in $Mg_{0.94}Al_{2.04}O_4$ synthetic spinel, by in situ powder neutron diffraction. Am Mineral (in press)

Poirier JP (1991) Introduction to the Physics of the Earth's Interior. Cambridge University Press, Cambridge, UK

Price GD, Putnis A, Agrell SO (1983) Wadsleyite, natural β-$(Mg,Fe)_2SiO_4$ from the Peace River meteorite. Can Mineral 21:29-35

Reid AF, Ringwood AE (1969) Six-coordinate silicon: high pressure strontium and barium aluminosilicates with the hollandite structure. J Solid State Chem 1:6-9

Ringwood AE (1962) Mineralogical constitution of the deep mantle. J Geophys Res 67:4005-4010

Ringwood AE, Major A (1966) Synthesis of Mg_2SiO_4-Fe_2SiO_4 spinel solid solution. Earth Planet Sci Lett 1:241-245

Ringwood AE, Major A (1967a) Some high pressure transformation of geophysical interest. Earth Planet Sci Lett 2:106-110

Ringwood AE, Major A (1967b) High-pressure reconnaissance investigations in the system Mg_2SiO_4-MgO-H_2O. Earth Planet Sci Lett 2:130-133

Ringwood AE, Reid AF, Wadsley AD (1967) High-pressure $KAlSi_3O_8$, an aluminosilicate with sixfold coordination. Acta Cryst 23:1093-1095

Ross NL, Shu JF, Hazen RM (1990) High-pressure crystal chemistry of stishovite. Am Mineral 75:739-747

Rossman GR, Smyth JR (1990) Hydroxyl contents of accessory minerals in mantle eclogites and related rocks. Am Mineral 75:775-780

Saxena SK, Dubrovinsky LS, Häggkvist P, Cerenius Y, Shen G, Mao HK (1995) Synchrotron x-ray study of iron at high pressure and temperature. Science 269:1703-1704

Saxena SK, Dubrovinsky LS, Lazor P, Cerenius Y, Haggkvist P, Hanfland M, Hu J (1996) Stability of perovskite ($MgSiO_3$) in the Earth's mantle. Science 274:1357-1359

Saxena SK, Shen G, Lazor P (1993) Experimental evidence for a new iron phase and implications for Earth's core. Science 260:1312-1314

Sclar CB, Carrison LC, Schwartz CM (1965a) High pressure synthesis and stability of a new hydronium bearing layer silicate in the system MgO-SiO_2-H_2O. Eos Trans Am Geophys Union 46:184

Sclar CB, Carrison LC, Schwartz CM (1965b) The system MgO-SiO_2-H_2O at high pressures, 25-130 Kb. and 375-1000°C (extended abstr). Basic Sci Div Am Ceram Soc Fall Mtg: 2-b-65F

Sclar CB, Carrison LC, Stewart OM (1967) High pressure synthesis of a new hydroxylated pyroxene in the system MgO-SiO_2-H_2O. Eos Trans Am Geophys Union 48:226

Serghiou G, Zerr A, Boehler R (1998) $(Mg,Fe)SiO_3$-perovskite stability under lower mantle conditions. Science 280:2093-2095

Shannon RD, Prewitt CT (1968) Effective ionic radii in oxides and fluorides. Acta Crystallogr B25:925-945

Sharp TG (1997) Natural occurrence of $MgSiO_3$-ilmenite and evidence for $MgSiO_3$-perovskite in a shocked L chondrite. Science 277:352-355

Sharp ZD, Hazen RM, Finger LW (1987) High-pressure crystal chemistry of monticellite. Am Mineral 72:748-755

Shen GY, Mao HK, Hemley RJ, Duffy TS, Rivers ML (1998) Melting and crystal structure of iron at high pressures and temperatures. Geophys Res Lett 25:373-376

Shieh SR, Mao HK, Hemley RJ, Ming LC (1998) Decomposition of phase D in the lower mantle and the fate of dense hydrous silicates in subducting slabs. Earth Planet Sci Lett 159:13-23

Shu J et al (1998) High-pressure phase transition in magnesiowüstite $(Fe_{1-x})O$. Eos Trans Am Geophys Union 79:S203

Skogby H, Bell DR, Rossman GR (1990) Hydroxide in pyroxene: variations in the natural environment. Am Mineral 75:764-774

Smyth JR (1987) β-Mg_2SiO_4: a potential host for water in the mantle? Am Mineral 72:1051-1055

Smyth JR (1994) A crystallographic model for hydrous wadsleyite (β-Mg_2SiO_4): an ocean in the Earth's interior? Am Mineral 79:1021-1024

Smyth JR, Kawamoto T, Jacobsen SD, Swope RJ, Hervig RL, Holloway JR (1997) Crystal structure of monoclinic hydrous wadsleyite [β-$(Mg,Fe)_2SiO_4$]. Am Mineral 82:270-275

Sugiyama M, Endo S, Koto K (1987) The crystal structure of stishovite under pressure up to 6 GPa. Mineral J (Japan) 13:455-466

Teter DM, Hemley RJ, Kresse G, Hafner J (1998) High pressure polymorphism in silica. Phys Rev Lett 80: 2145-2148

Tomioka N, Fujino K (1997) Natural (Mg,Fe)SiO$_3$-ilmenite and -perovskite in the Tenham meteorite. Science 277:1084-1086

Tsuchida Y, Yagi T (1989) A new, post-stishovite high-pressure polymorph of silica. Nature 340:217-220

Tsuchida Y, Yagi T (1990) New pressure-induced transformations of silica at room temperature. Nature 347:267-269

Wang SY, Sharma SK, Cooney TF (1993) Micro-raman and infrared spectral study of forsterite under high pressure. Am Mineral 78:469-476

Will G, Hoffbauer W, Hinze E, Lauterjung J (1986) The compression of forsterite up to 300 kbar measured with synchrotron radiation. Physica 139&140B:193-197

Woodland AB (1998) The orthorhombic to high-P monoclinic phase transition in Mg-Fe pyroxenes: can it produce a seismic discontinuity? Geophys Res Lett 25:1241-1244

Yagi T, Suzuki K, Akimoto S (1985) Static compression of wüstite (Fe$_{0.98}$O) to 120 GPa. J Geophys Res 90:8784-8788

Yamada H, Matsui Y, Ito E (1984) Crystal-chemical characterization of KAlSi$_3$O$_8$ with the hollandite structure. Mineral J (Japan) 12:29-34

Yamamoto K, Akimoto S-i (1974) High pressure and high temperature investigations in the system MgO-SiO$_2$-H$_2$O. J Solid State Chem 9:187-195

Yang H, Finger LW, Conrad PG, Prewitt CT, Hazen RM (1999) A new pyroxene structure type observed at high pressure: single-crystal x-ray diffraction and Raman spectroscopic studies of the Pbcn-P2$_1$cn phase transition in protopyroxene. Am Mineral (in press)

Yang H, Prewitt CT, Frost DJ (1997) Crystal structure of the dense hydrous magnesium silicate, phase D. Am Mineral 82:651-654

Zerr A, Reichmann H, Euler H, Boehler R (1993) Hydrostatic compression of γ-(Mg$_{0.6}$Fe$_{0.4}$)$_2$SiO$_4$ to 50.0 GPa. Phys Chem Minerals 19:507-509

Zha CS, Duffy TS, Downs RT, Mao HK, Hemley RJ (1996) Sound velocity and elasticity of single-crystal forsterite to 16 GPa. J Geophys Res 101:17535-17545

Zha CS, Duffy TS, Downs RT, Mao HK, Hemley RJ (1998) Brillouin scattering and x-ray diffraction of San Carlos olivine: direct pressure determination to 32 GPa. Earth Planet Sci Lett 159:25-33

Zhang J, Ko J, Hazen RM, Prewitt CT (1993) High-pressure crystal chemistry of KAlSi$_3$O$_8$ hollandite. Am Mineral 78:493-499

Zhang L, Ahsbahs H, Kutoglu A (1998) Hydrostatic compression and crystal structure of pyrope to 33 GPa. Phys Chem Minerals 25:301-307

Zou G, Mao HK, Bell PM, Virgo D (1980) High-pressure experiments on the iron oxide wüstite (Fe$_{1-x}$O). Carnegie Inst Washington Yearb 79:374-376

APPENDIX

Structural parameters and references for phases discussed in this chapter.

Magnesiowüstite, (Mg,Fe)O, Hazen (1976)

a, b, c (Å) α, β, γ (°) *Space group:*	4.2110	4.2110	4.2110	90	90	90	*Fm3m*
	x	*y*	*z*				
Mg,Fe	0	0	0				
O	1/2	1/2	1/2				

Perovskite, MgSiO₃, Horiuchi et al. (1987)

a, b, c (Å) α, β, γ (°) *Space group:*	4.7754	4.9292	6.8969	90	90	90	*Pbnm*
	x	*y*	*z*				
Mg	.5141	.5560	1/4				
Si	1/2	0	1/2				
O1	.1028	.4660	1/4				
O2	.1961	.2014	.5531				

Ilmenite, MgSiO₃, Horiuchi et al. (1982)

a, b, c (Å) α, β, γ (°) *Space group:*	4.7284	4.7284	13.5591	90	90	120	$R\bar{3}$
	x	*y*	*z*				
Si	0	0	.15768				
Mg	0	0	.35970				
O	.3214	.0361	.24077				

Stishovite, SiO₂, Ross et al. (1983)

a, b, c (Å) α, β, γ (°) *Space group:*	4.1801	4.1801	2.6678	90	90	90	*P4₂/mnm*
	x	*y*	*z*				
Si	0	0	0				
O	.3067	.3067	0				

Hypothetical post-stishovite SiO₂ (CaCl₂ struc.) at 50 GPa, Teter et al. (1998)

a, b, c (Å) α, β, γ (°) *Space group:*	3.9141	4.0218	2.5809	90	90	90	*Pnnm*
	x	*y*	*z*				
Si	0	0	0				
O	.2869	.3170	0				

Coesite, SiO₂, Levien and Prewitt (1981)

a, b, c (Å) α, β, γ (°) *Space group:*	7.1356	12.3692	7.1736	90	120.34	90	*C2/c*
	x	*y*	*z*				
Si1	.14035	.10833	.07230				
Si2	.50674	.15796	.54072				
O1	0	0	0				
O2	1/2	.11634	3/4				
O3	.2661	.12324	.9401				
O4	.3114	.10378	.3280				
O5	.0175	.21177	.4784				

Troilite, FeS I, Evans (1970)

a, b, c (Å) α, β, γ (°) *Space group:*	5.962	5.962	11.750	90	90	120	$P\bar{6}2c$
	x	*y*	*z*				
Fe	.3797	.0551	.1230				
S1	0	0	0				
S2	1/3	2/3	.0199				
S3	.6652	-.0031	1/4				

FeS III, Nelmes et al. (1998)

a, b, c (Å) α, β, γ (°) Space group:	8.1103	5.6666	6.4832	90	93.050	90	$P2_1/a$
	x	y	z				
Fe1	.0655	.7988	.0829				
Fe2	.2120	.2139	.1803				
Fe3	.4220	.7607	.4489				
S1	.1845	.0779	.8652				
S2	.5182	.9110	.7547				
S3	.3321	.4163	.5922				

Pyrite, FeS$_2$, Fujii et al. (1986)

a, b, c (Å) α, β, γ (°) Space group:	5.417	5.417	5.417	90	90	90	Pa3
	x	y	z				
Fe	0	0	0				
S	.3848	.3848	.3848				

Majorite, MgSiO$_3$, Angel et al. (1989)

a, b, c (Å) α, β, γ (°) Space group:	11.501	11.501	11.480	90	90	90	$I4_1/a$
	x	y	z				
Si1	0	1/4	3/8				
Si2	0	1/4	7/8				
Si3	.1249	.0065	.7544				
Mg1	.1253	.0112	.2587				
Mg2	0	1/4	.6258				
Mg3	0	0	1/2				
Si4	0	0	0				
O1	.0282	.0550	.6633				
O2	.0380	-.0471	.8562				
O3	.2195	.1023	.8021				
O4	.2150	-.0894	.7000				
O5	-.0588	.1617	.4680				
O6	-.1040	.2080	.7851				

Pyrope, Mg$_3$Al$_2$(SiO$_4$)$_3$, Pavese et al. (1995)

a, b, c (Å) α, β, γ (°) Space group:	11.4544	11.4544	11.4544	90	90	90	Ia3d
	x	y	z				
Si	3/8	0	1/4				
Al	0	0	0				
Mg	0	1/4	1/8				
O	.03295	.05032	.65334				

Ringwoodite, Mg$_2$SiO$_4$, Hazen et al. (1993)

a, b, c (Å) α, β, γ (°) Space group:	8.0709	8.0709	8.0709	90	90	90	Fd3m
	x	y	z				
Si	1/8	1/8	1/8				
Mg	1/2	1/2	1/2				
O	,2441065	,2441065	,2441065				

Wadsleyite, Mg_2SiO_4, Finger et al. (1993)

a, b, c (Å) α, β, γ (°) Space group:	5.6921	11.460	8.253	90	90	90	Imma
	x	y	z				
Mg1	0	0	0				
Mg2	0	1/4	.97013				
Mg3	1/4	.12636	1/4				
Si	0	.12008	.61659				
O1	0	1/4	.2187				
O2	0	1/4	.7168				
O3	0	.98946	.25562				
O4	.2608	.12286	.99291				

Phase D, $MgSi_2H_2O_6$

a, b, c (Å) α, β, γ (°) Space group:	4.7453	4.7453	4.3450	90	90	120	$P\bar{3}1m$
	x	y	z				
Mg	0	0	0				
Si	2/3	1/3	1/2				
O	.6327	0	.2716				
H	.536	0	.091				

Orthoenstatite, $MgSiO_3$, Ohashi (1984)

a, b, c (Å) α, β, γ (°) Space group:	18.225	8.813	5.180	90	90	90	Pbca
	x	y	z				
Mg1	.3758	.6538	.8660				
Mg2	.3768	.4869	.3589				
Si1	.2717	.3415	.0505				
Si2	.4735	.3374	.7980				
O1a	.1834	.3398	.0347				
O2a	.3111	.5023	.0433				
O3a	.3032	.2227	.8311				
O1b	.5624	.3402	.8001				
O2b	.4326	.4827	.6895				
O3b	.4474	.1952	.6039				

Clinoenstatite, $MgSiO_3$, Ohashi (1984)

a, b, c (Å) α, β, γ (°) Space group:	9.606	8.813	5.170	90	108.3	90	$P2_1/c$
	x	y	z				
Mg	.2511	.6533	.2177				
Mg	.2558	.0131	.2146				
Si	.0433	.3409	.2945				
Si	.5534	.8372	.2301				
O1	.8667	.3396	.1851				
O2	.1228	.5009	.3218				
O3	.1066	.2795	.6153				
O4	.3762	.8399	.1247				
O5	.6340	.9825	.3891				
O6	.6053	.6942	.4540				

Jadeite, NaAlSi$_2$O$_6$, Cameron et al. (1973)

a, b, c (Å) α, β, γ (°) Space group: 9.423	8.564	5.223	90	107.56	90	C2/c
	x	y	z			
Si	.2906	.0933	.2277			
Al	0	.9058	1/4			
Na	0	.3005	1/4			
O1	.1092	.0760	.1285			
O2	.3611	.2633	.2932			
O3	.3537	.0072	.0060			

Albite, NaAlSi$_3$O$_8$, Downs et al. (1994)

a, b, c (Å) α, β, γ (°) Space group: 8.1372	12.787	7.1574	94.245	116.605	87.809	C$\bar{1}$
	x	y	z			
Al$_1$0	.00887	.16835	.20845			
Si$_1$m	.00370	.82030	.23707			
Si$_2$0	.69161	.11020	.31483			
Si$_2$m	.68158	.88176	.36040			
Na	.26823	.98930	.14611			
O$_A$1	.00523	.13114	.96733			
O$_A$2	.59186	.99705	.28040			
O$_B$0	.81268	.10974	.19097			
O$_B$m	.82013	.85098	.25848			
O$_C$0	.01288	.30187	.27091			
O$_C$m	.02337	.69355	.22885			
O$_D$0	.20681	.10907	.38899			
O$_D$m	.18403	.86804	.43600			

Hollandite, KAlSi$_3$O$_8$, Zhang et al. (1993)

a, b, c (Å) α, β, γ (°) Space group: 9.315	9.315	2.723	90	90	120	I4/m
	x	y	z			
Si,Al	.3501	.1661	0			
K	0	0	1/2			
O1	.1526	.2036	0			
O2	.5406	.1648	0			

Quartz, SiO$_2$, Glinneman et al. (1992)

a, b, c (Å) α, β, γ (°) Space group: 4.921	4.921	5.4163	90	90	120	P3$_1$21
	x	y	z			
Si	.4698	0	0			
O	.4151	.2675	-.1194			

APPENDIX REFERENCES

Angel RJ, Finger LW, Hazen RM, Kanzaki M, Weidner DJ, Liebermann RC, Veblen DR (1989) Structure and twinning of single-crystal MgSiO₃ garnet synthesized at 17 GPa and 1800°C. Am Mineral 74:509-512

Cameron M, Sueno S, Prewitt CT, Papike JJ (1973) High-temperature crystal chemistry of acmite, diopside, hedenbergite, jadeite, spodumene, and ureyite. Am Mineral 58:594-618

Downs RT, Hazen RM, Finger LW (1994) The high-pressure crystal chemistry of low albite and the origin of the pressure dependency of Al-Si ordering. Am Mineral 79:1042-1052

Evans HT (1970) Lunar troilite: Crystallography. Science 167:621-623

Finger LW, Hazen RM, Zhang J, Ko J, Navrotsky A (1993) The effect of Fe on the crystal structure of wadsleyite β-(Mg₁₋ₓFeₓ)₂SiO₄, 0.00 < x < 0.40. Phys Chem Minerals 19:361-368

Fujii T, Yoshida A, Tanaka K, Marumo F, Noda Y (1986) High pressure compressibilities of pyrite and cattierite. Mineral J (Japan) 13:202-211

Glinnemann J, King HE Jr, Schulz H, Hahn Th, La Placa SJ, Dacol F (1992) Crystal structures of the low-temperature quartz-type phases of SiO₂ and GeO₂ at elevated pressure. Z Kristallogr 198:177-212

Hazen R M (1976) Effects of temperature and pressure on the cell dimension and x-ray temperature factors of periclase. Am Mineral 61:266-271

Hazen RM, Downs RT, Finger LW, Ko J (1993) Crystal chemistry of ferromagnesian silicate spinels: Evidence for Mg-Si disorder. Am Mineral 78:1320-1323.

Horiuchi H, Hirano M, Ito E, Matsui Y (1982) MgSiO₃ (ilmenite-type): Single crystal x-ray diffraction study. Am Mineral 67:788-793

Horiuchi H, Ito E, Weidner DJ (1987) Perovskite-type MgSiO₃: Single-crystal x-ray diffraction study. Am Mineral 72:357-360

Levien L, Prewitt CT, Weidner DJ (1980) Structure and elastic properties of quartz at pressure. Am Mineral 65:920-930

Nelmes RJ, McMahon MI, Belmonte SA, Allan DR, Gibbs MR, Parise JB (1998) High pressure structures of iron sulphide. *In:* Nakahara M (ed) Review of High-Pressure Science and Technology, Vol 7, p 202-204. Japan Society of High Pressure Science and Technology, Kyoto

Ohashi Y (1984) Polysynthetically-twinned structures of enstatite and wollastonite. Phys Chem Minerals 10:217-229

Pavese A, Artioli G, Prencipe M (1995) X-ray single crystal diffraction study of pyrope in the temperature range 30-973 K. Am Mineral 80:457-464

Ross N, Shu J-F, Hazen RM, Gasparik T (1990) High-pressure crystal chemistry of stishovite. Am Mineral 75:739-747

Takéuchi Y, Yamanaka T, Haga N, Hirano M (1984) High-temperature crystallography of olivines and spinels. *In:* I Sunagawa (ed) Materials Science of the Earth's Interior. p 191-231 Terra Scientific Publishing Company, Tokyo

Teter DM, Hemley RJ, Kresse G, Hafner J (1998) High pressure polymorphism in silica. Phys Rev Lett 80:2145-2148

Yang H, Prewitt CT, Frost DJ (1997) Crystal structure of the dense hydrous magnesium silicate, phase D. Am Mineral 82:651-654

Zhang J, Ko J, Hazen RM, Prewitt CT (1993) High-pressure crystal chemistry of KAlSi₃O₈ hollandite. Am Mineral 78:493-499

Chapter 10

THERMODYNAMICS OF HIGH PRESSURE PHASES

Alexandra Navrotsky

Thermochemistry Facility and Center for High Pressure Research
Departments of Chemical Engineering and Materials Science
University of California at Davis
Davis, California 95616

INTRODUCTION

In high pressure studies, as in other areas of mineralogy and petrology, thermo-dynamics plays a central role. The stability of mineral assemblages in the Earth's interior is described by their thermodynamic properties. Furthermore, to understand when and if different reservoirs in the Earth are not in equilibrium, one must first understand what equilibrium would be. It has been accepted for some time (Ringwood 1970, Jeanloz and Thompson 1983, Ito 1984) that the properties of the deep Earth are strongly influenced by phase transitions, olivine to beta phase to spinel near 410 km and spinel to perovskite plus magnesiowustite near 670 km receiving the greatest attention. In addition to defining regions of phase stability, the thermodynamic parameters define the width phase transition loops in the ferromagnesian silicates, and these thermodynamic widths are one of the factors determining the seismic width of the transition (Wood 1990, Bina and Helffrich 1994, Helffrich and Bina 1994).

The P-T slope of the transition is related, through the Clausius-Clapeyron equation

$$(dP/dT)_{equilibrium} = \Delta S / \Delta V \tag{1}$$

to the entropy of the transition and, therefore, to the sign and magnitude of the heat flow across the transition. This heat transfer is related in turn to the extent of slab penetration through the 670 km seismic discontinuity, as has been discussed in many recent mantle dynamics papers (Christensen and Yuen 1984, Tackley et al. 1993, Yuen et al. 1996, Weidner and Wang 1998). Heat flow associated with the movement of crystalline rocks (plastic flow regime), silicate melts, and hydrous fluids adds convective terms to heat transfer equations otherwise dominated by conduction. The heat capacity and heat of fusion of rocks must be known to model cooling history and evolution on a local, regional, or planetary scale (McKenzie and O'Nions 1988, Kojitani and Akaogi 1995). Thus fundamental thermodynamic parameters are clearly important to geophysics and geochemistry, and must be known accurately over the whole P-T range of the planet.

There have been a number of recent reviews of the thermodynamic properties of high pressure minerals, several attempts to produce internally consistent sets of thermodynamic data for the system $MgO-FeO-SiO_2$ at mantle pressures, and recom-mendations of "best" values of thermochemical parameters (Fei and Saxena 1986, Akaogi et al. 1989, Fei et al. 1990, Fei et al. 1991, Saxena et al. 1993, Fabrichnaya and Kuskov 1994, Fabrichnaya 1995, Saxena 1996). Rather than to generate another similar review, my purpose in this MSA tutorial is somewhat different. My goal is to describe, in part to the nonspecialist, what makes the gathering and evaluation of thermodynamic data for mineralogically important systems at high pressure, that is, in the range of about 3 to 30 GPa, a unique task. I summarize the methodologies available and their capabilities and limitations. I point out areas in which at least a first-order understanding of thermo-dynamic properties exists, other areas where some incomplete information is available,

0275-0279/98/0037-0010$05.00

and yet others in which we do not reliably know the phases likely to be stable, let alone their thermodynamic properties. I stress areas likely to develop significantly in the near future, and point out some unsolved problems which I consider important. To make the discussion concrete, but not exhaustive, I address several specific examples in detail.

FUNDAMENTAL RELATIONS

The driving force for a phase transition with increasing pressure is the decrease in volume associated with the transformation, as shown by the pressure derivative of the Gibbs free energy:

$$(\delta G/\delta P)_T = V \text{ or } (\delta \Delta G/\delta P)_T = \Delta V \tag{2}$$

A univariant phase boundary between a low pressure phase and a high pressure phase occurs when the decrease in free energy associated with this compaction just overcomes the initially unfavorable free energy of the high pressure phase, giving a locus of points in P-T space where $\Delta G = 0$. Therefore, along the phase boundary:

$$\Delta G = 0 = \Delta H^\circ - T\Delta S^\circ + \int_{1 \text{ bar}}^{P} \Delta V(P,T)dP \tag{3}$$

When two phases coexist, each of which forms a continuous solid solution, for example Mg-Fe substitution in the olivine (ol) and spinel (sp) polymorphs of $(Fe,Mg)_2SiO_4$, the partitioning of the elements between the two phases is given by the equality of chemical potentials, thus

$$\mu(P,T) \, Mg_2SiO_4(ol) = \mu(P,T) \, Mg_2SiO_4(sp) \tag{4}$$

and, simultaneously

$$\mu(P,T) \, Fe_2SiO_4(ol) = \mu(P,T) \, Fe_2SiO_4(sp) \tag{5}$$

These chemical potential terms are determined by the thermodynamics of the phase transitions in the end members and the thermodynamics of mixing in the solid solutions. This formalism and its application to mantle ferromagnesian minerals has been discussed in a number of papers (Akaogi et al. 1989, Fei et al. 1991, Fabrichnaya 1995, Saxena 1996) and in an earlier *Reviews in Mineralogy* volume (Navrotsky 1987). Here I make the point that enthalpic, entropic, and volumetric terms in these chemical potentials are often of comparable magnitude and must each be known to allow an accurate calculation of a binary two phase loop such as the olivine spinel transition.

Because high P-T phase transitions are often thought of as volumetrically driven, it is normal to consider a binary loop in pressure-composition space, with separate diagrams drawn at different temperatures to indicate how the transition loop moves with temperature. For melting, which is thought of as entropically driven, one normally draws a melting loop as a function of temperature and composition. But it is important to remember that a melting transition has a volume change as well as an entropy change, and a high pressure transition has an entropy change as well as a volume change, so that both any P-X or T-X diagram represents a section of a more general pressure-temperature-composition (P-T-X) surface. This point is especially important when one realizes that both pressure and temperature change with depth in the Earth, and that the region spanned by a phase transition in the mantle is neither isothermal nor isobaric.

Therefore, the basic question of the thermodynamicist is how to determine the variation of chemical potential as a function of temperature, pressure, and composition for each of the phases of interest. The following sections will describe experimental approaches applicable to high pressure materials.

THERMODYNAMIC DATA FROM HIGH PRESSURE PHASE EQUILIBRIA

Equation (3) implies that, if the phase boundary, the curve along which $\Delta G = 0$, is known accurately, the enthalpy, entropy, and volume changes for the reaction can be calculated. For transitions in the 5-30 GPa range, one must consider rather substantial uncertainties. Until recently, the uncertainties in temperature, caused by temperature gradients across a sample, were probably of the order of ± 30 K in multianvil experiment and ± 100-200 K in laser heated diamond cells. In general, the higher the temperature, the greater the uncertainty. An additional problem is the poorly known effect of pressure on thermocouple e.m.f. (Luth 1993). The problems of pyrometric measurements in the diamond cell add another level of uncertainty which is not always easy to assess (Heinz et al. 1991, Jeanloz and Kavner 1996). More recently, the development of tapered anvils, specially designed furnaces, and other improvements have allowed better control of both stress and temperature distribution in the multianvil (Weidner et al. 1992, Wang et al. 1998). There have been a number of improvements in the homogeneity and long-term stability of temperature distributions in the diamond cell. These include "externally heated" and "double sided laser heated" cells (Shen et al. 1996, Mao et al. 1998), laser feedback stabilization (Heinz et al. 1991), temperature distribution analysis (Manga and Jeanloz 1998), thermal pressure studies (Jeanloz and Kavner 1996), and deviatoric stress analysis (Meng et al. 1993). These improvements, coupled with the ability to make spectroscopic and/or diffraction measurements on a very small region of sample directly adjacent to a thermocouple, have reduced the uncertainty in temperature measurements. Nevertheless, it is unlikely that temperature is known with an accuracy better than about 1% of its absolute value, or ± 10 K at 1000 K and ± 20 K at 2000 K. Such temperature uncertainty translates into uncertainties in calculated free energy values shown in Table 1. Thus, if temperature uncertainties can be kept below about ± 20 K, the resulting uncertainty in ΔG for typical solid-solid transitions is less than ± 1 kJmol^{-1}.

It is generally regarded that pressure in diamond cell experiments, based on the ruby fluorescence scale, can be known to better than $\pm 1\%$, under quasi-hydrostatic conditions at pressures below 100 GPa and ambient temperature (Hemley and Mao 1997). Such hydrostatic conditions are, however, not easy to maintain at simultaneous high pressure and temperature. In multianvil experiments, using fixed point calibrations of pressure versus load, the pressure is assumed to be known with an accuracy of about $\pm 5\%$ (± 0.5 GPa at 10 GPa and ± 1.3 GPa at 25 GPa) (Weidner et al. 1992, Hemley and Mao 1997). The chapter by Duffy and Wang (this volume) further discusses techniques for pressure measurements in high P-T experiments. The effect of such uncertainties in pressure on the free energy change for a reaction are shown in Table 2. It is clear that uncertainties of the magnitude frequently encountered in experiments above 5 GPa at high temperature result in significant uncertainties in calculated free energies of reaction.

The accuracy with which the P-T slope of a phase transition can be known depends more on the internal consistency of pressure and temperature measurements than on their absolute accuracy. Thus, under favorable conditions and tight bracketing of reversals (see below), it is probably possible to determine dP/dT (when the P-T curve is a straight line) with an accuracy of about $\pm 10\%$, though $\pm 20\%$ to $\pm 50\%$ is probably more common. When the volume change for the reaction, ΔV, is known to better than $\pm 1\%$, the uncertainty in dP/dT translates almost directly into an uncertainty in the entropy change, ΔS, from Equation (1) (see Table 3).

Table 1. Effect of uncertainty in temperature on calculated
free energy from high pressure experiments*

$\Delta S°$ transition ($J K^{-1} mol^{-1}$)	Effect on $\Delta G°$ ($kJ mol^{-1}$) of an uncertainty in temperature of ΔT (K)	
	$\Delta T = 10$ K	$\Delta T = 100$ K
1	0.01	0.1
5	0.05	0.5
10	0.10	1.0
20	0.20	2.0

*Calculated as $\Delta T \Delta S$. Effects of temperature uncertainty on ΔV calculated through equation of state add further uncertainty, but that effect is generally smaller.

Table 2. Effect of uncertainty in pressure on calculated free
energy from high pressure experiments*

$\Delta V°$ transition $cm^3 mol^{-1}$	Effect on $\Delta G°$ ($kJ mol^{-1}$) of an uncertainty in pressure, ΔP (GPa)	
	$\Delta P = 1$ GPa	$\Delta P = 5$ GPa
1	1	5
5	5	25
10	10	50

*Calculated as $\Delta P \Delta V$. Effects of pressure uncertainty on ΔV calculated through equation of state add further uncertainties of smaller magnitude.

Table 3. Effect of uncertainty in Clausius-Clapeyron slope of a
transition on calculated entropy of transition*

ΔV transition $cm^3 mol^{-1}$	Effect on $\Delta S_{transition}$ (J/mol K) of an uncertainty in dP/dT of Δ (dP/dT) (GPa K^{-1})	
	Δ (dP/dT) = 10^{-4} GPa K^{-1}	Δ (dP/dT) = 10^{-5} GPa K^{-1}
1	0.1	1.0
5	0.5	5.0
10	1.0	10.0

*Assuming linear dP/dT, with ΔV = constant. Corrections arising from equation of state terms will add further small uncertainty.

Real curvature in a P-T plot comes from two sources: the variation of ∆V because of the difference in thermal expansivity and compressibility of the low and high pressure phases (or phase assemblages) and the variation of ∆S, arising from differences in the heat capacity of the low and high pressure assemblage. Apparent curvature can arise from problems of pressure and temperature calibration or measurement and from difficulties in attaining or retaining equilibrium (see below). For solid-solid transitions, I know of no cases in which P-T data unambiguously show a properly reversed curved phase boundary within one set of measurements over a limited temperature range (temperature varying by 300-500 K). Fitting data from various sources over a much larger temperature range (1000-2000 K) may require such curvature (see the example of the coesite-stishovite equilibrium below). Furthermore, as T→0 K, the third law of thermodynamics requires that phase boundaries flatten (dP/dT→0 as ∆S→0). Nevertheless, it is my opinion that there is a tendency in the literature to overfit solid-solid P-T boundaries with P- and T-dependent thermodynamic parameters which tend to cancel each other, and that, in most cases, a simple Clausius-Clapeyron slope, with ∆S and ∆V each constant, is both more robust and more honest. If one does use corrections for thermal expansion, compressibility, and heat capacity, all these corrections should be used simultaneously, rather than just one of them, because they tend to act in opposite directions and cancel each other (Akaogi et al. 1984).

The above discussion shows that issues of pressure and temperature calibration in high P-T experiments are crucial for obtaining reliable thermodynamic data. In comparing data from different laboratories, it is often hard to know whether the P,T points are based on identical, similar, or significantly different calibrations, and whether the experiments are done in the same way or with significant differences in cell assembly and procedure. An interlaboratory comparison ("round robin") using the same reaction done in different laboratories, as was done many years ago for piston-cylinder experiments in the 1-2 GPa range (Johannes et al. 1971) would be very useful for multianvil and diamond cell experiments. The calibration problem still requires considerable work.

A phase boundary is "reversed" when the low pressure (or low temperature) assemblage is converted to the high pressure (or high temperature) assemblage and then the high pressure (or high temperature) assemblage is converted back to the low pressure (or low temperature) assemblage, with the P-T conditions of this "bracket" as close together as possible. A synthesis reaction done in only one direction is called a "half bracket". In experiments above 5 GPa, half brackets or synthesis runs are the most common type of experiments, with the conversion of low pressure to high pressure polymorph generally being the direction of reaction. In older experiments, run durations were extremely short, usually of the order of a minute, but now, runs at relatively stable P-T conditions for hours are possible. Nevertheless, true "reversals" are technically difficult and seldom attempted. It is generally believed that equilibrium in most systems at 5-30 GPa is attained in a matter of minutes at temperatures of 1273 K or higher, and the general agreement between experimental P-T curves and those calculated based on calorimetric data for phase transitions involving silicate olivine, spinel, ilmenite, garnet, and perovskite phases (see below) suggests this is usually the case. However, some caution is needed, see discussion of the coesite-stishovite transition below.

Another issue relevant to high pressure equilibria is the production of "quench phases", materials of different structure formed upon the release of pressure and/or temperature. In situ studies, in which the phases are identified, and phase boundaries determined, by x-ray diffraction, spectroscopy, or other methods while the sample is held at high pressure and high temperature, are the only definitive way of determining whether quench phases are indeed a problem, although the inconsistency of observed phase boundaries on quenching and thermochemical data for the quench phases can strongly suggest that the phases brought back to ambient conditions are not the same as those at high P-T. For experiments involving melting, petrologists traditionally use textural evidence to identify quench phases, and such evidence is sometimes useful for solid-state transitions as well.

Metastable quench phases may be encountered under several conditions. If the high pressure structure can distort by a diffusionless (displacive or martensitic) transition to a slightly less dense and lower energy structure on pressure release, this may occur rapidly. An example is the formation of lithium niobate from perovskite by a diffusionless transformation (in contrast the formation of ilmenite from either lithium niobate or perovskite requires bond-breaking) (Ross et al. 1989, Leinenweber et al. 1991, Leinenweber et al. 1994). Another instance is the formation of α-PbO_2 phases from the fluorite structure of transition metal fluorides (Dandekar and Jamieson 1969, Jamieson 1970, Nagel and O'Keeffe 1971), of various dioxides (Syono and Akimoto 1968, Liu 1982, Ming and Manghnani 1982), and maybe even of SiO_2 (Hemley et al. 1994, Dubrovinsky et al. 1997). Yet another case is the formation of a $CaSi_2O_5$ phase with silicon in 5-fold coordination, possibly as a quench phase from a polymorph with all octahedral silicon (Angel et al. 1996, Angel 1997, Kudoh and Kanzaki 1998). If the high pressure phase is higher in free energy and more dense than an amorphous material, it can go "downhill" to the amorphous phase on pressure release, as well as on heating at atmospheric pressure. For this reason, and also on the basis of more microscopic considerations, metastability, pressure induced amorphization, and change in local coordination are closely interrelated (see Hemley et al. 1994 for a review of the case for silica). Thus $MgSiO_3$ perovskite (Ito et al. 1990) and SiO_2 stishovite (Holm et al. 1967) amorphize on heating to about 600 K, and $CaSiO_3$ perovskite (Hemley and Cohen 1992), although it can be brought down in pressure well below its stability field at about 12 GPa, transforms to glass at room temperature at pressures below about 3 GPa. Cryogenic quenching using liquid nitrogen in the multianvil may preserve otherwise unquenchable phases, as has been demonstrated for a high pressure polymorph of $Ca(OH)_2$ (Leinenweber et al. 1998). As one goes higher in pressure, the free energy driving forces for decomposition at ambient conditions increase, and amorphous materials become energetically more accessible from the high pressure phases. Thus it can be expected that quench phases become more of a problem with increasing pressure, making in situ methods of phase identification all the more desirable.

THERMOCHEMICAL STUDIES OF HIGH PRESSURE PHASES

Rather than determining $\Delta H°$ and $\Delta S°$ in Equation (3) from high P-T equilibria, one can measure the enthalpy and entropy change by calorimetric experiments, namely enthalpies of reaction and heat capacities, respectively. For a material without disorder, the standard entropy, $S°_T$, is related to the heat capacity, C_p, by

$$S°_T = \int_0^T (C_P/T)dT \qquad (6)$$

Thus to evaluate the entropy, the heat capacity must be known or estimated to temperatures approaching absolute zero. Adiabatic calorimetric measurements (Robie and Hemingway 1972) of heat capacity between liquid helium temperature and room temperature typically require several grams of sample; thus a recent study of synthetic coesite used 12.36 g of material synthesized by repeated 48 hour piston-cylinder runs at 3.2 GPa and 1073 K (Hemingway et al. 1998). Such synthesis is possible when several hundred mg can be made at a time, but making enough sample for cryogenic calorimetry of phases stable above 5 GPa is currently out of the question. Careful adiabatic C_p measurements can provide heat capacity data with an accuracy of $\pm0.5\%$ or better, and standard entropies at 298 K with an accuracy of $\pm0.5\%$.

Differential scanning calorimetry (DSC) (Callanan and Sullivan 1986) can determine heat capacities with an accuracy of ±1 to $\pm5\%$ (depending on the instrument and the care taken) in the range from liquid nitrogen temperature to about 1000 K, and with an accuracy of ±2 to $\pm10\%$ at 1000-1500 K. Samples of 30-50 mg are routine; samples as small as 10 mg have been used (Ashida et al. 1988). Thus this technique is suited for high pressure materials. A comparison of adiabatic calorimetric and DSC data for coesite is given in Figure 1.

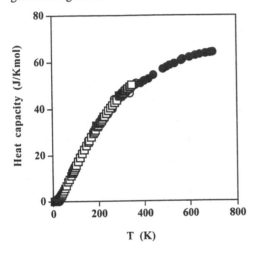

Figure 1. Heat capacity of coesite. Solid circles are DSC data of Akaogi et al. (1995); squares = adiabatic calorimetric data (Holm et al. 1967); circles = adiabatic calorimetric data (Hemingway et al. 1998); these latter two data sets are essentially identical and overlapping (see text).

The high temperature limit of DSC for high pressure samples is generally the temperature at which the high pressure phase begins to decompose, often between 600 and 800 K. Thus, for high pressure polymorphs, heat capacity data are often available over a limited range. At low temperature the data usually stop at a temperature too high (~150 K) for an appropriate extrapolation to T = 0 K according to a T^3 law (~20 K), and at high temperature, the data usually stop at 600-1000 K before lattice vibrations are fully excited and the Dulong and Petit limit of $C_v = 3nR$ is reached (1200-1400 K). Thus for such DSC data to be useful in thermodynamic modeling, sensible extrapolations to lower and higher temperature must be made. The extrapolation to 0 K and the calculation of S^0_{298} are greatly aided through the use of microscopic models of the vibrational density of states. These can come from simple but general descriptions of lattice vibrations arising from different bonding environments, as in the approach of Kieffer (Kieffer 1979, Akaogi et al. 1984) and others (Westrum and Komada 1986), from empirical estimates correlating entropy and polyhedra present in the structure (Robinson and Haas 1983), from more detailed lattice dynamical analysis (Wolf and Bukowinski 1987, Ghose et al. 1992) and atomistic simulations (Belonoshko 1994, Wentzcovitch et al. 1995). It is my opinion that combining such calculations with the measured heat capacities over a limited range of temperatures can both act as a test of the calculations and provide well-constrained standard entropies for high pressure phases. At high temperatures, one must be careful about the form of equations used to extrapolate C_p. Several of the polynomials used tend to show unrealistic curvature at temperatures outside their fitting range. It is probably safer to use a form such as that suggested by Saxena (Saxena 1996,

Dubrovinskaya et al. 1997) which ensures a linear behavior of heat capacity above the Dulong and Petit limit, that is

$$C_p = 3nR + AT \tag{7}$$

with A a moderately small constant. When taking heat capacity equations from the literature, the reader is cautioned to watch for pitfalls in extrapolation; I personally like to graph C_p over the entire temperature range I am interested in to make sure the function is sensibly behaved. In addition, phenomena associated with electronic transitions, order-disorder, strong anharmonicity, and premelting may further complicate the behavior of heat capacity at high temperature (Richet et al. 1994). It is not clear whether such effects become more or less important with pressure.

Heat capacity and entropy of course depend on pressure. However, if the conditions for equilibrium are written as in Equation (3), one does not have to explicitly consider this pressure dependence, since the effect of pressure is confined to the volume integral (the last term on the r.h.s.).

The enthalpy of reaction, ΔH° in Equation (3), needs to be known at one temperature. At other temperatures it can be calculated using the heat capacity difference between products and reactants, ΔCp.

$$\Delta H^\circ(T) = \Delta H^\circ(T_0) + \int_{T_0}^{T} \Delta C_p dT \tag{8}$$

The entropy of reaction can likewise be corrected as

$$\Delta S^\circ(T) = \Delta S^\circ(T_0) + \int_{T_0}^{T} (\Delta C_P/T) dT \tag{9}$$

These ΔC_P correction terms are generally significant at 1000-2000 K only when the enthalpy and entropy of reaction are small and when the two polymorphs (or assemblages) are sufficiently different in structure to have significantly different heat capacities.

$\Delta H^\circ(T_0)$ itself needs to be measured. The small amounts of sample quenched from P-T runs above 5 GPa (typically 1-25 mg) require a technique which uses only a few mg of sample. Solution calorimetry using molten lead borate as solvent is such a method and has been applied to a variety of high pressure materials (see Table 4). Recent developments (Navrotsky 1997) have enabled the utilization of samples as small as 1-2 mg. The drop-solution calorimetric method, suitable for high pressure samples which decompose upon heating, involves dropping the sample from room temperature into the molten oxide solvent at high temperature, normally 973 or 1073 K. The difference in enthalpy of drop solution between reactants and products gives the enthalpy of transformation at 298 K. The samples may be encapsulated or, as is now standard practice in Navrotsky's laboratory, they may be small pressed pellets of powder. Sample purity is critical to high temperature solution calorimetry (Liu et al. 1996, Navrotsky 1997). Particularly troublesome are impurities which oxidize (with large exothermic heat effects) during dissolution of the sample; these include graphite (from heater and/or capsule material, W. Re, or Mo (from thermocouples, lubricants, and dies) and traces of Fe, Cr, Ni (from pellet dies, pressure vessels, etc.). Thus a sample which is adequate for phase equilibrium or equation of state work may not be good enough for calorimetry. Calorimetric samples must be synthesized in sealed noble metal capsules and handled very carefully. Detailed study of the sample by microprobe, with emphasis on imaging and backscattered electron or specific element mapping for impurities, is essential to characterize good calorimetric samples. When appropriate care is taken, enthalpies of

Table 4. Summary of calorimetric studies of high pressure phases.

System	Phase	Reference
$CaSiO_3$	wollastonite, pseudowollastonite, high pressure phase	Charlu et al. 1978 Kiseleva et al. 1979
Al_2SiO_5	kyanite, andalusite, sillimanite	Anderson et al. 1977 Kiseleva et al. 1983
Mg_2SiO_4	olivine, beta phase, spinel	Akaogi et al. 1984, 1989 Kojitani & Akaogi 1994
Fe_2SiO_4	olivine spinel	Navrotsky et al. 1979 Akaogi et al. 1989 Kojitani & Akaogi 1994
Ni_2SiO_4	olivine spinel	Navrotsky et al. 1979
Co_2SiO_4	olivine, beta phase, spinel	Navrotsky et al. 1979
$NiAl_2O_4 - Ni_2SiO_4$	spinelloids	Akaogi & Navrotsky 1984a
$MgGa_2O_4 - Mg_2GeO_4$	spinelloids	Leinenweber & Navrotsky 1989
Mg_2SnO_4	spinel	Navrotsky & Kasper 1976
Co_2SnO_4	spinel	Navrotsky & Kasper 1976
SiO_2	quartz, coesite, stishovite	Holm et al. 1967 Akaogi & Navrotsky 1984b Akaogi et al. 1995 Liu et al. 1996 Hemingway et al. 1998
$MgSiO_3$	enstatite ilmenite	Brousse et al. 1984 Ito & Navrotsky 1985 Ashida et al. 1988
	garnet	Akaogi et al. 1987 Yusa et al. 1993
	perovskite	Ito et al. 1990 Akaogi & Ito 1993
$Mg_2SiO_4 - Fe_2SiO_4$	olivine, modified spinel, spinel	Akaogi et al. 1989 Kojitani & Akaogi 1994
$MnSiO_3$	pyroxenoid, pyroxene, garnet	Akaogi & Navrotsky 1985
$CaGeO_3$	pyroxenoid, garnet, perovskite	Ross et al. 1986
$CdGeO_3$	pyroxenoid, garnet, ilmenite, perovskite	Akaogi & Navrotsky 1987
$K_2Si_4O_9$	glass, sheet silicate, wadeite	Geisinger et al. 1987
Mg_2GeO_4	olivine, spinel	Ross & Navrotsky 1987
$MnTiO_3$	ilmenite, lithium niobate	Ko et al. 1989
GeO_2	quartz, rutile	Kume et al. 1989
ZrO_2	high P forms	Kume et al. 1989
$MgGeO_3$	orthopyroxene, clinopyroxene, ilmenite	Ross & Navrotsky 1988
$ZnSiO_3$	pyroxene, ilmenite	Leinenweber et al. 1989 Akaogi et al. 1990
$ZnGeO_3$	pyroxene, ilmenite	Leinenweber et al. 1989

transition can be determined with an accuracy of about $\pm2\%$. Recent work on high pressure dense hydrous magnesium silicates (DHMS) (Bose and Navrotsky 1998) has extended such calorimetric studies to high pressure hydrous phases (see below).

One limitation to the accurate determination of ΔS and ΔH from phase equilibria is the relatively small temperature range over which a phase boundary can be accurately reversed. At too low a temperature (typically below about 1200 K) reactions can become sluggish (see the quartz-coesite-stishovite example below) while above about 1700 K, temperature is harder to control and determine, or melting and/or reactions with capsule material may intervene. Combining calorimetric determination of ΔH with a ΔG value calculated from the P-T curve near the midpoint of the temperature range over which it has been determined provides a very effective way of constraining ΔS of high pressure phase transitions. This approach has proven very fruitful for a number of geophysically important transformations, including, in the $MgO-SiO_2$ system, quartz to coesite to stishovite (Liu et al. 1996), olivine (α) to wadsleyite (β) to ringwoodite (γ, spinel) (Akaogi et al. 1984, 1989), ilmenite to perovskite (Ito et al. 1990, Akaogi and Ito 1993), spinel to perovskite + periclase (Ito et al. 1990, Akaogi and Ito 1993), and in defining the stability field of the garnet phase in the $MgO-SiO_2-Al_2O_3$ system (Yusa et al. 1993, Akaogi et al. 1987).

ILLUSTRATIVE CASE STUDIES

This section uses three examples to illustrate both the success and the difficulty in extracting thermodynamic data from high pressure and thermochemical studies. The first example concerns phase equilibria in silica: the quartz-coesite and coesite-stishovite transitions. This illustrates the need for good samples, careful calorimetry, and attention to reversals in phase equilibria. The second concerns phase transitions in titanates, illustrating the problems associated with quench phases. The third example uses data for high pressure hydrous magnesium silicates to illustrate the importance of both thermochemical data and the equation of state.

Quartz-coesite-stishovite

Quartz, the ambient stable form of silica, transforms to coesite, a denser tetrahedral structure, near 2.5 GPa (Coes 1953). Coesite transforms to stishovite, the octahedrally coordinated rutile polymorph, near 9 GPa (Akimoto and Syono 1969). Quartz itself undergoes the thermodynamically complex (possibly both first and higher order components) α-β transition at 844 K; this transition shifts sharply to higher temperatures with increasing pressure (Cohen and Klement 1967). Stishovite decomposes on heating at atmospheric pressure to an amorphous phase energetically very similar to fused silica glass (Holm et al. 1967).

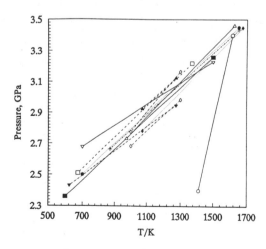

Figure 2. The quartz-coesite phase boundary. Two calculated boundaries are terminated by filled squares (Hemingway et al. 1998) and filled circles (Kuskov et al. 1991). The following symbols terminate curves representing experimental studies: Open diamonds = Mosenfelder et al. (1996); inverted open triangle = Bose and Ganguly (1995b); ◆ = Kosyakov and Ishbulatov (1984); ▼ = Bohlen and Boettcher (1982) plus Mirwald and Massonne (1980); ▲ = Akella (1979); ☐ = Kitahara and Kennedy (1964); and Δ = Boyd and England (1960). The line terminated with the open circles represents the α-β phase boundary for quartz derived by Cohen and Klement (1967). Graph and caption modified from Hemingway et al. (1998).

The quartz-coesite-stishovite transitions have been studied extensively by high pressure phase equilibria and calorimetry over the past 20 years, see Akaogi et al. (1995), Zhang et al. (1996), Liu et al. (1996), and Hemingway et al. (1998) for reviews. Most such studies were done between 700 and 1400 K. Figure 2 shows the results of various determinations of the quartz-coesite equilibrium. The determinations are in general agreement within a band of about 0.3 GPa in absolute pressure and most P-T slopes agree within about ±50%, including both experimental determinations and calculations from thermochemical data. Although one can discuss the specific details of various experiments in terms of tightness of reversals, pressure calibrations, non-hydrostatic effects, and other factors, it is probably fair to conclude that the range of values seen is a reflection of the current "state of the art" at present. It is interesting that the discrepancies between recent experimental determinations of the P-T boundary (for example Mosenfelder et al. (1996) versus Bose and Ganguly (1995a) are of the same magnitude as those of much older studies (for example Bohlen and Boettcher (1982) versus Akella (1979)).

The thermochemical calculation of this boundary is very sensitive to the parameters used because the enthalpy, entropy, and volume changes are small and the effects of thermal expansion, compressibility, and heat capacity are important. There have been several calorimetric determinations of the heat capacity of coesite, the entropy of the quartz-coesite transition (based on these heat capacity measurements), and the enthalpy of transformation (based on oxide melt solution calorimetry) (see Table 4). The early measurements of Holm et al. (1967) were done on a fine-grained natural coesite, and concern has been expressed that the high surface area and other imperfections could affect both the heat capacity and heat of solution (Akaogi et al. 1984). The heat capacity values determined by adiabatic calorimetry (Holm et al. 1967, Hemingway et al. 1998) agree well with each other and agree reasonably well with DSC values determined by Akaogi et al. (1995) (see Fig. 1). The natural coesite (Holm et al. 1967) did yield an enthalpy of transition different from that in two later studies on synthetic samples (Akaogi and Navrotsky 1984b, Akaogi et al. 1995). Those transition enthalpies on synthetic samples are consistent with the phase equilibrium data and entropy obtained from heat capacity data (Akaogi and Navrotsky 1984b, Akaogi et al. 1995, Hemingway et al. 1998).

The coesite-stishovite transition has produced some fairly contradictory experimental phase equilibrium data. A recent in situ x-ray diffraction study (Zhang et al. 1996) has identified much of the disagreement as arising from slow kinetics of transformation below about 1300 K, making tight bracketing of equilibrium impossible at low temperature (see Fig. 3). Zhang et al. (1996) proposed a phase diagram for silica (see Fig. 4) that is consistent with reversed equilibria and with thermochemical measurements.

The enthalpy of the coesite-stishovite transition has been determined several times (see Table 5). Early work by Holm et al. (1967) used natural samples, and, as discussed above, is suspect. A determination using synthetic coesite and stishovite (Akaogi and Navrotsky 1984b) had substantial correction terms for the graphite (from a graphite capsule) present in the stishovite sample. Because stishovite transforms to glass on heating, this graphite impurity could not be burned off. Two recent determinations (Akaogi et al. 1995, Liu et al. 1996) used very pure and well characterized samples. Using independently synthesized material, different calorimeters, and calorimetric techniques which differed in some details, they obtained similar values (see Table 5). These values appear consistent with the recent phase equilibria (Zhang et al. 1996). The difficulties in the early calorimetric studies and the extreme care taken in preparing and characterizing samples for the recent calorimetry illustrate the need for meticulous attention to sample purity for thermochemical studies.

Remaining uncertainties in the thermochemical data set for the silica polymorphs, some of which are pointed out by Hemingway et al. (1998), arise from the following

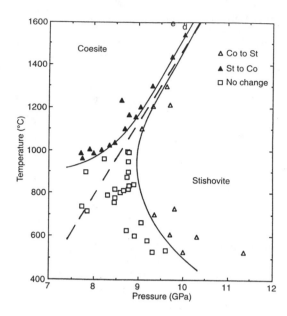

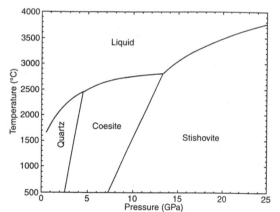

Figure 3. The coesite-stishovite trans-ition (Zhang et al. 1996). $\triangle$ = coesite $\rightarrow$ stishovite, $\blacktriangle$ = stishovite $\rightarrow$ coesite, $\square$ = no reaction. The curves show the kinetic phase boundary and indicate the difficulty in achieving bracketing of the reaction below about 1100°C. The boundary, P(GPa) = 4.65 + 0.0031T(K), consistent with both calorimetric meas-urements and the phase transformation data (Liu et al. 1996) is — — — —.

Figure 4. High pressure SiO$_2$ phase diagram proposed by Zhang et al. (1996).

factors: poor constraint on the heat capacity of stishovite at high temperature (where it can not be measured because of decomposition), the volumetric and heat capacity parameters for α-quartz at high P and T, and pressure calibration issues for both the piston-cylinder and the multianvil apparatus. It seems likely that the "best" thermo-chemical data for coesite and stishovite will continue to evolve slightly as these uncertainties are resolved in the future.

Phase transitions in titanates

MnTiO$_3$ and FeTiO$_3$ ilmenites transform to high pressure polymorphs (see Fig. 5). Initially indexed as having disordered ilmenite (corundum) structures (Ito and Matsui 1979, Syono et al. 1980), the materials quenched from high pressure and high temperature were later shown to have the lithium niobate structure (Ko and Prewitt 1988). When these lithium niobate materials were compressed at room temperature in a

Table 5. Thermodynamic parameters for quartz-coesite-stishovite phase transitions at 298 K*

ΔH (kJ mol⁻¹)	ΔS (J K⁻¹ mol⁻¹)	Reference
	quartz = coesite	
3.09 ± 1.0	-1.08	Hemingway et al. 1998
(5.06 ± 0.8)†	-0.96	Holm et al. 1967
3.08 ± 0.60		Akaogi et al. 1995
3.38 ± 0.60		Zhidikova et al. 1988
2.90	-2.96	Robie & Hemingway 1995
3.80	-0.96	Swamy et al. 1994
	coesite = stishovite	
(44.27 ± 1.42)	(-12.6 ± 1.0)	Holm et al. 1967
(48.95 ± 1.72)	(-4.2 ± 1.7)	Akaogi & Navrotsky 1984b
33.81 ± 1.06	-14.07 ± 0.70	Akaogi et al. 1995
29.85 ± 0.78	-16.73 ± 0.70	Liu et al. 1996
(36.5)	(-10.72)	Robie & Hemingway 1995
(42.90)	(-11.00)	Swamy et al. 1994

* Values, when measured by solution calorimetry near 970 K, have been corrected to 298 K by appropriate ΔC_P terms.
† Values in parentheses are considered less reliable.

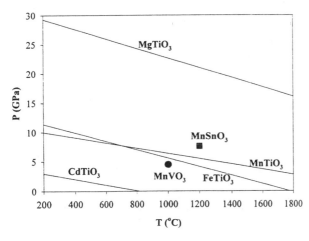

Figure 5. The ilmenite-perovskite transition in titanates and related materials. Lines represent measured phase boundaries, with calorimetric data and the reasoning discussed in the text indicating that the high pressure phases are perovskite rather than lithium niobate. . Points represent single (P,T) points of transition. From Navrotsky (1998).

diamond anvil cell (Ross et al. 1989, Leinenweber et al. 1991), they transformed to a perovskite structure at pressures far below those for the ilmenite-lithium niobate transition seen in multianvil quench experiments. This perovskite phase reverted to the lithium niobate on pressure release. It is now generally accepted that the perovskite is the stable structure at high P and T, and the lithium niobate forms by a diffusionless transition on pressure release and is a metastable quench product. Similar transitions were recently seen in $MgTiO_3$, but at higher pressures than for $MnTiO_3$ or $FeTiO_3$ (Linton et al. 1997). Again, the lithium niobate phase appears to be a quench product of the perovskite.

Intermediate compositions along the $FeTiO_3$-$MgTiO_3$ join also transform to a high pressure phase, which, though probably a perovskite in situ at high P and T, is recovered as a lithium niobate solid solution (Linton et al. 1997). This is the first documentation of complete and continuous substitution of Mg^{2+} by Fe^{2+} in the perovskite A-site.

Calorimetric studies have measured the enthalpy difference between ilmenite and perovskite in $CdTiO_3$ (Neil et al. 1971, Takayama-Muromachi and Navrotsky 1988) and between ilmenite and lithium niobate quench phase in $MnTiO_3$ (Ko et al. 1989), $FeTiO_3$ (Mehta et al. 1994, Linton et al. 1998), $MgTiO_3$ (Linton et al. 1998), and $MgTiO_3$-$FeTiO_3$ solid solutions (Linton et al. 1998). There is a linear increase in the $\Delta H°$ of transition between lithium niobate and ilmenite with increasing Mg content in $FeTiO_3$-$MgTiO_3$ (Linton et al. 1998). Similarly, $\Delta H°$ (ilmenite = lithium niobate) is a linear function of A-site cation size for a variety of titanates (see Fig. 6). This suggests that the size of the A-site cation dominates in the energetics.

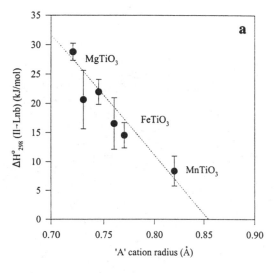

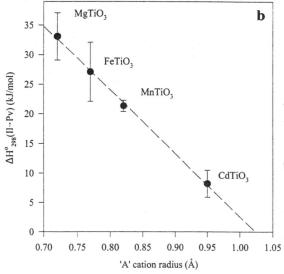

Figure 6. (a) Enthalpy of ilmenite → (metastable) lithium niobate transition versus A-site cation radius. Points between $MgTiO_3$ and $FeTiO_3$ represent $Mg_{1-x}Fe_xTiO_3$ solid solutions. Data from calorimetry from Linton et al. (1998) (b) Enthalpy of ilmenite → perovskite transition versus A-site cation radius. Data from phase equilibria from Linton et al. (1998). The combination of phase equilibrium boundaries (ilmenite → perovskite) and calorimetric study of quenched lithium niobate phases enables differentiation between stable phases (perovskite) and metastable quench products (lithium niobate) and the determination of thermodynamic properties for both.

Though in retrospect it is quite clear the lithium niobate form is a quench phase, this realization came slowly over a period of twenty years. The two crucial pieces of evidence were, first, the in situ observation of a transition at room temperature in the diamond anvil cell to a perovskite at a pressure much *lower* than that for the ilmenite-lithium niobate transition seen on quenching in multianvil runs (Ross et al. 1989, Leinenweber et al. 1991), and, second, the measurement of ΔH (ilmenite = lithium niobate) by calorimetry of $MnTiO_3$ (Ko et al. 1989) and $FeTiO_3$ (Mehta et al. 1994, Linton et al. 1998) which yielded values that could not be reconciled with the P-T slope observed from quench experiments. The first observation indicated that the lithium niobate phase was metastable, while the second observation strongly suggested that the high pressure phase defining the P-T slope observed in the multianvil quench experiments was not the lithium niobate form. These combined experiments, using different techniques, yielded a picture that would have been impossible to get from one set of data alone.

Dense hydrous magnesium silicates

Navrotsky and Bose (1995) presented a scenario that demonstrated that the interior of an older, slower, and therefore colder, subducting slab, can maintain pressure-temperature (P-T) regimes that provide an equilibrium pathway for descent of water to at least 600 km, without the need to consider kinetic overstepping of dehydration boundaries. However, the relevant phase diagram was based on estimated free energies, assumed equations of state, and preliminary calorimetric data. A recent paper (Bose and Navrotsky, 1998) presents calorimetric measurements of enthalpies of formation of well-characterized, synthetic samples of phase A [$Mg_7Si_2O_8(OH)_6$], talc [$Mg_3Si_4O_{10}(OH)_2$], and natural antigorite [$Mg_{48}Si_{34}O_{85}(OH)_{62}$], reversed phase equilibria on key reactions to 8 GPa, and a thermodynamic data set utilizing recent equations of state for phase A, talc, and antigorite that is consistent with the experimental results.

Table 6 lists the calorimetric results. Error bars represent two standard deviations of the mean. The enthalpy of formation of talc, from the oxides obtained from a thermochemical cycle, is -164.7 ± 2.3 kJ mol^{-1}. This corresponds to a standard heat of formation from the elements of -5897.9 ± 3.0 kJ mol^{-1}. The agreement with published values (kJ mol^{-1}), -5897.4 (Berman 1988), -5895.2 ± 3.7 (Holland and Powell 1990), -5900 ± 2 (Hemingway 1991), -5900.5 (Saxena et al. 1993); -5897.3 (Bose and Ganguly 1995a), is excellent.

Table 6. Thermochemical data for dense, high pressure magnesium silicates.

Phase	ΔH_{ds}*		$\Delta H_{oxides,\ 298}$	$\Delta H_{elem,\ 298}$
Talc	554 ± 0.5	(12)	-164.7	-5897.2
Phase A	831.0 ± 2.2	(5)	-225.3	-7114.6
Antigorite (A)[†]	8118.5 ± 69.2	(7)	-2246.4	-70942
Antigorite (B)[†]	8081.2 ± 48.7	(6)	-2209.1	-70905

Note: Units are in kJ mol^{-1}, all data from Bose & Navrotsky (1998).
 * Measured enthalpy of drop solution at 771°C in $2PbO \cdot B_2O_3$ solvent.
 Reported uncertainties are two standard deviations of the mean.
 Value in parentheses is number of experiments performed.
 † A and B represent two different samples of slightly different iron contents.

The enthalpy of formation of phase A is -225.3±2.2 kJ mol[-1] from the oxides or -7114.6±3.0 kJ mol[-1] from the elements. This had not been measured or estimated previously. For antigorite, the heat of formation from the oxides is -2246.4±0.9 kJ mol[-1] (sample 94NZ62), or, -2209.2±0.6 kJ mol[-1] (sample A3-JVC). These correspond to an enthalpy of formation from the elements of -70942±2 and -70905±2 kJ mol[-1] respectively. Optimized thermodynamic databases of Berman (1988) and Saxena et al. (1993) report -71364 and -71377 kJ mol[-1], respectively.

Results of the phase equilibria experiments are shown as points in Figure 7. The volume as a function of P and T was determined by indexing the diffraction lines to a single layer monoclinic unit cell for antigorite, and to a single layer pseudo-monoclinic unit cell for talc. A high-temperature Birch-Murnaghan equation of state (Wang et al. 1994) was used to fit the data and to determine the thermoelastic parameters: (K_T = bulk modulus, K' = its pressure derivative, _ = thermal expansivity) antigorite: $K_0 = 49.6±0.7$ GPa, $K' = 6.14±0.43$, $\alpha = 2.927×10^{-5}±0.073×10^{-5} K^{-1}$; talc: $K_0 = 35.5±0.3$ GPa, $K' = 12.1±0.4$, $\alpha = 2.16×10^{-5}±0.3×10^{-5} K^{-1}$. The in situ volumetric data for antigorite are shown in Figure 8.

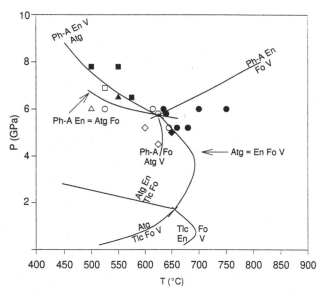

Figure 7. Experimental data of key reactions studied in the system MgO-SiO$_2$-H$_2$O (Bose and Navrotsky 1998). The open symbols represent stability of low P, low T assemblages, while the filled symbols represent stability of high P, high T assemblages. Squares represent Atg = Ph-A + En + V; circles represent Atg = En + Fo + V; triangles represent Atg + Fo = Ph-A + En; and diamonds represent Ph-A + Atg = Fo + V. The univariant reaction curves are computed using the self-consistent thermodynamic data set obtained by Bose and Navrotsky (1998). Reaction: Ph-A + En = Fo + V (Reaction 5) is from Luth (1995). Talc bearing reactions are from Bose and Ganguly (1995b). Atg = antigorite, En = enstatite, Fo = forsterite, Ph-A = phase A, Tlc = talc, V = vapor.

The new thermochemical data and equations of state were used to construct the phase diagram shown by solid curves in Figure 7. Thermodynamic properties for forsterite and enstatite are from Saxena et al. (1993), equation of state of phase A from Pawley et al. (1995), and for water from Belonoshko et al. (1992). All other date are from the work of Bose and Navrotsky (1998). The calculated positions of reactions all satisfy the phase equilibria, indicating internal consistency.

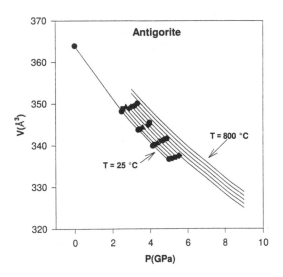

Figure 8. Variation of unit cell volume of antigorite as a function of pressure and temperature (Bose and Navrotsky 1998). Circles represent data at 25°C. The data above 25°C are isotherms from 100°C to 600°C at 100°C increments. The curves represent a fit to a high-temperature Birch-Murnaghan equation. The model fits are of correct functional form even when extrapolated beyond the experimental range; the calculated isotherms do not cross. Proper consideration of these equations of state is essential to the calculation of phase equilibria shown in Figure 7.

The maximum P, T stability of antigorite is at 5.8 GPa, 620°C. This is close in temperature, but about 0.6 GPa higher than that calculated by Pawley and Wood (1996). The calculated diagram clearly indicates that P-T paths exist during subduction of a cold slab that can lead to retention of a substantial amount of water in crystalline phases. Figure 7 also shows that small changes in P-T conditions can lead to very different sequences of phases and significantly change the fate of subducted water. The closely balanced thermodynamic properties of the various phase assemblages reflect not only small enthalpy and entropy values for pertinent reactions but also the very important role of the volumetric term. This stresses the importance of accurate equation of state data for both the low pressure layer silicates (talc and antigorite) and for the dense hydrous magnesium silicate (DHMS) phases, as well as for water itself. At high P and T (above 3 GPa and 500°C) the P-V-T properties of layer silicates are crucial to defining their stability fields. The appropriate formulation of the equation of state for such highly anisotropic materials remains a challenge from both an experimental and a theoretical point of view.

DIRECTIONS FOR FUTURE WORK

To obtain better thermodynamic data for high pressure phases, a number of technological issues must be addressed. Pressure calibration remains critical. The synthesis and equilibration of clean samples under well defined P-T conditions, with better controlled oxygen fugacities and water fugacities are essential. The equation of state of H_2O above 5 GPa needs better definition. P-V-T relations for complex and strongly anisotropic materials, such as layer silicates, need to be measured in situ and understood from a fundamental point of view. Samples of high purity need to be made for calorimetric studies and those studies must be conducted carefully.

For the MgO-FeO-SiO_2 system, the thermochemical and phase equilibrium data base has advanced considerably. The effect of other elements, especially Ca and Al, needs better thermodynamic definition. Thermodynamic descriptions of trace element incorporation are at present sketchy and largely empirical.

Silicate melts, including those containing volatiles (H_2O and CO_2) need much more systematic thermodynamic study at high P and T. Accurate phase equilibria, their rigorous thermodynamic analysis, and accurate measurement of melt densities at high P and T remain very worthwhile, albeit challenging, endeavors. A rigorous, accurate, and microscopically meaningful thermodynamic description of melts at high pressure is still a long way off.

CONCLUSIONS

At present, the combination of diamond cell, multianvil, and calorimetric techniques offer powerful tools for obtaining thermochemical properties of high pressure phases. Limitations still arise from a variety of sources: P-T calibration, attainment of equilibrium, identification of metastable quench phases, and the phase purity needed for solution calorimetric studies. The equation of state data are crucial for thermodynamic calculations. The combination of several experimental methodologies, followed by a rigorous modeling procedure, is far more powerful than the application of each technique separately.

ACKNOWLEDGMENTS

Much of the recent work presented here would not have been possible, both in terms of technique development and time-consuming measurements, without the support of CHiPR, the Center for High Pressure Research, and the NSF Science and Technology Center. I thank K.D. Grevel, R.J. Hemley, and M.C. Wilding for reviews of the manuscript.

REFERENCES

Akaogi M, Ito E (1993) Refinement of enthalpy measurement of MgSiO₃ perovskite and negative pressure-temperature slopes for perovskite-forming reactions. Geophys Res Let 20:1839-1842
Akaogi M, Ito E, Navrotsky A (1989) Olivine-modified spinel-spinel transitions in the system Mg₂SiO₄-Fe₂SiO₄: calorimetric measurements, thermochemical calculations, and geophysical application. J Geophys Res 94:15671-15685
Akaogi M, Navrotsky A (1984a) Calorimetric study of the stability of spinelloids in the system NiAl₂O₄-Ni₂SiO₄. Phys Chem Mineral 10:166-172
Akaogi M, Navrotsky A (1984b) The quartz-coesite-stishovite transformations: new calorimetric measurements and calculation of phase diagrams. Phys Earth Planet Inter 56:124-134
Akaogi M, Navrotsky A (1985) Calorimetric study of high pressure polymorphs of MnSiO₃. Phys Chem Mineral 12:317-323
Akaogi M, Navrotsky A (1987) Calorimetric study of high pressure phase transitions among the CdGeO₃ polymorphs (pyroxenoid, garnet, ilmenite, and perovskite structures). Phys Chem Mineral 14:435-440
Akaogi M, Navrotsky A, Yagi T, Akimoto S (1987) Pyroxene-garnet transformation: thermochemistry and elasticity of garnet solid solutions, and application to mantle models. *In:* M Manghnani, Y Syono (eds) High Pressure Research in Mineral Physics. p 251-260, American Geophysical Union, Washington, DC
Akaogi M, Ross NL, McMillan P, Navrotsky A (1984) The Mg₂SiO₄ polymorphs (olivine, modified spinel, and spinel)—thermodynamic properties from oxide melt calorimetry, phase relations, and models of lattice vibrations. Am Mineral 69:499-512
Akaogi M, Yusa H, Ito E, Yagi T, Suito K, Iiyama JT (1990) The ZnSiO₃ clinopyroxene-ilmenite transition: heat capacity, enthalpy of transition, and phase equilibria. Phys Chem Mineral 17:17-23
Akaogi M, Yusa H, Shiraishi K, Suzuki T (1995) Thermodynamic properties of α-quartz, coesite, and stishovite and equilibrium phase relations at high pressures and high temperatures. J Geophys Res 100:22337-22347
Akella J (1979) Quartz-coesite transition and the comparative friction measurements in the piston-cylinder apparatus using talc-alsimag-glass (TAG) and NaCl high pressure cells. N Jahrb Mineral Monatshefte 5:217-224
Akimoto S, Syono Y (1969) Coesite-stishovite transition. J Geophys Res 74:1653-1659
Anderson PAM, Newton RC, Kleppa OJ (1977) The enthalpy change of the andalusite-sillimanite reaction and the Al₂SiO₅ diagram. Am J Science 277:585-593

Angel RJ (1997) Transformation of fivefold-coordinated silicon to octahedral silicon in calcium silicate, $CaSi_2O_5$. Am Mineral 82:836-839

Angel RJ, Ross NL, Seifert F, Fliervoet TF (1996) Structural characterization of pentacoordinate silicon in a calcium silicate. Nature 384:441-444

Ashida T, Kume S, Ito E, Navrotsky A (1988) $MgSiO_3$ ilmenite: heat capacity, thermal expansivity, and enthalpy of transformation. Phys Chem Mineral 16:239-245

Belonoshko AB (1994) Molecular dynamics of silica at high pressures-equation of state, structure, and phase transitions. Geochim Cosmochim Acta 58:1557-1566

Belonoshko A, Shi PF, Saxena SK (1992) SUPERFLUID: a FORTRAN-77 program for the calculation of Gibbs free energy and the volumes of C-H-O-N-S-Ar mixtures. Compu Geosci 18:1267-1269

Berman RG (1988) Internally consistent thermodynamic data for minerals in the system Na_2O-K_2O-CaO-MgO-FeO-Fe_2O_3-Al_2O_3-SiO_2-TiO_2-H_2O-CO_2. J Petrol 29:445-552

Bina CR, Helffrich G (1994) Phase transition Clapeyron slopes and transition zone seismic discontinuity topography. J Geophys Res 99:15853-15860

Bohlen SR, Boettcher AL (1982) The quartz-coesite transformation: a precise determination and the effects of other components. J Geophys Res 87:7073-7078

Bose K, Ganguly J (1995a) Experimental and theoretical studies of stabilities of talc, antigorite and phase A at high pressures with applications to subduction processes. Earth Planet Sci Let 136:109-121

Bose K, Ganguly J (1995b) Quartz-coesite transition revisited: reversed experimental determination at 500-1200°C and retrieved thermodynamic properties. Am Mineral 80:231-238

Bose K, Navrotsky A (1998) Thermochemistry and phase equilibria of hydrous phases in the system MgO-SiO_2-H_2O: implications for volatile transport to the mantle. J Geophys Res 103:9713-9719

Boyd FR, England JL (1960) The quartz-coesite transition. J Geophys Res 65:749-756

Brousse C, Newton RC, Kleppa OJ (1984) Enthalpy of formation of forsterite, enstatite, akermanite, monticellite and merwinite at 1073 K determined by alkali borate solution calorimetry. Geochim Cosmochim Acta 48:1081-1088

Callanan JE, Sullivan SA (1986) Development of standard operating procedures for differential scanning calorimeters. Rev Sci Instrum 57:2584-2592

Charlu TV, Newton RC, Kleppa OJ (1978) Enthalpy of formation of some lime silicates by high-temperature solution calorimetry, with discussion of high pressure phase equilibria. Geochim Cosmochim Acta 42:367-375

Christensen UR, Yuen DA (1984) The interaction of a subducting lithospheric slab with a chemical or phase boundary. J Geophys Res 89:4389-4402

Coes L (1953) A new dense crystalline silica. Science 118:131-132

Cohen LH, Klement Jr WK (1967) High-low quartz inversion: determination to 35 kilobars. J Geophys Res 73:4245-4251

Dandekar DP, Jamieson JC (1969) Some high pressure phases of RX_2 fluorides. *In:* DB McWhan (ed) High Pressure Symposium Volume, Am Cryst Assoc

Dubrovinskaya NA, Dubrovinsky LS, Saxena SK (1997) Systematics of thermodynamic data on solids: thermochemical and pressure-volume-temperature properties of some minerals. Geochim Cosmochim Acta 61:4151-4158

Dubrovinsky LS, Saxena SK, Lazor P, Ahuja R, Eriksson O, Wills JM, Johansson B (1997) Experimental and theoretical identification of a new high-pressure phase of silica. Nature 388:362-365

Fabrichnaya OB (1995) Thermodynamic data for phases in the FeO-MgO-SiO_2 system and phase relations in the mantle transition zone. Phys Chem Mineral 22:323-332

Fabrichnaya OB, Kuskov OL (1994) Constitution of the Moon. 1. Assessment of thermodynamic properties and reliability of phase relation calculations in the FeO-MgO-$Al_2O_3$$SiO_2$ system. Phys Earth Planet Inter 83:175-196

Fei Y, Mao HK, Mysen BO (1991) Experimental determination of element partitioning and calculations in the FeO-MgO-Al_2O_3-SiO_2 system at high pressure and high temperature. J Geophys Res 96:2157-2169

Fei Y, Saxena S (1986) Thermochemical data for phase equilibria in the system Fe-Mg-Si-O at high pressure and temperature. Phys Chem Mineral 13:311-324

Fei Y, Saxena SK, Navrotsky A (1990) Internally consistent thermodynamic data and equilibrium phase relations for compounds in the system MgO-SiO_2 at high pressure and high temperature. J Geophys Res 95:6915-6928

Geisinger KL, Ross NL, McMillan P, Navrotsky A (1987) $K_2Si_4O_9$: energetics and vibrational spectra of glass, sheet silicate, and wadeite phases. Am Mineral 72:984-994

Ghose S, Choudhury N, Chaplot SL, Rao KR (1992) Phonon density of states and thermodynamic properties of minerals. *In:* SK Saxena (ed) Thermodynamic Data: Systematics and Estimations. Advances in Physical Geochemistry 10:283-314, Springer-Verlag, New York

Heinz DL, Sweeney JS, Miller P (1991) A laser heating system that stabilizes and controls the temperature: diamond anvil cell applications. Rev Sci Instrum 62:1568-1575

Helffrich G, Bina CR (1994) Frequency dependence of the visibility and depths of mantle seismic discontinuities. Geophys Res Let 21:2613-2616

Hemingway BS (1991) Thermodynamic properties of anthophyllite and talc: Corrections and discussion of calorimetric data. Am Mineral 76:1589-1596

Hemingway BS, Bohlen SR, Hankins WB, Westrum Jr EF, Kuskov OL (1998) Heat capacity and thermodynamic properties for coesite and jadeite, reexamination of the quartz-coesite equilibrium boundary. Am Mineral 83:409-418

Hemley RJ, Cohen RE (1992) Silicate perovskite. Ann Rev Earth Planet Sci 20:553-600

Hemley RJ, Mao, H-K (1997) Solids, static high-pressure effects in. Encycl Appl Phys 18:555-572

Hemley RJ, Prewitt CT, Kingma KJ (1994) High-pressure behavior of silica. In: PJ Heaney, CT Prewitt, GV Gibbs (eds) Silica: Physical Behavior, Geochemistry and Materials Applications. Rev Mineral 29:41-81

Holland TJB, Powell R (1990) An enlarged and updated internally consistent thermodynamic dataset with uncertainties and correlations: The system $K_2O-Na_2O-CaO-MgO-FeO-Fe_2O_3-Al_2O_3-TiO_2-SiO_2-C-H_2-O_2$. J Metamorph Geol 8:89-124

Holm JL, Kleppa OJ, Westrum EF (1967) Thermodynamics of polymorphic transformation in silica Thermal properties from 5 to 1070°K and pressure-temperature stability fields from coesite and stishovite. Geochim Cosmochim Acta 31:2289-2307

Ito E (1984) Ultra-high pressure phase relations of the system $MgO-FeO-SiO_2$ and their geophysical implications. In: Sunagawa I (ed) Materials Science of the Earth's Interior p 387-394, Terra Scientific, Tokyo

Ito E, Akaogi M, Topor L, Navrotsky A (1990) Negative P-T slopes for reactions forming $MgSiO_3$ perovskite confirmed by calorimetry. Science 249:1275-1278

Ito E, Matsui Y (1979) High-pressure transformations in silicates, germanates, and titanates with ABO_3 stoichiometry. Phys Chem Mineral 4:265-273

Ito E, Navrotsky A (1985) $MgSiO_3$ ilmenite: calorimetry, phase equilibria, and decomposition at atmospheric pressure. Am Mineral 70:1020-1026

Jamieson JC (1970) The phase behavior of simple compounds. Phys Earth Planet Inter 3:201-203

Jeanloz R, Kavner A (1996) Melting criteria and imaging spectroradiometry in laser-heated diamond-cell experiments. Phil Trans Roy Soc London 354:1279-1305

Jeanloz R, Thompson A (1983) Phase transitions and mantle discontinuities. Rev Geophys Space Phys 21:51-74

Johannes W, Bell M, Mao HK, Boettcher AL, Chipman DW, Hays JF, Newton RD, Seifert F (1971) An interlaboratory comparison of piston-cylinder pressure calibration using the albite-breakdown reaction. Contrib Mineral Petrol 32:24-38

Kieffer SW (1979) Thermodynamics and lattice vibrations of minerals: 2. Vibrational characteristics of silicates. Rev Geophys Space Phys 17:20-34

Kiseleva IA, Ogorodova LP, Topor ND, Chigareva OG (1979) A thermochemical study of the $CaO-MgO-SiO_2$ system. Geokhimiya 1811-1825

Kiseleva IA, Ostapenko GT, Ogorodova LP, Topor ND, Timoshkova LP (1983) Equilibrium between andalusite, kyanite, sillimanite and mullite (according to high temperature calorimetry data). Geokhimiya 9:1247-1256

Kitahara S, Kennedy GC (1964) The quartz-coesite transition. J Geophys Res 69:5395-5400

Ko J, Brown NE, Navrotsky A, Prewitt CT, Gasparik T (1989) Phase equilibrium and calorimetric study of transition of $MnTiO_3$ from the ilmenite to the lithium niobate structure and implications for the stability field of perovskite. Phys Chem Mineral 16:727-733

Ko J, Prewitt CT (1988) High-pressure phase transition in $MnTiO_3$ from the ilmenite to the $LiNbO_3$ structure. Phys Chem Mineral 15:355-362

Kojitani H, Akaogi M (1994) Calorimetric study of olivine solid solutions in the system $Mg_2SiO_4-Fe_2SiO_4$. Phys Chem Mineral 20:536-540

Kojitani H, Akaogi M (1995) Measurement of heat of fusion of model basalt in the system diopside-forsterite-anorthite. Geophys Res Let 22:2329-2332

Kosyakov AV, Ishbulatov RA (1984) Quartz-coesite transformation. In: OK Kuskov, AM Bychkov (eds) New Experimental Works in the Field of Geochemistry of Endogenic Processes. Geokhimiya no. 12 1926 (in Russian)

Kudoh Y, Kanzaki M (1998) Crystal chemical characteristics of α-$CaSi_2O_5$, a new high pressure calcium silicate with five-coordinated silicon synthesized at 1500°C and 10 GPa. Phys Chem Mineral 25:429-433

Kume S, Ohtaka O, Yamanaka T, Navrotsky A (1989) High pressure polymorphs in ceramics and minerals —GeO_2 and ZrO_2. *In:* MS Whittingham, S Bernasek, AJ Jacobson, A Navrotsky (eds) Reactivity of Solids p 285-287, North Holland, Amsterdam

Kuskov OL, Zhidikova AP, Semenov, YV, Babich, YV, Fabrichnaya OB (1991) Thermodynamics of silica polymorphism. Geokhimiya 8:1175-1185 (Geochem Int'l 1992:93-102)

Leinenweber K, McMillan PF, Navrotsky A (1989) Transition enthalpies and entropies of high pressure zinc metasilicates and zinc metagermanates. Phys Chem Mineral 16:799-808

Leinenweber K, Navrotsky A, (1989) Thermochemistry of phases in the system $MgGa_2O_4$-Mg_2GeO_4. Phys Chem Mineral 16:497-502

Leinenweber K, Schuelke U, Ekbundit S, McMillan PF (1998) Cryogenic recovery of unquenchable high-pressure samples using a multianvil device. *In:* M Manghnani, T Yagi (eds) Properties of Earth and Planetary Materials at High Pressure and Temperature. Geophysical Monograph Series 101:97-103, American Geophysical Union, Washington, DC

Leinenweber K, Utsumi W, Tsuchida Y, Yagi T, Kurita K (1991) Unquenchable high-pressure perovskite polymorphs of $MnSnO_3$ and $FeTiO_3$. Phys Chem Mineral 18:244-250

Leinenweber K, Wang Y, Yagi T, Hitoshi Y (1994) An unquenchable perovskite phase of $MgGeO_3$ and comparison with $MgSiO_3$ perovskite. Am Mineral 79:197-199

Linton JA, Fei Y, Navrotsky A (1997) Complete Fe-Mg solid solution in lithium niobate and perovskite structures in titanates at high pressures and temperature. Am Mineral 82:639-642

Linton JT, Fei Y, Navrotsky A (1998, in press) The $MgTiO_3$-$FeTiO_3$ join at high pressure and temperature. Am Mineral

Liu J, Topor L, Zhang J, Navrotsky A, Liebermann RC (1996) Calorimetric study of the coesite-stishovite transformation and calculation of the phase boundary. Phys Chem Mineral 23:11-16

Liu L-G (1982) High-pressure phase transformations of the dioxides: implications for structure of SiO_2 at high pressures. *In:* S Akimoto, MH Manghnani (eds) High-Pressure Research in Geophysics p 349-360, Center for Academic Publications, Tokyo

Luth RW (1993) Measurement and control of intensive parameters in experiments at high pressure in solid-media apparatus. *In:* RW Luth (ed) MAC Short Course on Experiments at High Pressure and Applications to the Earth's Mantle, p 15-37, Mineralogical Assoc Canada, Edmonton, Alberta

Luth RW (1995) Is phase A relevant to the Earth's mantle? Geochim Cosmochim Acta 59:679-682

McKenzie D, O'Nions RK (1988) The volume and composition of melt generated by extension of the lithosphere. J Petrol 32:1021-1091

Manga M, Jeanloz R (1998) Temperature distribution in the laser-heated diamond cell. *In:* M Manghnani, T Yagi (eds) Properties of Earth and Planetary Materials at High Pressure and Temperature. Geophysical Monograph Series 101:17-25, American Geophysical Union, Washington, DC

Mao H-K, Shen G, Hemley RJ, Duffy TS (1998) X ray diffraction with a double hot-plate laser-heated diamond cell. *In:* M Manghnani, T Yagi (eds) Properties of Earth and Planetary Materials at High Pressure and Temperature. Geophysical Monograph Series 101:27-34, American Geophysical Union, Washington, DC

Mehta A, Leinenweber K, Navrotsky A (1994) Calorimetric study of high pressure polymorphism in $FeTiO_3$: stability of the perovskite phase. Phys Chem Mineral 21:207-212

Meng Y, Weidner DJ, Fei Y (1993) Deviatoric stress in a quasi-hydrostatic diamond anvil cell: effect on the volume-based pressure calibration. Geophys Res Let 20:1147-1150

Ming LC, Manghnani MH (1982) High-pressure phase transformations in rutile-structured dioxides. *In:* S Akimoto, MH Manghnani (eds) High-Pressure Research in Geophysics, p 329-347, Center for Academic Publications, Tokyo

Mirwald PW, Massonne H-J (1980) The low-high quartz and quartz-coesite transition to 40 kbar between 600° and 1600°C and some reconnaissance data on the effect of $NaAlO_2$ component on the low quartz-coesite transition. J Geophys Res 85:6983-6990

Mosenfelder JL, Bohlen SR, Hankins WB (1996) "The quartz-coesite transition revisited" revisited. Geol Soc Am Abstracts with Programs Oct 28-31:A159-A160

Nagel L, O'Keeffe M (1971) Pressure and stress induced polymorphism of compounds with rutile structure. Mat Res Bull 6:1317-1320

Navrotsky A (1987) Models of crystalline solutions. *In:* HP Eugster, ISE Carmichael (eds) Thermodynamic Modeling of Geologic Systems: Minerals, Fluids, and Melts. Rev Mineral 17:35-69

Navrotsky A (1997) Progress and new directions in high temperature calorimetry revisited. Phys Chem Mineral 24:222-241

Navrotsky A (1998, in press) Energetics and crystal chemical systematics among ilmenite, lithium niobate, and perovskite structures. Chem Mater

Navrotsky A, Bose K (1995) Thermodynamic stability of hydrous silicates: Some observations and implications for water in the Earth, Venus and Mars. *In:* Volatiles in the Earth and Solar System. Conf Proc 341:221-228, AIP Press, New York

Navrotsky A, Kasper RB (1976) Spinel disproportiation at high pressure: calorimetric determination of enthalpy of formation of Mg_2SnO_4 and Co_2SnO_4 and some implications for silicates. Earth Planet Sci Let 31:247-255

Navrotsky A, Pintchovski FS, Akimoto S (1979) Calorimetric study of the stability of high pressure phases in the systems $CoO-SiO_2$ and $FeO-SiO_2$ and calculations of phase diagrams in $MO-SiO_2$ systems. Phys Earth Planet Inter 19:275-292

Neil JM, Navrotsky A, Kleppa OJ (1971) The enthalpy of the ilmenite-perovskite transformation in cadmium titanate. Inorg Chem 10:2076-2077

Pawley A, Wood BJ (1996) The low pressure stability of phase A, $Mg_7Si_2O_8(OH)_6$. Contrib Mineral Petrol 124:90-97

Pawley A, Redfern SAT, Wood BJ (1995) Thermal expansivities and compressibilities of hydrous phases in the system $MgO-SiO_2-H_2O$: talc, phase A and the 10-Å phase. Contrib Mineral Petrol 122:301-307

Richet P, Ingrin J, Mysen BO, Courtial P, Gillet P (1994) Premelting effects in minerals: an experimental study. Earth Planet Sci Let 121:589-600

Ringwood AE (1970) Phase transformations and the constitution of the mantle. Phys Earth Planet Inter 3:109-155

Robie RA, Hemingway BS (1972) Calorimeters for heat of solution and low-temperature heat capacity measurements. Geological Survey Professional Paper 755, US Govt Printing Office, Washington, DC

Robie RA, Hemingway BS (1995) Thermodynamic properties of minerals and related substances at 298.15 K and 1 bar (10^5 Pascals) pressure and at higher temperatures. US Geological Survey Bulletin 2131, US Govt Printing Office, Washington, DC

Robinson Jr GR, Haas Jr JL (1983) Heat capacity, relative enthalpy, and calorimetric entropy of silicate minerals: an empirical method of prediction. Am Mineral 68:541-553

Ross NL, Akaogi A, Navrotsky A, Susaki J, McMillan P (1986) Phase transitions among the $CaGeO_3$ polymorphs (wollastonite, garnet, and perovskite structures): studies by high pressure synthesis, high temperature calorimetry, and vibrational spectroscopy and calculation. J Geophys Res (Jamieson Memorial Vol) 91:4685-4698

Ross NL, Ko J, Prewitt CT (1989) A new phase transition in $MnTiO_3$:$LiNbO_3$-perovskite structure. Phys Chem Mineral 16:621-629

Ross NL, Navrotsky A (1987) The Mg_2GeO_4 olivine-spinel phase transition. Phys Chem Mineral 14:473-481

Ross NL, Navrotsky A (1988) Study of the $MgGeO_3$ polymorphs (orthopyroxene, clinopyroxene, and ilmenite structures) by calorimetry, spectroscopy, and phase equilibria. Am Mineral 73:1355-1365

Saxena SK (1996) Earth mineralogical model: Gibbs free energy minimization computation in the system $MgO-FeO-SiO_2$. Geochim Cosmochim Acta 60:2379-2395

Saxena SK, Chatterjee N, Fei Y, Shen G (1993) Thermodynamic Data on Oxides and Silicates: An Assessed Data Set Based on Thermochemistry and High Pressure Phase Equilibrium. Springer-Verlag, New York

Shen G, Mao HK, Hemley RJ (1996) Laser-heated diamond anvil cell technique: double-sided heating with multimode Nd:YAG laser. *In:* Advanced Materials '96, Proceedings of the 3rd NIRIM Int'l Symp on Advanced Materials, p 149-152 National Institute for Research in Inorganic Materials, Tsukuba, Japan

Swamy V, Saxena SK, Sundman B, Zhang J (1994) A thermodynamic assessment of silica phase diagram. J Geophys Res 99:11787-11794

Syono Y, Akimoto S (1968) High-pressure synthesis of fluorite-type lead dioxide. Mat Sci Res Bull 3:153-157

Syono Y, Yamauchi H, Ito A, Someya Y, Ito E, Matsui Y, Akaogi M, Akimoto S (1980) Magnetic properties of the disordered ilmenite $FeTiO_3$ II synthesized at very high pressure. *In:* H Watanabe, S Iida, Y Sugimoto (eds) Ferrites: Proc Int'l Conference, p 192-195 Tokyo, Japan

Takayama-Muromachi E, Navrotsky A (1988) Energetics of Compounds ($A^{3+}B^{4+}O_3$) with the perovskite structure. J Solid State Chem 72:244-256

Tackley PJ, Stevenson DJ, Glatzmaier GA, Schubert G (1993) Effects of an endothermic phase transition at 670 km depth in a spherical model of convection in the Earth's mantle. Nature 361:699-704

Wang Y, Getting IC, Weidner DJ, Vaughan MT (1998) Performance of tapered anvils in a DIA-type, cubic-anvil, high-pressure apparatus for X ray diffraction studies. *In:* M Manghnani, T Yagi (eds) Properties of Earth and Planetary Materials at High Pressure and Temperature. Geophysical Monograph Series 101:35-39, American Geophysical Union, Washington, DC

Wang Y, Weidner DJ, Liebermann RC, Zhao Y (1994) P-V-T equation of state of (Mg, Fe)SiO_3 perovskite: Constraints on the composition of the lower mantle. Phys Earth Planet Inter 83

Weidner DJ, Vaughan MT, Ko J, Wang Y, Liu X, Yeganeh-Haeri A, Pacalo RE, Zhao Y (1992) Characterization of stress, pressure, and temperature in SAM85, a DIA type high pressure apparatus. *In:* Syono Y, Manghnani (eds) High-Pressure Research: Applications to Earth and Planetary Sciences p 13-17, American Geophysical Union, Washington, DC

Weidner DJ, Wang Y (1998) Chemical- and Clapeyron-induced buoyancy at the 660 km discontinuity. J Geophys Res 103:7431-7441

Wentzcovitch RM, Ross NL, Price GD (1995) Ab initio study of $MgSiO_3$ and $CaSiO_3$ perovskites at lower-mantle pressures. Phys Earth Planet Inter 90:101-112

Westrum Jr EF, Komada N (1986) Progress in modeling heat capacity versus temperature morphology. Thermochim Acta 109:11-28

Wolf G, Bukowinski MST (1987) Theoretical study of the structural and thermodynamic properties of $MgSiO_3$ and $CaSiO_3$ perovskites: implications for lower mantle composition. *In:* MH Manghnani, Y Syono (eds) High Pressure Research in Mineral Physics. p 313-331, American Geophysical Union, Washington, DC

Wood BJ (1990) Postspinel transformations and the width of the 670-km discontinuity: a comment on "Postspinel transformations in the system Mg_2SiO_4-Fe_2SiO_4 and some geophysical implications" by E Ito and E Takahashi. J Geophys Res 95:12681-12685

Yuen DA, Cadek O, Van Keken P, Reuteler DM, Kyvalova H, Schroeder BA (1996) Combined results from mineral physics, tomography and mantle convection and their implications on global geodynamics. *In:* E Boschi, G Ekström, A Morelli (eds) Seismic Modelling of Earth Structure. p 463-505, Instituto Nazionale di Geofisica

Yusa H, Akaogi M, Ito E (1993) Calorimetric study of $MgSiO_3$ garnet and pyroxene: heat capacities, transition enthalpies and equilibrium phase relations in $MgSiO_3$ at high pressures and temperatures. J Geophys Res 98:6453-6460

Zhang J, Li B, Utsumi W, Liebermann RC (1996) In situ x-ray observations of the coesite-stishovite transition: reversed phase boundary and kinetics. Phys Chem Mineral 23:1-10

Zhidikova AP, Semenov YV, Babich YV (1988) Thermodynamics of polymorphic silica modifications according to calorimetric data. Terra Cognita 8:187-196

Chapter 11

SOLID SOLUTIONS AND ELEMENT PARTITIONING AT HIGH PRESSURES AND TEMPERATURES

Yingwei Fei

Geophysical Laboratory and Center for High Pressure Research
Carnegie Institution of Washington
5251 Broad Branch Road, N.W.
Washington, DC 20015

INTRODUCTION

Element partitioning data at high pressures and temperatures are crucial for understanding geological processes in Earth and planetary interiors and evolution of the planet. Experimental studies of major and minor element partitioning between solid phases, between solid and liquid, and between metallic and silicate liquids have provided a key link between the observed chemical signatures in natural rocks and the processes that produces these signatures. For example, experimentally determined element partitioning data have been used to estimate the pressure-temperature history of mantle-derived rocks (e.g. Boyd 1973, MacGregor 1974, Finnerty and Boyd 1987) and to infer the deep origin of diamond inclusions (e.g. Kesson and Fitz Gerald 1991, Harte et al. 1998). The homogeneous accretion model for the Earth has been tested based on observed siderophile element abundances in the upper mantle and experimentally determined metal-silicate partition coefficients of siderophile elements (e.g. Murthy 1991, Jones and Drake 1986, Walker et al. 1993, Li and Agee 1996). Partitioning data between silicate perovskite and silicate melt have been obtained to address the role of fractionation of silicate perovskite in a completely molten Earth (Kato et al. 1988a,b; Drake et al. 1993, McFarlane et al. 1994). Mg-Fe partitioning may be directly related to the magnesium-iron fractionation in the Earth at its primitive stage and determine the magnesium-iron balance throughout the Earth (Mao et al. 1982). Systematic determination of element partitioning, therefore, is essential in understanding the formation of the core, mantle and crust, and the equilibrium state of the core-mantle boundary.

Phase relations at high pressures and temperatures in chemical systems related to the Earth's mantle are of great interest because the seismic velocity discontinuities in the mantle may be accounted for by high-pressure phase transformations such as the olivine to wadsleyite and silicate spinel to silicate perovskite plus magnesiowüstite transitions. Most mantle minerals form solid solutions by cation substitutions, particularly by the Mg^{2+}-Fe^{2+}, Al^{3+}-Fe^{3+}, and $Mg^{2+}Si^{4+}$-$2Al^{3+}$ substitutions in the MgO-FeO-Fe_2O_3-Al_2O_3-SiO_2 system. To construct phase relations in a binary or a multicomponent system by thermodynamic modeling, two sets of data, the thermodynamic data of end-member phases and the mixing properties of solid solutions, are required. The mixing properties of the solid solutions are usually non-ideal. The excess free energy of a non-ideal mixing is dependent on composition, temperature, and pressure. Experimentally, the mixing properties can be determined from solution calorimetry measurements (e.g. Wood and Kleppa 1981, Geiger et al. 1987, Akaogi et al. 1989) or derived from element partitioning data between coexisting solid solutions (e.g. O'Neill and Wall 1987, Hackler and Wood 1989, Fei et al. 1991). Theoretically, non-ideal solutions are modeled by empirical or semi-empirical formulations which describe the excess free energy as a function of composition, temperature, and pressure (e.g. Guggenheim 1937, 1952; Thompson 1967).

0275-0279/98/0037-0011$05.00

In this paper, I review the methods of obtaining solution properties and summarize existing Mg-Fe partitioning data in the system $MgO-FeO-Al_2O_3-SiO_2$ at high pressures and temperatures. I also discuss the solid solutions involving MgSi-AlAl and $Al-Fe^{3+}$ substitutions in mantle minerals and provide a brief summary of trace element partitioning between silicate melts and mantle phases and siderophile element partitioning between solid (or liquid) metallic phase and silicate liquid at high pressures and temperatures. The review is also intended to provide guidance for the design of new partitioning experiments to obtain critical partitioning data that can more effectively constrain mantle phase relations and chemical differentiation models.

SOLID SOLUTION MODELS

In order to describe phase relations thermodynamically in pressure-temperature-composition (P-T-X_i) space, we need to model the excess free energy of mixing (G^{ex}) describing non-ideal behavior of a solid solution, as a function of P, T, and X_i. For an ideal solid solution, $G^{ex} = 0$, implying the activity coefficient (γ_i) to be 1. An ideal solid solution is then one whose free energy of mixing arises only from the random mixing of one mole of species. The excess free energy for most solid solutions is non-zero and is commonly expressed as a function of composition. Various solid solution models have been proposed, as discussed by Fei et al. (1986), Ganguly and Saxena (1987), and Navrotsky (1987). There are two groups of solution models. The first group originated from the Flory-Huggins model (Flory 1953), in which solutions are considered as athermal with zero excess enthalpy of mixing, commonly applied to mixtures of polymers. The later refined versions, such as the Wilson model (Wilson 1964), the quasi-chemical (Guggenheim 1952) and the non-random two-liquid (Renon and Prausnitz 1968) models, do involve enthalpy of mixing and have been reviewed by Acree (1984).

In the second group of models, the excess free energy is simply expressed as a function of the mole fraction. Guggenheim (1937) suggested that the excess energy G^{ex} for a binary solution can be expressed as a polynomial in mole fractions, X_1 and X_2.

$$G^{ex} = X_1X_2[A_0 + A_1(X_1 - X_2) + A_2(X_1 - X_2)^2 + ...], \tag{1}$$

where $A_{i\,(i = 0, 1, 2, ...)}$ are constants. The expression is often referred to as Redlich-Kister equation (Redlich and Kister 1948). The activity coefficients using the Redlich-Kister equation are given by

$$RT\ln\gamma_1 = X_2^2 [A_0 + A_1(3X_1 - X_2) + A_2(X_1 - X_2)(5X_1 - X_2) + ...] \tag{2}$$
$$RT\ln\gamma_2 = X_1^2 [A_0 + A_1(3X_2 - X_1) + A_2(X_2 - X_1)(5X_2 - X_1) + ...]. \tag{3}$$

In the geological and metallurgical literature, a two-parameter solution model, the so-called sub-regular or Margules formulation, has been used extensively for binary solutions. Its popularity in geochemistry is due to its simplicity of formulation; it is discussed in detail by Thompson (1967). In this model, the excess free energy is given by

$$G^{ex} = X_1X_2(X_2W_{12} + X_1W_{21}), \tag{4}$$

where W_{12} and W_{21} are adjustable constants, often referred to as the interaction parameters. When $W_{12} = W_{21}$, the solution is called a symmetric solution or a simple mixture with the formulation

$$G^{ex} = WX_1X_2 \tag{5}$$

The activity coefficients in the two-parameter Margules model are given by

$$RT\ln\gamma_1 = X_2^2 [W_{12} + 2X_1(W_{21} - W_{12})] \tag{6}$$

$$RTln\gamma_2 = X_1^2 [W_{21} + 2X_2(W_{12} - W_{21})]. \tag{7}$$

The simple mixture and the two-parameter Margules formulation are special cases of the polynomial Redlich-Kister equation with $A_0 = W$ in the simple mixture model and with $A_0 = (W_{21} + W_{12})/2$ and $A_1 = (W_{21} - W_{12})/2$ in the Margules model.

The interaction parameter W_{ij} can depend on T and P if the excess entropy (S^{ex}) and the excess volume (V^{ex}) are non-zero. It may be expressed as

$$W_{ij} = W_{ij}^H - TW_{ij}^S + (P - 1)W_{ij}^V, \tag{8}$$

where W_{ij}^H, W_{ij}^S, and W_{ij}^V represent the excess enthalpy, excess entropy, and excess volume contributions, respectively, to the interaction parameter W_{ij} ($= W_{ij}^G$).

The regular solution model was originally introduced by Hildebrand (1929) to describe mixtures with a non-zero excess enthalpy, but zero excess entropy and zero excess volume ($H^{ex} \neq 0$, $S^{ex} = V^{ex} = 0$). The formulation of the excess free energy for the regular solution is the same as that for the simple mixture (Eqn. 5) except that the interaction parameter, W, is independent of temperature and pressure.

The one-parameter simple mixture model and the two-parameter Margules formulation can be easily extended to ternary or multicomponent systems (e.g. Helffrich and Wood 1989) or adapted to treat a solid solution with more than one sublattice (e.g. Sundman and Ågren 1981). Such formulations have been popular in the geological literature because solid solutions in geological problems usually involve many components.

EXPERIMENTAL TECHNIQUES

There are two independent experimental data sources available for determining mixing properties, enthalpies of solution measured by solution calorimetry (e.g. Wood and Kleppa 1981, Akaogi et al. 1989), and element partitioning between the coexisting solid solutions. Element partitioning data at high pressures and temperatures have been obtained with the piston-cylinder apparatus (up to 4 GPa), the multi-anvil apparatus (up to 27 GPa), and the diamond-anvil cell technique (up to at least 136 GPa, the core-mantle boundary pressure). The combined use of the different techniques has allowed the collection of element partitioning data over a large range of pressure. The piston-cylinder apparatus, developed in the early 1960s (Boyd and England 1960), has been extensively used for phase equilibrium measurements under crustal and upper mantle conditions. High quality partitioning data have been obtained for many rock-forming systems (e.g. Kawasaki and Matsui 1983, Harley 1984, Lee and Ganguly 1988, Hackler and Wood 1989). The high quality of the data can be partly attributed to the unique quality of the piston-cylinder apparatus, which provides well calibrated pressure and temperature determination, large sample volume, and small temperature gradients. Elaborately designed reversal and time study experiments have been also possible because of readily available starting materials and relatively long duration of the experiments.

The multi-anvil apparatus was designed to achieve pressures beyond the piston-cylinder range so that phase equilibrium measurements can be made under mantle conditions (e.g. Ito et al. 1984, Wood and Rubie 1996, Katsura and Ito 1996, Martinez et al. 1997). Some of the limitations of the multi-anvil high-pressure technique, compared to the piston-cylinder, include large temperature gradients, small sample size, shorter run duration, and lack of availability of starting materials for reversal experiments. As a consequence, most multi-anvil experiments are of the synthesis type, rather than correctly designed exchange experiments.

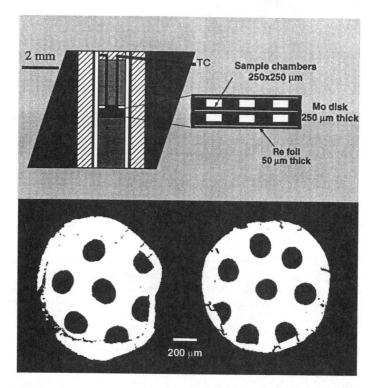

Figure 1. Experimental configuration of a multi-cell technique for ion-exchange experiments in the multi-anvil apparatus. A 10/5 assembly (Bertka and Fei 1997) is shown. Sample chamber consists of two 250-μm thick molybdenum disks separated by rhenium foils. The polished sections of the Mo disks (bottom) shows the 250-μm diameter sample cells in which different starting materials were loaded to be subjected to the same pressure and temperature conditions.

Fei and Bertka (1996) developed a multi-cell sample chamber technique for exchange experiments in the multi-anvil apparatus. Figure 1 shows the experimental configuration. The sample assembly consists of two 250-μm thick metal (molybdenum or rhenium) disks separated by rhenium foils. At least seven 250-μm diameter holes were drilled in each 1.5-mm diameter disk. Using this technique, starting materials with appropriate compositions loaded in the 250×250 μm sample cells were taken to identical high pressure and temperature conditions. A set of partitioning data over an entire compositional range of interest can be obtained in one experiment. Reversed experiments can also be designed by loading two disks of which one contains pre-synthesized high-pressure phase assemblages, as illustrated in Figure 2.

The laser-heated diamond-anvil cell technique has been used to obtain partitioning data to pressures higher than 26 GPa (e.g. Bell et al. 1979, Guyot et al. 1988, Fei et al. 1991, Kesson and Fitz Gerald 1991, Mao et al. 1997, Kesson et al. 1998). Because of short heating duration and large temperature gradients in conventional laser-heated diamond cell experiments, accurate equilibrium data for ion-exchange reactions have been typically hard to obtain. However, it is possible to obtain reliable Mg-Fe partitioning data between silicate perovskite (pv) and magnesiowüstite (mw) through the decomposition reaction, $(Mg,Fe)_2SiO_4 = (Mg,Fe)SiO_3$ (pv) + $(Mg,Fe)O$ (mw), at high pressure. In recent years, significant improvements in diamond cell laser-heating technique has been made, especially

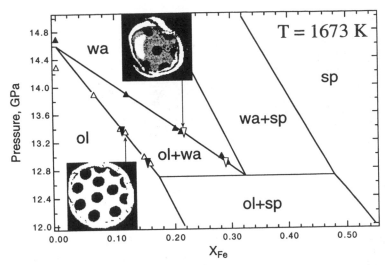

Figure 2. Phase relations at 1673 K in the system Mg_2SiO_4-Fe_2SiO_4 determined in the multi-anvil apparatus using the multi-cell technique (Fei and Bertka 1996). The open and solid triangles represent the compositions of the coexisting olivine and wadsleyite, respectively, when olivine solid solutions were used as the starting materials (bottom inset), whereas the reversed open and solid triangles represent the compositions of the coexisting wadsleyite and olivine, respectively, when pre-synthesized wadsleyite solid solutions were used as the starting materials (top inset).

by reducing temperature gradients and using a larger heating spot (30 μm) (Shen et al. 1996, Mao et al. 1997). A double-sided heating system with multimode Nd:YAG or YLF lasers can provide more uniform and constant temperatures (±50 K) in heated areas 30 to 50 μm wide to about 3500 K (Mao et al. 1997).

Mg-Fe^{2+} PARTITIONING DATA

Major element compositional models for the Earth's mantle can be effectively represented in the system MgO-FeO-Al_2O_3-SiO_2. Phase relations in this system at high pressures and temperatures have been extensively studied (e.g. Ito and Yamada 1982, Ito et al. 1984, Katsura and Ito 1989, Ito and Takahashi 1989, Fei et al. 1991, Wood and Rubie 1996). Solid solutions involving Mg-Fe^{2+} exchange include $(Mg,Fe)O$ magnesiowüstite, $(Mg,Fe)_2SiO_4$ polymorphs (olivine, wadsleyite, and spinel), $(Mg,Fe)SiO_3$ polymorphs (pyroxene, ilmenite, majorite, and perovskite), $(Mg,Fe)_3Al_2Si_3O_{12}$ garnet, and $(Mg,Fe)Al_2O_4$ aluminate spinel. Of all the Mg-Fe^{2+} solid solutions, magnesiowüstite, olivine, silicate spinel, pyroxene, garnet, and aluminate spinel form complete solid solutions and only magnesiowüstite is apparently stable over the entire mantle pressure range. Magnesiowüstite can form stable pairs of coexisting phases with $(Mg,Fe)_2SiO_4$ polymorphs (mw-ol, mw-wa, and mw-sp) and with $(Mg,Fe)SiO_3$ perovskite (mw-pv) (Fig. 3). It can therefore be used as a chemical probe (or reference) to study the Mg-Fe^{2+} distribution in the system (Fei et al. 1991). The solution properties of each solid solution may be derived from the element partitioning data by considering the distribution of an element between two solid solutions as an ion-exchange reaction. Other coexisting solid solutions in the system are olivine-wadsleyite (ol-wa), wadsleyite-spinel (wa-sp), olivine-spinel (ol-sp), ilmenite-majorite (ilm-maj), majorite-perovskite (maj-pv), ilmenite-perovskite (ilm-pv), olivine-pyroxene (ol-px), olivine-garnet (ol-gt), and garnet-pyroxene (gt-px). Table 1 summarizes the exchange reactions in the system.

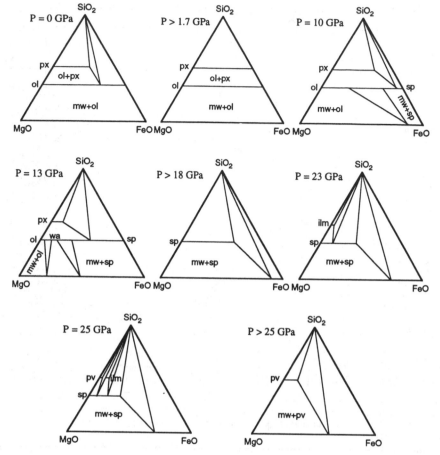

Figure 3. Phase relations and stability fields of the coexisting phases in the system MgO-FeO-SiO₂ at 1273 K and various pressures.

Magnesiowüstite-olivine (-wadsleyite, and -spinel)

Fei et al. (1991) determined the distribution of Mg and Fe between coexisting magnesiowüstite and olivine at a pressure of 2 GPa and temperatures of 1473 and 1723 K. The Mg-Fe distribution coefficient (K_D) is strongly dependent on the bulk iron content of the system. It varies from 0.28 at $X_{Fe}^{ol} = 0.05$ to 0.05 at $X_{Fe}^{ol} = 0.04$ (Fig. 4a). The distribution data also showed systematic variations with temperature; the higher K_D and the higher temperature for a given composition (Fig. 4a). Experiments were also performed at a temperature of 1723 K and pressures of 2, 4, and 9 GPa. The effect of pressure is relatively small. The general trend is that the K_D decreases with increasing pressure at constant temperature for a given composition.

(Mg,Fe)₂SiO₄ olivine transforms to its high-pressure polymorph wadsleyite (β-phase) at pressures between 12 and 15 GPa, depending on the FeO content and temperature. Wadsleyite has limited solid solution with the maximum FeO solubility in the structure to be about 26 mole%. At high FeO content, (Mg,Fe)₂SiO₄ olivine directly transforms to spinel structure. Therefore, magnesiowüstite coexists with wadsleyite up to $X_{Fe_2SiO_4} = 0.26$,

Table 1. Fe-Mg exchange reactions at high pressures and temperatures

No.	Reactions	K_D
1	Fe-mw + Mg-ol = Mg-mw + Fe-ol	$(X_{Fe}/X_{Mg})^{ol}/(X_{Fe}/X_{Mg})^{mw}$
2	Fe-mw + Mg-wa = Mg-mw + Fe-wa	$(X_{Fe}/X_{Mg})^{wa}/(X_{Fe}/X_{Mg})^{mw}$
3	Fe-mw + Mg-sp = Mg-mw + Fe-sp	$(X_{Fe}/X_{Mg})^{sp}/(X_{Fe}/X_{Mg})^{mw}$
4	Fe-mw + Mg-pv = Mg-mw + Fe-pv	$(X_{Fe}/X_{Mg})^{pv}/(X_{Fe}/X_{Mg})^{mw}$
5	Fe-mw + Mg-ilm = Mg-mw + Fe-ilm	$(X_{Fe}/X_{Mg})^{ilm}/(X_{Fe}/X_{Mg})^{mw}$
6	Fe-mw + Mg-maj = Mg-mw + Fe-maj	$(X_{Fe}/X_{Mg})^{maj}/(X_{Fe}/X_{Mg})^{mw}$
7	Fe-ilm + Mg-pv = Mg-ilm + Fe-pv	$(X_{Fe}/X_{Mg})^{pv}/(X_{Fe}/X_{Mg})^{ilm}$
8	Fe-ilm + Mg-maj = Mg-ilm + Fe-maj	$(X_{Fe}/X_{Mg})^{maj}/(X_{Fe}/X_{Mg})^{ilm}$
9	Fe-ilm + Mg-sp = Mg-ilm + Fe-sp	$(X_{Fe}/X_{Mg})^{sp}/(X_{Fe}/X_{Mg})^{ilm}$
10	Fe-ilm + Mg-wa = Mg-ilm + Fe-wa	$(X_{Fe}/X_{Mg})^{wa}/(X_{Fe}/X_{Mg})^{ilm}$
11	Fe-maj + Mg-pv = Mg-maj + Fe-pv	$(X_{Fe}/X_{Mg})^{pv}/(X_{Fe}/X_{Mg})^{maj}$
12	Fe-maj + Mg-sp = Mg-maj + Fe-sp	$(X_{Fe}/X_{Mg})^{sp}/(X_{Fe}/X_{Mg})^{maj}$
13	Fe-maj + Mg-wa = Mg-maj + Fe-wa	$(X_{Fe}/X_{Mg})^{wa}/(X_{Fe}/X_{Mg})^{maj}$
14	Fe-ol + Mg-wa = Mg-ol + Fe-wa	$(X_{Fe}/X_{Mg})^{wa}/(X_{Fe}/X_{Mg})^{ol}$
15	Fe-wa + Mg-sp = Mg-wa + Fe-sp	$(X_{Fe}/X_{Mg})^{sp}/(X_{Fe}/X_{Mg})^{wa}$
16	Fe-ol + Mg-sp = Mg-ol + Fe-sp	$(X_{Fe}/X_{Mg})^{sp}/(X_{Fe}/X_{Mg})^{ol}$
17	Fe-ol + Mg-px = Mg-ol + Fe-px	$(X_{Fe}/X_{Mg})^{px}/(X_{Fe}/X_{Mg})^{ol}$
18	Fe-ol + Mg-gt = Mg-ol + Fe-gt	$(X_{Fe}/X_{Mg})^{gt}/(X_{Fe}/X_{Mg})^{ol}$
19	Fe-gt + Mg-px = Mg-gt + Fe-px	$(X_{Fe}/X_{Mg})^{px}/(X_{Fe}/X_{Mg})^{gt}$
20	Fe-ol + Mg-Alsp = Mg-ol + Fe-Alsp	$(X_{Fe}/X_{Mg})^{a-sp}/(X_{Fe}/X_{Mg})^{ol}$

mw = magnesiowüstite; ol = olivine; wa = wadsleyite; sp = silicate spinel; pv = silicate perovskite; ilm = silicate ilmenite; maj = Al-free majorite; px = orthopyroxene; gt = garnet; Alsp = aluminate spinel.

whereas it coexists with spinel over the entire compositional range. The Mg-Fe distribution coefficient between coexisting magnesiowüstite and wadsleyite is about 0.33 at 15 GPa and 1773 K (Fei et al. 1991). Several investigators (Ito and Yamada 1982, Ito and Takahashi 1989, Fei et al. 1991, Bell et al. 1979, Wood and Rubie 1996, Akaogi et al. 1998) have determined Mg-Fe^{2+} partitioning data between coexisting magnesiowüstite and spinel at pressures between 15 and 25 GPa and temperatures between 1373 and 1873 K. In the experiments of Ito and Yamada (1982), Ito and Takahashi (1989), Akaogi et al. (1998), and Bell et al. (1979), both $(Mg,Fe)_2SiO_4$ olivine and $(Mg,Fe)SiO_3$ pyroxene were used as the starting materials to produce a spinel + magnesiowüstite + stishovite assemblage at high pressures and temperatures. The iron content of the spinel solid solution represents the maximum solubility of FeO in the spinel structure under a given pressure and temperature condition. Fei et al. (1991) and Wood and Rubie (1996) conducted the experiments using mixtures of $(Mg,Fe)_2SiO_4$ olivine and $(Mg,Fe)O$ magnesiowüstite as the starting materials. Therefore, the measured Mg-Fe distribution data were not limited to the compositions close to the three-phase field. The spinel-magnesiowüstite tie-line at or near the three-phase field gave a Mg-Fe distribution coefficient of 0.25(±0.10) over a compositional range from X_{Fe}^{sp} = 0.22 to 0.84 and a pressure range of 25-15 GPa (Fig. 4b). The tie-lines away from the three-phase field in the Mg-rich region gave higher Mg-Fe distribution coefficients.

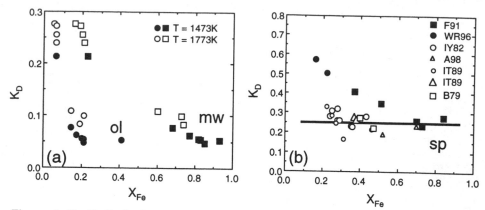

Figure 4. The Mg-Fe distribution coefficients between coexisting olivine and magnesiowüstite at 1473 and 1773 K as a function of iron content in olivine or in magnesiowüstite (a). Experimental data are from Fei et al. (1991). The coefficient between coexisting silicate spinel and magnesiowüstite as a function of iron content in spinel is shown in (b). F91, Fei et al. (1991); WR96, Wood and Rubie (1996); IY82, Ito and Yamada (1982); A98, Akaogi et al. (1998); IT89, Ito and Takahashi (1989); B79, Bell et al. (1979).

The coefficient increases with decreasing the bulk iron content (Fig. 4b), indicating that the tie-lines rotate towards an equilibrium composition with less iron partitioning into the magnesiowüstite solid solution in a more Mg-rich bulk composition.

Magnesiowüstite-perovskite

$(Mg,Fe)SiO_3$ perovskite is stable at pressures above 23 GPa and coexists with magnesiowüstite only in the Mg-rich region of the perovskite solid solution. The Mg-Fe distribution between coexisting magnesiowüstite and perovskite has been studied at high pressures and temperatures using the multi-anvil apparatus and the diamond-anvil cell. The reported Mg-Fe distribution coefficients, obtained at pressures below 26 GPa in the multi-anvil apparatus, range from 0.07 (Ito and Yamada 1982) to 0.35 (Katsura and Ito 1996) (Table 2). Ito et al. (1984) reported the distribution coefficient at 26 GPa and 1873 K which increases from 0.09 at $X_{Fe}^{pv} = 0.11$ to 0.15 at $X_{Fe}^{pv} = 0.03$. Katsura and Ito (1996) criticized the early work because the large uncertainties in composition analyses associated with the small grain size. They conducted experiments using $(Mg,Fe)_2SiO_4$ solid solutions mixed with B_2O_3 catalyst as the starting materials to enhance grain growth and the exchange reaction. The Mg-Fe distribution coefficient determined by Katsura and Ito (1996) is much higher than previously reported values. It increases from 0.16 at $X_{Fe}^{pv} = 0.14$ to 0.35 at $X_{Fe}^{pv} = 0.05$. Martinez et al. (1997) also reported K_D values of 0.23-0.26 based on analytical transmission electron microscopy measurements (Fig. 5).

The distribution coefficients between coexisting magnesiowüstite and perovskite at pressures above 25 GPa were determined by laser-heating $(Mg,Fe)_2SiO_4$ olivine starting materials in the diamond-anvil cell (Bell et al. 1979, Guyot et al. 1988, Fei et al. 1991, Kesson and Fitz Gerald 1991, Mao et al. 1997). Bell et al. (1979) and Fei et al. (1991) reported the average K_D values of 0.08 and 0.13 at about 30 GPa, respectively. These values are in a good agreement with the results obtained in the multi-anvil apparatus by Ito et al. (1984), despite a large temperature gradient in the laser-heated sample. Guyot et al. (1988) conducted experiments at pressures up to 75 GPa and showed sharp increase of K_D from 0.17 to 0.29 with increasing pressure from 25 to 40 GPa. Such a trend was not observed in the experiments of Kesson and Fitz Gerald (1991) who studied the Mg-Fe distribution coexisting magnesiowüstite and perovskite at pressures between 25 and 50 GPa.

Table 2. Experimental results on the Fe-Mg partitioning between two coexisting phases at high pressures and temperatures.

P (GPa)	T (K)	A+B	X_{Fe}^B	X_{Fe}^A	K_D	Ref.
2	1473	mw+ol	0.06	0.23	0.214	[1]
2	1473	mw+ol	0.14	0.68	0.077	[1]
2	1473	mw+ol	0.17	0.77	0.061	[1]
2	1473	mw+ol	0.20	0.82	0.055	[1]
2	1473	mw+ol	0.21	0.85	0.047	[1]
2	1473	mw+ol	0.21	0.85	0.047	[1]
2	1473	mw+ol	0.21	0.83	0.054	[1]
2	1473	mw+ol	0.41	0.93	0.052	[1]
2	1723	mw+ol	0.05	0.16	0.276	[1]
2	1723	mw+ol	0.06	0.19	0.272	[1]
2	1723	mw+ol	0.06	0.20	0.255	[1]
2	1723	mw+ol	0.06	0.21	0.240	[1]
2	1723	mw+ol	0.14	0.60	0.109	[1]
2	1723	mw+ol	0.21	0.73	0.098	[1]
2	1723	mw+ol	0.19	0.74	0.082	[1]
4	1723	mw+ol	0.06	0.22	0.226	[1]
4	1723	mw+ol	0.14	0.62	0.100	[1]
4	1723	mw+ol	0.20	0.71	0.102	[1]
4	1723	mw+ol	0.21	0.76	0.084	[1]
4	1723	mw+ol	0.41	0.89	0.086	[1]
9	1723	mw+ol	0.14	0.65	0.088	[1]
9	1723	mw+ol	0.14	0.66	0.084	[1]
9	1723	mw+ol	0.15	0.67	0.087	[1]
9	1723	mw+ol	0.20	0.78	0.070	[1]
9	1723	mw+ol	0.21	0.83	0.054	[1]
9	1723	mw+ol	0.47	0.93	0.067	[1]
15	1773	mw+wa	0.10	0.25	0.333	[1]
15	1773	mw+wa	0.12	0.32	0.290	[1]
15	1773	mw+wa	0.14	0.31	0.362	[1]
25	1373	mw+sp	0.24	0.53	0.280	[2]
25	1373	mw+sp	0.25	0.52	0.308	[2]
24.6	1373	mw+sp	0.28	0.55	0.318	[2]
24	1373	mw+sp	0.28	0.60	0.259	[2]
24	1373	mw+sp	0.29	0.61	0.261	[2]
23	1373	mw+sp	0.35	0.70	0.231	[2]
23	1373	mw+sp	0.36	0.71	0.230	[2]
22	1373	mw+sp	0.43	0.73	0.279	[2]
24.5	1373	mw+sp	0.22	0.46	0.331	[3]
24	1373	mw+sp	0.25	0.54	0.284	[3]
24	1373	mw+sp	0.27	0.60	0.247	[3]
23.2	1373	mw+sp	0.31	0.73	0.166	[3]
22.5	1373	mw+sp	0.46	0.80	0.213	[3]
22	1373	mw+sp	0.37	0.68	0.276	[3]
22	1873	mw+sp	0.38	0.70	0.263	[3]
20.3	1873	mw+sp	0.52	0.85	0.191	[4]
18.5	1873	mw+sp	0.70	0.91	0.231	[4]
20.4	1873	mw+sp	0.16	0.25	0.571	[5]
20.4	1873	mw+sp	0.22	0.36	0.501	[5]
15	1773	mw+sp	0.37	0.59	0.408	[1]
15	1773	mw+sp	0.51	0.75	0.347	[1]
15	1773	mw+sp	0.70	0.90	0.259	[1]
15	1773	mw+sp	0.73	0.92	0.235	[1]
15	1773	mw+sp	0.84	0.94	0.276	[1]
18-23	-	mw+sp	0.06	0.13	0.427	[6]
18-23	-	mw+sp	0.40	0.71	0.272	[6]
18-23	-	mw+sp	0.47	0.80	0.222	[6]
26	1873	mw+pv	0.030	0.170	0.151	[7]
26	1873	mw+pv	0.065	0.340	0.135	[7]
26	1873	mw+pv	0.064	0.340	0.133	[7]
26	1873	mw+pv	0.070	0.480	0.082	[7]
26	1873	mw+pv	0.110	0.580	0.090	[7]
26	1373	mw+pv	0.070	0.510	0.072	[2]
26	1373	mw+pv	0.095	0.470	0.118	[2]

P (GPa)	T (K)	A+B	X_{Fe}^B	X_{Fe}^A	K_D	Ref.
23	1900	mw+pv	0.053	0.137	0.353	[8]
23	1900	mw+pv	0.138	0.498	0.161	[8]
26	1573	mw+pv	0.053	0.174	0.266	[9]
26	1873	mw+pv	0.038	0.141	0.241	[9]
25	1373	mw+pv	0.070	0.450	0.092	[3]
25.5	1373	mw+pv	0.060	0.440	0.081	[3]
25-35	-	mw+pv	0.07	0.26	0.214	[6]
25-35	-	mw+pv	0.17	0.77	0.061	[6]
25-35	-	mw+pv	0.18	0.80	0.055	[6]
25-35	-	mw+pv	0.20	0.78	0.071	[6]
25-35	-	mw+pv	0.25	0.76	0.105	[6]
25	2500	mw+pv	-	-	0.167	[10]
30	2500	mw+pv	-	-	0.244	[10]
40	2500	mw+pv	-	-	0.286	[10]
50	2500	mw+pv	-	-	0.286	[10]
75	2500	mw+pv	-	-	0.286	[10]
26	~1673	mw+pv	0.05	0.27	0.142	[1]
26	~1673	mw+pv	0.07	0.38	0.123	[1]
26	~1673	mw+pv	0.11	0.48	0.134	[1]
25	-	mw+pv	-	-	0.345	[11]
30	-	mw+pv	-	-	0.158	[11]
35	-	mw+pv	-	-	0.326	[11]
40	-	mw+pv	-	-	0.248	[11]
50	-	mw+pv	-	-	0.361	[11]
70	-	mw+pv	-	-	0.500	[12]
135	-	mw+pv	-	-	0.420	[12]
29	1800	mw+pv	-	-	0.240	[13]
31	1800	mw+pv	-	-	0.110	[13]
32	1500	mw+pv	-	-	0.040	[13]
37	1500	mw+pv	-	-	0.120	[13]
40	1500	mw+pv	-	-	0.090	[13]
50	2000	mw+pv	-	-	0.290	[13]
24	1373	ilm+sp	0.13	0.11	1.200	[2]
20	2273	maj+sp	0.70	0.34	4.530	[14]
20	2273	maj+wa	0.06	0.08	0.734	[14]
14.4	1873	ol+wa	0.076	0.123	0.586	[15]
14	1873	ol+wa	0.053	0.093	0.546	[15]
13.7	1873	ol+wa	0.160	0.260	0.542	[15]
13.1	1473	ol+wa	0.079	0.141	0.523	[15]
12.2	1473	ol+wa	0.120	0.256	0.396	[15]
19.2	1873	wa+sp	0.081	0.126	0.611	[15]
16.5	1873	wa+sp	0.167	0.235	0.653	[15]
14.5	1873	wa+sp	0.218	0.326	0.576	[15]
13.5	1873	wa+sp	0.280	0.364	0.679	[15]
17.3	1473	wa+sp	0.043	0.074	0.562	[15]
13.9	1473	wa+sp	0.159	0.247	0.576	[15]
12.6	1473	wa+sp	0.240	0.354	0.576	[15]
5.7	1273	sp+ol	0.890	0.985	0.123	[16]
6.4	1273	sp+ol	0.660	0.910	0.192	[16]
6.9	1273	sp+ol	0.515	0.830	0.217	[16]
8.1	1273	sp+ol	0.385	0.750	0.209	[16]
9.8	1273	sp+ol	0.260	0.670	0.173	[16]
13.9	1673	ol+wa	0.065	0.120	0.510	[17]
13.6	1673	ol+wa	0.090	0.172	0.476	[17]
13.45	1673	ol+wa	0.106	0.193	0.496	[17]
13.4	1673	ol+wa	0.109	0.203	0.480	[17]
13.35	1673	ol+wa	0.118	0.212	0.497	[17]
13.14	1673	ol+wa	0.137	0.250	0.476	[17]
13	1673	ol+wa	0.149	0.283	0.444	[17]
12.9	1673	ol+wa	0.160	0.293	0.460	[17]
*13.35	1673	ol+wa	0.111	0.216	0.453	[17]
*12.9	1673	ol+wa	0.154	0.289	0.448	[17]

References:
[1] Fei et al. (1991)   [2] Ito & Yamada (1982)   [3] Ito & Takahashi (1989)
[4] Akaogi et al. (1998)   [5] Wood & Rubie (1996)   [6] Bell et al. (1979)
[7] Ito et al. (1984)   [8] Katsura & Ito (1996)   [9] Martinez et al. (1997)
[10] Guyot et al. (1988)   [11] Kesson & Fitz Gerald (1991)
[12] Kesson et al. (1998)   [13] Mao et al. (1997)   [14] Kato (1987)
[15] Katsura & Ito (1989)   [16] Nishizawa & Akimoto (1973)   [17] Fei & Bertka (1996)

The stars (*) indicate the reversal experiments in which pre-synthesized high-pressure phase was used as the starting material.

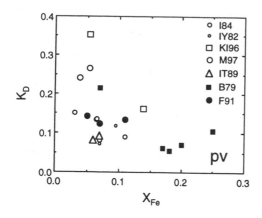

Figure 5. The Mg-Fe distribution coefficients between coexisting silicate perovskite and magnesiowüstite. I84, Ito et al. (1984);); IY82, Ito and Yamada (1982); KI96, Katsura and Ito (1996); M97, Martinez et al. (1997); IT89, Ito and Takahashi (1989); B79, Bell et al. (1979); F91, Fei et al. (1991).

Instead, they showed a systematic change in Mg numbers of coexisting magnesiowüstite and perovskite as a function of Mg number of the bulk composition, the Mg number of perovskite undergoes only a modest increase as the system becomes more magnesian, whereas that of magnesiowüstite increases from 61 to 90 when the Mg number of the system is increased from 73 to 92 (Kesson and Fitz Gerald 1991). Very recently, Mao et al. (1997) performed high-pressure diamond-cell experiments with the double-sided laser heated diamond cell technique (Shen et al. 1996). They demonstrated that the partitioning between magnesiowüstite and perovskite depends on the pressure, temperature, bulk Fe/Mg ratio, and ferric iron content; a systematic trend was found with K_D increasing with increasing pressure, increasing temperature, decreasing ferrous iron, and increasing ferric iron (Mao et al. 1997).

The $(Mg,Fe)SiO_3$ perovskite solid solution does not extend into the Fe-rich region. The maximum solubility of $FeSiO_3$ in the perovskite structure is defined by the reaction $(Mg,Fe)SiO_3$ (pv) = $(Mg,Fe)O$ (mw) + SiO_2 (st). The reaction has been studied by using laser-heating diamond-cell techniques (e.g. Yagi et al. 1979, Fei et al. 1991) and the multi-anvil apparatus (e.g. Ito and Yamada 1982, Ito et al. 1984, Ito and Takahashi 1989). There was a discrepancy in the determined maximum solubility of $FeSiO_3$ in the perovskite structure by the two difference techniques. The maximum solubility of $FeSiO_3$ was 11 mol % at 26 GPa and 1873 K, obtained by using the multi-anvil apparatus (e.g. Ito et al. 1984), whereas perovskites with 21 mol % $FeSiO_3$ were synthesized at similar pressure with the laser-heating diamond-anvil technique, according to Yagi et al. (1979). The discrepancy was re-examined by Fei et al. (1996) using both high-pressure techniques. They attributed the discrepancy to the large temperature gradient in the laser-heated samples. The data obtained with the multi-anvil apparatus showed that the maximum $FeSiO_3$ solubility in the perovskite increases from 0.05 at 1273 K to 0.12 at 2000 K at 26 GPa. Recent diamond-cell experiments up to 55 GPa (Mao et al. 1997) demonstrated that the maximum $FeSiO_3$ solubility increases rapidly with increasing pressure (Fig. 6). About 28 mol % $FeSiO_3$ can be dissolved into the perovskite structure at 50 GPa and 2000 K.

Ilmenite-perovskite (-majorite, -spinel, and -wadsleyite)

The $(Mg,Fe)SiO_3$ ilmenite solid solution is stable in only a narrow pressure interval (< 2 GPa). At very high temperature (>1900 K), the $(Mg,Fe)SiO_3$ majorite solid solution becomes stable (Kato 1986, Ohtani et al. 1991). The solubility of $FeSiO_3$ in the ilmenite structure is very limited ($X_{FeSiO_3} < 0.12$) (Ito and Yamada 1982). Because the two-phase

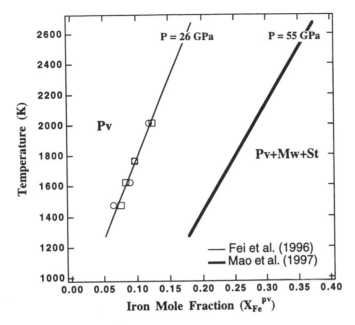

Figure 6. The maximum solubility of FeSiO₃ in the perovskite structure as a function of temperature and pressure. Experimental data at 26 GPa are from Fei et al. (1996), and are represented by the thin line. Data at higher pressures were obtained in the laser-heated diamond-anvil cell, represented by the heavy line (Mao et al. 1997).

fields of ilmenite-perovskite and ilmenite-majorite are small (Ito and Yamada 1982, Kato 1986, Ohtani et al. 1991), direct measurements of partitioning between the two coexisting phases are difficult. However, the phase relations can constrained by the phase boundary of the end-member and the univariant line. Kato (1986) studied the stability relation of the $(Mg,Fe)SiO_3$ majorite solid solution at 20 GPa and found two types of $(Mg,Fe)SiO_3$ majorites, non-cubic majorite for $X_{Fe} < 0.2$, and cubic majorite for $0.2 < X_{Fe} < 0.4$. However, cubic majorite was not observed at pressures between 16 and 24 GPa and at a temperature of 2073 K by Ohtani et al. (1991).

Silicate ilmenite coexists with spinel by the reaction $2(Mg,Fe)SiO_3$ (ilm) = $(Mg,Fe)_2SiO_4$ (sp) + SiO_2 (st). It can coexist with wadsleyite only in the very Mg-rich region ($X_{Fe} < 0.05$) at temperatures between 1800 and 2100 K. The Mg-Fe distribution coefficient between coexisting ilmenite and spinel $[(X_{Fe}/X_{Mg})^{sp}/(X_{Fe}/X_{Mg})^{ilm}]$ is about 1.20 at 24 GPa and 1373 K (Ito and Yamada 1982).

Majorite-perovskite (-spinel, and -wadsleyite)

The stability field of the $MgSiO_3$ end-member of the $(Mg,Fe)SiO_3$ majorite solid solution was determined by Sawamoto (1987), only stable at temperatures above 1900 K and pressures above 16 GPa. The solid solution, in principle, can coexist with perovskite, ilmenite, spinel, wadsleyite, or pyroxene solid solutions. Because of its extremely high P-T stability field, there are limited experimental data available for this system. Kato (1986) conducted experiments on $(Mg,Fe)SiO_3$ at 20 GPa and temperatures above 2073 K, and reported the Mg-Fe distribution coefficient between coexisting cubic majorite and spinel $[K_D = (X_{Fe}/X_{Mg})^{sp}/(X_{Fe}/X_{Mg})^{maj} = 4.53]$ and between coexisting non-cubic majorite and wadsleyite $[K_D = (X_{Fe}/X_{Mg})^{wa}/(X_{Fe}/X_{Mg})^{maj} = 0.73]$ at 20 GPa and 2273 K. Ohtani et al. (1991) reported that the maximum solubility of $FeSiO_3$ in non-cubic majorite is about 40 mol % at 2073 K, and decreases with increasing pressure.

Olivine-wadsleyite, wadsleyite-spinel, and olivine-spinel

The phase relations in the Mg_2SiO_4-Fe_2SiO_4 system at 1473 and 1873 K have been determined by Katsura and Ito (1989). The average Mg-Fe distribution coefficients between coexisting olivine and wadsleyite and between coexisting wadsleyite and spinel are 0.56(±0.02) and 0.64(±0.04), respectively, at 1873 K. The coefficients decrease slightly with decreasing temperature. The Mg-Fe distribution between coexisting olivine and spinel were studied by Akimoto and Fujisawa (1968) and Nishizawa and Akimoto (1973). The experimentally determined coefficient $[K_D = (X_{Fe}/X_{Mg})^{ol}/(X_{Fe}/X_{Mg})^{sp}]$ ranges from 0.123 to 0.217 at 1273 K (Nishizawa and Akimoto 1973). The three two-phase loops (ol-wa, wa-sp, and ol-sp) intercept to form the ol-wa-sp univariant line at pressures between 12 and 13.5 GPa, depending on the temperature. Thermodynamic calculations (e.g. Fei et al. 1991) showed that the phase equilibrium relations in the system could not be reproduced by assuming ideal solutions for the $(Mg,Fe)_2SiO_4$ polymorphs. Figure 7 shows the comparison between the calculated isothermal phase relations at 1873 K in the Mg_2SiO_4-Fe_2SiO_4 system with an ideal solution model and a two-parameter Margules solution model that best represents the experimental data. The data for the end-members and solution parameters used in the calculations are from Fei et al. (1991). The non-ideal solution behavior of the $(Mg,Fe)_2SiO_4$ polymorphs is consistent with the solution calorimetric measurements (Wood and Kleppa 1981, Akaogi et al. 1989) and the Mg-Fe partitioning data (Fei et al. 1991).

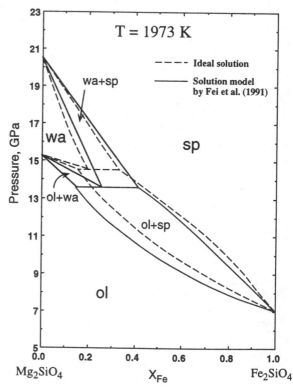

Figure 7. Comparison of the calculated isothermal phase relations at 1973 K in the system Mg_2SiO_4-Fe_2SiO_4 using an ideal solution model (dashed lines) and the solution model by Fei et al. (1991) (solid lines).

Fei and Bertka (1996, 1998) recently conducted multi-anvil experiments in the Mg_2SiO_4-Fe_2SiO_4 system at 1673 K, using the multi-cell sample chamber technique described above. The olivine-wadsleyite two-phase loop was precisely determined using both olivine and wadsleyite solid solutions as the starting materials. For example, they ran an experiment at 13.45 GPa and 1673 K using eight (Mg,Fe)$_2$SiO$_4$ olivine solid solutions with compositions from $(Mg_{0.92}Fe_{0.08})_2SiO_4$ to $(Mg_{0.78}Fe_{0.22})_2SiO_4$ (every 2 mol % Fe$_2$SiO$_4$ interval) as the starting materials, loaded into separate sample cells. The starting materials with compositions from $(Mg_{0.92}Fe_{0.08})_2SiO_4$ to $(Mg_{0.90}Fe_{0.10})_2SiO_4$ remained as a single olivine phase, whereas the starting materials with compositions from $(Mg_{0.80}Fe_{0.20})_2SiO_4$ to $(Mg_{0.78}Fe_{0.22})_2SiO_4$ transformed into the high-pressure phase wadsleyite. Two coexisting phases, olivine and wadsleyite, were observed in compositions between $(Mg_{0.90}Fe_{0.10})_2SiO_4$ and $(Mg_{0.80}Fe_{0.20})_2SiO_4$. The compositions of the coexisting olivine and wadsleyite phases at 13.45 GPa and 1673 K (Table 2) were determined based the microprobe analyses of the four samples in which olivine and wadsleyite coexisted. Reversal experiments were also performed at 12.9 and 13.35 GPa, using pre-synthesized wadsleyite solid solutions as the starting materials (Fig. 2). The determined Mg-Fe distribution coefficients between coexisting olivine and wadsleyite vary from 0.44 to 0.51, slightly smaller than the value at 1873 K determined by Katsura and Ito (1989) and consistent with the observed temperature-K_D trend.

Olivine-garnet (-pyroxene, and -aluminate spinel), and garnet-pyroxene

Many mineralogical reactions involving rock-forming minerals are used as geothermometers and geobarometers. In this study, I focus on the Mg-Fe distribution between coexisting olivine and orthopyroxene, between coexisting olivine and garnet, and between coexisting garnet and orthopyroxene. An extensive review of various geothermometric and geobarometric formulations of mineral reactions has been given by Ganguly and Saxena (1987).

The exchange of Mg and Fe between olivine and garnet has been extensively studied because of its geological application as a geothermometer (Kawasaki and Matsui 1977, O'Neill and Wood 1979, Hackler and Wood 1989). The Mg-Fe partitioning data were used to derive mixing properties of the Mg-Fe solid solutions involved in the exchange reaction. Because olivine forms several solid solution pairs with garnet, orthopyroxene, and aluminate spinel, its mixing properties derived from one pair of solid solutions must be consistent with the partitioning data in the other systems involving olivine. Wood and Kleppa (1981) determined the enthalpy of mixing of Fe-Mg olivine by solution calorimetry and showed some asymmetry in the mixing properties. However, phase equilibrium data in the systems involving Fe-Mg olivine (Nafziger and Muan 1967, Davidson and Mukhopadhyay 1984, O'Neill and Wall 1987, Bartholomew 1989) suggest that the olivine solid solution can be represented by a symmetric solution model with an interaction parameter W^{ol} value of between 7000 and 11000 J/mol (two-site basis) (Hackler and Wood 1989).

The Fe-Mg garnet is considered as an asymmetric solution (Ganguly and Saxena 1984, Geiger et al. 1987, Hackler and Wood 1989) and the Fe-Mg orthopyroxene as an ideal solution (Ganguly 1982, Ganguly and Saxena 1987). Hackler and Wood (1989) were able to reproduce the Fe-Mg partitioning data between olivine and garnet (O'Neill and Wood 1979, Hackler and Wood 1989) using a symmetric solution model for olivine with W^{ol} = 9000 J/mol (two-site basis) and an asymmetric solution model for garnet with W^{gt}_{FeMg} = 2080 J/mol and W^{gt}_{MgFe} = 6350 J/mol (three-site basis). Although the model of Hackler and Wood (1989) for garnet showed similar asymmetry in excess enthalpies as the solution calorimetry measurements of Geiger et al. (1987), the interaction parameters derived from the partitioning data are much smaller than those from the solution calorimetry measurements.

The partitioning of Fe and Mg between garnet and orthopyroxene has been studied up to 5 GPa and over a temperature range of 1248-1773 K (Kawasaki and Matsui 1983, Harley 1984, Lee and Ganguly 1988). The data suggest that the distribution coefficient K_D is essentially independent of Fe/Mg ratio (Kawasaki and Matsui 1983, Lee and Ganguly 1988). The $\ln K_D$ changes linearly with reciprocal temperature (Kawasaki and Matsui 1983, Harley 1984, Lee and Ganguly 1988). Hackler and Wood (1989) also showed that the partitioning of Fe and Mg between garnet and orthopyroxene at 2.6 GPa and 1473 K determined by Lee and Ganguly (1988) can be reproduced by an ideal solution model for orthopyroxene and the same solution model of garnet derived from the Fe-Mg partitioning data between olivine and garnet. However, the Fe-Mg partitioning data between olivine and orthopyroxene at 3 GPa and 1273 K determined by Matsui and Nishizawa (1974) were not reproduced by the same solution models.

Jamieson and Roeder (1984) determined the distribution of Mg and Fe between olivine and aluminate spinel at 1573 K. The distribution coefficient

$$K_D = (X_{Fe}/X_{Mg})^{Alsp}/(X_{Fe}/X_{Mg})^{ol}$$

varies between 0.94 and 1.23 and increases with $X_{Mg}{}^{Alsp}$. The same olivine solution model is applicable to this system and is used to represent the partitioning data (Jamieson and Roeder 1984). In addition to the Mg-Fe^{2+} exchange in the octahedral site, spinel solid solution is complicated by the disorder of the major cations (Fe^{2+}, Fe^{3+}, Mg^{2+}, Al^{3+}) between the octahedral and tetrahedral spinel sites. Solution models involving cation disorder have been discussed by O'Neill and Navrotsky (1984), O'Neill and Wall (1984), and Nell and Wood (1989).

MgSi-AlAl AND Al-Fe^{3+} SUBSTITUTIONS

The solid solution in garnet (or majorite) is complicated by a heterovalent substitution MgSi-AlAl in the octahedral sites in addition to the Mg-Fe exchange at high pressures and temperatures. For the iron-free system, majorite is characterized by a completely disordered distribution of Mg, Si, and Al in the two octahedral sites (sublattices) (Akaogi et al. 1987) representing the two end-members, $Mg_3[MgSi]^{oct}Si_3O_{12}$ (majorite) and $Mg_3[Al_2]^{oct}Si_3O_{12}$ (pyrope). A two-sublattice model (Sundman and Ågren 1981) is applicable to this system with Al^{3+} disordered in both sublattices (Fabrichnaya 1998). The majorite-pyrope solid solution can be approximated by an ideal mixing model, according to the solution calorimetric data (Akaogi et al. 1987). Al_2O_3 also dissolves into $MgSiO_3$ high-pressure phases with ilmenite and perovskite structures to form $(Mg,Al)(Al,Si)O_3$ ilmenite and perovskite solid solutions (Liu 1977, Weng et al. 1982, Kanzaki 1987, Irifune et al. 1996, Hirose and Fei 1998a). Kanzaki (1987), Irifune et al. (1996), and Hirose and Fei (1998a) determined the phase relations in the system $Mg_4Si_4O_{12}$ and $Mg_3Al_2Si_3O_{12}$ at 1273 K, 1773 K, and 2023 K, respectively. The stability field of ilmenite solid solution shrinks with increasing temperature, whereas the solubility of Al_2O_3 in the perovskite structure increases with increasing pressure. Ilmenite forms a complete solid solution over a narrow pressure interval (~ 2 GPa) at about 24 GPa and 1273 K (Kanzaki 1987). A complete solid solution of majorite forms at temperature above 1873 K (Akaogi et al. 1987, Gasparik 1992). Recent multi-anvil experiments using sintered diamonds as anvils (Ito et al. 1998) showed that pyrope transforms to orthorhombic perovskite at 37 GPa, indicating complete solid solution of perovskite above this pressure. Pyrope decomposes into orthorhombic perovskite and corundum at pressures between 26 GPa and 37 GPa (Irifune et al. 1996, Hirose and Fei 1998a). Figure 8 illustrates the change in phase relations in the system

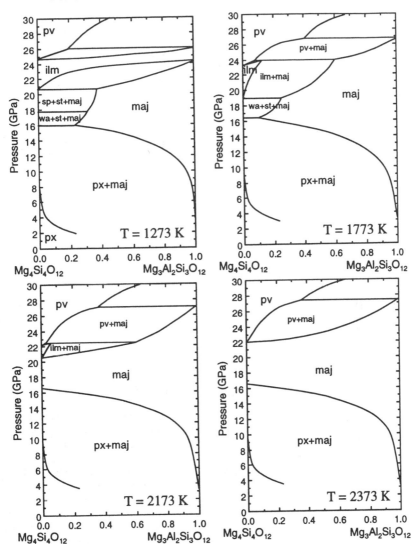

Figure 8. Phase relations in the system $Mg_4Si_4O_{12}$-$Mg_3Al_2Si_3O_{12}$ at various temperatures.

$Mg_4Si_4O_{12}$-$Mg_3Al_2Si_3O_{12}$ as a function of temperature.

Phase relations in the system $Fe_4Si_4O_{12}$-$Fe_3Al_2Si_3O_{12}$ have been determined at 1273 K up to 11 GPa (Akaogi and Akimoto 1977). The ferrosilite-almandine system forms limited solid solution by Fe^{2+}Si-AlAl substitution. The solubility limit of $Fe_4Si_4O_{12}$ in $Fe_3Al_2Si_3O_{12}$ garnet at 1273 K is about 40 mole% (Akaogi and Akimoto 1977). Almandine $(Fe_3Al_2Si_3O_{12})$ breaks down into FeO, Al_2O_3, and SiO_2 at about 20 GPa and 1700 K (Conrad 1998) and ferrosilite into Fe_2SiO_4 spinel plus stishovite at about 9 GPa (Akaogi and Akimoto 1977, Akimoto and Syono 1970).

In the $Mg_4Si_4O_{12}$-$Mg_3Al_2Si_3O_{12}$-$Fe_4Si_4O_{12}$-$Fe_3Al_2Si_3O_{12}$ quadrilateral, there are five binary solid solutions with Mg-Fe substitution, (Mg,Fe)SiO_3-pyroxene, -majorite, ilmenite, and -perovskite, and $(Mg,Fe)_3Al_2Si_3O_{12}$-garnet, and two binary solid solutions with MgSi-AlAl and FeSi-AlAl substitutions, $Mg_3(MgSi,AlAl)Si_3O_{12}$-majorite and $Fe_3(FeSi,AlAl)Si_3O_{12}$-majorite. Phase relations in each binary solid solution have been experimentally studied, as discussed above. However, no systematic study involving both Mg-Fe and Mg(Fe)Si-AlAl exchange has been undertaken yet. Wood and Rubie (1996) reported the Mg-Fe partitioning data between coexisting silicate perovskite and magnesiowüstite in Al_2O_3-bearing system, obtained from multi-anvil experiments. They found that Fe substitution in perovskite is strongly coupled to Al_2O_3 concentration. The difference in the Mg-Fe partitioning data between the Al_2O_3-bearing and the Al_2O_3-free systems may be attributed to the high Fe^{3+} content in the Al_2O_3-bearing system (Wood and Rubie 1996). Mössbauer study of silicate perovskite containing 3.3 mol % Al_2O_3, synthesized at 25 GPa and 1873 K in a multi-anvil apparatus, showed that approximately 50% of the iron is Fe^{3+} (McCammon 1997), suggesting that Fe^{3+} is an important component in the Al_2O_3-bearing system. The introduction of Fe^{3+} into the quadrilateral system would add further complexity to the solution model.

MINOR AND TRACE ELEMENT PARTITIONING

Minor element partitioning in the mantle-related systems has not been systematically studied at high pressures and temperatures. Scattered data are available from high-pressure experiments in which natural or synthetic mantle compositions were used as the starting materials. For example, experiments have been conducted in peridotitic compositions at pressures up to 25 GPa (e.g. Irifune 1987, Irifune and Ringwood 1987, Ito and Takahashi 1987, Takahashi 1986, Takahashi and Ito 1987, Zhang and Herzberg 1994, Herzberg and Zhang 1996, Fei and Bertka 1996, Kawamoto et al. 1996, Kawamoto and Holloway 1997, Fei and Bertka 1998), in basaltic compositions at pressures up to 28 GPa (e.g. Irifune and Ringwood 1993, Irifune et al. 1994, Yasuda et al. 1994, Ono and Yasuda 1996, Hirose et al. 1998), and in an iron-rich Martian mantle composition (Kamaya et al. 1993, Bertka and Fei 1997). Noticeably, minor elements such as Cr and Na partition preferentially into wadsleyite when olivine and wadsleyite coexist. The effect of minor element partitioning on the olivine-wadsleyite transition pressure and the width of the two-phase coexisting field is of great interest because it may provide an interpretation for the difference in the width of the 410-km mantle discontinuity derived from seismic observation and that predicted from phase equilibrium data (Fei and Bertka 1998). Recent high-pressure experiments have shown that the partitioning of H_2O between olivine and wadsleyite is about 1:10, strongly partitioning into wadsleyite (Inoue 1994, Young et al. 1993, Kohlstedt et al. 1996) and such behavior strongly affects the width of the two-phase coexisting field (Wood 1995).

It has been shown that ferric iron (Fe^{3+}) is readily accommodated in many high-pressure silicate phases such as $(Mg,Fe)_2SiO_4$ wadsleyite, $(Mg,Fe)_2SiO_4$ spinel, and $(Mg,Fe)SiO_3$ perovskite (Fei et al. 1992, 1994; McCammon et al. 1992, McCammon 1997). Preferential partitioning of Fe^{3+} into the high-pressure phase would significantly affect phase relations and the Mg-Fe^{2+} distribution between the coexisting phases at high pressures and temperatures. The apparent Mg-Fe distribution coefficient will also be affected by the Fe^{3+}-Al^{3+} coupling in an Al_2O_3-bearing system (Wood and Rubie 1996).

Trace element partitioning data between silicate melts and mantle phases at high pressure and temperature are crucial for understanding the processes of chemical differentiation in the Earth's interior. Partitioning data between silicate perovskite and silicate melt have been obtained at high pressure and temperature to address the existence of a magma ocean in the early history of the Earth and the role of fractionation of silicate perovskite in a completely molten Earth (Kato et al. 1988a,b; Drake et al. 1993, McFarlane et al. 1994, Gasparik and Drake 1995, Kato et al. 1996). Magnesiowüstite is the liquidus phase at pressures between 18 and at least 25 GPa in a peridotitic mantle composition (Herzberg and Zhang 1996, Hirose and Fei 1998b, Ohtani et al. 1998) and majorite garnet is another important liquidus phase at pressures from about 16.5 GPa to 23 GPa. The partitioning data between magnesiowüstite and silicate melt, and between majorite garnet and silicate melt, provide constraints on petrogenesis of komatiite and the depth of melt generation and segregation in mantle plumes (e.g. Kato et al. 1987, Ohtani et al. 1989, Drake et al. 1993, McFarlane et al. 1994).

It is technically challenging to obtain high-quality partitioning data of trace elements from high-pressure experimental charges. The existing trace element partitioning data between mantle phases and silicate melt were obtained by electron microprobe analysis with limited numbers of elements doped at high concentrations. It is desirable, however, to measure a large number of trace elements rapidly at natural abundances for trace element partitioning studies. The new generation of ion microprobe, with a high spatial resolution at ppb levels of detection limit, is an ideal tool for obtaining a comprehensive trace element partitioning database involving mantle phases. New trace element partitioning data between majorite garnet, silicate perovskite and silicate melt, obtained by ion microprobe analysis are starting emerge (Minarik et al. 1997, 1998).

The "excess siderophile element" problem of the Earth's upper mantle has long been recognized. Highly siderophile elements are more abundant in the upper mantle than predicted from the experimentally determined metal-silicate partition coefficients at ambient pressure and high temperatures (e.g. Ringwood 1966, Brett 1971, 1984; Jones and Drake 1986). Investigators have been attempting to reconcile the difference between the observed and calculated siderophile element abundances in the upper mantle by exploring the dependence of the metal-silicate partition coefficients on temperature, oxygen fugacity, compositions of metallic and silicate liquids, and pressure. Much of the research effort before 1993 on this subject was focused on temperature and oxygen fugacity effects (e.g. Jones and Drake 1986, Murthy 1991, Capobianco et al. 1993). Having recognized that the effect of temperature and oxygen fugacity alone on the metal-silicate partition coefficients cannot provide a complete resolution of the "excess siderophile element" problem of the Earth's mantle, several investigators have recently studied element partitioning between solid (or liquid) metallic phase and silicate liquid at high pressures and temperatures (Walker et al. 1993, Hillgren et al. 1994, Walter and Thibault 1995, Thibault and Walter 1995, Gessmann et al. 1995, Holzheid et al. 1995, Hillgren et al. 1996, Li and Agee 1996, Ohtani and Yurimoto 1996, Righter and Drake 1997, Ohtani et al. 1997). The partition coefficients for siderophile elements such as Ni, Co, Mo, W, and Re have complex relationships with pressure, temperature, oxygen fugacity, and the compositions of metallic and silicate liquids (Righter et al. 1997). By isolating the effect of different variables, Thibault and Walter (1995) and Li and Agee (1996) have demonstrated that the partition coefficients for Ni, Co, and Mo decrease with increasing pressure, whereas the coefficients for W and Re increase with increasing pressure (Thibault and Walter 1995, Righter and Drake 1997, Righter et al. 1997). The silicate melt compositional effect has also been clearly demonstrated by Walter and Thibault (1995) and Hillgren et al. (1996). The existing partitioning data seem to point to one conclusion that the absolute and relative abundances of the siderophile elements in the Earth's upper mantle could be consistent with metal-silicate equilibrium if core formation had taken place in a magma ocean with a depth of 750-1100 km (Li and Agee 1996, Righter and Drake 1997).

THERMODYNAMICS OF ION-EXCHANGE REACTIONS

For determining the properties of solid solutions from experimental data on the composition of coexisting phases, the distribution of a component (1 or 2) between two coexisting solid solutions (A and B) may be considered as an ion-exchange reaction:

$$1\text{-}A + 2\text{-}B = 2\text{-}A + 1\text{-}B, \tag{9}$$

where 1 and 2 are exchangeable components, such as Fe^{2+} and Mg^{2+} in Fe-Mg exchange reactions. At equilibrium,

$$G_r + RT \ln K_D + RT \ln[(\gamma_2/\gamma_1)^A/(\gamma_2/\gamma_1)^B] = 0, \tag{10}$$

where G_r is the Gibbs free energy of the reaction (9), calculated from thermodynamic data of the end-members, and is a function of only P and T. The distribution coefficient is defined as

$$K_D = (X_2/X_1)^A/(X_2/X_1)^B, \tag{11}$$

where X_1 and X_2 are the mole fractions, obtained from element partitioning experiments. The activity coefficient γ_i in a binary solution can be modeled, for example, using the two-parameter Margules model:

$$RT\ln\gamma_1^A = X_2^2[W_{12}^A + 2X_1(W_{21}^A - W_{12}^A)] \tag{12}$$
$$RT\ln\gamma_2^A = X_1^2[W_{21}^A + 2X_2(W_{12}^A - W_{21}^A)] \tag{13}$$
$$RT\ln\gamma_1^B = X_2^2[W_{12}^B + 2X_1(W_{21}^B - W_{12}^B)] \tag{14}$$
$$RT\ln\gamma_2^B = X_1^2[W_{21}^B + 2X_2(W_{12}^B - W_{21}^B)]. \tag{15}$$

By combining the Equations (10-15), the W parameters can be obtained from thermodynamic data for the end-members and element partitioning data for the ion-exchange reaction (9). However, the reaction (9) involves two solid solutions with two sets of adjustable W parameters, i.e. W_{12}^A, W_{21}^A, W_{12}^B, and W_{21}^B (two adjustable parameters even for the simple mixture where $W_{12}^A = W_{21}^A$, and $W_{12}^B = W_{21}^B$). Therefore, the solution for the W parameters may be not unique. One way of constraining the W parameters is to optimize the parameters by simultaneously considering multiple ion-exchange reactions involving a common solid solution. For example, magnesiowüstite coexists with olivine, wadsleyite, silicate spinel, or silicate perovskite solid solution. A set of solution parameters for those solid solutions can be obtained by simultaneously fitting the partitioning data. Similarly, the W parameters of the Mg-Fe exchange for olivine, garnet, orthopyroxene, and aluminate spinel solid solutions may be determined by fitting the Mg-Fe distribution data between coexisting olivine and orthopyroxene, between coexisting olivine and garnet, between coexisting olivine and aluminate spinel, and between coexisting garnet and orthopyroxene. In addition, the experimentally determined enthalpy of mixing by solution calorimetry can be used to further constrain the W parameters independently. Table 3 summarizes solution parameters of selected solid solutions.

The W parameters derived by different investigators may be different even if the same partitioning data were used because the thermodynamic data for the end-members used in the calculations may be different. The Gibbs free energy of the ion-exchange reaction (G_r), calculated from thermodynamic data of the end-members, plays an important role in determining the solution parameters through Equation (10). In general, the derived W solution parameters should be tied with a specific thermodynamic data base for self-consistency. For a limited solid solution, there is additional uncertainty in determining the W parameters because of fictive end-members whose thermodynamic data were estimated based on systematics. For example, the solubility of $FeSiO_3$ in the silicate perovskite structure is limited. The calculated maximum solubility, defined by the reaction $(Mg,Fe)SiO_3$ (pv) = $(Mg,Fe)O$ (mw) + SiO_2 (st), is affected by both the W parameters of the perovskite solid solution and the thermodynamic data of the fictive end-member $FeSiO_3$-perovskite used in the calculations. One can fit the solubility data to optimize the

Table 3. Selected *W* solution parameters of solid solutions derived from partitioning data.

Solid solution	$W_{Mg\text{-}Fe}$ J/mol	$W_{Fe\text{-}Mg}$ J/mol	*References*
(MgFe)O (mw)	16100	26300 - 5.56T	[1]
(MgFe)Si$_{0.5}$O$_2$ (ol)	4500 + 130P	6500 + 130P	[1]
(MgFe)Si$_{0.5}$O$_2$ (wa)	1000	2000	[1]
(MgFe)Si$_{0.5}$O$_2$ (sp)	3900 - 1.10T	3900	[1]
(MgFe)SiO$_3$ (pv)	4130 - 1.37T + 110P	-4050 - 2.45T + 150P	[1]
(MgFe)Si$_{0.5}$O$_2$ (ol)	4500	4500	[2]
(MgFe)Al$_{2/3}$SiO$_4$ (gt)	2117	693	[2]
(MgFe)SiO$_3$ (px)	0	0	[2]
(MgFe)Si$_{0.5}$O$_2$ (ol)	3500	3500	[3]
(MgFe)Si$_{0.5}$O$_2$ (ol)	5000	5000	[4]
(MgFe)Si$_{0.5}$O$_2$ (ol)	4140	4140	[5]
(MgFe)Al$_{2/3}$SiO$_4$ (gt)	815	815	[5]

P is in GPa, T in K.

mw = magnesiowüstite; ol = olivine; wa = wadsleyite; sp = silicate spinel; pv = silicate perovskite; gt = garnet; px = orthopyroxene.

References: [1] Fei et al. (1991) [2] Hackler and Wood (1989) [3] Davidson and Mukhopadhyay (1984) [4] O'Neill and Wall (1987) [5] O'Neill and Wood (1979).

thermodynamic data of the fictive end-member FeSiO$_3$-perovskite, assuming an ideal solution of perovskite (W = 0), or by optimizing the W parameters using the estimated thermodynamic data of FeSiO$_3$-perovskite based on systematics, or by optimizing both. The uncertainties associated with fictive end-members are also evident in the sublattice solution model which describes multicomponent mixtures with multiple crystallographic sites.

SUMMARY

Phase relations in the system MgO-FeO-SiO$_2$ at high pressures and temperatures are relatively well established through high-pressure experiments at least to conditions corresponding to the top portion of the lower mantle. The Mg-Fe partitioning data in this system are available over a wide pressure range. The solution models for magnesiowüstite, olivine, wadsleyite, spinel, and perovskite, derived from the partitioning data, are generally consistent with the phase equilibrium data at high pressures and temperatures (e.g. Fei et al. 1991). The solution properties of (Mg,Fe)SiO$_3$ ilmenite and majorite solid solutions are not well determined. Because (Mg,Fe)SiO$_3$ ilmenite forms limited solid solution, the perovskite-ilmenite-spinel+stishovite univariant line may provide the most useful information for constructing the phase diagram when it is combined with the determined phase relations for the MgSiO$_3$ end-member. The existence and stability field of cubic (Mg,Fe)SiO$_3$ majorite need to be confirmed.

In addition to the Mg-Fe exchange, the heterovalent substitution MgSi-AlAl becomes important when Al$_2$O$_3$ is introduced into the MgO-FeO-SiO$_2$ system. The existing experimental data only provide phase relations involving either the MgSi-AlAl substitution in the FeO-free system (Akaogi et al. 1987, Gasparik 1992, Kanzaki 1987, Irifune et al.

1996, Hirose and Fei 1998a) or the FeSi-AlAl substitution in the MgO-free system (Akaogi and Akimoto 1977). New experimental data in the $Mg_4Si_4O_{12}$-$Mg_3Al_2Si_3O_{12}$-$Fe_4Si_4O_{12}$-$Fe_3Al_2Si_3O_{12}$ quadrilateral system involving both Mg-Fe and Mg(Fe)Si-AlAl exchanges are required to define the system completely.

Partitioning experiments involving ferric iron (Fe^{3+}) require control of oxygen fugacity at high pressure and characterization of the Fe^{3+}/Fe^{2+} ratios in each coexisting phase. Quantitative determination of the distribution coefficients involving Fe^{3+} may rely on development of new techniques such as synchrotron Mössbauer spectroscopy (Hastings et al. 1991).

So far, all of the reported partitioning experiments involving high-pressure phases have been of the synthesis type. It is demonstrated that it is possible to design reversal experiments using pre-synthesized high-pressure phases as the starting materials and to obtain partitioning data efficiently by use of the multi-cell technique at high pressures and temperatures. The Mg-Fe distribution coefficient between coexisting magnesiowüstite and perovskite should be re-determined by correctly designed exchange experiments in which mixtures of magnesiowüstite and pre-synthesized perovskite are used as starting materials, and compared with the existing data derived from the decomposition of $(Mg,Fe)_2SiO_4$ olivine solid solution.

ACKNOWLEDGMENTS

This research was supported by NSF grants EAR-9418945 and EAR-96-28092, the National Science Foundation Center for High Pressure Research and the Carnegie Institution of Washington. I thank A. Navrotsky, R. J. Hemley, and C. M. Bertka for useful comments.

REFERENCES

Acree WE Jr (1984) Thermodynamic Properties of Nonelectrolyte Solutions. Academic Press, New York.
Akaogi M, Akimoto S (1977) Pyroxene-garnet solid solution equilibria in the system $Mg_4Si_4O_{12}$-$Mg_3Al_2Si_3O_{12}$ and $Fe_4Si_4O_{12}$-$Fe_3Al_2Si_3O_{12}$ at high pressures and temperatures. Phys Earth Planet Inter 15:90-106.
Akaogi M, Navrotsky A, Yagi T, Akimoto S (1987) Pyroxene-garnet transition: Thermochemistry and elasticity of garnet solid solutions, and application to a pyrolite mantle. *In:* MH Manghnani, Y Syono (eds) High Pressure Research in Mineral Physics. p 251-260, Terra Scientific, Tokyo—Am Geophys Union, Washington, DC.
Akaogi M, Ito E, Navrotsky A (1989) Olivine-modified spinel-spinel transitions in the system Mg_2SiO_4-Fe_2SiO_4: Calorimetric measurements, thermochemical calculation, and geophysical application. J Geophys Res 94:15671-15686.
Akaogi M, Kojitani H, Matsuzaka K, Suzuki T, Ito E (1998) Postspinel transformations in the system Mg_2SiO_4-Fe_2SiO_4: Element partitioning, calorimetry, and thermodynamic calculation. *In:* MH Manghnani, T Yagi (eds) Properties of Earth and Planetary Materials at High Pressure and Temperature. p 373-384, Am Geophys Union, Washington, DC.
Akimoto S (1987) High-pressure research in geophysics: past, present and future. *In:* MH Manghnani, Y Syono (eds) High Pressure Research in Mineral Physics. p 1-13, Terra Scientific, Tokyo—Am Geophys Union, Washington, DC.
Akimoto S, Fujisawa H (1968) Olivine-spinel solid solution equilibria in the system Mg_2SiO_4-Fe_2SiO_4. J Geophys Res 73:1467-1473.
Akimoto S, Syono Y (1970) High-pressure decomposition of the system $FeSiO_3$-$MgSiO_3$. Phys Earth Planet Inter 3:186-188.
Bartholomew PR (1989) Interpretation of the solution properties of Fe-Mg olivines and aqueous Fe-Mg chlorides from ion-exchange experiments. Am Mineral 74:37-49.
Bell PM, Yagi T, Mao HK (1979) Iron-magnesium distribution coefficients between spinel [$(Mg,Fe)_2SiO_4$], magnesiowüstite [$(Mg,Fe)O$], and perovskite [$(Mg,Fe)SiO_3$]. Carnegie Inst of Washington Yr Bk 78:618-621.

Bertka CM, Fei Y (1997) Mineralogy of Martian interior up to core-mantle boundary pressures. J Geophys Res 102:5251-5264.

Bertrand GL, Acree WE Jr, Burchfield T (1983) Thermodynamical excess properties of multicomponent systems: representation and estimation from binary mixing data. J Sol Chem 12:327-340.

Boyd FR (1973) A pyroxene geotherm. Geochim Cosmochim Acta 37:2533-2546.

Boyd FR, England JL (1960) Apparatus for phase-equilibrium measurements at pressures up to 50 kilobars and temperatures up to 1750°C. J Geophys Res 65:741-748.

Brett R (1971) The Earth's core: Speculations on its chemical equilibrium with the mantle. Geochim Cosmochim Acta 35:203-221.

Brett R (1984) Chemical equilibrium of the Earth's core and upper mantle. Geochim Cosmochim Acta 48:1183-1188.

Capobianco CJ, Jones JH, Drake MJ (1993) Metal-silicate thermochemistry at high temperature: Magma oceans and the 'excess siderophile element' problem of the Earth's upper mantle. J Geophys Res 98:5433-5443.

Conrad PG (1998) The stability of almandine at high pressures and temperatures. *In:* MH Manghnani, T Yagi (eds) Properties of Earth and Planetary Materials at High Pressure and Temperature. p 393-399, Am Geophys Union, Washington, DC.

Davidson PM, Mukhopadhyay DK (1984) Ca-Fe-Mg olivines: Phase relations and a solution model. Contrib Mineral Petrol 86:256-263.

Drake MJ, McFarlane EA, Gasparik T, Rubie DC (1993) Mg-perovskite/silicate melt and majorite garnet/silicate melt partition coefficients in the system CaO-MgO-SiO$_2$ at high temperatures and pressures. J Geophys Res Planets 98:5427-5431.

Fabrichnaya OB (1998) The phase relations in the FeO-MgO-Al$_2$O$_3$-SiO$_2$ system: Assessment of thermodynamic properties and phase equilibria at pressures up to 30 GPa. Geochim Cosmochim Acta, in press.

Fei Y, Bertka CM (1996) The α–β transition in systems relevant to the upper mantle. EOS (abstr), 77:649.

Fei Y, Bertka CM (1998) Chemical composition and mineralogy of the upper mantle, transition zone, and lower mantle. *In:* Y Fei, CB Bertka, BO Mysen (eds) Mantle Petrology: Field Observations and High-Pressure Experimentation, Special Publication in honor of F.R. Boyd, in press. Geochemical Society, Houston.

Fei Y, Mao HK, Mysen BO (1991) Experimental determination of element partitioning and calculation of phase relations in the MgO-FeO-SiO$_2$ system at high pressure and high temperature. J Geophys Res 96:2157-2169.

Fei Y, Mao HK, Shu J, Parthasathy G, Bassett WA (1992) Simultaneous high P-T x-ray diffraction study of β-(Mg,Fe)$_2$SiO$_4$ to 26 GPa and 900 K. J Geophys Res 97:4489-4495.

Fei Y, Saxena SK (1986) A thermochemical data base for phase equilibria in the system Fe-Mg-Si-O at high pressure and temperature. Phys Chem Mineral 13:311-324.

Fei Y, Saxena SK, Eriksson G (1986) Some binary and ternary silicate solution models. Contrib Mineral Petrol 94:221-229.

Fei Y, Virgo D, Mysen BO, Wang Y, Mao HK (1994) Temperature dependent electron delocalization in (Mg,Fe)SiO$_3$-perovskite. Am Mineral 79:826-837.

Fei Y, Wang Y, Finger LW (1996) Maximum solubility of FeO in (Mg,Fe)SiO$_3$-perovskite as a function of temperature at 26 GPa: Implication for FeO content in the lower mantle. J Geophys Res 101:11525-11530.

Finnerty AA, Boyd FR (1987) Thermobarometry for garnet peridotite xenoliths: a basis for upper mantle stratigraphy. *In:* PH Nixon (ed), Mantle Xenoliths. p 381-402, Wiley, New York.

Flory PJ (1953) Molecular configuration of polyelectrolytes. J Chem Phys 21:162-163.

Ganguly J (1982) Mg-Fe order-disorder of ferromagnesian silicates. II Thermodynamics, kinetics, and geological applications. *In:* SK Saxena (ed) Advances in Physical Geochemistry. p 58-99, Springer-Verlag, Berlin.

Ganguly J, Saxena SK (1987) Mixtures and Mineral Reactions, Springer-Verlag, Berlin.

Ganguly J, Saxena SK (1984) Mixing properties of aluminosilicate garnets: Constraints from natural and experimental data, and applications to geothermo-barometer. Am Mineral 69:88-97.

Gasparik T (1992) Melting experiments on the enstatite-pyrope joint at 80-152 kbar. J Geophys Res 97:15181-15188.

Gasparik T, Drake MJ (1995) Partitioning of elements among two silicate perovskites, superphase B, and volatile-bearing melt at 23 GPa and 1500-1600°C. Earth Planet Sci Lett 134:307-318.

Geiger CA, Newton RC, Kleppa OJ (1987) Enthalpy of mixing of synthetic almandine-grossular and almandine-pyrope garnets from high-temperature solution calorimetry. Geochim Cosmochim Acta 51:1755-1763.

Gessmann C, Rubie DC, O'Neill HST (1995) The effects of pressure, temperature, and oxygen fugacity on element partitioning between magnesiowüstite and liquid metal: Implications for the formation of Earth's core. Eur J Mineral 7:79.

Guggenheim EA (1937) Theoretical basis of Raoult's law, Trans Faraday Soc 33:151-159.

Guggenheim EA (1952) Mixtures, Clarendon Press, Oxford.

Guyot F, Madon M, Peyronneau J, Poirier JP (1988) X-ray microanalysis of high-pressure/high-temperature phase synthesized from natural olivine in a diamond-anvil cell. Earth Planet Sci Lett 90:52-64.

Hackler RT, Wood BJ (1989) Experimental determination of Fe and Mg exchange between garnet and olivine and estimation of Fe-Mg mixing properties in garnet. Am Mineral 74:994-999.

Harley S (1984) An experimental study of the partitioning of Fe and Mg between garnet and orthopyroxene. Contrib Miner Petrol 86:359-373.

Harte B, Harris JW, Hutchison MT, Watt GR, Wilding MC (1998) Lower mantle mineral associations in diamonds from Sao Luiz, Brazil. *In:* Y Fei, CB Bertka, BO Mysen (eds) Mantle Petrology: Field Observations and High-Pressure Experimentation, Special Publication in honor of F. R. Boyd, in press. Geochemical Society, Houston.

Hastings JB, Siddons DP, van Bürck U, Hollatz R, Bergmann U (1991) Mössbauer spectroscopy using synchrotron radiation. Phys Rev Lett 66:770-772.

Helffrich G, Wood BJ (1989) Subregular model for multicomponent solutions. Am Mineral 74:1016-1022.

Herzberg C, Zhang J (1996) Melting experiments on anhydrous peridotite KLB-1: Compositions of magmas in the upper mantle and transition zone. J Geophys Res 101:8271-8295.

Hildebrand JH (1929) Solubility, XII, Regular solutions. J Am Chem Soc 51:66-80.

Hillgren VJ, Drake MJ, Rubie DC (1994) High-pressure and high-temperature experiments on core-Mantle segregation in the accreting Earth. Science 264:1442-1445.

Hillgren VJ, Drake MJ, Rubie DC (1996) High pressure and high temperature metal-silicate partitioning of siderophile elements: the importance of silicate liquid composition. Geochim Cosmochim Acta 60:2257-2263.

Hirose K, Fei Y (1998a) Majorite-Perovskite Transformation in the system $MgSiO_3$-$Mg_3Al_2Si_3O_{12}$ at lower mantle pressures, submitted to Phys Earth Planet Inter.

Hirose K, Fei Y (1998b) Melting phase relations of mantle peridotite and MORB compositions in the lower mantle. EOS Trans, American Geophysical Union 79.

Hirose K, Fei Y, Ma Y, Mao HK (1998) Fate of subducted basaltic crust in the lower mantle, submitted to Nature.

Holzheid A, Rubie DC, O'Neill HSC, Palme H (1995) The influence of pressure on the metal-silicate partition coefficients for Ni and Co. Eur J Mineral 7:110.

Inoue T (1994) Effect of water on melting phase relations and melt composition in the system Mg_2SiO_4-$MgSiO_3$-H_2O up to 15 GPa. Phys Earth Planet Inter 85:237-263.

Irfune T (1987) An experimental investigation of the pyroxene-garnet transformation in a pyrolite composition and its bearing on the constitution of the mantle. Phys Earth Planet Inter 45:324-336.

Irfune T, Ringwood AE (1987) Phase transformations in primitive MORB and pyrolite compositions to 25 GPa and some geophysical implications *In:* MH Manghnani, Y Syono (eds) High Pressure Research in Mineral Physics. p 231-242, Terra Scientific, Tokyo—Am Geophys Union, Washington, DC.

Irfune T, Ringwood AE (1993) Phase transformations in subducted oceanic crust and buoyancy relationships at depths of 600-800 km in the mantle. Earth Planet Sci Lett 117:101-110.

Irfune T, Ringwood AE, Hibberson WO (1994) Subduction of continental crust and terrigenous and pelagic sediments: an experimental study. Earth Planet Sci Lett 126:351-368.

Irifune T, Koizumi T, Ando J (1996) An experimental study of the garnet-perovskite transformation in the system $MgSiO_3$-$Mg_3Al_2Si_3O_{12}$. Phys Earth Planet Inter 96:147-157.

Ito E, Takahashi E (1987) Melting of peridotite at uppermost lower-mantle conditions. Nature 328:514-517.

Ito E, Takahashi E (1989) Postspinel transformations in the system Mg_2SiO_4-Fe_2SiO_4 and some geophysical implications. J Geophys Res 94:10637-10646.

Ito E, Yamada H (1982) Stability relations of silicate spinel ilmenites and perovskites. *In:* S Akimoto, MH Manghnani (eds) High-Pressure Research in Geophysics. p 405-419, Center for Academic Publishing of Japan, Tokyo.

Ito E, Takahashi E, Matsui Y (1984) The mineralogy and chemistry of the lower mantle: an implication of the ultrahigh-pressure phase relations in the system MgO-FeO-SiO_2. Earth Planet Sci Lett 67:238-248.

Ito E, Kubo A, Katsura T, Akaogi M, Fujita T (1998) High-pressure transformation of pyrope $(Mg_3Al_2Si_3O_{12})$ in a sintered diamond cubic anvil assembly. Geophys Res Lett 25:821-824.

Jamieson HE, Roeder PL (1984) The distribution of Mg and Fe^{2+} between olivine and spinel at 1300°C. Am Mineral 69:283-291.

Jones JH, Drake MJ (1986) Geochemical constraints on core formation in the Earth. Nature 322:221-228

Kamaya N, Ohtani E, Kato T, and Onuma K (1993) High pressure phase transitions in a homogeneous model martian mantle. *In:* E Takahashi, R Jeanloz, D Rubie (eds) Evolution of the Earth and Planets, p.19-26, Am Geophys Union, Washington, DC.

Kanzaki M (1987) Ultrahigh-pressure phase relations in the system $MgSiO_3-Mg_3Al_2Si_3O_{12}$. Phys Earth Planet Inter 49:168-175.

Kato T (1986) Stability relation of $(Mg,Fe)SiO_3$ garnets, major constituents in the Earth's interior. Earth Planet Sci Lett 69:399-408.

Kato T, Irifune T, Ringwood AE (1987) Majorite partition behaviour and petrogenesis of Earth's upper mantle. Geophys Res Lett 14:1546-1549.

Kato T, Ringwood AE, Irifune T (1988a) Experimental determination of element partitioning between silicate perovskites, garnets and liquids: constraints on early differentiation of the mantle. Earth Planet Sci Lett 89:123-145.

Kato T, Ringwood AE, Irifune T (1988b) Constraints on element partition coefficients between $MgSiO_3$ perovskite and liquid determined by direct measurements. Earth Planet Sci Lett 90:65-68.

Kato T, Ohtani E, Ito Y, Onuma K (1996) Element partitioning between silicate perovskites and calcic ultrabasic melt. Phys Earth Planet Inter 96:201-207.

Katsura T, Ito E (1989) The system $Mg_2SiO_4-Fe_2SiO_4$ at high pressures and temperatures: Precise determination of stabilities of olivine, modified spinel, and spinel. J Geophy Res 94:15663-15670.

Katsura T, Ito E (1996) Determination of Fe-Mg partitioning between perovskite and magnesiowüstite. Geophys Res Lett 23:2005-2008.

Kawamoto T, Holloway JR (1997) Melting temperature and partial melt chemistry of H_2O-saturated mantle peridotite to 11 Gigapascals. Science 276:240-243.

Kawamoto T, Hervig RL, Holloway JR (1996) Experimental evidence for a hydrous transition zone in the early Earth's mantle. Earth Planet Sci Lett 142:587-592.

Kawasaki T, Matsui Y (1977) Partitioning of Fe^{2+} and Mg^{2+} between olivine and garnet. Earth Planet Sci Lett 37:159-166.

Kawasaki T, Matsui Y (1983) Thermodynamic analyses of equilibria involving olivine, orthopyroxene and garnet. Geochim Cosmochim Acta 47:1661-1679.

Kesson SE, Fitz Gerald JD (1991) Partitioning of MgO, FeO, NiO, MnO and Cr_2O_3 between magnesian silicate perovskite and magnesiowüstite: implications for the origin of inclusions in diamond and the composition of the lower mantle. Earth Planet Sci Lett 111:229-240.

Kesson SE, Fitz Gerald JD, Shelley JM (1998) Mineralogy and dynamics of a pyrolite lower mantle. Nature 393:252-255.

Kohlstedt DL, Keppler H, Rubie DC (1996) Solubility of water in the α, β and γ phases of $(Mg,Fe)_2SiO_4$. Contrib Mineral Petrol 123:345-357.

Lee HY, Ganguly J (1988) Equilibrium compositions of coexisting garnet and orthopyroxene: Experimental determinations in the system $FeO-MgO-Al_2O_3-SiO_2$, and applications. J Petrol 29:93-113.

Li J, Agee CB (1996) Geochemistry of mantle-core differentiation at high pressure. Nature 381:686-689.

Liu L (1977) The system enstatite-pyrope at high pressures and temperatures and the mineralogy of the Earth's mantle. Earth Planet Sci Lett 36:237-245.

MacGregor ID (1974) The system $MgO-Al_2O_3-SiO_2$:Solubility of Al_2O_3 in enstatite for spinel and garnet peridotite compositions. Am Mineral 59:110-119.

Mao HK, Bell PM, Yagi T (1982) Iron-magnesium fraction model for the Earth. *In:* S Akimoto, MH Manghnani (eds) High-Pressure Research in Geophysics. p 319-325, Am Geophys Union, Washington, DC.

Mao HK, Shen G, Hemley RJ (1997) Multivariable dependence of Fe-Mg partitioning in the lower mantle. Science 278:2098-2100.

Martinez I, Wang Y, Guyot F, Liebermann RC, Doukhan JC (1997) Microstructures and iron partitioning in $(Mg,Fe)SiO_3$ perovskite - $(Mg,Fe)O$ magnesiowüstite assemblages:An analytical transmission electron microscopy study. J Geophys Res 102:5265-5280.

Matsui Y, Nishizawa O (1974) Iron (II)-magnesium exchange between olivine and calcium-free pyroxene over a temperature range 800°C to 1300°C, Bull Soc Fr Mineral Crystallogr 97:122-130.

McCammon C (1997) Perovskite as a possible sink for ferric iron in the lower mantle. Nature 387:694-696.

McCammon CA, Rubie DC, Ross II CR, Seifert F, O'Neill HStC (1992) Mössbauer spectra of $^{57}Fe_{0.05}Mg_{0.95}SiO_3$ perovskite at 80 and 298 K. Am Mineral, 77:894-897.

McFarlane EA, Drake MJ, Rubie DC (1994) Element partitioning between Mg-perovskite, magnesiowüstite, and silicate melt at conditions of the Earth's mantle. Geochim. Cosmochim. Acta 58:5161-5172.

Minarik WG, Fei Y, Hauri E (1997) Trace element partitioning between majorite garnet, silicate perovskite and peridotite melt. EOS Trans, American Geophysical Union 78:F811-812.

Minarik WG, Hauri E, Fei Y (1998) Direct determination of trace-element partitioning between silicate perovskite and peridotite melt. EOS Trans American Geophysical Union 79.

Nafziger RH Muan A (1967) Equilibrium phase compositions and thermodynamic properties of olivines and pyroxenes in the system MgO-"FeO"-SiO$_2$. Am Mineral 52:1364-1385.

Nashizawa O, Akimoto S (1973) Partition of magnesium and iron between olivine and spinel, and between pyroxene and spinel. Contrib Mineral Petrol 41:217-230.

Navrotsky A (1987) Models of crystalline solutions. *In:* ISE Carmichael, HP Eugster (eds) Thermodynamic Modeling of Geological Materials: Minerals, Fluids and Melts. p 35-69, Mineral Soc Am, Washington, DC.

Nell J, Wood BJ (1989) Thermodynamic properties in a multicomponent solid solution involving cation disorder: Fe$_3$O$_4$-MgFe$_2$O$_4$-FeAl$_2$O$_4$-MgAl$_2$O$_4$ spinel. Am Mineral 74:1000-1015.

O'Neill HSC, Wood BJ (1979) An experimental study of Fe-Mg partitioning between garnet and olivine and its calibration as a geothermometer. Contrib Mineral Petrol 70:59-70.

O'Neill HSC, Wall VJ (1987) The olivine-orthopyroxene-spinel oxygen geobarometer, the nickel precipitation curve, and the oxygen fugacity of the Earth's upper mantle. J Petrol 28:1169-1191.

O'Neill HSC, Navrotsky A (1984) Cation distribution and thermodynamic properties of binary spinel solid solutions. Am Mineral 69:733-753.

Ohtani E, Kawabe I, Moriyama J, Nagata Y (1989) Partitioning of elements between majorite garnet and melt and implications for petrogenesis of komatiite. Contrib Mineral Petrol 103:263-269.

Ohtani E, Kagawa N, Fujino K (1991) Stability of majorite (Mg,Fe)SiO$_3$ at high pressures and 1800°C. Earth Planet Sci Lett 102:158-166.

Ohtani E, Yurimoto H (1996) Element partitioning between metallic liquid, magnesiowüstite, and silicate liquid at 20 GPa and 2500°C: A secondary ion mass spectrometric study. Geophys Res Lett 23:1993-1996.

Ohtani E, Yurimoto H, Seto S (1997) Element partitioning between metallic liquid, silicate liquid and lower-mantle minerals: implications for core formation of the Earth. Phys Earth Planet Inter 100:97-114.

Ohtani E, Moriwaki K, Kato T, Onuma K (1998) Melting and crystal-liquid partitioning in the system Mg$_2$SiO$_4$-Fe$_2$SiO$_4$ to 25 GPa. Phys Earth Planet Inter 107:75-82.

Ono S, Yasuda A (1996) Compositional change of majoritic garnet in a MORB composition from 7 to 17 GPa and 1400 to 1600°C. Phys Earth Planet Inter 96:171-179.

Redlich O, Kister AT (1948) Thermodynamics of non-electrolyte solutions, x-y-t relations in a binary system, Ind Eng Chem 40:314-345.

Renon H, Prausnitz JM (1968) Local compositions in thermodynamic excess functions for liquid mixtures. Am Inst Chem Eng J 14:135-144.

Righter K, Drake MJ (1997) Metal-silicate equilibrium in a homogeneously accreting earth: new results for Re. Earth Planet Sci Lett 146:541-553.

Righter K, Drake MJ, Yaxley G (1997) Prediction of siderophile element metal-silicate partition coefficients to 20 GPa and 2800°C: the effects of pressure, temperature, oxygen fugacity, and silicate and metallic melt compositions. Phys Earth Planet Inter 100:115-134.

Ringwood AE (1966) Chemical evolution of the terrestrial planets. Geochim Cosmochim Acta 30:41-104.

Sawamoto H (1987) Phase diagrams of MgSiO$_3$ at high pressures up to 24 GPa and temperatures up to 2200°C: Phase stability and properties of tetragonal garnet. *In:* MH Manghnani, Y Syono (eds) High Pressure Research in Mineral Physics. p 209-219, Terra Scientific, Tokyo—Am Geophys Union, Washington, DC.

Shen G, Mao HK, Hemley RJ (1996) Laser-heated diamond anvil cell technique: double-sided heating with multimode Nd: YAG laser, Proc ISAM'96, 149-152.

Sundman B, Ågren J (1981) A regular solution model for phases with several components and sublattices, suitable for computer applications. J Phys Chem Solids 42:297-301.

Takahashi E (1986) Melting of a dry peridotite KLB-1 up to 14 GPa: Implications on the origin of peridotitic upper mantle. J Geophys Res 91:9367-9382.

Takahashi E, Ito E (1987) Mineralogy of mantle peridotite along a model geotherm up to 700 km depth. *In:* MH Manghnani, Y Syono (eds) High Pressure Research in Mineral Physics. p 427-438, Terra Scientific, Tokyo—Am Geophys Union, Washington, DC.

Thibault Y, Walter MJ (1995) The Influence of pressure and temperature on the metal-silicate partition coefficients of nickel and cobalt in a model C1 chondrite and implications for metal segregation in a deep magma ocean. Geochim Cosmochim Acta 59:991-1002.

Thompson JB Jr (1967) Thermodynamic properties of simple solutions. *In:* PH Abelson (ed) Researches in Geochemistry. p 340-361, John Wiley, New York.

Walker D, Norby L, Jones JH (1993) Superheating effects on metal-silicate partitioning of siderophile elements. Science 262:1858-1861.

Walter MJ, Thibault Y (1995) Partitioning of tungsten and molybdenum between metallic liquid and silicate melt. Science 270:1186-1189.

Weng K, Mao HK, Bell PM (1982) Lattice parameters of the perovskite phases in the system $MgSiO_3$-$CaSiO_3$-Al_2O_3. Carnegie Institution of Washington Yearbook 81:273-277.

Wilson GM (1964) A new wxpression for the excess free energy of mixing. J Am Chem Soc 86:127-130.

Wood BJ (1995) The effect of H_2O on the 410-kilometer seismic discontinuity. Science 268:74-76.

Wood BJ, Kleppa OJ (1981) Thermochemistry of forsterite-fayalite olivine solutions. Geochim Cosmochim Acta 45:529-534.

Wood BJ, Rubie DC (1996) The effect of alumina on phase transformations at the 660-kilometer discontinuity from Fe-Mg partitioning experiments. Science 273:1522-1524.

Yagi T, Bell PM, Mao HK (1979) Phase relations in the system MgO-FeO-SiO_2 between 150 and 700 kbar at 1000°C. Carnegie Inst of Washington Yr Bk 78:614-618.

Yasuda A, Fujii T, Kurita K (1994) Melting phase relations of an anhydrous mid-ocean ridge basalt from 3 to 20 GPa: implications for the behavior of subducted oceanic crust in the mantle. J Geophys Res 99:9401-9414.

Young TE, Green II HW, Hofmeister AM, Walker D (1993) Infrared spectroscopic investigation of hydroxyl in β-$(Mg,Fe)_2SiO_4$ and coexisting olivine: implication for mantle evolution and dynamics. Phys Chem Mineral 19:409-422.

Yusa H, Akaogi M, Ito E (1993) Calorimetric study of $MgSiO_3$ garnet and pyroxene: heat capacities, transition enthalpies, and equilibrium phase relations in $MgSiO_3$ at high pressures and temperatures. J Geophys Res 98:6453-6460.

Zhang J, Herzberg C (1994) Melting experiments on anhydrous peridotite KLB-1 from 5.0 to 22.5 GPa. J Geophys Res 99:17729-17742.

Chapter 12

HIGH-PRESSURE MELTING OF DEEP MANTLE AND CORE MATERIALS

Guoyin Shen

Consortium for Advanced Radiation Sources, University of Chicago
and Advanced Photon Source, Argonne National Laboratory
9700 South Cass Avenue
Argonne, Illinois 60439

Dion L. Heinz

Department of the Geophysical Sciences and
James Franck Institute, University of Chicago
5734 South Ellis Avenue
Chicago, Illinois 60637

INTRODUCTION

Melting is the major force in chemical differentiation of the Earth. The concept is supported by evidence that early in our planet's history the core and mantle segregated from each other from a relatively homogenous body that condensed out of the solar nebula. The crust and upper and lower mantle then further chemically segregated. Melting experiments on Earth materials provide tight experimental constraints on the geotherm to great depth within the Earth. For example, the melting of silicate perovskite provides an upper bound on the temperature in the lower mantle (Heinz 1986), and the melting of iron places a constraint on the temperature at the inner-core/outer-core boundary (Jeanloz and Richter 1979, Boehler 1993). With the appropriate use of geophysical modeling the iron melting point can also be used to constrain the temperature at the core-mantle boundary. The process of melting is interesting in that the theories that describe melting have evolved from the exact description (i.e. the thermodynamic definition), to the believable approximations (e.g. Lindemann Law), and finally, to implementations of the exact criteria using modern computer science advances and complex approximations. On the other hand, experimental determination of melting has been possible for a very long time. It is only the technology used to achieve higher and higher temperatures and pressures that has changed over the last several decades.

We review very high-pressure melting of material believed to comprise the deep mantle and core of the Earth. After a review of some physical concepts and theoretical models of melting, we summarize the high-pressure melting of major deep mantle and core phases together with some experimental techniques. Data, both controversial and consistent, are evaluated and discussed in a historical context.

THEORY

Characteristics of melting

Melting is the transition from a solid to a liquid phase. The liquid state is characterized by the absence of long-range order and resistance to shear. Melting involves an increase in entropy, and a volume increase for most substances by a few percent. Thus, the mean interatomic distances increase generally by a few percent while the coordination number

0275-0279/98/0037-0012$05.00

changes by a few to 20 percent. A melt has a fractional coordination number, which can change continuously with pressures and temperatures. The microscopic structure of a melt is not well known; therefore, all melting theories developed thus far are imperfect.

For a single component, melting may take place congruently, with the composition of solid and liquid being the same (e.g. Fe, NaCl). Some compounds melt incongruently, decomposing into a liquid and another compound (e.g. enstatite at ambient pressure). For a multicomponent system, the melting phenomena are more complicated, and a phase diagram is usually required to describe the composition-temperature space at a given pressure. Generally, a multicomponent system may exhibit solid-solution melting, eutectic melting, or a mixture of both. Melting in the Earth obviously involves multicomponent systems (rocks). The equilibrium temperature of a multicomponent solid and its melt will always be lower than the melting point for the pure substances due to the entropy of mixing. The decrease in melting temperature depends on the nature and proportion of the components. At present, our knowledge of melting processes in the Earth is very primitive. Geophysicists usually use the melting curves of deep earth minerals and of iron to obtain constraints on the temperature profile of the Earth.

Thermodynamics of melting

The thermodynamic description of melting is relatively straightforward. Like all other phase transitions, the melting curve is defined when the Gibbs free energy of the solid and the liquid are equal. Melting is a first-order phase transition since it involves measurable discontinuities in volume and entropy, the first derivatives of the free energy. The slope of the melting curve is given by the Clausius-Clapeyron equation,

$$dT_m/dP = \Delta V_m/\Delta S_m \tag{1}$$

where $\Delta V_m = V_l - V_s$ is the melting volume at P and $\Delta S_m = S_l - S_s$ the melting entropy at P.

The entropy of melting is always positive, since the liquid is always more disordered than the solid. Melting volume is usually positive, and then, according to (1), results in a positive melting slope, i.e. the melting point increases as pressure increases. In general, the melting curve of a solid that undergoes no phase transition is convex toward the T-axis because the melt is more compressible than the solid. For materials with $\Delta V_m < 0$, the melting has a negative slope. In some cases, the melting curve may show a maximum at high pressures and usually ends at a triple point where another phase becomes stable. However, the melting curve of the low-pressure phase can be extrapolated in the stability domain of the high-pressure phase, thus leading to the so-called *amorphization* which occurs at high pressure (Mishima et al. 1984, Hemley et al. 1988).

Melting equations

There are a number of equations that have been used to extrapolate melting curves. In principle, the Clausius-Clapeyron equation yields a precise description of a melting curve, but in practice it has been difficult to use this equation because the details of the structure of a liquid are known with far less precision than are those of the solid. Historically semi-empirical models and theoretical models have been used to extrapolate low-pressure melting data to high pressures.

Simon equation. Simon melting law can be described by

$$P - P_0 = a \ [(T_m/T_0)^c - 1], \tag{2}$$

where P_0, T_0 are the reference pressure and temperature (e.g. ambient condition or a triple point), and a and c are parameters of material. The Simon equation has been fitted to the melting curve of many materials: solidified gases (Simon and Glatzel 1929), organic crystals (Babb 1963), and oxides and silicates (Bottinga 1985). Gilvarry (1956) showed that the Simon melting law is really a combination of the Murnaghan equation of state and the Lindemann law. Usually, the Simon equation cannot be used to extrapolate the melting curve outside the interval studied experimentally.

Kraut-Kennedy equation. The Kraut-Kennedy law shows that T_m is linear with $\Delta V/V_0$:

$$T_m = T_{m0} (1 + C \, \Delta V/V_0), \tag{3}$$

where T_m and $\Delta V/V_0$ are appropriate to the melting boundary and C is the parameter of the material. In the volume-temperature phase diagram, the melting curve obtained from Equation (3) is a straight line. The Kraut-Kennedy relation can be derived from the Lindemann law (Gilvarry 1966, Vaidya and Gopal 1966). The parameter C can be expressed as a function of the Grüneisen parameter: $C = 2 \, (\gamma-1/3)$. Thus we expect that the Kraut-Kennedy law is only applicable in the pressure range where Grüneisen parameter γ is insensitive to pressure. Indeed, the melting curves of ionic crystals and silicates are concave toward the $\Delta V/V_0$ axis, overestimating the melting temperature of minerals (Kennedy and Vaidya 1970).

Lindemann law. Lindemann (1910) assumed that melting occurs when the amplitude of the vibration of atoms reaches a value large enough so that the solid lattice is shaken apart, and in this theory the liquid plays no role. Gilvarry (1966) suggested that the root mean square amplitude of atomic vibrations at melting is a critical fraction f of the distance a of separation of nearest neighbor atoms:

$$<u^2> = f^2 a^2 . \tag{4}$$

If the vibrational frequencies of a solid are approximated with a Debye spectrum, the Lindemann law can be written:

$$T_m = (4\pi^2/9h^2) f^2 a^2 M \Theta_D^2 \tag{5}$$

where M is the mean mass of an atom, a is the mean interatomic spacing, Θ_D the Debye temperature, and h, k are, respectively, Plank's and Bolzmann's constants. Assuming $a = (V/N)^{1/3}$, N being Avogadro's number, Equation (5) can be expressed in terms of the atomic volume:

$$T_m = 0.00321 \, f^2 M V^{2/3} \Theta_D^2 \tag{6}$$

The Lindemann ratio f at the zero-pressure melting temperature is not exactly the same for all materials, but is sensitive to crystal structure and the nature of atomic forces. The Debye temperature can be expressed as a function of the elastic constants or determined from an inversion of heat capacity data. In most cases, the Lindemann law is applied semi-empirically, with f determined at one melting point; an equation of state is then used to extrapolate the melting point to high pressures (Mulargia 1986, Stacey 1977, Wolf and Jeanloz 1984). When no experimental melting point is available, a systematic study of substances with the same structure can lead to an estimation of one melting point (usually the zero-pressure point; Poirier 1989).

The Lindemann law holds for some classes of materials but not for other classes. Wolf and Jeanloz (1984), taking into account anharmonic corrections, found that the agreement between the curve predicted by the Lindemann law and experimental melting curves was poor for diopside, faylite, and pyrope, but agreed well for forsterite. Equation (6) shows that the Lindemann relation is based on the assumption that the frequency spectrum of a substance can be represented by a single number Θ_D. When a single value Θ_D does not suffice to describe thermoelastic properties, the Lindemann law appears to be invalid. Therefore, Anderson (1995) pointed out that the Lindemann law may be applied for dense oxides and silicates such as those identified for the lower mantle, but the upper mantle minerals may or may not obey the Lindemann law. Still, Poirier (1991) comments that "the Lindemann correlation, if applied to crystals of the same structure, may be the basis of fruitful systematics and reasonable extrapolations of the melting curve."

Molecular dynamics simulations

Computer simulations based on molecular dynamics have been used to study the structural, thermodynamic, and physical properties of many materials. The goal of these simulations is to reproduce the actual motion of a specified collection of molecules. Such numerical computations are widely applied for high-pressure melting. The molecular dynamics (MD) calculation rests on an interatomic potential either derived from ab initio quantum mechanical calculations or obtained empirically by fitting to observed properties. Additional details can be found in the chapter by Stixrude et al. (this volume). We will quote a few recent publications where the relevant references can be found.

Cohen and Gong (1994) performed large-scale MD simulations of clusters of MgO for melting up to 300 GPa, using the potential induced breathing (PIB) model, which allows the ions to behave in a nonrigid fashion. They found that, in the extreme pressure limit, melting is characterized only by dynamical changes such as onset of rapid diffusion, and not by local structural changes, since high pressure favors efficient packing of the liquid as well as the solid. Vocadlo and Price (1996) employed the supercell method to study melting of MgO to 150 GPa with a variety of potential models. Belonoshko and Dubrovinsky (1996a,b) studied the melting of MgO with a two-phase MD method (considering both solid and liquid phases), and found that the melting temperature is substantially lower than that predicted by a one-phase method (Vocadlo and Price 1996). The reported MD simulations (Belonoshko and Dubrovinsky 1996a,b; Cohen and Gong 1994, Cohen and Weitz 1998, Vocadlo and Price 1996) on MgO show that the calculated melting temperature is substantially higher than that determined experimentally by Zerr and Boehler (1994).

The recent MD simulation on Al_2O_3 (corundum) (Ahuja et al. 1998) produced a melting curve in agreement with experimental data (Shen and Lazor 1995) up to 25 GPa. MD studies of $MgSiO_3$ perovskite were carried out by Matsui and Price (1991), Kubicki and Lasaga (1992), Belonoshko (1994), Wentzcovitch et al. (1995), and Matsui (1998). The volume difference between $MgSiO_3$ perovskite and the liquid was found to decrease substantially with pressure, but a volume inversion between the two phases never occurs at high pressures in these studies, indicating a positive melting slope (dT_m/dP). MD has also been used for the simulation of crystals and liquids in silicate systems (e.g. Matsui 1998, Zhou and Miller 1997). Zhou and Miller (1997) proposed a method for direct determination of the slope of the liquidus at a given composition and temperature. They applied the method to the system Mg_2SiO_4-$MgSiO_3$ at low pressures and gave results in qualitative agreement with experimentally determined phase diagrams. For the MgO-$MgSiO_3$ system, Zhou and Miller (1997) calculated that the eutectic temperature is 280-560 K below the melting temperature of $MgSiO_3$ perovskite throughout the lower mantle.

EXPERIMENTAL TECHNIQUES

At pressures below about 25 GPa, different high pressure facilities have been used to study the effect pressure on melting relations; these include gas apparatus (Hollaway 1971), piston cylinder (Bell and Williams 1971), Bridgman anvil (Jamieson and Lawson

1962), belt apparatus (Hall 1958), and different kinds of multi-anvil presses (Kumazawa 1971, Wang et al. 1998). There are two major techniques generally used for ultrahigh pressure melting studies: shock-wave and internally heated diamond-cell methods. In this section, we briefly outline a few technical aspects of shock-wave and diamond-cell experiments.

Shock-wave experiments

In shock-wave experiments, the sample is subjected to high pressures and temperatures by dynamic processes. A shock wave is driven through a sample by means of an explosive charge or by the collision of a flying object. A number of techniques have been developed to measure the velocity of shock wave and the velocity to which the sample material has been accelerated. These measurements allow one to derive the density and pressure of the sample under loading (Brown and McQueen 1982).

A continuous Hugoniot curve in the diagram of density versus pressure usually means that the material undergoes no phase transition. In other words, an offset in the Hugoniot curve can be attributed to a phase transition of the sample. Because of the short time scales of shock experiments, possible overshoot and non-equilibrium are of concern. Hugoniot temperatures were often calculated (Brown and McQueen 1986). In recent years, advances have been made in the measurement of sample temperature during shock loading. For example, Holland and Ahrens (1998) employed a six-channel pyrometer and observed temperature decays in 250 ns. The technique and results are discussed further in Stixrude and Brown (this volume).

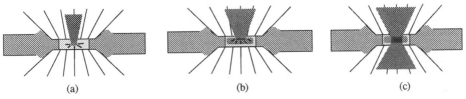

(a)   (b)   (c)

Figure 1. Typical sample configurations for laser heated diamond anvil cells: (a) low absorption sample; (b) opaque sample; (c) opaque sample with double-sided laser heating.

Laser-heated diamond cell

The generation and measurement of simultaneous high pressures and high temperatures in the laser-heated diamond anvil cell has undergone rapid development in recent years. Typical sample configurations for laser-heated diamond cells are shown in Figure 1. Major difficulties in laser heated diamond cell experiments are the characterization of samples at ultra high P-T conditions and the measurement of pressure and temperature. We take the opportunity to evaluate two aspects in these experiments: (1) the temperature measurement, and (2) the criteria used to define melting.

Temperature measurement. Temperature measurement is based on the collected radiation from thermal emission through the Planck radiation law:

$$I_\lambda = c_1\varepsilon(\lambda)\lambda^{-5}/\ [exp(c_2/\lambda T) - 1] \tag{7}$$

where I_λ is spectral intensity, λ is wavelength, T is temperature, $c_1 = 2\pi\ hc^2 = 3.7418 \times 10^{-16}\ Wm^2$, $c_2 = hc/k = 0.014388$ mK, and $\varepsilon(\lambda)$ is emissivity with $\varepsilon(\lambda) = 1$ for a blackbody.

In the Wien approximation, the irradiance from a grey body is given by the formula:

$$I_\lambda = c_1\varepsilon(\lambda)\lambda^{-5} / \exp(c_2/\lambda T) \tag{8}$$

This approximation is valid to within 4% for temperatures up to 6000 K (Heinz and Jeanloz 1987). The formula can be expressed by

$$J(\lambda) = \ln\varepsilon(\lambda) - w(\lambda)T^{-1} \tag{9}$$

where $J(\lambda) = \ln(I_\lambda \lambda^5/c_1)$ and $w(\lambda) = c_2/\lambda$, are two observable variables (Jeanloz and Heinz 1984). Since this equation is linear in T^{-1}, a linear least-squares fit can be used to solve for the temperature.

Temperature measurements by spectral radiometry have been validated by measuring melting points of a variety of materials at (close to) ambient pressure (Boehler et al. 1990, Heinz et al. 1991, Saxena et al. 1994, Shen et al. 1996). The biggest advantage of spectral radiometry is that temperatures can be measured independently from the absolute emissivity. However, materials are only approximately greybodies, since their emissivities are wavelength dependent. Moreover, temperatures estimated from the greybody approximation are more inaccurate at higher temperatures (Heinz 1993). Use of the available data on the wavelength-dependent emissivity at ambient pressure (de Vos 1954) gives about a 5% temperature correction downward at 3000 K and about 12% at 5000 K when compared to the results, assuming wavelength-independent emissivity. Unfortunately, such a correction cannot be accurately made because the wavelength dependence of emissivity is not currently known at high pressures and high temperatures, and it can be strongly dependent on the surface finish of a sample in the diamond cell. This is the limitation, or challenge, in the absolute accuracy of temperature measurement by spectral radiometry.

The temperatures reported with the laser-heated diamond cell have been obtained with the assumption of wavelength-independent emissivity, or a small given variation over the wavelength range measured for thermal emission. This consistent use of emissivity cannot, however, fully explain the large discrepancies existing in reported melting temperatures. Large errors could exist when there is a strong temperature gradient due to the chromatic aberration of a temperature measurement system. A reflecting objective was employed to reduce the system's chromatic aberration (Boehler et al. 1990, Jephcoat and Besedin 1996, Lazor et al. 1993, Sweeney and Heinz 1998). However, due to the strong dispersion of the diamond windows, the optical system cannot be totally free of chromatic aberration. The main source of error in the temperature estimate arises from the temperature gradient in a small hot spot. If temperature distribution is uniform, the uncertainty due to the chromatic aberration can be minimized by calibration. In order to check this, the melting temperature of platinum with wire heating was measured with a 6-mm thick silica glass plate between the platinum and the collecting lens (Shen et al. 1996). This glass created a severe chromatic aberration. The melting temperature was calculated to be 2380±25 K calibrated without this 6-mm thick glass. However, when a calibration was made with the glass, this yielded a melting temperature of 2060±25 K, near the literature value. Unfortunately, it is difficult to locate standards with known temperature gradients comparable in degree and size with the laser heating spot. Without such a calibration, the temperature profile obtained with an existing temperature gradient may not represent true values, due to the chromatic aberration. Therefore, it is important to have uniform heating on a large area of the sample to improve accuracy in temperature determination during laser-heating experiments. The temperature gradients have been greatly reduced by use of CO_2 lasers (Boehler 1992) and multimode YAG and YLF lasers (Shen et al. 1996), including combining beams having TEM_{00} and TEM_{01} modes (Shen et al. 1998a).

Melting criteria. Melting is thermodynamically defined as equilibrium between a solid and a liquid. When materials melt, their physical properties, such as density, viscosity, absorption properties, and electrical resistance, change dramatically. Such property changes are generally used for recognition of melting. Visual optical observation is a common way to determine whether melting has taken place. The temperature-resistance correlation is widely used as well, and it was used in internal wire-heated diamond cells to 43 GPa (Boehler 1986). The definition of *melt* in laser-heated diamond-cell experiments has been a subject of many articles (Anderson and Duba 1997, Boehler 1996, Duba 1994, Jeanloz and Kavner 1996, Jephcoat and Besedin 1996, Lazor and Saxena 1996). The lack of agreement in the literature on measured melting temperature indicates the difficulty in characterizing melting at extreme conditions in diamond cells.

Jeanloz and Kavner (1996) summarized five types of melting criteria, namely fluid flow, glass feature, quench texture, change in sample properties, and temperature-versus-laser-power correlation. They concluded that the most reliable criteria for determining melting inside the laser-heated diamond cell are fluid flow and quenched glass observations. The criterion of textural change (fluid flow) has been adopted by many studies on iron melting (Boehler 1993, Jephcoat and Besedin 1996, Williams et al. 1991). Fluid motion in the sample observed during heating is certainly an indication of a melt phase. However, it is less reliable for the determination of the onset of melting. Shen et al. (1993) pointed out that the textural change becomes less obvious as pressure increases; above 40 GPa only very occasional and small movement is typically observed, making it difficult to identify the onset of melting. The clear, continuous, fluid-like motion was observed only after the temperature was further increased by a few hundred degrees. Lazor and Saxena (1996) argued that the best criterion for recognition of melting is the one based on the combination of visual observation and laser power-temperature correlation, and believed that even in the absence of a visual observation, the latter alone could be used successfully. However, the underlying cause of the laser power-temperature correlation is still not well established. Other effects, such as changes in grain boundary or recrystallization, could play a role as well. When accompanied by beaded rims, melting is a likely candidate for the cause of the discontinuities in laser power-temperature correlation (Boehler 1993). The beaded rims alone could be interpreted as indicating a temperature above the melting point because of the time and size required to develop observable beads (Duba 1994).

The visual observation of fluid-like motion requires a high-resolution optical system to enable observation of a small texture change in a 1-2 μm area. The use of a long working-distance objective required for the diamond cell and heating optics limits the spatial resolution of the optical system, making the visual observation difficult and subjective. More objective criteria include laser power-temperature correlation (Boehler et al. 1990, Lazor et al. 1993), intensity discontinuity of reflected laser light (Boehler et al. 1990, Shen and Lazor 1995), thermal analysis (Sweeney and Heinz 1998), and x-ray diffraction (Shen et al. 1998b). The laser power-temperature correlation alone as a melting criterion requires further work in order to understand its underlying cause, and should be used together with visual observation (Boehler et al. 1990, Lazor and Saxena 1996). The measurement of reflected laser light from a heated area is not a good indicator of melting, as pointed out by Shen and Lazor (1995). A discontinuity in reflected intensity could be seen from any change in surface texture. The thermal analysis method is conceptually similar to the laser power-temperature correlation. Sweeney and Heinz (1998) found that melting temperatures determined by thermal analysis were exactly corroborated by quenched glass blobs on MgSiO₃ perovskite.

Recently, x-ray diffraction has been combined with laser-heated diamond cells and used for melting determination by x-ray diffraction (Mao et al. 1998, Shen et al. 1998b).

The appearance of diffraction peaks at a certain P-T condition clearly indicates the presence of a crystalline phase, because the double-sided laser heating technique provide essentially uniform heating over a region that is much larger (20-50 μm) than the x-ray beam size (10x10 μm). However, a simple loss of the diffraction peaks is not necessarily indicative of melting. While the small x-ray beam ensures the probing of a uniform high P-T region, the number of crystallites within the sampling region may be statistically insufficient for polycrystalline x-ray diffraction analysis. Possible crystal growth at high temperatures further compounds the problem. Several procedures were employed to minimize this effect in the Shen et al. (1998b) experiments. First, the iron sample was mixed with MgO (or Al_2O_3 or NaCl) powder in a ratio of approximately 1:1 by volume. Although the sample mixture helps to prevent coarse crystallinity, the preferred orientation effect; i.e. odd peak intensities or even missing diffraction peaks, was still observed at high temperature. Rotating the diamond cell around its axis allows the entire diffraction cone to be collected and averaged. With the laser beam coaxial to the rotating axis, the heating spot is unaffected by the rotation. In addition, efforts were made during experiments to bracket the melting temperatures by increasing and decreasing temperatures to observe the loss of diffraction peaks and the reappearance of crystalline peaks, respectively. Despite these efforts, Shen et al. (1998b) caution that the conclusive identification of melts is still limited by the energy-dispersive x-ray diffraction technique. Using an area detector with a monochromatic x-ray beam for a powder sample, or with a white x-ray beam for a single (a few) crystal(s), will significantly increase the diffraction statistics, and thus the reliability of melting determination. Direct measurement of the structure factor of the melt can also aid in determining the melting temperature.

MELTING OF DEEP MANTLE MINERALS

The melting temperature of the lower mantle is near the eutectic temperature in the $MgO-FeO-CaO-Al_2O_3-SiO_2$ system. Seismic observations show that the eutectic temperature must be above the mantle geotherm. Here we will discuss as examples the high-pressure melting of several end-member minerals: $MgSiO_3$ perovskite, $CaSiO_3$ perovskite, MgO, SiO_2, Al_2O_3. Table 1 summarizes the major experiments on the high-pressure melting of these minerals.

(Mg,Fe)SiO$_3$ perovskite

Iron-magnesium silicate perovskite is the dominant phase in the Earth's lower mantle (Knittle and Jeanloz 1987, Liu 1976, Mao et al. 1977). A large number of studies on high pressure melting can be found in the literature (Sweeney and Heinz 1998). Heinz and Jeanloz (1987b) performed the first systematic analysis of temperature measurement during laser heating. They identified two major issues that needed to be addressed in laser heating. The first is that the presence of temperature gradients (due to the mode structure of the heating laser) meant that some method of measuring the peak temperature was necessary. They also developed a tomographic technique to measure temperature gradients in the sample. The second major issue was the temperature fluctuations that were due to fluctuations in the heating laser power. This lead to their attempt at stabilizing the laser power using a feed back circuit. They used these newly developed techniques to measure the melting point of Bamble enstatite in the perovskite stability field.

Heinz and Jeanloz (1987a,b) noted several important effects during these experiments. The first is that Fe is mobile in the sample and can move via Soret diffusion. The second and possibly most important effect was that when the sample melted, it melted in a run-away manner. They attributed this to the partitioning of Fe preferentially into the melt, which absorbed more of the laser energy and led to the run-away character of the transition.

Table 1. A brief summary of melting experiments of mantle materials

Material	Reference	Comments	Melting criteria
Al_2O_3	Shen & Lazor 1995	Nd-YAG laser, metal heating plate, 25GPa	Laser power vs. temperature, YAG reflectivity
$CaSiO_3$	Shen & Lazor 1995	Nd-YAG laser, metal heating plate, 60 GPa	Laser power vs. temperature, YAG reflectivity
	Zerr et. al. 1997	CO_2 laser, 43 GPa	Glass
MgO	Zerr & Boehler 1994	CO_2 laser, 31.5 GPa	Glass
$(Mg,Fe)SiO_3$	Heinz & Jeanloz 1987a	Nd-YAG Laser, Bamble ensatite, 65 GPa	Glass blobs
	Knittle & Jeanloz 1989	Nd-YAG, Bamble ensatite, 96 GPa	Glass blobs
	Sweeney & Heinz 1993	Nd-YAG, Bamble ensatite, temperature stabilization, 94 GPa	Glass blobs and thermal analysis
	Zerr & Boehler 1993	CO_2 laser for Mg end member, and YAG for 10% iron, 62.5 GPa	Glass
	Shen & Lazor 1995	Nd-YAG laser, metal heating plate, 42 GPa	Laser power vs. temperature, YAG reflectivity
	Sweeney & Heinz 1998	Nd-YAG laser, Webster enstatite, 85 GPa	Glass blobs and thermal analysis
$(Mg,Fe)_2SiO_4$	Zerr & Boehler 1994	Nd-YLF laser, San Carlos olivine, 34 GPa	Glass
	Holland & Ahrens 1997	Shock wave	4300 ± 270 K @ 130 GPa
SiO_2	Shen & Lazor 1995	Nd-YAG laser, metal heating plate, 35GPa	Laser power vs. temperature, YAG reflectivity
	Lyzena et. al. 1983, Schitt & Ahrens 1989	Shock wave	4500 K @ 70 GPa 3000 ± 200 K @ 20-30 GPa

They used the appearance of glass as the melting criteria. The experiments reported measured the average temperature, which is determined by looking at the entire hot spot. Thus, the data needed to be corrected from average temperature to peak temperature, which is the highest temperature in the sample. Heinz and Jeanloz (1987a) used a correction from average temperature which was based upon the measurement of the temperature distributions in several independent experiments on the same composition using the tomographic technique developed in Jeanloz and Heinz (1984). They also measured the temperature profile across a partially molten spot and tried to determine the temperature where the melt and solid were coexisting. The major drawback of this experiment was the amount of time that it took to measure the temperature profile (of order 20-40 minutes). Heinz and Jeanloz (1987a) concluded that the melting temperature of $(Mg_{0.9}Fe_{0.1})SiO_3$ melted at approximately 3000 ± 500 K over the pressure range of 25-55 GPa. Heinz and Jeanloz (1987a) reported the pressure before heating, assuming that the thermal pressure was equal to the relaxation in the sample as it was heated.

In 1989 Knittle and Jeanloz (1989) followed up the work of Heinz and Jeanloz (1987a) by extending the experiments to 96 GPa. They detected a change in slope of the melting curve at 60 GPa, which they attributed to a change in the liquid structure at pressures above 60 GPa. They used a technique similar to that of Heinz and Jeanloz

(1987a), but did not make a fluctuation correction to their data. They did correct for the average to peak temperature for each experiment.

Heinz et al. (1991) developed a laser heating system that stabilized and controlled the temperature of the laser heated sample. They used a laser modulator that consisted of electro-opitcal liquid crystal cell which could effectively rotate the plane of polarization of the heating laser's beam, and a vertical polarizer that then rejected the horizontal component of the laser beam. This system allowed them to modulate the heating laser so that they could control the temperature of the sample by monitoring the brightness of the sample. The use of this system allowed Sweeney and Heinz (1993) to bypass the fluctuation correction that Heinz and Jeanloz (1987a) had used. Sweeney and Heinz (1993) still had to make a correction from the average temperature to the peak temperature since they did not measure the peak temperature directly. They arrived at a temperature that was slightly lower than the results of Heinz and Jeanloz (1987a) but still consistent with their results. Note that the actual temperatures reported in the paper by Sweeney and Heinz (1993) were the average temperatures that were measured. The correction to peak temperature where the melt first appears was not made in their figures.

Zerr and Boehler (1993) measured the melting point of $MgSiO_3$ using a CO_2 laser. The CO_2 laser allowed them to heat the Mg endmember without the presence of an absorber. The 10.6 µm laser beam is absorbed by the SiO_x polyhedra in the sample. The temperature gradient in their experiments (Zerr and Boehler 1993) were reduced in both the axial and radial direction. The reported melting temperatures are the last temperatures of the solid before the sample melted. They also measured the melting point of a $(Mg_{0.9}Fe_{0.1})SiO_3$ sample that they heated using a very stable Nd-YAG laser. As shown in Figure 2, their results differed significantly from those of Heinz and Jeanloz (1987a), Knittle and Jeanloz (1989) and Sweeney and Heinz (1993).

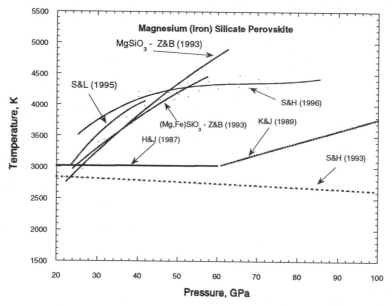

Figure 2. Melting of $(Mg,Fe)SiO_3$ perovskite. Only fitted lines to experimental data points are shown for clarity. H&J – Heinz and Jeanloz (1987); K&J – Knittle and Jeanloz (1989); S&H – Sweeney and Heinz (1993, 1998); Z&B – Zerr and Boehler (1993); S&L – Shen and Lazor (1995).

Shen and Lazor (1995) measured the melting temperature of $MgSiO_3$ using a metal plate beneath the sample to absorb the Nd-YAG laser beam and heat the sample. Melting was determined by plotting the laser power/sample temperature function and looking for the thermal anomaly associated with the fusion of materials. Their results agree within experimental uncertainties with those of Zerr and Boehler (1993) and the very recent data by Sweeney and Heinz (1998). In the experiments of Sweeney and Heinz (1998), melting was determined in situ by thermal analysis and corroborated by the appearance of glass in temperature quenched sample.

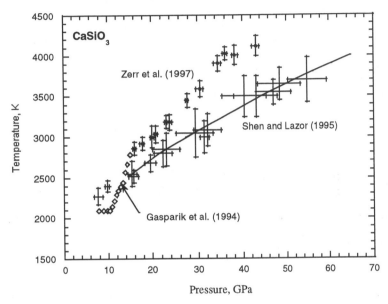

Figure 3. Melting of $CaSiO_3$ perovskite. Asolid line is the fit of the Simon melting formula (2) to the data of Shen and Lazor (1995) with $P_0 = 14$ GPa, $T_0 = 2490$ K, $a = 24$ GPa, and $c = 2.4$.

$CaSiO_3$ perovskite

The solubility of $CaSiO_3$ perovskite in $MgSiO_3$ phase is limited to a few percent (Irifune et al. 1989, Shen and Lazor 1995). $CaSiO_3$ perovskite is likely to exist in a separate phase in the lawer mantle. Gasparik et al. (1994) observed that the melting temperature for $CaSiO_3$ perovskite increased sharply by 360 K in the pressure interval of 13.5 - 15.2 GPa with multi-anvil apparatus. The first diamond cell data at pressures above 15 GPa by Shen and Lazor (1995) imply that the unusually deep dT_m/dP slope for $CaSiO_3$ perovskite (Gasparik et al. 1994) would flatten out at high pressures. The experiments of Shen and Lazor (1995) were carried out in a YAG laser heated diamond anvil cell with rhenium metal as an absorber of the laser light. A polished disc of the sample was in contact with the metal foil and heated by conduction. Melting was determined by plotting the laser power/sample temperature function and looking for the thermal anomaly associated with fusion of the material. Zerr et al (1997) measured melting temperatures of $CaSiO_3$ perovskite at 16-43 GPa by CO_2 laser heating in argon pressure media in diamond cells. The main criterion for melting in the Zerr et al (1997) experiments was a large increase in temperature due to an increase in absorption of the laser radiation, which is conceptually the same as the thermal anomaly in the laser power/sample temperature function (Shen and Lazor 1995). As shown in Figure 3, melting of Zerr et al (1997) shows

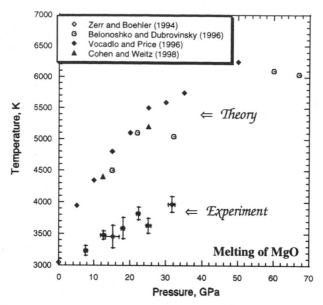

Figure 4. Melting of MgO. There is significant difference between the results of molecular dynamics simulations (Belonoshko and Dubrovinsky 1996, Vocadlo and Price 1996, Cohen ans Weitz 1998) and the experimental data by Zerr and Boehler (1994). The melting temperatures obtained by Cohen and Gong (1994) are higher than those of Cohen and Weitz (1998).

at higher temperatures than those of Shen and Lazor (1995). Zerr and Boehler (1994) obtained a melting slope (dT_m/dP) for $CaSiO_3$ perovskite that was almost identical to their previous data on $MgSiO_3$ perovskite (Zerr and Boehler 1993); in contrast Shen and Lazor (1995) showed that the melting slope of $MgSiO_3$ perovskite is higher than that of $CaSiO_3$ perovskite.

MgO

Magnesiowüstite, $(Mg,Fe)O$, is believed to be the second most abundant mineral in the lower mantle. The endmember MgO has been studied rather extensively by theory (Cohen and Gong 1994, Belonoshko and Dubrovinsky 1996b, Vocadlo and Price 1996, Cohen and Weitz 1998). Only one experimental study on high pressure melting has been done (Zerr and Boehler 1994). In contrast to the steep melting slope predicted by molecular calculations, Zerr and Boehler (1994) found that MgO has a flat melting curve with a slope at one atmosphere of 0.35 K/GPa, 3 to 5 times less than the theoretical prediction (Belonoshko and Dubrovinsky 1996b, Cohen and Gong 1994, Vocadlo and Price 1996, Cohen and Weitz 1998). Zerr and Boehler (1994) used the same techniques as for $(Mg,Fe)SiO_3$ perovskite and measured the melting of MgO to 31.5 GPa (Fig. 4). The melting curve of MgO is nearly parallel to that of FeO (Boehler 1992, Shen et al. 1993). The melting temperature of a magnesiowüstite, $(Mg_{0.85}Fe_{0.15})O$, is found to agree with the melting behavior of a solid solution system between MgO and FeO (Zerr and Boehler 1994).

SiO_2, Al_2O_3

Although free SiO_2 and Al_2O_3 most likely do not exist in substantial quantities in the lower mantle, they are constituents of silicate perovskites. Their melting data provide an endmember input for modeling the melting relations of the $FeO-CaO-Al_2O_3-MgO-SiO_2$ system, which represents the chemistry of the lower mantle.

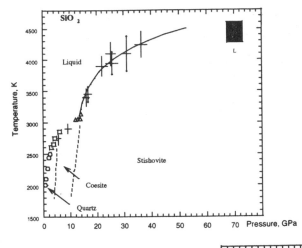

Figure 5. Melting of SiO_2.
○ = Jackson (1976)
□ = Kanzaki (1990)
△ = Zhang et al. (1993)
+ = Shen and Lazor (1995)
The size of bars represents the uncertainty in pressure and temperature. Shock-wave data for the melting of stishovite by Lyzenga et al. (1983) are also shown (L). A solid line is the fit of the Simon melting equation (2) to the data of Shen and Lazor (1995) with $P_0 = 13.6$ GPa, $T_0 = 3120$ K, $a = 1.5$ GPa, and $c = 9$. See also Dingwell, Chapter 13, this volume.

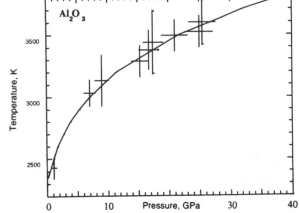

Figure 6. Melting of Al_2O_3. The results of Shen and Lazor (1995) are shown. The size of the symbols represents uncertainties in pressure and temperature. A solid line is the fit of the Simon melting equation (2) to the data with $P_0 = 0$ GPa, $T_0 = 2350$ K, $a = 1.75$ GPa, and $c = 6.5$.

At low pressures, melting of silica was reported by Jackson (1976), Kanzaki (1990) and Zhang et al. (1993). Shen and Lazor (1995) reported melting of stishovite at 15-37 GPa, using the same techniques as those used for $MgSiO_3$ perovskite. When extrapolating the data of Shen and Lazor (1995) to high pressures, they are in reasonable agreement with the melting point (T_m of 4500 K at 70 GPa) determined by shock-wave experiments (Lyzenga et al. 1983). A phase diagram is shown in Figure 5. A clear decrease in melting slope (dT_m/dP) can be seen for stishovite as pressure increases, indicating that the density of SiO_2 melt increases faster than that of the corresponding solid with increasing pressure.

Melting of corundum was studied by Shen and Lazor (1995) to 25.4 GPa for the first time (Fig. 6), using the metal-absorber technique. Even though it is unlikely to exist in the Earth in any substantial quantity as the pure oxide, the phase diagram of corundum is important because it is the endmember in Al_2O_3-containing systems. Technically, understanding the high-pressure melting of corundum is important, because in many melting experiments, corundum was used as thermal isolation material and pressure medium (e.g. Williams et al. 1991, Boehler 1993, Saxena et al. 1994). Melting temperatures of corundum were found to be higher than those of iron. Therefore, the use of corundum will not affect the determination of melting temperatures for iron unless there is a reaction at high pressure and temperatures.

MELTING OF CORE MATERIALS

Iron

Iron is believed to be the main component of the Earth's core based on a combination of evidence from geomagnetism (e.g. Jacobs 1987), seismology and high-pressure experimentation (e.g. Brown and McQueen 1986, Birch 1964), and cosmochemistry (Anders and Ebihara 1982). The phase diagram of iron has been studied extensively, because the melting temperature of iron can constrain the temperature at the inner core boundary and anchor the Earth's temperature profile.

At pressures below about 30 GPa, the phase diagram of iron is reasonably well known. At ambient pressure, the stable phase is α-Fe with a body-centered cubic (bcc) structure. At high temperature, it changes to γ-Fe, a face-centered cubic (fcc) structure, which transforms to a bcc δ-Fe phase at higher temperatures below the melting point (Strong et al. 1973). At high pressure, α-Fe transforms to ε-Fe with hexagonal-close-packed (hcp) structure. α-Fe to ε-Fe transformation has been studied in situ by x-ray (Akimoto et al. 1987, Bassett and Huang 1987). Wire-heated diamond anvil cells were used to study the ε-Fe to γ-Fe transformation (Boehler 1986, Mao et al. 1987). Recently, the transformation has been studied by synchrotron radiation (Funamori et al. 1996, Shen et al. 1998b). The experimental values of the α-ε-γ triple point are 11 GPa-763 K (Bundy 1965), 11.6 GPa-810 K (Boehler 1986), and 8.3 GPa-713 K (Akimoto et al. 1987).

Considerable controversy has surrounded the melting temperatures of iron at pressure over 30 GPa and between diamond-cell and shock-wave experiments (Boehler 1993, Brown and McQueen 1986, Williams et al. 1987, Yoo et al. 1993). Specifically, a range of P-T values had been reported for the ε-γ-liquid triple point. Moreover, the existence of new phases at high P-T conditions has been a subject of debate (Andrault et al. 1997, Boehler 1993, Dubrovinsky et al. 1998, Saxena et al. 1993, Shen et al. 1998b). The lack of agreement reflects the difficulties in the characterization of samples (e.g. unambiguous melting criteria) and in the measurement of pressure and temperature under extreme conditions. In this section, we briefly summarize the experiments to date. Table 2 gives a summary for the major experiments on the melting of iron.

Diamond-cell experiments. Liu and Bassett (1975) first studied melting of iron to 20 GPa with a wire-heated diamond anvil cell. Temperatures were measured by pyrometry and dependent on emissivity, which was estimated from the low-pressure data (Strong et al. 1973). The gasket was not introduced at that time, and large pressure gradients must have existed across the sample. Almost a decade later, high-pressure melting of iron was measured by Boehler (1986) to 43 GPa, using the same method but with new technologies in temperature measurement developed by that time (Boehler 1986, Jeanloz and Heinz 1984). The most significant improvement was the introduction of spectral pyrometry for temperature measurement. Temperatures were measured by collecting optical emission spectra over a large wavelength range from a hot wire, which were then fit to a blackbody radiation function. Knowledge of the absolute emissivity of the sample is not required; instead, the wavelength dependence of emissivity is needed to correct the measured intensities. Unfortunately, these properties for Earth materials are not known at high temperatures and pressures, if they are known at all. Constant emissivity within the 680-930 nm range was assumed in the Boehler (1986) study.

Mao et al. (1987) studied the ε-γ phase transition to 36 GPa using the wire-heated diamond cell with a different sample configuration. The slope of the γ-ε transition line was found to be 35 K/GPa by Mao et al. (1987), which is different from that obtained by Boehler (1986) (28.6 K/GPa). Williams et al. (1991) discussed the temperature gradient

Table 2. A brief summary of melting experiments of core materials

	Reference	Comments	Melting criteria
Fe	Liu & Basset 1975	Wire heated, melting, to 20 GPa	Resistance change
	Boehler 1986	Wire heated, melting, ε–γ, to 43 GPa	Resistance change
	Mao et al.. 1987	Wire heated, ε–γ, to 36 GPa	Resistance change
	Williams et al.. 1987	Laser heated, melting, to 105 GPa	Quench texture
	Boehler et al.. 1990	Laser heated, melting, to 120 GPa	Visual observation, reflectivity temperature vs laser power
	Boehler 1993	Laser heated, melting, new phase, γ–θ-*l* triple point, 100-200 GPa	Visual observation
	Shen et al.. 1993	Laser heated, melting, 50 GPa	Visual observation, temperature vs laser power
	Saxena et al. 1993	Laser heated, β phase, to 120 GPa	Temperature vs laser power
	Saxena et al. 1994	Laser heated, melting, to 150 GPa	Visual observation temperature vs laser power
	Yoo et al. 1995	Laser heated, x-ray, phase stability, to 110 GPa	X-ray diffraction
	Saxena et al. 1995	Laser heated, x-ray, dhcp	X-ray diffraction
	Jephcoat & Besedin 1996	Laser heated, melting, to 57 GPa	Visual observation
	Andraut et al. 1997	Laser heated, x-ray, orthorhombic structure, to 110 GPa	X-ray diffraction
	Dubrovinsky et al. 1998	wire heated, x-ray, dhcp, 68 GPa	X-ray diffraction
	Shen et al. 1998b	Double-sided laser heated, x-ray, melting, structure, ε–γ-l triple point, to 80 GPa	X-ray diffraction
	Brown & McQueen 1980, 1982	Shock wave	Sound velocity measurement, two discontinuities
	Brown & McQueen 1986	Shock wave	Sound velocity, Hugoniot temperature calculation
	Bass et al. 1987	Shock wave	Direct temperature measurement
	Yoo et al. 1993	Shock wave	Direct temperature measurement
	Gallagher & Ahrens 1994	Shock wave	Thermal conductivity measurement of widow materials
	Chen & Ahrens 1996	Shock wave	Shock wave on preheated iron
FeO	Knittle & Jeanloz 1991	Laser heated, melting, phase diagram, to 100 GPa, FeO, Fe-FeO	Quench texture
	Shen et al. 1993	Laser heated, melting, to 60 GPa, Fe, FeO	Visual observation
	Boehler 1992	Laser heated, melting, to 50 GPa Fe-FeO, Fe-FeS	Visual observation and laser power-temperature correlation
FeS	Williams & Jeanloz 1990	Laser heated, melting, to 100 GPa FeS, Fe-10%FeS	Visual observation, quench texture
	Boehler 1992	Laser heated, melting, to 50 GPa Fe-FeO, Fe-FeS	Visual observation and laser power-temperature correlation

across the diameter of a wire, and argued that the larger wire (20 μm) used in the Boehler (1986) experiment could result in lower temperature estimates when compared to the 10-μm wire used in the Mao et al. (1987) experiment. The resistive heating of wires in a diamond cell provided stable temperatures in large portions of the wire. Phase transitions and melting could be reliably detected by the change of electrical resistivity, which led Duba (1994) to conclude that resistive heating is "the most reliable measurement of melting of iron in the diamond-anvil cell." The low melting slope with the wire-heated diamond cell suggested a low-pressure ε-γ-liquid triple point, and a new phase was proposed by Boehler (1986) to explain the shock data at 200 GPa (Brown and McQueen 1986).

Williams et al. (1987) first reported high-pressure melting of iron to 105 GPa with the laser-heated diamond anvil cell. Their data were in disagreement with the those of Boehler (1986). The "cook and look" method was used to bracket the melting curve in Williams et al's (1987) experiments. The average temperatures measured by spectral radiometry were corrected to the peak values within the sample by measuring the temperature distribution across the laser-heated spot. Because of the extreme temperature gradients (~1000 K/μm) in the Williams et al. (1987) experiment, accurate measurement of the temperature distribution is not possible, as the correction for peak-versus-average temperature could introduce large errors. It is interesting to note that without such a correction the melting curve of Williams et al. (1987) would be identical to that of Boehler (1986) (see Fig. 12 in Williams et al. 1991). This leads to the question of proper correction for temperature gradients, both in laser-heating and wire-heating techniques.

Boehler et al. (1990) used a more powerful YAG laser to reduce the temperature gradient in a laser-heated spot. The reflecting objective was introduced to reduce the chromatic aberration in the temperature measurement optics. Melting of iron was determined in argon to 35 GPa, and continued in Al_2O_3 to 120 GPa. Melting temperatures obtained by Boehler et al. (1990) agreed well with (only slightly higher than) the data obtained with wire heating (Boehler 1986), but were significantly lower than those of Williams et al. (1987). Data on the wavelength dependence of tungsten emissivity were used in their temperature calculation, which resulted in temperature lower by about 200 K at 2500 K when compared to those calculated by assuming constant emissivity over the wavelength range. Temperatures were measured from a 3-μm area by locating a pinhole in front of the spectrograph's slit. No correction for peak-versus-average temperature was made in the Boehler et al. (1990) study. Melting was determined by observing convective motion at the onset of melting, which was found to be consistent with the observation of a discontinuity in the laser power-temperature correlation and a sharp drop of the intensity of reflected light from a hot spot (Boehler et al. 1990). Critical reviews of these melting criteria can be found in Duba (1994), Lazor and Saxena (1996), and Jeanloz and Kavner (1996).

These discrepancies led Shen et al. (1993) to measure high-pressure melting of iron with a laser-heated diamond cell system similar to that of Boehler et al. (1990). The iron sample was in an MgO medium and in direct contact with a polished corundum disc. Melting was detected in situ by watching the fluid-like motion in the molten sample illuminated by a defocused beam of Ar-laser and viewed by putting a 488 nm band-pass filter in front of a CCD camera. The first change in texture was considered indicative of melting. A large continuous change in texture was regarded as the point at which the melting temperature had been exceeded. When the laser was turned off after observing the first change in texture, no typical circular feature indicative of melting could be found. Shen et al. (1993) noted that the melting temperature would be overestimated substantially if the melting criterion were based on quench texture (Williams et al. 1987). At pressures higher than 40 GPa, the change in texture was found to be small, making visual observation very difficult. Shen et al. (1993) stopped at pressure about 150 GPa. Using the same laser-heated diamond-cell system and the sample geometry, Saxena et al. (1994) extended pressures to 1.5 GPa by adopting the laser power-temperature correlation as a melting criterion in addition to visual observation. The data of Shen et al. (1993) and Saxena et al.

(1994) are in agreement with those of Boehler et al. (1990), only at systematically slightly higher melting temperatures.

Analysis of the laser power-temperature correlation below the melting line gave some evidence of a possible new phase, called β-Fe by Saxena et al. (1993). Independently, Boehler (1993) reported melting of iron in an Al_2O_3 medium to 200 GPa. In his study, a more stable YLF laser and an intensified CCD detector were employed. Temperatures were measured from an area only 1 μm in diameter in the 550-800 nm spectral range, and no temperature correction was made. Possible diffraction effects on temperature measurements from a small area have been discussed (Jeanloz and Kavner 1996). Wavelength-independent emissivity was used in fitting to Planck's radiation function in the Boehler (1993) study. The iron sample was in direct contact with a polished disc of sapphire, in a surrounding matrix of ruby powder. Use of the sapphire disc reduced scattering of light, an important step for visual observation of melting. Another important modification is that the high-pressure cell was dried at 120°C in a vacuum oven and subsequently flushed with dry argon. The drying procedure ensured the removal of moisture and oxygen and avoided the possible reaction between sample and medium (Shen and Lazor 1995).

Based on the melting to 200 GPa (Boehler 1993), a triple point at 100 GPa and 2800 K was proposed based on the slope change of the melting curve. The slope of the ε-γ transition is well constrained by resistive heating experiments to over 40 GPa (Boehler 1986, Mao et al. 1987). Such a triple point requires a curved line of ε-γ transition above 50 GPa, which led Boehler (1993) to propose a new phase, based on the observation of small but reproducible changes in optical absorption, and changes in the slope of the laser power-temperature correlation. Jephcoat and Besedin (1996) measured the melting curve of argon and found that the melting points of iron and argon were equal at 47±1.0 GPa and 2750±200 K. Together with their observations at low pressures, they suggested that the melting curves of Boehler et al. (1990) and Shen et al. (1993) are lower bounds on the iron melting curve.

In the past few years, the laser-heated diamond cell has been combined with x-ray diffraction techniques that provides direct structural information (Andrault et al. 1997, Boehler et al. 1990, Mao et al. 1998, Saxena et al. 1995, Shen et al. 1998b, Yoo et al. 1995). Because x-ray diffraction is a measurement of a bulk property from a sampled volume, a uniform P-T condition within the whole volume is important in order to have a reliable result. Yoo et al. (1995) studied the phase diagram of iron to 130 GPa with a laser-heated diamond-cell system combined with an energy dispersive diffraction technique (beam line X17C, NSLS). The iron sample was contained in an Al_2O_3 medium and the low-power YAG laser in TEM_{00} mode was used to heat the sample from one end through a diamond. ε-Fe was found to be stable from 50 GPa to 110 GPa at high temperatures. Melting was determined by observing the disappearance of diffraction lines. Based on melting and solid phases, the ε-γ-liquid triple point was located at 2500 (±200) K and 50 GPa.

Using the same system, Saxena et al. (1995) reported a dhcp phase at about 38 GPa and temperatures between 1200 and 1500 K based on diffraction patterns of heated iron in a MgO medium. Andrault et al. (1997) reported a new phase with an orthorhombic structure in temperatures to 2375 K and pressures of 30 GPa-100 GPa. Angle-dispersive x-ray diffraction with an imaging plate was used by Andrault et al. (1997). The sample was placed in a rhenium gasket in an Al_2O_3 or SiO_2 pressure medium. The wire-heating technique was used again recently in an x-ray study of iron in argon to 48 GPa with a rotating anode source and an area detector (Siemens CCD) (Dubrovinsky et al. 1997), in Al_2O_3 to 68 GPa (a record-high pressure with wire heating), with synchrotron x-ray and

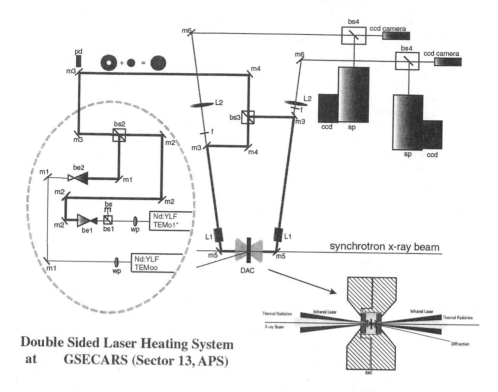

**Double Sided Laser Heating System
at   GSECARS (Sector 13, APS)**

Figure 7. Schematics of a double-sided laser heating system installed at GeoSoilEnviroCARS, Sector 13, at the Advanced Photon Source, Argonne National Laboratory. The heating laser is a combined beam of donut and gaussian beams. It is then split and directed to heat the sample in a diamond cell from both sides. The thermal radiation is collected by microscope optics from both sides for temperature measurement. X-ray measurements can be done *in situ* at high pressures and high temperatures. The x-ray beam is typically focused with Kirkpatrick-Baez mirrors.

energy dispersive diffraction (Dubrovinsky et al. 1998). Temperatures were estimated from the cell parameters of hcp iron. Both sets of results were interpreted in terms of the existence of a dhcp phase at temperatures above 1400 K and pressures above 35 GPa. These discrepancies reflect difficulties in combining laser-heated diamond cells with the x-ray diffraction technique.

Mao et al. (1998) pointed out that x-ray diffraction studies with laser heating could be improved by maximizing the heating spot size, x-ray intensity, spatial resolution, and diffraction accuracy, and minimizing the pressure gradients, temperature gradients (both radially and axially), temperature fluctuation, and sample differentiation. Based on a double-sided laser heating system (Shen et al. 1996), an integrated approach considering all these factors was performed with in situ x-ray diffraction measurements on a laser-heated sample in diamond cells (Mao et al. 1998, Shen et al. 1998b). A schematic of the system installed at the Advanced Photon Source is shown in Figure 7. Double-sided laser heating minimizes the axial temperature gradient in the laser-heated sample. Use of a multimode laser (Shen et al. 1996) or a laser system with both TEM_{00} mode (Gaussian) and TEM_{01} (donut) mode dramatically reduced the radial temperature gradient and increased the laser heating spot size at high pressure. A coaxial optical system provided an accurate alignment (within 2 μm) of the two laser-heating spots and their positions relative to the x-ray beam.

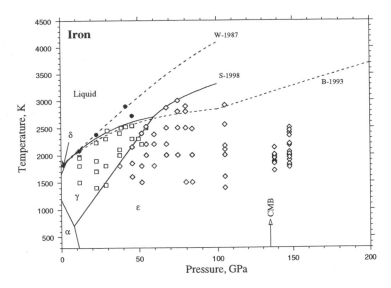

Figure 8. Phase diagram of iron to 200 GPa from diamond-cell experiments. Each symbol corresponds to one or more in situ x-ray diffraction measurements at a given P-T condition; data below 100 GPa are from Shen et al. (1998b), data above 100 GPa are from recent measurements from the Advanced Photon Source indicating the stability of the hcp phase (ε-Fe) to at least 155 GPa and 2500 K. Melting curves of W-1987 (Williams et al. 1987), B-1993 (Boehler 1993), and S-1998 (Shen et al. 1998b) are shown. Melting temperatures from Shen et al. (1993) to 50 GPa, Saxena et al. (1994) to 150 GPa, Jephcoat and Bedesin (1996) to 57 GPa are slightly higher than those of Boehler (1993); they are not shown here for clarity (see text). Phase boundaries are based on structure determinations by x-ray diffraction (Akimoto et al. 1987, Shen et al. 1998b).

The small x-ray beam size (<10 × 10 μm) improved the spatial resolution and ensured that the x-ray beam size was significantly smaller than the double-sided, laser-heated spot. By rotating the diamond cell around its axis, the diffraction statistics were improved when the samples suffered from strong preferred orientation, including growth of single crystals at high temperature.

High-pressure melting, phase transitions and structures of iron were studied to 84 GPa and 3500 K (Mao et al. 1998, Shen et al. 1998b). Recently, the pressure has been extended to 150 GPa. The phase diagram of iron to 200 GPa is shown in Figure 8. At pressures below 60 GPa, the lower bound on the melting curve was found to be close to those measured by Boehler et al. (1990) and Shen et al. (1993); however, at pressures above 60 GPa, x-ray data indicate melting at higher temperatures than reported in these studies, but still lower than the melting curve of Williams et al. (1987). The ε-γ-liquid triple point was determined as 60(±5) GPa and 2800(±200) K, based on data of the ε−γ phase transition and the observation of melting by in situ x-ray diffraction (Shen et al. 1998b). No solid phases other than ε-Fe and γ-Fe were observed in situ at high temperatures (>1000 K) and pressures to 150 GPa. However, the diffraction patterns of temperature-quenched products at high pressure can be fit to other structures such as dhcp. Melting was identified by the loss of crystalline diffraction peaks. As discussed in the section on melting criteria, care must be taken to identify melting by x-ray diffraction (Shen et al. 1998b). Further improvements in the use of x-ray diffraction for identifying melting should be possible with area detectors (e.g. CCD, image plate, multi-element solid state detectors), and by direct measurement of the structure factor of the melt.

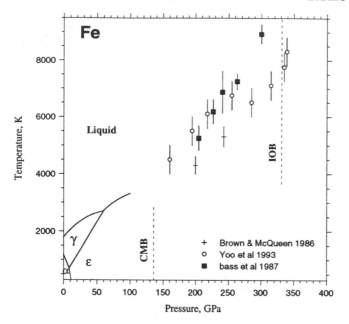

Figure 9. Summary of shock-wave together with the static pressure data for the phase diagram of iron (from Shen et al. 1998b).

Shock-wave experiments. The first shock-wave experiment on iron was done by Al'tshuler et al. (1962). By measuring the sound wave velocity along the Hugoniot, Brown and McQueen (1980, 1982) studied the melting of iron at core conditions. With a refinement in the temperature calculation made by assuming values for the specific heat and the Grüneisen parameters of iron, Brown and McQueen (1986) calculated the melting temperature at 243 GPa to be 5300±350 K and proposed a solid-solid phase transition at 200 GPa and 4200±350 K. Subsequently, temperature was measured directly in shock-wave experiments on iron to determine its melting temperature (Bass et al. 1987, Williams et al. 1987, Yoo et al. 1993). Temperature was determined by the radiance measured through the transparent window using the greybody approximation. Melting was defined by a discontinuity in the P-T path along the Hugoniot. This type of experiment requires a knowledge of the optical and thermal properties (especially thermal conductivity) of the window materials at extreme conditions. Lack of information could lead to large systematic errors, since the two materials must thermally equilibrate from their Hugoniot states. Gallagher and Ahrens (1994) reported a downward temperature correction of about 1000 K using a model that included the measured thermal conductivity of the window materials. Accordingly, they suggested that the adjustment to the measured temperatures (Bass et al. 1987, Williams et al. 1987, Yoo et al. 1993) would give good agreement with Brown and McQueen (1986) calculations. Recently, Chen and Ahrens (1996) performed shock-wave experiments on preheated iron to 1300°C and found melting at 74 GPa along the Hugoniot. Their results support the static diamond-cell experiments of Boehler (1993) and Saxena et al. (1994). Figure 9 summarizes the shock-wave studies of iron. Further discussion, including implications for the core, is provided by Stixrude and Brown (this volume).

FeO

Under ambient conditions, wüstite crystallizes in the NaCl (B1) structure with a nonstoichiometric formula of $Fe_{1-x}O$, where x is variable up to 0.12 (Hereafter $Fe_{1-x}O$ is designated as FeO). Above 17 GPa at 300 K, wüstite undergoes a displacive transition from B1 to a rhombohedral structure, and above 90 GPa at 600 K, a second transition to

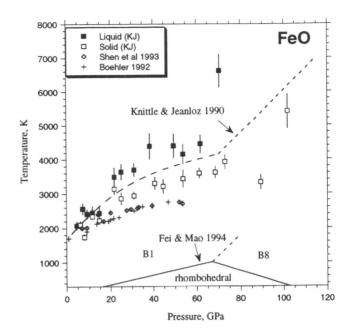

Figure 10. High pressure melting of FeO from laser-heating experiments. Phase boundaries were determined in resistance-heated diamond cells (Fei and Mao 1994).

the NiAs structure (Fei and Mao 1994, Mao et al. 1996). Melting of wüstite has been experimentally studied by Lindsley (1966) to 3 GPa with a piston-cylinder apparatus, and by Ringwood and Hibberson (1990) at a pressure of 16 GPa with a multi-anvil press. With the laser-heated diamond cell, the range of accessible pressures has been extended dramatically. These results are subject to the same concerns regarding temperature measurement and melting criteria as discussed above for iron. Here we briefly summarize experimental results for FeO (Fig. 10).

Knittle and Jeanloz (1991) measured the melting of FeO to 100 GPa. Their laser heating system is the same as the one used for the work on iron (Williams et al. 1991). Melting was determined by bracketing the lowest temperature measured in the liquid, and the highest temperature measured in the solid, at a given pressure. Above 70 GPa, a higher melting slope was obtained, consistent with their findings of a metallic phase transition by electrical resistivity measurements. Boehler (1992) measured the melting curve of FeO to 50 GPa, using a technique similar to that used for his iron melting experiments (Boehler et al. 1990). His data were in disagreement with Knittle and Jeanloz (1991). At 50 GPa, the melting temperature of Boehler is 1100 K lower than that of Knittle and Jeanloz (1991). Shen et al. (1993) measured the melting of wüstite to 60 GPa using a polished ruby chip as an insulating material. Their results agree closely with those of Boehler (1992), where an argon medium was used. Despite the discrepancies, all three sets of results indicate a crossover of the melting curves of Fe and FeO, above which pressure FeO melts at higher temperatures. The crossover occurs near 15 GPa and 2200 K in Shen et al. (1993), 20 GPa and 2200 K in Boehler (1992), and 5 GPa and 2000 K according to Knittle and Jeanloz (1991).

FeS

Stoichiometric FeS at ambient conditions (troilite, FeS-I) has a NiAs-type hexagonal structure. Above 3.4 GPa at 300 K, it transforms to a MnP structure (FeS-II) (King and Prewitt 1982), and above 7 GPa at 300 K to a monoclinic phase (FeS-III) (Fei et al. 1995,

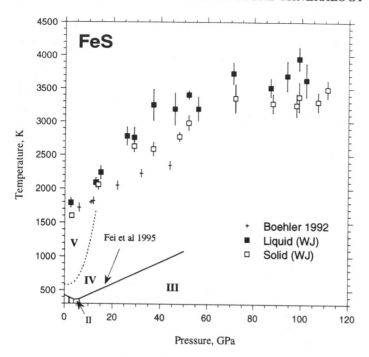

Figure 11. High pressure melting of FeS from laser-heating experiments. Phase boundaries are from Fei et al. (1995).

Kusaba and Syono 1998). High-pressure--high-temperature phases were reported with a NiAs-type superlattice structure (FeS-IV, FeS-V) (Fei et al. 1995). At ambient pressure, stoichiometric FeS melts slightly incongruently to Fe_7S_8-$Fe_{12}S_{13}$ and an iron-rich liquid (Jensen 1942). High-pressure experiments failed to resolve the presence of any incongruency (Boehler 1992, Williams and Jeanloz 1990). Melting temperatures reported by Williams and Jeanloz (1990) are higher than those reported by Boehler (1992). In the work of Williams and Jeanloz (1990), melting was determined by visual observation in situ, and quench texture observation with Al_2O_3 as a medium. In experiments by Boehler (1992), an argon medium was used and melting was recognized visually, together with the laser power-temperature correlation. The data of Williams and Jeanloz (1990) are consistent with estimates of melting temperatures of pyrrhotite ($Fe_{12}S_{13}$) at about 140 GPa in the shock-wave experiment (Brown et al. 1984). A phase diagram is shown in Figure 11.

Melting of iron alloy

As shown by Birch (1952), the density of the earth's outer core is about 10% too low to be pure iron, so the core must also contain some light elements. Candidate light elements are oxygen, sulfur, silicon, hydrogen, and carbon (Poirier 1994, Stixrude and Brown, this volume). At present, there is no particular constraint on the dominant light element in the core. Ternary and quaternary phase diagrams of Fe-light elements are needed to provide constraints on the nature of these light elements.

At ambient pressure, the addition of a small amount of light elements decreases the melting temperature of iron by up to a few hundred degrees. Data at high pressure are scarce and limited in the Fe-FeO-FeS system (Boehler 1992, Fei et al. 1997, Knittle and Jeanloz 1991, Ringwood and Hibberson 1990, Urakawa et al. 1987, Usselmann 1975, Williams et al. 1991). Eutectic melting was found in the Fe-FeS system by Usselman with the piston-cylinder apparatus, Fei et al. (1997) with a multi-anvil device, and Williams and

Jeanloz (1990) and Boehler (1996), both with laser-heated diamond cells. The eutectic temperatures were a few hundred degrees less than the melting temperatures of iron and FeS at pressures to 100 GPa. The eutectic melting temperatures were found by Boehler (1996) to increase faster with pressure than those of iron and FeS. The melting slope of Fe-FeO-FeS (a mixture of 70% Fe, 20% FeO, and 10% FeS) was reported with an even higher slope (Boehler 1996). In the Fe-FeO system, eutectic melting was reported by Ringwood and Hibberson (1990), using a multi-anvil press. In an experiment on the Fe-10% FeO system at 83 GPa, Knittle and Jeanloz (1991) found that the melting temperature was intermediate between melting points of iron and FeO, thus leading to the conclusion of a solid-solution between Fe and FeO. Boehler (1992) qualitatively agreed with the observation of Knittle and Jeanloz (1991), in that the melting point of iron was not significantly lowered in the presence of oxygen. He further pointed out the melting curves in the Fe-FeS-FeO system converge instead of diverge with pressure (Boehler 1996).

Several difficulties need to be addressed and overcome in melting experiments of multi-component systems at very high pressures (i.e. in diamond cells). (1) The laser heating spot is typically on the order of a few micrometers in these experiments, and non-uniform sample(s) on a scale of several micrometers could result in large experimental errors. A uniform heating spot of at least 20 μm in diameter is required for such a study (e.g. Boehler and Chopelas 1992, Shen et al. 1996). (2) Large temperature gradients in radial and axial directions could lead to compositional gradients (e.g. Soret diffusion), causing unmixing during heating and melting. (3) For opaque materials, the heat is generated at the sample-medium interface, which can cause a non-uniform sample layer. Such a thin layer makes the observation of eutectic melting or solidus difficult. These issues may be solved by the use of new techniques such as double-sided laser heating (Mao et al. 1998, Shen et al. 1996) and extensions of it.

CONCLUDING REMARKS

All models of the Earth's thermal history, chemical differentiation, and dynamical processes require knowledge of melting properties of candidate materials inside the Earth. Unfortunately, all melting theories so far are imperfect, in part because the microscopic structure of the melt is not well known. Our knowledge of high-pressure melting thus still largely relies on experiments. In the last fifteen years, constraints on the Earth's interior temperature have improved due to significant developments in high-pressure technologies needed for measuring melting properties of deep mantle and core materials. We have focused here on developments of the laser-heated diamond cell techniques. The use of spectral radiometry has greatly increased the accuracy of measuring temperature, measurements that have been validated by study of the melting points of a variety of materials at (or close to) ambient pressure (Boehler et al. 1990, Heinz et al. 1991, Saxena et al. 1994, Shen et al. 1996). With the use of proper feed-back systems, temperatures can be maintained to within 10 degrees at about 3000 K (Heinz et al. 1991), a precision sufficient to perform many types of high P-T experiments.

The lack of agreement in some previously measured melting temperatures reflects the difficulties in characterizing the state of materials under extreme P-T conditions. With recent technical advances many of these issues are being resolved. Production of large stable laser beams that interacts well with the sample open up whole new classes of experiments to be explored. Laser beams with total power of over 100 W have been used in diamond-cell experiments with CO_2 (Boehler and Chopelas 1992), YAG (Shen et al. 1996) and YLF lasers (Shen et al. 1998a). These systems allowing large heating spots, which in turn make the observation of the onset of the melting more reliable. The temperature gradient across the spot can now be dramatically reduced (Boehler and Chopelas 1992,

Mao et al. 1998) and can be further optimized by "building" a temperature distribution by mixing laser beams with different modes (Shen et al. 1998a). The uniform temperature distribution is also a key factor in accurately measuring the temperature (Shen et al. 1996). Very importantly, these powerful laser techniques open a new class of experiments aimed at measuring the high-pressure melting of multi-component system; e.g. the systematic investigation of the effect of light elements on the melting properties of iron and the melting of lower mantle rocks.

The future of static ultrahigh pressure melting studies is likely to be guided by the continued development of synchrotron radiation techniques. The high-brilliance synchrotron beam allows control of the x-ray beam to a size much smaller than the laser heating spot, but still with enough photon flux to measure accurately x-ray diffraction (Shen et al. 1998b) and to perform a variety of x-ray spectroscopies. For example, the double-sided heating technique (Shen et al. 1996, Mao et al. 1998) ensures a uniform temperature in the sampled volume. X-ray measurement provides an objective way of identifying melting, and an important extension of the visual observation methods used in previous studies. With the development of diffraction techniques giving sufficient statistics, melting can be determined by identifying the loss of diffraction peaks or by direct measurement of the structure factor of the melt. Since x-ray diffraction provides the means to give directly the structure and density of a material, accurate determination of the entire phase diagram, crystal structures, and equations of state to extreme P-T conditions, can now be obtained with these new techniques.

ACKNOWLEDGMENTS

We are grateful to R.J. Hemley and H.K. Mao for encouragement and useful discussions during the preparation of this review. We thank Jim Devine for preparation of figures and Sandy Heinz and Steve Gramsch for careful reading of the manuscript. This work was supported by the National Science Foundation and the W.M. Keck Foundation.

REFERENCES

Ahuja R, Belonoshko AB, Johansson B (1998) Melting and liquid structure of aluminum oxide using a molecular-dynamics simulation. Phys Rev E 57:1673-1676
Akimoto S, Suzuki T, Yagi T, Shimomura O (1987) Phase diagram of iron determined by high pressure/temperature x-ray diffraction using synchrotron radiation. *In:* MH Manghnani, Y Syono (eds) High Pressure Research in Mineral Physics. p 149-154 Terra Scientific, Tokyo/Am Geophys Union, Washington, DC
Al'tshuler LV, Bakanova AA, Trynin RF (1962) Shock adiabats and zero isotherms of seven metals at high pressures. Sov Phys JETP 15:65-74
Anders E, Ebihara M (1982) Solar system abundances of the elements. Geochim. Cosmochim Acta 46:2363-2380
Anderson OL (1995) Equation of State of Solids for Geophysics and Ceramic Science. Oxford Univ Press, Oxford, UK
Anderson OL, Duba AG (1997) Experimental melting curve of iron revisited. J Geophys Res 102:22659-22669
Andrault D, Fiquet G, Kunz M, Visocekas F, Hausermann D (1997) The orthorhombic structure of iron: An in situ study at high-temperature and high-pressure. Science 278:831-834
Babb SE (1963) Parameters in the Simon equation relating pressure and melting temperature. Rev Mod Phys 35:400-413
Bass JD, Svendsen B, Ahrens TJ (1987) The temperature of shock compressed iron. p 393-402 Am Geophys Union, Washington, DC
Bassett WA, Huang E (1987) Mechanism of the body centered cubic-hexagonal close-packed phase transition in iron. Science 238:780-783
Bell PM, Williams DW (1971) Pressure calibration in piston-cylinder apparatus at high temperature. p 195-215 Springer-Verlag, New York

Belonoshko AB (1994) Molecular dyanmics of MgSiO$_3$ perovskite at high pressures: equation of state, structure, and melting transition. Geochim Cosmochim 58:4039-4047

Belonoshko AB, Dubrovinsky LS (1996a) Molecular and lattice dynamics study of the MgO-SiO$_2$ system using a transferable interatomic potential. Geochim Cosmochim Acta 60:1645-1656

Belonoshko AB, Dubrovinsky LS (1996b) Molecular dynamics of NaCl (B1 and B2) and MgO (B1) melting: Two-phase simulation. Am Mineral 81:303-316

Birch F (1952) Elasticity and composition of earth's interior. J Geophys Res 57:227-286

Birch F (1964) Density and composition of the mantle and core. J Geophys Res 69:4377-4388

Boehler R (1986) The phase diagram of iron to 430 kbar. Geophys Res Lett 13:1153-1156

Boehler R (1992) Melting of Fe-FeO and Fe-FeS systems at high pressures: Constraints on core temperatures. Earth Planet Sci Lett 111:217-227

Boehler R (1993) Temperatures in the Earth's core from melting-point measurements of iron at high static pressures. Nature 363:534-536

Boehler R (1996) Melting of mantle and core materials at very high pressures. Phil Trans R Soc London A354:1265-1278

Boehler R, Chopelas A (1992) Phase transitions in a 500 kbar-3000 K gas apparatus. p 55-60 Am Geophys Union, Washington, DC

Boehler R, Von Bargen N, Chopelas A (1990) Melting, thermal expansion, and phase transitions of iron at high pressures. J Geophys Res 95:731-21,736

Bottinga Y (1985) On the isothermal compressibility of silicate liquids at high pressure. Earth Planet Sci Lett 74:350-360

Brown JM, Ahrens TJ, Shampine DL (1984) Hugoniot data for pyrrhotite and the Earth's core. J Geophys Res 89:6041-6048

Brown JM, McQueen RG (1980) Melting of iron under core conditions. Geophys Res Lett.7:533-536

Brown JM, McQueen RG (1982) The equation of state of iron and the Earth's core. *In:* S. Akimoto, MH Manghnani. High Pressure Research in Geophysics. p 611-623 Center for Academic Publications, Tokyo

Brown JM, McQueen RG (1986) Phase transitions, Gruneisen parameter and elasticity for shocked iron between 77 GPa and 400 GPa. J Geophys Res 91:7485-7494

Bundy FP (1965) Pressure-temperature phase diagram of iron to 200 kbar, 900 C. J Appl Phys 36:616-620

Chen GQ, Ahrens TJ (1996) High pressure melting of iron: new experiments and calculations. Phil Trans R Soc London 354:1251-1263

Cohen RE, Gong Z (1994) Melting and melt structure of MgO at high pressures. Physical Review B 50(17):12,301-12,311

Cohen RE, Weitz JS (1998) The melting curve and premelting of MgO. p 185-196 Am Geophys Union, Washington, DC

de Vos JC (1954) A new determination of the emissivity of tungsten ribbon. Physica 20:690-714

Duba AG (1994) Iron - what is melt? *In:* SC Schmidt, JW Shaner, GA Samara, M Ross (eds) High pressure Science and Technology – 1993 p 923-926 American Institute of Physics, New York

Dubrovinsky LS, Saxena SK, Lazor P (1997) X-ray study of iron with in situ heating at ultra high pressures. Geophys Res Lett 24:1835-1838

Dubrovinsky LS, Saxena SK, Lazor P (1998) Stability of β-phase: a new synchrotron x-ray study of heated iron at high pressure. Eur J Mineral 10:43-47

Fei Y, Bertka CM, Finger LW (1997) High pressure iron sulfur compound, Fe$_3$S$_2$, and melting relations in the Fe-FeS system. Science 275:1621-1623

Fei Y, Mao HK (1994) In situ determination of the NiAs phase of FeO at high pressure and high temperature. Science 266:1678-1680

Fei Y, Prewitt CT, Mao HK, Bertka CM (1995) Structure and density of FeS at high pressure and high temperature and the internal structure of Mars. Science 268:1892-1894

Funamori N, Yagi T, Uchida T (1996) High-pressure and high temperature in situ x-ray diffraction study of iron to above 30 GPa using MA8-type apparatus. Geophys Res Lett 23:953-956

Gallagher KG, Ahrens TJ (1994) First measurements of thermal conductivity in griceite and corundum at ultra high pressure and the melting point of iron. Eos Trans Am Geophys Union 75:653

Gasparik T, Wolf K, Smith CM (1994) Experimental determination of phase relations in the CaSiO$_3$ system from 8 to 15 GPa. Am Mineral 79:1219-1222

Gilvarry JJ (1956) The Lindemann and Gruneisen laws. Phys Rev 102:308-316

Gilvarry JJ (1966) Lindemann and Gruneisen laws and a melting law at high temperatures. Phys Rev Lett 16:1089-1091

Hall HT (1958) Some high pressure, high temperature apparatus design considerations: Equipment for use at 100,000 atmospheres and 3000°C. Rev Sci Instrum 29: 267-275

Heinz DL (1986) Melting Curve of Magnesian Silicate Perovskite. PhD Dissertation, University of California, Berkeley

Heinz DL, Jeanloz R. (1987) Temperature measurements in the laser heated diamond cell. *In:* HM Manghnani, Y Syono (eds) High Pressure Researches in Mineral Physics, p 113-127. Am Geophys Union, Washington, DC

Heinz DL, Sweeney JS, Miller P (1991) A laser heating system that stabilizes and controls the temperature: diamond anvil cell application. Rev Sci Instrum 62:1568-1575

Hemley RJ, Jephcoat AP, Mao HK, Ming LC (1988) Pressure induced amorphization of silica. Nature 334:52-54

Holland KG, Ahrens TJ (1998) Properties of LiF and Al_2O_3 to 240 GPa for metal shock temperature measurements. *In:* MH Manghnani, T Yagi (eds) Properties of Earth and Planetary Materials at High Pressure and Temperature, p 335-343 Am Geophys Union, Washington, DC

Holloway JR (1971) Internally heated pressure vessels. *In:* GC Ulmer (ed) Research Techniques for High Pressure and High Temperature, p 217-258 Springer-Verlag, New York

Irifune T, Susaki J, Yagi T, Sawamoto H. (1989) Phase transformations in diopside $CaMgSi_2O_6$ at pressures up to 25 GPa. Geophys Res Lett 16:187-190

Jackson I (1976) Melting of silica isotypes SiO_2, BeF_2 and GeO_2 at elevated pressures. Phys Earth Planet Inter 13:218-231

Jacobs JA (1987) Geomagnetism. Academic, New York

Jamieson JC, Lawson AW (1962) X-ray diffraction studies in the 100-kbar pressure range. J Appl Phys 33:776-780

Jeanloz R, Heinz DL (1984) Experiments at high temperature and pressure: laser heating through the diamond cell. J Physique 45:83-92

Jeanloz R, Kavner A (1996) Melting criteria and imaging spectroradiometry in laser-heated diamond-cell experiments. Phil Trans R Soc London A 354:1279-1305

Jeanloz R, Richter FM (1979) Convection, composition and the thermal state of the lower mantle. J Geophys Res 84:5497-5504

Jensen E (1942) Pyrrhotite: melting relations and composition. Am J Sci 240:695-709.

Jephcoat AP, Besedin SP (1996) Temperature measurement and melting determination in the laser-heated diamond-anvil cell. Phil Trans R Soc London A 354:1333-1360

Kanzaki M (1990) Melting of silica up to 7 GPa. J Am Ceram Soc 73:3706-3707

Kennedy GC, Vaidya SN (1970) The effect of pressure on the melting temperature of solids. J Geophys Res 75:1019-1022

King HE, Prewitt CT (1982) High pressure and high temperature polymorphism of iron sulfide (FeS). Acta Cryst B38:1877-1887

Knittle E, Jeanloz R (1987) Synthesis and equation of state of $(Mg,Fe)SiO_3$ perovskite to over 100 gigpascals. Science 235:668-670

Knittle E, Jeanloz R (1989) Melting curve of $(Mg,Fe)SiO_3$ perovskite to 96 GPa: evidence for a structural transition in lower mantle melts. Geophys Res Lett 16:421-424

Knittle E, Jeanloz R (1991) The high pressure phase diagram of $Fe_{0.94}O$: a possible constituent of the Earth's core. J Geophys Res 96:16169-16180

Kubicki JD, Lasaga AC (1992) Ab initio molecular dynamics simulations of melting in forsterite and $MgSiO_3$ perovskite. Am J Sci 292:153-183

Kumazawa M (1971) Multi-anvil sliding system -- A new mechanism of producing very high pressure in a large volume. High Temp-High Pres 3:243-260

Kusaba K, Syono Y (1998) Structures and phase equilibria of FeS under high pressure and temperature. *In:* MH Manghnani, T Yagi (eds) Properties of Earth and Planetary Materials at High Pressure and Temperature. p 297-305 Am Geophys Union, Washington, DC

Lazor P, Saxena SK (1996) Discussion comment on melting criteria and imaging spectroradiometry in laser-heated diamond-cell experiments (by R. Jeanloz and A. Kavner). Phil Trans R Soc London A 354:1307-1313

Lazor P, Shen G, Saxena SK (1993) Laser heated diamond anvil cell experiments at high pressure - melting curve of nickel up to 700 kbar. Phys Chem Minerals 20:86-90.

Lindemann FA (1910) Über die berechnung molekülarer eigenfrequenzen. Physikalisches Zeits 11:609-612

Lindsley DH (1966) Pressure-temperature relation in the system $FeO-SiO_2$. Carnegie Inst Wash Yearb 65:226-230

Liu L (1976) Orthorhombic perovskite phases observed in olivine, pyroxene, and garnet at high pressures and temperatures. Phys Earth Planet Inter 11:289-298

Liu LG, Basset WA (1975) The melting of iron up to 200 kbar. J Geophys Res 80:3777-3783

Lyzenga GA, Ahrens TJ, Mitchell AC (1983) Shock temperatures of SiO_2 and their geophysical implications. J Geophys Res 88:2431-2444

Mao HK, Bell PM, Hadidiacos C (1987) Experimental phase relations of iron to 360 kbar and 1400 C, determined in an internally heated diamond anvil apparatus. *In:* MH Manghnani, Y Syono (eds) High Pressure Research in Mineral Physics. p 135-138 Terra Scientific, Tokyo

Mao HK, Shen G, Hemley RJ, Duffy TS (1998) X-ray diffraction with a double hot plate laser heated diamond cell. *In:* MH Manghnani, T Yagi (eds) Properties of Earth and Planetary Materials at High Pressure and Temperature. p 27-34 Am Geophys Union, Washington, DC

Mao HK, Shu J, Fei Y, Hu J, Hemley RJ (1996) The wüstite enigma. Phys Earth Planet. Inter 96:135-145

Mao HK, Yagi T, Bell PM (1977) Mineralogy of the Earth's deep mantle: Quenching experiments of mineral compositions at high pressures and temperatures. Carnegie Inst Wash Yearb 76:502-504

Matsui M (1998) Computational modeling of crystals and liquids in the system $Na_2O-CaO-MgO-Al_2O_3-SiO_2$. *In:* MH Manghnani, T Yagi (eds) Properties of Earth and Planetary Materials at High Pressure and Temperature. p 145-151 Am Geophys Union, Washington, DC

Matsui M, Price GD (1991) Simulation of premelting behavior of $MgSiO_3$ perovskite at high pressure and temperatures. Nature 351:735-737

Mishima O, Calvert LD, Whalley E (1984) "Melting ice" at 77 K and 10 kbar: a new method of making amorphous solids. Nature 310:393-395

Mulargia F (1986) The physics of melting and temperatures in the earth's outer core. Quart J Roy Astr Soc 27:383-402

Poirier JP (1991) Introduction to the Physics of the Earth's Interior. p 264 Cambridge University Press, Cambridge

Poirier JP (1994) Light elements in the earth's outer core: A critical review. Phys Earth Planet Inter 85:319-337

Poirier J (1989) Lindemann law and the melting temperature of perovskites. Phys Earth Planet Inter 54:364-369

Ringwood AE, Hibberson W (1990) The system Fe-FeO revisited. Phys Chem Minerals 17:313-319

Saxena SK, Dubrovinsky LS, Haggkvist P (1995) X-ray evidence for the new phase b-iron at high pressure and high temperature. Geophys Res Lett 23:2441-2444

Saxena SK, Shen G, Lazor P (1993) Experimental evidence for a new iron phase and implications for Earth's core. Science 260:1312-1314

Saxena SK, Shen G, Lazor P (1994) Temperatures in Earth's core based on melting and phase transformation experiments on iron. Science 264:405-407

Shen G, Duffy T, Wang Y, Rivers M, Sutton S (1998a) Studies on materials under ultrahigh *P-T* at the Advanced Photon Source. *In:* RM Wentzcovitch, RJ Hemley, WJ Nellis, P Yu (eds) High Pressure Materials Research (in press) Materials Research Society, Warrendale, PA

Shen G, Lazor P (1995) Measurement of melting temperatures of some minerals under lower mantle pressures. J Geophys Res 100:17699-17713

Shen G, Lazor P, Saxena SK (1993) Melting of wüstite and iron up to pressures of 600 kbar. Phys Chem Minerals 20:91-96

Shen G, Mao HK, Hemley RJ (1996) Laser-heating diamond-anvil cell technique: Double-sided heating with multimode Nd:YAG laser. Advanced Materials '96—New Trends in High Pressure Research. p 149-152 NIRIM, Tsukuba, Japan

Shen G, Mao HK, Hemley RJ, Rivers ML (1998b) Melting and crystal structure of iron at high pressures. Geophy Res Lett 25:373-376

Simon F, Glatzel G (1929) Bemerkungen zur Schmelzdruckkurve. Z Anorg Allg Chem 178:309-316

Stacey FD (1977) Theory of melting: thermodynamics basis of Lindemann law. Austral J Phys 30:631-640

Strong HM, Tuft RE, Hannemann RE (1973) The iron fusion curve and the triple point. Metal Trans 4:2657-2661

Sweeney JS, Heinz DL (1993) Melting of iron-magnesium-silicate perovskite. Geophys Res Lett 20:855-858

Sweeney JS, Heinz DL (1998) Laser heating through a diamond anvil cell: melting at high pressure. *In:* MH Manghnani, T Yagi (eds) Properties of Earth and Planetary Materials at High Pressure and Temperature. p 197-213 Am Geophys Union, Washington, DC

Urakawa S, Kato M, Kumazawa M (1987) Experimental study of the phase relation in the system Fe-Ni-O-S up to 15 GPa. p 95-111 Terra, Tokyo

Usselmann TM (1975) Experimental approach to the state of the core: part 1. The liquidus relations of the Fe-rich portion of the Fe-Ni-S system from 30 to 100 kb. Am J Sci 275:278-290

Vaidya SN, Gopal ESR (1966) Melting law at high pressure. Phys Rev Lett 17:635-636

Vocadlo L, Price GD (1996) The melting of MgO—Computer calculations via molecular dynamics. Phys Chem Minerals 23:42-49

Wang Y, Getting IC, Weidner DJ, Vaughan MT (1998) Performance of tapered anvils in a DIA-type, cubic anvil, high pressure apparatus for x-ray diffraction studies. *In:* MH Manghnani, T Yagi (eds) Properties of Earth and Planetary Materials at High Pressure and Temperature, p 35-39 Am Geophys Union, Washington, DC

Wentzcovitch RM, Ross NL, Price GD (1995) Ab initio study of $MgSiO_3$ and $CaSiO_3$ perovskites at lower mantle pressures. Phys Earth Planet Inter 90:101-112

Williams Q, Jeanloz R (1990) Melting relations in the iron-sulfur system at ultra-high pressures: implications for the thermal state of the earth. J Geophys Res 95:19299-19310

Williams Q, Jeanloz R, Bass J, Svendsen B, Ahrens TJ (1987) The melting curve of iron to 250 GPa: a constaint on the temperature at the Earth's center. Science 236:181-182

Williams Q, Knittle E, Jeanloz R (1991) The high pressure melting curve of iron: A technical discussion. J Geophys Res 96:2171-2184

Wolf GH, Jeanloz R (1984) Lindemann melting law: Anharmonic correction and test of its validity for minerals. J Geophys Res 89:7821-7835

Yoo CS, Akella J, Compbell A, Mao HK, Hemley RJ (1995) Phase diagram of iron by in situ X-ray diffraction: implications for the Earth's core. Science 270:1473-1475

Yoo CS, Holmes NC, Ros M, Webb DJ, Pike C (1993) Shock temperatures and melting of iron at Earth core conditions. Phys Rev Lett 70:3931-3934

Zerr A, Boehler R (1993) Melting of $(Mg,Fe)SiO_3$-perovskite to 625 kilobars: indication of a high melting temperature in the lower mantle. Science 262:553-555

Zerr A, Boehler R (1994) Constraints on the melting temperture of the lower mantle from high-pressure experiments on MgO and magnesiowüstite. Nature 371:506-508

Zerr A, Serghiou G, Boehler R (1997) Melting of $CaSiO_3$ perovskite to 430 kbar and first in-situ measurements of lower mantle eutectic temperatures. Geophys Res Lett 24:909-912

Zhang J, Liebermann RC, Gasparik T, Herzberg CT, Fei Y (1993) Melting and subsolidus relations of SiO2 at 9-14 GPa. J Geophys Res 98:19785-19793

Zhou Y, Miller GH (1997) Constraints from molecular dynamics on the liquidus and solidus of the lower mantle. Geochim Cosmochim Acta 61:2957-2976

Chapter 13

MELT VISCOSITY AND DIFFUSION UNDER ELEVATED PRESSURES

Donald B. Dingwell

Bayerisches Geoinstitut
Universität Bayreuth
D-95440 Bayreuth, Germany

INTRODUCTION

The viscosity of liquids is a property of fundamental importance in the geosciences. The silicate melt phase has played and in most cases continues to play a key role in the chemical differentiation of the Earth's mantle and crust as well as in the condensed interiors of other planets. The mineralogical and geophysical consequences of these differentiation processes range from models of magma oceans in the history of the Earth's mantle to scenarios of extreme differentiation giving rise to magmatic ore deposits. The viscosity of iron-rich melts alloys is important for understanding both core formation and convection in the outer core. Behind these fields stands a growing body of applied and fundamental research on the subject of melt rheology. [*]

Melt viscosity has been well studied because of their materials applications, particularly for silicates. There are few useful boundaries to be drawn between the knowledge of melt rheology obtained geoscience and materials science lines of investigation. The formulation of the basic questions, the chemical compositions of interest, the range of viscosities and temperatures explored, the experimental methods employed and the empirical and theoretical models generated all overlap greatly. However, perhaps the single most important distinction that can be drawn here involves pressure. For example, the materials side of silicate melt research is understandably dominated by studies performed at ambient pressure (1 bar) where the processes and applications of interest invariably occur. In contrast, many geoscientific applications of melt research deal with elevated pressures. The range of pressure scales of natural processes in which silicate melts are involved ranges dynamically from volcanological applications at tens to hundreds of bars to core-mantle segregation at millions of bars. This extreme range of magnitudes means that the experimental approach to studying melt viscosities must explore a very broad range of measurement principles and technologies.

This chapter provides an overview of viscosity and diffusion of melts under pressure, focusing on studies of silicates. Much has been written on the viscosity and related properties of these melts. Melt properties have been addressed in previous *Reviews in Mineralogy* volumes, in particular Volume 24 (emphasis on equation of state), Volume 30

---

[*] It is of historical interest to note that amongst the earliest discussions of the pressure dependence of the viscosity of melts within the earth sciences was an analysis of the origin of basaltic magmas (Bowen 1928). Daly (1925) had proposed that the source of basaltic volcanism might lie in a layer of highly rigid basaltic glass at depth that generated basaltic magma via decompression and fluidizing of the magma. Early determinations of the pressure dependence of the viscosity of liquids was attempted by Bridgman (1926) who concluded that liquid viscosities generally did indeed increase with pressure. Those data appeared to support the geophysical feasibility of a glassy basaltic layer and this fallacious inference regarding the stability of rigid glass within the Earth was finally laid to rest in the 1970s by the observation that basaltic melt viscosity decreases with pressure (Kushiro 1976).

0275-0279/98/0037-0013$05.00

(emphasis on role of volatiles), and Volume 32 (emphasis on relaxation: Nicholls and Russell 1987, Carroll and Holloway 1994, Stebbins et al. 1995). Of particular value, as a background for the present discussion, are the treatments of relaxation (Dingwell 1995), viscoelasticity (Webb and Dingwell 1995) viscosity and configurational entropy (Richet and Bottinga 1995), and pressure dependence of structure and properties (Wolf and McMillan 1995). Their contents are not reworked here, rather, a prior knowledge of those contributions is assumed below. Additional information on shock-wave measurement of viscosities of liquids under very high P-T conditions can be found in Miller and Ahrens (1991). Finally, chapters by Bina and by Stixrude (this volume) provide discussions of core materials, including geophysical constraints on the outer core.

GENERAL CONSIDERATIONS

The vast body of knowledge on the properties of silicate melts at ambient pressure permits a fairly complete diagnostic overview of the phenomenology of melt viscosity. It cannot be overstated that sufficient analysis of this phenomenology is a necessary prerequisite for the design of experiments and interpretation of experimental results at elevated pressures. This is because the relative ease of sample access, manipulation and recovery from ambient temperature experiments ensures, in general, a higher degree of confidence in the results, better characterization of sample state, and the application of a much larger dynamic range of the controlling variables on melt viscosity than will be possible at elevated pressures. Thus the salient features of relevance to the design and interpretation of high pressure experiments are reviewed briefly here.

Newtonian considerations

The most fundamental assumption in the discussion of melt rheology relevant to the geosciences is the validity of the definition of a single-valued melt viscosity, independent of the applied or resultant strain rate. In consideration of most processes within the deep Earth we are concerned exclusively with the Newtonian response of the liquid or melt phase and therefore, the property of interest in most high pressure experimental investigations of melt rheology is the Newtonian viscosity. The limits on the validity of Newtonian rheology of silicate melts of geological interest have become very well defined, both theoretically and experimentally (Webb and Dingwell 1990, 1995). Silicate melts possess a viscoelastic behavior, which can be usefully approximated by a Maxwell body with a distribution of relaxation times, which generates non-exponential relaxation in time domain experiments or deviation from a Debye peak via a high frequency tail of the loss modulus. The relaxation mode controlling viscous flow in silicate melts has been linked quantitatively in temperature-time space to the exchange of Si and O atoms in the melt as expressed by diffusion experiments and by motional averaging in spectroscopic experiments. There is no evidence, either mechanical, or electrical, or enthalpic, for "downstream" relaxation modes at longer timescales due to complex formation. This point, of great significance in a number of areas of geophysics, such as comparison of frequency and time domain studies and interpretation of seismic monitoring of partially molten systems, has been generally accepted in the geo-scientific literature only in the past ten years with confusion existing as recently as the 1980s (Spera et al. 1983, Rivers and Carmichael 1987).

The onset of non-Newtonian viscosity is illustrated in two types of experiment in Figures 1 and 2. Time domain longitudinal dilatometric experiments reveal a regular and predictable onset of non-Newtonian rheology at approx. 2.5 log units of strain rate below the relaxational strain rate derived from the Maxwell equation

$$\tau = \eta/G = 1/\varepsilon_r \tag{1}$$

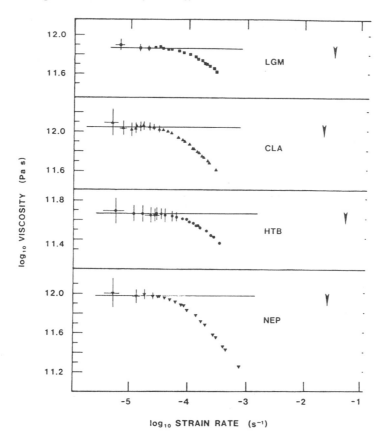

Figure 1. The onset of non-Newtonian rheology in geological melts as determined using fiber elongation viscometry. LGM = Little Glass Mountain Rhyolite; CLA = Crater Lake Andesite; HTB = Hawaiian tholeiitic basalt; NEP = nephelinite. The deviation from Newtonian behavior occurs ~3 log units of strain rate below thhe relaxational strain rate (signified by the arrowheads). Reproduced with permission from Webb and Dingwell (1990).

where η is the shear viscosity, G is the shear modulus, τ is the shear relaxation time and ε_r is the relaxational strain rate (Fig. 1).

Frequency domain studies of the same melt reveal a measurable values of an energy loss modulus at 2 log units of frequency below the relaxation frequency obtained from the Maxwell relation (Fig. 2).

Thus if the strain-rate in an experimental viscosity determination is well defined then the onset of non-Newtonian viscosity can usually be clearly avoided. This does not however mean that the limits are always easy to identify in high-pressure experimental determinations of melt viscosity. This is because the strain rate and the relaxation strain rate are not always well known. For this reason either a test of the Newtonian nature of melt viscosity at high pressure, or a clear fulfilling of the condition: strain rate (experimental) < 1000 strain rate (relaxational) is required.

Non-Arrhenian considerations

One of the most universal features of the viscosity of liquids is the fundamentally non-

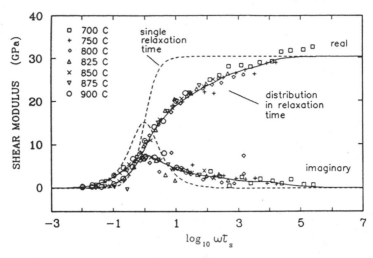

Figure 2. The onset of viscoelasticity in a rhyolitic melt as determined using low-frequency torsional deformation. The real and imaginary components of the shear modulus are plotted versus the log of the frequency-time product. The dashed lines indicate Debye behavior. The deviation from Debye behavior (perfect Maxwell body) towards a high frequency tail is evident. This represents the nonexponential nature of relaxation in silicate melts. Reproduced with permission from Webb (1992).

Arrhenian nature of its temperature-dependence. The terms "strong" and "fragile" have been introduced into the literature and successively quantified in various ways in order to parameterise the degree of curvature of the temperature-dependence of viscosity in an Arrhenian (log viscosity versus reciprocal absolute temperature) plot (Angell 1984). In comparison with most investigated inorganic liquids, silicate melts are relatively strong or Arrhenian. In fact, over the restricted temperature ranges defined by the phase equilibria of melting within the Earth, the Arrhenian approximation

$$\log \eta = a + b/T \tag{2}$$

can be tolerated for minor extrapolations. Unfortunately, the extremely large temperature dependence of the viscosity of silicate melts means that in general, non-Arrhenian models for the temperature-dependence of silicate melt viscosities must be generated in order to ensure the validity of extrapolation of larger temperature ranges. Nowhere is this more apparent than in explosive volcanism where vertical gradients in the viscosity of magma filling the conduit can range over 10 orders of magnitude (Dingwell 1998a). Thus a detailed inspection of the temperature dependence of liquid viscosities over the range of viscosities encountered from liquidus temperatures (as low as 10 Pa s) to the glass transition (ca. 10^{12} Pa s at typical dilatometric experimental timescales) reveals enormous departures from the Arrhenian approximation (Fig. 3) that are usually dealt with by empirical extensions of the Arrhenian relationship such as the Tamann-Vogel-Fulcher equation

$$\log \eta = a + b/(T-c) \tag{3}$$

In general, the degree of non-Arrhenian temperature dependence or fragility increases with melt basicity and thus the melts which are of most relevance to deep Earth processes can be expected to be strongly non-Arrhenian at high pressures. Added to this is the possibility, for which fragmentary evidence is presented below, that pressure may enhance the fragility of silicate melts. The primary consequence of all of this is that the description of the viscosity of a melt at high pressure must ultimately include data obtained at or near

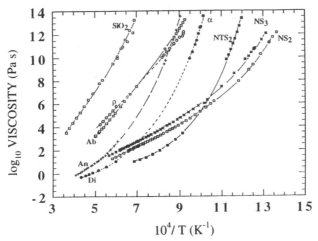

Figure 3. The non-Arrhenian nature of the temperature-dependence of the Newtonian viscosity of silicate melts as exhibited by several analog compositions of relevance to geological processes. The general trend towards more fragile behavior at lower absolute temperatures for a constant viscosity is evident. Ab = albite melt, Di = diopside melt, NTS_2 = sodium titanosilicate melt, NS_3 = sodium trisilicate, NS_2 = sodium disilicate Reproduced with permission from Richet and Bottinga (1995).

the temperature of interest for the application. Failing that possibility, safe extrapolation or even interpolation of viscosity in temperature will only be possible if a sufficiently broad range of viscosity is covered by experimentation. Due to the range of viscosities exhibited by such melts, this latter point generates the inevitability of combined or multiple experimental approaches to the construcion of viscosity-temperature relationships.

Geometrical considerations

The amorphous nature of liquids leads to the considerable simplification that the coefficients describing the elastic and viscous properties, the mechanical storage and loss moduli, are independent of sample orientation. This isotropic nature of silicate melts does not however preclude the distinction that must be made between the viscous coefficients of the two geometric components of deformation in the liquid state, the volume and shear viscosities, respectively. The shear and volume components represent the distinct response of the melt to deformation in the direction of and orthogonal to nearest neighbour atoms in the melt matrix. A longitudinal viscosity may be defined as a combination of shear and volume components analogously to the case for elastic moduli,

$$\eta_\lambda = \eta_v + (4/3)\, \eta_s \tag{4}$$

where η_λ, η_v and η_v are the longitudinal, volume and shear viscosities, respectively.

Experiments performed at ambient pressures have determined both shear and longitudinal viscosities. A comparison between shear viscosity data determined predominantly by concentric cylinder methods and longitudinal ultrasonic wave attenuation studies reveals a clearly defined proportionality between the longitudinal and shear viscosities. This proportionality yields, via Equation (1), the conclusion that the volume viscosity is similar in value to the shear viscosity. This equivalence of the coefficient of viscous response to changes in volume stress and shear stress is an importance link between alternative possibilities for the determination of viscosities at high pressure. In essence it means that the approach of metastable or stable molten systems to equilibrium density or volume as a result of a pressure change, is tightly linked to the shear relaxation and viscosity of the melt phase. In other words, volume relaxation, a process intrinsic to any experimentation in which a liquid is subjected to a pressure (volume stress) change, contains information on shear viscosity. It is difficult to overstate the importance of this connection for studies at high pressure. It essentially means that liquid PVT equation-of-state studies implicitly contain information on viscosity (see below).

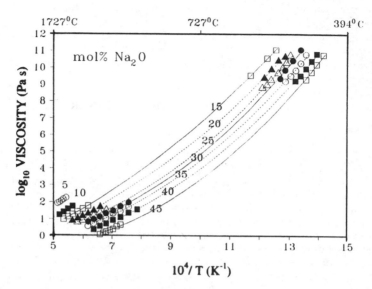

Figure 4. The variation of viscosity with composition in the SiO_2-Na_2O system. The strong and nonlinear decrease in viscosity with addition of Na_2O is evidence by comparison of the SiO_2 curve in Figure 3. The melt viscosity-temperature relationships become fragile with alkali addition. Reproduced with permission from Knoche et al. (1995).

Compositional considerations

Finally, in our list of considerations from experimental experiences at ambient pressure we must now make some points about expectations based on the compositional dependence of viscosity. We have already referred to the temperature-dependence of viscosity and its variation with composition above. The example of the system SiO_2-Na_2O illustrates several of the relevant features of the composition-dependence of silicate melts (Fig. 4). The endmember in this system, SiO_2, (see Fig. 3) possesses the highest viscosity at any given temperature of comparison of any silicate melt. The deviation from Arrhenian temperature dependence of the slightest for all silicate melts. The viscosity can be seen to decrease strongly and quite non-linearly with the addition of alkali oxide to silica. The general trend is towards a relatively fragile temperature dependence of viscosity which starts strongly and then decreases in intensity. The most alkali-rich of the melts illustrated here represent viscosity-temperature relationships which reflect the lower bounds of fluidity in geologically relevant melts. In addition their non-Arrhenian temperature dependence is typical of very basic melts of geoscientific relevance. This system illustrates clearly that the apparent activation energy of viscous flow that would be derived from viscosity determinations in the low viscosity range decreases strongly with the addition of alkali oxide. In contrast, in the high viscosity range the apparent activation energy increases considerably with the addition of alkali oxide to silica.

The extremely steep decrease in viscosity at very low alkali contents is typical of the behavior induced by the addition of a wide range of components to silica (Brückner 1971). Perhaps the most important example of this in the present context is the addition of water. Traces of water in the range of tens to hundreds of ppm of water can result in melt viscosity reduction in liquid silica of orders of magnitude. Given the general difficulty of excluding water from high pressure experiments this aspect must be emphasized. Characterization of water contents of high pressure viscosity measurements is highly advisable in distinguishing pressure-induced changes from water-induced changes. Fortunately, the effect of water on melt viscosity is most extreme in the high-viscosity rather than the low-viscosity range (e.g. Hess and Dingwell 1996). In the low viscosity range of many viscosity determinations at elevated pressures, the influence of water at these trace concentrations will be very moderate and much more easy to account for.

VISCOMETRY AT HIGH PRESSURE

Falling sphere viscometry

Modern experimental analysis of the influence of pressure on the viscosity of silicate melts of geoscientific interest began with the studies of Kushiro and coworkers in the late 1970s (Kushiro 1976, Kushiro et al. 1976, 1978a,b; see summary by Scarfe et al. 1987). He showed that the viscosity of several examples of relatively silicic melts decreased with viscosity up to pressures of 25 kbar. It should be noted that this trend had not been anticipated and in fact the assumption to that point in time had been that silicate melts would exhibit monotonously increasing viscosities with increasing pressure (see footnote on fisrt page of this chapter). Understandably, the observation of the opposite effect generated considerable interest, not only due to to the implications for melt and magma transport at depth within the Earth, but also because of the broader physico-chemical implications for liquids in general.

The bulk of data presently available on the influence of pressure on the viscosity of silicate melts has been obtained using the Stokesian settling or falling sphere method via Stokes settling law,

$$V = 2r^2\Delta\rho g/2\eta \tag{5}$$

where V is the settling viscosity, r is the sphere radius, $\Delta\rho$ is the density contrast between sphere and liquid, and g is the acceleration. As such, much of the discussion of the pressure dependence has been and continues to be restricted to the consideration of melts under conditions where the viscosities are less than 10^4 Pa s. Nevertheless a useful starting point for a discussion of these trends is provided by Figure 5 taken from a discussion of the influence of pressure on melt structure and properties by Wolf and McMillan (1995). The salient features of the viscosity trends observed for silicate melts include the following.

A restricted number of compositions permit comparison of aluminosilicate melts as a function of the compositional exchange operator $XAlSi_{-1}$. In the systems $Na_2O-Al_2O_3-SiO_2$ and $CaO-Al_2O_3-SiO_2$ the tectosilicate stoichiometry joins illustrate a shift from strong negative pressure dependence of viscosity at high SiO_2 contents to a weak negative pressure dependence at intermediate SiO_2 content. In the $CaO-Al_2O_3-SiO_2$ system the data illustrate a transition to a positive pressure dependence of viscosity as the SiO_2 is further reduced below 50 mol%.

Comparison of the available data with respect to the stoichiometrically estimated degree of polymerization of the melt (via the calculation of NBO/T, Mysen et al. 1982) also indicates a trend of decreasing intensity of viscosity decrease with increasing pressure as the SiO_2 decreases and is replaced by various oxide components which generate a

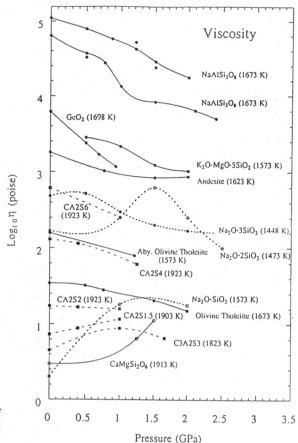

Figure 5. The pressure dependence of the viscosity of several analog melt compositions of relevance to geological processes.

CA2S6 = CaAl$_2$Si$_6$O$_{16}$
CA2S4 = CaAl$_2$Si$_4$O$_{12}$
CA2S2 = CaAl$_2$Si$_2$O$_8$
CA2S1.5 = CaAl$_2$Si$_{1.5}$O$_7$

Reproduced by permission from Wolf and McMillan (1995), where original references are listed.

depolymerization of the melt structure. Natural melt compositions approaching an ultrabasic chemical composition show very little influence of pressure on viscosity.

More recently, a considerable body of data on the viscosity of hydrous silicic melts has been developed (Dingwell 1987, Baker 1996, Schulze et al. 1996). In general, these studies have demonstrated, that under the conditions of pressure relevant to the transport of silicic magmas within the Earth's crust, the pressure dependence of granitic and other analog melts is very minor indeed. In fact in their recent summary and parameterization of this data base, Hess and Dingwell (1996) were able to set the pressure dependence of the viscosity of hydrous granitic liquids equal to zero for the pressure range up to 10 kbar and achieve a much improved fit to the entire available data set by the use of the first application of a non-Arrhenian model to these viscosity data over a very wide viscosity range via the relationship

$$\log \eta = [-3.545 + 0.833\ln(w)] + [9601 - 2368\ \ln(w)]/[T - (195.7 + 32.25\ \ln(w))] \quad (6)$$

where η is in Pa s, w is water content in wt %, and T is in K. This is consistent with the earlier observations of Kushiro (1978a) for hydrous andesitic melts as well.

Following the observation of opposing trends of pressure-dependence for anorthite and diopside liquids by Scarfe et al. (1987), the general nature of the transition from negative to positive pressure-dependence of viscosity in more depolmerised melts was investigated in detail by Brearley et al. (1986). These latter workers demonstrated that the transition from negative to positive dependence can occur for a given intermediate composition as a function of pressure, appearing as an minimum in the viscosity as a function of pressure. These results were later confirmed by Taniguchi (1992).

Recent developments in falling sphere methods

Experimental time constraints form the essential restrictions on the application of falling sphere viscometry to silicate melts at elevated pressures. If the viscosity of the melt phase is too low then the determination of experimental fall times using the traditional method of quenching and measurement of sphere position becomes impractical due to the high fall speed. Uncertainties in the sphere position versus time are contributed by inaccuracies in the exact start and stop times of sphere descent. For viscosities that are too high, the experimental problem of sample and experimental condition stability over excessive lengths of time becomes a problem. These two practical limits on falling sphere viscometry studies are responsible for the lower (10 Pa s) and upper (10⁵ Pa s) boundaries on traditional falling sphere investigations.

Persikov (1990) developed an in situ method for the determination of sphere descent by using a radioactive tracer (⁶⁰Co) in the falling sphere, together with a window in the pressure vessel, in order to record an effective sphere passage time. His method is schematically summarized in Figure 6.

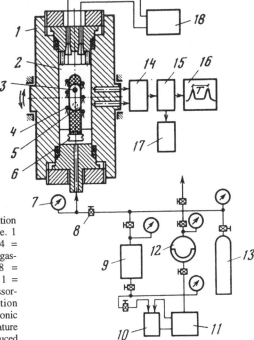

Figure 6. Schematic illustration of the radiation viscometer design for operation under gas pressure. 1 = vessel, 2 = heater, 3 = ⁶⁰Co-bearing sphere, 4 = thermocouples, 5 = capsule containing melt, 6 = gas-fluid separator vessel, 7 = pressure gauges, 8 = valves, 9 = oil-gas booster, 10 = oil reservoir, 11 = high pressure oil pump, 12 = oil-gas compressor-separator, 13 = gas supply, 14 = γ-scintillation counter, 15 = radiometric post, 16 = electronic recorder, 17 = digital printer, 18 = temperature measurement and automatic control panel. Reproduced with permission from Persikov et al. (1990).

Low viscosities. Lifting of the low viscosity restriction on falling sphere visco-metry has been sought in the avoidance of the need for quenching in such studies. Techniques for the in-situ or real-time monitoring of sphere position have been developed on the basis of shadowgraph imaging using high energy x-rays generated by synchrotrons (e.g. Kanzaki et al. 1987) and by monitoring the electrical conductivity of the charge during sphere descent (LeBlanc and Secco 1995). By reducing the effective viscosities measure-able using falling sphere methods, these methods have made viable the investi-gation of other, non-silicate melts under high pressures such as carbonates (Dobson et al. 1996) and sulfides (LeBlanc and Secco 1996).

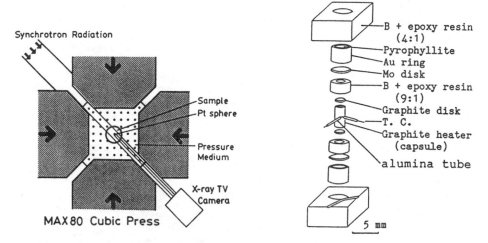

Figure 7 (left). The experimental design of the synchrotron-based MAX80 set-up for the in situ determination of viscosity using the falling sphere method at low viscosities. Sphere position is recorded by x-ray shadowgraphic means. Reproduced with permission from Kanzaki et al. (1987).

Figure 8 (right). The high pressure cell assembly used for the encapsulation and heating of the silicate melt samples studied by synchrotron-based falling sphere viscometry. The sample is surrounded by graphite. Reproduced with permission from Kanzaki et al. (1987).

The principle of the synchrotron-based methods initiated by Kanzaki et al. (1987) are illustrated schematically in Figures 7 and 8. The method relies on the in situ recording of sphere position during the fall using x-ray shadowgraph images obtained through the slits of a MAX80 multianvil press. This in situ method allows any temporal variations in the descent speed of the sphere to be identified. In this way the potential influence of temperature gradients on descent velocity or of delayed descent due to attachment of the sphere to the upper sample surface can be excluded by judicious selection of the frames of interest in a zone where the descent is constant and unhindered. These are considerable advantages over the quench-based applications of falling sphere viscometry and in fact in situ methods based on neutron absorption at lower pressures have been proposed (Winkler, pers. comm. 1997).

The electro-detection of LeBlanc and Secco (1995), illustrated in Figures 9 and 10, also circumvents the problems of attachment delay and temperature gradients by the determination of sphere passage time between two electrodes sensing bulk electrical conductivity. This method restricts the observation of the passage time to a region at the center of the capsule height where optimal thermal conditions should be obtained. It has been employed in a 1000 ton cubic anvil press and yields results for jadeite melt consistent with those obtained by Kushiro (1976) in the range of 2 to 3 GPa.

High viscosities. The high viscosity restriction on falling sphere viscometry has been addressed in recent years by the application of centrifuge-based techniques. Using

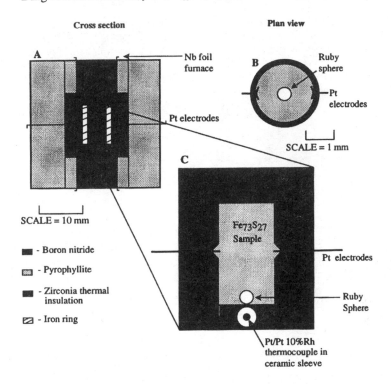

Figure 9. Experimental design for encapsulation of samples in the electro-detection method for determination of melt viscosity using the falling sphere method. Used with permission (LeBlanc and Secco 1995).

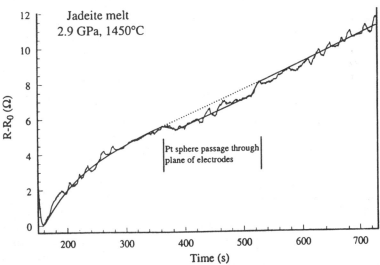

Figure 10. A plot of resistence versus time illustrating the measurement principle of the electro-detection method whereby the passage of the sphere is sensed as a temporal change in electrical conductivity as measured across two opposing electrodes placed centrally in the walls of the capsule. Reproduced with permission from LeBlanc and Secco (1995).

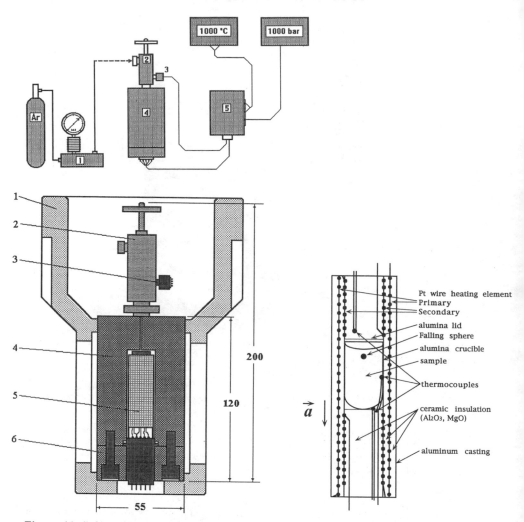

Figure 11 (left). Design of the centrifuge autoclave for the determination of melt viscosity using the falling sphere method under elevated pressure and acceleration. Upper diagram, 1 = gas compressor, 2 = valve, 3 = pressure transducer, 4 = autoclave, 5 = slip ring system. Lower diagram, 1 = autoclave/centrifuge adaptor, 2 = valve, 3 = pressure transducer, 4 = autoclave, 5 = internal heating, 6 = closure screws.

Figure 12 (right). Sample containment design for the centrifuge autoclave method for viscometry. Figures 11 and 12 reproduced with permission from Dorfman et al. (1996).

centrifuge accelerations up to 1000 g, the maximal viscosities obtainable using falling sphere techniques have been increased to 10^8 Pa s (Dorfman et al. 1996, 1997). The basic design of the heating system and autoclave used in the centrifuge falling sphere method are illustrated in Figures 11 and 12. The furnace is driven by a three-zone temperature control system to flatten out thermal gradients and is placed inside the autoclave to serve as the internal heating for what amounts to a miniature internally-heated pressure vessel. This autoclave sits opposite a counterweight in a laboratory centrifuge. A slip-ring system perimits the control of temperature and the monitoring of temperature and pressure during the the falling interval. For such studies at high viscosity, the quench uncertainties, which

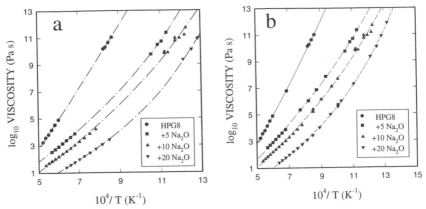

Figure 13. Comparison of viscosity data from the intermediate viscosity region, derived from the centrifuge autoclave method together with data from concentric cylinder methods at lower viscosity and micropenetration methods at higher viscosity. Reproduced with permission from Dorfman et al. (1996).

are minor to begin with, are further reduced via heating and cooling (quenching) without centrifuging. Samples are recovered after the quench and subjected to x-ray shadowgraph determination of sphere position. Figure 13 illustrates data obtained in the intermediate viscosity interval for some silicate melts using the centrifuge falling sphere method. The agreement with dilatometric determinations at lower temperatures and concentric cylinder determinations at higher temperatures is apparent. There have been recent developments in diamond-cell falling sphere techniques, although they have apparently not been applied to very high-temperature melts (King 1997).

Dilatometric viscometry

The high viscosity range of viscometry has long been underinvestigated in silicate melts of geological interest. There were a number of reasons for this. Many geoscientists were reluctant to combine viscosity measurements on equilibrium superliquidus melts with those on supercooled metastable subliquidus melts. Some reasons for this were, in principle, valid ones, such as the possibility of the transformation of the liquid over short or long times to other more stable phases (crystals or otherwise). Other reasons, such as the widely held impression that liquid structure and properties underwent discontinuous changes at the melting temperature or in the liquidus-solidus temperature range, were not valid concerns. It is now generally accepted that the temperature dependence of the structure and properties of silicate melts do not undergo changes induced by the crossing of the equilibrium crystallization temperature range provided the kinetics of crystal nucleation and growth can be outrun by the efficiency of the quench. As a consequence, strategies have been developed to synthesize supercooled liquid samples and subject them to property determinations (Dingwell 1993). These data, when combined with equilibrium property determinations at superliquidus temperatures, provide a much more complete description of the temperature-dependence of the liquid properties. Several techniques are well-suited to the investigation of the physical properties in the high viscosity range. In the case of studies of the viscosity of supercooled liquids at very high equilibrium viscosity values (in the range of 10^8 to 10^{12} Pa s) it is dilatometric studies of sample creep involving various geometries and deformation styles that have been most productive for viscometry.

At ambient pressure, the dilatometric experiments most widely applied for melts of geoscientific relevance are the micropenetration (Tauber and Arndt 1986, Hess 1996) and the cylinder compression (Neuville and Richet 1991) methods. The geometry of the

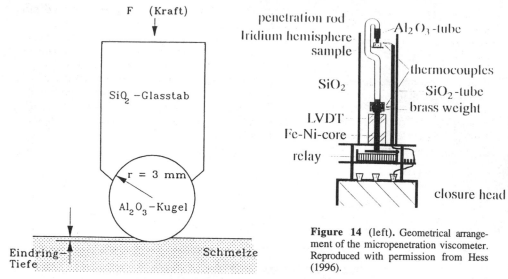

Figure 14 (left). Geometrical arrangement of the micropenetration viscometer. Reproduced with permission from Hess (1996).

Figure 15 (right). Schematic illustration of the high pressure dilatometer developed for use in micropenetration mode in a gas pressure vessel. Reproduced with permission from Gennaro et al. (1998).

micropenetration method is illustrated schematically in Figure 14. A hemispherical head or indenter under constant load is embedded from the surface of the sample into its volume. The indentation speed, under a given load, versus time corresponds to the viscosity of the melt. An example of an indentation curve and the resultant linearization is illustrated in Figure 14. Such ambient pressure dilatometric micropenetration viscometry is performed with sample and indenter support via silica glass and with a variable displacement trandsducer based on a vanadium metal core.

Adaptation of this method for use at elevated pressure has been initiated by Gennaro et al. (1998) using an inverted geometry of the sample/indenter assembly and the transducer (Fig. 15). The dilatometer developed is currently at the end of the testing stage and is capable of viscometry in the high viscosity range under the pressure range accessible by internally heated pressure vessels. Cylinder compression viscometry has also been extended recently to the range of internally heated gas pressure vessels by Schulze and Behrens (1998).

Glass transition determinations

One of the most important breakthroughs for the prospect of obtaining high pressure viscosity data in the near future concerns the determination of the glass transition. The quantitative link between the viscosity at the glass transition temperature and the effective relaxation time available for a given cooling rate (e.g. Stevenson et al. 1995) means that any observation of the glass transition in high pressure studies that is adequately quantified in timescale, temperature and pressure, can serve as a viscosity datum. This fact, together with the demonstrated equivalence of relaxation timescales for enthalpy, volume and shear stress relaxation, opens up a wide range of potential sensors for the glass transition that is solidly linked to viscosity. As described below, several of these options have been pursued already at elevated pressure but several more remain uninvestigated.

Volume relaxation. The relaxation of volume at the glass transition as a function of changing temperature has been demonstrated to coincide with relaxation time defining the shear viscosity of silicate melts (Dingwell 1995). As such, observations of volume or density relaxation during heating or cooling of a silicate melt yield viscosity data. The volume relaxation might be monitored in any number of ways. The simplest theoretical possibility might be optical sensing of the volume or thickness of a regular sample of highly viscous melt. An alternative might be the spectroscopic sensing of a volume-dependent parameter. In either case identification of the inflection in the temperature dependence of the volume, density, refractive index or any other volume-dependent spectroscopic parameter, together with knowledge of the experimental timescales (heating or cooling rates) could yield viscosity data. By systematic variation of the heating and/or cooling timescale, the activation energy of the process can be derived via the timescale dependence of the viscosity at the glass transition phenomenon.

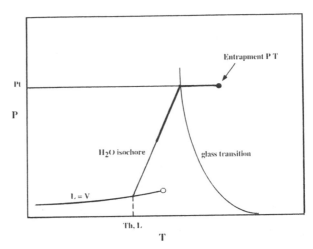

Figure 16. Schematic illustration of the obtainment of glass transition temperature data from heating stage analysis of the homogenization temperatures of fluid inclusions in glass.

The volume does not necessarily need to be measured directly as inferred above. Volume relations between phases can also be used to infer the glass transition. An example of such a technique is provided by the use of fluid inclusions in silicate melts by Romano et al. (1995). Figure 16 illustrates the principle they used to determine the glass transition of hydrous (water-saturated) silicic melts at elevated pressures. The densities of fluid inclusions trapped in glasses quenched isobarically from melts have densities which are defined by the known PVT equation of state of the fluid and the pressure and temperature of the glass transition for a given quench rate. By quenching a melt isobarically at a known rate, and subsequently determining the density (via the homogenization temperature) of the trapped fluid inclusions in the quenched glass, Romano et al. (1995) were able to determine glass transition temperatures for these fluid saturated systems at elevated pressures. Typical data from these studies are illustrated in Figure 17 where the mixed alkali effect which depresses the glass transition temperature (or isothermal viscosity) of intermediate compositions is apparent.

Such studies might be extended to much higher pressures by the judicious choice of saturating volatile with respect to its solubility and the phase diagram (especially the relative positioning of the triple point and the critical point) of the fluid (included) phase. Hydrous fluids will be problematic due to the high solubility of silicate components and lack of detailed equations of state for such systems but relatively pure candidates such as noble gases might success in extending the pressure range.

The possibility also exists that the surface or shape relaxation of included and/or

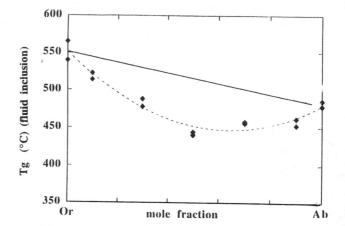

Figure 17. Glass transition temperature data obtained using the analysis of volume relaxation of synthetic fluid inclusions in glass. The mixed alakali effect is clearly visible. Figures 16 and 17 reproduced with permission from Romano et al. (1995).

including melt phases can be converted into viscosity data provided the surface tension can be estimated accurately enough. This too is made theoretically possible by the fundamental link between volume relaxation and viscosity at the glass transition.

Enthalpy relaxation. The link between enthalpy relaxation and the viscosity at the glass transition has also been well-defined and calibrated for silicate melts of geoscientific relevance (Stevenson et al. 1995) using the relationship

$$\log_{10} \eta_s \,(\text{at Tg}) \;=\; \text{constant} - \log_{10} q \;=\; \log_{10} \tau \,(\text{at Tg}) + \log_{10} G \qquad (7)$$

where η_s is the shear viscosity, q is the absolute value of the cooling rate (K/s), τ is the relaxation time and G is the shear modulus (Pa) and the value of the constant was provided by Stevenson et al. (1995) for silicic melts. Quite apart from being a simplification of enormous significance for the modelling of melt behavior in magma transport and volcanic eruptions, this link opens up another potential avenue for the investigation of melt viscosity at elevated pressures. The one study of the pressure dependence of the glass transition temperature of georelevant silicate melts using calorimetric means was conducted some 20 years ago by Rosenhauer et al. (1978). They demonstrated that the calorimetrically measured glass transition temperature of several silicate melts could be determined by simply designed DTA (differential thermal analysis) experiments in an internally heated pressure vessel. Their results, presented here in Figure 18, show that the glass transition temperature of albite melt decreases with increasing pressure whereas that of diopside melt increases. Simple DTA experiments should be readily possible at much higher pressures utilising solid and gas media encapsulation of samples.

Spectroscopic relaxation. One of the earliest studies to derive relaxation times, and thus, glass transition information, from high temperature silicate melts, was the ^{29}Si static NMR study of Liu et al. (1988). They observed motional averaging of the contributions of different Q species due to structural relaxation in the melt and inferred a strong link to the viscous flow process. The development of MAS techniques to higher temperature investigations of silicate melts holds out the hope that NMR-based detection of motional averaging may be capable of providing relaxation time data, and thus indirectly, viscosity data, at elevated temperatures for many silicate melt compositions (Stebbins 1995). Even more recently, efforts to develop in situ high temperature – high pressure probes for use in NMR spectroscopy are opening up the possibility of work at elevated pressure in molten silicate systems (Dupree, pers. comm., September 1998).

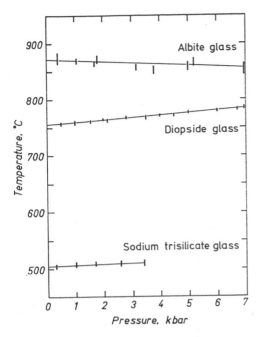

Figure 18. The pressure dependence of the calorimetric glass transition temperature obtained using differential thermal analysis in a gas pressure vessel. Reproduced with permission from Rosenhauer et al. (1979).

Many homogeneous equilibria involve reactions in the silicate melt phase whose kinetics are closely reflected in the relaxational timescale of resultant properties noted above such as enthalpy, volume and shear stress. Thus it should come as no surprise that the spectroscopic investigation of speciation relaxation in melts can also be used to derive viscosity data. The best investigated example of this is the innovative analysis of the relaxation of water species provided by Zhang et al. (1995). These authors performed time domain studies of the speciation of water bewteen hydroxyl groups and water molecules in a reaction which neccesarily involves the networkforming cations (Si,Al) in the melt. The approach to equilibrium was quantified as an effective or "apparent equilibration" time which was determined as a function of water content and temperature for silicic melts. An example of the data used to obtain these times is presented in Figure 19. The derived times were then compared by Zhang et al. (1998) with the stress relaxation times derived from the application of the Maxwell relation to the viscosity model of Hess and Dingwell (1996). This comparison, illustrated in Figure 20 displays a small absolute temperature offset (some 12 K) and a slope with a value of 1. This agreement between a viscosity model based largely on composite data from volume relaxation and viscometry and a spectroscopic study of water speciation illustrates that the spectrscopic investigation of speciation reactions that can be demonstrated, as in the case of water, to be linked to the kinetics of viscous flow, should provide ample opportunity for the determination of melt viscosities by spectroscopic means at much higher pressures.

The strong link provided between structural and property relaxation enables the use of in situ spectroscopic observations of reaction progress at constant temperature or of the temperature or pressure dependence of a homogeneous equilibrium to serve as a monitor of the glass transition. In situ infrared spectroscopic investigations of the temperature dependence of the reaction controlling the speciation of water dissolved in silicate melts provide just such an opportunity. The temperature dependence of the water speciation and the derived equilibrium constant are illustrated in Figure 21 from the study of Nowak and

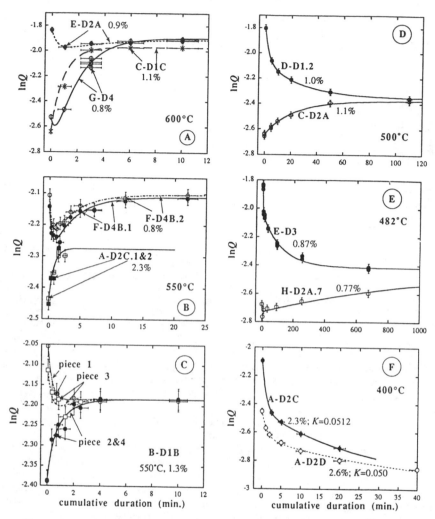

Figure 19. The approach of the water speciation reaction to equilibrium as determined in time series equilibration experiments on hydrous rhyolitic glasses. Reproduced with permission from Zhang et al. (1995).

Behrens (1995). The inflection marking the glass transition temperature is clearly visible. With knowledge of the heating or cooling rate for the transect of the glass transition, quantitative estimation of the viscosity corresponding to the glass transition temperature can be made (see below). Similar results were obtained by Shen and Keppler (1995).

The potential for spectroscopic detection of the glass transition temperature is enormous and it is hoped that such studies will be commonplace within a few years time. Shortening the timescale of observation of the speciation reactions by choosing alternate spectroscopies with higher frequencies of observation should allow for a considerable broadening of the range of observable viscosities using spectroscopic means, provided samples will remain in the metastable liquid state.

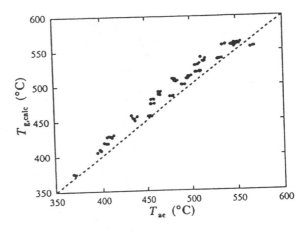

Figure 20. Comparison of the apparent equilibration temperature of the water speciation reaction with derived viscometrically derived glass transition temperatures for a series of hydrous granitic melts. The near agreement in temperatures is an indication of the equivalence of structural and property relaxation in these melts that lies behind the comparison presented below in Figure 22. Reproduced with permission from Zhang et al. (1998).

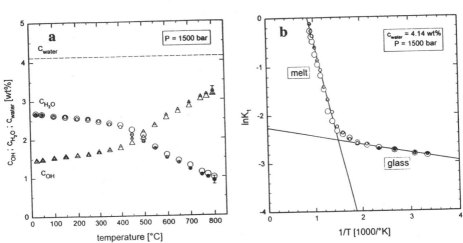

Figure 21. The temperature dependence of the concentrations of hydroxyl and molecular water, and the equilibrium constant defining their proportions, determined in situ using a sapphire anvil cell and infrared spectroscopy. Reproduced with permission from Nowak and Behrens (1995).

Electrical relaxation. An uninvestigated possibility for high pressure viscosity determination by indirect means concerns the possibility that the glass transition is accompanied by a strong enough inflection in the temperature dependence of the electrical conductivity. Most melts of geoscientific relevance are ionic conductors whose conductivity is determined by alkali mobility. Alkali-rich glasses do exhibit an inflection at the glass transition due to the inflection in the temperature dependence of the melt structure in going from a disequilibrium glass to a metastable liquid. This inflection could be potentially strong enough at elevated presures to permit the detection of the glass transition using electrical means in geometries that are already experimentally available.

Wave attenuation. Ultrasonic and low frequency torsional studies of the attenuation of shear, and longitudinal deformation waves in silicate melts have been recently reviewed (Webb and Dingwell 1995). The application of torsional methods in gas pressure vessels (Jackson 1986) and ultrasonic methods in piston cylinder and multianvil presses (Kung and

Rubie 1998) provide good porspects for the derivation of relatively low range viscosity data at considerably elevated pressures. The temperature dependence of the longitudinal and shear sound speeds has also been investigated in simple silicate melts from temperatures below the glass transition to above the liquidus using Brillouin scattering techniques (Askarpour et al. 1993). Clearly defined inflections in the temperature dependence of the V_p and V_s data signal the location of the glass transition, in situ (under elevated temperature and pressure) in these systems. Such methods, explicitly aimed at the determination of relaxation timescales for the melts under investigation are also a potentially rich source of data on the viscosities of silicate melts under high pressures.

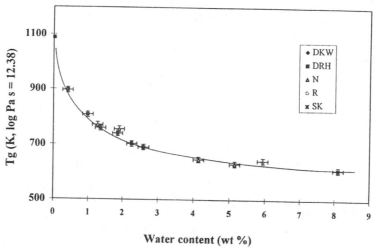

Water content (wt %)

Figure 22. The variation of the glass transition temperature of hydrous granitic melts with water content as obtained by volume, shear stress and structural relaxation studies. Reproduced with permission from Dingwell (1998), where original data sources are listed.

As mentioned above, the significance of glass transition studies for high pressure viscometry rests on two fundamental points. Firstly, the validity of the comparison of glass transition phenomenology and derived relaxation times from various structural and/or property investigations requires that the microscopic control on all of these measurements of relaxation is the same. That is to say that the various measurements all refer to a specific relaxation mode within the silicate melt structure. That this is true, as has been stated above, is emphasized in Figure 22, where various sources of data are included in an isoviscous comparison of the glass transition temperature of rhyolitic melt versus added water content where

$$Tg = 1059 - 83.47 \ln(C_{H_2O}) \tag{8}$$

and Tg is the glass transition at a viscosity of $10^{12.38}$ Pa s and C_{H_2O} is the water concentration in wt %. This comparison contains data derived from spectroscopic, volumetric and viscometric studies of water-bearing haplogranitic melts and illustrates convincingly that a simple function for the water concentration dependence of the glass transition unifies all of the observations. The parameterization of this data was in fact used as input into the viscosity model of Hess and Dingwell (1996) in order to fill out the region of high water content and high viscosity otherwise missing to date in viscometric studies of these important melt compositions. The resulting parameterization (Fig. 23). illustrates the importance of water in increasing the degree of non-Arrhenian behavior of silicate melts

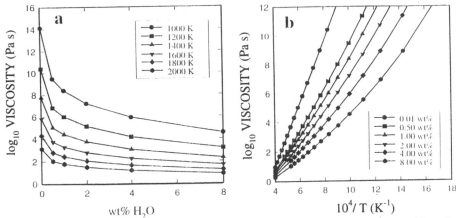

Figure 23. The temperature and water content dependence of the shear viscosity of haplogranitic melts. This parameterization was compiled using spectroscopic, volume and shear stress relaxation data. Reproduced with permission from Hess and Dingwell (1996).

whose anhydrous base composition behavior is very nearly Arrhenian. Secondly, the derived viscosity data is most useful in combination with higher temperature data obtained using one of the above described versions of the falling sphere method. And so we are reminded that, at high pressure, as at ambient pressure, complete descriptions of the temperature dependence of viscosity will rest on a combination of experimental methods, one or more for the high viscosity range and one or more for the low viscosity range.

Diffusivity studies

The demonstration of a negative pressure dependence of the viscosity of silicate melts was followed within a few years by the first simulations of the pressure dependence of component diffusivities in silicate melts using molecular dynamics (Angell et al. 1982). These studies were inspired, at least in part, by the possibility of an inverse proportionality between diffusivities and viscosity and the implications of the pressures dependence of simulated diffusivities for melt transport in the Earth and have, in turn, stimulated further work. In particular, these simulations predicted a maximum in the diffusivities of several components including Si. This would imply a minimum in melt diffusivity to reverse the "anomalous" negative pressure-dependence of the viscosity observed for silica-rich melts and has therefore stimluated further experimental work (see below). First, however, the link between diffusivities and viscosity must be critically evaluated.

For a wide range of inorganic liquids, an inverse proportionality between the diffusivities of specific components and the viscosity of the liquid is demonstrable. The Stokes-Einstein equation was formulated in order to explain this proportionality on a simple physical atomistic basis

$$\eta D = kT/\lambda \tag{9}$$

where D is the diffusivity, k is Boltzmann's constant and λ is an effective lengthscale or "jump distance" for diffusion. Where it can be demonstrated to be valid, this inverse relationship between the proportionality of diffusivities and viscosity offers a powerful opportunity for the estimation of melt viscosity by using diffusivity data in combination with Equation (3). This provides enormous experimental potential for viscosity determination because of the easing of experimental contraints for selfdiffusivity determinations in comparison with viscosity determination by, for example, the falling sphere method. The

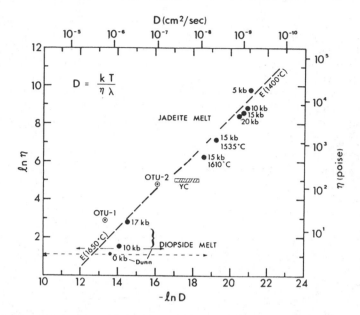

Figure 24. Oxygen diffusivity versus viscosity for a series of silicate melts. The correlation demonstrates the validity of the Stokes-Einstein approximation for silicate melts of relatively low viscosity. Reproduced with permission from Shimizu and Kushiro (1984).

determination of diffusivities can be accomplished by the detection of very short concentration profiles within a binary diffusion couple which in turn eases the difficulty of obtaining isothermal conditions over the scale of the experimental charge, a prime challenge in falling sphere determinations.

For the case of silicon diffusion, in the few cases where data are available, Equation (3) can be shown to be valid over a wide range of viscosities (Dingwell 1990). Thus experimental data on the self-diffusivity of silicon in melts probably yield reliable estimates of viscosity under all P-T-X conditions. For oxygen, the situation is somewhat more complex. The self-diffusivity of oxygen has also been linked to viscosity over a significant range of compositions at relatively low viscosities (Fig. 24; Shimizu and Kushiro 1984). This work has been complemented by relaxational spectroscopic data which suggest that oxygen diffusivity also correlates with viscous flow at much higher viscosities (Farnan and Stebbins 1990). Nevertheless, studies of oxygen gas permeation through glass melts at even higher viscosities (e.g. Yinnon and Cooper 1980, see review by Chakraborty 1995) illustrate clearly that at sufficiently low temperatures and high viscosities, the transport of oxygen within the melt can be detached or decoupled from the constraints of viscous flow. Thus the estimation of melt viscosity from oxygen diffusivity carries restrictions which distinguish it from the case for Si.

The situation may be generalised in terms of what may be usefully described as extrinsic and intrinsic diffusivities in silicate melts (see discussions by Dingwell 1990 and Chakraborty 1995). Extrinsic diffusivities refers to the condition where the timescale of viscous flow of the melt structure coincides with the inverse of the diffusive jump frequency of the cation of interest whereas intrinsic diffusivities refer to diffusivities which are completely detached from and unrelated to the viscous flow mechanism. Extrinsic diffusivities exhibit the inverse temperature-, composition- and pressure-dependence as viscosity and as such enable the use of Equation (9) for viscosity estimation. The diffusivities of trace or minor elements that conform to such behaviour can be seen to be extrinsically determined by the value of the viscosity, a property which, in the case of trace elements in depolymerised melts, can be seen to be essentially independent of their presence. Intrinsic diffusivities refer to the regime of component mobility where transport is independent of viscosity. Intrinsic diffusivities are invariably higher than extrinsic diffusivities and exhibit lower activation energies. Their pressure dependences often have opposite signs (e.g. Watson 1979 vs. Kushiro 1983). Intrinsic diffusivity data cannot be used to obtain viscosity data.

The diffusivities of further components of silicate melts may fall into one of three categories as a progressive function of cationic properties, or as a function of pressure or temperature. The three possibilities are, extrinsic behavior, intrinsic behavior and behavior transitional between the two. For the case of transitional behavior, the transition from intrinsic to extrinsic behavior occurs as a result of increasing temperature. Due to the independence of the pressure dependence of melt viscosity and of intrinsic diffusivities, the criterion for the onset of extrinsic behavior shifts with pressure. Thus many elements exhibiting intrinsic diffusivities at moderate pressures may exhibit extrinsic behavior at higher pressures for comparable temperatures.

The diffusivity behavior of most melt components, apart from Si, appears to be transitional. That is, at P-T-X conditions corresponding to high vicosities of the melt, the diffusivities are intrinsic, detached from any viscometric relevance, whereas at low viscosities they adopt extrinsic values as the temperature dependence of the intrinsic diffusivity timescale intersects that of the Maxwell relaxation time for viscous flow.

A final general point regarding the transition from intrinsic to extrinsic diffusive behavior is that the transition temperature depends both on the component-specific intrinsic diffusivity-temperature relationship and on the temperature-dependence of the viscosity of the melt. Thus the negative pressure-dependence of the viscosity of silica-rich melts, together with the negative pressure-dependence of many intrinsic component diffusivities, means that the two constraining curves for the transition temperature converge with increasing pressure in such systems. The consequence should be much lower transition temperatures and a much enhanced regime of extrinsic behavior at high temperatures. This may greatly increase the range of components of the melt whose diffusivities may be validly employed in obtaining viscosity data for high pressure melts.

In summary, Si duffusivity data may be used under all P-T-X conditions for the estimation of viscosity, oxygen diffusivities may be used at high temperatures and low viscosities, and other component diffusivities should only be used where extrinsic behavior can be demonstrated or confidently estimated. This means that, at the viscosities where falling sphere methods have been traditionally used in high pressure studies, and where geological and planetological aspects of melt transport are operative, Si and O appear to be extrinsic but other components may not be. Certainly, at the values of viscosity corresponding to several of the property relaxation studies described above, the behavior of most components can be considered to be intrinsic. Several studies now exist demonstrating the application of Stokes-Einstein or extrinsic diffusivity behavior to the estimation of viscosities at elevated pressures and a useful summary of the available data is provided by Poe and Rubie (1998) and reproduced here as Figure 25.

Fragility at high pressure. The importance of the non-Arrhenian temperature dependence of the viscosity has been outlined above using extensive data sets from low pressure studies. Those data sets incorporate determinations over ten orders of magnitude of viscosity and are therefore, of necessity composed of a combination of experimental

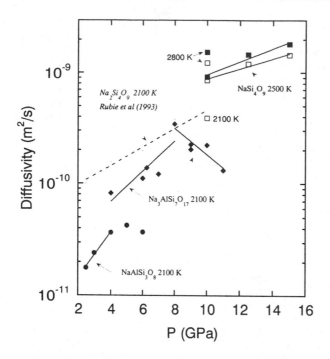

Figure 25. The pressure dependence of the diffusivity of oxygen and silicon in high pressure silicate melts. Reproduced with permission from Poe and Rubie (1998).

methods. How might we expect the fragility of silicate melts to change with pressure? The only melts approaching Arrhenian behavior at low pressure are those of tectosilicate stoichiometry with open, low density, three-dimensionally linked or "fully polymerized" tetrahedral structures. Such melts generally exhibit relatively high compressibilities and can be anticipated to adopt structures, including the distributions of T-O-T bond angles and energies, that are observed in more depolymerized structures at low pressure. Given that depolymerized melts exhibit marked fragility at low pressures it is tempting to speculate that polymerized melts might become significantly more fragile at high pressures. If so, one important experimental consequence is that the determination of viscosity becomes, at high pressure, the same composite challenge that it is at low pressure in that multiple methods will be needed for an adequate description over a reasonable range of viscosities.

Some pieces of evidence exist already to suggest that fragility is enhanced by pressure. The observation of decreasing viscosity and activation energy with increasing pressure, in falling sphere investigations of albite melt (Kushiro 1978b, Dingwell 1987) is one example. Similar behavior can be inferred using the diffusivity behavior of silicon and oxygen at even higher pressures (Poe et al. 1997). In principle, lower activation energies could be consistent with no increase in fragility. Data from the high viscosity regime near the glass transition are required for comparison to define the trend of fragility with increasing pressure. A good indication of the trend to expect is provided by a comparison of the high temperature, low-viscosity data of Dingwell (1987) with the low temperature, high-viscosity calorimetric glass transition data from Rosenhauer et al. (1979) which indicates increasing fragility with pressure. A further contribution to this matter is provided by molecular dynamics investigation utilising an ionic model for pure silica by Barrat et al. (1997). They conclude that SiO_2 exhibits increasing fragility with increasing pressure. The experimental test of this prediction should be given a high priority in future high pressure viscometry work.

Glass transition pressure and amorphization/vitrification. Much of the above discussion of the viscosity of silicate liquids has been augmented by the determination of melt viscosities in the metastable supercooled liquid regime. The kinetic limits on the fictive temperatures and resultant structures and properties that can be obtained from quenching such systems through the glass transition and into the glassy state are fairly well understood at low pressures. We have discussed almost exclusively, the glass transition as a function of temperature, and spoken in terms of glass transition temperatures for specific narrow ranges of cooling rate. For systems whose viscosity near the glass transition increases with pressure (e.g. $CaMgSi_2O_6$) can equally well define a glass transition pressure, where, during pressurization the metastable liquid will transform from a liquid to a glassy state. Such observations are indeed made in nonsilicate low viscosity systems (e.g. Wolf and McMillan 1995) but have not yet been explicitly performed on liquid silicate systems in static experiments.

Shock-wave studies of the equation of state of pre-heated molten silicate systems (Rigden et al. 1988, 1989), however, yield compressibility data that has been interpreted as the transition from the liquid to the glassy state (Dingwell and Webb 1990a). The kinetic interpretation of the compressibility kink reported by Rigden et al. (1988, 1989) for liquid silicates provides, if correct, an additional constraint on the pressure dependence of viscosity through the glass transition (cf. Miller et al. 1991).

The melting of metastable phases to the liquid state at temperatures much lower than that the melting temperature dictated by the global thermodynamics of the system represents a possibility to obtain a metastable liquid which is easy to visualize. Somewhat more complex is the possibility that this transformation can be shifted to such low temperatures, that the viscosity of the resultant "melt" is too high for viscous flow on the experimental timescale. High-pressure diamond cell experiments offer the possibility of investigating the behavior of highly unstable crystalline phases under large and rapidly applied pressure steps. Crystalline phases which become unstable due to such steps may then transform to the amorphous state at very low temperatures indeed. Amorphization (Mishima et al. 1984) is the term applied to the observation of a mineral-melt transformation under such (usually low temperature) conditions that the resultant "melt" is viscous enough to behave as a glass. Several systems exhibit this phenomenon when specific phases are taken outside of their stability fields as a function of increasing pressure (e.g. metastably compressed α-quartz and coesite; Hemley et al. 1988, 1994) and of decreasing pressure (e.g. decompression of $CaSiO_3$ perovskite, a high-pressure phase; Richet 1988).

Richet (1988) termed the decompression transformation of $CaSiO_3$ perovskite –"vitrification." The two terms, amorphization and vitrification, imply somewhat different meanings which raise the issue of the mechanism of the transformation. The term amorphization would appear to imply little mechanistically, simply the transformation from crystalline to amorphous material. The term vitrification is derived from a root implying the glassy state and might therefore be taken to imply that the transformation occurs directly from crystalline to glassy state. The sentences raise the question of the exact mechanism of these transitions and whether any viscous deformation involving the noncrystalline phase might be possible during the transformation. Careful observations of the transformation should reveal whether this is the case.

Besides their usefulness for mapping the metastable extensions of reaction curves in P-T space (see Fig. 26), such studies also provide constraints (lower bounds) on melt viscosities and offer the potential in future to generate glasses with enormously low fictive temperatures. The physical characterization of such glasses remains an important topic for the future.

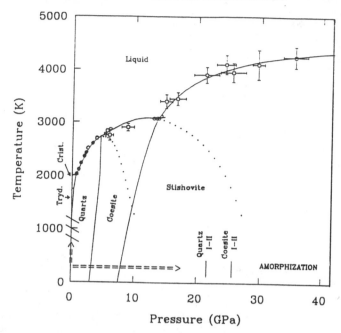

Figure 26. P-T phase diagram for SiO_2. The room-temperature amorphization pressures of α-quartz and coesite (horizontal arrow) are close to the metastable extensions of the equilibrium melting of these phases (Hemley et al. 1988, 1994). Transitions to metastable crystalline phases (quartz II and coesite II) occur at similar pressures. The temperature-induced amorphization of coesite (Richet 1988) is shown by the vertical arrow. After Hemley et al. (1994).

REFERENCES

Angell CA (1984) Strong and fragile liquids. *In:* Ngai KL and Wright GB (eds) Relaxation in Complex Systems. Office of Naval Research and Technical Information Service Lab, Arlington, VA 345 p
Angell CA, Cheeseman PA, Tamaddon S (1982) Pressure enhancement of ion mobilities in liquid silicates from computer simulation studies to 800 kilobars. Science 218:885-888
Askarpour V, Manghnani M, Richet P (1993) Elastic properties of diopside, anorthite and grossular glasses and liquids: a Brillouin scattering study up to 1400 K. J Geophy Res 98:17683-17689
Baker DR (1996) Granitic melt viscosities: empirical and configurational entropy models for their calculation. Am Mineral 81:126-134.
Barrat J-L, Badro J, Gillet P (1997) A strong to fragile transition in a model of liquid silica. Molecular Simulation 20:17-25.
Bowen NL (1929) The Evolution of the Igneous Rocks. Dover Publications, New York, 322 p
Brearley M, Dickenson JE, Scarfe CM (1986) Pressure dependence of melt viscosities on the join diopside-albite. Geochim Cosmochim Acta 50:2563-2570
Bridgman PW (1926) Proceedings of the Amer Acad Arts Sci 61:57-99
Brückner R (1971) Properties and structure of vitreous silica II. J Non-Cryst Sol 5:177-216
Carroll M, Holloway JR, eds (1994) Volatiles in Magmas. Rev Mineral 30, 517 p
Chakraborty S (1995) Diffusion in silicate melts. Rev Mineral 32:411-503
Daly RA (1925) Proc Am Phil Soc 64:283
Dickinson JE Jr, Scarfe CM, McMillan P (1990) Physical properties and structure of $K_2Si_4O_9$ melt quenched from pressures up to 2.4 Gpa. J Geophys Res 95:15675-15681
Dingwell DB (1987) Melt viscosities in the system $NaAlSi_3O_8$-H_2O-F_2O_1. *In:* Magmatic Processes: Physicochemical Principles. BO Mysen (ed) Geochemical Society Spec Publ 1:423-433
Dingwell DB (1990) Effects of structural relaxation on cationic tracer diffusion in silicate melts. Chem Geol 82:209-216
Dingwell DB (1993) Experimental strategies for the determination of granitic melt properties at low temperature. Chem Geol 108:19-30

Dingwell DB (1995) Relaxation in silicate melts: some applications in petrology. *In:* Stebbins JF, Dingwell DB, McMillan P (eds) Structure and Dynamics of Silicate Melts. Rev Mineral 32:21-66

Dingwell DB (1995) Viscosity and Anelasticity of melts and glasses. *In:* Ahrens TJ (ed) Mineral Physics and Crystallography—A Handbook of Physical Constants 2:209-217, Am Geophys Union, Washington, DC

Dingwell DB (1998) The glass transition in hydrous granitic melts. Phys Earth Planetary Inter 107:1-8

Dingwell DB, Webb SL (1989) Structural relaxation in silicate melts and non-Newtonian melt rheology in igneous processes. Phys Chem Minerals 16:508-516

Dingwell DB, Webb, SL (1990) Relaxation in silicate melts. Eur J Mineral 2:427-449

Dingwell DB, Pichavant M, Holtz H (1996) Experimental studies of boron in granitic melts. *In:* Grew E, Anovitz L (eds) Boron: Mineralogy, Petrology and Geochemistry. Rev Mineral 33:331-385

Dingwell DB, Romano C, Hess K-U (1996) The effect of water on the viscosity of a haplogranitic melt under P-T-X- conditions relevant to silicic volcanism. Contrib Mineral Petrol 124:19-28

Dingwell DB, Hess K-U, Romano C (1998) Extremely fluid behavior of hydrous peralkaline rhyolites Earth Planet Sci Lett 158:31-38

Dorfman A, Hess K-U, Dingwell DB (1996) Centrifuge-assisted falling sphere viscometry. Eur J Mineral 8:507-514

Dorfman A, Dingwell DB, Bagdassarov N (1997) A rotating autoclave for centrifuge studies: falling sphere viscometry. Eur J Mineral 9:345-350

Gennaro C, Dingwell DB, Pichavant M (1998) A dilatometer for high pressure viscometry. Eos Trans Am Geophys Union (in press)

Hemley RJ, Jephcoat AP, Mao HK, Ming LC, Manghnani, MH (1988) Pressure induced amophization of crystalline silica. Nature 334:52-54

Hemley RJ, Prewitt CT, Kingma KJ (1994) High-pressure behavior of silica. *In:* Heaney PJ, Prewitt CT, Gibbs GV (eds) Silica: Physical Behavior, Geochemistry, and Materials Applications. Rev Mineral 29:41-81

Hess K-U, Dingwell DB (1996) Viscosities of hydrous leucogranitic melts: a non-Arrhenian model. Am Mineral 81:1297-1300

Jackson I (1986) The laboratory study of seismic wave attenuation. *In:* BE Hobbs and HC Heard (eds) Mineral and Rock Deformation Studies. Paterson Vol p 11-23 Am Geophys Union, Washington, DC

Kanzaki M, Kurita K, Fujii T, Kato T, Shimomura O, Akimoto S (1987) A new technique to measure the viscosity and density of silicate melts at high pressure. *In:* Manghnani MH, Syono Y (eds) High Pressure Research in Mineral Physics, p 195-200 Am Geophys Union, Washington, DC

King HE Jr (1997) High pressure viscometry. *In:* Holzapfel WB, Isaac NS (eds) High-Pressure Techniques in Chemistry and Physics—A Practical Approach, p 122-150 Oxford University Press, New York

Knoche R, Dingwell DB, Seifert FA, Webb S (1994) Nonlinear properties of supercooled liquids in the Na_2O-SiO_2 system. Chem Geol 116:1-16

Kung J, Rubie D (1998) Measurement of elastic velocities for mantle phases in the multianvil press. Terra Abstracts 10:32

Kushiro I (1976) Changes in viscosity and structure of melt of $NaAlSi_2O_6$ composition at high pressures. J Goephy Res 81:6347-6350

Kushiro I (1978a) Density and viscosity of hydrous calkalkaline andesite magma at high pressures. Carnegie Inst Wash Yrbk 77:675-677

Kushiro I (1978b) Viscosity and structural changes of albite ($NaAlSi_3O_8$) melt at high pressures. Earth Planet Sci Lett 41:87-90

Kushiro I, Yoder HS, Mysen BO (1976) Viscosities of basalt and andesite melts at high pressures. J Geophy Res 81:6351-6356

Miller GH, Ahrens TJ (1991) Shock-wave measurement. Rev Mod Phys 63:910-948

Miller GH, Stolper EM, Ahrens TJ (1991) The equation of state of a molten komatiite—1. Shock wave compression to 36 GPa. J Geophys Res 96:11831-11848

Mysen BO, Virgo D, Seifert FA (1982) The structure of silicate melts: implications for chemical and physical properties of natural magma. Rev Geophy 20:353-383

Nicholls J, Russell JK (1990) Modern Methods of Igneous Petrology. Rev Mineral 24, 314 p

Neuville D, Richet P (1991) Viscosity and mixing in molten (Ca,Mg) pyroxenes and garnets. Geochim Cosmochim Acta 55:1011-1019

Nowak M, Behrens H (1995) The speciation of water in haplogranitic glasses and melts determined by in situ near infrared spectroscopy. Geochim Cosmochim Acta 59:3445-3450

Persikov ES, Zharikov VA, Bukhtiyarov PG, Polsk´oy SF (1990) The effect of volatiles on the properties of magmatic melts. Eur J Mineral 2:621-642

Poe B, Rubie D (1998) Transport properties of silicate melts at high pressure. *In:* Aoki H, Syono Y, Hemley RJ (eds) Physic Meets Mineralogy (in press)

Poe B, McMillan P, Rubie DC, Chakraborty S, Yarger J, Diefenbacher J (1997) Silicon and oxygen selfdiffusivities in silicate liquids measured to 15 Gigapascals and 2800 Kelvin. Science 276:1245-1248

Richet P (1984) Viscosity and configurational entropy of silicate melts. Geochim Cosmochim Acta 48:471-484

Richet P (1988) Superheating, melting and vitrification through decompression of high pressure minerals. Nature 331:56-58

Rigden S, Ahrens T, Stolper EM (1988) Shock compression of molten silicates: results for a model basalt composition. J Geophy Res 93:367-382

Rigden S, Ahrens T, Stolper EM (1989) High-pressure equations of state of molten anorthite and diopside. J Geophy Res 94:9508-9522

Rivers ML, Carmichael ISE (1987) Ultrasonic studies of silicate melts. J Geophys Res 92:9247-9270

Romano C, Dingwell DB, Sterner SM (1994) Kinetics of quenching of hydrous feldspathic melts: quantification using synthetic fluid inclusions. Am Mineral 79:1125-1134

Rosenhauer M, Scarfe CM, Virgo D (1979) Pressure dependence of the glass transition in glasses of diopside, albite and sodium trisilicate composition. Carnegie Inst Wash Yrbk 78:556-559

Sato H, Manghnani M (1984) Ultrasonic measurements of V_p and Q_p: relaxation spectrum of complex modulus on basalt melts. Phys Earth Planet Inter 41:18-33

Scarfe CM, Mysen BO, Virgo DL (1987) Pressure dependence of the viscosity of silicate melts. Geochem Soc Spec Publ 1:59-67

Schulze F, Behrens H, Holtz F, Roux J, Johannes W (1996) The influence of water on the viscosity of a haplogranitic melt. Am Mineral 81:1155-1165

Schulze F, Behrens H (1998) Viscosity determination by creep deformation of viscous melts at pressures up to 300 MPa: first results for hydrous melts. Terra Nova Abstr Suppl EMPG Orléans, France

Shaw HR (1963) Obsidian-H_2O viscosities at 1000 and 2000 bars in the temperature range 700 to 900°C. J Geophys Res 68:6337-6343

Shaw HR (1972) Viscosities of magmatic silicate liquids: an empirical method of prediction. Am J Sci 272:870-889

Shen A, Keppler H (1995) Infrared spectroscopy of hydrous silicate melts to 1000°C and 10 kbars: direct observation of water speciation in a diamond anvil cell. Am Mineral 80:1335-1338.

Shimizu N, Kushiro I (1984) Diffusivity of oxygen in jadeite and diopside melts at high pressures. Geochim Cosmochim Acta 48:1295-1303

Spera, FJ, Borgia A, Strimple J (1988) Rheology of melts and magmatic suspensions—1. Design and calibration of concentric cylinder viscometer with application to rhyolitic magma. J Geophy Res 93:10273-10294.

Stebbins JF (1995) Dynamics and structure of silicate and oxide melts: nuclear magnetic resonance studies. Rev Mineral 32:191-246

Stebbins JF, McMillan PW, Dingwell DB (eds) (1995) Structure and Dynamics of Silicate Melts. Rev Mineral 32, 616p

Stevenson R, Dingwell DB, Webb SL, Bagdassarov NL (1995) The equivalence of enthalpy and shear relaxation in rhyolitic obsidians and quantification of the liquid-glass transition in volcanic processes. J Volc Geotherm Res 68:297-306

Taniguchi H (1992) Entropy dependence of viscosity and the glass transition temperature of melts in the system diopside-anorthite. Contrib Mineral Petrol 109:295-303

Tauber P, Arndt J (1986) Viscosity-temperature relationship of liquid diopside. Phys Earth Planet Inter 43:97-103.

Watson EB (1979) Calcium diffusion in a simple silicate melt to 30 kbar. Geochim. Cosmochim. Acta 43:313-322.

Webb SL (1992) Low frequency shear and structural relaxation in rhyolite melt. Phys Chem Minerals 19:240-245

Webb SL, Dingwell DB (1990a) The onset of non-newtonian rheology of silicate melts. A fiber elongation study. Phys Chem Minerals 17:125-132

Webb SL, Dingwell DB (1990b) Non-newtonian rheology of igneous melts at high stresses and strain rates: experimental results for rhyolite, andesite, basalt, and nephelinite. J Geophys Res 95:695-701

Webb SL, Dingwell DB (1995) Viscoelasticity. Rev Mineral 32:95-118

White BS, Montana A (1990) The effect of H_2O and CO_2 on the viscosity of sanidine liquid at high pressures. J Geophy Res 95:15683-15693

Wolf G, McMillan P (1995) Pressure effects on silicate melt properties. Rev Mineral 32:505-561

Yinnon H, Cooper AR Jr (1980) Oxygen diffusion in multicomponent glass forming silicates. Phys Chem Glasses 21:204-211

Zhang Y, Stolper EM, Ihinger PD (1995) Kinetics of the reaction $H_2O + O = 2$ OH in rhyolitic and albitic glasses: preliminary results. Am Mineral 80:593-612

Chapter 14

PRESSURE-VOLUME-TEMPERATURE EQUATIONS OF STATE

Thomas S. Duffy

Department of Geosciences
Princeton University
Princeton, New Jersey 08544

Yanbin Wang

Consortium for Advanced Radiation Sources
The University of Chicago
5640 S. Ellis Avenue
Chicago, Illinois 60637

INTRODUCTION

Recent advances in experimental and theoretical techniques have resulted in major improvements in our understanding of the physical and chemical properties of minerals and other materials at high pressures (P) and temperatures (T). Experiments using the diamond anvil cell (DAC), multi-anvil press (MAP), and shock compression are yielding increasingly precise measurements on minerals at ever higher pressures and temperatures. Experimentally accessible P-T conditions now span the range of values encountered in terrestrial planet interiors, and reach deep into the interior of the giant planets. At the same time, advances in computational power and quantum statistical techniques are now yielding accurate predictions of crystal structures and bonding properties under these extreme conditions.

Together with the crystal structure, the equation of state (EOS) is the most fundamental parameter obtained from high-pressure investigations. The equation of state of a system describes the relationships among the thermodynamic variables volume (V), pressure, and temperature (or energy (E)). As such, it reflects the underlying atomic structure and chemical bonding of the components involved. The application of pressure is of prime importance for EOS studies because pressure varies the interatomic spacing, resulting in large changes in structure, bonding, and electronic configurations.

The P-V-T equation of state is one of the most critical parameters for understanding the behavior of minerals under the ultrahigh pressure conditions encountered in planets and their satellites. The equation of state forms a link between the observable physical properties of a planet (seismic velocity structure, moment of inertia, etc.), the P-T conditions within the planet, and the atomic structure of the constituent materials. The equation of state is also necessary to assess the stability of mineral assemblages under very high P-T conditions.

This review focuses on recent experimental advances in the determination of P-V-T equations of state of minerals and metals. First, the basic formalism of the equation of state under both ambient and high-temperature conditions is reviewed. In the second section, experimental methods for measuring equations of state are described and recent results are summarized. In the final section the accuracy of P-V-T equations of state is examined in more detail, focusing on P-V-T standards and highlighting areas of focus for future work.

0275-0279/98/0037-0014$05.00

The *P-V-T* equation of state is of fundamental importance not only in mineralogy, but in a variety of other fields including condensed matter physics, astrophysics, and materials science. For more detailed treatments of topics covered in this review, a number of references can be consulted on high-pressure equations of state (Eliezer and Ricci 1991), shock compression (McQueen et al. 1970), high-pressure experimental techniques (Eremets 1996), mineral elasticity (Liebermann, this volume) and theoretical descriptions of the equation of state (Cohen and Stixrude, this volume; Ross 1993).

EQUATIONS OF STATE

We are interested in developing *P-V-T* equations of state that can accurately describe the behavior of solids under pressures and temperatures encountered in the terrestrial planets ($P < 360$ GPa, $T < 7000$ K). We assume the Helmholtz free energy, F, of a solid can be divided into independent static lattice, vibrational, and electronic components:

$$F(V,T) = F_c(V) + F_{vib}(V,T) + F_{el}(V,T) \tag{1}$$

where $F_c(V)$ is the cold (0 Kelvin) compression curve that depends only on volume, $F_{vib}(V,T)$ is the contribution due to atomic vibrations, and $F_{el}(V,T)$ is the contribution due to thermal excitation of electrons. While the static lattice energy is the largest part of the total energy, it is the vibrational energy that usually determines the stable high-temperature crystal structure and contributes nonnegligibly to the total pressure and energy.

The pressure is given by the derivative of energy with respect to volume:

$$P = -\left(\frac{\partial F}{\partial V}\right)_T \tag{2}$$

and hence can be written as:

$$P(V,T) = P_c(V) + P_{vib}(V,T) + P_{el}(V,T) \tag{3}$$

where $P_c(V)$ is the 0 K pressure, and P_{vib} and P_{el} are the vibrational and electron contributions to the pressure, respectively.

For most geological materials in the *P-T* range under consideration, the contribution of the electrons can be neglected. However, this is not necessarily always the case. Ultra-high pressure shock wave experiments on metals represent one exception (e.g. Holmes et al. 1989). Nevertheless with this simplification, the pressure can be written as:

$$P(V,T) = P_c(V) + P_{th}(V,T) \tag{4}$$

Here P_c indicates a reference curve which could be the 0 K isotherm, but is more commonly the room temperature isotherm which is more readily measured experimentally. In this case, the thermal pressure, $P_{th}(V,T)$ represents the difference in pressure between the high *P-T* state and the reference curve.

Higher order derivatives of the energy function yield the isothermal bulk modulus, K_T, and its first pressure derivative, K_T':

$$K_T = -V\left(\frac{\partial P}{\partial V}\right)_T \tag{5}$$

$$K_T' = \left(\frac{\partial K_T}{\partial P}\right)_T \tag{6}$$

Isothermal equation of state

To develop a *P-V-T* equation of state, it is first necessary to specify the form of $P_c(V)$. There is no unambiguous choice, as a large number of semi-empirical equations of state exist, none of which can be rigorously justified on theoretical grounds. For mineralogical applications, the third-order Birch-Murnaghan equation is widely used. This equation was developed by Birch (1947) who expanded the strain energy, *F*, in a Taylor's series in strain, *f*:

$$F = a_0 + a_1 f + a_2 f^2 + a_3 f^3 + a_4 f^4 + ... \qquad (7)$$

The main issues involve choice of the appropriate measure of finite strain and convergence of the series. Empirically, it has been demonstrated that a Eulerian strain measure provides a more strongly convergent series than Lagrangian strain. At very high values of strain, however, the convergence may no longer be reliable.

Using the Eulerian strain measure and retaining terms up to third order in strain yields the Birch-Murnaghan equation:

$$P = \frac{3K_{T0}}{2}\left[\left(\frac{V_0}{V}\right)^{7/3} - \left(\frac{V_0}{V}\right)^{5/3}\right]\left\{1 - \frac{3}{4}\left(4 - K_{T0}'\right)\left[\left(\frac{V_0}{V}\right)^{2/3} - 1\right]\right\} \qquad (8)$$

where the subscript *0* refers to ambient pressure conditions.

While the Birch-Murnaghan equation is the most widely used by mineralogists, many other equations of state have also been developed. It was recently proposed that solids possessing a variety of bond types (metallic, ionic, covalent, van der Waals) can, despite differing interatomic interactions, be described by a universal *P-V* equation of state of the form (Vinet et al. 1986, Vinet et al. 1989):

$$P(x) = \frac{3K_{T0}(1 - X)}{X^2}\exp\left[\frac{3}{2}\left(K_{T0}' - 1\right)(1 - X)\right] \qquad (9)$$

where $X = (V/V_0)^{1/3}$ is the linear compression. This equation arises from an expression for the cohesive energy that varies only as a function of normalized interatomic separation. A single expression can describe a range of bonding types because the form of the *P-V* relation in compression is dominated by repulsive interactions for all classes of solids.

For materials under very strong compression (i.e. volatiles in the deep earth, interiors of the giant planets), equations of state based on an exponential repulsive potential, such as that of Vinet, are superior to finite strain theories (Hemley et al. 1990, Loubeyre et al. 1996). Effective intermolecular pair potentials can also be used to model interatomic forces and determine the equation of state (Ross 1993, Duffy et al. 1994). A variety of other simple equations of state applicable to broad classes of solids have been proposed (e.g. Kumari and Dass 1990, Holzapfel 1991, 1996; Taravillo et al. 1996). A recent comparison of different equations of state to very high pressure (>1000 GPa) supports the use of the Vinet equation (Hama and Suito 1996). However, for compressions found in the terrestrial planets, the Birch-Murnaghan equation is normally adequate.

Thermal equation of state

High-temperature Birch-Murnaghan equation. One simple, but ad hoc, approach for evaluating *P-V-T* data is to develop a high-temperature form of the third-order Birch-Murnaghan equation. In this case, Equation (8) is used, but the zero-pressure parameters (V_0, K_{T0}, K_{T0}') all refer to an initial high-temperature state. In other words, the material is first raised to high temperature, then compressed along the isotherm of interest. High-temperature values for the bulk modulus and volume are given by:

$$K_{T0}(T) = K_{T0}(T_0) + \left(\frac{\partial K_T}{\partial T}\right)_P (T - T_0) \tag{10}$$

$$V_0(T) = V_0(T_0) \exp \int_{T_0}^{T} \alpha(T) dT \tag{11}$$

where $\alpha(T)$ is the ambient-pressure thermal expansion coefficient which can normally be represented simply as:

$$\alpha(T) = a + bT - \frac{c}{T^2} \tag{12}$$

The high-temperature Birch-Murnaghan equation has been extensively used in fitting multi-anvil and diamond cell data (e.g. Utsumi et al. 1998). The main limitation of the approach is that higher order and mixed P-T derivatives are neglected. For example, it is usually assumed that the pressure derivative of the bulk modulus, K_{T0}', does not vary with temperature. Recently, the mixed P-T derivative of the adiabatic bulk modulus was determined for MgO with an uncertainty of ±40% from acoustic velocity measurements at simultaneous high pressure and temperature (Chen et al. 1998). Least-squares fits to P-V-T data can be used to determine the parameters in (10)-(12) (e.g. Wang et al. 1996). The ambient pressure bulk modulus and its pressure derivative are often fixed in the analysis.

Thermal pressure equation of state. In this method, the thermal pressure is constrained using thermodynamic data that is often measurable in the laboratory (Anderson 1984). Taking the derivative of Equation (4) with respect to temperature at constant volume yields:

$$\left(\frac{\partial P}{\partial T}\right)_V = \left(\frac{\partial P_{th}}{\partial T}\right)_V \tag{13}$$

Combining this with the thermodynamic identity:

$$\left(\frac{\partial P}{\partial T}\right)_V = \alpha K_T \tag{14}$$

yields:

$$P_{th} = \int_{T_0}^{T} \alpha K_T dT \tag{15}$$

High-temperature data indicate that the product αK_T is, in general, temperature dependent, but becomes nearly constant for many materials above the Debye temperature (Anderson et al. 1991). In the event that αK_T is also independent of volume, the thermal pressure becomes linear in T (Anderson 1984):

$$P_{th} = \alpha K_T (T - T_0) \tag{16}$$

In the more general case where αK_T depends on both volume and temperature, the thermal pressure can be expressed as (Anderson et al. 1989, Jackson and Rigden 1996):

$$P_{th} = \int_{T_0}^{T} \alpha K_T dT + (\partial K_T / \partial T)_V \left[-\ln(V / V_0)(T - T_0) + \int_{T_0}^{T} \int_{T_0}^{T} \alpha dT dT \right] \tag{17}$$

Jackson and Rigden (1996) discuss the implementation of (17) in fitting *P-V-T* data through a thorough analysis of both previously published and synthetic data sets. They considered the sensitivity of the fitted results to uncertainties in the data and various assumptions that may be made in the analysis (e.g. the location of the principal isotherm). They conclude that using this approach, values of $(\partial K_T / \partial T)_P$ and $\partial^2 K_T / \partial P \partial T$ can be resolved from high-quality *P-V-T* data with precisions of 10% and 50%, respectively.

Lattice dynamics. In the lattice dynamics approach, the crystal is viewed as composed of a collection of harmonic oscillators, and the Helmholtz free energy is obtained by summing over all the normal mode vibrational frequencies using lattice dynamics theory. A simple but successful form for characterization of the lattice vibrational modes is the Debye model which treats the solid as a continuous medium and parameterizes the vibrational spectrum in terms of a single characteristic temperature. This corresponds to a single nondispersive acoustic phonon branch representation of the vibrational modes.

The thermal pressure is determined using the Mie-Gruneisen equation:

$$P(V,T) - P_r(V,T_r) = \frac{\gamma}{V} \left[E(V,T) - E_r(V,T_r) \right] \tag{18}$$

where the subscript *r* represents a reference state which is often chosen to be the 300 K isotherm. The Gruneisen parameter, γ, is assumed to be a function of volume only:

$$\gamma = \gamma_0 \left(\frac{V}{V_0} \right)^q \tag{19}$$

The parameter *q* is usually taken to be 1 implying γV = constant. Support for such a relationship has been obtained from shock experiments on porous materials and measured sound velocities along the Hugoniot (McQueen 1991). Using the Debye model, the thermal energy is then given by:

$$E_{th} = \frac{V}{\gamma} P_{th} = 3nRTD(x) \tag{20}$$

where *n* is the number of atoms per formula unit, *R* is the gas constant, and

$$x = \frac{\Theta(V)}{T} \tag{21}$$

where $\Theta(V)$ is the Debye temperature and $D(x)$ is the Debye function given by:

$$D(x) = \frac{3}{x^3} \int_0^x \frac{z^3 dz}{e^z - 1} \tag{22}$$

The Debye temperature, Θ, is assumed to be independent of *T* and a function of volume only according to:

$$\frac{d \log \Theta}{d \log V} = -\gamma \tag{23}$$

The specific heat at constant volume, C_V, can be obtained from the temperature derivative of the energy at fixed volume:

$$C_V = 3nk\left(4D(x) - \frac{3x}{e^x - 1}\right) \tag{24}$$

An effective value of Θ_0 can be obtained by fitting heat capacity data at ambient pressure or from expressions using the measured acoustic velocities.

An analysis of the implementation of this method for fitting P-V-T data has also been provided by Jackson and Rigden (1996). The different methods for fitting P-V-T data generally yield comparable results for a given data set (Fei et al. 1992, Wang et al. 1994, Jackson and Rigden 1996). It has been argued that the lattice dynamics approach is preferable because it more accurately represents the temperature dependence of α and allows for internally consistent conversion between thermodynamic derivatives at constant temperature and constant entropy (Jackson and Rigden 1996).

The Debye model, however, relies on an oversimplified density of states which can be extended with a classical model of statistical thermodynamics giving the following quasi-harmonic vibrational free energy which includes a simple anharmonic correction (e.g. Wallace 1972):

$$F_{vib} = \int\left[\frac{h\upsilon}{2} + kT\ln\left(1 - \exp\left(-\frac{h\upsilon}{kT}\right)\right) + akT^2\right]g(\upsilon)d\upsilon \tag{25}$$

where the integration is over the vibrational density of states, $g(\upsilon)$, with υ being the vibrational frequency, and h and k are Planck's constant. and Boltzmann's constant, respectively. A Kieffer-type density of states can be easily established based on vibrational spectroscopic data (Guyot et al. 1996). The anharmonic term $a(\upsilon)$ is defined by:

$$a(\upsilon) = \left(\frac{\partial \ln \upsilon}{\partial T}\right)_V \tag{26}$$

(Gillet et al. 1991). Through the thermodynamic identity

$$\left(\frac{\partial \ln \upsilon}{\partial T}\right)_V = \left(\frac{\partial \ln \upsilon}{\partial T}\right)_P + \alpha K_T\left(\frac{\partial \ln \upsilon}{\partial P}\right)_T \tag{27}$$

$a(\upsilon)$ can be determined experimentally using spectroscopic techniques (IR and Raman) with α and K_T determined from P-V-T measurements.

Although the thermal pressure is insensitive to the details of the density of states, Guyot et al. (1996) have shown that vibrational free energy of the Debye model underestimates the entropy of forsterite by a few percent and that a more detailed density of states and its pressure and temperature dependence (and thus $a(\upsilon)$) helps in resolving the discrepancy with the experimental entropy data.

EXPERIMENTAL DETERMINATION OF
EQUATIONS OF STATE

Shock compression

Shock compression is the most established technique for investigating P-V-T equations of state. Due to the large, irreversible energy deposition inherent in the shock process, dynamic compression necessarily produces simultaneous high pressure and high

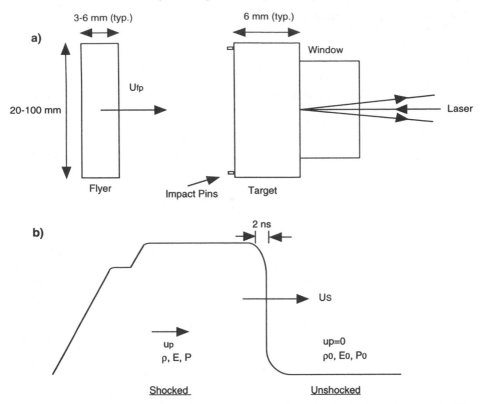

Figure 1. (a) Schematic illustration of a shock wave experiment. U_{fp} is the flyer plate velocity. The impact pins are used to trigger recording instrumentation and establish the impact time. The laser probe at the target-window interface is used to detect the shock wave arrival and measure the particle velocity. (b) Schematic profile of a shock pulse traveling through a material. A typical pulse width is 1 µs and the risetime of the shock front is typically 2 ns. Unloading of the high-pressure state is due to rarefactions arising from the edges of the sample or the flyer rear surface.

temperature states. More importantly, the pressure can be obtained directly from the conservation laws governing the flow, and independent pressure calibration is not required. For this reason, nearly all high-pressure-temperature experiments are ultimately tied to shock standards for pressure calibration. However, the shock method has a number of disadvantages besides the high cost and time-consuming nature of the experiments. The most significant limitation is that shock techniques do not provide direct information on the crystal structure existing in the high *P-T* state.

Shock compressed states are normally achieved by impact of a high-velocity flat plate (flyer plate) into a second, stationary plate (target) (Fig. 1a). The flyer plate is accelerated down a one- or two-stage gun barrel and impacts the target with a velocity in the range of 0.5-7.0 km/s. Impact with the target produces one-dimensional, steady shock pulse propagating through both the flyer and sample (Fig. 1b). The shock front travels supersonically relative to the uncompressed medium with a shock velocity, U_S. The material behind the shock front is accelerated to a particle velocity, u_p.

Depending on the impact velocity and materials under study, pressures of up to several hundred GPa can be achieved with light gas guns. Recently, equations of state have been measured at pressures between 1000-4000 GPa using laser driven shocks (Cauble et al. 1998), although the equation of state determination is not as precise as for gas gun experiments.

Application of the principles of conservation of mass, momentum, and energy across the shock front yield the Rankine-Hugoniot equations which relate the kinematic variables of the flow (U_S, u_p) to the thermodynamic variables of interest—pressure (or longitudinal stress), volume, and the specific internal energy, E:

$$P = \frac{U_s u_p}{V_0} \tag{28}$$

$$\frac{V_0}{V} = \frac{U_S}{U_S - u_p} \tag{29}$$

$$E - E_0 = \frac{1}{2} u_p{}^2 = \frac{1}{2}(P + P_0)(V_0 - V) \tag{30}$$

In the last equation, the first part of the equality means that the material achieves an increase in internal energy (per unit mass) that exactly equals the kinetic energy per unit mass. Thus, impact energy is exactly divided between kinetic energy imparted to the target and internal energy that goes into heating the target. In a typical shock experiment, the particle and shock velocities are measured by one of a variety of techniques (e.g. fast streak camera, laser interferometer) and the Rankine-Hugoniot equations are then used to constrain the thermodynamic state.

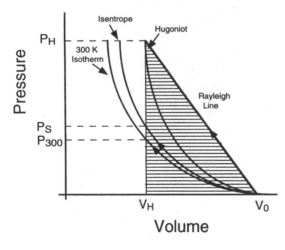

Figure 2. Schematic illustration of the relationships among the isotherm, isentrope, and Hugoniot centered at 300 K. The shaded region beneath the Rayleigh line represents the specific internal energy increase under shock compression. P_H and V_H are the pressure and specific volume achieved during shock compression. P_S and P_{300} are the pressures corresponding to V_H along the isentrope and isotherm, respectively.

Shock states are represented by a Hugoniot which is the locus of final states achieved by shock compression in P-V space (Fig. 2). It is not the thermodynamic path followed by the material. The thermodynamic path is a straight line from the initial volume to the final volume (called the Rayleigh line). The specific internal energy increase under shock compression is given by the area under the triangle defined by the Rayleigh line (shaded region in Fig. 2). In contrast, the area under the principal isentrope represents the reversible work done by isentropic compression to a given volume. The difference between this area and the area under the Raleigh line is the amount of irreversible work that is available for heating the target. Thus, the shock process can result in very strong heating of the sample.

Empirically, it has been observed that over large ranges of compression many materials can be described by a linear relationship between shock velocity and particle velocity. This relation is known as the shock wave equation of state:

$$U_S = c_0 + s u_p \tag{31}$$

The intercept, c_0, is equal to the bulk sound velocity which is the limiting value for a weak shock:

$$c_0^2 = V_0 K_{0S} \tag{32}$$

where K_{0S} is the ambient-pressure adiabatic bulk modulus. The slope, s, is related to the pressure derivative of the adiabatic bulk modulus:

$$s = \frac{K_{0S}' + 1}{4} \tag{33}$$

Jeanloz (1989) compared the shock wave equation of state to the third-order Birch-Murnaghan equation by expanding each as a function of strain and comparing terms. He found empirically that the two equations of state are nearly identical to leading order in finite strain.

A break in the slope of the U_S-u_p curve can be identified as a phase transition occurring under shock compression (McQueen et al. 1970). Many phase transitions in minerals were first identified through shock wave experiments. For silicates, the transition pressure under shock loading is often significantly higher than found in corresponding static high-pressure experiments. This is a consequence of kinetic hindering of the transformation under the short time scales of the shock experiment. In addition, questions remain about the exact nature of the high-pressure phases achieved in dynamic experiments as direct examination of the material in its high-pressure state is not possible.

Extensive numerical compendiums of shock wave data for rocks, minerals, and other materials are available (e.g. Marsh 1980, Ahrens and Johnson 1995) as are detailed descriptions of experimental techniques (Ahrens 1987). and the fundamental physics of shock compression (e.g. Zel'dovich and Raizer 1966, Migault 1998).

Shock wave equation of state determinations are based on the use of standards. A standard is a material whose equation of state has been determined in such a way that the use of the EOS of another material is not required. This can be achieved by taking advantage of symmetry constraints imposed when the impactor and target are composed of the same material. A number of materials (e.g. 2024 Al, Fe, Cu, Pt, Ta) have been qualified as standards and these have been extensively cross-checked (e.g. McQueen et al. 1970). However, the possibility of systematic error remains. One such source of systematic error is material rigidity which may be important at pressures where shock heating is insufficient to melt or significantly soften the material (McQueen et al. 1970). For example, significant shear stresses have been observed in molybdenum at pressures of 10-80 GPa, even when shock compressed from an initial temperature of 1673 K (Duffy and Ahrens 1994).

Temperatures achieved during shock compression cannot be directly determined by application of the Rankine-Hugoniot equations. To calculate the temperature, the Hugoniot energy equation is combined with the thermodynamic law:

$$dE = TdS - PdV = C_V(dT + T\frac{\gamma}{V}dV) - PdV \tag{34}$$

to yield the following differential equation that can be solved for the shock temperature, T_H:

$$dT_H = \frac{(V_0 - V)}{2C_V} dP_H + \left[\frac{(P_H - P_0)}{2C_V} - \frac{\gamma}{V}T_H\right]dV \tag{35}$$

where the subscript H refers to Hugoniot conditions. The primary uncertainty in the calculated temperature arises from C_V. The uncertainty in the Gruneisen parameter only weakly affects the calculated temperature. The specific heat used in the temperature calculation can range from a simple constant value of $3R$, where R is the gas constant, to values calculated from Debye theory (Eqn. 24) to models incorporating significant electronic contributions. Figure 3 shows calculated temperatures along the Hugoniot for NaCl, MgO, Au, and Pt where C_V has been calculated using the Debye model. It is notable that most multi-anvil and laser-heated diamond cell experiments are carried out at temperatures that are much higher than achieved in shock compression experiments on the in situ standards.

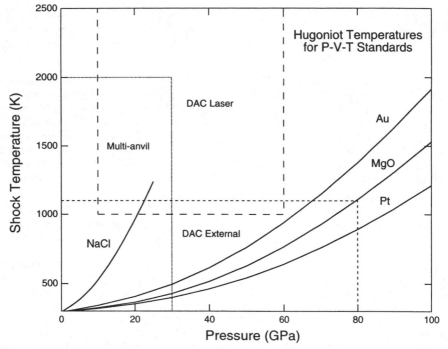

Figure 3. Calculated Hugoniot temperatures in materials used as static P-V-T standards. The short, medium, and long dashed boxes represent the regimes in which high P-T diffraction experiments have been reported to date for the multi-anvil press, externally heated diamond cell, and laser heated diamond cell, respectively.

Optical pyrometry has been used to directly measure shock temperatures of transparent (Kormer et al. 1965, Ahrens 1987) and opaque materials (Bass et al. 1987). The measurement of temperatures in opaque materials (metals) is more complicated as the results depend on the transmission properties of the window in contact with the sample, as well as the relative thermal diffusivities of the window and sample. Nevertheless, temperature measurements have provided much insight into the thermodynamic properties and melting behavior of shock compressed materials.

An important application of shock experiments is the determination of isentropes and isotherms. To determine the principal adiabat, the Mie-Gruneisen Equation (18), with the Hugoniot as the reference curve, is combined with the Hugoniot energy Equation (30) to yield:

$$P_H\left[1 - \frac{\gamma}{2V}(V_0 - V)\right] = P_S - \frac{\gamma}{V}(E_S - E_0) \tag{36}$$

The energy along the isentrope is given by:

$$E_S = E_0 - \int_{V_0}^{V} P_S dV \tag{37}$$

These expressions can be evaluated iteratively, or if a form for the isentrope assumed (e.g. Eqn. 8 or 9), the integral can be evaluated directly to determine $P_S(V)$. Once the adiabat is determined, the corresponding 300 K isotherm can be found by accounting for the thermal energy difference between the adiabat and isotherm using the Mie-Gruneisen equation again:

$$P_{300}(V) = P_S(V) - \frac{\gamma}{V}\left[E_S(T,V) - E_{300}(300,V)\right] \tag{38}$$

The thermal energies along the adiabat and isotherm at volume, V, can be obtained from the Debye model (19)-(23). An additional application of the Mie-Gruneisen equation allows for calculation of isotherms away from room temperature by accounting for energy differences between the isotherm of interest and the 300 K curve. Alternatively, the Hugoniot temperature (35) can be calculated at each volume, and the thermal components of pressure (18) and energy (20) subtracted to obtain pressures and energies along the 0 K compression curve. Isotherms at temperatures other than 0 K can then be found by adding back the necessary components of thermal pressure and energy (Hixson and Fritz 1992).

Multi-anvil press

Multi-anvil presses provide sample volumes that are on the order of 1 mm^3, which is several orders of magnitude greater than in the DAC, and offer several technical advantages in high-pressure research: (1) the relatively large sample volume provides robust counting statistics so that data collection time is short, (2) the encapsulated sample is placed inside an electric resistance heater; resistive heating provides stable temperatures for hours or even days, (3) high temperature is an efficient way to minimize effects of non-hydrostatic stresses in the sample, (4) pressure and temperature gradients are small (order of 0.1 GPa/mm and 10 K/mm, respectively, or less) and well characterized, and (5) with careful design in anvil geometry, the ram-load vs pressure hysteresis can be minimized and thus a wide pressure and temperature range can be covered in a single experiment. The main disadvantage for the MAP is that pressure and temperature range is limited compared to the laser-heated diamond cell. MAP researchers are pushing the limits constantly. In the last 10 years, the maximum pressure range has been tripled, from ~10 GPa to more than 30 GPa, thanks largely to the use of sintered diamond as an anvil material and the use of synchrotron radiation sources (Kato et al. 1992, Shimomura et al. 1992, Kondo et al. 1993).

Almost all multi-anvil presses consist of a uniaxial hydraulic ram compressing a set of guide blocks (pressure tooling), which are thrust blocks that transform the uniaxial load into three dimensions, pushing a number of anvils toward the center, generating high pressures on the sample. There are various types of pressure tooling and anvil

Table 1. Types of multi-anvil apparatus used with synchrotron x-ray diffraction.

Type of apparatus	Anvil geometry	Cell geometry	Anvil material and P,T range	Direction of diffraction vector	Reference(s)
DIA	Six anvils with square tips	cube	WC: 15 GPa, 2000C SDC: 20+ GPa, 1200C	Vertical	Shimomura et al. (1985)
MA-8 or 6/8	Eight second-stage cubes compressed by six first-stage wedges	octahedron	WC: 30 GPa, 2000+C	Inclined 35.3° from ram force direction (T-Cup) or subparallel to load direction (Walker module)	T-Cup: Vaughan et al. (1998) Walker module: Clark (1996)
6/8 in DIA	MA8 second stage assembly compressed in a large DIA	octahedron	WC: 30 GPa, 2000+C SDC: 40+GPa, 2000+C	Horizontal, with first stages notched	Shimomura et al. (1992)
Paris-Edinburgh	Opposed anvils with semi-spherical recess in each anvil	sphere	WC: 15 GPa, 1500C SDC: 30 GPa, 1200C	Through anvil gaps	Khvostantsev (1984) Besson et al. (1992)

configurations. Table 1 lists the most commonly used anvil configurations and their potential P, T range for diffraction applications. For earlier MAPs, see reviews by Ohtani et al. (1979), Shimomura et al. (1985), Akimoto (1987), and Vaughan (1994); for recent developments, see Shimomura et al. (1992) (6/8 in DIA) and Vaughan et al. (1998) (T-Cup).

The earliest equation of state measurements were carried out using the DIA apparatus. Yagi (1978) studied several alkali halides to 9 GPa and 1073 K, using a laboratory x-ray source. Prior to the mid 80's, however, few P-V-T studies were carried out on important mantle minerals with multi-anvil presses. This is mainly because the multi-anvil press utilizes solid media for pressure transmission that are x-ray absorbing. Laboratory x-ray sources such as the rotating anode do not provide sufficient x-ray brightness. The development of the DIA apparatus MAX-80 and 90 at the Photon Factory (Japan; Shimomura et al. 1985) and SAM-85 at NSLS (USA; Weidner et al. 1992) at synchrotron sources signified a major advance in multi-anvil technology. The dramatic improvement in experimental efficiency and data quality (Shimomura et al. 1985, Yagi 1985) has since placed multi-anvil presses in the forefront in P-V-T equation of state studies. Most of the major mantle minerals have been studied so far: MgO (Utsumi et al. 1998), diopside (Zhao et al. 1998), jadeite (Zhao et al. 1997), olivine (Meng et al. 1994, Guyot et al. 1996), wadsleyite (Meng et al. 1994), ringwoodite (Meng et al. 1994), majorite garnet (Wang et al. 1998), $CaSiO_3$ perovskite (Wang et al. 1996), and $(Mg,Fe)SiO_3$ perovskite (Wang et al. 1994, Funamori et al. 1996). A representative energy-dispersive diffraction pattern for $CaSiO_3$ perovskite at 11.7 GPa and 1172 K is shown in Figure 4a. Table 2 lists the data published to date.

There are many important issues to be addressed in obtaining high quality P-V-T data. Here we summarize some of the key issues (see also Wang et al. 1998).

1. Effects of non-hydrostatic stress were recognized very early on as a major source of errors (Sato et al. 1975). Liquid cells (sample embedded in a methanol/ethanol mixture) were developed for room-temperature studies. However, as the methanol/ethanol mixture freezes above 10 GPa, no truly hydrostatic condition could be achieved at higher pressures. As understanding of strength of materials developed, it became clear that high temperature is perhaps the best way to eliminate non-hydrostatic stress.

Non-hydrostatic stresses can affect the quality of P-V-T data in several ways:

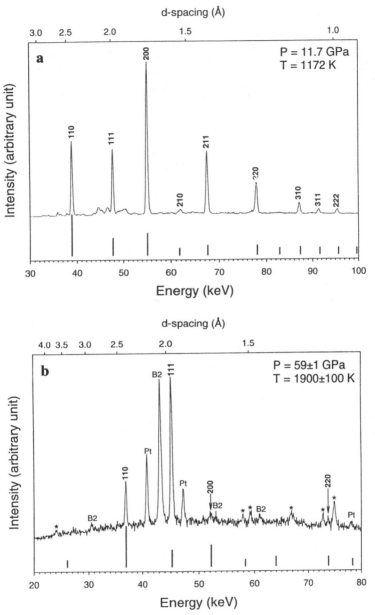

Figure 4. Representative energy dispersive diffraction patterns for $CaSiO_3$ perovskite from in situ high *P-T* experiments in (a) the multi-anvil press at 11.7 GPa and 1172 K (Wang et al. 1996) and (b) diamond anvil cell at 59 GPa and 1900 K (S.-H. Shim, G. Shen, and T.S. Duffy, unpublished data). Calculated patterns are shown at the bottom of each figure. In (a), a total of 11 diffraction peaks are observed. Some weak peaks from lower pressures phases (Ca_2SiO_4 and $CaSi_2O_5$) are present near the (111) peak. In (b) the lines labeled with *hkl* values are for $CaSiO_3$ and those labeled B2 and Pt are for the high-pressure phase of NaCl and platinum, respectively. Weak lines indicated with an asterisk are fluorescence lines. The difference in relative intensities between the two samples is due to preferred orientation.

Table 2. P-V-T Equation of state data from multi-anvil and diamond cell experiments.

Material	P_{max} (GPa)	T_{max} (K)	K_{T0} (GPa)	K_{T0}'	$(\partial K_T/\partial T)_P$ (GPa/K)	Ref.
MgSiO$_3$ (oen)	4.5	1000	102.8(2)	10.2(12)	-0.037(5)	Zhao et al., 1995
NaAlSi$_2$O$_6$ (jd)	8.2	1280	125(4)	5.0*	-0.017(5)	Zhao et al., 1997a
CaMgSi$_2$O$_6$ (di)	8.2	1280	109(4)	4.8(6)	-0.021(4)	Zhao et al., 1998
α-Mg$_2$SiO$_4$	7.6	1019	127.4*	4.8*	-0.021(2)	Meng et al., 1993b
β-(Mg,Fe)$_2$SiO$_4$	26	900	174(3)	4.0*	-0.027	Fei et al., 1992a
β-Mg$_2$SiO$_4$	7.6	872	172.6*	4.8*	-0.027(5)	Meng et al., 1993b
γ-Mg$_2$SiO$_4$	9.9	1122	182.6*	5.0*	-0.028(3)	Meng et al., 1993b
γ-Mg$_2$SiO$_4$	30	700	182(2)	4.0(3)	-0.027(5)	Meng et al., 1994
(Mg,Fe)SiO$_3$ (pv)	30	900	261(4)	4.0*	-0.063(5)	Mao et al., 1991
MgSiO$_3$ (pv)	11	1300	261*	4.0*	-0.023(11)	Wang et al., 1994
MgSiO$_3$ (pv)	20	773	261*	4.0*	-0.02	Utsumi et al., 1995
MgSiO$_3$ (pv)	29	2000	261*	4.0*	-0.028(17)	Funamori et al., 1996
MgSiO$_3$ (pv)	57	2500	261*	4.0*	-0.027(5)	Fiquet et al., 1998
CaSiO$_3$ (pv)	13	1600	232(8)	4.8(3)	-0.033(8)	Wang et al., 1996
Py$_{62}$Mj$_{38}$	11	1163	160(3)	4.9(5)	-0.020(1)	Wang et al., 1998c
(Mg,Fe)O	30	800	157*	4.0*	-0.027(3)	Fei et al., 1992b
MgO	10	1673	153(3)	4.0*	-0.034(3)	Utsumi et al., 1998
Mg(OH)$_2$	80	600	54.3(15)	4.7(2)	-0.018(8)	Fei and Mao, 1993
Mg(OH)$_2$	11	873	39.6(14)	6.7(7)	-0.011(2)	Xia et al., 1998
MgCO$_3$	8.6	1285	103(1)	4.0*	-0.021(2)	Zhang et al., 1997
(Fe,Mg)CO$_3$	8.9	1073	112(1)	4.0*	-0.026(2)	Zhang et al., 1998
FeCO$_3$	8.9	1073	117(1)	4.0*	-0.031(3)	Zhang et al., 1998
FeSi	9.5	1081	172(3)	4.0*	-0.043(8)	Guyot et al., 1997
h-BN	9	1280	17.6(8)	19.5(34)	-0.0069(8)	Zhao et al., 1997b

Abbreviations: oen - orthoenstatite; jd - jadeite; di -diopside; pv-perovskite; py - pyrope; mj - majorite; h-BN - hexagonal boron nitride, * - fixed value. Numbers in parentheses are one standard deviation uncertainties in the last digit(s).

1.1. Strength of pressure media around the sample. In a solid medium cell, a macroscopic non-hydrostatic stress field may be present as a boundary condition imposed on the sample, causing relative shifts in diffraction peak positions. As the x-ray data are obtained with a given diffraction vector, only the lattice planes that are perpendicular to that vector (to satisfy Bragg's law) are sampled in a powdered specimen. The results are therefore biased (Weidner et al. 1992). For example, in the DIA apparatus, the cell usually generates a stress field with the unique maximum (or minimum) principal stress component (σ_1) vertical and the other two components about equal (σ_3). The diffraction condition is so selective that only the lattice planes close to normal to the principal stress σ_1 can be "seen". If one uses an in situ standard to measure the pressure, the apparent pressure (P_{app}) is different from the real pressure (P_{real}) in the following manner:

$$P_{real} - P_{app} = \frac{K}{\mu}(\sigma_1 - \sigma_3) \tag{39}$$

where K and μ are bulk and shear moduli of the pressure standard (Weidner et al. 1992). The magnitude of the differential stress, ($\sigma_1 - \sigma_3$), is ultimately limited by the yield strength of the material. Large uncertainties in pressure can exist because the ratio (K/μ) can be large (e.g. about 2 for NaCl and 6 for Au).

1.2. Elastic anisotropy of the sample introduces a microscopic non-hydrostatic stress field at the grain-to-grain level in a powdered sample (Weidner et al. 1994). The main effect of this highly heterogeneous stress field is broadening of the diffraction peaks which causes errors in determining peak positions, even if the sample is under a perfectly hydrostatic boundary condition.

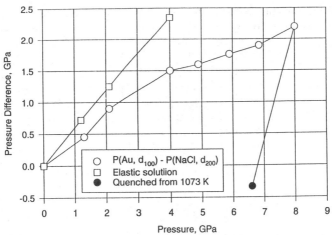

Figure 5. Pressure difference in a two-phase sample (Au dispersed in NaCl). Circles are the measured pressure difference between NaCl and Au with increasing sample pressure (based on the NaCl scale) at room temperature. The last data point is after heating to 1073 K and then quenching to room temperature. Squares are predicted values using the elastic solution (Eqns. 40-43).

1.3. Often the sample is mixed with another material (the pressure standard). As a multi-phase sample is compressed, the elasticity mismatch between the two solid phases generates a local shear stress field around the dispersed second-phase grains. This can be observed in an experiment where Au powder is dispersed in a NaCl matrix. When compressed at room temperature, equations of state for both NaCl (Decker 1971) and Au (Anderson et al. 1989) yielded remarkably different pressures, with the gold pressure 2 GPa higher than that of NaCl at 8 GPa (see Fig. 5). Wang et al (1998b) examined this effect by considering a spherical inclusion in a infinite matrix under a hydrostatic boundary condition P_0. Classic Lamé solutions in a spherically symmetric case give:

$$P_i = P_0 \left(1 + \frac{4\mu_m}{3K_m}\right) \bigg/ \left(1 + \frac{4\mu_m}{3K_i}\right) \tag{40}$$

$$P_m = P_0 - 4\mu_m B \left(\frac{a}{r}\right)^3 \tag{41}$$

$$B = \left(\frac{P_0}{3K_m}\right)\left(\frac{K_m}{K_i} - 1\right) \bigg/ \left(1 + \frac{4\mu_m}{3K_i}\right) \tag{42}$$

where subscript i indicates inclusion and m matrix, a is the diameter of the inclusion, and r is the distance from the center of the inclusion. The differential stress normal to the inclusion surface is

$$\sigma_1 - \sigma_3 = -6\mu_m B \left(\frac{a}{r}\right)^3 \tag{43}$$

Figure 5 shows the pressure difference, $P_i - P_m$, for an Au inclusion in NaCl matrix (open squares) as predicted by the above elastic solution, as compared with experimental data. Obviously the elastic solution fails when yielding occurs; but this qualitative example provides some insights. Equations (40)-(43) all involve the shear modulus of the matrix (μ_m). In case of $\mu_m = 0$, $P_i = P_m$ and $(\sigma_1 - \sigma_3) = 0$.

All of these problems can be easily solved with heating. As the yield strength of any material decreases with increasing temperature, $(\sigma_1 - \sigma_3)$ in Equation (39) becomes negligible and peak broadening disappears. It can also be seen in Figure 5 that after heating to 1073 K, the gold pressure becomes much closer to NaCl pressure (confirming that the discrepancy at room temperature is not due to equations of state of the two materials).

2. Pressure and temperature gradients. Pressure gradients are ultimately controlled by the strength of the sample which again can be minimized by heating. Special furnace designs can be used to reduce temperature gradients (e.g. Takahashi et al. 1982). More importantly, with synchrotron sources, the x-ray beam can be collimated to a cross section of $100 \times 100 \ \mu m^2$ or less (which is necessary for maintaining accuracy in the diffraction geometry), and data collection is performed adjacent to the thermocouple. Virtually no temperature gradient is present in the diffracting volume.

3. P-T path. Earlier P-V-T measurements were carried out either along certain isotherms or at random P-T points. Isothermal compression is a useful technique but time consuming (generally one temperature per experiment). Special anvil designs can reduce the pressure-load hysteresis significantly, allowing a wide P-T range to be covered by a single experiment (e.g. Wang et al. 1998b). The optimal P-T path is to compress to the highest pressure and then heat to the maximum temperature. Data collection is performed at certain predetermined temperatures from the maximum down, in order to minimize the magnitude of non-hydrostatic stress. Once the sample is cooled down to room temperature, the pressure is lowered by decreasing the ram load, the sample is heated again to the highest point and the process repeated. A uniform P-T grid can be generated in a single experiment that can be treated as a number of isothermal curves (Wang et al. 1998c).

With these techniques, very accurate P-V-T data can be obtained. Uncertainties as low as ±0.02% or 200 ppm (1σ) have been achieved (Wang et al. 1998c, Utsumi et al 1998), based on peak position data alone. An example of a P-V-T equation of state fit to the high-temperature Birch-Murnaghan equation of state on a pyrope-majorite solid solution is shown in Figure 6 (Wang et al. 1998c).

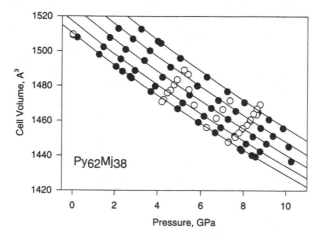

Figure 6. P-V-T data for $Py_{62}Mj_{38}$ garnet (Wang et al. 1998). The solid circles are data at 300, 473, 673, 873, and 1073 K, respectively (curves shows fits to a high-temperature Birch-Murnaghan equation at these temperatures). Open circles are from an additional run at temperatures of 523, 723, 923, and 1123 K. Uncertainties of each measurement (about 0.02%) are much smaller than the size of the symbols.

4. Data analysis techniques. Usually, peak position information is extracted from each x-ray spectrum and unit cell volume is then determined by the positions of the peaks. Problems arise when significant peak overlap occurs. Minerals with low symmetry belong to this category. Even for some high symmetry minerals, e.g. tetragonal majorite, peak overlap can cause considerable difficulty in determining the cell volume accurately (Wang et al. 1998c). Zhao et al. (1997a, 1998) extended the Rietveld technique used for monochromatic diffraction data to energy dispersive data to get more accurate lattice parameters and volumes. Intensity information can be ignored, and least-squares fit is performed based on all peak positions and background (LeBail fit). This technique reduces the uncertainty of the results significantly; for example, uncertainties on the order of 100 ppm can be achieved for monoclinic minerals such as diopside and jadeite (Zhao et al. 1997a, 1998).

5. In order to obtain high accuracy *P-V-T* data, the sample should be examined in the widest possible *P-T* range. Thus the sample is often under *P-T* conditions where the phase of interest is thermodynamically metastable. Any slight alteration in state can cause irreversible changes in unit cell volume, resulting in an erroneous equation of state. Although some minerals exhibit large irreversible changes that are easily detected (e.g. $(Mg,Fe)SiO_3$ perovskite: Wang et al. 1994; $CaSiO_3$ perovskite: Wang et al. 1996), others show very subtle changes (e.g. pyrope-majorite: Wang et al. 1998c). Post-mortem studies (IR, Raman, TEM, microprobe etc.) are essential to be sure that the measured properties are reversible.

6. An outstanding problem is the pressure effects on thermocouple emf. It is difficult to measure directly the effects, although some studies have attempted to look at emf changes of one type of thermocouple versus another (Getting and Kennedy 1970). New techniques such as epithermal neutron resonance broadening (Clark et al. 1996) may help resolve this issue in the future.

Diamond anvil cell

The diamond anvil cell combined with x-ray diffraction techniques also provides a powerful means for determining the equation of state properties of materials at very high pressures and temperatures. Among the advantages of the diamond cell are its simplicity of operation and the transparency of the diamonds to a wide range of electromagnetic radiation including x-rays and visible light. Very high pressures (in excess of 100 GPa) can be readily achieved with this device. However, the diamond cell has the disadvantage that the compressed sample volume is small (~0.001 mm^3). High temperature experiments are also complicated by the potential for large temperature gradients and non-hydrostatic pressure distributions.

P-V-T equation of state measurements in the diamond anvil cell have been reported using both resistive and laser heating techniques. Combined high pressure and temperature measurements (to 10 GPa and 200°C) in the diamond cell were first reported more than 30 years ago (Bassett and Takahashi 1965). Technical aspects involved in generating and measuring high temperatures using resistive heating methods have been discussed elsewhere (Adams and Christy 1992). In the internal heating technique, electrical current is passed directly through the sample thereby heating it to temperatures as high as 2000 K (e.g. Liu and Bassett 1975). The main disadvantages of internal heating are that the sample must be electrically conducting, and that relatively large temperature gradients may exist in the sample chamber.

More uniform temperature fields can be created using an external heater placed outside the sample chamber. The temperature gradient is minimized because the entire cell is heated in this case. In one study (Mao et al. 1991), a Mao-Bell diamond cell made from inconel was used together with an external platinum-wire resistance heater placed over the cylinder of the cell. Simultaneous high pressures and temperatures to 30 GPa and 900 K were obtained with this arrangement. The peak achievable temperature can be increased at the cost of a larger temperature gradient by including an inner molybdenum wire resistance heater close to the diamonds in addition to the outer furnace. Nonetheless, the maximum achievable temperature is limited. The modifications and tradeoffs necessary to achieve increasingly high temperature are discussed by Mao and Hemley (1996). The practical temperature limit for this technique is estimated to be 1600 K. Currently, pressures as high as 100 GPa have been achieved at 1100 K with this method (Fei and Mao 1994).

The resistively heated diamond cell can be combined with x-ray diffraction measurements using the energy dispersive technique at a synchrotron x-ray facility to determine the cell volume as a function of pressure and temperature. Pressures are usually determined from the lattice parameter of an internal standard (e.g. gold, sodium chloride). The temperature can be measured to within a few degrees using a thermocouple. It has been reported in these experiments that pressure decreases with increasing temperature at a rate of about 5 GPa/100 K (Fei et al. 1992b), as a result of relaxation of the diamond cell. A detailed description of the cell and experimental procedure has been provided by Fei (1996).

The external heating technique has been used by Y. Fei and colleagues to measure P-V-T properties of a variety of silicates, oxides, and hydroxides to pressures up to 30 or more GPa and temperatures to 900 K. As shown in Table 2, the technique has been applied to $(Mg,Fe)SiO_3$-perovskite (Mao et al. 1991), β-$(Mg_{0.84}Fe_{0.16})_2SiO_4$ (Fei et al. 1992a), γ-Mg_2SiO_4 (Meng et al. 1994), $(Mg_{0.6}Fe_{0.4})O$ (Fei et al. 1992b), $Mg(OH)_2$ (Fei and Mao 1993). More recently, the high-pressure hydrous silicate phase D has been studied to pressures of 50 GPa and 1200 K (Frost and Fei 1998).

Other types of diamond cells for simultaneous high P-T experiments have also been described. A cell optimized for hydrothermal studies has been developed by Bassett for use up to 10 GPa and 1473 K (Bassett et al. 1993). This cell has been used to determine the anisotropic thermal expansion coefficients of calcite at 1 GPa and 773 K (Wu et al. 1995). A Merrill-Bassett cell has also been developed that can be used with a vacuum furnace to 1000 K and 20 GPa (Schiferl 1987, Schiferl et al. 1987). An important advantage of this cell is that it can be used for single-crystal diffraction experiments. This P-V-T equation of state of a single-crystal orthoenstatite was successfully studied to 4.5 GPa and 1000 K with this system (Zhao et al. 1995). (Table 2). A high-temperature diamond cell suitable for x-ray diffraction experiments on polycrystalline materials to 30 GPa and 773 K has also been reported by Kikegawa (1987)

As with the multi-anvil press, non-hydrostatic pressure conditions may exist inside the diamond anvil cell at high pressures (e.g. Kinsland and Bassett 1977, Wilburn et al. 1978, Singh 1993), which could influence P-V-T equation of state determinations. The variation of deviatoric stress in a quasi-hydrostatic diamond cell sample have been investigated as a function of pressure and temperature (Meng et al. 1993a). In this study, a sample of gold was compressed in a neon pressure medium. Small deviatoric stresses were detectable at 15-30 GPa, but rapidly disappeared upon heating at a temperature of 650 K. The presence of deviatoric stresses will affect both the pressure calibration and the magnitude of the apparent thermal pressure. In the case of brucite, $Mg(OH)_2$, there are large differences in thermoelastic parameters determined in the diamond cell (Fei and Mao 1993) and multi-anvil press (Xia et al. 1998) (Table 2). These differences may reflect the presence of non-hydrostatic stresses in the diamond cell experiments which did not use a pressure medium. Room-temperature DAC EOS measurements for brucite under quasi-hydrostatic conditions in the DAC (Duffy et al. 1995b) yield a bulk modulus that is more than 20% lower than obtained under non-hydrostatic conditions (Fei and Mao 1993) and in good agreement with the multi-anvil press data (Xia et al. 1998).

To achieve the higher pressure and temperature states needed to reproduce the conditions of the Earth's lower mantle and core, it is necessary to use laser heating techniques with the diamond anvil cell. Laser heating was first used with a diamond cell in 1974 (Ming and Bassett 1974), and the technique has evolved considerably since. Technical aspects of laser heating remain controversial, however, and a number of publications discussing the relative merits of different laser heating and temperature measurement techniques are available (e.g. Boehler and Chopelas 1991, Jephcoat and Besedin 1996, Jeanloz and Kavner 1996). The main experimental challenge in using laser heating to determine *P-V-T* equations of state is to create a uniform pressure and temperature distribution in the laser heated spot. Numerical calculations of the temperature distribution in the laser heated sample chamber provide some constraints on the temperature gradients that may exist for different sample geometries (Manga and Jeanloz 1998, Morishima and Yusa 1998).

Laser heating is free of the pressure and temperature restrictions that exist in the externally heated cell since the heating is localized and the support apparatus of the diamond cell is not appreciably weakened. Temperatures above 5000 K have been achieved at pressures over 100 GPa with this method. However, if the volume of the sample is constrained, a thermal pressure is generated. It is therefore necessary to measure the pressure in situ, and not rely on measurements of the pressure in a cold region of the sample, or before and after heating. The magnitude of the thermal pressure effect depends on the extent to which the volume of the heated sample is constrained, and hence may depend on the particular sample environment. Two endmember cases involve heating at constant pressure and heating at constant volume. In the former case, the thermal pressure is zero while it is a maximum in the latter case. The magnitude of potential thermal pressure effects was first treated theoretically by Heinz (1990).

The first high-pressure high-temperature diffraction study using synchrotron radiation on a laser heated sample was carried out by Boehler et al. (1989) on γ-Fe. Results to 17 GPa and 1900 K were used to constrain the pressure dependence of the thermal expansivity of this material (Boehler et al. 1990). The use of laser heating for high accuracy *P-V-T* determinations has been enhanced by a series of technical advances. Among these advances is the development of third-generation synchrotron beamlines optimized for high-pressure experiments (Hausermann and Hanfland 1996, Rivers et al. 1998). The use of undulator insertion devices together with Kirkpatrick-Baez focusing optics has greatly increased the intensity of the x-ray beam impinging on the sample, leading to reduction in data collection time and increasing the accessible pressure range. Figure 4b shows an example of an x-ray diffraction pattern from the laser heated diamond cell. In this experiment, the sample consisted of a mixture $CaSiO_3$ perovskite and platinum (with a sodium chloride insulation layer) heated to 1900 K at 59 GPa. Improved laser heating techniques have also been developed including the adoption of double-side heating and flat-topped multimode YAG and YLF lasers (Shen et al. 1996, Mao et al. 1998). An example of a temperature profile from the laser-heated diamond cell is shown in Figure 7. Alternatively, a high-powered CO_2 laser can be used to reduce radial temperature gradients (Boehler and Chopelas 1991). Angle dispersive x-ray diffraction techniques with imaging plates can be used to record higher resolution diffraction patterns during laser heating than can be achieved with energy dispersive techniques (Andrault et al. 1997).

The combination of laser heating with third generation synchrotron radiation has

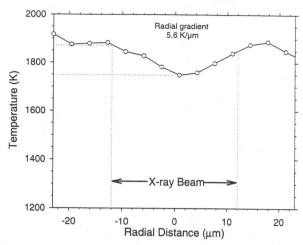

Figure 7. Temperature profile across the laser heated spot in a diamond anvil experiment at 55 GPa. The sample consisted of a homogeneous mixture of $MgSiO_3$ perovskite and platinum. Double-sided heating was carried out using the TEM_{01} mode of a Nd:YLF laser (S.-H. Shim, G. Shen, and T. S. Duffy, unpublished data). In this experiment, a 25-μm diameter x-ray beam was used yielding a total temperature variation of ±70 K across the heated spot.

resulted in a number of studies examining phase boundaries and stability of mineral phases. However, still only relatively few studies have directly addressed the P-V-T equation of state under these conditions. Thermal pressure effects in the laser-heated diamond cell were directly investigated by Fiquet et al. (1996) in experiments on MgO to 16 GPa and 3000 K using an argon medium to provide insulation from the diamonds. The authors calculated an anomalously large value of the Anderson-Gruneisen parameter (δ_T = 25-30) based on pressures measured by ruby fluorescence before and after heating. The unexpectedly large value for this parameter was attributed to a deviation from constant pressure conditions. More recently, Kavner et al. (1998), using platinum as an in situ pressure marker, reported large decreases in pressure as MgO samples were heated above 2000 K. These studies suggest that neither constant pressure nor constant volume conditions apply in the sample chamber during laser heating, and emphasize the importance of measuring pressure in situ.

In the highest P-T conditions yet achieved for static equation of state determination, Fiquet et al. (1998) reported P-V-T equation of state data for silicate perovskite to 57 GPa and 2500 K using Pt as an internal pressure standard. The results of this experiment were consistent with earlier multi-anvil studies (Wang et al. 1994, Funamori et al. 1996a) but disagree with lower P-T diamond cell experiments (Mao et al. 1991, Stixrude et al. 1992). Figure 8 compares the lattice parameters determined for $MgSiO_3$ perovskite at ambient temperature and 2000 (±100) K from multi-anvil and diamond cell studies. These data are suggestive of a possible change in slope in the pressure dependence of the b and c lattice parameters above 25 GPa at 300 K. The implications of P-V-T equation of state results for $(Mg,Fe)SiO_3$, $CaSiO_3$ and $(Mg,Fe)O$ for lower mantle mineralogy have been discussed in a number of recent publications (e.g. Wang and Weidner 1996, Fiquet et al. 1998, Hama and Suito 1998, Jackson 1998). The general consensus of these studies is that upper mantle mineralogies provide a good match to the seismic properties of the Earth's lower mantle and a change in bulk mantle composition at 670-km depth is not required by the data.

Table 2 summarizes results from static P-V-T studies to date. Figure 9 shows the

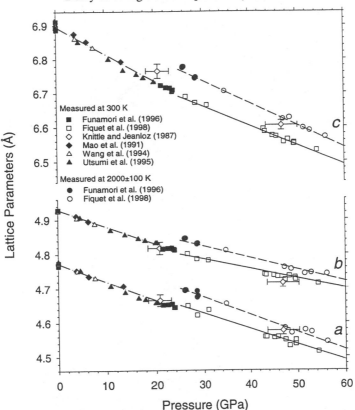

Figure 8. Lattice parameters for MgSiO₃ perovskite from multi-anvil and diamond cell experiments at room temperature and near 2000 K. The solid lines show fits to the room temperature lattice parameters above 25 GPa, while the dash-dot line shows the fit below 25 GPa. Dashed lines show fits to measured lattice parameters at 2000 ± 100 K and pressures above 25 GPa.

resulting values of the temperature derivative of the bulk modulus plotted as a function of K_T. While there is considerable scatter in the results, two points are worth noting. First, $(\partial K_T/\partial T)_P$ appears to depend on K_T either weakly or not at all. Second, values of $(\partial K_T/\partial T)_P$ for silicates tend to fall in a relatively narrow range of $(\partial K_T/\partial T)_P = -0.025\pm0.01$ GPa/K. Exceptions to this latter observation include orthoenstatite, which is well-known to have anomalous elastic properties, and one of the five reported values for silicate perovskite (Mao et al. 1991). The large temperature derivative obtained for this sample has been explained as a consequence of irreversible changes occurring during heating of the metastable Fe-bearing perovskite (Wang et al. 1994).

ACCURACY OF P-V-T EQUATIONS OF STATE

Pressure calibration at room temperature

Reliable pressure calibration is essential for all equation of state determinations. In the diamond anvil cell, the ruby fluorescence technique, which relates the wavelength shift of the ruby R1 fluorescence peak to pressure, serves as a fast and convenient secondary pressure scale (Piermarini et al. 1975, Mao et al. 1978) at pressures up to and even above 100 GPa. The ruby scale was first calibrated against x-ray diffraction data for NaCl to 20

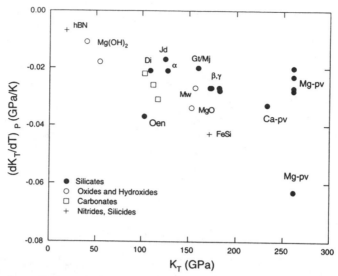

Figure 9. Temperature derivative of isothermal bulk modulus versus the bulk modulus from *P-V-T* equation of state measurements in the diamond anvil cell and multi-anvil press See Table 2 for data and description of abbreviations used.

GPa (Piermarini et al. 1975) using a semi-empirical equation of state for NaCl (Decker 1971). At higher pressures, the ruby scale is based on comparison of room-temperature isotherms of metallic standards from shock data with x-ray measurements under nonhydrostatic (Mao et al. 1978) and quasi-hydrostatic conditions (Mao et al. 1986). While the precision of the ruby scale is high, the overall accuracy is limited to about 3% (Mao et al. 1978). The principal uncertainties regarding the absolute accuracy of the ruby scale include: (1) experimental uncertainties in dynamic and static compression data used for calibration, (2) neglect of contributions due to dynamic and static shear strengths, and (3) uncertainties in the thermal correction from Hugoniot states to the static isotherm. In high-pressure x-ray diffraction experiments using a diamond cell or multi-anvil press, pressure is often directly determined from a calibrated marker material (e.g. NaCl, MgO, Au, Pt) included in the sample chamber. The accuracy of the pressure scale from in situ markers is comparable to that of the ruby scale, as both are ultimately traceable to reduced shock wave isotherms. The achievement of very high precision (0.1%), however, in pressure determination, has been reported up to 9 GPa by very precise volume measurements of quartz and CaF_2 for use as internal standards in hydrostatic single-crystal experiments (Angel et al. 1997).

Intercomparisons between shock and static equations of state can form the basis of tests of the internal consistency of different standards used for pressure calibration. For example, Ross et al. (1986) compared the room temperature equation of state of solid argon from diamond cell data using the ruby fluorescence technique with a theoretical isotherm obtained by reduction of shock data to 80 GPa. The good agreement between the resulting curves implies that the reduction of shock data to an isotherm is accurate even when shock temperatures are extremely high. Similarly, Greene et al. (1994) demonstrated the consistency of diamond cell data for aluminum (using Pt as an internal pressure standard) to 220 GPa with shock data for the same material. This demonstrates that the shock and static equations of state of Al and Pt are mutually consistent to at least this pressure.

It has long been known that precise measurements of elastic wave velocities and volume, together with knowledge of some other thermodynamic properties, are sufficient for direct determination of pressure in a high-pressure device (Smith and Lawson 1954). The feasibility of combining elasticity and x-ray diffraction probes as a function of P, V, and T to constrain the pressure without reference to a secondary scale has been examined recently (Spetzler and Yoneda 1993). Direct calculation of pressures have recently been carried out for the first time in experiments on San Carlos olivine to 32 GPa (Zha et al. 1998). In this study, measurements of the adiabatic bulk modulus by Brillouin spectroscopy were combined with volumes determined by x-ray diffraction to compute the pressure by using the following expression derived from the equation for the isothermal bulk modulus (Eqn. 5):

$$P = P_0 - \int_{V_0}^{V} \frac{K_S dV}{V(1 + \alpha\gamma T)} \tag{44}$$

together with the relationship, $K_S = K_T(1 + \alpha\gamma T)$.

The necessary thermodynamic parameters (α, γ) are known for San Carlos olivine. A comparison of the pressures determined by (44) with independent ruby fluorescence measurements is shown in Figure 10. The ruby fluorescence pressures generally agree with those by the Brillouin/diffraction technique, but the ruby pressures tend to lie at slightly higher pressures. A third-order Birch-Murnaghan equation of state fit to the ruby measurements produces pressures that exceed the mean Brillouin determination by about 2% between 10-30 GPa. The San Carlos olivine measurements and the ruby fluorescence scale are consistent within their mutual uncertainties. The precision of the Brillouin/diffraction measurements is ±2%, but this could be reduced in the future by higher precision bulk modulus determinations. The recent development of simultaneous high P-T ultrasonic elasticity measurements combined with synchrotron x-ray diffraction also offers the opportunity for direct pressure determination (Chen et al. 1998; Liebermann, this volume).

High temperature standards

At pressures to ~30 GPa, NaCl (Decker 1971, Birch 1986) is the most commonly used internal standard, particularly in multi-anvil experiments (e.g. Utsumi et al. 1998). Other widely used standards include Au (Jamieson et al. 1982, Heinz and Jeanloz 1984, Anderson et al. 1989), MgO (Jamieson et al. 1982, Anderson et al. 1993), and Pt (Holmes et al. 1989). The ruby fluorescence scale is less useful at high temperature (above 700 K) because of strong line broadening and uncertainties in calibration (Yen and Nicol 1992). A number of alternative fluorescence standards have been proposed. Sm-doped $Y_3Al_5O_{12}$ has been calibrated to 25 GPa and 820 K (Hess and Schiferl 1992). An advantage of this material is that, unlike ruby, the fluorescence wavelength shift is essentially independent of temperature. Another alternative is $Sm^{2+}:SrB_4O_7$ which was initially examined to 673 K (Lacam and Chateau 1989). This material can serve as an alternate to ruby even at ambient temperature as there is reported to be strong luminescence up to 130 GPa (Datchi et al. 1997). In addition there is little thermal broadening up to 900 K. Raman frequencies of ^{13}C and ^{12}C diamond chips can also be used as a precise P-T sensor to 25 GPa and 1200 K (Schiferl et al. 1997).

In a landmark paper, Jamieson et al. (1982) examined a number of candidates for a primary x-ray standard for P-V-T studies, emphasizing diamond cell applications. The criteria by which standards were evaluated were the following: (1) chemical inertness, (2) ready availability, (3) experimental data in the P-T range of interest, (4) large compressibility, and (5) simple x-ray diffraction pattern. To these we could also add the

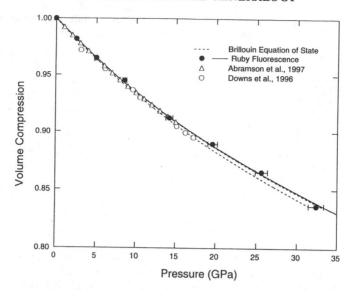

Figure 10. Volume compression data for San Carlos olivine (Zha et al. 1998). Filled symbols show x-ray volume measurements with pressure determined by ruby fluorescence (± 3% uncertainty). The solid line is a third-order finite strain fit to this data. The dashed lines show the upper and lower bound of the pressure determined by Brillouin/diffraction data as discussed in the text. The upper limit of the Brillouin determination nearly overlaps the finite strain fit from ruby data. Also shown are previous data for San Carlos olivine (Abramson et al. 1997) and forsterite (Downs et al. 1996) based on the ruby pressure scale.

additional requirements of (6) a wide P-T stability range, (7) a high atomic number, (8) low shear strength, and (9) internal consistency among shock, static, and ultrasonic data.

Jamieson et al. (1982) concluded that gold (Au) was the material that best satisfied the list of 5 requirements, and it is compatible with (6) through (9) as well. Platinum was also considered, but deemed less suitable primarily because of its low compressibility, which renders it less precise for pressure measurements. MgO was rejected because of inconsistency between available shock and static data.

For the multi-anvil press, NaCl satisfies all criteria except (7) which is not a serious disadvantage due to the larger sample volumes available in this device. While the P-T range of NaCl in the B1 structure is limited from the point of view of the diamond cell, it is adequate for most of current multi-anvil press experiments. However, NaCl may be affected by grain growth at high temperature. The reliability of the Decker equation of state for NaCl has been extensively tested. Birch (1978, 1986) examined available ultrasonic, volumetric, and shock compression data for NaCl to produce isotherms to 30 GPa and 773 K. He also constrained the behavior of a number of thermodynamic functions over this range. Birch's equation of state is in good agreement with that of Decker. Theoretical studies (Feldman et al. 1988, Hemley and Gordon 1985) also provide support for the Decker EOS, as do experimental intercomparisons using different pressure standards (Funamori et al. 1996).

In the case of MgO, the discrepancy between shock and static data has now been largely resolved as it has become clear that MgO exhibits relatively high shear strength under non-hydrostatic compression (Duffy et al. 1995a). Recently, the P-V-T equation of state of MgO was determined in the multi-anvil press to 10 GPa and 1673 K (Utsumi et al. 1998). New ultrasonic data have just become available for MgO up to 8 GPa and 1600 K

(Chen et al. 1998). In view of the fast growing *P-T* range, MgO may become an important pressure standard for the MAP above the B1-B2 transition pressure for NaCl.

Since Jamieson et al.'s (1982) work, two new *P-V-T* EOSs have been developed for gold (Heinz and Jeanloz 1984, Anderson et al. 1989). In their study of the thermoelastic properties of MgSiO$_3$ perovskite to 30 GPa and 2000 K, Funamori et al. (1996) used several different pressure standards (MgO, Au, NaCl) and different equations of state to provide cross-checks over the *P-T* range studied. In general, the equations of state tested (Anderson et al. 1989, Birch 1986, Decker 1971, Heinz and Jeanloz 1984, Jamieson et al. 1982) were self-consistent with the notable exception of the equation of state of Au of Jamieson et al. (1982) . The source of the discrepancy for this equation of state lies in the choice for the volume dependence of the Gruneisen parameter. Jamieson et al. (1982) assumed that γ/V = constant for gold. However, the other equations of state for gold adopt values of *q* ranging from 1.7 (Heinz and Jeanloz 1984) to 2.5 (Anderson et al. 1989) which are more consistent with thermodynamic data for this material.

It is possible to constrain both pressure and temperature from simultaneous measurements of the volumes of two internal standards (Jamieson et al. 1982). This requires a very high accuracy in the volume measurement to avoid large errors in the temperature determination. This approach is also useful as a means of cross-checking *P-T* standards if an independent temperature measurement (by thermocouple, for example) is available. Measurements on gold and NaCl to 7 GPa and 873 K demonstrated that feasibility of this approach and revealed evidence for a small, but systematic discrepancy between equations of state for the two materials (Yagi et al. 1985) which could be related to non-hydrostatic stresses in the experiments or uncertainty in Jamieson's equation of state for gold. More recently, measurements of NaCl and MgO with volume precisions of 0.07% and 0.02%, respectively were used to simultaneously constrain the pressure and temperature with a precision of about 1% at about 8 GPa and 1200 K (Utsumi et al. 1998). A similar analysis was also carried out for a mixture of hexagonal BN and NaCl at 7 GPa and 670 K (Zhao et al. 1997b). Using the diamond anvil cell, Ming et al. (1983) performed simultaneous x-ray diffraction experiments on an NaCl-Au mixture to 10.2 GPa and 723 K using a thermocouple for temperature determination. The pressures from Au (using Jamieson's equation of state) were in agreement with NaCl pressures to within 2.7% at 10.2 GPa and there was no clear systematic difference between pressures determined by the two scales. Intercomparisons between Au and tungsten (Schiferl et al. 1987) and Au and MgO (Ming et al. 1984), however, suggest that pressures produced by the gold scale of Jamieson et al. are too high. As discussed above, this is likely to be related to the volume dependence of *q* used in the analysis of Jamieson et al. (1982).

Consistency of EOS determinations by different techniques

In assessing the validity of equations of state, it is important to check the consistency of equations of state determined by different experimental techniques. EOS parameters can be determined by four separate techniques: (ultrasonic or acousto-optic) sound velocity measurements, shock compression, static compression, and first-principles theory. Each method has its own strengths and weaknesses. Sound velocity measurements, for example, are typically limited to low pressures. The reliability of static compression data depends on the *P* and *T* calibration, and could be biased by non-hydrostatic stresses. Shock compression suffers from temperature uncertainties, and could also be biased by shear stresses. First-principles theory is limited by the practical necessity of simplifying to make the problem tractable.

A comparison of shock, static, and ultrasonic equations of state for gold is shown in Figure 11. While static and shock data are in good agreement, ultrasonic values for K_T'

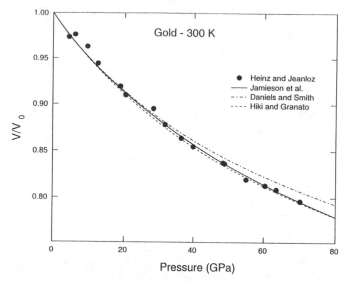

Figure 11. Equations of state of gold from static compression (filled symbols), shock (solid line), and ultrasonic data (dashed lines). Curves from ultrasonic data are calculated using Equation (8).

range from 5.2 to 6.4 (Daniels and Smith 1958, Hiki and Granato, 1966, Golding et al. 1967). At high pressures, this leads to an uncertainty of order 10% in pressure, depending on which equation of state is used. Furthermore, the parameter q, which controls the magnitude of the thermal correction to the shock data, has been determined on the basis of thermodynamic relations involving K_T' (Heinz and Jeanloz 1984, Anderson et al. 1989). The uncertainty in K_T' from static compression data (5.5 ± 0.8) precludes using this data to distinguish between ultrasonic values (Heinz and Jeanloz 1984). Improved determinations K_T' from high-pressure sound velocity measurements would better constrain the thermal pressure and equation of state of this material.

There has also been debate regarding the correct functional form for specifying the P-V relation (i.e. the reliability of the third-order Birch-Murnaghan equation). For example, Greene et al. (1994) argue that the Birch-Murnaghan equation is unreliable based on their analysis of static compression and reduced shock data for aluminum to 220 GPa ($V/V_0 = 0.5$). Fitting the data to a Birch-Murnaghan equation yields $K_T' = 4.1$-4.3, which they argue is inconsistent with ultrasonic data for this material ($K_T' = 5.1$) which should provide a thermodynamically correct value of K_T', since it is based on direct measurements of elastic constants as a function of pressure. The problem with this analysis is that the accuracy of the ultrasonic determinations of K_T' appears to be insufficient to be used as a criterion to distinguish amongst equations of state. In fact, published values of K_T' for aluminum from ultrasonic data cover a much wider range ($K_T' = 4.4$ - 5.2) than considered by Greene et al. (1994). As discussed above, similar problems are encountered when cross-comparing equations of state of gold determined by different techniques (Fig. 11).

Figure 12 provides a broader test of the consistency of equations of state determined by ultrasonic elasticity and shock compression, two techniques which are completely independent. As shown in Figure 12a, bulk moduli obtained from shock isotherms of a number of metals are in excellent agreement (within a few per cent) with ultrasonic elasticity values. On the other hand, the agreement between the pressure derivatives of the bulk modulus for shock and ultrasonic data is poor (Figure 12b). On the whole, the

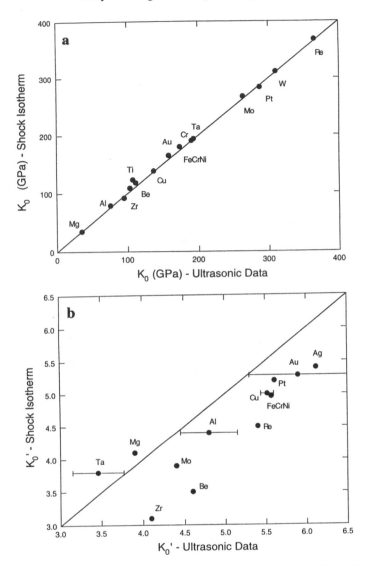

Figure 12. Comparison of (a) bulk modulus and (b) pressure derivative of bulk modulus for metals determined from reduced shock wave data and from high-pressure ultrasonic elasticity measurements. In cases where multiple ultrasonic measurements are available, the mean value is plotted and error bars show the range of measured values.

ultrasonic determinations of K_T' are typically about 10% larger than values from shock data. However, when ultrasonic values of K_T' reported by different laboratories for the same material are compared, there can be considerable variation, suggesting that the experimental precision of current ultrasonic data set may be poor (see Fig. 12b). The lack of consistency between thermodynamic parameters determined by these two techniques should be a source of concern. The recent advances in elasticity measurements described in the chapter by Liebermann may offer an opportunity to resolve this discrepancy.

SUMMARY

The thermoelastic properties of minerals play a crucial role in understanding the mineralogy, structure, and dynamics of the terrestrial planets. Major advances in mineral thermoelasticity have occurred over the last seven years due to the development of new techniques for accurate experimental determinations of P-V-T equations of state in both the multi-anvil press and the diamond cell. Many of the major mantle minerals have been studied to temperatures up to 1000 K or more and pressures of 30 GPa or higher. While significant progress has been made, a number of key issues relating to the pressure and temperature generation and characterization in both devices warrant continued attention. A thorough understanding of the pressure and temperature distributions, especially the effects of non-hydrostatic stresses and temperature gradients, is of the highest importance. As x-ray diffraction techniques progress and lead to higher accuracy in volume determinations, it becomes increasingly important to better characterize the pressure standards and thermal equations of state used in these experiments. We expect that there will be continued progress in these areas, and that increasingly accurate measurements will soon be regularly performed under the P-T conditions of the lower mantle.

ACKNOWLEDGMENTS

We thank R.C. Liebermann, R.J. Hemley, S.-H. Shim, and A. Kavner for valuable comments and S.-H. Shim for assistance in figure preparation. This research was supported by the NSF. Portions of this work were performed at GeoSoilEnviroCARS (GSECARS), Sector 13, Advanced Photon Source at Argonne National Laboratory. GSECARS is supported by the National Science Foundation–Earth Sciences, Department of Energy–Geosciences, W.M. Keck Foundation, and the United States Department of Agriculture. Use of the Advanced Photon Source was supported by the U.S. Department of Energy, Basic Energy Sciences, Office of Energy Research, under Contract No. W-31-109-Eng-38. Experimental results were also obtained at beamline X17 of the National Synchrotron Light Source, Brookhaven National Laboratory.

REFERENCES

Abramson EH, Brown JM, Slutsky LJ, Zaug J (1997) The elastic constants of San Carlos olivine to 17 GPa. J Geophys Res 102:12,253-12,264

Adams DM, Christy AG (1992) Materials for high temperature diamond anvil cells. High Pressure Res 8:685-689

Ahrens TJ (1987) Shock wave techniques for geophysics and planetary physics. In: CG Sammis and TL Henyey (eds) Methods of Experimental Physics. p 185-235 Academic, San Diego, CA

Ahrens TJ, Johnson ML (1995) Shock wave data for minerals. In: TJ Ahrens (ed) Mineral Physics and Crystallography: A Handbook of Physical Constants. p 143-184 Am Geophys Union, Washington, DC

Akimoto S (1987) High-pressure research in geophysics: Past, present and future. In: MH Manghnani, Y Syono (eds) High-Pressure Research in Mineral Physics. p 1-13 Am Geophys Union, Washington, DC

Anderson OL (1984) A universal thermal equation of state. J Geodyn 1:185-214

Anderson OL, Isaak DG, Oda H (1991) Thermoelastic parameters for 6 minerals at high temperature. J Geophys Res 96:18037-18046

Anderson OL, Isaak DG, Yamamoto S (1989) Anharmonicity and the equation of state for gold. J Appl Phys 65:1534-1543

Anderson OL, Oda H, Chopelas A, Isaak DG (1993) A thermodynamic theory of the Gruneisen ratio at extreme conditions: MgO as an example. Phys Chem Mineral 19:369-380

Andrault D, Fiquet G, Kunz M, Visocekas F, Hausermann D (1997) The orthorhombic structure of iron: An in situ study at high-temperature and high-pressure. Science 278:831-834

Angel RJ, Allan DR, Milletich R, Finger LW (1997) The use of quartz as an internal pressure standard in high-pressure crystallography. J Appl Cryst 30:461-466

Bass JD, Svendsen B, Ahrens TJ (1987) The temperature of shock compressed iron. *In:* MH Manghnani, Y Syono (eds) High-Pressure Research in Mineral Physics. p 393-402 Am Geophys Union, Washington, DC

Bassett WA, Shen AH, Bucknum M, Chou IM (1993) A new diamond anvil cell for hydrothermal studies to 2.5 GPa and from -190°C to 1200°C. Rev Sci Instrum 64:2340-2345

Bassett WA, Takahashi T (1965) Silver iodide polymorphs. Am Mineral 50:1576-1594

Besson JM, Hamel G, Grima T, Nelmes RJ, Loveday JS, Hull S, Hausermann D (1992) A large volume pressure cell for high temperatures. High Pressure Res 8:625-630

Birch F (1947) Finite elastic strain of cubic crystals. Phys Rev 71:809-824

Birch F (1978) Finite strain isotherms and velocities for single-crystal and polycrystalline NaCl at high pressures and 300 K. J Geophys Res 83:1257-1268

Birch F (1986) Equation of state and thermodynamic parameters of NaCl to 300 kbar in the high-temperature domain. J Geophys Res 91:4949-4954

Boehler R, Bargen NV, Chopelas A (1990) Melting, thermal expansion, and phase transition of iron at high pressures. J Geophys Res 95:21731-21736

Boehler R, Besson JM, Nicol M, Nielsen M, Itie JP, Weill G, Johnson S, Grey F (1989) x-ray diffraction of Gamma-Fe at high-temperatures and pressures. J Appl Phys 65:1795-1797

Boehler R, Chopelas A (1991) A new approach to laser heating in high pressure mineral physics. Geophys Res Lett 18:1147-1150

Cauble R, Perry TS, Bach DR, Budil KS, Hammel BA, Collins GW, Gold DM, Dunn J, Celliers P, DaSilva LB, Foord ME, Wallace RJ, Stewart RE, Woolsey NC (1998) Absolute equation-of-state data in the 10-40 Mbar (1-4 TPa) regime. Phys Rev Lett 80:1248-1251

Chen G, Liebermann RC, Weidner DJ (1998) Elasticity of single crystal MgO to 8 Gigapascals and 1600 Kelvin. Science 280:1913-1916

Clark, SM (1996) A new energy-dispersive powder diffraction facility at the SRS. Nucl Instrum Meth A 381:161-168

Clark SM, Jones RL, te-Lindert R, Walker D, Johnson MC, Fowler P (1996) The determination of the Seebeck coefficients of some common thermocouple wires using epithermal neutron resonance broadening. International Union of Crystallography, XVIIth Congress, Abstracts C-541

Daniels WB, Smith CS (1958) Pressure derivatives of the elastic constants of copper, silver, and gold to 100,000 bars. Phys Rev 111:713-721

Datchi F, LeToullec R, Loubeyre P (1997) Improved calibration of the $SrB_4O_7:Sm^{2+}$ optical pressure gauge: Advantages at very high pressures and temperatures. J Appl Phys 81:3333-3339

Decker DL (1971) High-pressure equation of state for NaCl, KCl, and CsCl. J Appl Phys 42:3239-3244

Downs RT, Zha CS, Duffy TS, Finger LW (1996) The equation of state of forsterite to 17.2 GPa and effects of pressure media. Am Mineral 81:51-55

Duffy TS, Ahrens TJ (1994) Dynamic response of molybdenum shock compressed at 1400°C. J Appl Phys 76:835-842

Duffy TS, Hemley RJ, Mao HK (1995a) Equation of state and shear strength at multimegabar pressures: Magneisum oxide to 227 GPa. Phys Rev Lett 74:1371-1374

Duffy TS, Shu J, Mao HK, Hemley RJ (1995b) Single-crystal x-ray diffraction of brucite to 14 GPa. Phys Chem Minerals 22:277-281

Duffy TS, Vos, WL, Zha CS, Hemley RJ, Mao HK (1994) Sound velocities in dense hydrogen and the interior of Jupiter. Science 263:1590-1593

Eliezer S, Ricci RA (eds) (1991) High-Pressure Equations of State: Theory and Applications. North Holland, New York

Eremets MI (1996) High Pressure Experimental Methods. Oxford University Press, New York

Fei Y (1996) Crystal chemistry of FeO at high pressure and temperature. *In:* MD Dyar, C MacCammon, MW Schaefer (eds) Mineral Spectroscopy: A Tribute to Roger Burns. p 243-254 Spec Pub Geochemical Soc

Fei Y, Mao HK (1993) Static compression of $Mg(OH)_2$ to 78 GPa at high temperature and constraints on the equation of state of fluid H_2O. J Geophys Res 98:11875-11884

Fei Y, Mao HK (1994) In situ determination of the NiAs phase of FeO at high pressure and temperature. Science 266:1678-1680

Fei Y, Mao HK, Shu J, Parthasarathy G, Bassett WA (1992a) Simultaneous high-P, high-T x-ray diffraction study of β-$(Mg,Fe)_2SiO_4$ to 26 GPa and 900 K. J Geophys Res 97:4489-4495

Fei Y, Mao HK, Shu JF, Hu J (1992b) P-V-T equation of state of magnesiowustite ($Mg_{0.6}Fe_{0.4}$). Phys Chem Minerals 18:416-422

Feldman JL, Mehl MJ, Boyer LL, Chen NC (1988) Ab initio calculations of static lattice properties for NaCl and a test of the Decker equation of state. Phys Rev B 37:4784-4787

Fiquet G, Andrault D, Dewaele A, Charpin T, Kunz M, Hausermann D (1998) P-V-T equation of state of MgSiO₃ perovskite. Phys Earth Planet Inter 105:21-31

Fiquet G, Andrault D, Itie JP, Gillet P, Richet P (1996) x-ray diffraction of periclase in a laser-heated diamond-anvil cell. Phys Earth Planet Inter 95:1-17

Frost DJ, Fei Y (1998) Stability of phase D at high pressure and high temperature. J Geophys Res 103:7463-7474

Funamori N, Yagi T, Utsumi W, Kondo T, Uchida T, Funamori M (1996) Thermoelastic properties of MgSiO₃ perovskite determined by in situ x-ray observations up to 30 GPa and 2000 K. J Geophys Res 101:8257-8269

Getting IC, Kennedy GC (1970) Effect of pressure on the emf of chromel-alumel and platinum-platinum 10% rhodium thermocouples. J Appl Phys 41:4552-4562

Gillet P, Richet P, Guyot F, Fiquet G (1991) High-temperature thermodynamic properties of forsterite. J Geophys Res 96:11,805-11,816

Golding BS, Moss SC, Averbach BL (1967) Composition and pressure dependence of the elastic constants of gold-nickel alloys. Phys Rev 158:637-646

Greene RG, Luo H, Ruoff AL (1994) Al as a simple solid: High pressure study to 220 GPa (2.2 Mbar). Phys Rev Lett 73:2075-2078

Guyot F, Wang YB, Gillet P, Ricard Y (1996) Quasi-harmonic computations of thermodynamic parameters of olivines at high-pressure and high-temperature: A comparison with experimental data. Phys Earth Planet Inter 98:17-12

Guyot F, Zhang JH, Martinez I, Matas J, Ricard Y, Javoy M (1997) P-V-T measurements of iron silicide (ε–FeSi): Implications for silicate-metal interactions in the early Earth. Eur J of Mineral 9:277-285

Hama J, Suito K (1996) The search for a universal equation of state correct up to very high pressures. J Phys—Cond Matter 8:67-81

Hama J, Suito K (1998) Equation of state of MgSiO₃ perovskite and its thermoelastic properties under lower mantle conditions. J Geophys Res 103:7443-7462

Hausermann D, Hanfland M (1996) Optics and beamlines for high-pressure research at the European Synchrotron Radiation Facility. High PressureRes 14:223-234

Heinz DL (1990) Thermal pressure in the laser-heated diamond anvil cell. Geophys Res Lett 17:1161-1164

Heinz DL, Jeanloz R (1984) The equation of state of the gold calibration standard. J Appl Phys 55:885-893

Hemley RJ, Gordon RG (1985) Theoretical study of solid NaF and NaCl at high pressures and temperatures. J Geophys Res 90:7803-7813

Hemley RJ, Mao HK, Finger LW, Jephcoat AP, Hazen RM, Zha CS (1990) Equation of state of solid hydrogen and deuterium from single-crystal x-ray diffraction data to 26.5 GPa. Phys Rev B 42:6458-6470

Hess NJ, Schiferl D (1992) Comparison of the pressure-induced frequency shift of Sm-YAG to the ruby and nitrogen vibron pressure scales from 6 K to 820 K and 0 to 25 GPa and suggestions for use as a high-temperature pressure calibrant. J Appl Phys 71:2082-2086

Hiki Y, Granato AV (1966) Anharmonicity in the noble metals; Higher order elastic constants. Phys Rev 144:411-419

Hixson RS, Fritz JN (1992) Shock compression of tungsten and molybdenum. J Appl Phys 71:1721-1728

Holmes NC, Moriarity JA, Gathers GR, Nellis WJ (1989) The equation of state of platinum to 660 GPa (6.6 Mbar). J Appl Phys 66:2962-2967

Holzapfel WB (1991) Equations of state for ideal and real solids under strong compression. Europhys Lett 16:67-72

Holzapfel WB (1996) Physics of solids under strong compression. Rep Prog Phys 59:29-90

Jackson I, Rigden SM (1996) Analysis of P-V-T data: Constaints on the thermoelastic properties of high-pressure minerals. Phys Earth Planet Inter 96:85-112

Jackson I (1998) Elasticity, composition and temperature of the Earth's lower mantle: a reappraisal. Geophys J Int'l 134:291-311

Jamieson JC, Fritz JN, Manghnani MH (1982) Pressure measurement at high temperature in x-ray diffraction studies: Gold as a primary standard. *In:* S Akimoto, MH Manghnani (eds) High Pressure Research in Geophysics. p 27-47 Center for Academic Publications, Tokyo

Jeanloz R (1989) Shock wave equation of state and finite strain theory. J Geophys Res 94:5873-5886

Jeanloz R, Kavner A (1996) Melting criteria and imaging spectroradiometry in laser-heated diamond-cell experiments. Phil Trans R Soc London A 354:1279-1305

Jephcoat AP, Besedin SP (1996) Temperature measurement and melting determination in the laser-heated diamond-anvil cell. Phil Trans R Soc Lond A 354:1333-1360

Kato T, Ohtani E, Kamaya N, Shimomura O, Kikegawa T (1992) Double-stage multi-anvil system with sintered diamond anvil for x-ray diffraction experiment at high pressures and temperatures. *In:* Y Syono,

MH Manghnani (eds) High Pressure Research in Mineral Physics: Application to Earth and Planetary Scientces. p 33-36 Am Geophys Union, Washington, DC

Kavner A, Duffy TS, Heinz DL, Shen G (1998) Thermoelastic behavior of platinum and MgO in the laser-heated diamond anvil cell. Eos Trans Am Geophys Union Spr Mtg Suppl S163

Khvostantsev LG (1984) A verkh-niz (up-down) toroid device for generation of high pressure. High Temp High Press 16:165-169

Kikegawa T (1987) X-ray diamond anvil press for structural studies at high pressures and high temperatures. *In:* MH Manghnani, Y Syono (eds) High-Pressure Research in Mineral Physics. p 61-68 Am Geophys Union, Washington, DC

Kinsland GL, Bassett WA (1977) Strength of MgO and NaCl polycrystals to confinining pressures of 250 kbar at 25°C. J Appl Phys 48:978-985

Kondo T, Sawamoto H, Yoneda A, Kato M, Matsumuro A, Yagi T (1993) Ultrahigh-pressure and high-temperature generation by use of the MA8 system with sintered diamond anvils. High Temp High Pressure 25:105-112

Kormer SB, Sinitsyn MV, Kirillov GA, Urlin VD (1965) Experimental determination of temperature in shock-compressed NaCl and KCl and of their melting curves at pressures up to 700 kbar. Sov Phys JETP 21:811-819

Kumari M, Dass N (1990) An equation of state applied to sodium-chloride and cesium- chloride at high-pressures and high-temperatures. J Phys-Cond Matter 2:3219-3229

Lacam A, Chateau C (1989) High-pressure measurements at moderate temperatures in a diamond anvil cell with a new optical sensor- $SrB_4O_7:Sm^{2+}$. J Appl Phys 66:366-372

Liu L, Bassett WA (1975) The melting of iron up to 200 kbar. J Geophys Res 80:3777- 3782

Loubeyre P, LeToullec R, Hausermann D, Hanfland M, Hemley RJ, Mao HK, Finger LW (1996) x-ray diffraction and equation of state of hydrogen at megabar pressures. Nature 383:702-704

Manga M, Jeanloz R (1998) Temperature distriubtion in the laser-heated diamond cell. *In:* MH Manghnani, T Yagi (eds) Properties of Earth and Planetary Materials at High Pressure and Temperature. p 17-26 Am Geophys Union, Washington, DC

Mao HK, Bell PM, Shaner JW, Steinberg DJ (1978) Specific volume measurements of Cu, Mo, Pd, and Ag and calibration of the ruby R1 fluorescence pressure gauge from 0.06 to 1 MBar. J Appl Phys 49:3276-3283

Mao HK, Hemley RJ (1996) Experimental studies of Earth's deep interior: Accuracy and versatility of diamond cells. Phil Trans R Soc Lond A 354:1-18

Mao HK, Hemley RJ, Fei Y, Shu JF, Chen LC, Jephcoat AP, Wu Y, Bassett WA (1991) Effect of pressure, temperature, and composition on lattice parameters and density of $(Fe,Mg)SiO_3$ perovskites to 30 GPa. J Geophys Res 96:8069-8079

Mao HK, Shen G, Hemley RJ, Duffy TS (1998) x-ray diffraction with a double hot-plate laser-heated diamond cell. *In:* MH Manghnani, T Yagi (eds) Properties of Earth and Planetary Materials at High Pressure and Temperature. p 27-34 Am Geophys Union, Washington, DC

Mao HK, Xu J, Bell PM (1986) Calibration of the ruby pressure gauge to 800 kbar under quasi-hydrostatic conditions. J Geophys Res 91:4673-4677

Marsh SP (1980) LASL Shock Hugoniot Data. Univ of California Press, Berkeley, CA

McQueen RG (1991) Shock waves in condensed media: Their properties and the equation of state of materials derived from them. *In:* S Eliezer and RA Ricci (eds) High-Pressure Equations of State: Theory and Applications. North-Holland, Amsterdam

McQueen RG, Marsh SP, Taylor JW, Fritz JN, Carter WJ (1970) The equation of state of solids from shock wave studies. *In:* R Kinslow (ed) High-velocity Impact Phenomena. p 294-419 Academic, San Diego, CA

Meng Y, Fei Y, Weidner DJ, Gwanmesia GD, Hu J (1994) Hydrostatic compression of gamma-Mg_2SiO_4 to mantle pressures and 700 K: Thermal equation of state and related thermoelastic properties. Phys Chem Minerals 21:407-412

Meng Y, Weidner DJ, Fei YW (1993a) Deviatoric stress in a quasi-hydrostatic diamond anvil cell: Effect on the volume-based pressure calibration. Geophys Res Lett 20:1147-1150

Meng Y, Weidner DJ, Gwanmesia GD, Liebermann RC, Vaughan MT, Wang Y, Leinenweber K, Pacalo RE, Yeganeh-Haeri A, Zhao Y (1993b) In situ high P-T x-ray diffraction studies on 3 polymorphs (α, β, γ) of Mg_2SiO_4. J Geophys Res 98:22199-22207

Migault A (1998) Concepts of shock waves. *In:* D Benest, C Froeschle (eds) Impacts on Earth. p 79-112 Springer-Verlag, New York

Ming LC, Bassett WA (1974) Laser heating in the diamond anvil press up to 2000°C sustained and 3000°C pulsed at pressures up to 260 kbars. Rev Sci Instrum 45:1115- 1118

Ming LC, Manghnani MH, Balogh J, Qadri SB, Skelton EF, Jamieson JC (1983) Gold as a reliable internal pressure calibrant at high temperatures. J Appl Phys 54:4390- 4397

Ming LC, Manghnani MH, Balogh J, Qadri SB, Skelton EF, Webb AW, Jamieson JC (1984) Static P-T-V measurements on MgO: Comparison with shock wave data. *In:* JR Asay, RA Graham, GK Straub (eds) Shock Waves in Condensed Matter 1983. p 57-60 North Holland, Amsterdam

Morishima H, Yusa H (1998) Numerical calculations of the temperature distribution and the cooling speed in the laser-heated diamond anvil cell. J Appl Phys 83:4572-4577

Ohtani A, Onodera A, Kawai N (1979) Pressure apparatus of split-octahedron type for x- ray diffraction studies. Rev Sci Instrum 50:308-315

Piermarini GJ, Barnett JD, Forman RA (1975) Calibration of the pressure dependence of the R1 ruby fluorescence line to 195 kbar. J Appl Phys 46:2774-2780

Rivers ML, Duffy TS, Wang Y, Eng PJ, Sutton SR, Shen G (1998) A new facility for high-pressure research at the Advanced Photon Source. *In:* MH Manghnani, T Yagi (eds) Properties of Earth and Planetary Materials at High Pressure and Temperature. p 79-88 Am Geophys Union, Washington, DC

Ross M (1993) High pressure equations of state: Theory and applications. *In:* R Winter, J Jonas (eds) High Pressure Chemistry, Biochemistry, and Materials Science. p 1-41 Kluwer Academic Press, New York

Ross M, Mao HK, Bell PM, Xu JA (1986) The equation of state of dense argon: A comparison of shock and static studies. J Chem Phys 85:1028-1033

Sato Y, Yagi T, Ida Y, Akimoto S (1975) Hysteresis in the pressure-volume relation and stress inhomogeneity in composite material. High Temp High Pressure 7:315-323

Schiferl D (1987) Temperature compensated high-temperature high-pressure Merrill-Bassett diamond anvil cell. Rev Sci Inst 58:1316-1317

Schiferl D, Fritz JN, Katz AI, Schaefer M, Skelton EF, Qadri SB, Ming LC, Manghnani MH (1987) Very high temperature diamond-anvil cell for x-ray diffraction: Application to the comparison of the gold and tungsten high-temperature-high-pressure internal standards. *In:* MH Manghnani, Y Syono (eds) High-Pressure Research in Mineral Physics. p 75-83 Am Geophys Union, Washington, DC

Schiferl D, Nicol M, Zaug JM, Sharma SK, Cooney TF, Wang SY, Anthony TR, Fleischer JF (1997) The diamond $^{13}C/^{12}C$ isotope Raman pressure sensor system for high-temperature/pressure diamond-anvil cells with reactive samples. J Appl Phys 82:3256-3265

Shen G, Mao HK, Hemley RJ (1996) Laser-heating diamond anvil cell technique: Double-sided heating with multimode Nd:YAG laser. *In:* Advanced Materials '96—New Trends in High-Pressure Research. p 149-152

Shimomura O, Yamaoka S, Yagi T, Wakatsuki M, Tsuji K, Fukunaga O, Kawamura H, Aoki K, Akimoto S (1985) Multi-anvil type high pressure apparatus for synchrotron radiation. *In:* S Minomura (ed) Solid State Physics Under Pressure. p 351-356, KTK Scientific Publishers, Tokyo

Shimomura O, Utsumi W, Taniguchi T, Kikegawa T, Nagashima T (1992) A new high pressure and high temperature apparatus with the sintered diamond anvil for synchrotron radiation use. *In:* Y Syono, MH Manghnani (eds) High Pressure Research in Mineral Physics: Application to Earth and Planetary Sciences. p 3-12 Am Geophys Union, Washington, DC

Singh AK (1993) The lattice strains in a specimen (cubic system) compressed nonhydrostatically in an opposed anvil device. J Appl Phys 73:4278-4286

Smith AH, Lawson AW (1954) The velocity of sound in water as a function of temperature and pressure. J Chem Phys 22:351-359

Spetzler HA, Yoneda A (1993) Performance of the complete travel-time equation of state at simultaneous high pressure and temperature. Pure Appl Geophys 141:379-392

Stixrude L, Hemley RJ, Fei Y, Mao HK (1992) Thermoelasticity of silicate perovskite and magnesiowustite and stratification of the Earth's mantle. Science 257:1099-1101

Takahashi E, Yamada H, Ito E (1982) An ultrahigh-pressure furnace assembly to 100 kbar and 1500°C with minimum temperature uncertainty. Geophys Res Lett 9:805-807

Taravillo M, Baonza VG, Nunez J, Caceres M (1996) Simple equation of state for solids under compression. Phys Rev B 54:7034-7045

Utsumi W, Funamori N, Yagi T, Ito E, Kikegawa T, Shimomura O (1995) Thermal expansivity of $MgSiO_3$ perovskite under high pressures up to 20 GPa Geophys Res Lett 22:1005-1008

Utsumi W, Weidner DJ, Liebermann RC (1998) Volume measurement of MgO at high pressures and temperatures. *In:* MH Manghnani, T Yagi (eds) Properties of Earth and Planetary Materials at High Pressure and Temperature. p 327-334 Am Geophys Union, Washington, DC

Vaughan MT (1993) In situ diffraction using synchrotron radiation at high P and T in a multi-anvil device. *In:* RW Luth (ed) Short Course Handbook on Experiments at High Pressure and Applications to the Earth's Mantle. p 95-130 Mineralogical Society of Canada, Edmonton, Alberta

Vaughan MT, Weidner DJ, Wang Y, Chen JH, Koleda CC, Getting IC (1998) T-CUP: A new high-pressure apparatus for x-ray studies. *In:* M Nakamura (ed) Rev High Pressure Sci Tech 7:1520-1522

Vinet P, Ferrante J, Smith JR, Rose JH (1986) A universal equation of state for solids. J Phys C 19:L467-L473

Vinet P, Rose JH, Ferrante J, Smith JR (1989) Universal features of the equation of state of solids. J Phys Cond Matter 1:1941-1963

Wallace DC (1972) Thermodynamics of Crystals. John Wiley and Sons, New York

Wang Y, Getting IC, Weidner DJ, Vaughan MT (1998a) Performance of tapered anivls in a DIA-type, cubic-anvil, high-pressure apparatus for x-ray diffraction studies. *In:* MH Manghnani, T Yagi (eds) Properties of Earth and Planetary Materials at High Pressure and Temperature. p 35-39 Am Geophys Union, Washington, DC

Wang Y, Weidner DJ, Meng Y (1998b) Advances in equation-of-state measurements in SAM-85. *In:* MH Manghnani, T Yagi (eds) Properties of Earth and Planetary Materials at High Pressure and Temperature. p 365-372 Am Geophys Union, Washington, DC

Wang Y, Weidner DJ, Zhang J, Gwanmesia GD, Liebermann RC (1998c) Thermal equation of state of garnets along the pyrope-majorite join. Phys Earth Planet Inter 105:59-71

Wang Y, Weidner DJ (1996) dμ/dT of the the lower mantle. Pure Appl Geophys 146:533-549

Wang Y, Weidner DJ, Guyot F (1996) Thermal equation of state of CaSiO₃ perovskite. J Geophys Res 101:661-672

Wang Y, Weidner DJ, Liebermann RC, Zhao YS (1994) P-V-T equation of state of (Mg,Fe)SiO₃ perovskite- Constraints on composition of the lower mantle. Phys Earth Planet Inter 83:13-40

Weidner DJ, Wang Y, Vaughan MT (1994) Yield strength at high pressure and temperature. Geophys Res Lett 21:753-756

Weidner DJ, Vaughan MT, Ko J, Wang Y, Liu X, Yeganeh-Haeri A, Pacalo RE, Zhao Y (1992) Characterization of stress, pressure, and temperature in SAM85, a DIA type high pressure apparatus. *In:* Y Syono, MH Manghnani (eds) High Pressure Research: Application to Earth and Planetary Science. p 13-18 Am Geophys Union, Washington, DC

Wilburn DR, Bassett WA, Sato Y, Akimoto S (1978) X-ray diffraction compression studies of hematite under hydrostatic, isothermal conditions. J Geophys Res 83:3509-3512

Wu TC, Shen AH, Weathers RS, Bassett WA, Chou IM (1995) Anisotropic thermal- expansion of calcite at high pressures—an in situ x-ray diffraction study in a hydrothermal diamond anvil cell. Am Mineral 80:941-946

Xia X, Weidner DJ, Zhao H (1998) Equation of state of brucite: Single-crystal Brillouin spectroscopy and polycrystalline pressure-volume-temperature measurement. Am Mineral 83:68-74

Yagi T (1978) Experimental determination of thermal expansivity of several alkali halides. J Phys Chem Solids 39:563-571

Yagi T, Shimomura O, Yamaoka S, Takemura K, Akimoto S (1985) Precise measurement of compressibility of gold at room temperature and high temperature. *In:* S Minomura (ed) Solid State Physics Under High Pressure: Recent Advance with Anvil Device. p 363-368 KTK Scientific Publishers, Tokyo

Yen J, Nicol M (1992) Temperature-dependence of the ruby luminescence method for measuring high-pressures. J Appl Phys 72:5535-5538

Zel'dovich YB, Raizer YP (1966) Physics of Shock Waves and High Temperature Phenomena. Academic Press, New York

Zha CS, Duffy TS, Downs RT, Mao HK, Hemley RJ (1998) Brillouin scattering and X- ray diffraction of San Carlos olivine: Direct pressure determination to 32 GPa. Earth Planet Sci Lett 159:25-34

Zhang J, Martinez I, Guyot F, Gillet P, Saxena SK (1997) x-ray diffraction study of magnesite at high pressure and high temperature. Phys Chem Minerals 24:122-130

Zhang J, Martinez I, Guyot F, Reeder RJ (1998) Effects of Mg²⁺ substitution in calcite-structured carbonates: Thermoelastic properties. Am Mineral 83:280-287

Zhao Y, Von Dreele RB, Zhang J, Weidner DJ (1998) Thermoelastic equation of state of monoclinic pyroxene: CaMgSi₂O₆. Rev High Pres Sci Tech 7:25-27

Zhao YS, Von Dreele RB, Shankland TJ, Weidner DJ, Zhang JZ, Wang YB, Gasparik T (1997a) Thermoelastic equation of state of jadeite: NaAlSi₂O₆: An energy-dispersive Rietveld refinement study of low symmetry and multiple phases diffraction. Geophys Res Lett 24:5-8

Zhao YS, Von Dreele RB, Weidner DJ, Schiferl D (1997b) P-V-T data of hexagonal boron nitride hBN and determination of pressure and temperature using thermoelastic equations of state of multiple phases. High Pressure Research 15:369-386

Zhao YS, Schiferl D, Shankland TJ (1995) A high P-T single crystal x-ray diffraction study of thermoelasticity of MgSiO₃ orthoenstatite. Phys Chem Minerals 22:393-398

Chapter 15

ELASTICITY AT HIGH PRESSURES AND TEMPERATURES

Robert C. Liebermann

Center for High Pressure Research
and Department of Geosciences
State University of New York at Stony Brook
Stony Brook, New York 11794

Baosheng Li

Center for High Pressure Research
And Mineral Physics Institute
State University of New York at Stony Brook
Stony Brook, New York 11794

INTRODUCTION

Seismological investigations provide the primary source of information about the properties and processes of the Earth's interior, especially for depths greater than a few hundreds of kilometers (i.e. depths below which rock samples have not yet reached the Earth's surface). Seismic models of the variation of compressional and shear wave velocities and densities presumably reflect radial and lateral variations of chemical composition, mineralogy, pressure and temperature. Successful interpretation of these seismic models in terms of the variables above requires experimental and theoretical information on the elasticity of deep Earth materials under the elevated conditions that characterize the Earth's interior.

This chapter focusses on experimental studies to measure the elasticity of deep Earth materials at high pressures and/or temperatures. After introducing some basic equations of elasticity, we assess the state of the art in such studies, with emphasis on the very significant progress in the past decade in the range of pressures and temperatures over which these experiments can be conducted, and then concentrate on the application of ultrasonic interferometric techniques in multi-anvil apparatus to measure wave velocities using both polycrystalline and single-crystal specimens. In doing so, we utilize several important recent data on the elasticity of deep Earth materials and illustrate some of the implications of these new data for mantle composition, lateral temperature variation, and anisotropy of seismic wave velocities. For additional details on the latter topics, the reader is referred to Liebermann (1998).

This chapter is complimentary to that of Stixrude, Cohen and Hemley and of Gillet, Hemley and McMillan (both in this volume), and to the excellent article by Jackson and Rigden (1998) in the Ringwood memorial volume on *The Earth's Mantle* on "Composition and temperature of the Earth's mantle: Seismological models interpretated through experimental studies of Earth materials."

BASIC ELASTICITY THEORY

In the limit of infinitesimal strain that governs wave propagation in solid and liquid media, Hooke's law implies a linear relation between each component of the strain tensor ε_{kl} and each component of the stress tensor σ_{ij}; the coefficients of these linear relationships are the elastic "constants"/stiffnesses/moduli, C_{ijkl} and Hooke's law becomes:

0275-0279/98/0037-0015$05.00

$$\sigma_{ij} = C_{ijkl}\varepsilon_{kl}$$

where i, j, k, and l = 1, 2, 3 and repeated indices imply summation over all values of the indices. There are thus 81 components to the fourth-order C_{ijkl} tensor, which reduces to 36 components due to the symmetry of the stress and strain tensors.

In the Voigt notation, the tensor C_{ijkl} components are reduced to matrix C_{ij} components using the following translation of indices:

Tensor:	11	22	33	23	32	13	31	12	21
Matrix:	1	2	3	4	4	5	5	6	6

and Hooke's law becomes:

$$\sigma_i = C_{ij}\varepsilon_j$$

where i,j = 1,2,3,4,5,6. C_{ij} has 36 independent components; energy and work considerations reduce this to 21 for the least symmetric triclinic system. As the symmetry increases, the number of independent elastic moduli C_{ij} decreases to 9 (orthorhombic-olivine), 7 (trigonal-corundum), 5 (tetragonal-rutile), and 3 (cubic-spinel), as seen in Table 1.

By applying Newton's laws of motion $(\partial\sigma_{ij}/x_j) = \rho\ (\partial^2 u_i/\partial t^2)$, substituting the σ_{ij}-ε_{kl} relationship, and utilizing Gauss' theorem, we obtain the "Christofel" equations:

$$\rho V^2 u_{0i} = C_{ijkl}n_j n_l u_{0k}$$

where ρ is density, V velocity, u the particle displacement, and n_j, n_l are the wavefront normal directions. These equations were the basis of the "long-wave" approximation used by Born to develop the elastic moduli/constants from lattice theory. For any general direction, there are 3 mutually perpendicular displacements (u_1, u_2, u_3), but these represent neither longitudinal/compressional or transverse/shear waves. In specific "pure mode" directions, the wavefront normals n are normal (for P waves) or parallel (for S waves).

For polycrystalline aggregates of minerals, the elastic properties have been customarily averaged by the Voigt, Reuss, or Hill averaging schemes (e.g. O.L. Anderson 1965). The Voigt (K_V and G_V) average assumes uniformity of strain; i.e. that the grains fit together well, but that discontinuities in stress occur across the grain boundaries. The Reuss (K_R and G_R) average assumes uniformity of stress; i.e. that the stress is uniform throughout the polycrystalline agrgregate, but the grains will not fit together well.

For cubic symmetry, these averages of the bulk and shear moduli are:

$$K_V = K_R = 1/3\ (C_{11} + 2\ C_{12})$$

$$G_V = 1/5\ (2\ C' + 3\ C_{44})$$

$$GR = 5\ C'C_{44}/(2\ C_{44} + 3\ C')$$

where $C' = 1/2\ (C_{11} - C_{12})$.

More exact or constraining estimates of the isotropic averages of the elastic properties of isotropic aggregates can be obtained via the Hashin-Shtrkman averages (e.g. Watt et al. 1976, 1988). These bounds are designated A and B and are related to the single crystal C_{ij} as follows for materials of cubic symmetry:

$$K_A = K_B = 1/3\ (C_{11} + 2\ C_{12}),\ \text{unique as in the Voigt-Reuss averages.}$$

$$G_A = C' + 3\ [5/(C_{44} - C') - 4\ \beta_1]^{-1}$$

$$G_B = C_{44} + 2 [5/(C' - C_{44}) - 6 \beta_2]^{-1}$$

These are the most restrictive bounds on the elasticity of monomineralic aggregates that are possilble without more specific assumptions or knowledge of the geometrical distribution of crystallites.

Table 1. Single crystal elastic moduli for various crystal symmetries
(following Truell et al. 1969 and Nye 1957).

General, or triclinic symmetry (21 elastic moduli C_{ij} — example: feldspars):

$$\begin{Vmatrix} \sigma_{11} \\ \sigma_{22} \\ \sigma_{33} \\ \sigma_{23} \\ \sigma_{31} \\ \sigma_{12} \end{Vmatrix} = \begin{Vmatrix} C_{11} & C_{12} & C_{13} & C_{14} & C_{15} & C_{16} \\ C_{12} & C_{22} & C_{23} & C_{24} & C_{25} & C_{26} \\ C_{13} & C_{23} & C_{33} & C_{34} & C_{35} & C_{36} \\ C_{14} & C_{24} & C_{34} & C_{44} & C_{45} & C_{46} \\ C_{15} & C_{25} & C_{35} & C_{45} & C_{55} & C_{56} \\ C_{16} & C_{26} & C_{36} & C_{46} & C_{56} & C_{66} \end{Vmatrix} \begin{Vmatrix} \varepsilon_{11} \\ \varepsilon_{22} \\ \varepsilon_{33} \\ \varepsilon_{23} \\ \varepsilon_{31} \\ \varepsilon_{12} \end{Vmatrix}$$

As the symmetry increases, the number of independent C_{ij} decreases, as seen in the following tables of the C_{ij} matrix:

Monoclinic symmetry (13 elastic moduli C_{ij} — example: clinopyroxene):

$$\begin{matrix} C_{11} & C_{12} & C_{13} & -- & C_{15} & -- \\ C_{12} & C_{22} & C_{23} & -- & C_{25} & -- \\ C_{13} & C_{23} & C_{33} & -- & C_{35} & -- \\ -- & -- & -- & C_{44} & -- & C_{46} \\ C_{15} & C_{25} & C_{35} & -- & C_{55} & -- \\ -- & -- & -- & C_{46} & C_{56} & C_{66} \end{matrix}$$

Orthorhombic symmetry (9 elastic moduli C_{ij} — example: olivine):

$$\begin{matrix} C_{11} & C_{12} & C_{13} & -- & -- & -- \\ C_{12} & C_{11} & C_{23} & -- & -- & -- \\ C_{13} & C_{23} & C_{33} & -- & -- & -- \\ -- & -- & -- & C_{44} & -- & -- \\ -- & -- & -- & -- & C_{55} & -- \\ -- & -- & -- & -- & -- & C_{66} \end{matrix}$$

Trigonal symmetry (7 elastic moduli C_{ij} — example: corundum):

$$\begin{matrix} C_{11} & C_{12} & C_{13} & C_{14} & (-C_{25}) & -- \\ C_{12} & C_{11} & C_{13} & (-C_{14}) & C_{25} & -- \\ C_{13} & C_{13} & C_{33} & -- & -- & -- \\ -- & -- & -- & C_{44} & -- & (-C_{25}) \\ -- & -- & -- & -- & C_{44} & C_{14} \\ -- & -- & -- & -- & -- & 1/2(C_{11}-C_{12}) \end{matrix}$$

Trigonal symmetry (6 elastic moduli C_{ij} — example: quartz):

$$\begin{matrix} C^{11} & C_{12} & C_{13} & C_{14} & -- & -- \\ C_{12} & C_{11} & C_{13} & (-C_{14}) & -- & -- \\ C_{13} & C_{13} & C_{33} & -- & -- & -- \\ -- & -- & -- & C_{44} & -- & -- \\ -- & -- & -- & -- & C_{44} & C_{14} \\ -- & -- & -- & -- & -- & 1/2(C_{11}-C_{12}) \end{matrix}$$

Table 1, continued

Tetragonal symmetry (6 elastic moduli C_{ij} — example: rutile):

C_{11}	C_{12}	C_{13}	--	--	C_{16}
C_{12}	C_{11}	C_{13}	--	--	$(-C_{16})$
C_{13}	C_{13}	C_{33}	--	--	--
--	--	--	C_{44}	--	--
C_{16}	--	--	--	C_{44}	--
$(-C_{16})$	--	--	--	--	$1/2(C_{11}-C_{12})$

Hexagonal symmetry (5 elastic moduli C_{ij} — example: beryl):

C_{11}	C_{12}	C_{13}	--	--	--
C_{12}	C_{11}	C_{13}	--	--	--
C_{13}	C_{13}	C_{33}	--	--	--
--	--	--	C_{44}	--	--
--	--	--	--	C_{44}	--
--	--	--	--	--	$1/2(C_{11}-C_{12})$

Cubic symmetry (3 elastic moduli C_{ij} — example: spinel):

C_{11}	C_{12}	C_{12}	--	--	--
C_{12}	C_{11}	C_{12}	--	--	--
C_{12}	C_{12}	C_{11}	--	--	--
--	--	--	C_{44}	--	--
--	--	--	--	C_{44}	--
--	--	--	--	--	$1/2(C_{11}-C_{12})$

In the cubic system (e.g. garnets, periclase), there are three elastic moduli, C_{11}, C_{12}, and C_{44} (note that this system is optically isotropic, but elastically isotropic). For isotropic elastic solids, the additional condition that $2\,C_{44} = (C_{11} - C_{12})$ reduces the number of independent C_{ij} to 2. It is customary in geophysics to utilize the adiabatic bulk modulus K_S and shear modulus G which are related to the C_{ij} as follows:

$$K_S = 1/3\,(C_{11} + 2\,C_{12}) = \rho\,(V_P^2 - 4/3\,V_S^2)$$

$$G = C_{44} = 1/2\,(C_{11} - C_{12}) = \rho\,V_S^2$$

where V_P and V_S are, respectively, the velocities of compressional and shear waves, and ρ is the density. The adiabatic bulk modulus K_S is related to the isothermal bulk modulus K_T measured in static compression studies by:

$$K_S = K_T\,(1 + T\alpha^2 K_S/\rho C_P)$$

where α = volume thermal expansion, and C_P = specific heat at constant pressure.

Other parameters discussed in this chapter and important in discussions of mantle geophysics include:

[...all at P = 1 bar (0.0001 GPa)]

K_0 = adiabatic bulk modulus K_S	G_0 = shear modulus G
$K_0' = (\partial K_S/\partial P)_T$	$G_0' = (\partial G/\partial P)_T$
$K_0'' = (\partial^2 K_S/\partial P^2)_T$	$G_0'' = (\partial^2 G/\partial P^2)_T$

For the purposes of this discussion, we assume that the pressure and temperature derivatives of K_S and K_T are identical, but this is clearly not so from a thermodynamical point of view (e.g. Anderson et al. 1968).

STATE OF THE ART IN LABORATORY STUDIES
OF ELASTICITY OF MINERALS

The elasticity of solids at elevated pressures and temperatures can be determined by techniques utilizing static or shock-wave compression (at zero frequency) or dynamical acoustic vibration (to frequencies as high as 10^{13} Hz). The former provide data on the volume as a function of pressure and temperature over a wider P and T range than can currently be achieved for acoustic experiments. However, the compression studies only indirectly measure the elastic bulk modulus and its derivatives and provide no information on the shear elasticity of solids. The reader is referred to the tutorial review by Duffy and Wang (1998, this volume) for an authoritative summary of the state of the art in compression studies and to Knittle (1995) for a tabulation of extant data.

This chapter will be restricted to studies by acoustic techniques to characterize the elasticity of minerals (i.e. measurements of the elastic bulk (K_S) and shear (G) moduli and the velocities of compressional and shear waves, V_P and V_S). These techniques include those based on natural resonance vibrations, phonon-photon scattering phenomena, and ultrasonic wave propagation for both single-crystal and polycrystalline specimens. The emphasis here will be on progress in the past decade in experiments at elevated pressures and/or temperatures; see also earlier reviews of technological developments using acoustic techniques by Anderson and Liebermann (1968), Schreiber, Anderson and Soga (1973), Weidner (1987), Li et al. (1998a), Parise et al. (1998), and Figure 9.3 and Table 9.1 of Jackson and Rigden (1998), and data compilations by Anderson et al. (1968), Bass (1995), Anderson and Isaak (1995).

As illustrated in Figure 1a, most of the studies prior to 1988 were conducted versus pressure at room temperature (maximum P = 8 GPa (Kinoshita et al. (1979)) or versus temperature at room pressure (maximum T = 1800 K (Goto et al. 1988, 1989). A notable exception was the work of Spetzler and colleagues (Spetzler 1970, Spetzler et al. 1972a) on MgO and NaCl to 0.8 GPa and 800 K; see also later studies by Isaak and Graham (1976) on a natural garnet to 0.5 GPa and 473 K.

When compared in Figure 1a with the geotherm, it is clear that those experimental conditions fell far short of those achieved in the Earth's mantle. Progress in the past decade has made conditions approaching those for the transition zone (at depths greater than 400 kilometers) achievable.

Progress in high-temperature experiments at ambient pressure and
high-pressure experiments at room temperature

In the UCLA laboratory, Anderson and Isaak and their colleagues have continued to utilize the technique of resonant ulrasound spectroscopy (RUS, sometimes also called Rectangular Paralelliped Resonance-RPR or Resonant Sphere Technique-RST) to study the elastic behavior of minerals to extreme temperatures at room pressure (see summary of extant data in Anderson and Isaak 1995). Isaak et al. (1989a,b) utilized RUS to study the elasticity of single crystals of Mg_2SiO_4-forsterite and MgO to 1700-1800 K. The application of this technique to controlled atmospheres was demonstrated by the experiments on iron-bearing olivines (Isaak 1992). Recently, Isaak and colleagues have demonstrated the feasibility of using RST for accurate measurements of the shear modes and shear modulus at high pressures for fused silica (Isaak et al. 1998); extension to

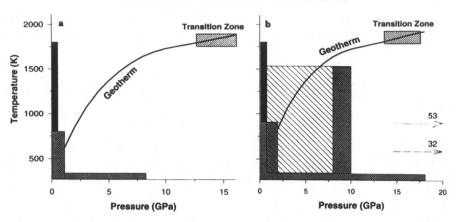

Figure 1. (a) Pressures and temperatures achievable for acoustic experiments in the laboratory in 1988 as compared with the Earth's geotherm to depths of the transition zone (410-660 km). Based on data from: Rectangular paralleliped resonance (RPR; also known as resonance ultrasound spectroscopy, RUS): T = 1800 K at P = 1 bar—Goto et al. (1988, 1989). Ultrasonic interferometry: P = 8 GPa at T = 300 K—Kinoshita et al. (1979); P = 0.8 GPa and T = 800 K—Spetzler (1970; see also Spetzler et al. 1972a).
(b) Pressures and temperatures achievable for laboratory acoustic experiments in 1998. Sources of data: Resonance ultrasound spectroscopy—RUS (also known as Rectangular paralleliped resonance—RPR, and Resonant Sphere Technique—RST); T = 1800 K at P = 1 bar—Goto et al. (1988, 1989; Isaak et al. 1989a,b; Isaak 1992) for oxides and silicates. Brillouin spectroscopy: T = 300 K and P = 32 GPa (Zha et al. 1998b) for San Carlos olivine, P = 58 GPa (Zha et al. 1994) for SiO_2 glass, P = 60 GPa (Zha et al. 1997) for MgO. Impulsive stimulated scattering: T = 300 K and P = 20 GPa (Chai et al. 1997a) for a natural garnet. Ultrasonics: P = 18 GPa at T = 300 K—Li et al. (1996b) for polycrystalline wadsleyite; P = 18 GPa at T = 300 K for iron—data of Baosheng Li in Mao et al. (1998); P = 2 GPa and T = 900 K—Niesler and Fisher (1992); P = 10 GPa and T = 1573 K—Li et al. (1998a) for polycrystalline forsterite; P = 8 GPa and T = 1573 K—Chen et al. (1998a) for single-crystal MgO in conjunction with synchrotron x-radiation of both sample and NaCl pressure standard. Diagrams courtesy of Ganglin Chen, as modified by Jun Liu.

compressional modes and moduli is still plagued by complications with both the theory and experiments resulting from the interaction of the specimen vibrations with the pressurizing medium.

Using other spectroscopic techniques, there has been dramatic progress in extending the pressures attainable in room temperature experiments beyond the 4 GPa the pioneering study of Bassett et al. (1982). Using laser-induced phonon spectroscopy in the form of impulsive stimulated scattering (ISS) in a diamond-anvil cell (Brown et al. 1989), the University of Washington group has measured the single-crystal elastic moduli (C_{ij}) of natural specimens of mantle minerals (olivine, orthopyroxene and garnet) to pressures as high as 20 GPa (Zaug et al. 1992, 1993; Chai et al. 1997 a,b; Abramson et al. 1997).

The Geophysical Laboratory group has extended the pressures achievable in Brillouin scattering experiments using single-crystal specimens of Mg_2SiO_4 forsterite and wadsleyite to 16 GPa (Duffy et al. 1995a, Zha et al. 1996, 1997, 1998a), San Carlos olivine to 32 GPa (Zha et al. 1998b) and MgO to 60 GPa (Zha et al. 1997), and for SiO_2 glass to 58 GPa (Zha et al. 1994). The University of Illinois laboratory has obtained Brillouin spectroscopic data for synthetic pyrope and MgO single crystals to 20 GPa (Sinogeikin et al. 1998).

Using a new method for obtaining acoustic velocities to very high pressures from the acoustic modes found in the vibrational sidebands of fluorescing transition-metal cations doped into the lattice, Chopelas (1996; see also Chopelas et al. 1996), has measured some of the elastic moduli and their pressure derivatives for MgO, $MgAl_2O_4$, Al_2O_3, and pyrope single crystals to pressures as high as 62 GPa. There are some remaining discrepancies between the data obtained by this technique and the other acoustic techniques; disadvantages of this technique are that it does not determine individual elastic moduli and is only applicable to those materials in which a fluorescent ion can be introduced.

Using ultrasonic interferometric techniques and multi-anvil, high-pressure apparatus, experiments have been performed in a number of Japanese laboratories on both single-crystal (e.g. Yoneda 1990, Yoneda and Morioka 1992) and polycrystalline (e.g. Fujisawa 1998) specimens to pressures of 8 and 14 GPa, respectively. Similar work in our laboratory to 18 GPa is summarized in detail below; this work utilizes the interferometric techniques that Jackson and his colleagues at the Australian National University have used in a piston-cylinder apparatus with a fluid pressure medium to study both single-crystal and polycrystalline specimens to 3 GPa (e.g. Jackson and Niesler 1982, Webb 1989, Niesler and Jackson 1989, Rigden et al. 1992)

Within the past year, a new method has been developed at the Geophysical Laboratory for estimating the single-crystal elasticity from x-ray diffraction of polycrystalline samples under non-hydrostatic compression (Singh et al. 1998). This technique has been utilized to study the elasticity and rheology of iron to 211 GPa (Mao et al. 1998) and of gold and rhenium to 37 GPa (Duffy et al. 1998).

Progress in experiments at simultaneous high pressures and temperatures

All of the above high-pressure spectroscopic and ultrasonic experiments were performed at room temperature and utilized the ruby pressure scale or fixed transition points (e.g. Bi, ZnTe, ZnS) for pressure calibration. Several efforts are underway to extend these ultrasonic measurements to simultaneous high pressures and temperatures. In Boulder and Bayreuth, these experiments have been extended to gigahertz frequencies (following Spetzler et al. 1993) in both gas pressure vessels (to 0.04 GPa and 350 K by Chen et al. 1996b) and diamond-anvil cells (to 4.4 GPa and 500 K by Shen et al. 1998; see also Spetzler et al. 1996). This technique has the advantage of requiring only very small specimens (as small as 60 micron in length), but to date, has not been successfully applied to measure shear wave velocities (Shen et al. 1998, Reichmann et al. 1998).

In both the Bayreuth and Stony Brook laboratories, programs have been pursued over the past three years to extend the ultrasonic measurements in the megahertz frequency range to simultaneous high pressures and temperatures (see below and Niesler and Fisher 1992, Knoche et al. 1998, and Li et al. 1998a papers for progress reports on these experiments). Although these experiments demonstrated the feasibility of performing measurements of both compressional and shear wave velocities to pressures above 10 GPa and simultaneous temperatures above 1300 K, it was not possible in these experiments to have a direct determination of pressure at high temperature (note that this was also true in the gigahertz experiments of Shen et al. 1998).

Over the past two years, the ultrasonics team at Stony Brook has succeeded in adapting the multi-anvil ultrasonic techniques for use in a DIA-type, cubic-anvil apparatus (SAM 85) installed on the superconducting wiggler beamline at the National Synchrotron Light Source of the Brookhaven National Laboratory (e.g. Liebermann et al. 1998, Parise et al. 1998). We describe these technological developments in detail below. These developments enable in situ ultrasonic and x-ray measurements to be performed simultaneously to pressures of 8 GPa and temperatures of 1600 K on both single-crystal (Chen et al. 1998) and polycrystalline (Li et al. 1998b, Sinelnikov et al. 1998) specimens.

As the result of these technological advances, the elastic properties of mantle minerals may now be investigated experimentally by a variety of acoustic techniques to pressures and temperatures approaching those of the transition zone of the Earth's mantle (see Fig. 1b).

ULTRASONIC INTERFEROMETRY IN MULTI-ANVIL APPARATUS

Following on the earlier work in Japan (e.g. Kinoshita et al. 1979, Fukizawa and Kinoshita 1982, Sasakura et al. 1989, 1990; Yoneda 1990, Yoneda and Morioka 1992), our laboratory has been working for the past eight years to adapt the techniques of ultrasonic interferometry for the determination of elastic wave velocities of polycrystalline and single-crystal specimens in multi-anvil apparatus at high pressures and temperatures. This section summarizes our progress in this endeavor (see also Gwanmesia and Liebermann 1992, Rigden et al. 1992, Gwanmesia et al. 1993, Li et al. 1996a, 1998a, Liebermann et al. 1998).

Hot-pressing of polycrystalline aggregates: Experiments to 3 GPa

An essential feature of our experimental approach has been to hot-press the optimum quality polycrystalline specimen of high-pressure phases of mantle minerals. This approach grew out of work done in the laboratory of Ringwood at the Australian National University (ANU) in the 1970s (e.g. Liebermann et al. 1975). To be suitable for such ultrasonic measurements, the hot-pressed specimens must be free of pores and cracks and preferred orientation of grains. Furthermore, the high-pressure phase should have uniform chemical composition and crystal structure and be sufficiently fine-grained so that the ultrasonic experiments can be performed at high enough frequencies (e.g. 10-70 MHz) to minimize dispersion caused by diffraction effects from grain boundaries and energy reflection from the side walls of the specimen. We characterize these hot-pressed specimens precisely by x-ray, optical, scanning-transmission electron microscopy and also by ultrasonic techniques at P = 1 bar and T = 300 K. The early series of these specimens (the olivine, wadsleyite and ringwoodite phases of Mg_2SiO_4, pyrope-majorite garnets, stishovite) were fabricated by in a 2000-ton uniaxial split-sphere apparatus (USSA-2000—Fig. 2). (see description in Liebermann and Wang 1992, Gwanmesia and Liebermann 1992) and their elastic wave travel times measured as a function of hydrostatic pressure to 3 GPa using the phase comparison technique of ultrasonic interferometry in a piston-cylinder apparatus in collaboration with Sally Rigden and Ian Jackson of the ANU (e.g. Gwanmesia et al. 1990a,b; Rigden et al. 1991, 1992; Rigden et al. 1994, Li et al. 1996c).

Example of stishovite. The study of the high-pressure form of SiO_2 known as stishovite presented two difficulties not encountered in the previous syntheses of the high-pressure phases of silicates. First, stishovite is much denser (4.29 g/cm³) than its low-pressure polymorphs quartz (62%) or coesite (45%) and amorphous or fused silica (95%); as a consequence of these large volume reductions, the hot-pressed specimens are likely to be smaller in dimension. Second, stishovite is extremely anisotropic in its elastic wave velocities (31% for P waves and 60% for S waves); thus even a small degree of preferred orientation within the polycrystal would prduce a directional dependence of its velocitis and their pressure derivatives.

To solve the problem of densely packing the starting material, a solid fused quartz rod (3-mm diameter) was used as a starting material; these rod usually weigh about 50 mg, which is 45% greater than for amorphous silica powder. The recovered specimens were about 2.5 mm long and 2.5 mm in diameter, and most importantly, the specimens exhibited good cylindrical shape with flat ends; x-ray spectra demonstrate that they are fully transformed to stishovite within 30 minutes at 1323 K.

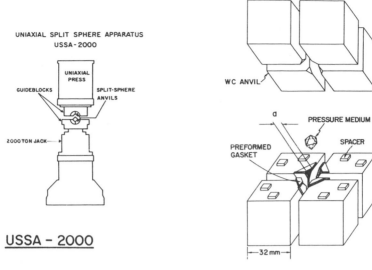

UNIAXIAL SPLIT SPHERE APPARATUS
USSA-2000

USSA - 2000

UPPER GUIDEBLOCK

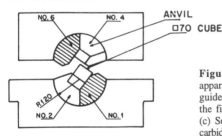

LOWER GUIDEBLOCK

Figure 2. Schematic for 2000-ton uniaxial split-sphere apparatus (USSA-2000). (a) Hydraulic press with guideblocks an split-sphere anvils in place. (b) Details of the first-stage spherical steel anvils in the guideblocks. (c) Second-stage with the assembly of eight tungsten carbide cubic anvils of truncation edge length (TEL = a) which compress an octahedral pressure medium.

Following the procedures developed by Gwanmesia et al. (1990a; see also Gwanmesia and Liebermann 1992), we utilized the cell assembly shown in Figure 3 and several P-T-t paths (Fig. 4). In all cases, the cell pressure is first increased slowly, after which the temperature is increased to the desired value. After maintaining the sample under the peak P-T conditions for a suitable time interval (15-60 min.), the pressure is decreased slowly. The cell temperature follows one of three paths: (1) Path Q—The sample is quenched to room temperature in a few seconds and the pressure then decreased at about 1 GPa/hr; (2) Path R—Decompression and cooling occur simultaneously in the stability field of stishovite to room temperature before final decompression to ambient conditions; (3) Path S1—The pressure and temperature initially decrease simultaneously in the stability field of stishovite, at 673 K, the temperature is kept approximately constant and the pressure is released slowly at a rate of about 0.5 GPa/hour (annealing period) to about 4 GPa, then the pressure and temperature are again decreased simultaneously to ambient conditions at 1.0 GPa/hour and 50 K/hour; Path S2—Temperature was decreased rapidly to 673 K at peak pressure and then decompression followed the S1 path.

The annealing temperature in paths S1 and S2 is important to preserve the high-pressure phase and to release the intergranular stresses. Higher annealing temperatures will result in transforming back to the low-pressure phase. One sample was simultaneously decompressed and cooled from P = 11 GPa and T = 1373 K, the recovered product was

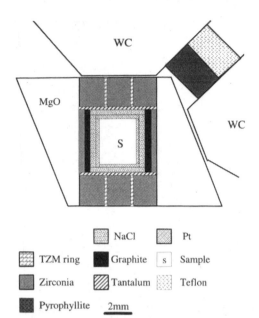

NaCl   Pt

TZM ring   Graphite   s   Sample

Zirconia   Tantalum   Teflon

Pyrophyllite   2mm

Figure 3. The cell diagram used for hot-pressing polycrystals of stishovite and other high-pressure phases (following Li et al. 1996c).

found to be coesite with trace of stishovite; this is because the run conditions entered the coesite stability field at relatively high temperature (>823 K).

Diffraction spectra from the end of the specimen are compared with JCPDS spectra of stishovite (card #150026) from 20 to 70 degrees in two theta; all the observed diffraction peaks match those of stishovite. In some cases, spectra were taken from both ends of the specimens to check its homogeneity. When this was done, very similar spectra were observed.

Figure 4. Comparison of the P-T-t paths explored in the hot-pressing experiments for poly-crystalline stishovite (following Li et al. 1996c).

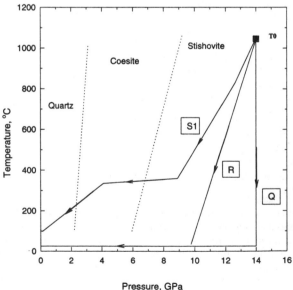

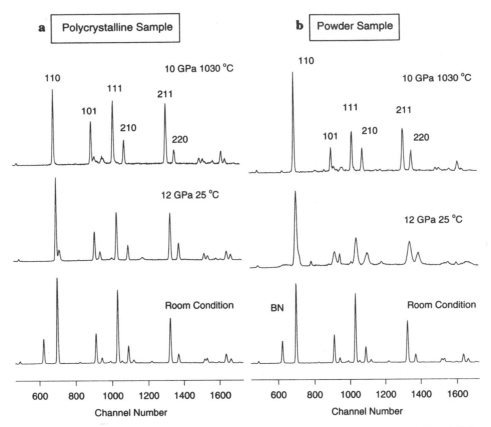

Figure 5. Comparison of x-ray diffraction spectra at room conditions with those at high P and T for stishovite. (a) polycrystalline sample; (b) powder sample (following Li et al. 1996c).

Bulk densities of the specimens were determined using Archimedes immersion method and carbon tetrachloride as the weighing fluid. All of the fully-transformed specimens have bulk densities within 1% of the theoretical x-ray density for stishovite (4.287 g/cm³), including the specimen #1533 (4.274 g/cm³) used for subsequent ultrasonic studies at high pressure (Li et al. 1996c).

SEM photographs were taken from the broken surfaces along the axial direction reveals that most grains are less than 7 microns, have excellent crystal faces and some exhibit triple junction boundaries. No significant variation in grain size was observed from the center to the boundaries of the specimen. The textural homogeneity of these specimens suggests that both the temperature and pressure gradients in the sample chamber are small.

Internal residual strain may affect the elastic wave velocity measurements for polycrystalline samples as noted by Spetzler et al. (1972b). Recently, Weidner et al. (1994) have developed a new technique based on the peak width of the x-ray diffraction peaks to determine the microscopic yield strength using synchrotron radiation in conjunction with a DIA-type cubic anvil apparatus (SAM 85) installed in the superconducting wiggler beamline (X17B1) of the National Synchrotron Light Source at the Brookhaven National Laboratory. We have used this new technique to study powder and polycrystalline specimens of stishovite (Figs. 5 and 6). The two specimens were simultaneously compressed (in different parts of the same sample chamber) up to P = 12 GPa and then heated to T = 1303 K and monitored with in situ x-ray diffraction. Both the powder and polycrystalline samples were surrounded by BN as the pressure transmitting medium with NaCl as the pressure standard. As the powder sample is compressed at room temperature, the individual diffraction peaks broaden as a consequence of the compacting process (Fig. 5b), indicating an increase in microscopic deviatoric stress. When heated under high pressure, these peaks sharpen above 673 K and approach their initial widths. In contrast, as the polycrystalline specimen is compressed to 12 GPa and subsequently heated to 1303 K (Fig. 5a), the diffraction peaks exhibit no broadening or sharpening. These data indicate that the microscopic deviatoric stress experienced by the individual grains in the polycrystal under high pressure and temperature is extremely low. Consequently, we conclude that the single-crystal grains within this polycrystalline aggregate are well-equilibrated (at least at 10 GPa and 1273 K), and the hot-pressed specimen is free of residual strain; see Figure 6.

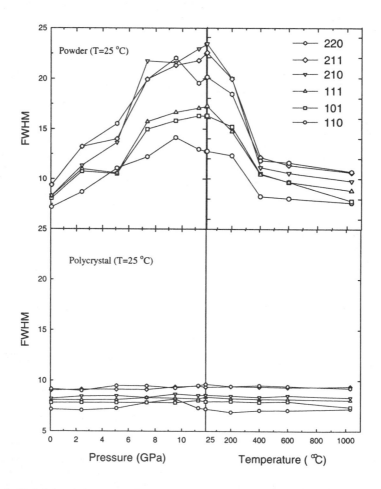

Figure 6. Variations of the x-ray diffraction peak widths (FWHM) with pressure and temperature for powdered and polycrystalline stishovite samples (following Li et al. 1996c).

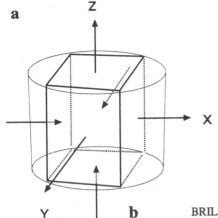

a

Figure 7. (a) Schematic diagram for the P-wave velocity measurements in the axial (Z) and radial (X and Y) directions in cylindrical specimens of polycrystalline stishovite (following Li et al. 1996c). (b) Acoustic wave velocities in single-crystal stishovite as a function of crystallographic direction from the study of Weidner et al. (1982). Comparison of Hashin-Shtrikman averages of single-crystal elastic velocities with those measured in polycrystalline stishovite by Li et al . (1996c).

b

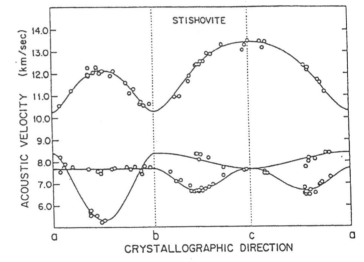

BRILLOUIN SPECTROSCOPY: SiO$_2$ - STISHOVITE

Anisotropy of Stishovite

P waves ~ 31%   S waves ~ 60%

Isotrpic Velocities:

	Hashin-Shtrikman	Voigt-Reuss-Hill
V$_P$ (km /sec)	11.93 (5)	11.92 (24)
V$_S$ (km /sec)	7.18 (4)	7.16 (20)

P wave velocities were measured using the pulse transmission technique (e.g. Liebermann et al. 1975) in different directions in the sample (Fig. 7a). The observed velocities in the axial and radial directions are identical within the experimental uncertainty. These results indicate that the possibility of having preferred orientation in these specimens is extremely low. Precise velocity measurements using an ultrasonic interferometer were

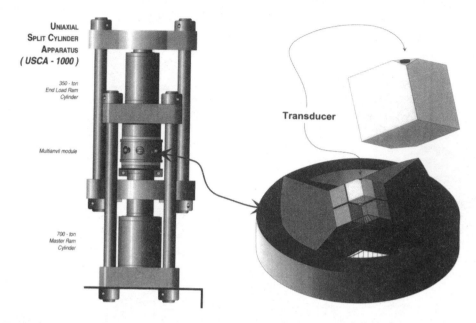

Figure 8. (a) The 1000-ton uniaxial split-cylinder (USCA-1000) apparatus used for acoustic experiments, consisting of a modified Kennedy-Getting type press and a Walker-type cylindrical multi-anvil module. (b) Exploded view of the containment ring of the Walker module and three of the removable split cylinders with the second stage WC cubes. The blow-up is the cube used as the buffer rod in the acoustic experiments, also showing the position of the ultrasonic transducer (following Li et al. 1996a).

conducted at the Australian National University at ambient pressure and temperature conditions, yielding $V_p = 11.78 \pm 0.11$ km/sec, $V_s = 7.12 \pm 0.07$ km/sec. As shown in Figure 7b, these results agree within the mutual uncertainty with those calculated from the Hashin-Shtrikman average of the single-crystal elastic moduli determined using Brillouin spectroscopy by Weidner et al. (1982).

In summary, the application of these high pressure techniques to hot-pressed polycrystalline stishovite is successful; the specimens using fused quartz rods as starting material and recovered along P-T path S1 are homogeneous, fine-grained (~7 mm), well crystallized with equilibrium texture, low porosity (less than 1%), free of preferred orientation and in agreement with the Hashin-Shtrikman average of the single-crystal velocities for both P and S waves. In the following section, we describe in detail the measurement of the pressure dependence of the elastic wave velocities in multi-anvil apparatus.

Ultrasonics in uniaxial split-cylinder apparatus at room temperature

The implementation of ultrasonic interferometric measurements using the 1000-ton uniaxial split-cylinder apparatus (USCA-1000) is shown schematically in Figure 8 and has been described in detail by Li et al. (1996a). Diagonally opposite corners of one cube are truncated to yield lapped surfaces on which the transducer and sample are mounted. This cube thus serves as the buffer rod to transmit the acoustic signals to and from the sample. **Figure 9** is a cross section of the octahedral cell assembly and the buffer rod cube designed to perform acoustic velocity measurements. The sample is surrounded by lead on the sides (0.3-mm wall thickness) and bottom (2.0-mm-thick disk), thereby providing a pseudohydrostatic pressure medium which protects the sample from cracking at high

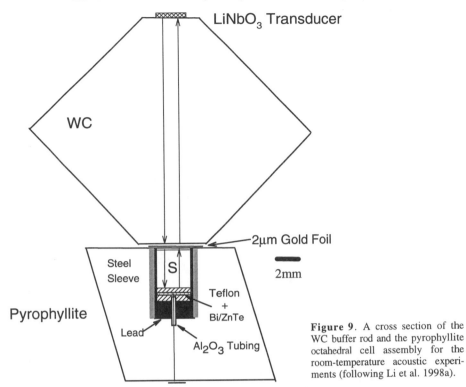

LiNbO₃ Transducer

WC

Steel
Sleeve

S

Pyrophyllite

Lead

2µm Gold Foil

2mm

Teflon
+
Bi/ZnTe

Al₂O₃ Tubing

Figure 9. A cross section of the WC buffer rod and the pyrophyllite octahedral cell assembly for the room-temperature acoustic experiments (following Li et al. 1998a).

pressures. A teflon disk (~2 mm thick) is placed at the end of the sample to enhance the ultrasonic signals by increasing the mismatch between the sample and the backup material (Fig. 9). The sample and surrounding lead are inserted into a steel sleeve (6.0 mm long, OD 4.4 mm, ID 3.4 mm). Pressure sensors (Bi and/or ZnTe) are placed in the teflon disk next to the sample as shown in Figure 9. Therefore, the pressure scale is obtained in each individual run through observation of the phase transformations in Bi (I-II 2.55GPa, III-V 7.7 GPa) and/or ZnTe (at 9.6 GPa and 12.0 GPa) (Lloyd 1971, Kusaba et al. 1993). The reproducibility of the cell pressure for the same specimen is better than 3% for given oil pressure (or ram force).

The pressure gradient in the sample region was investigated by enclosing Bi in AgCl at the center of the cell assembly and at the surface in contact with the WC anvil in two different runs, and determined to be less than 0.2 GPa/mm for our acoustic experiments.

The acoustic signals are generated and received using disk-shaped (3.2-mm diameter), 40 MHz LiNbO₃ transducers (36°Y-cut for compressional wave and 41°X-cut for shear waves). A spring-loaded sliding pin is placed in the vertical gap between the bottom three anvils to make contact with one electrode of the transducer while the buffer rod cube serves as the electrical ground. At elevated pressure, the transducer remains stress-free since it is located in the gap between the first-stage anvils and the second-stage cubes, allowing precise travel-time measurements over a wide frequency range (20 to 70 MHz).

The ultrasonic phase-comparison method implemented on the Australian Scientific Instruments Ultrasonic Interferometer employed in our laboratories has been described in detail in previous studies (Rigden et al. 1988, Niesler and Jackson 1989, Rigden et al. 1992, Li et al. 1996a). The output from a continuous wave source is gated to produce a

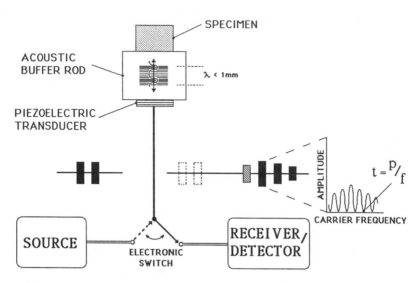

Figure 10. Schematic diagram of the method of phase comparison ultrasonic interferometry (following Rigden et al. 1992).

pair of phase-coherent, high frequency pulses (Fig. 10). These pulses are applied to the transducer which is bonded to the buffer rod. The elastic waves generated by each pulse are reflected and transmitted at the buffer rod/sample interface, and the transmitted portion reverberates inside the sample, resulting in a series of 'sample' echoes following the buffer rod echo (Fig. 11). If the applied pulses are separated by the apparent two-way travel time through the sample, the first buffer echo from the second applied source pulse will superimpose with the first sample echo from the first source pulse. As the carrier frequency is varied, alternate constructive and destructive interferences between the superimposed signals will occur, resulting in a series of maxima and minima on the amplitude spectrum modulated by the transducer response envelope. Frequencies for pth and $(p+n)$th interference extrema, f_p and f_{p+n} can be used to estimate the apparent travel time by $t'_{est} = n/(f_{p+n} - f_p)$; then the p value is calculated from $p = f_p t'_{est}$ and the closest half or integral value is therefore assigned to frequency f_p and all remaining extrema can be assigned sequentially. In practice, the interference minima are normally used to reduce the travel-time data because they are sharper than the maxima. For the situation in which the amplitude ratio of the first buffer rod echo and the first sample echo (B1/S1) is very different from unity, the perturbation to travel time from the transducer response envelope has to be taken into consideration, especially for interference maxima (see Jackson et al. 1981, Niesler and Jackson 1989).

Polycrystalline Lucalox alumina. An example of the acoustic signals is shown in Figure 12 for P waves in a polycrystalline specimen of alumina (General Electric brand Lucalox). Three successive sample echoes are clearly visible. Overlap and interference (Fig. 12b) of these sample echoes with that from the buffer rod enables the determination of the travel time at high frequencies with a precision of 1.6×10^{-3} (Fig. 12c); this high acoustic quality and precision of the travel time determination are maintained as pressure is increased.

The dispersion of the travel times from 20 to 40 MHz ($\sim$1%) is much more pronounced than that at frequencies higher than 40 MHz in which range the travel times are

TRANSDUCER

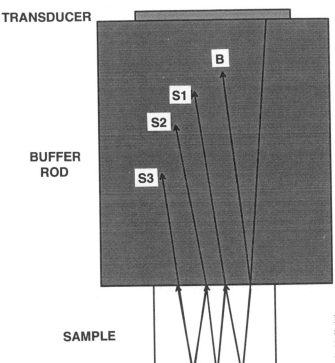

**BUFFER
ROD**

SAMPLE

Figure 11. Schematic diagram showing the acoustic ray paths of incident pulse, buffer rod echo (B), and 1st and 2nd echoes in the sample (S1, S2).

insensitive to the frequency. This characteristic of the dispersion persists at all pressures in our experiments. This observed dispersion in the frequency range lower than 40 MHz is not intrinsic to the sample (i.e. change of velocity with frequency). Instead, the dispersion is largely caused by departures from the assumption of wave propagation in an infinite medium, due to the small size of the sample as has been suggested by Rigden et al. (1992). Small samples are likely to be effected by the sidewall reflections from the sample at certain low frequencies. Comparison of the results of this study with measurements for samples of the same material but larger diameters (I. Jackson, pers. comm. 1996) show that the dispersion in the frequency range 20-50 MHz decreases from 1% to 0.3% as the sample diameter increases from 2.9 to 7 mm. Further measurements in the frequency range of 120-150 MHz on both polycrystalline alumina and forsterite samples (diameter 2.7-2.9 mm) show a very good agreement (better than 10^{-3}) with measurements at frequency range 40-70 MHz; this confirms that the travel times at low-frequency range are influenced by the experimental configuration, and that reliable travel times can be obtained at the high-frequency range. The uncertainties in these measured travel times are thus estimated to be about 0.3%.

The bulk (K_S) and shear (G) moduli for Lucalox alumina as calculated from the measured P and S wave velocities are compared with single-crystal data of Gieske and Barsch (1968) in Figure 13. The bulk modulus from our study is about ~1% higher than the Hashin-Shtrikman bounds calculated from single-crystal data. The shear modulus is almost identical in the pressure range of this study. Both K_S and G show linear pressure dependences to 10 GPa and first-order polynomial fits yield $(\partial K_S/\partial P)_T = 4.4$ and $(\partial G/\partial P)_T = 1.8$ which are in good accord with the data from the single-crystal data of Gieske and Barsch (1968) and from polycrystalline specimens of Schreiber and Anderson (1966).

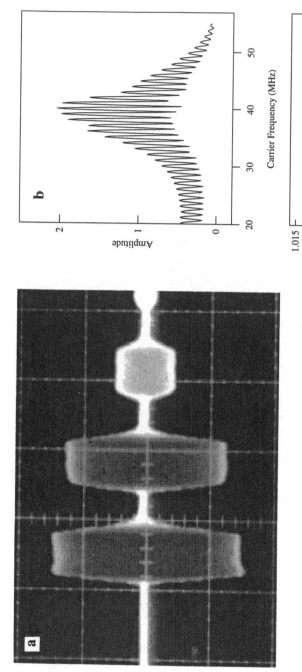

Figure 12. (a) P wave echo trains for Lucalox alumina on a WC buffer rod with V9 bond at 45 MHz. The first echo is the reflection at the WC cube/sample interface (B1) and the second, third, and fourth echoes are the first (S1), second (S2) and third (S3) reflections from the far end of the sample (following Li et al. 1996a). (b) overlap and interference of the sample echoes with that from the buffer rod permits determination of travel time at high frequencies (c).

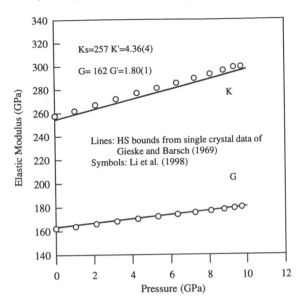

Figure 13. The bulk (K$_S$) and shear (G) moduli for Lucalox alumina calculated from the measured compressional and shear wave velocities. Symbols are results from this study and the lines are Hashin-Shtrikman bounds calculated from single crystal data of Gieske and Barsch (1968) and their extrapolations (following Li et al. 1998a).

The "sample friendly" cell assembly enables us to perform multiple measurements on a single sample. Upon recovery from each high-pressure experiment, these specimens were carefully examined, including measurement of dimensions, flatness and wave speeds at room conditions. No noticeable changes in dimensions have been observed even after multiple measurements. Sometimes, minor repolishing of the sample surface is performed to remove residual gold foil. The results of bench-top velocity measurements are reversible. The overall reproducibility in travel time for duplicate high pressure runs is about 0.3% (see Li et al. 1996a, Fig. 9).

Polycrystalline olivine and wadsleyite. Polycrystalline specimens of both the olivine and wadleyite (ß phase) polymorphs of Mg$_2$SiO$_4$ were hot-pressed in the USSA-2000 apparatus and studied at pressures to 13 GPa in the USCA-1000 apparatus using the above techniques (Li et al. 1996b). The P and S wave velocities for the olivine polymorph (forsterite) are shown in Figure 14 along with data from other laboratories. The P and S wave velocities from our new measurements increase more slowly than the finite strain extrapolations of low-pressure data of Kumazawa and Anderson (1969) (0-0.2 GPa) and Graham and Barsch (1969) (0-1.0 GPa), differences reaching about 1% for P wave and ~2% for S wave at 12.5 GPa (see Li et al. 1996b, Fig. 1). However, as seen in Figure 14, they agree with the recent single-crystal data from ultrasonic (0-6 GPa) and Brillouin scattering (3.1-16.2 GPa) studies. Considering the difference in acoustic techniques and source and form of specimens, there is remarkable consistency in these data, especially for P < 10 GPa. At higher pressures, the Brillouin scattering data exhibit larger scatter (especially in the compressional wave), but the Duffy et al. (1995) /Zha et al. (1996) data at 15-16 GPa agree with the extrapolation of our data and those of Yoneda and Morioka (1992). The velocities at room conditions were measured and compared with the velocities before and after each high-pressure run; a reproducibility better than 0.2% is obtained. These olivine and wadsleyite data measured at pressures corresponding the those at depths of 410 km are critical in attempts to infer the olivine content of the upper mantle from the magnitude of the seismic discontinuity at this depth (e.g. Duffy et al. 1995, Li et al. 1996b).

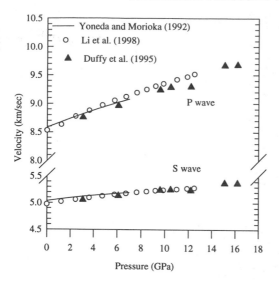

Figure 14. P and S wave velocities measured in this study and comparison with single crystal data from ultrasonic and Brillouin scattering techniques. The open circles are the experimental results from this study, the solid triangles are the Brillouin scattering data of Duffy et al. (1995), and the solid lines are calculated from ultrasonic measurements on single crystal forsterite of Yoneda and Morioka (1992) (following Li et al. 1998a).

Single crystals of forsterite and San Carlos olivine. These ultrasonic techniques in multianvil apparatus have also been successfully applied to study the elasticity of olivine single crystals to 13 GPa (see Chen et al. 1996a). The goals of these measurements were to demonstrate the feasibility of performing such experiments on single-crystal samples and to re-examine the unusual behavior of the shear modulus C_{55} versus pressure in San Carlos olivine (Zaug et al. 1993).

Figure 15a shows the acoustic echo pattern observed at 10 GPa for the S-waves (polarized in [001]) in forsterite using a 40 MHz $LiNbO_3$ transducer. The first echo is the reflection from the end of the WC cube and the ensuing echoes are internal reflections within the sample; one can see three sample reflections. In Figure 15b, we illustrate the sharpness of the interference patterns between the WC buffer rod echo and the first sample echo for the S-waves in San Carlos olivine at 13 GPa. The travel time versus frequency data based on the amplitude data in Figure 15b are shown in Figure 15c; from 15 to 30 MHz the travel time increases and then from 30-60 MHz attains a steady value ($\pm 0.15\%$). A demodulation algorithm has been applied to the interference amplitude data to remove the amplitude trend effected by the electronic components and the transducer response (the resultant amplitude spectrum is shown in the inset in Fig. 15b; for details, see Spetzler et al. 1993).

We obtained travel-time data for the C_{55} mode in forsterite and for the C_{22} and C_{55} modes in San Carlos olivine and used these data to calculate the elastic moduli versus pressure (Chen et al. 1996a). As an example, our C_{55} data for forsterite agree well with the earlier ultrasonic study of Yoneda and Morioka (1992) to 6 GPa and the recent Brillouin scattering data of Duffy et al. (1995; see also Zha et al. 1996) to 16 GPa (Fig. 16). Most significantly, C_{55} is a linear function of pressure in all three studies.

The data of Chen et al. (1996a) for C_{55} in San Carlos also increase linearly with pressure, unlike the earlier work of Zaug et al. (1993). Subsequent measurements by impulsive stimulated scattering (Abramson et al. 1997) and by Brillouin scattering (Zha et al. 1998b) have extended these acoustic measurements on San Carlos olivine to higher pressures, but with lingering debate about the non-linearity of the modulus-pressure behavior.

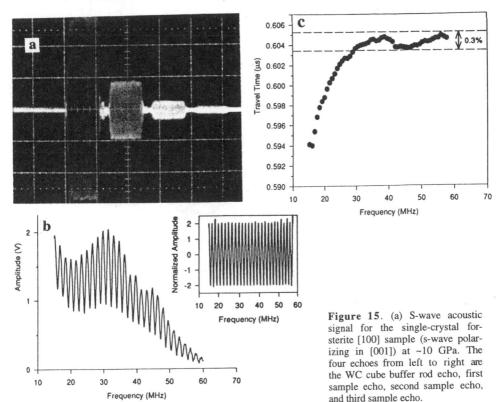

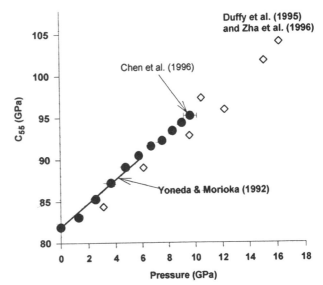

Figure 15. (a) S-wave acoustic signal for the single-crystal forsterite [100] sample (s-wave polarizing in [001]) at ~10 GPa. The four echoes from left to right are the WC cube buffer rod echo, first sample echo, second sample echo, and third sample echo. (b) S-wave interference pattern from the single-crystal San Carlos olivine [100] sample (s-wave polarizing in [001]) at ~13.5 GPa. The inset shows the same interference data after applying a demodu-lation procedure to the raw data. (c) Travel time data from the reduced interference data shown in the inset of (b) (data of Chen et al. 1996a, following Li et al. 1998a).

Figure 16. Shear elastic modulus C_{55} versus pressure for single crystals of synthetic forsterite (Mg_2SiO_4). ● = data from Chen et al. (1996); diamonds: data from Duffy et al. (1995) and Zha et al. (1996); line from Yoneda and Morioka (1992).

Polycrystalline iron. Velocity measurements on the epsilon phase of iron have been conducted at 16-17 GPa and room temperature using a modified cell assembly with a 10 mm octahedron and 5 mm truncation edge length of Toshiba WC cubes. A glass buffer rod was inserted into the octahedron to extend the sample into the center of the cell. ZnS was embedded in a teflon disk adjacent to the sample and used as an in situ pressure marker. Alpha-phase iron rod (2.0 mm diameter) was used as a starting material and the epsilon phase was produced by transformation at high pressure. Travel-time measurements were made at 16-17 GPa to ensure fully transformation from the alpha to the epsilon phase. Previous P-V compression data were used to obtain the sample length correction for the epsilon phase at these conditions (Mao et al. 1990). We obtained a result of $V_P = 6.9 \pm 2$ km/sec and $V_S = 3.5 \pm 1$ km/sec for epsilon iron. These measurements demonstrate the feasibility of meauring velocities of unquenchable phases at high pressures. The consistency with the back-extrapololated data of Mao et al. (1998) also provide support to the measurements of elastic wave velocities at core pressures by the non-hydrostatic x-ray diffraction technique described above (see Fig. 2 of Mao et al. 1998).

Ultrasonics in USCA-1000 apparatus at simultaneous high P and T

In the previous section, we have presented new data for elastic wave velocities in millimeter-sized polycrystalline and single-crystal specimens using ultrasonic interferometry in a multianvil apparatus. These measurements eliminate the necessity and the potential uncertainties in a five-fold extrapolation in pressure to reach conditions of the mantle transition zone. However, without direct measurements of elastic wave velocities at simultaneous high pressure and temperature, comparisons with seismic models are still fraught with ambiguity. For instance, attempts to match the olivine-beta phase transformation with the observed seismic discontinuity at 410 km depth to infer the olivine content of the mantle have resulted in widely scattered results, ranging from 27-44% (Duffy et al. 1995) to 45-65% (Gwanmesia et al. 1990b), even when new measurements of $(\partial K_T / \partial T)_P$ for the beta phase by Fei et al. (1992) and Meng et al. (1993) are incorporated. Obviously, the solution to this dilemma is to directly measure velocities of relevant mantle phases at the P and T conditions of the Earth's transition zone. Such experimental work has been initiated in our laboratory and others (Li et al. 1998a, Knoche et al. 1998). In this section, we illustrate implementation of such experiments in a multi-anvil apparatus.

The cell assembly for the USCA-1000 multianvil apparatus (Fig. 9) has been modified to enable velocity measurements at simultaneous high pressures and temperatures to be made by inserting a cylindrical graphite furnace with caps at both ends into an octahedron of semi-sintered MgO (Fig. 17). The furnace is surrounded by a sleeve and end plugs of zirconia which serve as thermal insulators to surround the sample; two MgO washers fill the remaining space inside the furnace. Holes in the zirconia end plugs accommodate an acoustic buffer rod and thermocouple feedthroughs. On one end, an alumina rod (Coors brand 998—3.2-mm diameter, 3.9-mm length) serves as another acoustic buffer rod; it is surrounded by metal foil (such as Pt or Cu) to provide electrical contact between the WC cube and the graphite furnace. On the other end of the cell, a piece of two-hole mullite tubing (1.7-mm diameter) contains the thermocouple wires (W3%Re/W25%Re); a Pt (or TZM) ring outside the mullite serves as the other furnace electrode. To protect the sample, the thermocouple junction does not touch the sample, but instead is located at the bottom of the NaCl cup in a position symmetrical to that of the sample/buffer rod interface. The high thermal conductivity of NaCl should reduce the temperature gradient across the 1.5-mm long sample. From measurements in a hot-pressing cell of similar design, we estimate the temperature gradient in the sample to be ~15 K/mm (Gwanmesia and Liebermann 1992).

Polycrystalline forsterite. For our pilot studies at high P and T, we used two

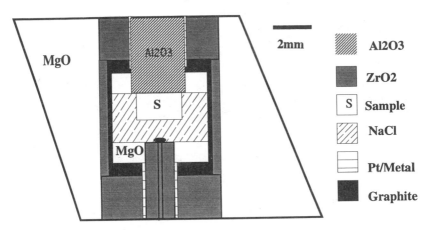

Figure 17. Modified cell assembly for simultaneous high-pressure and high-temperature velocity measurements. W3%Re/W25%Re thermocouples are used (following Li et al. 1998a).

polycrystalline specimens of forsterite which were hot-pressed in either the USSA-2000 USCA-1000 apparatus. Figure 18 shows an example of the S wave signals at 2 GPa and 673 K. The three strong echoes are the reflections from the ends of the WC cube, the alumina buffer rod, and the sample. The weak reflections following the sample echo are another cycle of the buffer rod and sample reflections plus possibly reflections from the bottom of the NaCl cup.

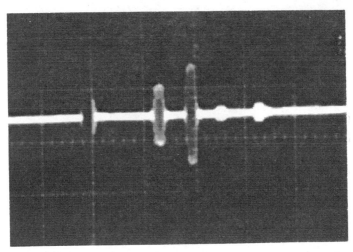

Figure 18. The S wave acoustic signals at 2 GPa and 673 K for a polycrystalline forsterite sample. Three strong echoes from left to right are from the first WC cube, first alumina buffer rod, and the first sample echoes. The weak signals are the second alumina buffer rod, sample and NaCl echoes (following Li et al. 1998a).

Travel-time measurements have been performed along a specially designed P-T path to peak P and T conditions of 10 GPa and 1273 K (see path A→B→C→D→E in Fig. 19). No dimensional changes are observed in our specimens after the acoustic experiments, and the polished surfaces remain optically flat. The cell pressure at the peak oil pressure (400 bars)

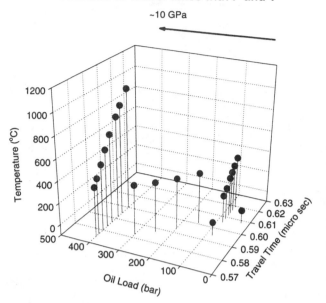

Figure 19. S wave travel times measured along specially designed P-T path. (A) Room conditions; A-B: pressurizing to ~2 GPa at room T; (B-C) preheating to ~400°C at ~2 GPa; C-D: further pressuring to ~10 GPa at temperature close to 400°C; (D-E) heating at constant load to peak temperature of 1000°C (following Li et al. 1998a).

of the experiment was calibrated by the observation of the coesite/stishovite phase transformations according to the revised phase boundary by Zhang et al. (1996) and Liu et al. (1996) following the same P-T path as discussed above. Phase transformations were obtained at about 1523 K which constrains the cell pressure to be 9.6 GPa for oil pressures of 400 bar (~300 tons).

Ultrasonics in multi-anvil apparatus at simultaneous high P and T in conjunction with synchrotron x-radiation

In the section above, we described our first experiments on ultrasonic velocities in polycrystalline olivine at simultaneous temperatures of 1300 K and pressures of 10 GPa (Li et al. 1998a; see also Knoche et al. 1998). While these experiments dramatically expanded the range of P-T space which can be explored in ultrasonic studies of the elasticity of minerals and ceramics, they are subject to limitations, including: (1) Uncertainty in cell pressures; calibration at high temperature is difficult due to the influence of the P-T-t path utilized in the experiment and the necessity of relying previous calibrations based on phase transitions in materials (e.g. Zhang et al. 1996). (2) Length changes of the specimen at high P and T must be known to convert the observed travel times to velocities; at high temperature, this requires knowledge of the volume of the specimen.

In this section, we describe recent progress in our laboratory to couple the advances in sound velocity techniques in multi-anvil apparatus with advances in large-volume, high-pressure studies at the National Synchrotron Light Source (NSLS) of the Brookhaven National Laboratory (Weidner et al. 1992). We illustrate this progress with new data on polycrystalline specimens of the wadsleyite (ß) phase of Mg_2SiO_4 (Li et al. 1998b,c) and $MgSiO_3$-perovskite (Sinelnikov et al. 1998), as well as single crystals of MgO (Chen et al. 1998) conducted at conditions as high as pressures of 8 GPa and temperatures of 1600 K.

The ultrasonic techniques developed for the USCA-1000 apparatus have been adapted for use in a DIA-type, cubic-anvil apparatus (SAM 85) and installed on the

P-V-Vp-Vs-T in DIA-type Apparatus (SAM85)

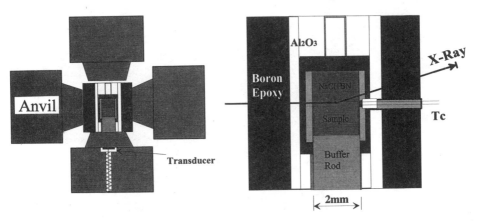

Figure 20. Schematic diagram of the cubic anvil apparatus (SAM 85) and cell assembly utilized for simultaneous ultra-sonic and x-ray experiments in studies by Li et al. (1998b), Chen et al. (1998a), and Sinel-nikov et al. (1998a).

superconducting wiggler beamline (X17B) at NSLS (Weidner et al. 1992). In Figure 20, we show the acoustic piezoelectric transducer-tungsten carbide anvil arrangment and the high-temperature cell assembly used in SAM 85. The LiNbO$_3$ transducers (40 MHz, 36° Y-cut for compressional wave and 41° X-cut for shear waves) are mounted onto the backside of the WC anvil (Toshiba Grade F) with Aremco Crystalbond, and connected to the interferometer by coaxial cables. The WC anvil serves as an acoustic buffer rod to transmit the high-frequency signal (20 to 70 MHz) into the cell assembly.

The polycrystalline or single-crystal sample is centered within the cubic cell assembly (Fig. 20) and is surrounded by a boron nitride sleeve. The acoustic signal into the sample via another buffer rod of either fused silica glass or alumina ceramic. The NaCl disc serves two important purposes: (1) a pseudo-hydrostatic pressure environment for the sample; and (2) a pressure standard at room and high temperature using the Decker et al. (1972) equation of state. A tungsten rhenium thermocouple monitors the temperature; temperatures are reproducible within 40 K for the same power to the graphite furnace. x-ray spectra of both the polycrystalline specimen and the NaCl medium which surrounds it are monitored continuously (Fig. 21); the former provides PVT data to compliment the velocity measurements and the latter the pressure standard; the sharp diffraction spectrum from the polycrystalline sample is typical of those observed throughout the range of pressures and temperatures of our experiments for those synthesized in multi-anvil apparatus in our laboratory and demonstrates the well-sintered nature of these samples (Gwanmesia and Liebermann 1992, Gwanmesia et al. 1993, Li et al. 1996c).

These experiments provide direct measurements of both the velocities (V_p and V_s) and the volume (V, and thus the length) of samples at high P and T; consequently, we call these P-V-V_p-V_s-T, or P-V^3-T, experiments. Such experiments can be used to provide modifed absolute pressure scales as well as offering the promise of improving our ability to interpret seismic models of the Earth's interior.

Polycrystalline forsterite and wadsleyite. An excellent example of the comprehensive nature of these experiments is provided by the work on polycrystalline specimens of the wadsleyite (or ß) phase of Mg$_2$SiO$_4$. (Li et al. 1998b,c). A representative

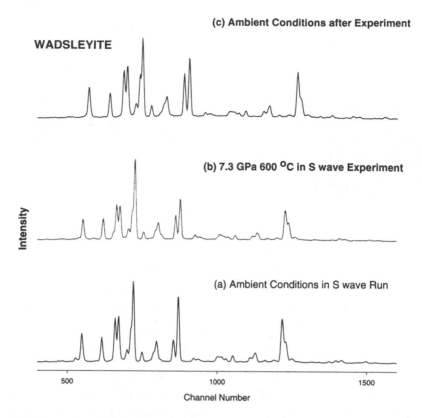

Figure 21. Comparison of x-ray diffraction spectra in S wave experiments at (a) ambient conditions; (b) 7.3 GPa and 873 K; (c) ambient conditions after recovery from high P-T experiment.

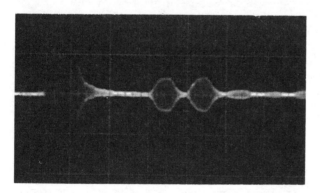

Figure 22. Acoustic echo train for P waves in polycrystalline wadsleyite at 4 GPa and 573 K. The first echo is from the WC anvil, the second from the glass buffer rod, and the third from the polycrystalline sample (see Fig. 20 for reference).

acoustic echo pattern is shown in Figure 22. The experimental P-T paths for the separate S and P wave experiments is shown in Figure 23. In the S wave run, the sample was first pressurized at room temperature to the peak value of ~7 GPa, and then heated to 873 K at constant ram load. During heating, the pressure initially increased slightly and then decreased above 473 due to relaxation of the non-hydrostatic stresses in the solid-media cell assembly (Fig. 20). In the P wave experiment, we performed a pre-heating to 573 K at 3.8

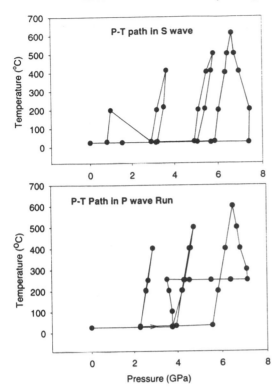

Figure 23. P-T-t paths for acoustic experiments on polycrystalline wadsleyite using the SAM 85 apparatus (following Li et al. 1998c).

GPa before further pressurization to ~7 GPa; from the improvement in the signal to noise ratio of the acoustic signals, we conclude that such pre-heating reduces the risk of damage to the sample when it is exposed to high P and T conditions. After reaching peak P and T conditions, both P and S wave experiments were continued by a series of cooling/heating cycles accompanied by gradual decompression (see Fig. 23). The samples recovered from each experiment were examined in detail; no changes in dimension reflecting permanent deformation were observed, nor were any macroscopic cracks evident. Moreover, the well-polished surfaces of the sample were maintained, thus allowing the samples to be used for subsequent experiments without re-polishing.

In Figure 24, the cell volumes of the wadsleyite sample are plotted versus pressure for various temperatures to 873 K. These data can be analyzed independently to obtain the parameters of the P-V-T equation of state of wadsleyite (as in the Fei et al. 1992 and Meng et al. 1993 experiments); they also provide direct measurement of the change of length of the polycrystalline sample under the assumption that its compression and expansion is isotropic and uniform.

Travel-time data for the S wave experiments over the entire range of P and T explored is illustrated in Figure 25 and converted to shear velocities in Figure 26 using the length determinations above. With these P-V^3-T data, Li et al. (1998b,c) have provided direct measurements of the adiabatic bulk (K_S) and shear (G) moduli and their first pressure and temperature derivatives.

For the wadsleyite study, the experimental range of P and T was limited by the thermodynamic stability of this high-pressure polymorph (which is metastable below pressures of ~13 GPa). Consequently, the temperatures were kept to less than 873 K to

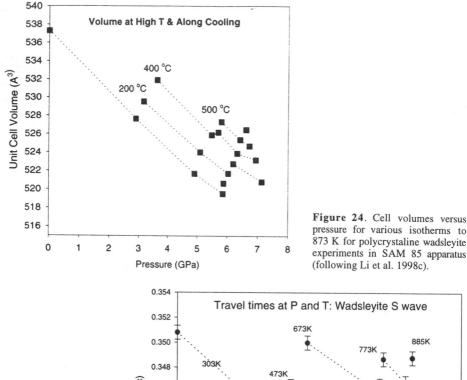

Figure 24. Cell volumes versus pressure for various isotherms to 873 K for polycrystaline wadsleyite experiments in SAM 85 apparatus (following Li et al. 1998c).

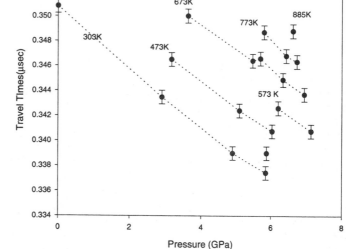

Figure 25. Travel-times of S waves vs. pressure for various isotherms to 873 K for polycrystalline wadsleyite experiments in SAM 85 apparatus (following Li et al. 1998c).

prevent back-transformation to the olivine phase.

Single-crystal MgO. The SAM 85 apparatus is, however, capable of much higher temperatures in this pressure range. This is illustrated in Figure 27 which shows the P-T field examined in the ultrasonics study of single-crystal MgO by Chen et al. (1998): temperatures approaching 1600 K at pressures up to 8 GPa were attained. By combining the travel-time data for three acoustic modes in single-crystal MgO with the high-precision x-ray diffraction data of Utsumi et al. (1998) for polycrystalline MgO obtained in the same SAM 85 apparatus over a comparable wide P-T range, Chen et al. (1998) were able to determine the cross pressure and temperature derivatives of the elastic moduli $(\partial^2 C_{ij}/\partial P \partial T)$. Their results indicate that the effect of cross pressure and temperature

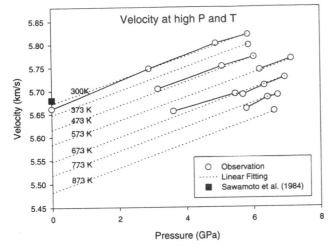

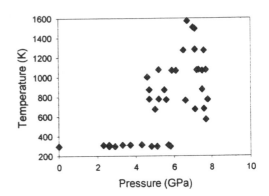

Figure 26. Shear wave velocities versus pressure for various isotherms to 873 K for polycrystalline wadsleyite experiments in SAM 85 apparatus (following Li et al. 1998c).

Pressure-Temperature Coverage for the MgO Ultrasonic Experiment

Figure 27. Pressure and temperature range of ultrasonic experiments on the elasticity of single crystal MgO by Chen et al. (1998a), using the SAM 85 apparatus installed on the superconducting wiggler beamline at the National Synchrotron Light Source at the Brookhaven National Laboratory.

dependence on the behavior of C_{11} and C_{44} is very different. Where as the cross derivative $(\partial^2 C_{11}/\partial P \partial T)$ is about 10^{-3}/K, the cross derivative for the C_{44} mode $(\partial^2 C_{44}/\partial P \partial T)$ is an order of magnitude smaller in absolute value. As a consequence, there is excellent agreement between the new high-pressure data of Chen et al. (1998) and the previous high-temperature data of Isaak et al. (1989b) after correction for the effect of pressure (see Fig. 28). The high P-T data of Chen et al. (1998a) also can be used to evaluate the evolution of the elastic anisotropy of the cubic symmetry phase of MgO; this anisotropy decreases with increasing pressure at ambient temperature, but then increases as temperature is increased at high pressure.

SUMMARY

We have reported on the state-of-the-art in experimental techniques to determine the elasticity of deep Earth materials at elevated pressure sand temperatures. Advances over the past 10 years make it possible to conduct laboratory studies at conditions approaching those of the transition zone (depths of 400 to 700 km). In this paper, we have summarized these technological advances, and discussed in detail the progress in the application of ultrasonic interferometric techniques in multi-anvil, high-pressure apparatus to measure elastic wave

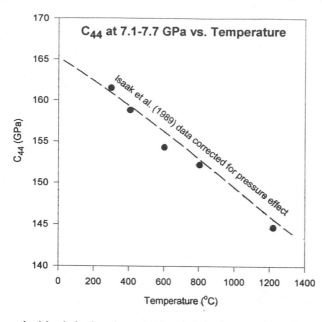

Figure 28. Elastic modulus C_{44} of single-crystal MgO as a function of temperature at 7.1-7.7 GPa in comparison with room pressure data of Isaak et al.(1989) (following Chen et al. 1998a).

velocities in both polycrystalline and single-crystal specimens.

ACKNOWLEDGMENTS

It is a pleasure to thank our colleagues and collaborators who have participated in the evolution of the Stony Brook High Pressure Laboratory's ultrasonics program over the past 7 years: Gabriel Gwanmesia for initiating the hot-pressing of polycrystals at high pressures and temperatures, Ian Jackson and Tibor Gasparik for assistance in the adaptation of ultrasonic interferometry techniques to multi-anvil apparatus, Ganglin Chen for leading the interfacing of the ultrasonic experiments with the synchrotron x-rays, and graduate students Joseph Cooke, Lucy Flesch, Jun Liu and Yegor Sinelnikov for their participation. We are especially grateful to the SAM 85 team of Jiuhua Chen, Michael Vaughan and Donald Weidner for providing us with the facilities and technical support to perform the experiments at the NSLS of Brookhaven National Laboratory. Their efforts have led to the technological developments that have made possible the new generation of ultrasonic measurements of the elasticity of mantle minerals at elevated conditions of pressure and temperature. These high-pressure experiments were conducted in the Stony Brook High Pressure Laboratory, which is jointly supported by the State University of New York at Stony Brook and the NSF Science and Technology Center for High Pressure Research (CHiPR: EAR 89-20239). This research is also supported by NSF grants to RCL (EAR 93-04502 and 96-14612). This is Mineral Physics Institute contribution number 251.

REFERENCES

Abramson EH, Brown JM, Slutsky LJ, Zaug J (1997) The elastic constants of San Carlos olivine to 17 GPa. J Geophys Res 102:12,253-12,263
Anderson OL (1965) Determination and some uses of isotropic elastic constants of polycrystalline aggregates, using single-crystal data. *In:* Physical Acoustics IIIB:43-95 WP Mason (ed) Academic Press, New York
Anderson OL, Liebermann RC (1968) Sound velocities in rocks and minerals: Experimental methods extrapolations to very high pressures: results. *In:* Physical Acoustics IVB:330-472 WP Mason (ed)

Academic Press

Anderson OL, Schreiber E. Liebermann RC, Soga N (1968) Some elastic constant data on minerals relevant to geophysics. Rev Geophys 6:491-524

Anderson OL and. Isaak DG (1995) Elastic constants of Mantle Minerals at High Temperature. *In:* Mineral Physics and Crystallography: A Handbook of Physical Constants. TJ Ahrens (ed) Am Geophys Union, Washington, DC, p 64-80

Bass JD (1995) Elasticity of minerals, glasses, and melts. *In:* Mineral Physics and Crystallography: A Handbook of Physical Constants. TJ Ahrens (ed) Am Geophys Union, Washington, DC, p 46-63

Bassett WA, Shimizu H, Brody EM (1982) Pressure dependence of elastic modulit of forsterite by Brillouin Scattering in the diamond cell. p 115-124 *In:* High-Pressure Research in Geophysics. S Akimoto, MH Manghnani (eds) Center for Academic Publications, Tokyo and D Reidel, Boston

Brown JM, Slutsky LJ, Nelson KA, Cheng L-T (1989) Single-crystal elastic constants for San Carlos peridot: An application of impulsive stimulated scattering. J Geophys Res 95:9485-9492

Chai M, Brown JM, Slutsky LJ (1997a) The elastic constants of the pyrope-grossular-almandine garnet to 20 GPa. Geophys Res Lett 24:523-526

Chai M Brown JM, Slutsky LJ (1997b) The elastic constants of an aluminous orthopyroxene to 12.5 GPa. J Geophys Res 102:14,779-14,785

Chen G, Li B, Liebermann RC (1996a) Selected elastic moduli of single-crystal olivines from ultrasonic experiments in mantle pressures. Science 272:979-980

Chen G, Yoneda A, Getting IC, Spetzler HA (1996b) Cross pressure and temperature derivatives of selected elastic moduli for olivine from gigahertz ultrasonic interferometry. J Geophys Res 101:25,161-25,171

Chen G, Liebermann RC, Weidner DJ (1998) Elasticity of single-crystal MgO to 8 Gigapascals and 1600 Kelvin. Science 280:1913-1916

Chopelas A (1996a) The fluorescence sideband method for obtaining acoustic velocities at high compressions: Application to MgO and $MgAl_2O_4$. Phys Chem Minerals 23:25-37

Chopelas A, Reichmann HJ, Zhang L (1996) Sound velocities of fived minerals to mantle pressure determined by the sideband fluorescence methods. *In:* Mineral Spectroscopy: A Tribute to Roger G. Burns. MD Dyar, C McCammon, MW Schaefer (eds) Geochemical Soc Spec Publ No 5:229-242

Duffy TS, Anderson DL (1989) Seismic velocities in mantle minerals and the mineralogy of the upper mantle. J Geophys Res 94:1895-1912

Duffy TS, Zha C-s, Downs RT, Mao H-k, Hemley RJ (1995a) Elasticity of forsterite to 16 GPa and the composition of the upper mantle. Nature 378:170-173

Duffy TS, Shen G, Heinz DL, Ma Y, Hemley RJ, Singh AK (1998) Lattice strains in gold and rhenium under non-hydrostatic compression. Proc Materials Res Soc 1997 Fall Meeting (in press)

Fei Y, Mao H-k, Shu J, Parthasarathy G, Bassett WA, Ko J (1992) Simultaneous high-P and high-T x-ray diffraction study of β-$(Mg,Fe)_2SiO_4$ to 26 GPa and 900 K. J Geophys Res 97:4489-4495

Fujisawa H (1998) Elastic wave velocities of forsterite and its ß-spinel form and chemical boundary hypothesis for the 410-km discontinuity. J Geophys Res:103:9591-9608

Fukizawa A, Kinoshita H (1982) Shear wave velocity jump at the olivine-spinel transformation in Fe_2SiO_4 by ultrasonic measurements in situ. J Phys Earth 30:245-253

Gieske JH, Barsch GR (1968) Pressure dependence of the elastic constants of single-crystalline aluminum oxide. Phys Stat Sol 29:121-131

Goto T, Anderson OL (1988) Apparatus for measuring elastic constants of single crystals by a resonance technique up to 1825 K. Rev Sci Instrum 59:1405-1408

Goto T, Anderson OL, Ohno I, Yamamoto S (1989) Elastic constants of corundum up to 1825 K. J Geophys Res 94:7588-7602

Graham EK Jr, Barsch GR (1969) Elastic constants of single-crystal forsterite as a function of temperature and pressure. J Geophys Res 74:5949-5959

Gwanmesia GD, Liebermann RC, Guyot F (1990a) Hot-pressing and characterization of polycrystals of β-Mg_2SiO_4 for acoustic velocity measurements. Geophys Res Lett 17:1331-1334

Gwanmesia GD, Rigden SM, Jackson I, Liebermann RC (1990b) Pressure dependence of elastic wave velocity for ß-Mg_2SiO_4, and the composition of the Earth's mantle. Science 250:794-797

Gwanmesia GD, Liebermann RC (1992) Polycrystals of high-pressure phases of mantle minerals: Hot-pressing and characterization of physical properties, High Pressure Research: Application to Earth and Planetary Sciences. Y Syono, MH Manghnani (eds) Am Geophys Union, Washington, DC, p 117-135

Gwanmesia GD, Li B, Liebermann RC (1993) Hot pressing of polycrystals of high-pressure phases of mantle minerals in multi-anvil apparatus. Pure Applied Geophys 141:467-484

Isaak DG (1992) High-temperature elasticity of iron-bearing olivines. J Geophys Res 97:1871-1885

Isaak DG, Graham EK (1976) The elastic properties of an almandine-spessartine garnet and elasticity in the garnet solid solution series. J Geophys Res 81:2483-2489

Isaak DG, Anderson OL, Goto T (1989a) Elasticity of single-crystal forsterite measured to 1700 K. J Geophys Res 94:5895-5906

Isaak DG, Anderson OL, Goto T (1989b) Measured elastic moduli of single-crystal MgO up to 1800 K. Phys Chem Minerals 16:704-713

Isaak DG, Carnes JD, Anderson OL (1998) Elasticity of fused silica spheres under pressure using resonant ultrasound spectroscopy (RST) measurements at high pressure for fused silica. J Acoust Soc (in press)

Jackson I, Niesler H (1982) The elasticity of periclase to 3 GPa and some geophysical implications. p 93-113. *In:* High-Pressure Research in Geophysics. S Akimoto, MH Manghnani (eds) Center for Academic Publications, Tokyo and D Reidel, Boston

Jackson I, Rigden SM (1996) Analysis of P-V-T data: Constraints on the thermoelastic properties of high-pressure minerals. Phys Earth Planet Inter 96:85-112

Jackson I, Rigden SM (1998) Composition and temperature of the Earth's mantle: Seismological models interpreted through experimental studies of Earth materials. *In:* The Earth's Mantle, Composition, Structure and Evolution. I Jackson (ed) Cambridge University Press, p 405-460

Jackson I Niesler H, Weidner DJ (1981) Explicit correction of ultrasonically determined elastic wave velocities for transducer-bond phase shift. J Geophys Res 86:3736-3748

Karki BB (1997) Structure and elasticity of MgO at high pressure. Am Mineral 82:51-60

Kinoshita H, Hamaya N, Fujisawa H (1979) Elastic properties of single-crystal NaCl under high pressures to 80 kbar. J Phys Earth 27:337-350

Knittle E (1995) Static compression measurements of equations of state. *In:* Mineral Physics and Crystallography: A Handbook of Physical Constants. TJ Ahrens (ed) Am Geophys Union, Washington, DC, p 98-142

Knoche R, Webb SL, Rubie DC (1998) Measurements of acoustic wave velocities at P-T conditions of the Earth's mantle. *In:* Properties of Earth and Planetary Materials at High Pressure and Temperature. Geophys Monogr 101:119-128 Am Geophys Union, Washington, DC

Kumazawa M, Anderson OL (1969) Elastic moduli, pressure derivatives, and temperature derivatives, of single-crystal olivine and single-crystal forsterite. J Geophys Res 74:5961-5972

Kusaba K, Galoisy L, Wang Y, Vaughan MT, Weidner DJ (1993) Determination of phase transition pressures of ZnTe under quasi-hydrostatic conditions. *In:* Experimental Techniques in Mineral and Rock Physics. *In:* RC Liebermann, CH Sondergeld (eds) PAGEOPH 141:644-652

Li B, Jackson I, Gasparik T, Liebermann RC (1996a) Elastic wave velocity measurement in multi-anvil apparatus to 10 GPa using ultrasonic interferometry. Phys Earth Planet Inter 98:79-91

Li B, Gwanmesia GD, Liebermann RC (1996a) Sound velocities of olivine and beta polymorphs of Mg_2SiO_4 at Earth's transition zone pressures. Geophys Res Lett 23:2259-2262

Li B, Rigden SM, Liebermann RC (1996c) Elasticity of stishovite at high pressure. Phys Earth Planet Inter 96:113-127

Li B, Chen G, Gwanmesia GD, Liebermann RC (1998a) Sound velocity measurements at mantle transition zone conditions of pressure and temperature using ultrasonic interferometry in a multianvil apparatus. *In:* Properties of Earth and Planetary Materials at High Pressure and Temperature. Geophys Monogr 101:41-61 Am Geophys Union, Washington, DC

Li B, Liebermann RC, Weidner DJ (1998b) Elastic moduli of wadsleyite (ß-Mg_2SiO_4) to 7 gigapascals and 873 Kelvin. Science 281:675-677

Li B, Liu J, Flesch L, Liebermann RC, Weidner DJ (1998c) P-V-V_p-V_s-T measurements on wadsleyite to 7 GPa and 873 K: Implications for the 410-km seismic discontinuity. J Geophys Res (submitted)

Liebermann RC (1998) Elasticity of mantle minerals: Experimental studies. Chapter 7 in Mineral Physics and Seismic Tomography. S Karato (ed) Am Geophys Union Monograph (in preparation)

Liebermann RC, Wang Y (1992) Characterization of sample environment in a uniaxial split-sphere apparatus. *In:* High Pressure Research: Application to Earth and Planetary Sciences Y Syono, MH Manghnani (eds) Terra Scientific Publishing and Am Geophys Union, Tokyo and Washington, DC, p 19-31

Liebermann RC, Ringwood AE, Mayson DJ, Major A (1975) Hot-pressing of polycrystalline aggregate at very high pressure for ultrasonic measurements. *In:* Proc 4th Conf on High Pressure. J Osugi (ed) Physico-Chemical Soc Japan, Tokyo, p 495-502

Liebermann RC, Chen G, Li B, Gwanmesia GD, Chen J, Vaughan MT, Weidner DJ (1998) Sound velocity measurements in oxides and silicates at simultaneous high pressures and temperatures using ultrasonic techniques in multi-anvil apparatus in conjunction with synchrotron x-radiation determination of equation of state. Rev High Pressure Sci Technol 7:75-78

Liu J, Topor L, Zhang J, Navrotsky A, Liebermann RC (1996) Calorimetric study of coesite stishovite transformation and calculation of the phase boundary. Phys Chem Minerals 23:11-16

Lloyd EC (1971) Accurate Characterization of the high pressure environment, Natl Bur Standards Spec Publ 326:1-3

McSkimin HJ (1950) Ultrasonic measurement techniques applicable to small solid specimens. J Acoustic Soc Am 22:413-418

Mao H-k, Chen LC, Hemley RJ, Jephcoat AP, Wu Y, Bassett WA (1989) Stability and equation of state of $CaSiO_3$-perovskite to 134 GPa. J Geophys Res 94:17,889-17,894

Mao H-k, Wu Y Chen LC, Shu JF, Jephcoat AP (1990) Static compression of iron to 300 GPA and $Fe_{0.8}Ni_{0.2}$ alloy to 260 GPa: Implications for composition of the core. J Geophys Res 95:21,737-21,742

Mao H-k, Shu J, Shen G, Hemley RJ, Li B, Singh AK (1998) Elasticity and rheology of iron above GPa: Constraints on the Earth' s inner core. Nature (submitted)

Meng Y, Weidner DJ, Gwanmesia GD Liebermann RC, Vaughan MT, Wang Y, Leinenweber K, Pacalo RE, Yeganeh-Haeri A, Zhao Y (1993) In-situ high P-T x-ray diffraction studies on three polymorphs (a, b, g) of Mg_2SiO_4. J Geophys Res 98:23199-23207

Niesler H, Jackson I (1989) Pressure derivatives of elastic wave velocities from ultrasonic interferometric measurements on jacketed polycrystals. J Acoustic Soc Am 86:1573-1585

Niesler H, Fisher G (1992) Technique for ultrasonic mesurement of samples at 2.0 GPa and 600°C. High Temperature-High Pressure 24:65-74

Nye JF (1957) Physical Properties of Crystals: Their Representation by Tensors and Matrices. Clarendon Press, Oxford, 322 p

Parise JB, Weidner DJ, Chen J, Liebermann RC, Chen G (1998) In situ studies of the properties of materials under high-pressure and temperature conditions using multi-anvil apparatus and synchrotron x-rays. Ann Rev Mater Sci 28:349-374

Reichmann HJ, Angel RJ, Spetzler H, Bassett WA (1998) Ultrasonic interferometry and x-ray measurements on MgO in a new diamond anvil cell. Am Mineral (in press)

Rigden SM, Gwanmesia GD, FitzGerald JD, Jackson I, Liebermann RC (1991) spinel elasticity and seismic structure of the transition zone of the mantle. Nature 354:143-145

Rigden SM, Gwanmesia GD, Jackson I, Liebermann RC (1992) Progress in high-pressure ultrasonic interferometry, the pressure dependence of elasticity of Mg_2SiO_4 polymorphs and constraints on the composition of the transition zone of the Earth's mantle. *In:* High Pressure Research: Application to Earth and Planetary Sciences Y Syono, MH Manghnani (eds) Terra Scientific Publishing and Am Geophys Union, Tokyo and Washington, DC, p 167-182

Rigden SM, Gwanmesia GD, Liebermann RC (1994) Elastic wave velocities of pyrope-majorite garnet to 3 Gpa. Phys Earth Planet Inter 86:35-44

Sasakura T, Suito K, Fujisawa H (1989) Measurement of ultrsonic wave velocities in fused quartz under hydrostatic pressures up to 6.0 Gpa. Proc XIth Int'l Conf, Int'l Assoc Advancement of High Pressure Science and Technology (AIRAPT) 2:60-72, Kiev Naukova Dumka, USSR

Sasakura T, Yoneda H, Suito K, Fujisawa H (1990) Variations of the elastic constants of InSb near the covalent-metallic transition. High Pressure Res 4:318-320

Schreiber E, Anderson OL (1966) Pressure derivatives of the sound velocities of polycrystalline alumina. J Am Ceramic Soc 49:184-190

Schreiber E, Anderson OL, Soga N (1973) Elastic Constants and Their Measurement. McGraw-Hill, New York, p 196

Shen AH, Reichmann H-J, Chen G, Angel RJ, Bassett WA, Spetzler H (1998) GHz ultrasonic interferometry in a diamond anvil cell: P-wave velocities in periclase to 4.4 GPa and 207°C. *In:* Properties of Earth and Planetary Materials at High Pressure and Temperature Geophysical Monograph 101:71-77, Am Geophys Union, Washington, DC

Sinelnikov YD, Chen G, Neuville DR, Vaughan MT, Liebermann RC (1998a) Ultrasonic shear velocities of $MgSiO_3$-perovskite at 8 GPa and 800 K and lower mantle composition. Science 281:677-679

Singh AK, Mao H-k, Shu J, Hemley RJ (1998) Estimation of single-crystal elastic moduli from polycrystalline x-ray diffraction at high pressure: Application to FeO and iron. Phys Rev Lett 80:2157-2160

Sinogeikin SV, Bass JD (1998) Single-crystal elasticity of synthetic pyrope and MgO by Brillouin scattering to 20 Gpa. Eos Trans Am Geophys Union 79:S163

Spetzler HA (1970) Equation of state of polycrystalline and single-crystal MgO to 8 kilobars and 800K. J Geophys Res 75:2073-2087

Spetzler HA, Sammis CG, O'Connell RJ (1972) Equation of state of NaCl: Ultrasonic measurements to 8 Kbar and 800°C and static lattice theory. J Phys Chem Solids 33:1727-1750

Spetzler H, Schreiber E, O'Connell RJ (1972b) Effect of stress-induced anisotropy and porosity on elastic properties of polycrystals. J Geophys Res 77:4938-4944

Spetzler HA, Chen G, Whitehead S, Getting IC (1993) A new ultrasonic interferometer for the determination of equation of state parameters of sub-millimeter single crystals. Pure Applied Geophys 141:341-377

Spetzler HA, Shen A, Chen G, Herrmannsdoerfer G, Schulze H, Weigel R (1996) Ultrasonic measurements in a diamond anvil cell. Phys Earth Planet Inter 98:93-99

Truell R, Elbaum C, Chick BB (1969) Ultrasonic Methods in Solid State Physics. Academic Press, New York, 464 p

Utsumi W, Weidner DJ, Liebermann RC (1998) Volume measurement of MgO at high pressures and temperatures. *In:* Properties of Earth and Planetary Materials at High Pressure and Temperature. Geophys Monogr 101:327-333 MH Manghnani, T Yagi (eds) Am Geophys Union

Watt JP (1988) Elastic properties of polycrystalline minerals: Comparison of theory and experiment. Phys Chem Minerals 15:579-587

Watt JP, Davies GF, O'Connell RJ (1976) The elastic properties of composite materials. Rev Geophys Space Phys 14:541-563

Webb S (1989) The elasticity of the upper mantle orthosilicates olivine and garnet to 3 GPa. Phys Chem Minerals 16:684-692

Weidner DJ (1987) Elastic properties of rocks and minerals. *In:* Methods of Experimental Physics, 24A: 1-30, Academic Press, New York

Weidner DJ, Bass JD, Ringwood AE, Sinclair E (1982) The single-crystal elastic moduli of stishovite. J Geophys Res 87:4740-4746

Weidner DJ, Vaughan MT, Ko J, Wang Y, Liu X, Yeganeh-Haeri A, Pacalo RE, Zhao Y (1992) Characterization of stress, pressure, and temperature in SAM85, a DIA type high Pressure apparatus. *In:* High Pressure Research: Application to Earth and Planetary Sciences Y Syono, MH Manghnani (eds) Terra Scientific Publishing and Am Geophys Union, Tokyo and Washington, DC, p 13-17

Weidner DJ, Wang Y, Vaughan MT (1994) Yield strength at high pressure and temperature. Geophys Res Lett 21:753-756

Yoneda, A (1990) Pressure derivatives of elastic constants of single-crystal MgO and $MgAl_2O_4$. J Phys Earth 38:19-55

Yoneda A, Morioka M (1992) Pressure derivatives of elastic constants of single-crystal forsterite. *In:* High Pressure Research: Application to Earth and Planetary Sciences. Y Syono, MH Manghnani (eds) Terra Scientific Publishing and Am Geophys Union, Tokyo and Washington, DC, p 207-214

Zaug J Abramson E, Brown JM, Slutsky LJ (1992) Elastic constants, equations of state and thermal diffusivity at high pressure. *In:* High Pressure Research: Application to Earth and Planetary Sciences. Y Syono, MH Manghnani (eds) Terra Scientific Publishing and Am Geophys Union, Tokyo and Washington, DC, p 157-166

Zaug JM, Abramson EH, Brown JM, Slutsky LJ (1993) Sound velocities in olivine at earth mantle pressures, Science 260:1487-1489

Zha C-s, Hemley RJ, Mao H-k, Duffy TS, Meade C (1994) Acousti c velocities and refractive index of SiO_2 glass to 57.5 GPa by Brillouin scattering. Phys Rev B50:13105-13112

Zha C-s, Duffy TS, Downs RT, Mao H-k, Hemley RJ (1996) Sound velocity and elasticity of single-crystal forsterite to 16 GPa. J Geophys Res 101:17,535-17,545

Zha C-s, Mao H-k, Hemley RJ, Duffy TS (1997) Elasticity measurement and equation of state for MgO to 60 Gpa. Eos Trans Am Geophys Union 78:F752

Zha C-s, Duffy TS, Mao H-k, Downs RT, Hemley RJ, Weidner DJ (1997) Single-crystal elasticity of ß-Mg_2SiO_4, to the pressure of the 410 km seismic discontinuity of the Earth's mantle. Earth Planet Sci Lett 147:E9-E15

Zha C-s, Duffy TS, Downs RT, Mao H-k, Hemley RJ, Weidner DJ (1998a) Single-crystal elasticity of the and of Mg_2SiO_4 polymorphs at high pressure. *In:* Properties of Earth and Planetary Materials at High Pressure and Temperature. Geophysical Monogr 101:9-16 Am Geophys Union, Washington, DC

Zha C-s, Duffy TS, Downs RT, Mao H-k, Hemley RJ (1998b) Brillouin scattering and x-ray diffraction of San Carlos olivine: direct pressure determination to 32 Gpa. Earth Planet Sci Lett 159:25-33

Zhang J, Li B, Utsumi W, Liebermann RC (1996) In situ x-ray observations of the coesite-stishovite transition: reversed phase boundary and kinetics. Phys Chem Minerals 23:1-10

Chapter 16

RHEOLOGICAL STUDIES AT HIGH PRESSURE

Donald J. Weidner

*Center for High Pressure Research
and Department of Geoscience
State University of New York at Stony Brook
Stony Brook, New York 11794*

INTRODUCTION

The Earth is a dynamic planet. Earthquakes, volcanoes, plate motions, glacial rebound all attest to movements within the Earth. Our ability to predict stress fields in subduction zones, chemical mixing within the mantle, time scales of mantle convection all rest on our understanding of the manner in which Earth materials respond to stress. Indeed, the flow that enables Earth dynamics occurs in the mantle, a region that is predominantly in the solid state. Regions that transmit shear waves, often with little or no attenuation, must contribute to the motion of plates, offering the daunting task to the experimentalist of evaluating the nature of solid-state deformation within the time scale of the laboratory framework and inferring the salient characteristics that define Earth movements that span billions of years. Indeed this task includes defining the role of mantle pressures and for Earth minerals that are mostly stable only at high pressures.

Rheology, in its broadest sense, encompasses the entire response of a material to stress (Poirier 1985). This response can be purely elastic, anelastic, or plastic. Elastic responses include changes in shape, length, and volume that are both instantaneous and reversible. Anelastic responses are reversible—that is, the deformation returns to zero if the stress becomes zero—but the response is not instantaneous. There is a time lag between applied stress and deformation. Such processes dissipate energy, give rise to a finite Q (or quality factor) and result in dispersive acoustic propagation—or frequency dependent acoustic velocities. Plastic deformation is neither instantaneous nor reversible. Permanent deformation results from applied stresses. Generally, plastic deformation accompanies non-hydrostatic stress and requires a threshold value of stress to initiate the process. If no threshold value is required, then the material is considered viscous. In the simplest case, such as a fluid, the response to stress is instantaneous but not reversible.

This chapter focuses on plastic deformation of solids and the experimental advances of the last decade to explore these properties at pressures in excess of one GPa. Elastic responses at elevated pressures and temperatures are discussed by Duffy and Wang (this volume) and Liebermann and Li (this volume). In the following, we use the term rheological properties to indicate those related to permanent deformation of a solid.

It has been long recognized that the theoretical strength of a crystalline solid, one composed of a regular periodic, and undisturbed, array of atoms, far exceeds the experimentally determined strength. Indeed, it has been the imperfections of the atomic arrangement, the dislocations, that have been implicated in the reduction of strength. With a mechanical driving force, these imperfections are mobilized, causing a change in shape of the material, and when the driving force diminishes, these imperfections remain in their new configuration with the associated deformation in tact. Thus, the plastic response of a solid to applied stress depends on a wide array of variables. In addition to the normal 'state' variables of pressure, temperature, and crystal structure, these plastic properties

0275-0279/98/0037-0016$05.00

depend on the distribution and types of defects in the solid, the ability to exchange atoms, ions, or chemical complexes with the environment, as well as the shape and size of the grains. The history of the material becomes important. Thus, the rheological properties can no longer be considered state variables in the sense that we use this term in classical thermodynamics. Processes are irreversible, in that entropy is created during deformation. In this regard, the concepts of irreversible thermodynamics are embraced. The goal of achieving 'steady state' replaces that of 'equilibrium'. In the experimental laboratory, neither of these ideals is often realized. In fact, they may not be attained in the Earth. Nonetheless, the system becomes better defined and the processes contributing to the flow of a solid more easily identified and quantified if the steady state can be achieved.

Solids have several possible mechanisms of mobilizing the crystalline defects in order to accommodate the applied stress. Generally one of these mechanisms is dominant at any pressure and temperature. The goal of the experimentalist is to characterize the stress-strain history along with the operative mechanism of deformation. With this data in hand, one may then approximate the behavior of each mechanism to the Earth's time scale and stress levels by extrapolation. The key elements of a comprehensive experiment include a continuous measure of the deforming stress, a measure of the plastic strain and its time dependence, and an assessment of the deformation mechanism that is operative during the experiment. These measurements need to be carried out in a chemically controlled environment with the appropriate crystal structure of the target material in the pressure and temperature regime in question. In low pressure experiments, typical stress levels of 10s of MPas and strain rates of 10^{-7} sec^{-1} can be achieved. In the high pressure systems discussed here, such low levels are not yet achieved. We are still at the point where we can conduct a less than perfect experiment at high pressure. For example, multi-anvil systems are ideally designed to produce large (25 GPa) pressures hydrostatically. Introducing a deforming stress on a sample after it has been pressurized and heated to the desired state is still a difficult task. Measuring the magnitude of this deforming stress is even more difficult. However, a great deal of progress has been made and directions for future progress have been identified.

After a review of important physical concepts and constraints, we will examine several approaches to understanding the rheology of Earth materials at elevated pressure and temperature. Most of the studies do not yet meet the ideal. Still many insights have already been obtained.

PHYSICAL CONSTRAINTS

Stress

Central to a quantification of rheology is the fundamental understanding of the description and limitations of stress and strain. Generally, stress refers to the nature of the force field acting on the material and strain defines the deformation of the material. The material is viewed as a continuum in which stress and strain are evaluated at every point in the continuum. Here we do not consider the atomistic nature of the material, however, point defects or dislocations can be the source of the stress field, but the descriptions become valid only at some finite distance from these defects. Stress and strain must be treated as tensors, yet much of tensor theory is not necessary to present the fundamental principles. We simplify the analysis by restricting the description to a Cartesian coordinate system whose axial directions are labeled by subscripts of 1, 2, or 3. In this notation, the position of a point in the material is given by the vector from the origin of the coordinate system to the point as $\mathbf{x} = \{x_1, x_2, x_3\}$.

Stress is force per unit area acting at a point in the sample. To visualize this consider a

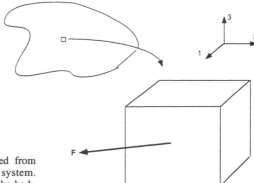

Figure 1. Small cube of material removed from sample with edges parallel to coordinate system. Force vector indicates the effect of the rest of the body across the face of the cube.

small cube removed from the sample around the point in question. If we make the edges of the cube parallel to the coordinate system as illustrated in Figure 1, we define the stress acting at the point as the force that the rest of the body exerts on each cube face divided by the area of the face. The force acting on each face is a vector quantity, $\mathbf{F}$, and the orientation of each face can be defined by a vector, namely its normal, $\mathbf{n}$, a vector of unit length pointing outward from the cube and perpendicular to the surface. The definition of stress needs to reflect both the orientation of the force and the orientation of the surface, requiring stress to be a second rank tensor, $\sigma = \{\sigma_{ij}\}$, where both i and j are 1, 2, or 3. Or:

$$\sigma_{ij} = lim_{V \to 0} \frac{F_i n_j}{A_j} \tag{1}$$

where A_j is the area of the face and V is the volume of the cube. This definition reflects the need for the forces on opposite faces of the cube to have opposite signs (as the normals will have opposite signs) in order that there be no unbalanced forces acting on the cube. The sign of the stress is the product of the sign of the force component and the sign of the outward normal yielding tension as positive and compression, negative. The further requirement that there be no torque on the cube leads to:

$$\sigma_{ij} = \sigma_{ji} \tag{2}$$

thus, even though the notation allows for 9 stress components, only 6 are independent.

Stress, which defines the force field within a material is related to a surface force per unit area, $\mathbf{S}$, applied at the surface with an outward normal, $\mathbf{n}$, by:

$$\sum_{j=1}^{3} \sigma_{ij} n_j = S_i \tag{3}$$

where the stress is evaluated just inside the body at the surface and Equation (3) represents three equations, one for each component of the vector, $\mathbf{S}$. Equation (3) allows us to relate the force applied to the surface of a material to the stresses that are generated. The stress in the body also includes contributions from 'internal' sources such as dislocations. Stress, being a point property, can vary within a body, but the requirement that the body is not accelerating leads to the constitutive relation:

$$\sum_{j=1}^{3} \frac{\partial \sigma_{ij}}{\partial x_j} = 0 \tag{4}$$

in the absence of any body forces, which are forces such as gravity that act on the volume of the material. Again, this represents three equations, that is for i = 1, 2, and 3.

The value of stress is indexed to a particular coordinate system. The values of the same stress field referenced to a different coordinate system follow the rotation rules for tensors. That is:

$$\sigma_{ij} = \sum_{m'=1}^{3} \sum_{n'=1}^{3} a_{im'} a_{jn'} \sigma_{m'n'} \tag{5}$$

where primes indicate a different coordinate system than the unprimed and the direction cosines between the axes is given by $a_{im'}$. Even though the individual terms of the stress tensor change in magnitude with differently oriented coordinate system, there are some features that remain constant on rotation. An important one is the trace, or the sum of the diagonal terms. Pressure is simply given by:

$$P = -1/3 \ \mathrm{tr}\{\sigma\} \tag{6}$$

and is thus well defined independently of the orientation of the reference coordinate system. The minus sign reflects the convention that pressure is considered positive for compression and stress is positive for tension.

The deviatoric stress is the total stress with 1/3 the trace subtracted from each of the diagonal terms. Thus, the trace of the deviatoric stress tensor is zero and the deviatoric stress reflects shape altering stresses with no pressure component included. Deviatoric stress is indeed the driving force field for flow.

For any stress field, there exists a coordinate system where the off-diagonal terms are zero and the diagonal terms are called the principal stresses. If the stress is the deviatoric stress, then the sum of the principal stresses is zero. Thus, stress can be represented in any coordinate system by six components, or it can be represented by the principal stresses and the orientation of this special coordinate system. In addition, a purely deviatoric stress can be represented in a coordinate system where all of the diagonal terms are zero and the off diagonal terms are non zero. This coordinate system indicates the shear stresses operative within the system and the orientation of the coordinate system reflects the orientation of the maximum shear stress.

Strain

Strain is a measure of the deformation of the material. As with stress, strain can be described by a tensor and is a point property. Deformation is a relational property, and requires a definition of an undeformed state. We define the position of a point in the undeformed body by **x** and the position of the same point after deformation is located at **y**. The displacement of the point between these two reference states is given by:

$$\mathbf{u} = \mathbf{y} - \mathbf{x} \tag{7}$$

using the same coordinate system to measure the position of the point both before and after the deformation. We have considerable freedom in choosing the reference state. Usually, we do so in order to reflect the phenomenon that we are considering. If we are evaluating the compressibility, then we may prefer to define the reference system to be that occurring at zero pressure. If we are considering dislocation flow in a solid at elevated pressure, then the reference state may be best served by a hypothetical body where there are no dislocations but evaluated at the pressure of interest. Ultimately, we want a strain metric, that reflects the deformation of the body. It should relate the undeformed state to the deformed state, but should be zero for displacements that translate or rotate the body. Since it is a point property, we should expect that it should reflect the positional changes in the local neighborhood of the point, hence involving gradients of displacement. A definition that meets all of these requirements is given by:

$$\gamma_{ij} = \frac{1}{2}\left[\frac{\partial u_i}{\partial y_j} + \frac{\partial u_j}{\partial y_i} - \sum_{k=1}^{3} \frac{\partial u_k}{\partial y_i}\frac{\partial u_k}{\partial y_j} \right] \tag{8}$$

This metric is called the Eulerian strain tensor. While this definition is adequate, it is not unique. Here the strain is referenced to the final position of the points. A Lagrangian description references the deformation to the initial position (x) with a similar, but slightly different equation defining strain. In the event of small displacements, the quadratic terms after the summation symbol become vanishingly small and all definitions converge to the more usual infinitesimal strain tensor. Here, we will simply refer to strain by the symbol, ε, unless we specifically need to distinguish Eulerian and Lagrangian strain.
From the definition,

$$\varepsilon_{ij} = \varepsilon_{ji} \tag{9}$$

In the small displacement limit, the diagonal terms reflect linear strains in directions parallel to the coordinate axes, and the off-diagonal terms indicate shear distortions within the plane defined by the coordinates of the subscripts (i.e. ε_{12} represents distortion in the plane containing the 1 and 2 coordinate axes).

As with stress, strain is a tensor and inherits all of the mathematical tools available to tensors. This includes coordinate system rotations illustrated in Equation (5), the ability to define a coordinate system with vanishing off-diagonals, and with invariants such as the trace. For small deformations, the trace is simply the volumetric strain.

Strain does not describe the interaction of the body with the rest of the world as directly as does stress through forces by virtue of Equation (3). Generally, strain need never be continuous across interfaces, rather displacement defines the interaction between two objects across an interface. Displacement can be continuous and its gradient, which is related to strain, can be discontinuous.

Material properties

The relation between force and deformation is a property of the material. In the elastic case for infinitesimal deformation, this becomes Hooke's law, given as:

$$\sigma_{ij} = \sum_{k=1}^{3}\sum_{l=1}^{3} c_{ijkl}\varepsilon_{kl} \tag{10}$$

where c_{ijkl} are the elastic stiffness coefficients. This relation is fundamental to wave propagation, and the elastic moduli are often determined from acoustic velocities as discussed by Liebermann and Li (this volume). The material is usually elastic even for finite volume changes associated with hydrostatic pressure. Then, using the Eulerian definition of strain, one obtains the Birch-Murnaghan relation discussed by Duffy and Wang (this volume) to describe the pressure-volume relationship.

Plastic deformation can be generally parameterized using the following:

$$\dot{\varepsilon} = \frac{\partial \varepsilon}{\partial t} = A\sigma^n \, exp\left\{\frac{-(Q*+PV*)}{RT}\right\} \tag{11}$$

where the dot indicates the time derivative, ε represents a measure of the deviatoric strain, σ is a measure of the deviatoric stress, A is a constant that may depend on grain size, $Q*$ represents an activation energy, P is pressure, $V*$ is an activation volume, R is the gas constant, and T is absolute temperature. The exponent, n, embodies the non-linear dependence of the stress-strain rate relationship. Ashby (1972) introduced a representation of the deformation mechanism as illustrated in Figure 2 for spinel (MgAl$_2$O$_4$). These diagrams indicate the shear stress vs temperature for several strain rates. The map is divided into regions that are defined by the dominant flow mechanism. The deformation map applies to a fixed grain size as some of the mechanisms, such as grain boundary diffusion, may be grain size dependent.

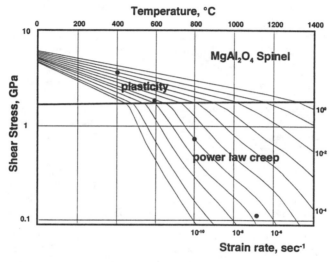

Figure 2. Deformation mechanism map for spinel (MgAl$_2$O$_4$), from Weidner et al. (1998). Stress-temperature relationships are indicated for specific strain rates.

Equation (11) can be used as an empirical relation for most of the regions of the deformation map. In particular the effective value of n can distinguish between competing mechanisms. In the deformation diagram, the value, n, reflects the number of decades of strain rate per decade of stress. However, the physical basis of Equation (11) is most appropriate to the power law creep region. In the low temperature – high stress regime, dislocation glide generally governs the plastic deformation and obstacles such as impurities or other dislocations limit the strain rate. In this regime, the stress levels must overcome a threshold value, termed the athermal stress, for flow to proceed. Frost and Ashby (1982) suggest a rate-equation for plasticity limited by such lattice resistance as:

$$\dot{\varepsilon} = A\sigma^2 \, exp\left\{\frac{-Q*}{RT}\left[1 - (\sigma/\tau)^{3/4}\right]^{4/3}\right\} \tag{12}$$

where τ is the athermal stress. When the deviatoric stress is lower than τ, flow will not occur unless the thermally activated processes of the power law creep regime are available.

Work hardening can occur in this region as the dislocations interact with each other and with impurities as the amount of deformation increases resulting in an increase in τ. Thus, the strength becomes dependent on the history of deformation and the dislocation density should increase with deformation.

In the power law region, where dislocation recovery mechanisms are operative, n assumes a value of 3-5. Through recrystallization or dislocation climb, the dislocations are able to overcome obstacles and continue to deform the sample. These recovery processes are often controlled by diffusion of ions or complexes. They result in a stress level that is very dependent on the rate of deformation. Experimental characterizations of these diffusion processes can provide information about the activation energy and volume for the flow law itself since flow is limited by such short range diffusion.

For regions where grain boundary or inter-crystalline diffusion is responsible for the deformation of the material, n obtains the value of 1. In this region the material behaves as a Newtonian fluid with a fixed viscosity. Since diffusion flow is sensitive to the diffusion path length, this mechanism depends on the grain size, smaller grains expanding the region where diffusion controls the deformation. It is expected that either power law creep or diffusional flow control the deformation processes in the most of the mantle, while the dislocation glide region may be important in the cold subducting slab.

The salient parameters in Equation (11), such as activation volume and energy, can be linked to the defect mobility in the material (Jaoul 1990, Karato 1989) allowing diffusion measurements to help constrain rheological properties (Jaoul and Raterron 1994, Bejina et al. 1997) as well as theoretical calculations (Ita and Cohen 1998). An in depth discussion of the physics of crystal deformation that leads to constitutive relations such as Equation (11) is beyond the scope of this work and readers are referred to Poirier (1985), Evans and Kohlstedt (1995), Weertman and Weertman (1975), and Hirth and Lothe (1982) for more information.

RECOVERY EXPERIMENTS

Rheology experiments on most rock forming minerals have been explored using single crystals, polycrystals, and natural samples at crustal pressures and temperatures over the past several decades. Since 1980, there are over 150 entries in the GeoRef index for olivine rheology alone. The traditional experiment consists of applying a uniaxial load to a cylindrical sample. Most commonly, the strain rate is held constant as the force on the piston is monitored with time and total deformation. A steady-state condition is sought in which the stress remains independent of total deformation. Microstructures are examined with transmission electron microscopy to determine the slip systems and identify the recovery mechanisms. In steady-state, the microstructure is uniform in the sample with a relatively constant and reproducible dislocation density. Pressure can be applied to the system using a gas such as that designed by Paterson (Paterson 1970) to pressures of a few hundred MPa, or with a solid in a Griggs type device (Griggs 1967) to pressure as high as 3 GPa. These systems rely on the ability to design a low friction interface between the pressurizing medium and the piston, with the force on the piston being generated by a hydraulic system that is independent of the pressurizing system. With these systems, a wide range of experimental conditions have been explored. Effects of partial melting, chemical environment, volatile activity, temperature and strain rate have been delineated.

In order to examine the effects of pressure at mantle conditions and to explore the properties of the high pressure phases that constitute the bulk of the Earth, higher pressures are required. Both the diamond anvil cell and the multi-anvil high pressure devices offer the capability of generating pressures that reach into the lower mantle. The diamond anvil cell can provide pressure to the center of the Earth, however, temperature is more difficult to generate and maintain. Externally heated diamond anvil cells reach 1200K (Fei et al. 1992) and laser heated diamond cells can extend the temperature range to several thousand degrees. While the laser spot size tends to be very small (a few microns), new techniques employing multi-mode laser heating, illuminating both sides of the sample, hold the potential for more uniform temperatures over dimensions of 10 microns or so (Mao et al. 1998a). The diamond anvil cell loads the sample along a single axis. While the sample is usually contained in a gasket to help provide a hydrostatic pressure field, a solid sample still experiences a stress field with a uniaxial component parallel to the loading direction. While these systems have severe limitations to explore the full range of rheological properties, they offer the capability of applying deviatoric stress at very high pressure. Multi-anvil systems can generate pressures to 25 GPa with fairly uniform temperatures up to 3000K on millimeter sized samples. The force field generated by these systems is generally of cubic symmetry and are thus inherently hydrostatic. By building a cell that has axial symmetry, it is possible to apply a uniaxial stress to the sample, however, with only a single loading system, it is difficult to introduce the deviatoric stress after the sample has been brought to the pressure and temperature region of interest. Further difficulties are encountered in measuring the stress and strain as a function of time. Recent use of synchrotron x-radiation offers the possibility of providing such information. In the following, the successes to date are reviewed, with special emphasis on the potential use of synchrotron radiation in these measurements.

Multi-anvil studies. The 6-8 double stage multi-anvil high pressure system has become widely used to generate pressures up to 25 GPa, sufficient to replicate the environment at the top of the lower mantle. The second stage, illustrated in Figure 3, consists of 8 cubic anvils with the inner corners truncated into triangles. The second stage is compressed in a cubic cavity formed by the first stage anvils by a force that is applied by a single ram. An octahedral pressure-transmitting media is compressed by the triangular faces of the cube as the 8 second stage anvils are forced closer together. Temperature can be generated by passing an electrical current through an electrically conducting cylindrical furnace within the octahedron and measured with a thermocouple. The sample is contained within the furnace. These high pressures can be generated with tungsten carbide anvils since gaskets and extruding pressure medium support the large stress gradients generated in the anvils. Thus, these systems must be inherently inefficient in generating pressure, since a significant amount of the force that is applied by the external ram must be used to support the anvils themselves. Therefore, it is difficult, a priori, to calculate the pressure in the sample from knowledge of the loading force and the area of the sample chamber.

Deviatoric stress can be generated by making the elastic and flow properties of the pressure medium anisotropic. Some pioneering attempts have been made to access this deviatoric stress to define rheological properties and deformation mechanisms at high pressure and temperature (Wang et al. 1988, Green et al. 1990, Liebermann and Wang 1992, Bussod et al. 1993, Yamazaki et al. 1996, Karato and Rubie 1997, Ando et al. 1997, Durham and Rubie 1998). However, it is even more difficult to estimate the deviatoric stress in the sample than it is to estimate the pressure since the whole system is responding to the force generated by this single ram. Wang et al. (1988) propose using the dislocation density in an olivine sample as a piezometer. Previous calibrations of the steady-state dislocation density of olivine (Kohlstedt et al. 1976) are used for this purpose. Bussod et al. (1993) increase pressure and temperature to the desired conditions and then further advance the ram, monitoring ram displacement, to define time dependent strain in the sample and define the activation volume. Durham and Rubie (1998) discuss the difficulty in estimating the stress owing to uncertainty in gasket deformation during the pumping process.

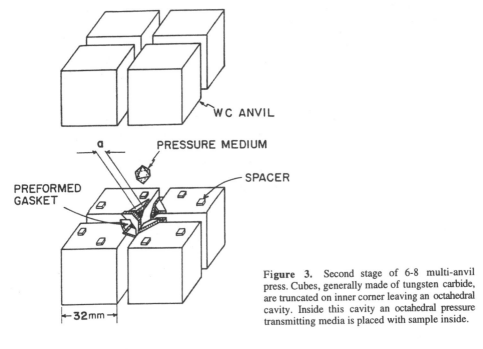

Figure 3. Second stage of 6-8 multi-anvil press. Cubes, generally made of tungsten carbide, are truncated on inner corner leaving an octahedral cavity. Inside this cavity an octahedral pressure transmitting media is placed with sample inside.

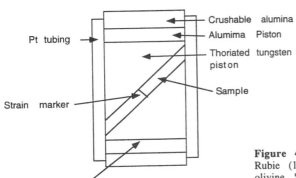

Figure 4. Cell assembly use by Karato and Rubie (1997) for inducing large strains in olivine. Sample is about 200 μm thick and oriented 45° to the stiff axis of the cell.

Karato and Rubie (1997) propose a complex cell assembly illustrated in Figure 4, which is designed to produce large strains and provide the deviatoric stress to the sample while the sample is at high pressure and temperature. The sample is a thin disc about 200 microns thick sandwiched between two hard pistons that are truncated at 45° to their axes. The disc is prepared in two halves with a strain marker between the pieces and oriented perpendicular to the piston truncations. Sample flow is indicated by rotation of the marker. The deviatoric stress is kept low during initial compaction by using crushable alumina at the ends of the piston. Once the porosity is compressed from the end-plugs, further compression will have a deviatoric component with compression along the axis of the anvils as a result of the elastic anisotropy of the cell. Samples are recovered corresponding to differing run times at temperature to monitor the strain as a function of time. Deviatoric stress is estimated from the maximum resulting strain of the sample assuming that this strain is initially stored elastically in the piston. The data are analyzed assuming that this is a relaxation experiment, with the total length of piston and sample remaining fixed during the experiment. Perhaps the most limiting assumption in this experiment is that the piston is held in a frictionless manner. The rotation of the strain maker is assumed to reflect only the response of the sample to the pressure, but the pistons cannot advance in the cell if there are significant forces holding them in position. It is important to determine if the resulting relaxation rate is limited by the sample or by the interaction of the cell with the piston. Despite such possible problems, this experiment represents a major step forward in harnessing the stress field of the multi-anvil system to produce deformation of the sample.

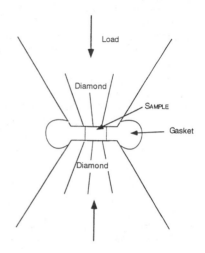

Figure 5. Schematic of a diamond anvil cell.

Diamond anvil cell studies. The diamond anvil cell, in contrast to the multi-anvil devices, creates pressure that is naturally non-hydrostatic since pressure is generated by pushing the two diamonds against each other with the sample in the middle. Hydrostatic pressure is routinely generated in this environment using a fluid along with the sample inside a gasket. Still it is difficult to create hydrostatic pressure for pressures beyond that required to solidify the pressure medium which is about 10 GPa at room temperature. A typical diamond cell configuration is illustrated in Figure 5. A small hole inside a metal gasket provides a sample chamber. As the force is applied to the diamonds, both the sample and sample chamber deform and compress. Pressure is often monitored by the shift in fluorescence of ruby. If the sample is stronger than the gasket, the magnitude of the deviatoric stress is limited by the strength of the sample. In a series of papers, Meade and Jeanloz 1988a, Meade and Jeanloz 1988b, Meade and Jeanloz 1990 capitalize on this phenomena to determine the yield strength of several important materials at very high pressures. The determination of the magnitude of the deviatoric stress follows from measuring the pressure gradient across the diameter of the sample (Sung et al. 1977). Using Equation (4) in a coordinate system with the 1 axis perpendicular to the diamond surface, we deduce:

$$\frac{\partial P}{\partial x_2} = \frac{\partial \sigma_{12}}{\partial x_1} \tag{13}$$

With the symmetry constraint that the shear stress vanish on the mid-plane of the sample, then the magnitude of the maximum shear stress is defined by the pressure gradient and the thickness of the deformed sample. The first of these is determined by measuring the fluorescence shift of the ruby line for many small ruby crystals distributed throughout the sample. The thickness is measured on the recovered sample. Since these experiments are carried out at room temperature, the deformation mechanism should be in the high-stress low-temperature regime where sample strength is the primary rheological characterization with very little time dependence of this property. Thus, such experiments directly sample important deformation characteristics of the specimen. Methods using x-radiation to measure deviatoric stress in the diamond anvil cell are also available and discussed in a latter section.

Chai and Brown (1996) demonstrate that the two ruby fluorescence lines that are normally used for pressure calibration have different shifts when a single crystal is stressed non-hydrostatically. Using this splitting as an in situ piezometer, Chai et al. (1998) prepare a layered diamond anvil cell sample with a single crystal ruby on a layer of single crystal olivine. The sample is surrounded with an argon pressure medium and the geometry is controlled so that the diamonds contact the sample stack as pressure is elevated. In this manner they are able to create and measure a uniaxial stress in the olivine while the sample is at elevated pressure. Since the yield strength of olivine is less than that of the ruby, they are able to determine the strength of the olivine sample. Transmission electron microscopy of the recovered olivine then reveals the slip systems that were responsible for the yielding of olivine.

This class of diamond anvil cell experiments have been restricted to room temperature measurements. Extension to high temperature is possible using externally heated diamond anvil cells. As the sample moves into high temperature creep, it will become important to monitor the time dependence of both the stress and the strain in order to define the parameters of the flow law. Laser heating is less promising for these classes of diamond cell experiments because of the large lateral temperature gradients over the entire sample that result from the focused laser beam.

IN SITU MEASUREMENTS WITH SYNCHROTRON X-RAYS

Quantitative rheology experiments at elevated temperature and pressure require accurate measurements of the time dependent stress-strain relation. Recovery experiments that are aimed at capturing the state of the sample at different times during deformation require several experimental runs that are quenched and recovered at different times during the system evolution. Slight differences in cell preparation can create differences in the stress state that are interpreted as the time dependent behavior. Estimates of the stress from the mechanical properties of the cell introduces large uncertainties as this behavior is poorly established. While the results provide insights into the rheological behavior of the sample, they must be regarded as qualitative. Quantitative flow laws demand in situ time dependent monitoring of stress and strain as pressure and temperature are controlled. The last decade has witnessed a remarkable expansion of the ability to use x-rays to monitor samples at high pressure and temperature. We are reaching the point where it is possible to obtain an x-ray diffraction pattern of a sample at elevated pressure and temperature of comparable quality to the best that can be obtained at ambient conditions. Furthermore, the time scale to gather such an image of the sample is routinely of the order of 30 seconds. These advances have resulted from the development of high-energy, high-flux x-rays from synchrotron sources merged with high pressure technology. Deviatoric stress, once considered a nuisance in these experiments is now being harnessed, quantified, and interpreted in terms of flow properties of the material. This field is still in its infancy; resolution of stress is limited, strain metrics need better definition, experimental protocols need refining. The advent of third generation synchrotrons at APS (at Argonne National Laboratory), ESRF (in Grenoble), and SPring8 (in Kobe) provide hope of further developing the capabilities as well as the continuing advances at NSLS (Brookhaven National Labs).

Synchrotron radiation. X-ray diffraction provides a measure of the separation of crystalline planes within a solid following the well known Bragg's law:

$$n\lambda = 2d \sin\theta \qquad\qquad (14)$$

where n is an integer, λ is the x-ray wave length, 2θ is the scattering angle, and d is the spacing between lattice planes for the crystallites oriented with the normals to these planes at an angle of 90 - θ from the incident x-ray direction and within the scattering plane. Diffraction experiments record scattered x-ray intensity as a function of either θ for a fixed wavelength (angle dispersive) or energy (proportional to $1/\lambda$) at a fixed θ when the incident x-ray source has a white spectrum (energy dispersive). Peaks occur in these recordings when Bragg's law is satisfied.

The combination of high-flux, high-energy, and small divergence of x-rays generated by synchrotrons have a proven potential for use with high pressure experiments that place sever restrictions on x-ray experiments as the x-ray beam must be able to enter the sample and the diffracted beam must survive to the detector. X-rays are absorbed by all materials between the source and the detector. Generally, the absorption of x-rays increases with the atomic number of the material and decreases with increased energy (shorter wavelength) of the x-ray. Typical cells for multi-anvil devices, with an MgO pressure medium will remove x-rays with energies lower than about 20 kev. If amorphous boron is used as the pressure transmitting medium, this low energy cut off is about 15 kev. Diamond anvil cells have a similar, but slightly lower threshold. Accessing the sample through a Be gasket allows the diamond anvil studies to lower the energy cut-off to 4 kev (Hemley et al. 1997, Mao et al. 1998b). A sample jacketed in a platinum container requires energies in excess of 40 kev for diffraction experiments. Since the sample in any of these experiments is relatively small, a well collimated beam is required and the low divergence of the synchrotron beam can provide a spatially well-defined x-ray beam at large distances from the beam-defining slits. Figure 6 illustrates the flux as a function of energy for several x-ray sources. Beam-line X17 at NSLS, which is generated by a super-conducting wiggler magnet inserted into a straight section of the electron beam satisfies the energy criteria of these cell materials. In a similar manner, all beam lines at the APS meets the energy-flux combination to produce a quality source for high pressure studies.

Energy dispersive systems take full advantage of the total x-ray flux offered by these sources. The high flux of x-rays for energies up to 100 kev yields a wide region where diffraction peaks can be generated. The detection electronics enable the entire spectrum to be collected simultaneously at a fixed θ. Angle dispersive analyses require that the incident beam be filtered by a monochromator which removes x-rays at all energies except a very narrow band at the desired wavelength. This greatly reduces the x-ray through-put and increases the data gathering time if the detector is required to scan θ. This loss can be partially recovered if an area detector, such as film or an imaging plate, is used that simultaneously collects data over a range of θ. In general, higher resolution and more reliable peak intensities are obtained from angle dispersive data, while collimators on energy dispersive detectors provide a better definition of sample position and remove diffraction peaks from the material surrounding the sample.

The portion of the specimen that is sampled by the x-ray beam is defined by the intersection of the incident x-ray beam and that visible through the detector collimator. Typical multi-anvil systems limit the incident beam to 100×200 microns with similar

X-ray Spectra for APS and NSLS

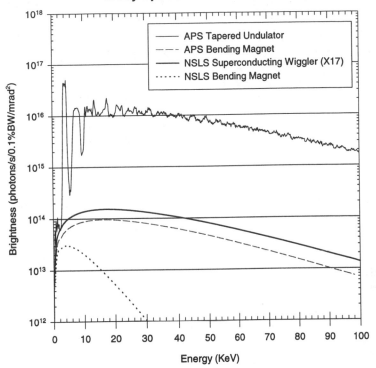

Figure 6. X-ray flux as a function of energy for several synchrotron sources.

dimensions on the detector. Different portions of the sample volume are examined by moving the press on a high precision table. Diamond anvil cells illuminate a much smaller area of sample by focusing the x-ray beam to a spot size of the order of 10 microns. This allows the x-rays to sample a portion of the sample with a relatively uniform temperature even if it generated by a laser. The smaller sampling volume requires a smaller grain size to obtain good powder statistics.

High pressure equipment. In principle, the metrics discussed in the following section for defining stress and strain can be equally applied to multi-anvil systems and to diamond cell systems. Most of the examples will be from multi-anvil devices using the NSLS superconducting wiggler source. Two systems are currently available. One is a small version of the 6-8 apparatus discussed above, the T-cup (Vaughan et al. 1998). This guide block is illustrated in Figure 7. The main departure from a standard 6-8 includes the smaller size, designed to work with a 200 ton hydraulic loading system, and the cut-out of metal from the first stage to allow x-rays to enter and exit the sample volume through the gaps in the second stage anvils. The second and more standard x-ray system is a DIA apparatus (Weidner et al. 1992), a cubic anvil system. Pressure is applied in the DIA by advancing 6 anvils simultaneously onto the sample volume as illustrated in Figure 8. X-rays enter and leave the sample through the gaps between anvils. The T-cup operates to pressures of 24 GPa compared to 13 GPa for the DIA but with a loss in sample volume.

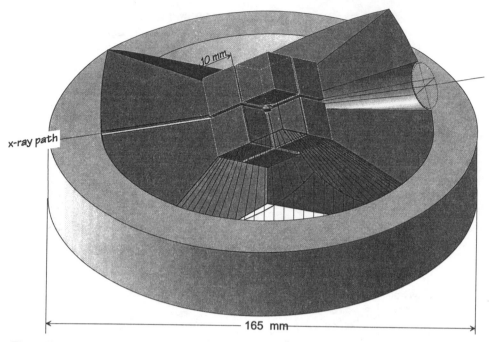

Figure 7. The Stony Brook T-cup guide block. Small version of the standard 6-8 guide block, but with x-ray access to the sample.

Micro-scale stress

Deviatoric stress generally accompanies hydrostatic pressure in all high pressure devices unless each individual grain is surrounded by a fluid medium that has no strength. This stress field can be uniform throughout the sample, with little or no variation from grain to grain, or it can be heterogeneous, varying in magnitude and orientation from grain to grain or even point to point as long as Equation (4) is satisfied. Uniform stresses arise from anisotropies of the elastic and plastic properties of the pressurizing medium or from the symmetry of the loading system. The diamond anvil cell is an example of the latter as the forces are applied to the sample through a uniaxial loading geometry. A heterogeneous stress field results from heterogeneities within the sample such as a two phase mixture or randomly oriented elastically anisotropic grains. A sample that is a comprised of loosely packed grains will experience deviatoric stresses as it is compressed owing to the void space between grains and the point contacts of the grains. Microscopic stresses will occur even if the sample is externally loaded hydrostatically Theories regarding stress induced by thermal expansion in a polyphase aggregate made of anisotropic grains is applicable to the current problem. Studies such as those by Walsh (1973), Kreher (1990), Tvergaard and Hutchinson (1988), and many others allow one to estimate properties of the microscopic deviatoric stress for particular material properties. These studies reveal that the microscopic stress varies within the individual grains, but as Evans and Clarke (1980) point out, the grain size should not affect the amplitude of the microscopic stress field.

It is convenient to distinguish these scales relative to the scale of the x-ray sampling volume. We use the term, macroscopic stress, to indicate the deviatoric stress field that is uniform over the sampling volume and microscopic stress to indicate the stress field that is

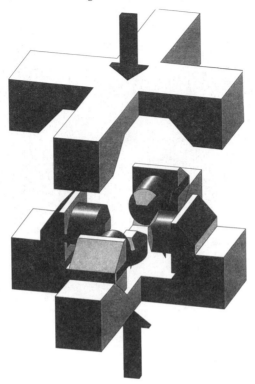

Figure 8. DIA type guide block. This has been the most common x-ray compatible multi-anvil system.

heterogeneous in this volume as they will have different manifestations in the diffracted x-ray signal and they have different expressions of the plastic strain field. Macroscopic stress, as we discuss latter, will result in a distortion of the Debye rings for all diffraction lines and will introduce different 'strains' in each diffraction peak for elastically anisotropic materials. Plastic strain will be manifest as a change in shape of the entire sample. Microscopic stress broadens the diffraction peak with the resulting plastic strain occurring at the expense of elastic strain of the sample. In this regard, a deformation experiment that utilizes microscopic stress is a relaxation type of experiment in that the volume of the sample remains quite constant as the stress relaxes by converting the elastic strain to plastic strain. For macroscopic stress experiments, the loading system, including the pressure medium, will strain with the sample. In this section, we discuss how we can measure and utilize microscopic stress for defining rheological properties.

Figure 9 illustrates three x-ray diffraction spectra, corresponding to different conditions, for $MgAl_2O_4$ spinel. The sample consisted of a loose polycrystalline powder with grain size of a few microns, packed into the cell assembly shown in Figure 10 in a layer about 0.5 mm thick along with a second layer consisting of NaCl powder for pressure calibration, all in a cylinder of 1-mm diameter inside a boron nitride sleeve which was contained in a cylindrical graphite furnace. The room pressure pattern was taken before compression, while the 10 GPa pattern was obtained after compression at room temperature. The center of the peaks moved to higher energy (smaller d-spacing) as the peaks broaden considerably. The third pattern is taken after the sample had been heated over a few hours. This high temperature pattern also demonstrates shifts of the peaks

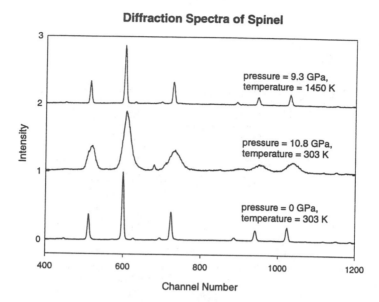

Figure 9. X-ray diffrac-ion signals for spinel at ambient conditions, at elevated pressure before heat-ing and at high pressure and temperature. Peak widths increase dramatically with pressure but diminish with heating.

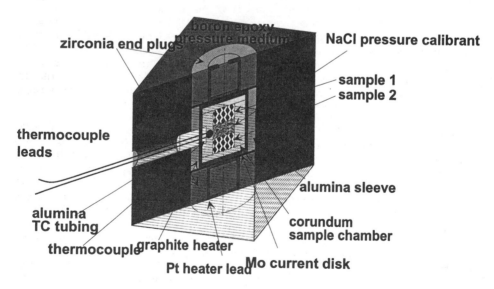

Figure 10. Sample cell assembly for the DIA apparatus.

owing to the elevated pressure and temperature, but the peaks have become similar in width to the initial pattern before compression. The broadening of the peaks is the result of microscopic stress in the sample while the narrowing occurs as the stress relaxes due to plastic deformation of the grains. Rheological information can be derived from the quantification of this process.

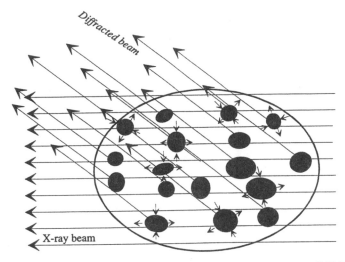

Figure 11. Schematic of x-ray diffraction from grains in a randomly varying stress field. Each grain will provide a peak with a slightly different *d*-spacing, thus altering the shape of the diffracted peak.

Model peak shape. Each diffraction peak is the sum of diffraction from only a subset of grains within the specimen. These are the grains that have the particular orientation which aligns the specific set of lattice planes with the diffraction vector. Figure 11 illustrates this picture. The solid grains represent the grains that are in diffracting condition. The small arrows around each grain represent the deviatoric stress field at the position of these grains. The effect of this microscopic deviatoric stress field will be the broadening of x-ray diffraction lines and the amount of line broadening is determined by the distribution of longitudinal strain parallel to the diffraction vector. The x-ray signal will reflect only the elastic strain, since plastic strain will leave the lattice planes unperturbed. The shape of the diffracted peak is defined by the magnitude distribution of the principal stresses and the orientation distribution of the stress field.

Each grain that contributes to the diffraction peak is sampling the stress field at the location of the grain or subgrain. We model the manner that the diffraction signal will reflect this stress field by expressing the stress in terms of the principal stress components and the orientation of the principal stress relative to the grain. Assuming a random distribution of the orientation of the stress field, we can derive the distribution of lattice spacings for the particular diffraction peak and all possible principal stress values. These calculations have been done numerically by generating a uniform distribution of orientations of the stress field, evaluating each stress field in a coordinate system defined by the orientation of the grain, and calculating the lattice spacing of the diffracting planes. We need to know the single crystal elastic properties of the material to relate the stress to the strain. Figure 12 illustrates the predicted distribution of lattice spacings of the (400) diffraction peak for spinel for a random distribution of orientation of a stress field whose principal stress magnitude is $(-\sigma,0,0)$—note that this is uniaxial compression. The resulting peak is highly asymmetric, with a maximum corresponding to an increase in lattice spacing and a scale length given by $S_{1111}\sigma$, where S_{1111} is an elastic compliance that is equal to one over Young's modulus for an isotropic material. The surprising result that compression actually leads to an expansion of the lattice spacing for many grains can be understood by the fact that there are many more orientations where the uniaxial stress acts perpendicular to

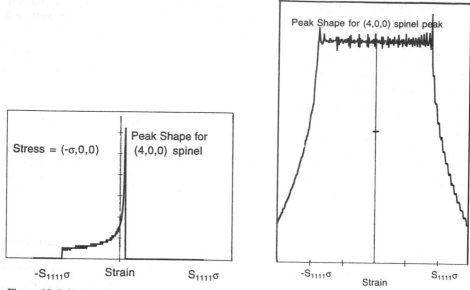

Figure 12 (left). Distribution of lattice spacing for the spinel (400) peak where the local deviatoric stress field has a magnitude of $(-\sigma,0,0)$ but is randomly oriented.

Figure 13 (right). Distribution of lattice spacing for the spinel (400) peak where the local deviatoric stress field has a magnitude of $(\sigma-x/3,-\sigma-x/3,2x/3)$ where $\sigma > x > -\sigma$, but is randomly oriented.

the diffraction vector—causing expansion by a Poisson ratio process—than there are parallel to the diffraction vector, which would cause compression. As the sample is compressed, the high stress points that are supporting the load will yield first and the low stress points that are being protected from the load will experience an increase in deviatoric stress. A terminal situation will obtain when all points in the sample are at the strength limit. In this case the principal stress at each point will have the same maximum shear stress. The principal stress for the purely deviatoric stress field will then be $(-\sigma-x/3,\sigma-x/3,2x/3)$ where $-\sigma < x < \sigma$, and the maximum shear stress is σ. A randomly oriented stress field for this system will produce the distribution of lattice spacings for the (400) spinel peak illustrated in Figure 13. This peak is flat-topped and symmetric with a full width at half height given by σS_{1111}. A diffraction peak will be given by the distribution of lattice spacing peak convolved with the instrument response. The actual situation for a compressed sample probably includes a distribution of σ that is dictated by the rheological properties and the geometry of the individual grains. These models suggest that a measure of deviatoric stress can be obtained from the strain as measured by the full width at half height multiplied by Young's modulus of the solid after removing the instrument response from the observed peak width. In the example given here, this would yield 2σ.

Other contributions to peak width. The above example applies to a polycrystalline sample under hydrostatic compression. Each grain is at the same pressure with a superimposed randomly oriented deviatoric stress. If the macroscopic stress has a deviatoric stress component, then it is possible to elastically broaden the peaks because of the macroscopic field. This results from the possibility that the Poisson's ratio of the crystallites may be anisotropic and that this anisotropy interacts with the macroscopic deviatoric stress field acting perpendicular to the diffraction vector. For multi-anvil systems, the symmetry of the macroscopic stress field should reflect the symmetry of the cell assembly. This generally has a unique axis, defined by the cylindrical sample chamber, nearly parallel to the diffraction vector. A macroscopic stress field with this geometry will not broaden the peaks. In contrast, for a diamond anvil cell, the diffraction vector is usually perpendicular to the load axis. A macroscopic stress in this case can broaden diffraction peaks that correspond to an anisotropic Poisson ratio. However, peaks such as (200) for cubic crystal symmetries should not experience peak broadening as the Poisson's ratio should be isotropic for this orientation.

Peak broadening will occur when the coherence length of the sample becomes small. Reduction in grain size, internal fracturing, onset of amorphization, high density of stacking faults, can all break up the periodicity of the lattice and thereby reduce the coherence length. The broadening due to small grain size is difficult to separate from strain broadening for any single diffraction peak, but the contributions of these two sources can be separated if several peaks are analyzed. Gerward et al. (1976) demonstrated that, for energy dispersive data, the observed peak width, W_o is given by:

$$W_o^2 = W_s^2 + W_d^2 + W_i^2 \tag{15}$$

where the subscript s is the strain contribution, d indicates the coherence length contribution, and i is the instrument contribution. The instrument must be calibrated as a function of energy. This can be done with diffraction standards and with the sample before compression. The broadening due to strain is linearly proportional to energy and the broadening due to coherence length is independent of energy. This yields:

$$W_o^2 - W_i^2 = A + BE^2 \tag{16}$$

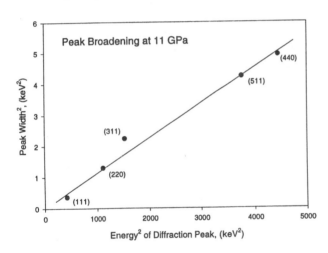

Figure 14. Energy dependence of peak broadening for spinel at high pressure. Slope of a straight line reflects the strain contribution to broadening, and the intercept reflects the grain size broadening (from Weidner et al. 1998).

where A is the contribution from grain size and B is the contribution from strain. A plot of the left hand side of Equation (16) against energy squared should yield a straight line whose slope yields the strain term and whose intercept at zero energy yields the grain size effect. Figure 14 gives such a plot for several diffraction peaks from the spinel sample after compression and before heating. Clearly, in this case, the peak broadening is a result of strain and not grain size reduction on loading.

In the above analysis, we have assumed that every point in the sample is at the same hydrostatic pressure and that the peak broadening is due to variations in the deviatoric

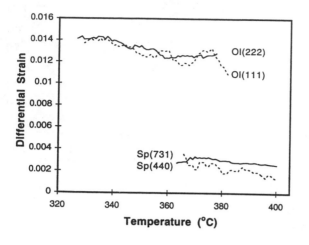

Figure 15. Peak strains as a function of time for a fayalite sample transforming to the spinel structure. The compressed fayalite sample experienced peak broadening during compression, the spinel phase forms strain free (after Chen et al. 1998b).

stress field. Point to point variations in the hydrostatic pressure can also cause peak broadening. However, as discussed above, variations in hydrostatic pressure must be accompanied with shear stresses of comparable magnitude since:

$$\frac{\partial P}{\partial x_1} = \frac{\partial \sigma_{12}}{\partial x_2} + \frac{\partial \sigma_{13}}{\partial x_3} \tag{17}$$

and the length scales involved in the pressure variations must be comparable to the length scales of the shear stress variations. Still it is interesting to determine whether the peak broadening results from heterogeneous pressure or heterogeneous deviatoric stress. A recent study by Chen et al. (1998b) provides some insight. They monitored the time dependence of the phase transition of fayalite transforming to the spinel structure. Pressure was increased into the spinel field and monochromatic x-rays were continuously recorded as a function of time over a 50-K temperature range using an imaging plate. During the recording a totally olivine sample completely transformed to the spinel phase. The peak strains as a function of time are illustrated in Figure 15. The olivine peaks are considerably broadened owing to the stress heterogeneity, but the spinel peaks are very sharp, indicating a uniform hydrostatic pressure in the spinel grains. Because of the transformation, the spinel phase can change shape and grow free of deviatoric stress, thus minimizing its free energy. If, however, there is a spatial variation of hydrostatic pressure in the olivine phase, then the spinel phase should reflect this variation. The absence of peak broadening in the spinel phase suggest that this is not the case and that the peak broadening of the olivine phase is due to heterogeneities in the deviatoric stress field.

Effects of elastic anisotropy. The above analysis provides a scaling that relates the observes strain broadening to the magnitude of the deviatoric stress field as:

$$\sigma = \varepsilon / S_{1111} \tag{18}$$

Most crystals are elastically anisotropic yielding a different S_{1111} for each diffraction peak. If each grain is sampling the local stress field, then we expect that the diffraction peaks would reflect the crystalline elastic anisotropy. On the other hand, each grain may simply deform to fit the local shape. In this case, the grain will sample the local strain field. These two extremes represent the Reuss and Voight models that are used to describe the local stress-strain field in a polycrystalline aggregate, with the expectation that a real situation will fall somewhere in between these two limits. Spinel is extremely anisotropic even

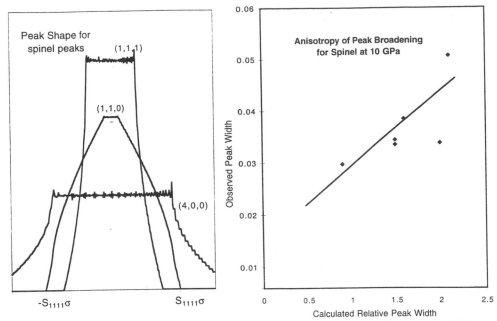

Figure 16 (left). Calculated peak shape for several spinel peaks where the local deviatoric stress field has a magnitude of (σ-x/3,-σ-x/3,2x/3) where σ > x > -σ, but is randomly oriented.

Figure 17. (right) Comparison of observed peak width with calculated peak width. Differences in the calculations between peaks reflects the elastic anisotropy.

though it has cubic symmetry. Figure 16 shows the calculated peak shape for several diffraction peaks following the methods outlined above. All peaks are normalized to S_{1111} and thus reflect the relative strains expected if they are sampling the local stress field. There is a variation of a factor of 2 in peak widths from the (111) peak to the (400). Figure 17 illustrates the observed peak strain for the spinel sample after cold compression compared to these calculations. The fit would be horizontal if each peak samples the strain field and should have a 0 intercept if they sample the local stress field. We see that they are indeed between these two limits indicating that the solid occupies an intermediate state between the Reuss and Voight bounds. Still the error is small to simply use Equation (18) to estimate stress and to use the scatter in the stress estimate among the peaks to suggest the uncertainty.

Dislocation contributions. A primary mechanism of plastic flow is the movement of dislocations within the solid. Some dislocations are always present within the solid, others are created at grain boundaries, at defects, and even homogeneously under deviatoric stress. The dislocation is a line in the crystal lattice marking the boundary between a slipped area on a crystallographic plane (by the Burger's vector, **b**) and the unslipped area. The material flows as the dislocation line moves through the solid. The distortion of the lattice is manifest as strain and stress within the solid—internally generated. This stress field contributes to the total stress field at any point in the solid. It will both contribute to the stress field measured with peak broadening and to the motivating force for moving other dislocations. Indeed, the process of annealing is simply the plastic response of the solid to this stress field by moving dislocations, eventually out of the material. Thus, the flow of the material in response to both the externally generated stress field and the internally generated stress field informs us of the rheological properties. The stress fields of the dislocations can lock other dislocations from moving. This is the case for work hardening of materials. Thermally activated recovery processes are required to aid the dislocation motion and release this stress. Even in the case of work hardening, the association of plastic flow with the volume average of the stress, as deduced from peak broadening, provides a quantitative measure of the work hardened strength.

The contribution of dislocations to the broadened diffraction peak can be estimated from the strain associated with the dislocation (see for example Hull and Bacon 1984). The strain field associated with a screw dislocation as:

$$\varepsilon_{11} = \varepsilon_{22} = \varepsilon_{33} = \varepsilon_{12} = 0$$

$$\varepsilon_{13} = -\frac{b}{4\pi}\frac{\sin\phi}{r} \tag{19}$$

$$\varepsilon_{23} = \frac{b}{4\pi}\frac{\cos\phi}{r}$$

where the 3 axis is parallel to the line of the dislocation, r is distance from the dislocation, ϕ is the angle in the 1-2 plane from the plane that has been displaced, and b is the amount of the displacement (the magnitude of the Burger's vector). This strain field can be expressed as principal strains as follows:

$$\varepsilon_{11} = -\varepsilon_{33} = \frac{b}{4\pi r} \tag{20}$$

$$\varepsilon_{ij} = 0, ij \neq 11, 33$$

By considering a random distribution of dislocation orientation relative to a particular set of lattice planes, we evaluate the contribution of this strain to peak broadening at a fixed r and integrate this effect, weighted by the volume of material at r, over r. Integrating from r = b to r = 100 b, we produce Figure 18 as the dislocation-broadened diffraction peak. The limit of the integration should be defined by the dislocation density. This example would be appropriate for dislocations that are separated by r = 200 b. For a value of b = 5 A, this corresponds to a dislocation density of $10^{10}/cm^2$. The full width at half maximum of strain is 0.002. Higher dislocation densities effectively reduce the volume per cent of material that is at low strain. A dislocation density of $10^{12}/cm^2$ is required to increase the volume average strain to 0.02. In the peak width analysis, a value of 0.002 is at the current resolution limit.

Edge dislocation strain field is more complicated than that for a screw dislocation. However, the strain energy density of an edge dislocation is a simple scalar multiple of

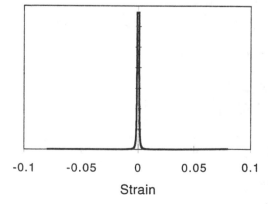

Figure 18. Diffraction peak shape of a solid with a screw dislocation density of $10^{10}/cm^2$. Full width at half maximum is 0.002 for this peak.

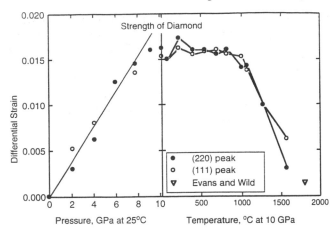

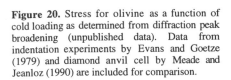

Figure 19. Strain for a diamond sample (after Weidner et al. 1994b) as a function of loading at room temperature, and then temperature as pressure remains constant. The point labeled Evans and Wild is from a 3-point bending study of diamond strength at room pressure by Evans and Wild (1965).

1/(1 - ν) (or about 3/2) times that of a screw dislocation where ν is the Poisson's ratio. Thus the strain distribution for an edge dislocation should be similar to that of a screw dislocation with the amplitude of the curve increased by a constant scaling factor.

The relative importance of the stress field generated by dislocations and that resulting from the compression of an heterogeneous sample will vary during the experiment. Figure 19 (Weidner et al. 1994b) illustrates the strain determined from peak broadening of a diamond sample as a function of pressure on cold loading, and then as a function of temperature as pressure is held constant. Strain increases linearly with pressure over the entire compression portion to the final pressure of about 10 GPa. This linearity implies that the strain is totally elastic (i.e. stress is proportional to strain). On heating, there is no change in strain up to temperatures of 1000°C. Above this temperature, strain diminishes with further heating. Dislocations are probably most abundantly generated in concert with plastic deformation. However, their motion is effective in reducing the stress even though they may contribute to the observed stress field. Continued heating further reduces the observed stress field even though the dislocation density probably continues to increase as dislocation recovery processes are still not dominating the flow. Most oxides and silicates that we have studied depart from the linear loading curve as illustrated for in Figure 20.

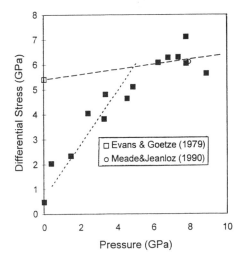

Figure 20. Stress for olivine as a function of cold loading as determined from diffraction peak broadening (unpublished data). Data from indentation experiments by Evans and Goetze (1979) and diamond anvil cell by Meade and Jeanloz (1990) are included for comparison.

Deviatoric stress initially increases linearly with pressure, but then the stress saturates and increases at a slower rate or remains constant with further loading. This indicates that the deviatoric stress exceeds the yield strength of the solid and it has begun to deform. Also shown in Figure 20 are room temperature yield strengths obtained from indentation experiments at room pressure (Evans and Goetze 1979) and diamond anvil cell measurements at high pressure (Meade and Jeanloz 1990). These three measurements involve very different methods of determining yield strength, yet provide quantitatively similar results.

Relaxation experiments. The above discussion outlines the tools that are available for extracting rheological information from monitoring peak broadening with environmental conditions and time. The x-ray pattern defines the deviatoric elastic strain in the sample at any time. This strain allows us to determine both the stress and the plastic strain. A quantitative measure of the deviatoric stress comes from the elastic relation with the deviatoric strain. The plastic strain is deduced from the relation:

$$\varepsilon_{tot} = \varepsilon_{pl} + \varepsilon_{el} \tag{21}$$

where ε_{pl} represents the total non-elastic portion of strain and ε_{el} is the elastic portion. As the sample temperature is changed and the strength lowered then, since the experimental conditions maintain ε_{tot} constant, the plastic component of strain increases at the expense of the elastic portion. Thus, we can estimate the total plastic strain as the deficit between elastic and total strain. This generally amounts to about 1-5% of plastic strain during these measurements. The strain rate of the deformation process is then given by:

$$\partial\varepsilon_{pl}/\partial t = -\partial\varepsilon_{el}/\partial t \tag{22}$$

This is, thus, a relaxation type of rheological experiment (Rutter et al. 1978). As Rutter et al. (1978) point out, this class of experiments is useful to define the empirical parameters of Equation (11) from information relating to the first time derivative of the observations, and to define properties of the dynamics of dislocation motion such as the athermal stress, τ, of Equation (12), from second derivatives. Assuming the power flow law of Equation (11), at any temperature:

$$\partial\varepsilon_{pl}/\partial t = k \cdot \sigma^n \tag{23}$$

or since:

$$\partial\varepsilon_{pl}/\partial t = -\partial\varepsilon_{el}/\partial t \propto -\partial\sigma/\partial t \tag{24}$$

then:

$$\partial\sigma/\partial t = k' \cdot \sigma^n \tag{25}$$

yielding:

$$d\sigma/\sigma^n = k' \cdot dt \tag{26}$$

or

$$(1-n) \cdot \ln(\sigma) = \ln(t+c) + k'' \tag{27}$$

where c is a constant of integration. Thus, if a plot of either log stress or log strain with log time yields a straight line then the slope of the line is $1/(1-n)$.

Weidner et al. (1998) report the results of such analysis for a sample of $MgAl_2SiO_4$ in the spinel structure. A powder sample of spinel was loaded into the pressure chamber and compressed to 11 GPa in SAM85. Then the sample was heated to 400°C in a few seconds. Diffraction data were collected continuously for a period of several minutes, each spectra

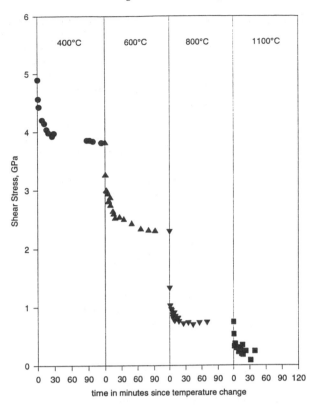

Figure 21. Time dependence of the differential stress in spinel at different temperatures (from Weidner et al. 1998). Data are from the x-ray diffraction peak broadening for spectra collected in 30 sec. Temperature changes were made in less than 10 sec.

was collected for 30 seconds. Data were collected for about 2 hours at this temperature. Then the temperature was raised to 600°C in 1 second and the same procedure was used to collect several spectra. This was repeated at 800°C and 1100°C. A second sample was run in a similar manner. The difference is that the temperature was raised to 600°C immediately from room temperature and the temperature and pressure were maintained for three days.

Time dependence of differential stress is illustrated in Figure 21 for the different temperatures of the first data set. Here the shear stress, $[\sigma_3 - \sigma_1]/2$, is given as a function of time in minutes. At each temperature there is an initial fast relaxation that slows down with time. The strain rate derived from the slope of the equivalent strain vs time curve ranges from 10^{-4} to 10^{-7}/sec. The data given in terms of strain are plotted as a function of time on a log-log plot in Figure 22. The lines are straight suggesting a well defined value for n at each temperature. Lines are drawn on the figure with slopes corresponding to different values of n. The lower temperature data have large values of n, but by 1100°C the value of n decreases to about 3, a value that is consistent with a power law creep mechanism. Figure 23 shows the 3-day relaxation experiment at 600°C. Again the data define a straight line in a log-log representation. The slope and magnitude are consistent for the two different experiments.

Spinel has been studied in several different experiments in the past. Frost and Ashby (1982) summarize the rheological properties of spinel with a deformation mechanism map

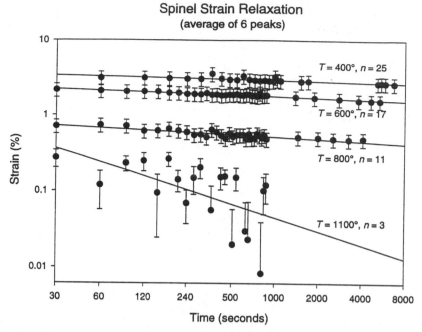

Figure 22. Strain as a function of time for spinel samples at different temperatures (from Weidner et al. 1998).

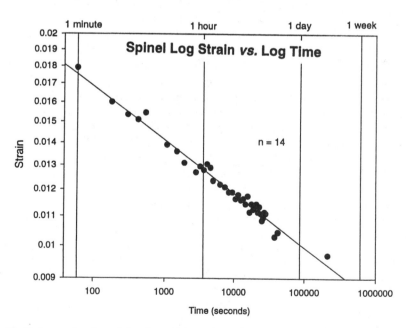

Figure 23. Strain as a function of time for spinel samples at 600°C for a 3-day experiment (from Weidner et al. 1998).

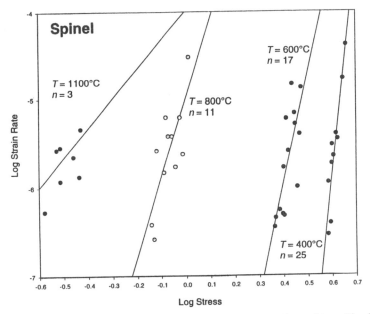

Figure 24. Log strain rate vs log stress for spinel at different temperatures (same data as Fig. 22). Slopes indicate effective exponent of a power law relationship.

as illustrated in Figure 2. The low temperature regime is characterized by high stress and very high values of the effective n. As temperature increases, the strain rate lines fan out, consistent with lower values of n and lower values of stress. We plot our data for shear stress as a function of temperature on the Frost and Ashby deformation map. Indeed our data follow the deformation region corresponding to our inferred strain rate very precisely. Our effective n values are quite consistent with those deduced from this diagram.

Rutter et al. (1978) suggests that, in the plastic regime, which would correspond to these data at 800°C and lower, that this projection should be concave upward asymptotically approaching the athermal stress with increased time. A straight line either implies insufficient time to relax to the athermal stress, or dislocation recovery processes are operative. The data do not allow a distinction between these possibilities. The fact that the relaxation continues for several days suggest that recovery processes are operative and that the flow is entering the power-law creep regime.

Strain rate can be derived from the time dependence of the observations, yielding a direct comparison with stress. Figure 24 illustrates the results for these spinel data in a log-log plot. In the single experiment strain rates from 10^{-7} to 10^{-4}/sec are observed for several different temperatures with a straight line relationship with stress whose slope yields an effective power in a power-law relationship. The calculated slope is in agreement with that deduced from the log stress - log time relations. The strain rate-stress curves for the three-day experiment are illustrated in Figure 25, where strain rates reach values close to 10^{-9}/sec. Still the values and slopes are internally consistent.

Microscopic deviatoric stress analyses discussed above provide a straight forward manner to define both stress and strain with a simple measurement. The deviatoric stress is generated as a polycrystalline specimen is compressed. The plasticity regime can be qualitatively identified by the value of n determined from the relaxation experiment. High

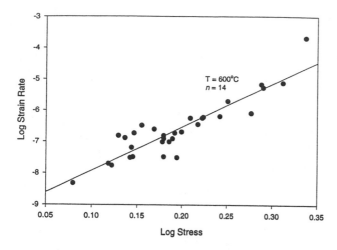

Figure 25. Log strain rate versus log stress for three-day experiment for spinel at 600°C. Strain rates approach values of 10^{-9}/sec.

values (>6) indicate a dislocation glide regime, intermediate values (2 > n > 6) indicate dislocation recovery mechanisms, while a value of 1 would indicate a diffusion deformation mechanism. Chen et al. (1998a) have used this methodology to study the effect of water weakening in polymorphs of olivine. They conclude that olivine is drastically weakened by 400°C in the dislocation glide regime with only a minor amount of water, while both wadsleyite and ringwoodite are considerably stronger than olivine in this regime and that even a few percent water, which is incorporated into the crystal structure, has little effect on the rheology in this regime. At the highest temperature (1000°C) there is some indication that the hydrous ringwoodite phase is slightly weakened relative to the dry sample.

Peak broadening is particularly suited for high values of deviatoric stress. Even diamond can be deformed in this system (Weidner et al. 1994b). It can be a particularly valuable tool for studying the transition from dislocation glide to the power-law creep regimes. An understanding of this regime will be useful for defining flow characteristics in the cold subducting lithosphere. Lower stress limits of about 100 MPa have been encountered. At this point the strain broadening of the x-ray peak is small compared to the inherent instrument width of the peak. Future improvements may come by using monochromatic sources which have higher resolution.

A further limitation comes from the fact that stress is generated on compression due to grain-grain interactions. After the stress relaxes owing to flow, it cannot be regenerated. This limits the experimental range to relatively small strains—a few percent. In this range, it is difficult to isolate transient creep from steady state creep.

Macro-scale stress

Stresses that are uniform on the scale of the x-ray sampling volume can be generated with either anisotropic loading of the cell, such as can be deliberately introduced in the diamond anvil system, or by anisotropic properties of the pressure medium that is loaded by a multi-anvil system. Macroscopic stresses can be present with no microscopic stress field. In this case the diffraction peaks will remain sharp, and information about the stress field will be contained in the positions of the diffracted peaks. The material in the stress field will strain in response to the stress as given by Equation (10). Or, rewriting this equation in terms of the elastic compliance tensor, **s**:

$$\varepsilon_{ij} = \sum_{k=1}^{3}\sum_{l=1}^{3} s_{ijkl}\sigma_{kl} \tag{28}$$

The position of the diffracted x-ray records the spacing of the specific lattice planes that are oriented parallel to the diffraction vector of the x-ray, which bisects the incident and scattered vectors. The macroscopic strain for a particular diffraction vector is thus:

$$\varepsilon_{11}(hkl) = \delta d_{hkl}/d_{hkl} \tag{29}$$

defining the direction, 1, parallel to the diffraction vector. There are two possible ways to determine the stress field from these equations. The first is to measure the strain for any diffraction peak while varying the orientation of the diffraction vector by an angle, ϕ, about the incident x-ray direction (see Fig. 26). This is done by changing the orientation of the detector. For each orientation, the strain measure of Equation (29) reflects the strain in the subset of grains that are oriented in diffracting condition (the subset changes with ϕ) and the strains are sampling different orientations of the stress field and elastic moduli. For example, a simple uniaxial stress will expand the lattice planes that are oriented perpendicular to the load direction and contract those parallel to it. Thus, the Debye rings, that are the powder diffraction rings that form circles around an incident monochromatic beam, will become distorted in response to the stress field. If the sample can be further rotated to allow all possible incident x-ray directions, then it is possible to map out the entire stress field. Geometric restrictions of the x-ray optics and high pressure apparatus often severely limits the orientations for x-ray access to the sample. A priori knowledge of the symmetry and orientation of the stress field must then be used. Kinsland and Bassett (1977) made use of this stress metric to measure high pressure strength of MgO and NaCl. The experiments were done in a diamond anvil cell with the x-ray beam entering the cell in the plane perpendicular to the diamond axis. Thus, the Debye rings sampled the stress field parallel and perpendicular to the loading direction.

A second type of measure of the deviatoric stress field can be made using diffraction for a single sample orientation if the a priori information about the stress field informs us that the principal stresses are of cylindrical symmetry and oriented either parallel or perpendicular to the diffraction vector (Singh and Kennedy 1974, Singh 1993, Singh and Balasingh 1994, Weidner et al. 1992, Weidner et al. 1994a). If the sample is elastically anisotropic, then each diffraction line will strain by a different amount. Following Equation (28) this becomes:

$$\varepsilon_{11} = s_{1111}\sigma_{11} + s_{1122}\sigma_{22} + s_{1133}\sigma_{33} \tag{30}$$

Figure 26. Illustration of Debye rings for diffracted x-rays. A macroscopic stress field will distort the circles.

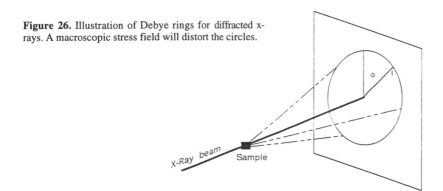

where the elastic moduli are evaluated in a coordinate system with the l direction parallel to the diffraction vector. With elastic properties that are direction dependent, as even cubic crystals exhibit, the coefficients of the stress terms will be multiplied by different constants for each diffraction peak. Using 2 or more diffraction peaks enable a determination of the axial and radial stresses. The diamond anvil cell standard x-ray geometry is with the x-ray beam incident through the diamond and the diffracted beam through the other diamond. This places the diffraction vector parallel to a radial stress axis. The standard multi-anvil geometry is with the diffraction vector parallel to the axis of the cylindrical sample chamber and hence the axial axis of the stress field. Both geometries can be used in Equation (30). This calculation must make the assumption that the polycrystalline stress-strain field is of the Reuss type. That is the grain samples the local stress field. If the grains sample the local strain field, then all grains will exhibit the same strain and no information can be gained about the deviatoric stress. Thus, we expect that the stress estimated in this fashion is a lower bound and may be of the order of half of the actual deviatoric stress.

With the introduction of beryllium gaskets, diamond anvil studies are able to utilize the first of these methods to characterize the deviatoric stress field in the diamond anvil cell as demonstrated by Hemley et al. (1997). The results of using the deformation of the entire Debye ring in conjunction with analyzing the strain fields of each diffraction line allow Singh et al. (1998) to estimate the elastic anisotropy of the sample, yielding information about the elastic constant tensor. Mao et al. (1998b) have further inverted this type of data to obtain strength at very high pressures in iron.

The use of macroscopic stress for defining flow laws is still in the development stage. Strain metrics, that can monitor plastic strain of the sample as a function of time are not yet developed for use with x-ray sources. Monitoring of sample length or strain markers such as that used by Karato and Rubie (1997) hold a possible application for in situ, time dependent stress-strain determinations. As these develop, it may become possible to derive the necessary data for large strains, assuring the data correspond to steady state processes and not just transient phenomena. As the next generation of experiments develop from these initial efforts, increased resolution, larger strains, lower stresses may soon become accessible to the experimental arsenal for defining rheological properties of Earth materials in a wide pressure-temperature space, and for all phases that exist within the Earth.

ACKNOWLEDGMENTS

The work of this paper was inspired and supported by many co-workers. Yaqin Xu, Frederic Bejina, Yujin Wu, Juhua Chen, Mike Vaughan all provided insights and fruitful discussions. Mike Vaughan helped with production and assembly of the figures. Jerry Hastings and Pete Siddons helped to provide reliable x-rays at NSLS on X17. This paper was supported by CHiPR (NSF EAR92-19245). Comments on the manuscript by R. Hemley and J.-P. Poirier were helpful. Thanks to all. This is MPI publication No. 239.

REFERENCES

Ando J, T Irifune, T Takeshita, K Fujino (1997) Evaluation of the non-hydrostatic stress produced in a multi-anvil high pressure apparatus. Phys Chem Minerals 24:139-148
Ashby MF (1972) A first report on deformation-mechanism maps. Acta Metall 20:887-897
Bejina F, P Raterron, J Zhang, O Jaoul, RC Liebermann (1997) Activation volume of silicon diffusion in San Carlos olivine. Geophys Res Lett 24:2597-2600
Bussod GY, T Katsura, DC Rubie (1993) The large volume multi-anvil press as a high P-T deformation apparatus. Pure Applied Geophys 141:579-599
Chai M, JM Brown (1996) Effects of static non-hydrostatic stress on the R lines of ruby single crystals. Geophys Res Lett 23:3539-3542

Chai M, JM Brown, Y Wang (1998) Yield strength, slip systems and deformation induced phase transitions of San Carlos olivine up to the transition zone pressure at room temperature. *In:* MH Manghnani, T Yagi (eds) Properties of Earth and Planetary Materials at High Pressure and Temperature 101:483-493 Am Geophys Union, Washington, DC

Chen J, T Inoue, DJ Weidner, Y Wu, MT Vaughan (1998a) Strength and water weakening of mantle minerals, olivine, wadsleyite, and ringwoodite. Geophys Res Lett 25:575-578

Chen J, DJ Weidner, MT Vaughan, R Li, JB Parise, CC Koleda, KJ Baldwin (1998b) Time resolved diffraction measurements with an imaging plate at high pressure and temperature. Rev High Pressure Sci Technol 7:272-274

Durham WB, DC Rubie (1998) Can the multianvil apparatus really be used for high-pressure deformation experiments? In: MH Manghnani, T Yagi (eds) Properties of Earth and Planetary Materials at High Pressure and Temperature 101:63-70 Am Geophys Union, Washington, DC

Evans AG, DR Clarke (1980) Residual stresses and microcracking induced by thermal contraction inhomogeneity. In: DPH Hasselman, RA Heller (eds) Thermal Stresses in Severe Environments. Plenum Press, p 629-649

Evans B, C Goetze (1979) The temperature variation of hardness of olivine and its implication for polycrystalline yield stress. J Geophys Res 84:5505-5524

Evans B, DL Kohlstedt (1995) Rheology of rocks. *In:*T Ahrens (eds) Rock Physics and Phase Relations: A Handbook of Physical Constants 3:148-165 Am Geophys Union, Washington, DC

Evans T, RK Wild (1965) Three point bending of single crystal diamond. Philos Mag 12:479

Fei Y, HK Mao, J Shu, J Hu (1992) P-V-T equation of state of magnesiowustite $(Mg_{0.6}Fe_{0.4})O$. Phys Chem Minerals 18:416-422

Frost HJ, MF Ashby (1982) Deformation-Mechanism Maps, The Plasticity and Creep of Metals and Ceramics. Pergamon Press, Oxford

Gerward L, S Morup, H Topsoe (1976) Particle size and strain broadening in energy-dispersive x-ray powder patterns. J App Physics 47:822-825

Green HW, TE Young, D Walker, CH Scholz (1990) Anticrack-associated faulting at very high pressure in natural olivine. Nature 348:720-722

Griggs DT (1967) Hydrolytic weakening of quartz and other silicates. Geophys J Royal Astr Soc 14:19-31

Hemley RJ, HK Mao, G Shen, J Badro, P Gillet, M Hanfland, D Hausermann (1997) X-ray imaging of stress and strain of diamond, iron, and tungsten at megabar pressures. Science 276:1242-1245

Hirth JP, J Lothe (1982) Theory of Dislocations, 2nd Edn. John Wiley & Sons, New York

Hull D, DJ Bacon (1984) Introduction to dislocations. Pergamon Press, Oxford

Ita J, RE Cohen (1998) Diffusion in MgO at high pressure: implications for lower mantle rheology. Geophys Res Lett 25:1095-1098

Jaoul O (1990) Multicomponent diffusion and creep in olivine. J Geophys Res 1990:17631-17642

Jaoul O, P Raterron (1994) High-temperature deformation of diopside crystal: 3. Influence of pO_2 and SiO_2 precipitation. J Geophys Res 99:9423-9439

Karato S-i (1989) Defects and plastic deformation in olivine. *In:* S-i Karato , M Toriumi (eds) Rheology of solids and of the Earth. Oxford Univ Press, p 176-208

Karato S-i, DC Rubie (1997) Toward an experimental study of deep mantle rheology: a new multianvil sample assembly for deformation studies under high pressures and temperatures. J Geophys Res B, Solid Earth and Planets 102:20,111-20,122

Kinsland GL, W Bassett (1977) Strength of MgO and NaCl polycrystals to confining pressures of 250 Kbar at 25°C. J Appl Phys 48:978-985

Kohlstedt DL, C Goetze, WB Durham (1976) Experimental deformation of single crystal olivine with application to flow in the mantle. *In:* RGJ Strens (eds) The physics and chemistry of minerals and rocks. John Wiley, New York, p 35-49

Kreher W (1990) Residual stresses and stored elastic energy of composites and polycrystals. J Mech Phys Solids 38:115-128

Liebermann RC, Y Wang (1992) Characterization of sample environment in a uniaxial split-sphere apparatus. *In:*Y Syono , MH Manghnani (eds) High Pressure Research: Application to Earth and Planetary Sciences, Geophys Monogr 67:19-31. Terra Scientific Publishing and Am Geophys Union, Tokyo and Washington, DC

Mao HK, G Shen, RJ Hemley, TS Duffy (1998a) X-ray diffraction with a double hot-plate laser-heated diamond cell. *In:*MH Manghnani, T Yagi (eds) Properties of Earth and Planetary Materials at High Pressure and Temperature 101:27-34. Am Geophys Union, Washington, DC

Mao HK, J Shu, G Shen, RJ Hemley, B Li, AK Singh (1998b) Elasticity and rheology of iron above 200 GPa adn the nature of the earth's inner core. Nature (in press)

Meade C, R Jeanloz (1988a) The yield strength of B1 sand B2 phases of NaCl. J Geophys Res 93:3270-3274

Meade C, R Jeanloz (1988b) The yield strength of MgO to 40 GPa. J Geophys Res 93:3261-3269

Meade C, R Jeanloz (1990) The strength of mantle silicates at high pressures and room temperature: implications for the viscosity of the mantle. Nature 348:533-535

Paterson MS (1970) A high temperature high pressure apparatus for rock deformation. Int'l J Rock Mech Min Sci 7:517-526

Poirier J-P (1985) Creep of Crystals: High-Temperature Deformation Processes in Metals, Ceramics and Minerals. Cambridge University Press, Cambridge, UK

Rutter EH, BK Atkinson, DH Mainprice (1978) On the use of the stress relaxation testing method in studies of the mechanical behavior of geological materials. Geophys J R Astr Soc 55:155-170

Singh AK (1993) The lattice strains in a specimen (cubic symmetry) compressed nonhydrostatically in an opposed anvil device. J Appl Phys 73:4278-4286

Singh AK, C Balasingh (1994) The lattice strains in a specimen (hexagonal system) compressed nonhydrostatically in an opposed anvil high pressure setup. J Appl Phys 75:4956-4962

Singh AK, GC Kennedy (1974) Uniaxial stress component in tungsten carbide anvil high-pressure x-ray cameras. J Appl Physics 45:4686-4691

Singh AK, HK Mao, J Shu, RJ Hemley (1998) Estimation of single-crystal elastic moduli from polycrystalline x-ray diffraction at high pressure: Applications to FeO and iron. Phys Rev Lett 80:2157-2160

Sung CM, C Goetze, HK Mao (1977) Pressure distribution in the diamond anvil press and the shear strength of fayalite. Rev Sci Instrum 48:1386-1391

Tvergaard V, JW Hutchinson (1988) Microcracking in ceramics induced by thermal expansion or elastic anisotropy. J Am Ceram Soc 71:157-166

Vaughan MT, DJ Weidner, YB Wang, JH Chen, CC Koleda, IC Getting (1998) T-cup: A new high-pressure apparatus for x-ray studies. Rev High Pressure Sci Technol 7:1520-1522

Walsh JB (1973) Theoretical bounds for thermal expansion, specific heat, and strain energy due to internal stress. J Geophys Res 78:7637-7646

Wang Y, RC Liebermann, JN Boland (1988) Olivine as an in situ piezometer in large-volume, high-pressure apparatus. Phys Chem Minerals 15:493-497

Weertman J, JR Weertman (1975) High temperature creep of rock and mantle viscosity. Ann Rev Earth Planet Sci 3:293-315

Weidner DJ, MT Vaughan, J Ko, Y Wang, X Liu, A Yeganeh-Haeri, RE Pacalo, Y Zhao (1992) Characterization of stress, pressure, and temperature in SAM85, A DIA type high pressure apparatus. *In:* Y Syono, MH Manghnani (eds) High-Pressure Research: Application to Earth and Planetary Sciences, Geophys Monogr 67:13-17. Terra Scientific Publishing Co and Am Geophys Union, Tokyo and Washington, DC

Weidner DJ, Y Wang, MT Vaughan (1994a) Deviatoric stress measurements at high pressure and temperature. *In:* SC Schmidt et al (eds) High-Pressure Science and Technology (Proc 1993 AIRAPT Conf), p 1025-1028

Weidner,DJ, Y Wang, MT Vaughan (1994b) Strength of diamond. Science 266:419-422

Weidner DJ, YB Wang, G Chen, J Ando (1998) Rheology measurements at high pressure and temperature. *In:* MH Manghnani, T Yagi (eds) Properties of Earth and Planetary Materials at High Pressure and Temperature 101:473-480 Am Geophys Union, Washington, DC

Yamazaki D, T Kato, E Ohtani (1996) Preferred orientation of olivine in stress field under ultra-high pressures. J Seismol Soc Japan 49:39-53

Chapter 17

VIBRATIONAL PROPERTIES
AT HIGH PRESSURES AND TEMPERATURES

Philippe Gillet

Institut Universitaire de France, Laboratoire de Sciences de la Terre
Ecole Normale Supérieure de Lyon
46 allée d'Italie, 69364
Lyon cedex 07 France

Russell J. Hemley

Geophysical Laboratory and Center for High Pressure Research
Carnegie Institution of Washington
5251 Broad Branch Road NW
Washington, DC 20015

Paul F. McMillan

Department of Chemistry and Biochemistry
and Center for Solid State Science
Arizona State University
Tempe, Arizona 85287

INTRODUCTION

The vibrational dynamics of minerals determine many of their fundamental properties, such as heat capacity and the lattice thermal expansion. The vibrational properties are governed by the same balance of interatomic forces that determine the mineral structure, so that vibrational spectroscopy is a useful and direct probe of changes in mineral structure and bonding at high pressures and temperatures, including phase transitions likely to occur deep within the Earth. These spectroscopic techniques are thus highly complementary to direct structural studies using x-ray and neutron diffraction. The diffraction techniques indirectly give vibrational information through the averaged displacements of atoms determined during structural refinements and are described by temperature (e.g. Debye-Waller) factors. These dynamical effects can also often be determined from vibrational data or calculated by the methods of lattice dynamics to provide a further useful complementarity between the vibrational and structural studies (Gillet et al. 1993). Vibrational spectroscopy is particularly useful in the study of second order displacive phase transitions associated with "soft mode" behavior, in which the vanishing restoring force to the set of atomic displacements associated with the "soft" vibrational mode drives the phase transition (Scott 1974, McMillan 1985, Dove 1993, Navrotsky 1994). In general, vibrational spectra are very sensitive to subtle changes in local or long-range symmetry, and detailed vibrational studies often permit elucidation of temperature- or pressure-induced phase transition mechanisms through an identification of the microscopic order parameters (Salje 1990, Dove 1993).

The vibrational spectra of minerals are conveniently measured in the laboratory by non-destructive optical techniques, primarily infrared and Raman spectroscopies (McMillan and Hofmeister 1988). These techniques are readily coupled with high-pressure devices

0275-0279/98/0037-0017$05.00

such as diamond cells, as well as with high-temperature furnaces or laser heating methods, to yield information on mineral dynamics over the entire range of P-T conditions of the deep interior of the Earth and other planets (Hemley et al. 1987, Sharma 1989, McMillan and Wolf 1995, Gillet 1996). The interatomic forces that determine the vibrational dynamics also determine the mineral elasticity, which can be readily probed in situ at high P and T by Brillouin scattering (Liebermann and Li, Chapter 15) or other laser spectroscopic techniques, such as impulsive stimulated scattering (Zaug et al. 1992). This technique can also be used to study the thermal conductivity of minerals (Chai et al. 1996).

The vibrational spectrum of a given mineral is highly characteristic of that phase, and infrared and Raman spectroscopies are useful techniques for phase identification, even in petrologic thin section (Mao et al. 1987, Gillet and Goffé 1988, McMillan 1989, McMillan et al. 1996a,b). This property has proved useful both in studies of natural rocks and for examination of experimental run products containing high-pressure minerals. The vibrational spectra also reveal the presence of molecular species contained within the crystal, such as trace quantities of OH in mantle minerals (Bell 1992, Bell and Rossman 1992, Rossman 1996). The mineral composition of the Earth's crust and mantle is now generally well established (Anderson 1989, Poirier 1991, Gillet 1995). Vibrational spectroscopy on both natural samples and samples recovered from high pressure experiments has played a key role in establishing the mineral assemblages and their phase relations (McMillan et al. 1996a,b).

A primary focus of current research has now moved toward studying the vibrational spectra of minerals in situ at high pressures and at high temperatures, to extract structural and thermodynamic information relevant to the deep Earth. Such studies have become possible mainly due to significant technical advances over the past 10 to 15 years, that permit high quality in situ measurements to be made under simultaneous high P-T conditions. The experiments complement advances in theoretical techniques that allow the prediction of vibrational dynamics of minerals with increasing accuracy (Bukowinski 1994, Chizmeshya et al. 1994, Stixrude et al., Chapter 19, this volume).

In addition to providing a means for phase identification and structural characterization, obtaining the vibrational spectra of minerals has been particularly important for constraining and developing our understanding of theoretical models of the thermodynamic and bonding properties of minerals calculated with empirical or ab initio methods, which are used to predict mineral stability and properties under various P-T conditions (Tossell and Vaughan 1992, Navrotsky 1994). Vibrational spectra are also used to construct and constrain models for the heat capacity, vibrational entropy and equations of state of minerals (Anderson 1995). In situ studies of the vibrational properties of minerals at high temperature and high pressure reveal immediately and directly the intrinsic anharmonic nature of the vibrations. The anharmonicity of vibrational modes obtained from spectroscopic measurements at high temperature and at high pressure can be quantified and incorporated in the thermodynamic calculations. These issues are described and discussed below.

The vibrational properties of high-pressure minerals have been recently reviewed by McMillan et al. (1996b). In this chapter, we give an overview of approaches used to obtain vibrational spectra of minerals in situ under high pressure, high temperature, or simultaneous high P-T conditions relevant to the deep Earth. The principle focus is on Raman, infrared, and Brillouin spectroscopies, but other new methods are also presented. We also provide examples of the methodology used to extract information on the thermodynamic and thermoelastic properties of minerals from the vibrational data, in concert with measurements of structural parameters and elastic properties at high P-T

conditions, using simple mineralogical systems as examples. Finally, we summarize some recent high *P-T* results on minerals, and identify critical areas where further work is needed to understand the mineralogical basis of the properties and processes of deep planetary interiors.

BASIC CONSIDERATIONS

Mineral lattice vibrations

Atoms in crystals oscillate about their equilibrium positions, and since they are periodically arranged and interact with their neighbors, the vibrations take the form of atomic displacement waves traveling through the whole lattice (Born and Huang 1954, McMillan 1985, Dove 1993). These waves can be either longitudinal (L) when the atom displacements lie parallel to the wave propagation direction, or transverse (T) when the displacements are perpendicular to the wave propagation direction. For most minerals, the lattice vibrations contain both T and L components: in cases of high symmetry, the L and T displacements constitute separate vibrational modes.

Within a given (primitive) crystallographic unit cell, the oscillation frequency of atoms (ν) about their equilibrium positions is determined by the relative positions of the atoms within the unit cell, their masses, and the bonding or repulsive forces between them. In solid minerals, the vibrational frequencies are also dependent upon the phase relationship between the atomic displacements in neighboring unit cells, which determines the wavelength (λ) of the resulting lattice vibrational wave. The wave vector is given by $q = 2\pi/\lambda$, in the direction of propagation of the vibrational wave. A vibrational mode in a crystal is described by its frequency (ν_i), for a given wave vector q. This is a classical description of crystal lattice dynamics. In a quantum mechanical view, the vibrational excitations in crystals are described in terms of *phonons*, which correspond to units of vibrational energy transfer between vibrational states. The phonons possess properties analogous to particles ("quasi-particles"), particularly in terms of their momentum (or "quasi-momentum", because the phonons do not possess true mass) exchange and scattering phenomena (Reissland 1973). These properties are important in determining the thermal conductivity of non-metallic minerals (Wallace 1972, Reissland 1973, Dove 1993, Chai et al. 1996).

The dependence of lattice vibrational mode frequencies, or phonon energies, on the wave vector constitutes the phenomenon of *dispersion* of the lattice vibrations, and the mode frequencies are grouped into *branches* ($\nu_i(q)$) for vibrations with similar patterns of atomic displacement within a given unit cell, but with different phase relationships between adjacent unit cells (McMillan 1985, Dove 1993). The slopes of the $\nu(q)$ relation for the three branches with lowest frequency (termed the *acoustic* branches, because of their role in the propagation of sound waves through the mineral) in the limit of long wavelength (i.e. as $q \rightarrow 0$) determine the transverse and longitudinal sound speeds in the mineral, and hence its elastic constants. The other branches give rise to the *optic* vibrational modes, because it is these that interact with light. If there are n atoms in the primitive unit cell, there are $3n$ phonon branches. Because three of these are acoustic branches, there are $3n-3$ optic branches, and hence $3n-3$ optical vibrations in the limit of long wavelength (i.e. as $q \rightarrow 0$, at the center of the first *Brillouin zone*, in reciprocal space).

The optical vibrations correspond to specific displacement patterns of atoms within the crystal unit cell. If the mineral is a simple metal or an ionic crystal, such as Fe or MgO, the atomic displacements are simply described in terms of Cartesian displacements of the atoms or ions about their equilibrium positions. If the mineral contains identifiable "molecular" units, such as the silicate (SiO_4^{4-}) tetrahedra in olivine or the trigonal planar carbonate groups in magnesite ($MgCO_3$), some of the vibrational modes can be described in terms of the "internal" vibrations of the molecular units (Fig. 1). The remainder of the vibrational modes constitute the "external" vibrations, which involve motions of the counterions (e.g. Mg^{2+} for forsterite and magnesite), and hindered translations and rotations of the "molecular" units (Fig. 1).

For a given vibrational mode, at $q = 0$, the atomic displacements are identical (i.e. they are in phase) from one unit cell to another. The vibrations are then described simply in terms of the symmetry species of the symmetry point group for the unit cell. The number and type of each symmetry species, as well as their spectroscopic activity, are readily enumerated using the technique of factor group analysis (McMillan 1985, McMillan and Hess 1988). For magnesite, for example, the symmetries of the 30 vibrational modes obtained from factor group analysis are:

$$\Gamma = 1A_{1g} + 4E_g \qquad \textit{Raman active}$$
$$+ 4A_{2u} + 5E_u \qquad \textit{IR active}$$
$$+ 1A_{2u} + 1E_u \qquad \textit{acoustic}$$
$$+ 3A_{2g} + 2A_{1u} \qquad \textit{inactive}$$

and the related atomic motions are shown in Figure 1. The A_{2g} and A_{1u} modes are spectroscopically inactive because they do not give rise to any interaction with the absorbed light that leads to an infrared or Raman signal. The activity of vibrational modes are determined by quantum mechanical selection rules for the vibrational transitions (McMillan 1985, McMillan and Hess 1988). These selection rules are dictated by the interactions between the oscillating electric field of the light absorbed by the mineral and the change in permanent dipole moment or in that induced by the light interaction with the vibrating atoms.

Overview of spectroscopic techniques

Optical spectroscopy (i.e. those using UV-visible-IR light) probes excitations in materials in the long wavelength limit, at $q \sim 0$, because the wavelength of this radiation is long compared with the lattice spacings in the crystal (McMillan 1985, McMillan and Hofmeister 1988, Dove 1993). Infrared absorption or reflection spectroscopy involves direct interaction between the radiation in the infrared region of the spectrum and vibrational modes with the same energies, via a resonance phenomenon. In a conventional infrared absorption experiment, broad-band IR radiation is passed through the mineral and the intensity of the transmitted light is measured as a function of the wavelength (McMillan and Hofmeister 1988). Missing intensities at given wavelengths in the transmission spectrum correspond to the energies of vibrational transitions. The electric field of the incoming IR light interacts directly with the oscillating dipoles induced by the lattice vibrations. A vibrational mode is thus IR-active only if it induces a change in the dipole moment of a bond. Asymmetric vibrations give in general strong IR absorption bands because they induce larger changes in dipole moments than more symmetric vibrations (see for example the IR v_3 mode in magnesite, described below).

Most laboratory studies utilize "conventional" IR sources, in particular, a "globar" (heated ceramic) source for the mid-IR (500-8000 cm^{-1}) or a Hg emission lamp for low frequency studies in the "far" infrared (10-500 cm^{-1}). These generally give good results when coupled with the Fourier transform (FTIR) technique. Recently, infrared spectroscopic techniques that involve the use of intense synchrotron infrared radiation as a source, which can have 10^2-10^4 times the flux of a conventional source, have been developed. The high flux makes the technique particularly useful for high-pressure studies and microspectroscopy (Reffner et al. 1994, Meade et al. 1994, Carr et al. 1995, Lu et al. 1998).

ν_1 ν_2

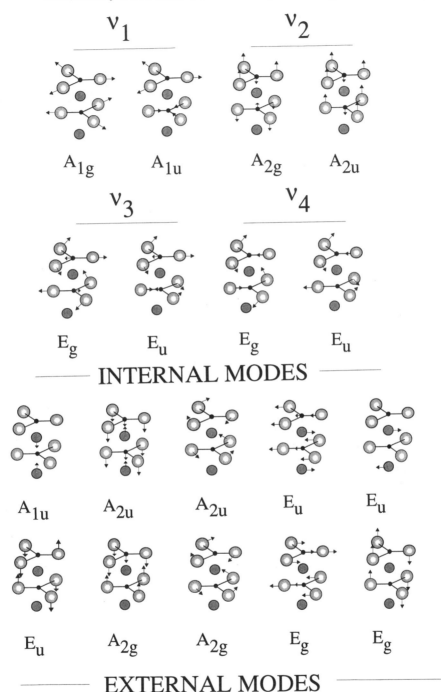

A_{1g} A_{1u} A_{2g} A_{2u}

ν_3 ν_4

E_g E_u E_g E_u

INTERNAL MODES

A_{1u} A_{2u} A_{2u} E_u E_u

E_u A_{2g} A_{2g} E_g E_g

EXTERNAL MODES

Figure 1. Atomic displacements associated with the 27 optical vibrations of magnesite (after White 1974). The 3 acoustic modes are not shown. The letters and subscripts (e.g., E_g) refer to symmetry species and are numbered according to classical group theory.

Infrared *reflectivity* is measured by reflection of the IR beam from a sample surface, and the reflectance spectrum gives a rich array of information on the optical and dielectric properties as well as the vibrational behavior of the sample. In particular, the lower and upper limits of the reflectance band associated with the IR-active mode give the frequencies of the transverse and longitudinal components associated with that vibration [called transverse optic (TO) and longitudinal optic (LO)] (McMillan and Hofmeister 1988). The reflectance technique is easily applied at ambient temperature and high temperatures; although it is less well adapted to in situ studies at high pressure in the diamond cell, vibrational reflectance spectra above 200 GPa have been measured using synchrotron radiation (Hemley et al. 1996, Goncharov et al. 1996). The IR absorption spectrum is measured via transmission of the infrared beam through a thin plate of sample, or through a suspension of finely powdered sample held in a transparent medium. This technique is readily carried out in the diamond cell. In general, with a well collimated beam and high-quality sample, IR absorption peaks correspond to the transverse optic (TO) vibrational modes only; however, the transmission spectrum can contain some component of reflectance, and features due to LO vibrations can be observed (Hofmeister 1996).

In a typical Raman scattering experiment, visible light from a laser source passes through the sample. About 10^{-3} of the incoming light is scattered by the crystal atoms. Most of this scattered light exits at the same wavelength (elastic Rayleigh scattering). About 10^{-6} of the light interacts inelastically with the sample, inducing transitions between vibrational states, so that the scattered light contains components at a different energy from that of the incoming radiation (Raman scattering) (Long 1976, McMillan and Hofmeister 1988). The energy difference between the incoming light and the inelastically scattered light corresponds to the energy of the vibrational transitions. From a microscopic point of view, the electrical field of the incoming light creates an instantaneous dipole moment by deforming the electronic clouds of the bonded atoms. The nuclei tend to move in the direction of the new position of the electrons. When the nuclear displacements correspond to those of a crystal vibrational mode, then Raman scattering occurs. The inelastically scattered light is analyzed with a spectrometer and the Raman spectrum consists of peaks shifted in energy from the Rayleigh line (i.e. the energy of the incident beam). The positions of the Raman peaks relative to the Rayleigh line provide the frequencies of the Raman-active vibrations (usually expressed in cm^{-1}; Fig. 2). In general, the most symmetric modes give rise to strong Raman peaks because they induce the largest change in bond polarizability (Long 1977, McMillan 1985, McMillan and Hofmeister 1988).

The general features of IR and Raman spectra of minerals are illustrated using magnesite ($MgCO_3$) as an example (Fig. 3). This phase has been shown to be extremely stable under ultrahigh pressure and temperature conditions (Gillet 1993, Biellmann et al. 1993), and it may act as a source or sink for carbon-bearing species in the mantle. The primitive unit cell contains two $MgCO_3$ formula units, so that $n = 5$. There are thus $3n-3 = 27$ optic vibrations, which have the symmetry species described in the previous section. The internal vibrations of the CO_3^{2-} groups are readily identified in the Raman and IR spectra of the crystals (Figs. 2 and 3). These modes lie at high frequencies (between 600 and 1500 cm^{-1}) (White 1974). The remaining external modes which involve translation and rotation of the CO_3 groups and vibrations of the Mg atoms lie between 80 and 450 cm^{-1}. Spectra taken of isotopically substituted species can be useful in identifying vibrational modes (Fig. 4), and also in thermodynamic calculations of isotopic fractionation between different phases, using the vibrational data (Gillet et al. 1996).

Brillouin spectroscopy is an optical technique related to Raman scattering in which the monochromatic incident light is scattered by the three (two transverse and one longitudinal) acoustic branches at very low frequency. Brillouin scattering is used to measure the

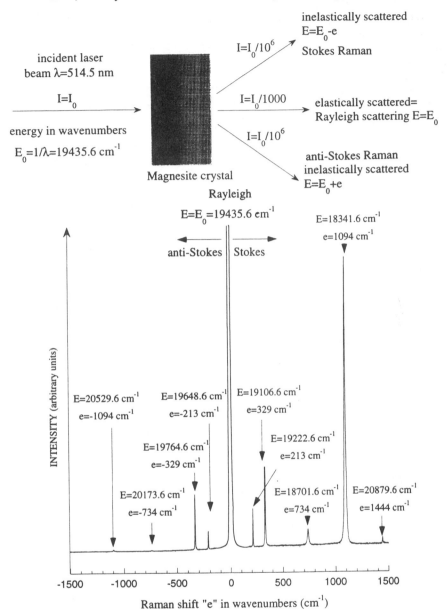

Figure 2. The Raman scattering process. In this example, the incident beam of an Ar laser ($\lambda = 514.5$ nm corresponding to an energy (in wavenumber units) of 19435.6 cm^{-1}) is scattered by a magnesite single crystal. $1/10^3$ of the incident beam intensity is elastically scattered at the same energy and corresponds to Rayleigh scattering. $1/10^6$ of the incident ligth is scattered inelastically and is accompanied by positive (anti-Stokes scattering) or negative (Stokes scattering) energy changes. These energy modifications result from transitions in vibrational energies related to the interaction of the incident beam with crystal vibrations. The intense peak observed on the Stokes side at an energy of 18431 cm^{-1} corresponds to a Stokes shift of 1094 cm^{-1}. A similar peak with less intensity is observed on at 20529.6 cm^{-1}, with an ant-Stokes shift of -1094 cm^{-1}. The Stokes peaks are more intense than the anti-Stokes peaks because the ground state is more populated than higher vibrational states.

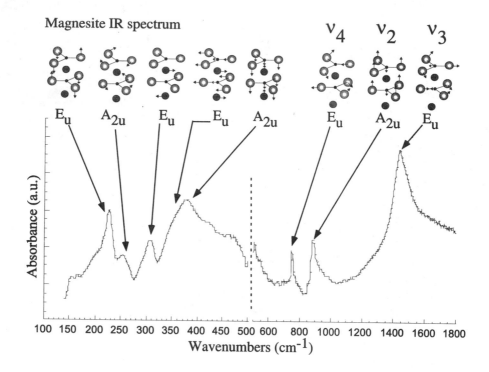

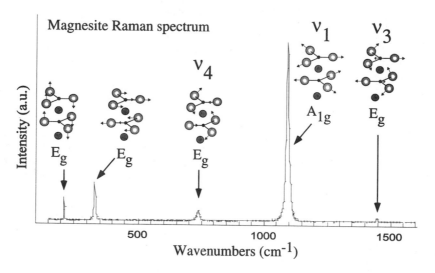

Figure 3. (a) Raman spectrum of magnesite. Unpolarized spectrum recorded with a micro-Raman spectrometer. The atomic motions involved in the Raman-active vibrations are outlined. The symmetric mode v_1 gives rise to a strong Raman peak because the associated vibration induces a large change in bond polarizability. (b) IR absorption spectrum of magnesite obtained on a cold-pressed powder (Grzechnik et al. 1998). The asymmetric mode v_3 gives rise to a strong IR absorption band because of the large dipole moment change in the dipole moment of the CO_3 group.

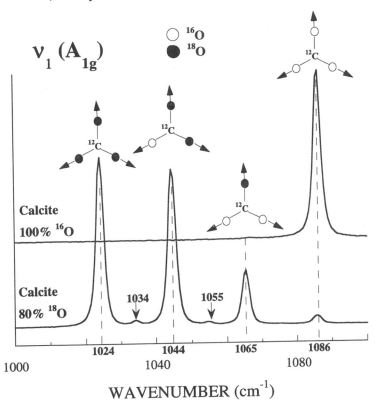

Figure 4. Raman spectrum of the symmetric strectching vibration of the CO_3 group in natural calcite containing mostly ^{16}O compared to that of a synthetic calcite sample containing 80% ^{18}O and 20% ^{16}O (after Gillet et al. (1996)) The incomplete substitution of ^{18}O leads to the presence of four types of CO_3 group within the calcite structure. The change of symmetry of the CO_3 groups associated with this substitution leads to a change in the selection rules and four peaks instead of one are observed in the ^{18}O-rich sample. The frequencies of the symmetric stretch in the substituted sample decrease with increasing reduced mass of the oxygen ions in the CO_3 group.

velocity of acoustic waves travelling through the mineral and thus the related elastic moduli. With the availability of highly sensitive detectors and spectrometers, both Raman and Brillouin scattering spectroscopies are readily carried out in situ at high temperatures and at high pressures, in the diamond anvil cell. Another related technique recently applied to mineralogy for studying acoustic phonons uses impulsive stimulated scattering (Zaug et al. 1992). This technique has also been also used to measure the thermal conductivity of minerals (Chai et al. 1996). Liebermann and Li (Chapter 15, this volume) provide a review of other techniques used to study elastic properties of minerals, and their applications to deep Earth materials.

Effects of dispersion

Because of the $q~0$ "selection rule", IR and Raman spectra of minerals are usually quite simple and are easy to interpret in terms of the number of modes predicted to appear by symmetry analysis. This leads to the great power of IR and Raman spectroscopy in detailed studies of changes in mineral structure and bonding at high pressure and temperature, through analysis of the frequencies, linewidths, and shapes of the zone center vibrational modes. However, determination of thermodynamic properties such as the heat capacity or vibrational contribution to the entropy requires knowledge of the full dispersion curves, or their sum throughout the entire Brillouin zone: this is the vibrational density of states [VDOS, or $g(v)$] (Wallace 1972, Kieffer 1985). For minerals at ambient pressure, this information has been obtained from inelastic neutron scattering experiments, (Ghose 1988) which generally requires large samples. Recent advances in "large volume" cell design (Mao and Hemley, Chapter 1, this volume) have now enabled inelastic neutron scattering studies to be carried out in situ at high temperatures to pressures in the several tens of GPa range (Parise et al. 1994, Schwoerer-Böhning et al. 1996). Measurements on Fe_3Pt performed to 7 GPa reveal the pressure dependence of the transverse acoustic branch frequencies (Schwoerer-Bohning et al. 1996). The results indicate the potential for studying core-alloy materials at high pressures; however, it is difficult to envisage this technique attaining the megabar pressure regime in the near future.

Recently, the availability of high brightness x-ray synchrotron sources has rendered possible measurement of phonon dispersion relations and $g(v)$ via inelastic x-ray scattering (Sette et al. 1998, Schwoerer-Böhning et al. 1998a,b; Alp et al. 1998). Although the sample size requirements may remain too large for diamond cell applications, this technique might provide a convenient way for measurement of $g(v)$ relations at high pressure in the near future. Phonon dispersion relations can also be obtained via nuclear resonant forward scattering (Alp et al. 1998, Hemley et al., Chapter 18), which holds great promise for $g(v)$ in situ measurements at high-pressure .

Alternative methods to overcome the problem posed by the limited sampling of the vibrational modes across the Brillouin zone by IR and Raman experiments include (1) supplementing experimental data with calculations of the full mineral lattice dynamics (Sangster 1970, Bukowinski 1994, Stixrude et al., Chapter 19, this volume), or (2) constructing model $g(v)$ functions constrained by the observed spectra (Kieffer 1985, Gillet 1996, Gillet et al. 1996). Both of these methods have now been applied with some success to enable geophysically useful calculations of mineral thermodynamic properties to be made based on the spectroscopic data obtained in situ at high pressure and high temperature.

IN SITU HIGH-PRESSURE MEASUREMENTS

Diamond cells

The diamond anvil cell is the most versatile apparatus for carrying out in situ spectroscopic measurements at very high static pressures (Mao and Hemley, Chapter 1, this volume). This utility is related to the great strength of diamond, coupled with its high transparency throughout wide portions of the electromagnetic spectrum. Several review papers provide the basic aspects on diamond-cell technologies and their use for vibrational and other spectroscopic measurements at high pressures (Jayaraman 1983, Hemley et al. 1987, Gillet 1996, Le Toullec et al. 1988, Chervin et al. 1992). The general principle of optical measurements of samples in the diamond cell is illustrated in Figure 5. The material is placed in the hole (usually 10-200 μm in diameter) of a gasket squeezed between two gem-quality diamonds. Small chips of ruby are used for measuring pressure by excitation of their fluorescence (Mao et al. 1986). The gasket hole is filled with a pressure-transmitting medium (soft solid, liquid, or rare gas). Laser and IR beams can be transmitted through the diamonds and interact with samples under pressure. Most of the vibrational spectroscopic experiments carried out on minerals at ambient conditions can be performed on samples in a diamond cell (Fig. 5) using conventional spectroscopic instrumentation and by adapting the optical path of the incoming light to the geometrical constraints of the cell. However, some requirements are necessary for the sample preparation, for the diamonds, and for the choice of the pressure-transmitting medium.

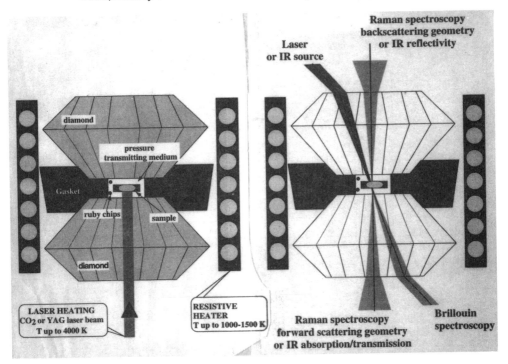

Figure 5. (a) High *P-T* optical spectroscopy in the diamond cell. The sample is placed in the hole (e.g. 10-200 μm in diameter) of a metallic gasket squeezed between two gem-quality diamonds. Small chips of ruby are used for measuring pressure by excitation of their fluorescence. The gasket hole is filled with a pressure-transmitting medium (soft solid, liquid, or rare gas). High temperatures are achieved either with small resistive heaters surrounding the diamonds or by focussing on the sample through the diamonds the beam of a high power IR laser (YAG or CO₂). Absorption of the beam by the sample permits heating up to 4000 K at pressures up to at least 135 GPa. (b) Various electromagnetic beams (visible, IR) can be directed on the sample through the diamonds for direct probing of the sample under sustained high pressures and temperatures. The samples can also be studied at ambient conditions after pressure and temperature release.

Diamonds must be correctly choosen for their transparency to visible and IR light to carry out IR, Raman, or Brillouin spectroscopy at elevated pressures (Ferraro 1984, Chervin 1998). Pure diamond has no first-order IR spectrum, but has a first-order Raman spectrum consisting of a single sharp peak at 1332 cm^{-1}, and a second order Raman spectrum arising from overtone vibrations which extends from 2200 cm^{-1} to 2800 cm^{-1} (Fig. 6). Most natural diamonds contain impurities, particularly nitrogen, which lead to fluorescence in the visible region of the spectrum, giving rise to a background signal in Raman spectra (Fig. 6). This fluorescence signal can be so strong that it masks the weak Raman signal of the the sample contained between the diamonds, so that diamonds (type Ia) must be carefully selected for the spectroscopic experiment. In some cases type Ia diamonds also exhibit sharp luminescence bands which can be observed during Raman spectroscopic experiments (Fig. 6). Type Ia diamonds strongly absorb IR radiation at 1000-1400 cm^{-1} due to N impurities (Fig. 6), which precludes their use for IR measurements in this range. Type IIa diamonds are best suited for high-pressure IR experiments. All diamonds have an intrinsic second-order phonon absorption at 1700-2700 cm^{-1}; however, recent work has shown that excellent spectra can be measured through this range by the use of small (<2 mm) diamonds (Struzhkin et al. 1997).

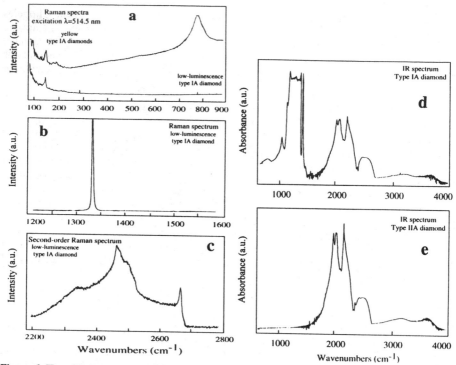

Figure 6. IR and Raman spectra of diamond (from Chervin 1998). (a) Comparison of a highly-fluorescent yellow type-Ia diamond and a low-luminescence type Ia diamond. The spectra were obtained with the 514.5 nm beam of an Ar laser. The fluorescent diamond exhibits a strong background fluorescence as well as a large peak centered around 800 cm^{-1}.which position changes with the wavelength of the excitation laser. (b) First-order spectrum of diamond. The peak at 1333 cm^{-1} corresponds to the C-C stretching vibration (T_{2g} symmetry). (c) Second-order Raman spectrum of a low-luminescence type Ia diamond. (d) Absorption IR spectrum of a polished gem-quality platelet of a type Ia diamond. (e) Similar spectrum for a type IIa diamond.

High-pressure Raman spectroscopy

In the past two decades new generations of Raman spectrometers have appeared, especially micro-Raman spectrometers. These permit focusing the exciting laser beam on the sample and the collection of the Raman signal through the objective of an optical microscope (Turrell and Corset 1996). Microscope objectives with high numerical aperture (typically >0.7) focus the laser beam into a very small sample volume (on the order of 3-10 μm^3) and efficiently collect the light scattered by this volume. Special care must be taken with some microscope objectives, as their coatings may give rise to Raman or narrow fluorescence bands which may obscure the Raman signal of the studied samples. Refinements in micro-Raman spectroscopy, including the use of confocal optics and new scattering geometries, revolutionized the study of minerals at high pressure in diamond cells (Hemley et al. 1987). The use of confocal optics in the light path permits substantial reduction in extraneous background radiation, such as fluorescence from the diamonds or blackbody radiation from furnace components, because the signal is obtained primarily from the sample volume illuminated by the incident laser optics   (Dhamelincourt et al. 1993, McMillan and Wolf 1995). Significant lowering of the recording time and greatly improved signal/noise ratio are now achieved with charge-coupled-device (CCD) detector systems. Although useful spectroscopic studies can be carried out using scanning spectrometers and photomultiplier detection in the diamond cell (Chopelas 1990, Chopelas and Boehler 1992), this is difficult at very high temperature, or for weakly scattering samples at high pressure (Hemley et al. 1987, McMillan and Wolf 1995). Multi-channel Raman spectroscopy with sensitive (now using back-thinned detectors) CCD detectors has revolutionized in situ studies at high pressure and high temperature of weakly scattering samples, including opaque samples and even metals.

Several other major advances in the basic instrumentation for Raman spectroscopy have occurred in recent years. Most of this has been the result of the development of holographic filters, conceptually simple optical systems that provide exceedingly fine discrimination at a prescribed wavelength and thereby provide excellent scattered light rejection that is essential for Raman spectroscopy (Kim et al. 1993). The invention and application of these devices are obviating the need for double and triple spectrographs, thereby allowing higher light throughput as well as much simpler (and inexpensive) instrumentation to be used. A second generation of "supernotch" filters has now been developed which provide even better performance (Schoen et al. 1993), and these are now in routine use for high pressure optical studies of deep Earth materials. More recently, holographic transmission gratings have been developed which can replace conventional reflecting gratings; systems employing these gratings have few optical components and have high light throughput (Arns 1995) (although they are somewhat less flexible). This technology continues to evolve, with important implications for high-pressure mineral physics.

High-pressure Raman spectroscopy at room temperature. High-pressure Raman spectra of Earth materials can now be obtained routinely at room temperature over the entire pressure range of the Earth's mantle (e.g. to >135 GPa) with diamond cells. With care, measurements can be carried out to significantly higher pressure, as demonstrated by studies of compressed molecular systems and diamond to >250 GPa (Mao and Hemley 1994). Measurements in this range are typically complicated by the appearance of stress-induced fluorescence from natural diamonds (Mao and Hemley 1991). Recent work has shown that this background signal is suppressed with the use of ultrapure synthetic single-crystal diamond anvils (Goncharov et al. 1998). Recent results for selected high-pressure minerals are reviewed later in this chapter.

The diamond cell is ideally suited for micro-Raman spectroscopy, which permits precise control over the region within the sample examined, and gives excellent signal/noise ratio. The use of confocal optics is of prime importance for discriminating the Raman signal from background fluorescence of diamonds. Finally, it should be mentioned that all the transmitting pressure media used in all high-pressure experiments have Raman spectra which may interfere with that of the sample. This is obviously the case for organic fluids, such as alcohol mixtures or hydrocarbons which have first-order Raman bands in the 900-1300 cm^{-1} region (Lemos and Camatgo 1990), but also occurs for the rare gases and soft salts which have no first order Raman spectrum, but posses second-order Raman spectra below 300 cm^{-1}. The samples can be of various types: polished single crystals, sintered polycrystals or powders dispersed in the pressure-transmitting medium. With improvements in the sensitivity of detectors, it should be pointed out that high-pressure Raman scattering measurements can also be performed on samples with high optical absorption or reflectivity, including metals (e.g. Massey et al. 1990, Olijnyk 1992).

High P-T Raman spectroscopy. The heating of samples pressurized in diamond cells has permitted major advances in our understanding of the behavior of minerals under the extreme pressure and temperature conditions especially relevant to the Earth's upper and lower mantle. Two different approaches are used for the heating of samples in the diamond cell: resistive heating and laser heating. Heating in the diamond cell (in principle up to ~1500 K) can be achieved by placing small resistance furnaces surrounding the diamond anvils (Fig. 5). This technique has been successfully used in high-pressure synchrotron x-ray diffraction experiments, for example to determine high temperature equations of state of mantle silicates, iron and sulfides (Duffy and Wang, Chapter 14, this volume). There have been a few attempts to use this technique to record vibrational spectra of Earth materials at simultaneous high pressure and temperature (Arashi 1987, Kraft et al. 1991, Farber and Williams 1996, Cooney et al. 1992, Wolf and McMillan 1995). These measurements have generally been limited to temperatures below 1100 K because of (1) mechanical instability of the cell during heating and thus pressure variations, (2) graphitization of diamonds and (3) need for using a pressure calibrant that remains robust at high temperature (i.e. an alternative such as Sm-YAG because ruby fluorescence vanishes above 800 K). It is expected that this technique, which permits precise temperature control and measurement, can be considered a "workhorse" for high-pressure measurements in the lower temperature range.

Laser heating constitutes an alternative method for heating samples compressed in diamond cells (Mao and Hemley, Chapter 1). An infrared CO_2 laser technique has been developed and used to obtain high quality Raman spectra under simultaneous high P-T conditions up to 30 GPa and 2000 K (Gillet et al. 1993) (Fig. 7). Type IIa diamonds, selected for low fluorescence and transparency to the 10.6 μm wavelength of the laser, are needed for such studies. The beam of a CO_2 laser is focused through a ZnSe plano-convex lens on the pressurized sample. Hot spots, 30-100 μm diameter-sized, are created on the sample. A spectrometer arrangement can be designed to permit simultaneous recording of the Raman spectra, the ruby fluorescence and the thermal emission of the sample. A microscope with a long-working distance objective permits focusing the excitation laser beam to a 2-μm spot and collection of the Raman signal in the backscattered direction (Fig. 7). The presence of a confocal pinhole before the spectrometer entrance ensures a sampling of the same 2-3 μm-sized zone for both the Raman and thermal emission signals. The imaging system allows one to illuminate ruby chips placed at the edge of the gasket hole and to collect their fluorescence spectrum via optical fibers. Optical fibers are also used to collect the thermal emission from a 4 μm diameter zone of the heated spot (Fig. 7). During local heating of the sample, the ruby, the diamonds and the gasket remain cold (T<600 K). A similar arrangement has been used for synchrotron x-ray diffraction (Fiquet et al. 1996 1998). Phase transitions in $CaTiO_3$ and the high P-T stability of $MgCO_3$ have been studied using the combined CO_2 laser heating/Raman spectroscopy technique (Gillet 1993, Gillet et al. 1993) (Fig. 8).

Raman spectroscopy at simultaneous high P-T conditions with either external or laser heating techniques is still in its infancy. In the past, the accuracy of temperature measurements in conventional single-sided laser heating has been much discussed (Fiquet et al. 1996, Heinz and Jeanloz 1987, Boehler and Chopelas 1992, Mao et al. 1998, Manga and Jeanloz 1998). Because of the development of thermal pressure on localized heating (Heinz 1990, Fiquet et al. 1996, Dewaele et al. 1998, Andrault et al. 1998), corrections must be made if the pressure is measured in the unheated part of the sample assembly. Rapid advances are being made to resolve instrumental problems associated with the technique. For example, the double-sided laser heating with complementary mode-tuned lasers has produced stable and uniform heating within a sample (Shen and Heinz, Chapter 12, this volume). It is likely that a variety of optical spectroscopic measurements on mineral systems at simultaneous high P-T conditions using externally- and laser-heated diamond cells will become nearly routine in the next few years.

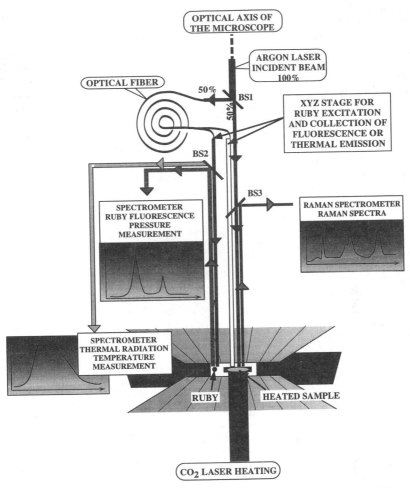

Figure 7. Optical set-up for collecting Raman spectra at simultaneous high pressures and high temperature using a laser heated diamond anvil cell (Gillet 1996). BS denotes beamsplitters (see text).

High-pressure, low-temperature Raman spectroscopy. Measurements at high pressure and low temperature provide fundamental information on physical properties, such as the origin of low-temperature thermal expansion and heat capacity. Raman measurements at high pressures and at temperatures between 4 K and 320 K can be performed with diamond cells inserted in a crysotat (Jayraman 1983, Chervin et al.,1992). For instance, Gillet et al. (1998) used a membrane-type diamond anvil cell (Chervin et al. 1992) for measuring the Raman spectrun of $MgSiO_3$-perovskite between room pressure and 25 GPa and 11 K and 320 K (Fig. 9). The cell was installed in a helium flow cryostat allowing temperature to be varied between 320 K and 4 K. The cryostat was coupled to a XY Dilor Raman microspectrometer equiped with a CCD detector. The spectrometer was used in backscattering geometry. The exciting laser beam was focused down to 5 μm on the sample through an optical access in the cryostat with a long working distance objective. The backscattered Raman light was collected through the same objective. Temperatures were measured using two thermocouples soldered on to the stainless steel gasket.

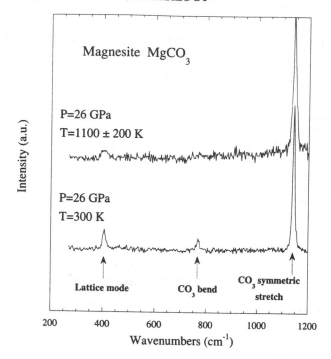

Figure 8. High-pressure Raman spectra of MgCO₃ (magnesite) recorded at 26 GPa at room temperature and at 1100 K using a laser-heated diamond cell (after Gillet 1993).

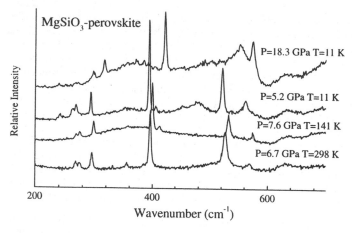

Figure 9. Raman spectra of MgSiO₃-perovskite at high pressures and low temperatures. The measurements were performed with a diamond anvil cell placed in a cryostat and coupled to a micro-Raman spectrometer (Gillet et al. 1998).

High-pressure infrared spectroscopy

In infrared spectroscopy, it is convenient to distinguish between the near IR (10000-4000 cm⁻¹ or 1-2.5 μm), the mid-IR (4000-400 cm⁻¹ or 2.5-25 μm) and the far-IR (400-10 cm⁻¹ or 25-1000 μm) regions. As most of the vibrational frequencies of minerals lie between 10 and 4000 cm⁻¹, only mid- and far-IR experiments will be discussed in this chapter. Far-IR measurements are typically much more difficult to carry out than mid-IR experiments even at room conditions, due to limitations in source intensity and detector sensitivity. Most in situ high pressure IR measurements on minerals to date have been carried out in the mid-IR region, although some far-IR studies have been reported (Ferraro

1984, Hofmeister et al. 1989, Grzechnik et al. 1998). However, far-IR measurements have been performed in diamond cells down to a few cm^{-1} (Challener and Thompson 1986).

Infrared measurements in the diamond cell require that both the diamonds and the pressure-transmitting medium must be transparent to IR radiation in the wavelength range of interest. Because of the strong impurity-related mid-IR absorptions of type I diamonds in the mid-IR (Fig. 6), this requires the use of type IIa diamonds for IR spectroscopy in diamond cells in this spectral region (Ferraro 1984). However, even "transparent" diamonds may permit transmission of only ~20-30% down to even a few per cent of the incident IR beam, in the "window regions" between strong absorptions (Fig. 6). An exciting recent development in diamond cell technology as a whole is the advent of ultrapure synthetic single-crystal diamond as well as the prospect of "designer diamond anvils" grown by chemical vapor deposition (Israel and Vohra 1998). These diamonds can be grown with essentially no impurities, in principle to any desired size; the spectroscopic window regions become determined entirely by the intrinsic second-order diamond absorption bands near 2000 cm^{-1}. Even these effects can be minimized by use of thin anvils, as mentioned above.

Because of the strong IR absorption coefficient of the principal bands of the largely ionic minerals, mineral plates must be thinned to less than a few μm in width for single crystal IR absorption measurements (Hofmeister 1987, McMillan and Hofmeister 1988, Hofmeister et al. 1989). For this reason, most diamond cell studies have been carried out as transmission experiments on powdered samples (i.e. particle sizes on the order of a few μm) dispersed in a "transparent" medium. Typical supporting matrix materials for powder IR spectroscopy of solids include the alkali halides KBr or CsI, which have their principal IR absorption bands at low frequency. The high frequency edge of KBr absorption occurs near 450 cm^{-1}, and that for CsI lies near 100 cm^{-1} (Ferraro 1984). It is convenient that these are soft materials that can be used as reasonably hydrostatic pressure-transmitting media over a wide range. In a clean IR absorption experiment, only the transverse optic (TO) components of the IR-active modes would appear in the absorption spectrum: however, due to the nature of the sample and optics, features due to longitudinal (LO) components can appear in the spectrum (McMillan and Hofmeister 1988). The difference in frequency between TO and LO components is termed the TO-LO splitting, and scales with the magnitude of the dipole moment change during the vibration (and hence also with the intensity of the IR absorption). Weak IR modes have small TO-LO splitting, and the peak maximum in an IR transmission experiment lies close to the TO mode frequency. Strongly active IR modes, such as Si-O stretching vibrations, or the single IR-active mode of MgO, have TO-LO splittings that can be on the order of several hundred cm^{-1}. In this case, the peak maximum in the powder transmission experiment can lie substantially above that of the TO mode frequency and the peak shape be can considerably distorted, depending on the component of reflectance in the transmission spectrum (McMillan and Hofmeister 1988).

Sample preparation. Absorption experiments may be prepared in the form of a thin film, which can be obtained by cold sintering of a powder between two diamonds, or by ion milling a single crystal (Hofmeister et al. 1987 1989). The appropriate thickness depends on the studied mineral, and on the strength of the modes of interest. The usual thickness is between 5 and 30 μm. The sample is then placed in the gasket hole of the diamond cell on one diamond face and covered with the pressure-transmitting medium, or in some cases sandwiched between two films of the medium. The latter type of loading provides better hydrostatic conditions (see later in this chapter). The resulting spectra on such samples show absorption bands at the TO frequencies and only limited LO components and reflection features between the LO and TO component. Spectra of stishovite are shown as an example in Figure 10.

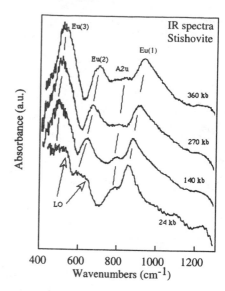

Figure 10. High-pressure IR spectra of stishovite (after Hofmeister 1996). The spectra were obtained from a cold-pressed powder of stishovite in a diamond cell.

Alternatively, the sample may be prepared in the form of a powder (0.1 to 2 μm sized particles) dispersed in the pressure medium (polyethylene, KBr or CsI, depending on the wavelength range of interest). The mixing ratio between the sample and the pressure-transmitting medium is approximately 1:10, but this can be adjusted to enhance different features in the spectrum. As noted earlier, the resulting transmission spectra are quite complex and can be difficult to interpret reliably. When the incident IR beam strikes a sample particle it is partly absorbed and partly reflected, depending upon the optical characteristics of the sample and the medium. The relative proportion of transmitted and reflected light depends on the angle of incidence of the IR beam on the particle, and upon the refractive index match of the sample and the surrounding matrix, both of which change as a function of pressure, and both of which vary rapidly with wavelength in the IR region of strong absorption. The resulting spectrum thus generally contains both pure absorption bands giving rise to transmission minima with symmetric shape and absorption maxima at the TO frequencies, and bands with substantial reflection components forming broad features with maximum absorption between the TO and LO frequencies (McMillan 1985, McMillan and Hofmeister 1988, Hofmeister et al. 1989, Hofmeister 1987, Grzechnik et al. 1998). These issues have recently been demonstrated and discussed by Hofmeister (1996) for the case of SiO_2 stishovite, including changes in the relative sampling of the LO and TO components with increasing pressure.

Synchrotron IR. As discussed above, high-energy synchrotron radiation sources generate high flux over a very broad region of the electromagnetic spectrum, including the infrared. Because of the significant enhancement in our ability to probe microscopic samples provided by synchrotron sources, the source is ideally suited to studies of materials under extreme pressures. When coupled with an FTIR interferometer and special microscopes for high-pressure cells, up to five orders of magnitude in sensitivity are gained relative to a grating system used for high-pressure IR measurements, and up to three orders of magnitude relative to FTIR measurements with a conventional source. Although enhanced flux can also be achieved with new infrared laser techniques (see below), the combination of very broad spectral distribution, ease of interfacing with a conventional FTIR instrument, and pulsed time structure illustrate important advantages of the synchrotron technique.

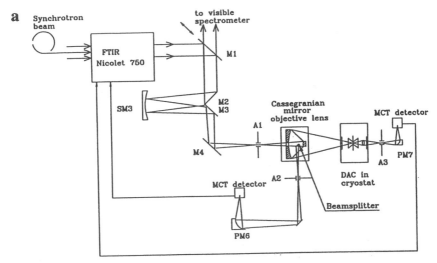

Figure 11a. Synchrotron infrared spectroscopy. Schematic of the optical setup at the National Synchrotron Light Source, Brookhaven National Laboratory, beamline U2B (Hemley et al. 1998; Lu et al. 1998). The focused and collimated synchrotron IR beam passes through a Fourier transform infrared spectrometer (FTIR) and directed either to a commercial IR microscope (not shown) or to a custom-designed, long working-distance microscope. The microscope contains Cassegrain-type reflecting objectives that focus the IR beam to the specimen for both absorption and reflection measurements on small samples, either free-standing or in high-pressure cells, cryostats, or furnaces. The system is integrated with additional instrumentation (not shown) that allows spectral characterization of the sample by Raman and UV-visible absorption, reflectivity, and fluorescence spectroscopy. The mirrors and pinholes are labeled by M and A, respectively (PM parabolic; SM spherical mirror).

Recent work has demonstrated the utility of synchrotron radiation for high-pressure IR spectroscopy. The technique permits study of microscopic specimens down to the diffraction limit of the light (Reffner et al. 1994, Carr et al. 1995, Hemley et al. 1998). The intense flux of a synchrotron beam increases the brightness of the collimated beam by 2 to 4 orders of magnitude over conventional globar sources. With proper optics (Fig. 11), a synchrotron beam can provide a diffraction-limited spot size in the principal wavelength range from the far- to near-IR. The overall improvement provides the opportunity for microspectroscopy with greatly improved sensitivity and spatial resolution; it can also be used for time-dependent spectroscopy. The improvement in spatial resolution effectively converts IR spectroscopy into a structural microprobe given the fact that distinct symmetries of mineral polymorphs exhibit distinct IR bands.

High-pressure acoustic spectroscopies

We consider two high-pressure spectroscopic techniques for measuring acoustic vibra-tions in minerals–Brillouin scattering and impulsive stimulated scattering. Application of Brillouin scattering to minerals was pioneered by Weidner (1975), and the first measurements carried out at very high pressures were performed by Bassett and Brody (1977). The technique is illustrated schematically in Figure 12. The technique involves scattering off thermally-generated acoustic waves propagating in a material. The incident monochromatic laser beam is scattered by these modulations and undergoes a change in wavelength directly related to the velocity of acoustic phonons. This change is expressed by the relative shift in wavenumber from the elastically scattered light (the Rayleigh line). The Brillouin frequency shift (Δv) is related to the acoustic velocity v by

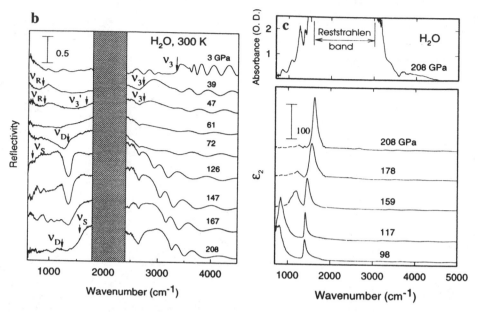

Figure 11b. Synchrotron IR reflectivity spectra of H_2O ice to 208 GPa (300 K). The interference pattern observed in the lowest pressure measurement can be used to determine the index of refraction (see Chapter 18). In reflectivity vibrational modes are identified by inflections (labeled v_i). The region near 2000 cm^{-1} is that of second-order absorption by diamond. (c) The imaginary part (ε_2) of the dielectric function ($\varepsilon = \varepsilon_1 + i \varepsilon_2$) determined from a Kramers-Kronig transform of the reflectivity spectra (Goncharov et al. 1996). The corresponding absorption spectrum at 208 GPa showing the strong absorption in the region between the LO and TO peaks (Reststrahlen band). Above 60 GPa, ice is non-molecular (Chapter 18).

$$v = \Delta v \lambda_0 / \sin\theta \tag{1}$$

where λ_0 is the wavelength of the incident beam and θ the angle between the incident beam and the scattered light. The acoustic velocity is a function of the direction of propagation of the acoustic phonon in the crystal. Orientation of the incident laser beam with respect to crystallographic axes permits direct exploration of the orientational dependence of the velocity and a full detemination of the tensor of elastic constants (Cummins and Schoen 1972). Applications to elasticity of deep Earth materials are examined further by Liebermann and Li (Chapter 15, this volume). Brillouin scattering also provides information on mechanisms of phase transitions, acoustic attenuation, and piezoelectricity (Pine 1981).

Advances in laser technologies and the development of non-linear optical methods have given rise to 'stimulated' spectroscopic techniques that are being used for probing minerals under pressure. As mentioned above, a particularly important one for high-pressure mineralogy is impulsive scattering spectroscopy (ISS). The technique provides similar information to Brillouin scattering but differs in a fundamental way by operating in the time, rather than the frequency, domain (Brown et al. 1988 1989). In ISS, the output from a laser is split into two beams and recombined to form a standing wave grating; a second laser is scattered from the grating (Fig. 13). In addition to information on acoustic wave velocities, it provides information on relaxation processes such as thermal diffusivity not available from the techniques described above (Chai et al. 1996). Results for San Carlos olivine are compared with the Brillouin scattering measurements in Figure 12e.

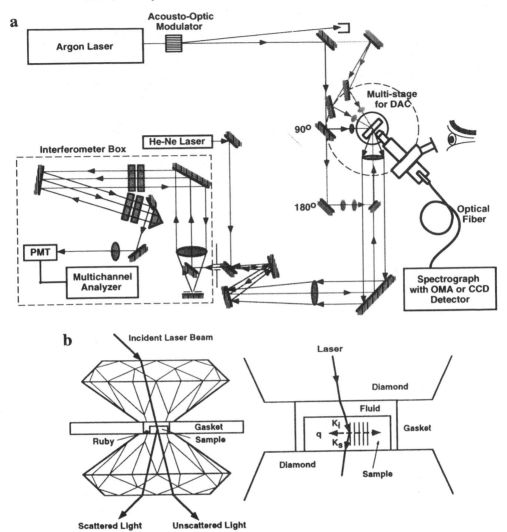

Figure 12. High-pressure Brillouin scattering spectroscopy. (a) Schematic of high-pressure single-crystal Brillouin scattering setup. The beam of an Ar laser is modulated to reduce spurious scattering and focused on the high-pressure sample in the diamond cell. A multistage is used, which allows different scattering geometries as well as rotation of the cell about two axes (χ and ω). The scattered light is collected and analyzed using a 6-pass tandem Fabry-Perot interferometer connected to a photomultiplier tube and multichannel analyzer. (b) Detail showing showing the scattering geometry. k_i = wavevector of the incident photon, k_s = wavevector of the scattered photon; q = acoustic phonon propagation (from Zha et al. 1996). **Continued on the following page:**
(c) Brillouin data for forsterite at high pressures (from Zha et al. 1996). Brillouin spectrum at 16.2 GPa measured in a He medium (room temperature). R = Rayleigh peak; D = diamond peak; He = peak from the pressure medium; S = S wave of the sample; P = P wave from the sample. (d) Variation of the compressional and shear velocity as a function of the direction of propagation in the crystal (measured at 6.1 GPa). (e) Elastic moduli (stiffness coefficients C_{ij}) of forsterite as a function of pressure determined from the Brillouin measurements. The solid lines are fits to the Brillouin data. The dashed lines are the results of impulsive scattering spectroscopy (ISS) measurements on $(Mg,Fe)_2SiO_4$ (San Carlos olivine), as discussed in the following figure (from Zha et al. 1996).

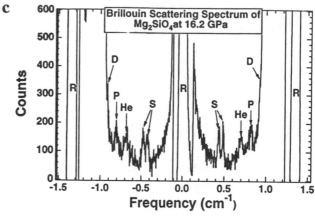

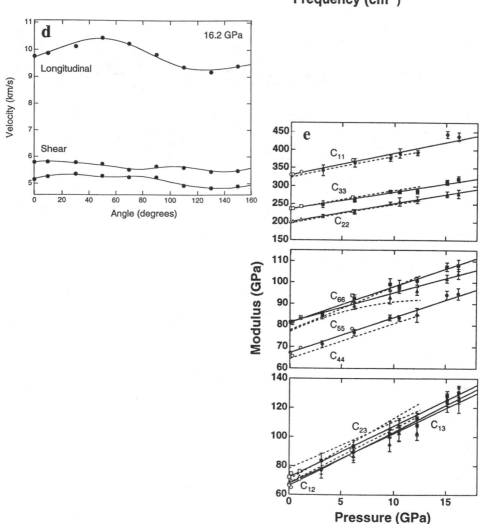

Other optical techniques

A growing number of other types of stimulated spectroscopies are being used for in situ high-pressure measurements, although primarily for studies in condensed matter physics. In hyper-Raman scattering, the sample is irradiated with a high-power laser, and hyper-Raman lines are found at frequencies $2v_o \pm v_i$, where the v_i are the vibrational frequencies. The hyper-Raman effect arises from non-linear optical effects (Long 1977). Hyper-Raman spectroscopy has the potential to be extremely useful in high pressure studies, because IR-active modes which are silent in the normal Raman spectrum appear in the hyper-Raman spectrum due to the different selection rules (Long 1977). Hyper-Raman scattering could be extremely useful to follow the behavior of the IR-active modes in silicate perovskites, as has been demonstrated for $SrTiO_3$ (Inoue et al. 1985). The hyper-Raman technique has also been applied to crystals with the rocksalt and rutile structures (Vogt and Neumann 1979, Denisov et al. 1980b), as well as calcite (Polivanov and Sayakhov 1981, Denisov et al. 1981, 1982).

Methods for generating tunable laser radiation throughout the IR continue to evolve. This includes various diode and color center lasers, which have been used in high-pressure studies over selected regions of the infrared, but mainly for applications in condensed matter physics (Cui et al. 1995). Near-IR techniques based on optical parametric amplifier technology have recently been developed for micro-spectroscopy (Wang et al. 1998). Such techniques can be readily extended to high-pressure studies.

Vibrations can also be excited in combination with electrons. High-pressure *vibronic* spectra, the frequencies and intensities of coupled electronic and vibrational excitations, reveal the pressure dependence of both the electronic and vibrational levels in the system, which can be used to constrain acoustic velocities. Chopelas and Nicol (1982) showed that the side bands present on the V^{2+} or Cr^{3+} luminescence bands for these ions implanted in MgO contained information on the vibrational density of states of the host oxide crystal, due to vibronic coupling between the electronic excitation of the transition-metal ion and the lattice vibrations of MgO. The technique has been applied to a number of other materials (e.g. Zhang and Chopelas 1994). Vibronic spectra reveal details about the electronic structure (see Hemley et al., Chapter 18, this volume). For example, coupling of phonons to 'crystal field' excitations under pressure can be studied (Goncharov et al. 1994).

Dynamical properties from other methods

In principle, structure refinements in x-ray and neutron diffraction can reveal information on the pressure variation of the average vibrational dynamics through the pressure effect on thermal parameters. This has not been widely applied because of the difficulty of measuring a sufficient number of reflections in a high-pressure cell, and the problem of obtaining accurate intensities (by carrying out accurate absorption corrections). King and Finger (1978) performed careful measurements on NaCl and extracted accurate thermal parameters. Good agreement was obtained with the results of simple exponential plus coulomb potential calculations. It is expected that there will be more extensive use of this approach with the continued development and application of the synchrotron x-ray diffraction techniques for crystal structure refinements.

Inelastic neutron scattering can be used to excite vibrations; it requires high energies and the measurement of low momentum transfer. Given the intensity of current neutron sources, generally large samples are required. Nevertheless, high-pressure studies have been carried out on a growing number of minerals under pressure, including iron oxide (S. Klotz, private communication). New x-ray spectroscopic methods also provide measurements of dynamics. Phonon dispersion relations can also be obtained via nuclear resonant forward scattering (Alp et al. 1998; Hemley et al., Chapter 18, this volume).

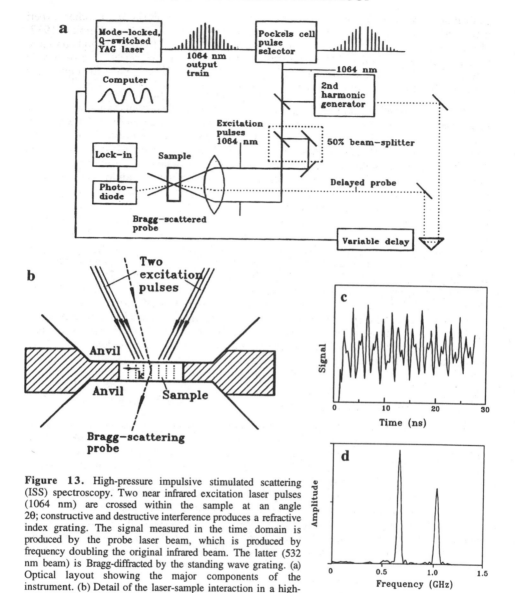

Figure 13. High-pressure impulsive stimulated scattering (ISS) spectroscopy. Two near infrared excitation laser pulses (1064 nm) are crossed within the sample at an angle 2θ; constructive and destructive interference produces a refractive index grating. The signal measured in the time domain is produced by the probe laser beam, which is produced by frequency doubling the original infrared beam. The latter (532 nm beam) is Bragg-diffracted by the standing wave grating. (a) Optical layout showing the major components of the instrument. (b) Detail of the laser-sample interaction in a high-pressure cell. (c) Diffraction signal as a function of probe delay for the probe pulse from an 80-mm thick crystal of San Carlos olivine. (d) Fourier transform of the signal showing the two acoustic modes. The pressure dependence of the elastic moduli determined from the acoustic velocities are shown in Figure 12e (after Brown et al. 1989).

Characterization of stress states

Vibrational spectroscopy plays an important role in characterizing the state of stress in a sample. First, various vibrational modes have been calibrated against other standards to provide a useful measure of hydrostatic pressure in a sample. For example, the pressure dependence of the Raman mode of in ^{13}C diamond (see below) has been used for calibration in high *P-T* experiments (Frantz and Mysen, unpublished). In other experiments, vibrational modes in hydrogen and quartz have been used (e.g. Hemley 1987, Hemley et al. 1996). More detailed information about the state of stress in high-pressure samples can also be obtained. Many transformations in minerals are sensitive to deviatoric stresses, so it is of prime importance to consider the nature of the pressure conditions generated by different types of pressure-transmitting media.

Purely hydrostatic conditions prevail if the pressure medium remains liquid or gaseous under pressure. This is the case for the 4:1 methanol:ethanol mixture up to 10 GPa and for the 16:3:1 methanol:ethanol:water mixture up to 15 GPa (Fujishiro and Namura 1994, Piermarini et al. 1973). Above these pressures vitrification of the mixtures occurs and "quasi-hydrostatic conditions" prevail, with typical pressure gradients on the order of 1-2 GPa over a 200-μm hole at 20-30 GPa. Similar quasi-hydrostatic conditions are obtained with soft solids like KBr, NaCl or CsI. Rare gases (argon, neon and helium) provide quasi-hydrostatic conditions above their freezing point, and the lighter ones provide the best media. Meng et al. (1993) have recently shown by x-ray diffraction that neon becomes non-hydrostatic above 15 GPa and that deviatoric stresses on the order of 0.15 GPa develop in compressed samples (Bell and Mao 1981). Helium and hydrogen remain the most hydrostatic and can be used into the megabar range. The degree of hydrostaticity can be inferred from broadening of the linewidth of the ruby fluorescence lines used for measuring pressure, by calibrating the change in separation of the R_1 and R_2 ruby fluorescence lines (Chai and Brown 1996), or by measuring pressure differences within the gasket hole (e.g. assuming the hydrostatic pressure shift in fluoresence peaks or vibrational modes). Estimates of the shear stress (σ) acting on the sample parallel to the diamond culet can be obtained through the following relation (Sotin et al. 1985, Gillet et al. 1988, Meade and Jeanloz 1988):

$$\sigma = \frac{h}{2}\left(\frac{\partial P}{\partial r}\right) \qquad (2)$$

where h is the thickness of the gasket and ($\partial P/\partial r$) the pressure gradient across the gasket hole. Using typical values of $h = 60$ μm and ($\partial P/\partial r$) = 1 GPa/100 μm one finds $\sigma = 0.3$ GPa, in good agreement with the value measured by x-ray diffraction by Meng et al. (1993). In general, direct contact of the sample between the diamonds with no pressure-transmitting medium leads to non-hydrostatic conditions. In runs performed with pressure media, care must also be taken to avoid bridging between sample and the diamonds, which creates large deviatoric stresses. Very thin polished samples (e.g. 20 μm down to just a few μm, depending on the pressure) are required if hydrostatic conditions are to be approached. When the sample is in powdered form a low sample to pressure medium ratio is needed to avoid intergranular bridging. Moderate heating of the sample to a few hundred degrees in an appropriate pressure medium can be used to relax the deviatoric stresses induced by compression at room temperature.

The presence of non-hydrostatic stresses can cause splitting of otherwise degenerate vibrational modes (Eggert et al. 1995), can result in enhanced TO-LO splitting of Raman modes in non-centrosymmetric materials such as SiO_2 quartz (Fig. 14) and $AlPO_4$ (Hemley 1987, Jayaraman et al. 1987, Gillet et al. 1995), and can modify the mean pressure of phase transitions as demonstrated for $CaAlSi_2O_8$ anorthite (Daniel et al. 1997), or otherwise change the pressure-induced frequency shifts of the vibrational modes. On the other hand, if the non-hydrostatic conditions are quantified, the measurements thus provide important additional information on material behavior. For example, techniques have recently been developed to produce, control, and measure uniform non-hydrostatic conditions in polycrystalline samples at very high pressures (to >200 GPa) to determine elastic moduli,

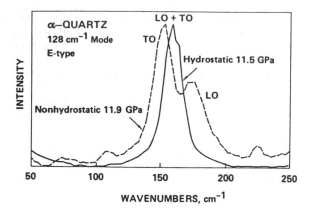

Figure 14. Raman measurement of the *E*-symmetry mode at 128 cm⁻¹ in SiO_2 quartz on compression in a diamond cell. During hydrostatic compression achieved with a soft pressure medium surrounding the sample, the LO-TO compounds of this vibrational mode are at the same frequency. When the sample is directly squeezed between the two diamond anvils (i.e. under highly non-hydrostatic conditions) one observes a strong LO-TO splitting of this mode (from Hemley 1987).

texture, and rheology by radial x-ray diffraction (Hemley et al. 1997, Singh et al. 1998, Mao et al. 1998).

High-temperature vibrational spectroscopy

For completeness, we also point out that high-temperature Raman and IR spectroscopy on minerals and silicate melts at ambient pressure has been carried out for many years using simple resistance furnaces (Sakurai and Sato 1971, Gervais and Piriou 1975, Gervais et al. 1973, reviewed by McMillan and Wolf 1995). Miniaturized heating stages can be easily adapted on micro-Raman spectrometers for obtaining Raman spectra up to 1300 K (Gillet et al. 1993). Mysen and Frantz (1992) developed a simple type of furnace that is easily adaptable for micro-Raman and micro-IR spectroscopy. It consists of a metallic wire heated by the Joule effect. Richet et al. (1993) have improved this technique using Pt/10%Ir or pure Ir wires. Temperatures up to 2700 K can be obtained and the set-up is readily adapted to synchrotron x-ray sources, and for both Raman or Brillouin spectroscopic experiments (Richet et al. 1993, Vo-Than et al. 1996). The problems encountered during high-temperature Raman measurement as well as the data corrections (black body emission from the sample, temperature- and frequency-dependence of the first- and second-order Raman scattering) have been discussed in detail by Daniel et al. (1995), McMillan and Wolf (1995) and Gillet (1996). Examples of measurements up to 2000 K on minerals can be found for instance in Richet et al. (1993) for alumina, Daniel et al. (1995) for anorthite, Gillet et al. (1997) for forsterite and Sharma (1989) for enstatite. Grzechnik and McMillan (1998) used the Ir wire- heating technique to obtain IR transmission spectra of OH groups in SiO_2 glass to nearly 2000 K. Raman spectroscopy has also been carried out on SiO_2 glass and melt to near 2400 K at room pressure, using the CO_2 laser heating technique (1988).

PHASE TRANSFORMATIONS

SiO_2 system as an example

Because the number, type, and spectroscopic activity of vibrational modes of a given mineral are governed by selection rules imposed by symmetry, phase transitions which involve symmetry changes are easily observed by vibrational spectroscopic techniques from changes in the number or relative intensity of bands in the IR and Raman spectra. Further, the frequencies of the vibrational modes depend intimately upon the bond lengths and bond angles within the mineral. A careful examination of the pressure dependency of the vibrational frequencies thus provides an "indirect" tool for probing the response of crystalline structures to increasing pressure, complementary to diffraction measurements.

Finally, in some cases the microscopic atomic movements associated with a phase transition can be directly inferred from vibrational data because the atomic displacements associated with the normal modes of vibration are known. To date, there have been relatively few infrared and Raman spectroscopic studies of important minerals under the high *P-T* conditions of the Earth's crust and mantle, and this represents a field that is now ripe for exploration using the newly-developed techniques described in the previous section. All of the points described above can, however, be illustrated by using the simplest mineralogical system, SiO_2, as a model, which has been the object of several recent studies.

The stability fields and the crystal chemical properties at high temperature and high pressure of the SiO_2 polymorphs (α-quartz, β-quartz, cristobalite, trydimite, coesite and stishovite), as well as its melt and glasses, have been extensively studied because these phases represent archetypal framework structures which provide insights on the intrinsic behaviour of the Si-O bond. In the past decade, the database on these phases has been markedly increased through in situ measurements, including new determinations of the phase stability fields and equations of state, utilizing powder and single-crystal x-ray diffraction, and vibrational spectroscopies (Hemley et al. 1994 1998). The phase diagram of SiO_2 is given by Shen and Heinz (Chapter 12, this volume) and Dingwell (Chapter 13, this volume). Vibrational spectroscopy has proven to be an essential technique for studying the structural behaviour of both crystalline and amorphous phases of SiO_2 at high pressures and high temperatures; new structural variants have been discovered during room-temperature compression, and structural and configurational changes in SiO_2 glasses and melts at high temperatures have been assesed (Hemley et al. 1994).

SiO_2 polymorphs

Quartz, cristobalite, trydimite, coesite, and stishovite are the known thermodynamically stable phases which can be quenched to room pressure and temperature conditions. Both quartz and cristobalite have non-quenchable β forms at high temperature: the α-β transformations of these phases have been studied extensively by vibrational spectroscopy (McMillan and Hofmeister 1988). Raman spectroscopy performed at simultaneous high pressure and high temperature using a laser heated diamond cell can be used to readily detect in situ the transformations between these phases (Fig. 15).

Metastable high-pressure transitions in SiO_2, including both crystalline-crystalline transitions and amorphization, were examined in detail by Hemley (1987). In this study, Raman spectroscopy was essential for detailing the effects of non-hydrostatic or uniaxial stress conditions on the high-pressure behavior (Fig. 14). A crystalline-crystalline phase transition was observed in coesite near 25 GPa and room temperature, just before the onset of pressure-induced amorphization in this material. Kingma et al. (1993a,b) found a crystalline-crystalline transition in α-quartz metastably compressed to 21 GPa, preceeding or accompanying pressure-induced amorphization (Fig. 16). Cristobalite also undergoes metastable transformations at 1.2 and 11 GPa before transforming to an amorphous phase above 30 GPa (Palmer et al. 1994, Yahagi et al. 1994). All the reported crystal-crystal transitions are found to be reversible upon decompression. Similar transformations have also been reported for chemical analogues of α quartz and cristobalite, such as $AlPO_4$ (Gillet et al. 1995).

Post-stishovite phases merit special attention. In the high-pressure phase stishovite, Si is in six-fold coordination, in contrast to the other known polymorphs in which Si is in four-fold coordination. Many attempts have been made to identify possible crystalline phases of silica denser than stishovite. A number of theoretical calculations, as well as comparisons to related systems, have predicted transitions of stishovite to various denser

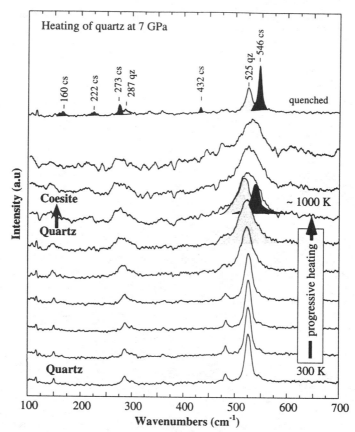

Figure 15. Quartz-coesite transition monitored by *in situ* Raman spectroscopy. A single crystal of quartz 20 μm thick and 150 μm wide, pressurized at 7 GPa in a diamond anvil cell, is progressively heated with a CO_2 laser. Progressive heating is accompanied by band broadening and frequency shifts toward lower wavenumbers. Heating during a few minutes at around 1000 ± 100 K leads to a partial transformation of quartz into coesite outlined by the appearance of two bands in the most intense band. Shutting off the laser power leads to instantaneous quenching at 7 GPa. The quenched sample shows Raman bands characteristic of both polymorphs.

phases at ultrahigh pressures. Stishovite, having the rutile structure, can transform above 50 GPa to a $CaCl_2$ (Cohen 1992), α-PbO_2, α-$PbCl_2$, or a ZrO_2 structure (Cohen 1994, Haines et al. 1996, Karki et al. 1997, Belonoshko et al. 1996, Dubrovinsky et al. 1997). Among the theoretically predicted phases, the SiO_2 phase with the $CaCl_2$ structure is now well established (Tsuchida and Yagi 1989, Kingma et al. 1995).

The Raman spectrum of stishovite shows all four modes expected from symmetry analysis: A_{1g} at 753 cm^{-1}, B_{2g} at 967 cm^{-1}, E_g at 589 cm^{-1}, and B_{1g} at 231 cm^{-1} (Hemley et al. 1986). The behavior of the low-lying B_{1g} mode is particularly interesting, because its frequency decreases with increasing pressure (Fig. 17) (Hemley 1987). The atomic displacements associated with this mode correspond to those required to transform stishovite from the rutile into the calcium chloride structure (Hemley 1987, Nagel and O'Keeffe 1971). Kingma et al. (1995) studied in detail by in situ Raman spectroscopy at high pressure and room temperature the transition of stishovite from the rutile to the $CaCl_2$ structure. They demonstrated that the transition is reversible upon decompression, as

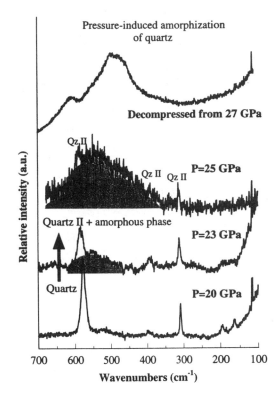

Figure 16. Metastable transition and pressure-induced amorphization of quartz above 20 GPa observed by Raman spectroscopy. Up to 20 GPa the spectrum is characteristic of quartz. Between 20 and 23 GPa, birefringent lamellae appear throughout the crystal (Kingma et al. 1993a); band broadening as well as disapearance of some bands is related to the transformation of quartz into a new variant called quartz II. At 25 GPa, a major band can be observed in spectra recorded in some places of the sample, which resembles that of SiO_2 glass compressed at similar pressures (Hemley et al. 1986). The 25 GPa spectrum has been corrected for the strong fluorescence background systematically observed at the onset of amorphization, which might be due to the creation of defects within the sample. The spectrum of the decompressed material indicates that the sample is predominately amorphous.

expected for a second order displacive phase transition, associated with a soft mode. Total energy calculations as well as determination of the soft mode frequency as a function of pressure predict the transition to occur near 50 GPa (Cohen 1992), in excellent agreement with the experimental observation (Kingma et al. 1995) (Fig. 17). The predicted splitting of the E_g mode of stishovite into the expected Raman-active modes B_{3g} and B_{2g} of the $CaCl_2$ structure is observed. Moreover, the B_{1g} mode which softens during compression was also predicted to become a hard mode in the $CaCl_2$ structure. This work indicates that free SiO_2 would exist in the $CaCl_2$ structure below 1200-1500 km within the Earth's lower mantle, depending on the *P-T* slope of the transition, which has not yet been determined experimentally.

A further transition may occur (to α-PbO_2 type structure) near 80 GPa (Dubrovinsky et al. 1997, Teter et al. 1998). Evidence for this transition has been obtained very recently from analysis of a strongly shocked meteorite (El Ghoresy et al. 1998). It should also be mentioned here that other forms of SiO_2 have been "produced" in numerical simulations. For instance, a fully five-coordinated form of silica can be produced if quartz is compressed under highly non-hydrostatic pressure conditions above 20 GPa (Badro et al. 1997, Badro et al. 1996).

Amorphization

The formation of structurally disordered, metastable crystalline phases during room temperature compression seems to be a general feature associated with pressure-induced amorphization for most of the known SiO_2 polymorphs. Figure 16 shows how in situ Raman spectroscopy reveals the passage of crystalline α-quartz to a new crystalline polymorph ("quartz II") and finally to an amorphous material characterized by a strong decrease in Raman intensity and the growth of broad glass-like bands similar to those observed in silica glass compressed to the same pressure (Hemley et al. 1986) (Fig. 16). On compression, the driving force appears to be the higher density allowed by an amorphous (or disordered) state of the material relative to that of the initial phase (e.g. a low-density crystal). A metastable transformation to another crystalline phase (which may be disordered or poorly crystallized) may represent the first step of this densification. In SiO_2, initial compression is achieved by closure of the Si-O-Si angles (Hazen et al. 1989). Further collapse of these angles cannot occur due to oxygen-oxygen and silicon-oxygen repulsions as revealed by the limiting values reached in the frequencies of Raman- and IR-active modes associated with the bending of the Si-O-Si bonds (Hemley 1987, Williams et al. 1993, Wolf et al. 1992). Further compression is achieved by pressure-driven disordering that leads ultimately to an amorphous solid. High-pressure IR spectra on quartz and coesite show that amorphization at the highest pressure (e.g. >25 GPa) is accompanied by an increase in Si coordination from four- to six-fold (Williams et al. 1993).

Skinner and Fahey (1963) observed that the high-pressure phases stishovite and coesite amorphize when heated at moderate temperatures at room pressure. This phenomenon has been studied by high-temperature Raman spectroscopy (Grimsditch et al. 1994, Gillet et al. 1990). Broad bands are observed to appear upon heating due to amorphization of the high-pressure phase into a glass dominated by tetrahedrally coordinated Si (Fig. 18). Stishovite first transforms to an amorph akin to a densified glass which then relaxes upon heating to a normal SiO_2 glass (Grimsditch et al. 1994) (Fig. 17). Temperature-induced amorphization was also observed by Raman spectroscopy during room-pressure heating of $MgSiO_3$-perovskite (Durben and Wolf 1992) and $MgSiO_3$-ilmenite (Reynard and Rubie 1996). Like melting, amorphization of crystals could thus result from either super-pressing crystals at low temperature well beyond their thermodynamic stability fields, or by heating high-pressure phases that have been decompressed metastably to ambient conditions. Although these are metastable transformations, it is possible to understand both pressure- and temperature-induced amorphization in the standard thermodynamic sense; i.e. in terms of volume-driven and entropy-driven processes. Thermodynamic continuity between the high-temperature liquid and the amorphous solid further implies a connection between the onset of amorphization and the metastable extensions of the melting curves (Mishima et al. 1984, Hemley et al. 1988, Richet 1988). This last point is supported by recent molecular dynamics simulations of SiO_2 (Badro et al. 1998).

Following the early studies by Hemley (1987) on SiO_2 polymorphs, in situ vibrational spectroscopy has proven to be a convenient technique for investigating pressure-induced amorphization in several other minerals, including feldspars (Daniel et al. 1997), zeolites (Gillet et al. 1996), micas (Faust and Knittle 1994), wollastonite (Serghiou and Hammack 1993), berlinite (Gillet et al. 1995), serpentine, and $Ca(OH)_2$ (Meade et al. 1992) and GeO_2 (Wolf et al. 1992). Richet and Gillet (1997) have recently reviewed pressure-induced amorphization for the SiO_2 polymorphs as well as for other minerals.

Future work

The study of the SiO_2 system by vibrational spectroscopic techniques has shown that these tools are extremely powerful for detecting previously unsuspected phase transitions in mineral systems. Much work remains in characterizing the full vibrational dynamics of the new SiO_2 phases. The IR spectrum of the $CaCl_2$-type structure of SiO_2 has not been measured, nor have IR and Raman spectra of the α-PbO_2-type phase. More phases are likely to be discovered at conditions of high temperature, pressure, and stress (Badro et al. 1996, 1997; Teter et al. 1998). Attempts to synthesize the predicted five-fold coordinated

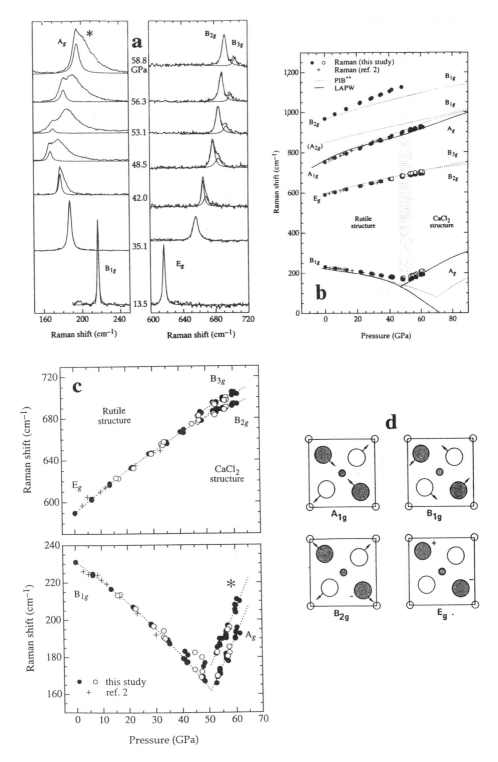

Figure 17 (opposite page). Rutile- to $CaCl_2$-type phase transition in stishovite (after Kingma et al. 1995). (a) High-pressure Raman spectra showing the negative shift of the B_{1g} mode with increasing pressure up to the phase transition at around 50 GPa. Also shown is the splitting of the E_g mode of stishovite into the B_{2g} and B_{3g} modes active in the $CaCl_2$ structure. (b) Pressure dependence of the Raman mode frequencies for stishovite and $CaCl_2$-type phase. (c) Detail of the pressure shift for the low-frequency soft mode. The frequencies are compared to those calculated from an *ab initio* model (PIB) and first-principles-theory (LAPW) (see Stixrude et al., Chapter 19). (d) Atomic motion associated with the Raman active vibrations of stishovite. The movements associated with the soft B_{1g} mode are those required to transform stishovite to the $CaCl_2$ structure.

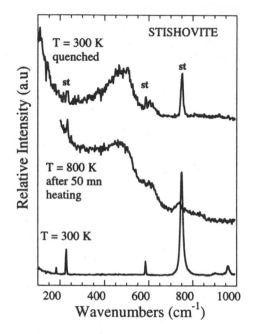

Figure 18. Temperature-induced amorphization of stishovite at room pressure observed by *in situ* Raman spectroscopy. Features marked 'st' denote peaks of untransformed stishovite.

phase of SiO_2 and to obtain its Raman and IR spectra would also be useful for identifying SiO_5 groups within other more complex silicate structures (Angel et al. 1996, Gautron et al. 1997). Finally, data on the acoustic phonons of all the SiO_2 polymorphs are scarce, so that important data relevant to the mineralogy and seismological properties of the Earth's lower mantle are missing. The means for carrying out such studies has now become routine with recent developments in high-temperature diamond-cell technology, coupled with Raman and infrared spectroscopy and synchrotron x-ray diffraction. The work remains to be carried out for other phases important in mantle and lower crustal mineralogy.

P-V-T DEPENDENCE OF VIBRATIONAL FREQUENCIES

It is usually observed that the frequencies of the vibrational modes are sensitive to pressure and/or temperature changes, due to anharmonic effects, which has important consequences for their thermodynamic properties. In order to perform a detailed analysis of these dependencies, one needs accurate measurements of the phonon frequencies along with the unit cell volume over as wide a pressure and temperature range as possible. The recent improvements in both x-ray diffraction and vibrational spectroscopic techniques at extreme *P-T* conditions now permit such a detailed analysis.

The detailed analysis of pressure and temperature effects on vibrational modes of solids is well described by phonon physics (Wallace 1972, Reissland 1973). To introduce the study of pressure and temperature effects on vibrational mode frequencies in minerals, we take a simple approach to define classically used parameters, such as the Grüneisen parameter, and to provide a link between these and structural changes occurring in minerals during pressurization and/or heating. We first discuss the case of the simplest minerals with the covalent diamond and ionic NaCl (or MgO) structures, and then extend the argument to more complex minerals, using magnesite as an example.

Grüneisen parameters

The Grüneisen parameter is widely used to describe the response of materials to high pressures. There are many macroscopic and microscopic definitions of this parameter in common use (Poirier 1991). The original definition applied to the lattice modes of simple solids. For a vibrational mode of frequency v_i one defines a *mode*-Grüneisen parameter by

$$\gamma_i = -\frac{d\ln(v_i)}{d\ln(V)} \tag{3}$$

where V is the volume. For spectroscopic measurements performed at fixed temperature T and varying pressures we can define an isothermal mode-Grüneisen parameter γ_{iT}. It is expressed as a function of the crystal bulk modulus K_T and the measured frequency shift by

$$\gamma_{iT} = \frac{K_T}{v_i}\left(\frac{\partial v_i}{\partial P}\right)_T \tag{4}$$

This equation defines γ_{iT} for any vibrational mode, and most studies at high pressures and fixed temperature use this definition for all observable frequencies. The mode-Grüneisen parameters can be related to the *thermal* Grüneisen parameter γ_{th}

$$\gamma_{th} = \frac{\alpha K_T V}{C_V} \tag{5}$$

where $\alpha = \frac{1}{V}\left(\frac{\partial V}{\partial T}\right)_P$ is the coefficient of volumetric thermal expansion and C_V is the specific heat at constant volume. γ_{th} can be written as

$$\gamma_{th} = \frac{\sum \gamma_{iT} C_{Vi}}{\sum C_{Vi}} \tag{6}$$

where C_{Vi} is the contribution of the ith vibrational mode to the total specific heat. When all the γ_{iT} are equal one has $\gamma_{iT} = \gamma_{th}$.

For spectroscopic measurements performed at fixed P and variable T one can define an *isobaric* mode-Grüneisen parameter γ_{iP} at constant pressure

$$\gamma_{iP} = -\frac{1}{\alpha v_i}\left(\frac{\partial v_i}{\partial T}\right)_P \tag{7}$$

The volume dependence of the two Grüneisen parameters is defined by the second Grüneisen parameters:

$$q_{iT} = \left(\frac{\partial \ln \gamma_{iT}}{\partial \ln V}\right)_T \tag{8}$$

and

$$q_{iP} = \left(\frac{\partial \ln \gamma_{iP}}{\partial \ln V} \right)_P \tag{9}$$

An independent determination of both the temperature dependence $[v_i(T)_P]$ and the pressure dependence $[v_i(P)_T]$ of a given phonon frequency permits a separation of the temperature dependence into its two components: (a) the implicit volume-driven (quasi-harmonic) contribution produced by thermal expansion and (b) the explicit phonon-excitation (intrinsic anharmonicity) contribution, due to phonon-phonon interactions (Gervais and Piriou 1975, Peercy and Morosin 1973, Gillet et al. 1989). The intrinsic mode anharmonicity can be quantified with a parameter a_i, defined by

$$a_i = \left(\frac{\partial \ln v_i}{\partial T} \right)_V = \alpha(\gamma_{iT} - \gamma_{iP}) \tag{10}$$

Finally, the volume dependence of the a_i parameters is defined by a parameter, m_i:

$$m_i = \frac{\partial a_i}{\partial \ln V} \tag{11}$$

The derivation and use of these parameters is illustrated using the simple minerals diamond and NaCl (MgO) as examples.

Example of diamond

The vibrational spectrum of cubic diamond is very simple: only a single triply degenerate mode is expected at the Brillouin zone center:

$$\Gamma_{vib} = T_{2g}(R)$$

This gives rise to a single first order peak in the Raman spectrum at 1332 cm^{-1}. Since this mode is not polar, it is not IR-active, and it does not exhibit LO-TO splitting. The Raman spectrum of diamond has been studied to ~40 GPa at room temperature (Hanfland et al. 1985, Boppart et al. 1985, Muinov et al. 1994). The frequency shift has a constant slope up to 40 GPa. Zouboulis and Grimsditch (1991) have measured Raman scattering in diamond at high temperatures, to 1900 K, at room pressure. The Raman mode exhibits a linear frequency decrease in the 300 K-1900 K temperature range (Fig. 19), as is often the case in crystals which do not undergo any phase transition.

To compute the various Grüneisen and anharmonic parameters defined in the previous section, one needs the variation of molar volume or that of K_T and α with pressure and temperature respectively. The variation of K_T with pressure at a given temperature can be inferred from x-ray diffraction measurements of the molar volume variations as a function of pressure; these data are often fitted by a Birch-Murnaghan equation of state (Anderson 1995):

$$P = \frac{3}{2} K_{0T} \left(\left(\frac{V_0}{V} \right)^{\frac{7}{3}} - \left(\frac{V_0}{V} \right)^{\frac{5}{3}} \right) \left(1 + \frac{3}{4} \left(K'_{0T} - 4 \right) \left(\left(\frac{V_0}{V} \right)^{\frac{2}{3}} - 1 \right) \right) \tag{12}$$

where V_0 is the molar volume at room pressure and temperature T, V is the molar volume at P and T, K_{0T} is the isothermal bulk modulus at room pressure and temperature T and K_{0T}' its pressure derivative $K'_{0T} = \left(\frac{\partial K_{0T}}{\partial P} \right)_T$. For diamond, $K_{0T} = 442$ GPa and $K_{0T}' = 4$.

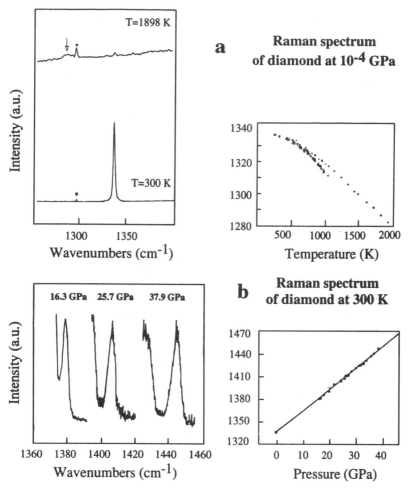

Figure 19. Effect of pressure and temperature on the Raman-active mode of diamond. (a) Spectra recorded at room pressure and high temperature (after Zouboulis and Grimsditch 1991) and corresponding temperature shifts. The peak labelled with a dot represents a plasma line. (b) Spectra recorded at room temperature and high pressure (after Hanfland et al. 1985) and corresponding shifts.

The thermal expansion coefficient for diamond is known at least up to 1700 K (see Reeber and Wang 1996).

For the Raman-active mode of diamond, one obtains the following results either from the $\ln(v_i)$-$\ln(V)$ analysis or by the use of Eqs. (4) and (7), which take into account the pressure or temperature dependence of K_T and α. γ_{iT} at 300 K increases from 0.96 at room pressure to 1.2 at 40 GPa. Similar values for γ_{iT} are obtained for the acoustic modes (McSkimin and Andreatch 1972) and all lie very close to the value of $\gamma_{th} = 1$. γ_{iP} at room pressures decreases from 2.53 to 1.36 between 300 and 2000 K (Fig. 20). At ambient conditions $q_{iT} = -3.2$. For a quasi-harmonic phonon, (i.e. when only volume variations are responsible for the frequency shifts) γ_{iT} and γ_{iP} should be equal. The observed difference between these parameters demonstrates the role of phonon-phonon interactions in determining the vibrational properties of diamond, in addition to pure volume (quasi-

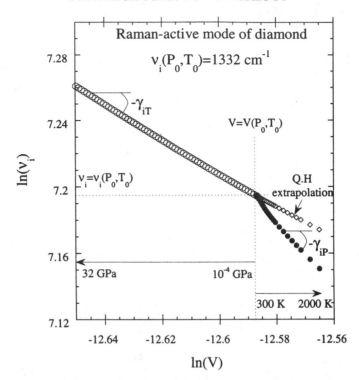

Figure 20. Logarithmic plot of the frequency of the Raman-active mode of diamond as a function of volume. The volume variations are inferred from thermal expansion and compressibility data. The slope of this curve gives the γ_{iT} and γ_{iP} mode-Grüneisen parameters. The difference between the quasi-harmonic extrapolation of high-pressure data in the high-temperature region and the measured values in this region are a signature of intrinsic anharmonicity; i.e., pure temperature dependence of the frequency shift (see text).

harmonic) effects that could be predicted from a knowledge of the compressibility and thermal expansion.

In general, intrinsic anharmonic effects can be directly observed graphically in $\ln(\nu_i)$-$\ln(V)$ plots in the domains over which pressure- and temperature-induced effects on normal mode frequencies overlap. The existence of such effects shows up as a different slope for the $\ln(\nu_i)$-$\ln(V)$ plots at constant temperature and at constant pressure, respectively. The intrinsic anharmonic behavior is revealed in such plots by extrapolating the high-pressure room-temperature relation toward the high-temperature room-pressure region. If the mode behaves quasi-harmonically, then the extrapolation of the high- pressure data on one hand, and the experimental high-temperature data on the other, should match (within experimental precision), since only volume would affect the vibrational frequencies. Significant differences between the two curves would provide a definite signature of intrinsic anhar-monicity of the mode.

A simple relation, obtained by integration of Equation (8), can be used to extrapolate the high-pressure data:

$$\ln(\nu_i(P_0,T)) = \ln(\nu_i(P_0,T_0)) - [(\gamma/q)((V(P_0,T)/V(P_0,T_0))^q-1)] \tag{13}$$

where $\gamma_{iT} = \gamma_{iP}$ and $q = q_{iT}$ are assumed to be constant over the volume range investigated. For diamond the high-temperature frequencies of the Raman-active mode do not lie on the quasi-harmonic extrapolation of the high pressure data, implying significant intrinsic anharmonicity for this mode. Integration of Equation (10) gives

$$ln(v_i(V,T))-ln(v_i(V,T_0))=\int_{T0}^{T1}a_idT = \Delta ln\, v_{ith} \tag{14}$$

The purely thermal contribution to the vibrational frequency $\Delta ln v_{ith}$ is obtained by substracting the actual frequency from that expected for a quasi-harmonic behaviour. As observed in Fig. 20, $\Delta ln v_{ith}$ decreases linearly with T suggesting that Equation (14) can be approximated by

$$\Delta ln(v_{th})=a_i\Delta T \tag{15}$$

This result shows that the frequency of the diamond Raman mode cannot be solely a function of volume, as has been suggested previously (Liu 1993). The intrinsic mode-anharmonic parameter a_i is found to be constant over a large temperature range with a value close to -1.4 10^{-5} K^{-1}.

Relationship between frequency and structural changes. Diamond is a monoatomic mineral held together by covalent bonds of only one type and it can be treated in the spirit of the original Grüneisen model, in which bulk properties are simply related to bond length changes. The Raman frequency of diamond at ambient conditions is determined by the first neighbor covalent C-C bond stretching force constant and the second neighbor C...C repulsive interactions (Guth et al. 1990). In a first approximation, these can be subsumed into an "effective" C-C bond stretching force constant (k), and we can consider the vibrational frequency change on compression as due to changing this constant. The Raman-active mode of diamond involves only C-C bond stretching and its frequency v_i is related to the effective force constant k by

$$v_i=B_ik^{1/2} \tag{16}$$

A change in frequency is thus directly related to a change of k with compression. The corresponding mode-Grüneisen parameter can be written (Sherman 1980)

$$\gamma_{iT}=-\left(\frac{Ar^3}{B_ik^{1/2}}\right)\left(\frac{\partial B_ik^{1/2}}{\partial Ar^3}\right)_T=-\left(\frac{r}{6k}\right)\left(\frac{\partial k}{\partial r}\right)_T \tag{17}$$

where the unit cell volume V (or molar volume) has been replaced by $V = Ar^3$, r being the C-C bond length. It can be estimated that a change of 1% in the C-C bond length causes a 6% change in the effective force constant, for a covalent compound (Herzberg 1950). The mode-Grüneisen parameter is thus approximately equal to unity, consistent with that inferred from the spectroscopic measurements.

Example of NaCl and MgO

A similar approach can be followed for simple diatomic minerals like MgO, which is isostructural with NaCl. Following Sherman (1980), one can relate the mode-Grüneisen parameter to properties of the interatomic potential. For these ionic crystals, one may neglect bond-bending interactions, and the equilibrium distance r_0 between atoms is determined by the balance between a long-range attractive force and a short-range repulsive force. The associated interatomic potential $E(r)$ can be written

$$E(r)=-\frac{A}{(r_0+x)^m}+\frac{B}{(r_0+x)^n} \tag{18}$$

with n > m and x representing a deviation from the equilibrium position. The constants A, B, m and n are determined from experimental data including lattice energy and bulk modulus. The resulting $E(r)$ curve is asymmetric and the atoms oscillate nonlinearly (i.e. anharmonically) around r_0. The force constant is obtained by the second derivative of E with respect to r. Neglecting the small terms one gets

$$\frac{\partial^2 E}{\partial x^2} \approx k_0 \left[1 - (n+m+3)\frac{x}{r_0} \right] \tag{19}$$

where k_0 is the force constant when $x = 0$. This relation gives the variation of k with x. For small changes in equilibrium spacing the force constant varies according to

$$\frac{dk}{dr} = -\frac{k_0}{r_0}(m+n+3) \tag{20}$$

Substituing this expression into Equation (19) gives a new expression for γ_{iT} for small values of x,

$$\gamma_{iT} = \frac{n+m+3}{6} \tag{21}$$

Development of $\dfrac{\partial^2 E}{\partial r^2}$ to higher orders leads to the volume dependence of γ_{iT} (i.e. q_{iT}) and one obtains

$$q_{iT} = \frac{n+m+nm+1}{3(n+m+3)} \tag{22}$$

For ionic solids n ≈ 1 and m ≈ 9. The expected mode-Grüneisen parameter is thus close to 2.2, in good agreement with the γ_{iT} measured for the TO IR modes of a large number of alkali halides including NaCl (Ferraro 1984, Hofmeister 1997b). The theoretical value of q_{iT} is ~0.5 which is also close to that obtained from the experimental data (~0.9).

In the case of simple minerals, it thus possible to relate with a pair interaction potential the frequency shifts of the vibrational modes to bond length changes as the volume changes. This is no longer likely to be the case for more complex minerals, where internal structural rearrangements can occur to accomodate the response to preaaure or temperature. γ_{iT} is related to changes in the force constant and is therefore a measure of the anharmonic nature of the interaction potential, determined within the quasi-harmonic model of crystal lattice vibrations. The a_i parameter measures a different type of anharmonicity which is present at constant volume, and which results from phonon-phonon coupling perhaps along with electronic redistributions in the crystal (Wallace 1972, Reissland 1973, Dove 1993).

Example of magnesite

The structure of magnesite consists of layers of Mg atoms that alternate with layers of CO_3 groups (Fig. 21). The Mg atoms are octahedrally coordinated to 6 oxygen atoms. The structure can be seen as a polyhedral linkage of corner sharing MgO_6 octahedra and trigonal CO_3 units. As already mentioned the internal vibrations of the CO_3 groups are easily identified in the Raman and IR spectra and lie at higher frequencies than the lattice modes. The effect of pressure on both the IR and Raman spectra of magnesite have been measured to 40 GPa (Gillet 1993, Grzechnik et al. 1998, Williams et al. 1993, Gillet et al. 1993). Gillet et al. (1993) also recorded the Raman spectrum up to 800 K at room pressure.

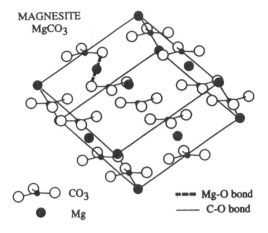

MAGNESITE
MgCO₃

○─●─○ CO₃

● Mg

▪▪▪ Mg-O bond

── C-O bond

Figure 21. Structure of MgCO₃ magnesite showing the position of CO₃ groups and MgO₆ octahedra. During heating, the thermal expansion is mostly accomodated by changes in the Mg-O bond length and only minor modifications of the C-O bond length. Similarly, most of the compression is accomodated by the Mg-O bond (Ross 1997, Markgraf and Reeder 1985).

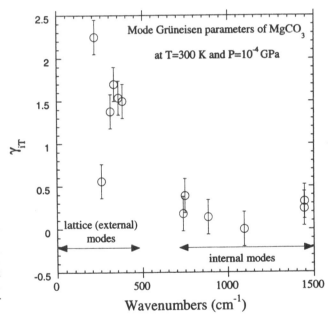

Figure 22. Mode-Grüneisen parameters γ_{iT} of the IR- and Raman-active modes of MgCO₃ magnes-ite. They have been calculated from the pressure-induced shifts reported by Gillet (1993) and Grzechnik et al. (1998) using a bulk modulus K_T = 112 GPa

The Grüneisen parameters γ_{iT} of selected Raman and IR modes of magnesite are shown in Figure 22. It can be noticed that the values related to the internal modes of the CO₃ units are significantly smaller than those of the external modes. A first consequence of this observation concerns the estimation of the macroscopic Grüneisen parameter γ_{th} from averaged values of the γ_{iT} using either Equation (6) or a simple arithmetic mean value. For magnesite one has $\gamma_{th} = 1.4$ from Equation (5). It is clear that the calculated values of γ_{th} will strongly depend on the type (internal vs external) and the number of modes included in the calculation. When only a small fraction of the active vibrational modes are used, the discrepancy can be very large [see Williams et al. (1993) for the SiO₂ polymorphs]. For magnesite, 10 of the 12 internal modes have been tracked at high pressures and their γ_{iT} values range between 0.15 and 0.4. 12 of the 15 internal optical modes have also been studied at high pressures and the γ_{iT} values range between 0.12 and 1.54. A simple average of all the γ_{iT} leads to 0.8-0.9, while that normalized by the specific heat relation [Equation (6)] gives 1.28. If only the internal modes are used in the calculation a very large under-estimation of γ_{th} results.

When pressure is applied to magnesite it is the MgO_6 octahedra which undergo most of the compression, the CO_3 groups being much less affected (Ross 1997). Although it is mathematically valid to use the classical definition of the mode-Grüneisen parameter (using the crystal unit cell volume) for all modes, it appears physically more reasonable to correlate the pressure (and temperature)-induced frequency shifts with specific changes in bond length, polyhedral volumes and bond angle. This contrasting behaviour of the CO_3 groups and MgO octahedra of magnesite during compression is reflected in the relative frequency shifts, $\frac{1}{v}\left(\frac{\partial v}{\partial P}\right)_T$, which are higher for the external than for the internal modes.

A similar observation can be made for the the relative frequency shifts, $\frac{1}{v}\left(\frac{\partial v}{\partial T}\right)_P$, which are smaller for the internal modes with respect to the external modes, in good agreement with the high temperature x-ray structural refinements which indicate a very small change of the C-O bond length and a strong variation of the Mg-O bond length (Markgraf and Reeder 1985).

Other correlations between frequency shifts and structural changes

Several attempts have been made to relate spectroscopic properties (frequencies and mode-Grüneisen parameters) and functionals of the interatomic potentials in solids or their equations of state (Liu 1992, Hofmeister 1991a,b). Hofmeister (1991a,b) used a model developed by Brout (1959) that relates the observed vibrational frequencies and their pressure derivatives to the bulk modulus and its pressure derivative of a given compound. Gillet et al. (1992) have shown that there is a correlation between the mineral bulk moduli and the mean frequency shifts of their vibrational modes with pressure. A correlation was also inferred between the temperature-induced mean frequency shifts and the bulk coefficient of thermal expansion. Reynard and Gillet (unpublished data) have further investigated these correlations and shown that they also hold for internal modes of molecular groups in minerals (CO_3 groups in carbonates and SiO_4 groups in silicates, for example) and they thus provide insights on polyhedral compressibility and thermal expansion.

CALCULATING THERMODYNAMIC PROPERTIES

A useful application of vibrational spectroscopy for geophysics is in the internally consistent calculation of the thermodynamic (specific heat, entropy) and thermoelastic (elastic constants, equation of state) of minerals at extreme pressure and temperature conditions. Calorimetric determinations of the specific heat and heat content of high-pressure mineral phases have been made at low pressure (see Navrotsky, Chapter 10, this volume). Direct measurement of these thermodynamic properties at high pressures is not straightforward. Moreover, even measurements on high-pressure minerals at ambient conditions are often complicated by the minute amounts available and because they are metastable at ambient pressure and high temperature. For instance, the most abundant mineral stable in the Earth's lower mantle, $(Mg,Fe)SiO_3$ perovskite, transforms first to an amorphous phase and then to enstatite upon heating above 600 K (Durben and Wolf 1992) making the detemination of its specific heat difficult at high temperature in the laboratory. The formation of the amorphous phase inside the metastable perovskite crystal also causes an anomalously large apparent thermal expansion to be measured (Durben and Wolf 1992; Wang et al. 1994). It is therefore useful to supplement available thermodynamic data on high-pressure minerals with calculations of these properties from vibrational data. These calculations also provide information on the statistical mechanical origin of measured behavior as well as tests of consistency of various measurements.

Specific volume measurements are readily carried out under simultaneous high *P-T* conditions up to 30 GPa and 2000 K using multianvil presses (Funamori et al. 1996) and up to 100 GPa and 3000 K with laser-heated diamond cells (e.g. Fiquet et al. 1998, Mao and Hemley, Chapter 1, this volume). Again, independent determinations of equations of state derived from vibrational data provide information on the origin of high *P-T* behavior as well as tests of consistency. The methodology is illustrated using forsterite as an example of an important mantle mineral for which a large data set already exists.

Vibrational density of states

Inelastic neutron scattering (INS) experiments, coupled with theoretical calculations, currently provide the most complete picture of the vibrational density of states [VDOS $g(v)$] of crystals at the present time (Ghose 1988, Dove 1993). Such data are scarce for high-pressure minerals because large quantities of sample are required for the measurements. Further, the INS experiment can not yet be applied to minerals at the very high pressures corresponding to the lower mantle and transition zone. As already noted above, synchrotron-based techniques such as inelastic x-ray scattering may hold some promise for work at high pressure, but these are in their infancy. The only techniques capable of providing information on the full VDOS under the complete range of *P-T* conditions within the Earth are theoretical calculations, either ab initio or empirical. Despite spectacular recent advances both in the accuracy and efficiency of these calculations and the complexity of the minerals that can be studied (Tossell and Vaughan 1992, Bukowinski 1994), they cannot yet be applied routinely and reliably to all Earth materials.

For the purpose of thermodynamic calculations, approximate simple descriptions of the VDOS such as the Debye model (ED) (Ita and Stixrude 1992, Jackson and Rigden 1996) or the Kieffer model (EK) (Kieffer 1979c, 1985) are often used to bypass these problems for high-pressure minerals. In the ED model, the frequency distribution is greatly oversimplified for most minerals (Fig. 23) and the model parameters (Debye temperature and Grüneisen parameter) are generally obtained by fitting the experimentally measured equation of state and heat capacity data. The EK model incorporates additional details of the frequency spectrum of the mineral (Fig 23) by including Raman, IR and acoustic data. Such models have now been used extensively for the prediction of thermodynamic and thermoelastic properties of high-pressure minerals (Kieffer 1979c, 1985; McMillan 1985, Chopelas 1990, 1996; Gillet et al. 1996, McMillan et al. 1996).

Statistical thermodynamics

From a knowledge of the VDOS, the entropy at a given volume and temperature can be calculated (Wallace 1972). For a given vibrational frequency $v_i(V,T)$ the corresponding entropy S_i is given by

$$S_i(V,T)=k_B\left[-ln\left(1-exp\left(-\frac{hv_i}{k_BT}\right)\right)+\frac{hv_i}{k_BT}\frac{1}{\left(exp\left(\frac{hv_i}{k_BT}\right)-1\right)}\right] \tag{23}$$

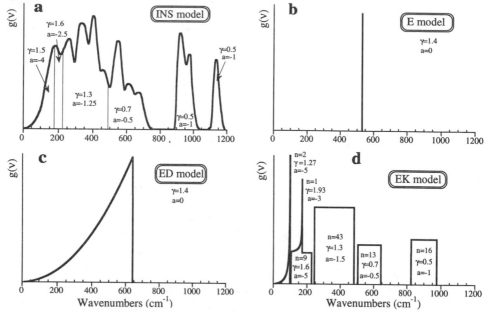

Figure 23. Vibrational density of states (VDOS) of Mg_2SiO_4 forsterite. (a) Neutron scattering measurements provide the best approach for obtaining the most accurate description of the frequency distribution of the vibrational modes of a crystal (INS model). Such measurements are available for forsterite (Ghose, 1988). (b) Einstein model (E model): the whole VDOS is represented by a single oscillator representing all 84 vibrational modes of forsterite. (c) Debye model (ED model): all the modes are assumed to be acoustic and follow a simple distribution starting from 0 up to a cut-off frequency inferred from acoustic measurements. (d) Kieffer model (EK model: the 3 acoustic modes follow a Debye model. The optical modes are placed in continua or in single Einstein oscillators if some particular modes can be assigned in the Raman and IR spectra. In all the models the γ_{iT} and a_i values for different frequency ranges are given.

The contribution of a given vibration to the specific heat at constant volume (C_{vi}) is given by:

$$C_{Vi}=T\frac{\partial S_i}{\partial T}\bigg|_V \tag{24}$$

and its contribution to the thermal pressure P_{th}^i is given by:

$$P_{th}^i(V,T)-P_{th}^i(V,T_0)=\int_{T_0}^{T}\frac{\partial S_i}{\partial V}\bigg|_{T'}dT' \tag{25}$$

with

$$\frac{\partial S_i}{\partial V}\bigg|_T=k_B\frac{\gamma_{iT}}{V}\left(\frac{hv}{k_BT}\right)^2\frac{exp\left(\frac{hv}{k_BT}\right)}{\left(exp\left(\frac{hv}{k_BT}\right)-1\right)^2} \tag{26}$$

The crystal entropy S, specific heat C_v and thermal pressure P_{th} are obtained by summation of Equations (23), (24), (25) over the whole VDOS. The total pressure and thus the high-temperature equation of state at various V and T conditions is obtained from

$$P(V,T)=P(V,T_0)+P_{th}(V,T)-P_{th}(V,T_0) \qquad (27)$$

$P(V,T_0)$ is calculated from room-temperature elastic data using a finite-strain equation of state (Anderson 1995):

$$P(V,T_0)=\frac{3}{2}K_0\left[\left(\frac{V_0}{V}\right)^{7/3}-\left(\frac{V_0}{V}\right)^{5/3}\right]\left[1+\frac{3}{4}(K_0'-4)\left(\left(\frac{V_0}{V}\right)^{2/3}-1\right)\right] \qquad (28)$$

Finally, the specific heat at constant pressure is obtained from

$$C_p = C_v + \alpha^2 V K_T T \qquad (29)$$

where α is the coefficient of volume thermal expansion obtained from the calculated values of $V(T,P)$.

In Equation (23), the vibrational frequencies are needed as functions of V and temperature T in order to derive Equations (24) and (25). The frequency variation with V and T can be written as a Taylor series for $d\ln(v_i)$ as:

$$d\ln v_i=\left(\frac{\partial \ln v_i}{\partial \ln V}\right)_{V0,T0}d\ln V+\left(\frac{\partial \ln v_i}{\partial T}\right)_{V0,T0}dT+\frac{1}{2}\left(\frac{\partial}{\partial \ln V}\frac{\partial \ln v_i}{\partial \ln V}\right)_{V0,T0}d^2\ln V$$
$$+\left(\frac{\partial}{\partial T}\frac{\partial \ln v_i}{\partial \ln V}\right)_{V0,T0}dTd\ln V+\frac{1}{2}\left(\frac{\partial}{\partial T}\frac{\partial \ln v_i}{\partial T}\right)_{V0,T0}d^2T \qquad (30)$$

Integration of Equation (30) leads to

$$\ln\frac{v_i}{v_i^0}=-\gamma_{iT}^0\ln\frac{V}{V_0}+a_i^0(T-T_0)-\frac{1}{2}\gamma_{iT}^0 q_{iT}^0\ln^2\frac{V}{V_0}+m_i^0(T-T_0)\ln\frac{V}{V_0}+\frac{1}{2}m_i^{'0}(T^2-T_0^2) \qquad (31)$$

where $v_i^0 = v_i(V,T_0)$, $v_i =v_i(V,T)$, $m_i^0=\left(\frac{\partial a_i}{\partial \ln V}\right)_{V0,T0}$, $m_i^{'0}=\left(\frac{\partial a_i}{\partial T}\right)_{V0,T0}$,

$$\gamma_{iT}^0=\left(\frac{\partial \ln v_i}{\partial \ln V}\right)_{V0,T0},\qquad q_{iT}^0=\left(\frac{\partial \ln \gamma_{Ti}}{\partial \ln V}\right)_{V0,T0}.$$

All of these parameters can be derived from spectroscopic measurements performed at high pressures and high temperatures, as described above. Usually the calculations of S and C_v are carried out within the harmonic (i.e. with v_i independent of pressure and temperature) or quasi-harmonic ($\gamma_{iT} = \gamma_{iP}$ and $a_i = 0$) approximations. However, it has been pointed out that the effects of intrinsic mode anharmonicity (i.e., the variations in mode frequencies with temperature at constant volume) become important at high temperatures (Gillet et al. 1989, 1991). Mode-anharmonicity parameters a_i are evaluated by separately measuring vibrational mode frequencies at high temperature and at high pressure using the techniques outlined previously.

Application to forsterite

The mineral forsterite provides a good example for demonstrating a calculation of thermodynamic properties from the VDOS. The measured VDOS of forsterite is compared with various simplified models in Figure 23. Analysis of the vibrational spectra for this mineral indicate that it is necessary to take into account intrinsic mode anharmonicity in the calculation of the specific heat and entropy at temperatures in excess of 1000 K (Fig. 24). Forsterite can be considered as a "standard" mineral for this purpose because most of its thermodynamic and thermoelastic properties have been measured over a wide range of

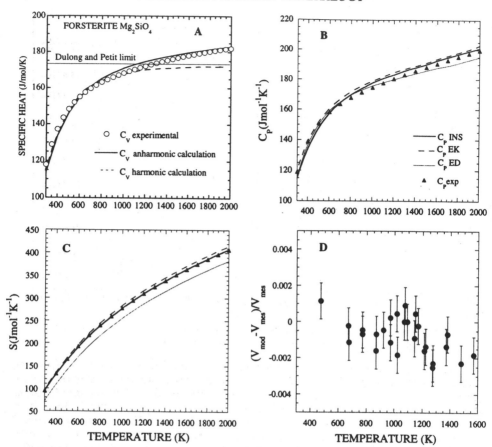

Figure 24. Thermodynamic properties of forsterite. (a) Specific heat at constant volume. The dots represent the values of C_v obtained from relation $C_v = C_v + \alpha^2 V K_T T$, where C_v is the measured specific heat at constant pressure (from Gillet et al., 1991), α the measured coefficient of thermal expansion (Bouhifd et al. 1996), K_T the measured bulk modulus (Isaak et al. 1989) and V the molar volume. The measured C_v crosses the harmonic Dulong and Petit limit above 1200 K. Also shown are the values of C_v calculated under harmonic assumptions with the different VDOS shown on Figure 2. The calculated C_v values tend toward the Dulong and Petit limit at high temperature and do not account for the observed excess C_v above 1200 K. (b) Comparison between the measured specific heat at constant pressure and that calculated with the three VDOS models outlined in Figure 23. Intrinsic mode anharmonicity has been taken into account in the calculations. (c) Comparison between measured entropy and that inferred from vibrational calculations including anharmonic behaviour of the vibrational modes. (d) Differences between the molar volumes calculated with the EK model of Figure 23 and values measured up to 8 GPa and 1300 K by x-ray diffraction (after Guyot et al. 1996). The agreement between both calculated and measured values is excellent and the difference never exceeds ±0.5%. Notice also that the Debye model is less efficient than other models of the VDOS for reproducing the measured values of the specific heat and entropy.

pressure and temperature conditions. The effect of pressure and temperature on the vibrational modes of forsterite is known up to 50 GPa at room temperature (Hofmeister et al. 1989, Chopelas 1990, Durben et al. 1993) and to 2000 K at room pressure (Gillet et al. 1991, 1997). All the parameters needed to calculate frequencies as a function of volume are thus available (Fig. 25). As discussed by Guyot et al. (1996) a Debye model is not nearly sufficient to describe the VDOS: a Kieffer-type model or the measured VDOS itself are required for thermodynamic calculations (Fig. 23) (Rao et al. 1988).

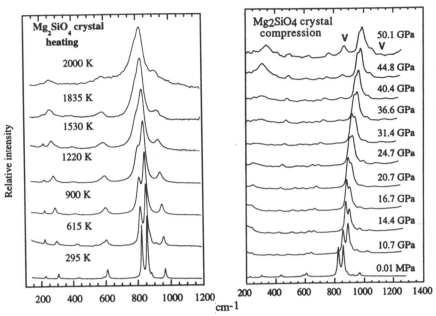

Figure 25. (a) Raman spectra of forsterite at high-pressure and room temperature (from Durben et al. 1993) and at high temperature and room pressure (from Gillet et al. 1997).

The degree of departure of the high temperature specific heat from the limit of Dulong and Petit (DP), namely $3nR$ (= 174 J/mol-K for forsterite) is a classical measure of intrinsic anharmonicity (Cynn et al. 1998). In Figure 24, we have plotted C_v obtained from Equation (29). C_p has been measured up to 1900 K (Gillet et al. 1991) and α and K_T have been measured up to 2000 K (Isaak et al. 1989, Bouhifd et al. 1996). The DP limit is exceeded at about 1300 K, and the excess C_v reaches 5% at 2000 K. C_v can be calculated with the VDOS given in Figure 23. Under the assumption that the crystal is harmonic or quasi-harmonic, the calculated C_v tends toward the DP at high temperature and does not account for the actual value of C_v. However, taking into account the volume dependence and intrinsic anharmonicity of the vibrations brings the calculated C_v to within 1% of the values obtained from calorimetric measurements with Equation 32. Similar good agreement between measured and calculated entropy is only obtained when anharmonicity is taken into account.

The vibrational model can also be used to predict the variation of volume as a function of pressure and temperature. Figure 24 shows the very good agreement between measured and calculated values of the molar volume of forsterite at various *P-T* conditions. In summary, the example shows that knowledge of the pressure and temperature dependence of vibrational mode frequencies of minerals can be used to carry out accurate determination of their fundamental thermodynamic and thermoelastic properties under the extreme conditions at depth within the Earth.

SELECTED HIGH *P-T* VIBRATIONAL SPECTRA

The database on the vibrational properties of minerals has increased rapidly in the past ten years. The status of vibrational spectroscopy of mantle minerals has recently been reviewed by McMillan et al. (1996a). The purpose of the section below is to give a brief survey of current data with a special emphasis on those minerals for which vibrational

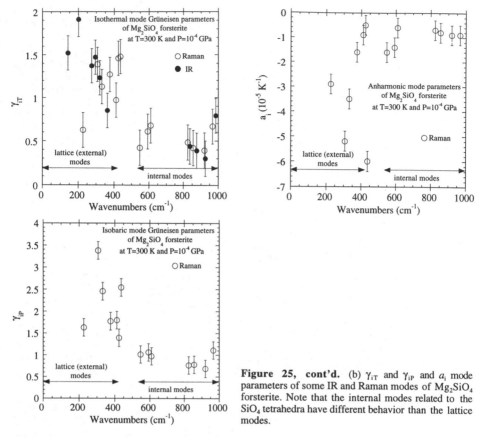

Figure 25, cont'd. (b) γ_{iT} and γ_{iP} and a_i mode parameters of some IR and Raman modes of Mg_2SiO_4 forsterite. Note that the internal modes related to the SiO_4 tetrahedra have different behavior than the lattice modes.

properties at high pressures and temperatures are useful for understanding the structure, composition and dynamics of the Earth's mantle. An overview of mantle mineralogy is provided in Chapters 2-8 of this volume.

$(Mg,Fe)_2SiO_4$ polymorphs

$(Mg,Fe)_2SiO_4$ olivine. Olivine has been thoroughly studied by vibrational spectroscopies. The pressure dependences at room temperature of the Raman, mid-IR and far-IR spectra have been measured in several studies for the end-members forsterite and fayalite, as well as for some intermediate compositions (Besson et al. 1982, Dietrich and Arndt 1982, Gillet et al. 1988, Hofmeister et al. 1989, Chopelas 1990, Williams et al. 1990, Gillet et al. 1991, Durben et al. 1993, Wang et al. 1993, Hofmeister 1997a). All the data show that the high-frequency modes (>500 cm⁻¹), related to the internal vibrations of the isolated SiO_4 units of the olivine structure, have smaller relative pressure-induced shifts, $\frac{1}{\nu}\left(\frac{\partial \nu}{\partial T}\right)_P$, or equivalently, smaller γ_{iT} parameters, than the lattice modes which involve vibrations of the MgO_6 octahedra (Fig. 22). This effect is clearly related to the larger incompressibility of the Si-O bond compared to that of the Mg-O bond as shown by high pressure x-ray diffraction studies (Kudoh and Takéuchi 1985). Gillet et al. (1997) have reported the Raman spectra of forsterite from 77 K to 2000 K and discused the intrinsic anharmonicity of the Raman-active modes. The relative temperature-induced

shifts, $\frac{1}{v}\left(\frac{\partial v}{\partial T}\right)_P$, are likewise smaller for the internal modes than for the lattice modes, indicating a very small expansion of the Si-O bond relative to the Mg-O bond upon heating (Fig. 25). Chopelas (1990), Gillet et al. (1991), and Guyot et al. (1996) have used high-pressure IR and Raman data to predict the thermodynamic and thermoelastic properties of forsterite over a wide P-T range, in excellent agreement with available experiment data.

On room temperature compression of forsterite above 40 GPa, new bands appearing near 750 cm^{-1} and 960 cm^{-1} in the Raman spectrum give evidence for the formation of Si-O-Si linkages (i.e. isolated Si_2O_7 units) within the structure, perhaps accompanied by an increase in the silicon coordination (Durben et al. 1993). Williams et al. (1990) have demonstrated from IR spectra that Fe_2SiO_4 fayalite undergoes pressure-induced amorphization when compressed at ambient temperature, consistent with x-ray investigation (Richard and Richet 1990). The former workers also interpreted changes in intensity of a band in the 600-800 cm^{-1} region above 35 GPa as related to the formation of six-coordinated silicon species.

The elastic constants and sound velocities of forsterite and San Carlos olivine $((Mg_{0.9},Fe_{0.1})SiO_4)$ have been measured up to 32 GPa using Brillouin scattering and impulsive stimulated scattering (Zha et al. 1998, 1996; Zaug et al. 1993, Abramson et al. 1997) (Fig. 12). These studies show that the shear elastic moduli in olivine are sensitive to the Fe content. A non-linear change of the shear constants above 20 GPa is observed for San Carlos olivine, which could lead to an eventual lattice instability perhaps associated with the observed pressure-induced amorphization at room temperature (Williams et al. 1990, Guyot and Reynard 1992, Andrault et al. 1995). Further vibrational studies in situ temperatures of $(Mg,Fe)_2SiO_4$ at high pressures and high will be required to understand and characterize the compression mechanisms and phase transformations of olivine in the upper mantle and transition zone.

An area that continues to deserve further attention is the exploration of "pre-melting" behavior of forsterite and other mantle minerals (Richet et al. 1994). The pre-melting regime is characterized by a rapid, anomalous increase in heat capacity and entropy below the thermodynamic melting point (Richet and Fiquet 1991). The origin of the phenomenon has been studied in detail using in situ Raman spectroscopy at high temperature for $(Ca,Mg)GeO_4$ olivines, and has been related to M_1-M_2 disordering, as well as to extreme anharmonicity in several sets of modes (Fiquet et al. 1992, Richet et al. 1994). This behaviour has now been observed to occur for silicate olivine, pyroxene and garnet, and it has obvious implications for rheology of high temperature mineral assemblages in the mantle, as well as for the kinetics of phase transformations.

β-(Mg,Fe)$_2$SiO$_4$. Vibrational spectra of the β polymorph of Mg_2SiO_4 were first reported by Akaogi et al. (1984). The Raman spectrum for this phase was later obtained by McMillan and Akaogi (1987), Chopelas (1991) and Reynard et al. (1996). In all of these studies, a strong peak was observed at 723 cm^{-1}, due to the Si-O-Si linkage vibrations in this phase. Williams et al. (1986) measured the powder IR spectrum of β-Mg_2SiO_4 to 27 GPa at room temperature, and Cynn and Hofmeister (1994) have obtained mid- and far-IR spectra of a sample with composition $(Mg_{0.9}Fe_{0.1})SiO_4$, to ~23 GPa. The spectrum of the Fe-containing sample is less well-resolved, perhaps due to Fe/Mg disorder. Reynard et al. (1996) measured the temperature shifts of the Raman peaks of β-Mg_2SiO_4 up to 800-900 K (the back-transformation temperature), and combined the data with the pressure shifts measured by Chopelas (1991) to determine intrinsic mode-anharmonicity parameters. They also observed the appearance of new peaks in the Raman spectrum between 800 K and 1000 K, consistent with the formation of a defective spinelloïd intermediate phase during the back-transformation of β-Mg_2SiO_4 to forsterite. Chopelas (1991) observed changes in the pressure shifts at 9.2 GPa and perhaps also at 17 GPa, which might indicate a structural transformation in β-Mg_2SiO_4. Similar changes in slope are apparent in the IR data taken at high pressure (Cynn and Hofmeister 1994). Liu et al. (1994) have also recorded the Raman spectrum of β-Mg_2SiO_4 up to 18 GPa. Chopelas (1991) used Kieffer's model (1979) to propose a thermodynamic database for this β-Mg_2SiO_4. Finally, Zha et al. (1997, 1998) measured the Brillouin spectrum of β-Mg_2SiO_4 up to 14 GPa and proposed a complete set of elastic constants. All elastic moduli vary linearly with pressure up to 14 GPa. Comparison with similar data for forsterite enable the direct determination of the velocity jump expected at 410 km in a forsterite-rich mantle.

γ-$(Mg,Fe)_2SiO_4$. This phase is cubic and has a spinel structure. Its Raman and IR spectra are known quite well at ambient conditions (Jeanloz 1980, Akaogi et al. 1984, Guyot et al. 1986, McMillan and Hofmeister 1988). The pressure shifts of the Raman peaks of γ-$(Mg,Fe)_2SiO_4$ were measured to 20 GPa (Liu et al. 1994, Chopelas et al. 1994). Chopelas et al. (1994) used their data for quasiharmonic calculations of heat capacity and entropy and proposed a calculation of the phase diagram of the $(Mg,Fe)_2SiO_4$ polymorphs.

$(Mg,Fe)SiO_3$ polymorphs

$(Mg,Fe)SiO_3$ pyroxene. In comparison with $(Mg,Fe)_2SiO_4$, much less work has been done on the vibrational spectroscopy of pyroxenes, either at ambient or mantle conditions (McMillan et al. 1996). The $(Mg,Fe)SiO_3$ rich-enstatite phase shows complex polymorphism as a function of pressure and temperature and of iron content (Yang and Ghose 1994). The orthorhombic phase (*Pbca*) is stable to 8 GPa and 1300 K. Above this temperature, there is a displacive transformation to protoenstatite (*Pbcm*) at ambient pressure for pure $MgSiO_3$, but Fe-containing phases transform to *C2/c* structures. At pressures above 8 GPa, orthopyroxene (*Pbca*) transforms into a high-density clinoenstatite structure (Angel et al. 1992, Pacalo and Gasparik 1990) of *C2/c* symmetry which reverts upon decompression to another orthorhombic phase of *P21a* symmetry.

Chopelas and Boehler (1992) have studied the pressure dependence of the Raman modes of the *Pbca* and *C2/c* polymorphs. The spectrum of the clinonenstatite polymorph is similar to that of diopside at the same pressure, consistent with the *C2/c* symmetry (Angel et al. 1992). Dietrich and Arndt (1982) have investigated the infrared spectrum of natural orthopyroxene at lower pressures, to 5 GPa, up to 250°C. Sharma (1989) and Ghose et al. (1994) have investigated the effect of temperature on the Raman spectra of $MgSiO_3$ pyroxene phases. The Raman and infrared spectra of $CaMgSi_2O_6$ (diopside) have been obtained at ambient conditions, and a Raman spectrum at 16 GPa is presented by Chopelas and Boehler (1992). The elastic constants and sound velocities of $(Mg_{0.9},Fe_{0.1})SiO_3$ have been measured up to 12 GPa using impulsive stimulated scattering (Chai et al. 1998).

$(Mg,Fe)SiO_3$ ilmenite. There have been several IR and Raman studies of this phase at ambient conditions (Ross and McMillan 1984, McMillan and Ross 1987, Madon and Price 1989, Hofmeister and Ito 1992, Reynard and Rubie 1996). Liu et al. (1994) recorded the principal Raman peaks of $MgSiO_3$-ilmenite to 30 GPa. Reynard and Rubie (1996) carried out a more complete study, in which accurate pressure shifts were obtained for nine Raman modes, with $(\partial v_i/\partial P)_T$ ranging from 1.7 to 3.7 cm^{-1}/GPa. These authors also measured the temperature shifts of the Raman peaks [$(\partial v_i/\partial T)_P$ ranging between -0.015 and -0.029 cm^{-1}/K], from which the intrinsic mode-anharmonicity parameters have been calclualted. These a_i parameters are very small, indicating that $MgSiO_3$-ilmenite behaves nearly quasiharmonically over the temperature range studied. Because the anharmonicity parameters for garnets are much larger (Gillet et al. 1992), it has been proposed that intrinsic mode anharmonicity corrections might play a role in determining the slope of

garnet-ilmenite phase boundaries (Reynard and Rubie 1996, Reynard and Guyot 1994).

(Mg,Fe)SiO₃ garnet and other garnets. X-ray diffraction, IR and Raman studies of pure $MgSiO_3$ garnet indicate that this phase has a slight tetragonal distortion from cubic symmetry when quenched to ambient conditions (McMillan et al. 1989). At high temperatures and pressures, there is extensive solid solution in a cubic garnet phase between pyrope $(Mg_3Al_2Si_3O_{12})$ and the majorite $MgSiO_3$ end-member (Akaogi et al. 1987, Kanzaki 1987). Up to 20% mole pyrope content, the quenched samples are tetragonal and cubic for higher Al contents (Heinemann et al. 1997). In addition, a majorite garnet phase, containing octahedrally coordinated silicon, with composition in $(Mg,Fe,Ca)_3(Si,Al,Fe)_2Si_3O_{12}$, is likely to be present within the transition zone of the Earth's mantle. Rauch et al. (1996) and Manghnani et al. (1998) have measured the Raman spectrum of the tetragonal end-member of $MgSiO_3$, up to 20 and 30 GPa, respectively, and observed a linear variation of the modes frequencies with increasing pressure. Rauch et al. (1996) have also reported a reversible change in the Raman spectrum of a 0.94 $MgSiO_3$:0.6 $Mg_3Al_2(SiO_4)_3$ sample between 5.1 and 8.5 GPa and suggest a possible transiton from a tetragonal to an orthorhombic structure. Manghnani et al. (1998) do not report a similar transition in 0.8$MgSiO_3$:0.2$Mg_3Al_2(SiO_4)_3$ and 0.5$MgSiO_3$:0.5$Mg_3Al_2(SiO_4)_3$ samples. Liu et al. (1995) have recorded the high frequency (>800 cm⁻¹) Raman spectrum of a garnet with composition 0.9$MgSiO_3$:0.1Al_2O_3, to 13.6 GPa at room temperature, and 875 K at room pressure.

Garnets in the $(Ca,Fe,Mg)_3(Al,Fe)_2Si_3O_{12}$ (pyrope-grossular-almandine-spessartine) system are important phases in the upper mantle. There have been several systematic infrared and Raman studies of these minerals (McMillan et al. 1996, Hofmeister and Chopelas 1991). Hofmeister and Chopelas (1991) used the vibrational spectra of pyrope and grossular to construct model density-of-states functions, $g(\omega)$, for calculation of the specific heat and entropy of these phases. Raman and IR spectra for cubic garnets have been recorded at high pressure and high temperature (Gillet et al. 1992, Dietrich and Arndt 1982, Mernagh and Liu 1990). Gillet et al. (1992) have used these data for pyrope, grossular and almandine to calculate intrinsic mode-anharmonicity parameters, showing that a relatively large (~3%) anharmonic correction to the heat capacity would be necessary.

The elasticity and sound velocity of different majoritic garnets have been studied at ambient conditions by Brillouin spectroscopy. Sinogeikin et al. (1997) have shown that the elastic properties of natural majorite $(Mg_{0.75}Fe_{0.21}Ca_{0.01})SiO_3$ from the Catherwood meteorite are indistinguishable from those of the pure Mg end-member (Bass and Kanzaki 1990). The elasticity of different cubic garnets have been also obtained from Brillouin measurements at ambient conditions (O'Neill et al. 1989, 1991; Bass 1989). Chai et al. (1997) measured the pressure dependence of the elastic constants of a natural pyrope-grossular-almandine garnet to 20 GPa by implusive stimulated scattering. The results are consistent with very recent high-pressure Brillouin scattering data of garnets over a similar pressure range (Conrad et al. 1998).

(Mg,Fe)SiO₃ perovskite. Because of its dominant role in the Earth's lower mantle, the physical properties of silicate perovskite $(MgSiO_3)$ have been investigated in numerous recent studies. This material has been the subject of various vibrational spectroscopic investigations. Williams et al. (1987) recorded the first Raman spectrum of $MgSiO_3$ perovskite at ambient conditions; then Hemley et al. (1989), Chopelas and Boehler (1992), Liu et al. (1994), Chopelas and Boehler (1992) and Chopelas (1996) measured the pressure-induced shifts of some Raman-active modes between ambient and 65 GPa at room temperature. Durben and Wolf (1992) and Gillet et al. (1996) investigated the temperature dependence at room pressure of the Raman spectra of $MgSiO_3$ perovskite from 77 to 600 K

and 300 K respectively. More recently, Gillet et al. (1998) (Fig. 9) have recorded the Raman spectra of $MgSiO_3$ perovskite at simultaneous high-pressure (up to 25 GPa) and low-temperature conditions (between 11 K and 320 K). This allowed determination of several mode frequencies (v_i) as a function of both pressure and temperature. The cross-derivatives $\partial^2 v_i / \partial T \partial P$ of vibrational frequencies were measured for the first time. These parameters are negative for the two lowest-frequency modes at 250 and 255 cm^{-1} (about -6 $\times 10^{-4}$ cm^{-1}GPa^{-1}K^{-1}) and positive for the other modes (+3 $\times 10^{-4}$ to +5 $\times 10^{-4}$ cm^{-1}GPa^{-1}K^{-1}. The anharmonicity of these modes is relatively small, indicating that no soft-mode behavior is expected over a wide P-T regime. IR data measured under ambient and high pressures have also been measured (Madon and Price 1989, Williams et al. 1987, Lu et al. 1994). Such data have been used by various authors for deriving the entropy, specific heat, thermal pressure, and equation of state, as well as various thermoelastic parameters, such the Grüneisen parameter of $MgSiO_3$ perovskite at mantle P-T conditions (Gillet et al. 1998, Chopelas 1996, Gillet et al. 1996, Hemley et al. 1989, Lu et al. 1994).

SiO_2

We have already discussed the use of vibrational spectroscopy in the study of phase transitions in the SiO_2 system. The effect of pressure and temperature on the IR and Raman-active modes of the principal polymorphs has been reviewed recently by Hemley et al. (1994) and Dolino et al. (1994).

The vibrational properties of this mineral have been extensively studied at high pressures and high temperatures (Dolino et al. 1994). Most of the IR- and Raman-active modes have been tracked up to the pressure of amorphization (Wong et al. 1986, Hemley 1987, Jayaraman et al. 1987, Williams et al. 1993) and also at high temperatures (Gervais and Piriou 1975, Gillet et al. 1990, Castex and Madon 1995). The Raman and IR spectra of a metastable high-pressure phase (quartz II) have been presented by Kingma et al. (1993b) and Williams et al. (1993), respectively. The effect of pressure on many of the Raman modes and a few IR modes of coesite are known (Hemley 1987, Williams et al. 1993). Gillet et al. (1990) and Liu et al. (1997) measured the effect of temperature on the Raman spectrum from which mode- anharmonic parameters can be inferred. The thermodynamic properties of this mineral have been evaluated from vibrational modeling (Gillet et al. 1990, Akaogi et al. 1995). The elastic constants at ambient conditions have been obtained by Brillouin spectroscopy (Weidner and Carleton 1977).

Williams et al. (1993) measured the pressure dependence of two peaks in the powder mid-IR transmission spectrum of stishovite up to 36 GPa. Two bands were followed to approximately 36 GPa, with pressure shifts of 2.4 and 3.3 cm^{-1}/GPa, respectively. More recently, Hofmeister (1996) carried out similar measurements in both mid and far-IR and has followed the pressure dependency of five IR modes. The pressure induced shifts of the Raman modes have been been studied up to 50 GPa where the transition to the $CaCl_2$–type structure occurs (Hemley 1987, Kingma et al. 1995) (Fig. 17). Gillet et al. (1990) and Liu et al. (1997) measured the temperature dependence of the Raman active modes of stishovite at ambient pressure from which intrinsic mode- anharmonic parameters can be evaluated. The heat capacity and entropy of this phase has been calculated using vibrational modeling (Gillet et al. 1990, Akaogi et al. 1995). The elastic constants of stishovite have been measured at ambient conditions by Weidner et al. (1982).

Simple oxides

There have been a large number of theoretical studies of the lattice dynamics of MgO under lower mantle conditions (Hemley et al. 1985, Agnon and Bukowinski 1990, Isaak et al. 1993). MgO has the cubic NaCl (B1) structure, which has one triply degenerate

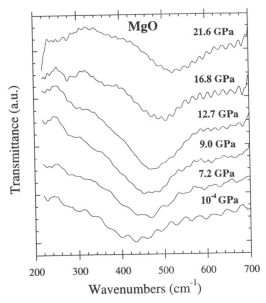

Figure 26. Absorption IR spectra of MgO at high pressures. The sample is a cold-pressed disk of MgO powder compressed in a diamond cell in between two CsI disks. From Grzechnik et al. (in preparation).

IR-active mode and no first-order Raman spectrum. The IR spectrum at ambient conditions is well understood (Piriou 1964, Ferraro 1984). The infrared reflectance spectrum of MgO has been studied at high temperature and room pressure (Piriou 1974, Piriou and Cabannes 1968). Recently, measurement of the far-IR spectrum of MgO up to 22 GPa at room temperature (Fig. 26) has permitted the first direct determination of the microscopic Grüneisen parameter γ_{iT} of this mineral (Grzechnik et al, in preparation). The values obtained (1.5-1.6) are close to that of γ_{th} (1.52), as expected for such a simple compound. The mode-Grüneisen parameters have also been constrained from vibronic optical spectra (Chopelas and Nicol 1982). There is a close resemblance between the VDOS deduced from these data and the VDOS obtained from neutron scattering measurements (Sangster et al. 1970). The pressure dependence of the sidebands modes of Cr^{2+} doped MgO crystals has been measured up to 37 GPa (Chopelas 1992, 1996; Chopelas and Nicol 1982) providing and independent detemination of the mode-Grüneisen parameters of the vibrational modes of MgO. The same authors used these data for MgO to calculate thermodynamic and thermoelastic properties of this phase within the quasiharmonic approximation, up to 40 GPa (Chopelas 1996).

Iron oxides

Vibrational Raman spectra, including both first- and second-order bands, of Fe_2O_3 have been measured under pressure (Massey et al. 1990). The assignment to vibrational transitions (distinguished from magnetic excitations) was made possible by ^{17}O isotope substitution.

Hydrous phases

The water content of the Earth's mantle is estimated to range between 100 and 500 ppm (Thompson 1992, Jambon 1994). Several major questions are actively debated (see also Mysen et al., Chapter 3, this volume): (1) How is this amount of water stored in the mantle? In a free hydrous fluid, accomodated as hydrogen defects in nominally anhydrous minerals or in special hydrous phases? (2) How does the presence of water affect the physical and chemical properties of minerals (elasticity, thermal and electrical conductivity,

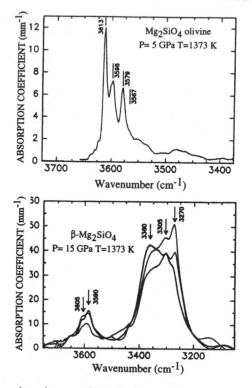

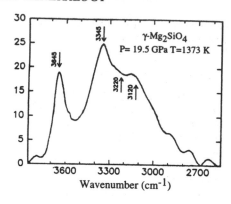

Figure 27. Absorbtion IR spectra in the O-H stretching region for the three polymorphs of Mg_2SiO_4 (after Kohlstedt et al. 1996). Samples were equilibrated at *P-T* conditions indicated in the figures in the presence of water.

viscosity, etc.)? (3) Is there a distinct possibility of detecting the presence of water in the mantle by geophysical prospecting, using the effect of water on the physical properties of deep Earth minerals?

Hydrous minerals such as micas, amphiboles, humites and dense hydrous magnesium silicates (DHMS = phase A, B, C, D, E and F) have been regarded as potential hosts for water in the mantle (Thompson 1992, Kanzaki 1991, Prewitt and Finger 1992, Gasparik 1993) but their stability seems to be restricted to H_2O-rich regions or cold regions such as subducting slabs (Thompson 1992, Bell 1992). More recently, it has been suggested that the nominally anhydrous mantle minerals are the most likely storage sites of hydrogen (and thus water) in the mantle. In that case, hydrogen is often located in specific crystallographic sites or as point defects (Kohlstedt et al. 1996).

IR and Raman spectroscopies permit a clear identification of hydrous species in minerals either in the form of H_2O molecules or OH (Fig. 27). This identification relies on the study of characteristic vibrations of these species: O-H stretching, H-O-H bending, or their overtones and combinations (Rossman 1988). Molecular H_2O has three IR- and Raman-active vibrational modes: symmetric stretch (v_1 at 3652 cm^{-1}), asymmetric stretch (v_3 at 3756 cm^{-1}) and bend (v_2 at 1595 cm^{-1}). The isolated OH group has a single stretching mode at 3735 cm^{-1}. In condensed phases, like silicates, the OH stretching frequency is affected by hydrogen bonding due to the polarity of the O-H bond. Attraction of OH to any adjacent negative center such as oxygen to which they are not bound, lowers the frequency of the O-H stretching vibration by a few hundred of wavenumbers. Quantitative determination of the OH concentration is achieved by carrying out absorbance measurements in the frequency region of the OH stretching mode on samples of different thickness (Rossman 1988, Paterson 1982). The pleochroic scheme of IR absorption bands obtained on well-oriented single crystals using polarized IR beam imposes strong constraints on the orientation of the OH dipole within the crystal structure (Beran and Putnis 1983, Libowitzky and Beran 1995).

Nominally anhydrous mantle minerals

Upper mantle silicates. In the past, vibrational spectroscopy has been extensively used to characterize the trace hydrogen content of olivine, pyroxene, and garnets, and other accessory minerals from natural samples extracted from the mantle by magmas (Bell 1992, Rossman 1988, 1996, Bell and Rossman 1992, Rossman and Smyth 1990). IR absorption measurements on small gem-quality samples have shown that, besides H_2O-bearing fluid inclusions, OH groups are also present in the crystal structure. The amount of OH is highly variable from sample to sample, but some general trends are observed. More generally, larger concentrations of OH are observed in clinopyroxenes (up to 1000 ppm equivalent H_2O) than in garnets (up to 200 ppm equivalent H_2O) and olivine (150 ppm equivalent H_2O). The solubility of water in the form of OH species has been experimentally studied in olivine (Kohlstedt et al. 1996, Young et al. 1993) and pyroxene (Ingrin et al. 1995, Hercule and Ingrin 1997). For olivine there is a strong increase in solubility with pressure, and 1200 ppm H_2O can be stored at 12 GPa.

Transition zone and lower mantle silicates. Only majoritic-garnet samples have been so far clearly identified as inclusions in diamonds (Moore and Gurney 1985). The β and γ phases of $(Mg,Fe)_2SiO_4$ have never been observed in samples derived from the mantle. There are also some diamond inclusions which are likely to represent former high-pressure phases from the lower mantle (now back-transformed to their low pressure polymorphs) like $(Mg,Fe)SiO_3$ and $CaSiO_3$ perovskites. IR or Raman studies apparently have not yet been reported to determine whether or not these natural samples of the mantle contain OH groups. Nevertheless, following the prediction of Smyth (1987) that β-$(Mg,Fe)_2SiO_4$ can host large amounts of water in the form of OH species, numerous experiments have been carried out to investigate the possibility of storing OH species in these anhydrous silicates (McMillan and Akaogi 1987, Kohlstedt et al. 1996, Young et al. 1993, Inoue 1994, Inoue et al. 1995). In fact, most of these minerals do actually accept OH groups in their structures as inferred from IR absorption experiments. Up to 3% equivalent H_2O can be introduced at high pressures in both β-Mg_2SiO_4 and γ-$(Mg,Fe)_2SiO_4$ (Fig. 27). Meade et al. (1994) have used infrared synchrotron radiation to measure the OH content of small single crystals (40 μm dimensions) of $MgSiO_3$-perovskite synthetized in the presence of water. The absorbance spectra indicate that at least 50 to 100 ppm equivalent H_2O can be stored in perovskite under typical lower mantle pressure and temperature. Lu et al. (1994) have obtained similar spectra using conventional IR spectroscopy. Pawley et al. (1993) have also shown by IR spectroscopy that SiO_2 stishovite can accept hydrogen in defect sites within the structure and that the OH concentration increases with the trace Al content of the samples.

Cynn and Hofmeister (1994) have measured the effect of pressure on the IR spectrum of hydrous β-Mg_2SiO_4 up to 23 GPa. The two most intense OH stretching vibrations have been studied only up to 12 GPa. The band at 3710 cm^{-1} has a strong positive shift with increasing pressure and that at 3329 cm^{-1} has a strongly negative one. These frequency changes can either be explained by variations in the O-H bond length upon compression or by no change in the O-H bond length but rather a change in the distance between the O atom of the O-H bond and adjacent O anions of the SiO_4 and MgO_6 polyhedra. Liu et al. (1998) have measured the effect of pressure and temperature on the Raman spectrum of hydrous β-Mg_2SiO_4. The introduction of OH species in the β-Mg_2SiO_4 structure appears to shift most of the Raman peaks toward lower frequencies. This has important implications for the calculations of the thermodynamic and thermoelastic properties of this hydrated crystal. The two peaks corresponding to the OH stretching modes have opposite behavior upon heating and compression. Inoue et al. (1998) have used Brillouin spectroscopy to measure the elastic constants and thus the compressibility of γ-Mg_2SiO_4 single crystals containing 2.2% equivalent H_2O. The isentropic bulk and shear moduli are 16% and 10% lower than those of "dry" γ-Mg_2SiO_4.

Dense hydrous magnesium silicates

High pressure hydrous minerals in the system MgO-SiO_2-H_2O are potential hosts for water in particular regions of the mantle such as subduction zones. Some of these phases (chondrodite, clinohumite, phases A-B-C-D-E-F) (Prewitt and Finger 1992) have been studied by vibrational spectroscopy at room conditions but also at high pressures. The IR and Raman spectra of chondrodite, $Mg_5Si_2O_8(OH)_2$; phase A, $Mg_7Si_2O_8(OH)_6$; phase B, $Mg_{24}{}^{VI}Si_2{}^{IV}Si_6O_{38}(OH)_4$; and phase E, $^{VI+IV}Mg_2SiO_4{\cdot}2H_2O$; are known at room conditions (Williams 1992, Cynn et al. 1996, Faust and Williams 1996, Liu et al. 1997). The OH stretching region of these phases is quite complex due to the presence of different protonated sites. The frequencies of the OH stretching vibrations are as low as 3350 cm^{-1}, which indicates a large degree of hydrogen bonding. IR and Raman spectra provide strong constraints on the localization of the H atoms within the crystal structures (Cynn et al. 1996). The behavior of these OH stretching vibrations upon pressure increase is complex, reflecting subtle structural changes. In chondrotite, phase B and phase E as already observed in β-Mg_2SiO_4, some of these modes exhibit positive shifts with pressure while others have negative ones. These opposite variations can be rationalized by variable changes in the length and thus strength of the hydrogen OH••••O bond [see discussions in Cynn et al. (1996) and Faust and Williams (1996)].

Brucite

The high-pressure behavior of brucite [$Mg(OH)_2$] is of interest because this mineral serves as a useful analog for complex hydrogen-bearing silicates of the crust and mantle. High-pressure Raman measurements showed evidence for a distortion of the structure under pressure (Duffy et al. 1995). Recently, IR absorption spectra of brucite single crystals have been measured using synchrotron radiation (Lu et al. 1998). Measurements were done at the U2B NSLS beamline using a FTIR spectrometer and diamond cells (Fig. 11). Samples several micrometers in thickness were loaded into the pressure chamber with ruby (as pressure gauge) and KBr or NaCl (as pressure-transmitting media). The IR spectrum of brucite in the mid-IR consists of an O-H stretching fundamental and combination bands involving this mode and low-frequency lattice vibrations (Fig. 28). The O-H stretching mode is only weakly active in the geometry used (direction of propagation is perpendicular to the basal plane of the crystals) because the transition moment direction is parallel to the wave vector of light. At 3.7 GPa, a new band appears at lower frequencies, and it gains intensity and softens with further increase in pressure. The initial O-H fundamental broadens under pressure and gradually looses intensity. On pressure release, the initial IR spectrum is restored at ambient pressure, but some hysteresis is observed and another narrow band is seen in the vicinity of the main fundamental. The observed phenomena are interpreted as due to a phase transition, which involves displacements of hydrogen atoms from their original axial sites, as proposed on the basis of recent Raman and neutron diffraction measurements (Parise et al. 1994, Duffy et al. 1995).

CONCLUSIONS AND OUTLOOK

Vibrational spectroscopy provides a powerful family of techniques for probing the dynamical behavior of minerals at high pressure and at high temperature, including

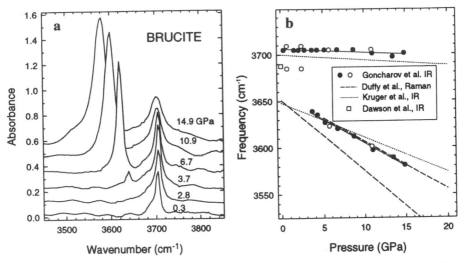

Figure 28. (a) IR absorption spectra of brucite as a function of pressure, courtesy of Goncharov et al. (to be published). (b) Pressure dependence of O-H stretching mode frequencies (Lu et al. 1998) and comparison to previous IR and Raman measurements (Dawson et al. 1973, Kruger et al. 1989, Duffy et al. 1995).

crystalline phase transitions and melting phenomena. Essential parameters for calculating thermodynamic properties and equations of state of minerals at depth can be extracted from the spectroscopic data to extend the database to the *P-T* regime where direct thermodynamic measurements are difficult or not currently possible. The advent and refinement of both external and laser heating techniques in diamond cells make it possible to collect vibrational spectroscopic data over the entire *P-T* range of the Earth. To date, only a few minerals have been studied in this way, and all major mantle and lower crustal minerals are candidates for future study. It will be particularly important to study minerals containing hydrous species, either as trace OH incorporated as impurities in their nominally anhydrous structure, or as an intregal part of the crystal structure. There has been no mention of molten phases in this review, despite that fact that melting is an important phenomenon within the mantle: only a very few studies have focused on the structure, dynamics and properties of silicate melts at very high pressures (see review by Wolf and McMillan 1995), and this represents an area for significant future effort using vibrational spectroscopy at high pressures and high temperatures. The high sensitivity of new techniques is likely to make it possible to measure with optical techniques the high-pressure vibrational properties of metals of relevant to the core.

   The techniques of high-pressure Raman, infrared, and Brillouin spectroscopies continue to be developed with advances in detector, spectrometer, and anvil technology. New methods enabled by advances in both synchrotron radiation and laser methods complement these more standard techniques. In particular, methods for obtaining information on phonon dispersion and densities of states at high pressure and high temperature are particularly exciting. Moreover, the development of a new generation of brighter radiation sources based on the free-electron laser is likely to provide still higher accuracy and sensitivity in the study of the vibrational dynamics of minerals under extreme *P-T* conditions. The future is thus bright for obtaining a more detailed understanding of the fundamental dynamical character of deep Earth materials, and how this behavior helps to control large-scale properties and processes of the Earth and planets.

ACKNOWLEDGMENTS

We thank all our colleagues—particularly F. Guyot, B. Reynard, I. Daniel, G. Fiquet, Y. Ricard, H.K. Mao, G.H. Wolf, A. Grzechnik—for continuous and fruitful collaborations and shared scientific experiences. PG and RJH warmly thank J. Matas and S. Gramsch for their generous and cheeful help during the preparation of the manuscript. We all owe special thanks to Paul Ribbe, *Reviews in Mineralogy* Series Editor, for his untiring efforts to make these volumes possible. This work was supported by INSU and by the NSF.

REFERENCES

Abramson EH, Brown JM, Slutsky LJ, Zaug J (1997) The elastic constants of San Carlos olivine to 17 GPa. J Geophys Res 102:12253-12264

Aines RD, Rossman GR (1984) Water in minerals? A peak in the infrared. J Geophys Res 89:4059-4071

Agnon A, Bukowinski MST (1990) Thermodynamic and elastic properties of a many-body model for simple oxides. Phys Rev B 41:7755-7766

Akaogi M, Navrotsky A, Yagi T, Akimoto SI (1987) Pyroxene-garnet transformations: thermochemistry and elasticity of garnet solid solutions, and application to a pyrolite mantle. *In:* MH Manghnani, Y Syono (eds) High Pressure Research in Mineral Physics: Applications to Earth and Planetary Sciences. p 251-260 Terra Scientific/American Geophysical Union, Tokyo/Washington, DC

Akaogi M, Ross NL, McMillan PF, Navrotsky A (1984) The Mg_2SiO_4 polymorphs (olivine, modified spinel, spinel)—thermodynamic properties from oxide melt solution calorimetry, phase relations, and models of lattice vibrations. Am Mineral 69:499-512

Akaogi M, Yusa H, Shiraishi K, Suzuki T (1995) Thermodynamic properties of a-quartz, coesite and stishovite and equilibrium phase relations at high pressures and high temperatures. J Geophys Res 100:22337-22347

Alp EE, Sturhahn W, Toellner T, Lee P, Schwoerer-Böhning M, Hu M, Hession P, Sutter J, Abbamonte P (1998) Advances in high-energy-resolution x-ray scattering at beamline 3-ID. Adv Photon Source Res (available from Argonne Nat'l Laboratory) 1:9-14

Anderson DL (1989) Theory of the Earth. Blackwell, Boston

Anderson OL (1995) Equations of State of Solids for Geophysics and Ceramic Science. Oxford University Press, New York

Andrault D, Bouhifd MA, Itié JP, Richet P (1995) Compression and amorphization of $(Mg,Fe)_2SiO_4$ olivines: an x-ray diffraction study up to 70 GPa. Phys Chem Minerals 22:99-107

Andrault D, Fiquet G, Itié JP, Richet P, Gillet P, Haüsermann D, Hanfland M (1998) Thermal pressure in the laser-heated diamond-anvil cell: An x-ray diffraction study. Eur J Min 10 (in press)

Angel RJ, Chopelas A, Ross NL (1992) Stability of high-density clinoenstatite at upper-mentle pressures. Nature 358:322-324

Angel RJ, Ross NL, Seifert F, Fliervoet TF (1996) Structural characterization of pentacoordinate silicon in a calcium silicate. Nature 384:441-444

Arashi H (1987) Raman spectroscopic studies at high temperatures and high pressures: application to determination of P-T diagram of ZrO_2. *In:* MH Manghnani, Y Syono (eds) High Pressure Research in Mineral Physics, p 335-339 Terra Scientific/American Geophysical Union, Tokyo/Washington, DC

Arns JA (1995) Holographic transmission gratings improve spectroscopy and ultrafast laser performance. SPIE Proc 2404:174-181

Badro J, Barrat JL, Gillet P (1996) Numerical simulation of a-quartz under nonhydrostatic compression: memory glass and five-coordinated crystalline phases. Phys Rev Lett 76:772-775

Badro J, Gillet P, Barrat JL (1998) Melting and pressure-induced amorphization of quartz. Europhys Lett 42:643-648

Badro J, Teter DM, Downs RT, Gillet P, Hemley RJ, Barrat JL (1997) Theoretical study of a novel five-coordinated silica polymorph. Phys Rev B 56:5797-5806

Bass JD (1989) Elasticity of grossular and spessartine garnets by Brillouin spectroscopy. J Geophys Res 94:7621-7628

Bass JD, Kanzaki M (1990) Elasticity of a majorite-pyrope solid solution. Geophys Res Lett 17:1989-1992

Bassett WA, Brody EM (1977) Brillouin scattering: a new way to measure elastic moduli at high pressures. *In:* MH Manghnani, S Akimoto (eds) High Pressure Research—Applications in Geophysics. p 519-532 Academic, New York

Bell DR (1992) Water in mantle minerals. Nature 357:646-647

Bell DR, Rossman GR (1992) The distribution of hydroxyl in garnets from the subcontinental mantle of southern Africa. Contrib Mineral Petrol 111:161-178

Bell DR, Rossman GR (1992) Water in the Earth's mantle: the role of anhydrous minerals. Science 255:1391-1397

Bell PM, Mao HK (1981) Degree of hydrostaticity in He, Ne, and Ar pressure-transmitting media. Carnegie Inst Washington Yearb 80:404-406

Belonoshko AB, Dubrovinsky LS, Dubrovinsky NA (1996) A new high-pressure silica phase obtained by molecular dynamics. Am Mineral 81:785-788

Beran A, Putnis A (1983) A model for the OH position in olivine derived from IR spectroscopy. Phys Chem Minerals 9:57-60

Besson JM, Pinceaux JP, Anastopoulos C, Velde B (1982) Raman spectra of olivine up to 65 kbar. J Geophy Res 87:10773-10775

Biellmann C, Gillet P, Guyot F, Peyronneau J, Reynard B (1993) Experimental evidence for carbonate stability in the Earth's lower mantle. Earth Planet Sci Lett 118:31-41

Boehler R, Chopelas A (1992) Phase transitions in a 500 kbar-3000 K gas apparatus. *In:* Y Syono, MH Manghnani (eds) High Pressure Research in Minerals Physics: Applications to Earth and Planetary Sciences. p 55-67 Terra Scientific/American Geophysical Union, Tokyo/Washington, DC

Boppart H, van Straaten J, Silvera IF (1985) Raman spectra of diamond at high pressure. Phys Rev B 32:1423-1425

Born M, Huang K (1954) Dynamical Theory of Crystal Lattices. Clarendon Press, Oxford

Bouhifd MA, Andrault D, Fiquet G, Richet P (1996) Thermal expansion of forsterite up to the melting point. Geophys Res Lett 23:1143-1146

Brout R (1959) Sum rule for lattice vibrations in ionic crystals. Phys Rev 113:43-44

Brown JM, Slutsky LJ, Nelson KA, Cheng L-T (1988) Velocity of sound and equation of state for methanol and ethanol in diamond-anvil cell. Science 241:65-67

Brown JM, Slutsky LJ, Nelson KA, Cheng L-T (1989) Single-crystal elastic constants for San Carlos peridot: an application of impulsive stimulated scattering. J Geophys Res 94:9485-9492

Bukowinski MST (1994) Quantum geophysics. Ann Rev Earth Planet 22:167-205

Carr GL, Hanfland M, Williams GP (1995) Midinfrared beamline at the National synchrotron Light source port U2B. Rev Sci Instrum 66:1643-1645

Castex J, Madon M (1995) Test of the vibrational modelling for the lambda-type transitions: Application to the alpha-beta quartz transition. Phys Chem Minerals 22:1-10

Chai M, Brown JM, Slutsky LJ (1996) Thermal diffusivity of mantle minerals. Phys Chem Minerals 23:470-475

Chai M, Brown JM, Slutsky LJ (1998) The elastic constants of Kilbourne Hole orthopyroxene to 12.5 GPa. J Geophys Res 102:14779-14785

Chai M, Brown JM (1996) Effetcs of static non-hydrostatic stress on the R lines of ruby single crystals. Geophys Res Lett 23:3539-3542

Chai M, Brown JM (1997) The elastic constants of a pyrope-grossular-almandine garnet to 20 GPa. Geophys Res Lett 24:523-526

Challener WA, Thompson JD (1986) Far-infrared spectroscopy in diamond anvil cells. Appl. Spectrosc. 40:298-303

Chervin JC (1998) Céramiques et matériaux pour l'optique. *In:* P Boissinot, P Langlois, JP Michel (eds) Matériaux et joints d'étanchéité pour les hautes pressions. p 55-80 Lavoisier, Paris

Chervin JC, Canny B, Gauthier M, Pruzan P (1992) Micro-Raman at variable low-temperature and very high pressure. Rev Sci Instrum 64:203-206

Chizmeshya A, LaViolette R, Wolf GH (1994) Variational charge relaxation in ionic crystals: an efficient treatment of statics and dynamics. Phys Rev 50:15559-15574

Chopelas A (1990) Thermochemical properties of forsterite at mantle pressures derived from vibrational spectroscopy. Phys Chem Mineral 17:149-156

Chopelas A (1991) Thermal properties of $\beta-Mg_2SiO_4$ at mantle pressures derived from vibrational spectroscopy: implications for the mantle at 400 km depth. J Geophys Res 96:11817-11829

Chopelas A (1992) Sound velocities of MgO to very high compression. Earth Planet Sci Lett 114:195-202

Chopelas A (1996) Thermal expansivity of lower mantle phases MgO and $MgSiO_3$ perovskite at high pressure derived from vibrational spectroscopy. Phys Earth Planet Inter 98:3-15

Chopelas A, Boehler R, Ko T (1994) Thermodynamics and behaviour of $\gamma-Mg_2SiO_4$ at high pressure: implications for Mg_2SiO_4 phase equilibrium. Phys Chem Minerals 21:351-359

Chopelas A, Boehler R (1992) Raman spectroscopy of high pressure $MgSiO_3$ phases synthesized in a CO_2 laser heated diamond anvil cell: perovskite and clinopyroxene. *In:* Y Syono, MH Manghnani (eds) High Pressure Research in Mineral Physics: Applications to Earth and Planetary Sciences. p 101-108 Terra Scientific/American Geophysical Union, Tokyo/Washington, DC

Chopelas A, Nicol MF (1982) Pressure dependence to 100 kbar of the phonons of MgO at 90 and 295 K. J Geophys Res 87:8591-8597

Cohen RE (1992) First-principles predictions of elasticity and phase transitions in high pressure SiO_2 and geophysical applications. *In:* Y Syono, MH Manghnani (eds) High Pressure Research in Mineral Physics: Applications to Earth and Planetary Sciences. p 425-431 Terra Scientific/American Geophysical Union, Tokyo/Washington, DC

Cohen RE (1994) First-principle theory of crystalline SiO_2. *In:* PJ Heaney, CT Prewitt, GV Gibbs (eds) Silica—Physical Behavior, Geochemistry and Materials Applications. Rev Mineral 29:369-402,

Conrad PG, Zha CS, Mao HK, Hemley RJ (1998) The high-pressure elastic moduli of pyrope, grossular, and andradite. Am Mineral (in press)

Cooney T, Sharma SK, Schiferl D (1992) Raman study of forsterite under simultaneous high pressure and high temperature. Abstract book of 29th International Geological Congress, Kyoto Japan 3:688

Cui, L, Chen NH, Silvera IF (1995) Infrared properties of ortho and mixed crystals of solid deuterium at megabar pressures and the question of metallization in the hydrogens Phys Rev Lett 74:4011-4014

Cummins HZ, Schoen PE (1972) Linear scattering from thermal fluctuations. *In:* FT Arrechi, EO Schulz-Dubois (eds) Laser Handbook. p 1029-1076, North Holland

Cynn H, Hofmeister AM, Burnley PC (1996) Thermodynamic properties and hydrogen speciation from vibrational spectra of dense hydrous magnesium silicates. Phys Chem Minerals 23:361-376

Cynn H, Isaak DG, Anderson OL (1998) Elastic properties of forsterite at high pressure obtained from high temperature database. *In:* MH Manghnani, T Yagi (eds) Properties of Earth and Planetary materials at High Pressure and Temperature, p 345-355 American Geophysical Union, Washington, DC

Cynn H, Hofmeister AM (1994) High-pressure IR spectra of lattice modes and OH vibrations in Fe-bearing wadsleyite. J Geophys Res 99:17717-17727

Daniel I, Gillet P, McMillan PF, Wolf G, Verhelst-Voorhees M (1997) High-pressure behaviour of anorthite. J Geophys Res 102:10313-10325

Daniel I, Gillet P, McMillan PF, Poe BT (1995) In situ high-temperature Raman spectroscopic studies of aluminosilicates liquids. Phys Chem Minerals 22:74-86

Daniel I, Gillet P, McMillan PF, Richet P (1995) An in-situ high-temperature structural study of stable and metastable $CaAl_2Si_2O_8$ polymorphs. Min Mag 59:25-33

Dawson P, Hadfield CD, Wilkinson GR (1973) The polarized infra-red and Raman spectra of $Mg(OH)_2$ and $Ca(OH)_2$. J Chem Solids 34:1217-1225

Dewaele A, Fiquet G, Gillet P (1998) Temperature and pressure distribution in the laser-heated diamond anvil cell. Rev Sci Instrum 69:2421-2426

Dhamelincourt P, Barbillat J, Delhaye M (1993) Laser confocal Raman microspectrometry. Spect Eur 5:16-26

Dietrich P, Arndt J (1982) Effects of pressure and temperature on the physical behavior of mantle-relevant olivine, orthopyroxene, and garnet: 2. Infrared absorption and microscopic Gruneisen parameters. *In:* W Schreyer (eds) High Pressure Research in Geoscience, p 307-319 E Schweizerbart'sche Verlags-binchhandlung

Dolino G, Valade M (1994) Lattice dynamical behaviour of anhydrous silica. *In:* PJ Heaney, CT Prewitt, GV Gibbs (eds) Silica—Physical Behavior, Geochemistry and Materials Applications. Rev Mineral 29:403-432

Dove, MT (1993) Introduction to Lattice Dynamics. Cambridge University Press

Dubrovinsky LS, Saxena SK, Lazor P, Ahuja R, Eriksson O, Wills JM, Johansson B (1997) Experimental and theoretical identification of a new high-pressure phase of silica. Nature 388:362-365

Duffy TS, Meade C, Fei Y, Mao HK, Hemley RJ (1995) High-pressure phase transition of brucite, $Mg(OH)_2$. Am Mineral 80:222-230

Durben DJ, McMillan PF, Wolf GH (1993) Raman study of the high pressure behavior of forsterite Mg_2SiO_4 crystal and glass. Am Mineral 78:1143-1148

Durben DJ, Wolf GH (1992) High-temperature behavior of metastable $MgSiO_3$ perovskite: a Raman spectroscopic study. Am Mineral 77:890-893

Exarhos GJ, Frydrych WS, Walrafen GE, Fisher M, Pugh E, Garofalini SH (1988) Vibrational spectra of silica near 2400°K: measurement and molecular dynamics simulation. In RJH Clark, DA Long (eds) Proc 11th International Conferenc on Raman Spectroscopy, p 503-504. John Wiley and Sons, New York

Eggert JH, Hemley RJ, Mao HK (1995) Comparison of the one and two-phonon spectra in stressed diamond. Bull Am Phys Soc 40:148

El Ghoresy A, Dubrovinsky L, Saxena SK, Sharp TS (1998) A new poststishovite silicon dioxide polymorph with the baddelyite structure (zircon oxide) in the SNC meteorite Shergotty: evidence for extreme shock pressure. Meteoritics Planet Sci 33:A45

Farber DL, Williams Q (1996) An in situ Raman spectroscopic study of $Na_2Si_2O_5$ at high pressures and temperatures: Structures of compressed liquids and glasses. Am Mineral 81:273-283

Faust J, Knittle E (1994) The equation of state, amorphization, and high-pressure phase diagram of muscovite. J Geophys Res 99:19785-19792

Faust J, Williams Q (1996) Infrared spectra of phase B at high pressures: Hydroxyl bonding under compression. Geophys Res Lett 23:427-430

Ferraro JR (1984) Vibrational Spectroscopy at High External Pressures. The Diamond Anvil Cell. Academic Press, Orlando

Finger LW, King H (1978) A revised method of operation of the single-crystal diamond cell and refinement of the structure of NaCl at 32 kbar. Am Mineral 63:337-342

Fiquet G, Gillet Ph, Richet P (1992) Anharmonicity and high-temperature heat capacity of crystals: the examples of Ca_2GeO_4, Mg_2GeO_4 and $CaMgGeO_4$ olivines. Phys Chem Minerals 18:469-479

Fiquet G, Andrault D, Dewaele A, Charpin T, Kunz M, Hausermann D (1998) P-V-T equation of state of $MgSiO_3$ perovskite. Phys Earth Planet Inter 105:21-32

Fiquet G, Andrault D, Itié JP, Gillet P, Richet P (1996) X-ray diffraction of periclase in a laser-heated diamond anvil cell. Phys Earth Planet Inter 95:1-17

Fujishiro I, N Mura Y (1994) Measurement of the viscoelastic properties of lubricants under high pressure by DAC. High Pressure Liquids and Solutions, p 149-167 Current Japanese Materials Research, Tokyo

Funamori N, Yagi T, Utsumi W, Kondo T, Uchida T, Funamori M (1996) Thermoelastic properties of $MgSiO_3$ perovskite determined by in situ X ray observations up to 30 GPa and 2000 K. J Geophys Res 101:8257-8269

Gasparik T (1993) The Role of volatiles in the transition zone. J Geophys Res 98:4287-4299

Gautron L, Fitz Gerald JD, Kesson SE, Eggleton RA, Irifune T (1997) Hexagonal Ba-ferrite: a good model for the crystal structure of a new high-pressure phase $CaAl_4Si_2O_{11}$. Phys Earth Planet Inter 102:223-230

Gervais F, Piriou B, Cabannes F (1973) Anharmonicity in silicate crystals: temperature dependense of A_u type vibrational modes in $ZrSiO_4$ and $LiAlSi_2O_6$. J. Phys Chem Solids 34:1785-1796

Gervais F, Piriou B, Cabannes F (1973) Anharmonicity of infrared vibration modes in the nesosilicate Be_2SiO_4. Phys Stat Sol B 55:143-154

Gervais F, Piriou B (1975) Temperature depedence of transverse and longitudinal optic modes in the α and β phases of quartz. Phys Rev B 11:3944-3950

Ghose S (1988) Inelastic neutron scattering. *In:* F Hawthorne (eds) Spectroscopic Methods in Mineralogy and Geology. Rev Mineral 18:161-192

Ghose S, Choudury N, Pal Chowdhury C, Sharma SK (1994) Lattice dynamics and Raman spectroscopy of protoenstatite $Mg_2Si_2O_6$. Phys Chem Minerals 20:469-471

Gillet P (1993) Stability of magnesite ($MgCO_3$) at mantle pressure and temperature conditions. A Raman spectroscopic study. Am Mineral 78:1328-1331

Gillet P (1995) Mineral physics, mantle mineralogy and mantle dynamics. C R Acad Sci Ser II 320:341-356

Gillet P (1996) Raman spectroscopy at high pressure and high temperature. Phase transitions and thermodynamic properties of minerals. Phys Chem Minerals 23:263-275

Gillet P, Badro J, Varel B, McMillan P (1995) High pressure behaviour of $AlPO_4$. Amorphization and the memory glass effect revisited. Phys Rev B 51:11262-11269

Gillet P, Biellmann C, Reynard B, McMillan PF (1993) Raman spectroscopic studies of carbonates. Part I: High-pressure and high-temperature behaviour of calcite, magnesite, dolomite, aragonite. Phys Chem Minerals 20:1-18

Gillet P, Daniel I, Guyot F, Matas J, Chervin JC (1998) A thermodynamic model for MgSiO3 perovskite derived from the pressure, temperature and volume dependence of the Raman mode frequencies. J Geophys Res (in press)

Gillet P, Daniel I, Guyot F (1997) Anharmonic properties of Mg_2SiO_4-forsterite measured from the volume dependence of the Raman spectrum. Eur J Mineral 9:255-262

Gillet P, Fiquet G, Daniel I, Reynard B (1993) Raman spectroscopy at mantle pressure and temperature conditions. Experimental set-up and the example of $CaTiO_3$ perovskite. Geophys Res Lett 20:1931-1934

Gillet P, Fiquet G, Malézieux JM, Geiger C (1992) High-pressure and high-temperature Raman spectroscopy of end-member garnets: pyrope, grossular and andradite. Eur J Mineral 4:651-664

Gillet P, Guyot F, Malezieux JM (1989) High-pressure and high-temperature Raman spectroscopy of Ca_2GeO_4: some insights on anharmonicity. Phys Earth Planet Inter 58:141-154

Gillet P, Guyot F, Wang Y (1996) Microscopic anharmonicity and the equation of state of $MgSiO_3$-perovskite. Geophys Res Lett 23:3043-3046

Gillet P, Le Cléach A, Madon M (1990) High-temperature Raman spectroscopy of the SiO_2 and GeO_2 polymorphs: anharmonicity and thermodynamic properties at high-temperature. J Geophys Res 95:21635-21655

Gillet P, Malezieux JM, Dhamelincourt MC (1988) Microraman multichannel spectroscopy up to 2.5 GPa using a sapphire-anvil cell: experimental set-up and some applications. Bull Minéral 111:1-15

Gillet P, Malézieux JM, Itié JP (1996) Phase changes and amorphization of zeolites at high pressures. The case of natrolite and mesolite. Am Mineral 81:651-657

Gillet P, McMillan P, Schott J, Badro J, Grzechnik A (1996) Thermodynamic properties and isotopic fractionation of calcite from vibrational spectroscopy of ^{18}O-substituted calcite. Geochim Cosmochim Acta 60:3471-3485

Gillet P, Richet P, Guyot F, Fiquet G (1991) High-temperature thermodynamic properties of forsterite. J Geophys Res 96:11805-11816

Gillet P, Goffé B (1988) On the significance of aragonite occurrences in the Western Alps. Contrib Mineral Petrol 99:70-81

Grimsditch M, Popova S, Brazkhin VV, Voloshin RN (1994) Temperature-induced amorphization of SiO_2 stishovite. Phys Rev B 50:12984-12986

Grzechnik A, McMillan PF (1998) Temperature dependence of the OH⁻ absorption in SiO_2 glass and melt to 1975 K. Am Mineral 83:331-338

Goncharov AF, Struzhkin VV, Ruf T, Syassen K (1994) High-pressure Raman study of the coupling of crystal field excitations to phonons in Nd-containing cuprates. Phys Rev B 50:13841-13844

Grzechnik A, McMillan PF (1998) Temperature dependence of the OH⁻ absorption in SiO_2 glass and melt to 1975 K. Am Mineral (submitted)

Grzechnik A, Simon P, Gillet P, McMillan PF (1998) An infrared study of $MgCO_3$ at high pressure. Physica B (submitted)

Guth JR, Hess AC, McMillan PF, Petuskey WT (1990) An ab initio valence force field for diamond. J Phys C Solid State Phys 2:8007-8014

Guyot F, Boyer H, Madon M, Velde B, Poirier JP (1986) Comparison of the Raman microprobe spectra of $(Mg,Fe)_2SiO_4$ and Mg_2GeO_4 with olivine and spinel structure. Phys Chem Minerals 13:91-95

Guyot F, Wang Y, Gillet P, Ricard Y (1996) Experimental measurements of high-pressure high-temperature volumes of olivines by synchrotron x-ray diffraction. Comparison with quasi-harmonic computations. Phys Earth Planet Inter 98:17-29

Guyot F, Reynard B (1992) Pressure-induced structural modifications and amorphization in olivine compounds. Chem Geol 96:411-420

Haines J, Léger JM, Schulte O (1996) Pa3 modified fluorite-type structure in metal dioxides at high pressures. Science 271:629-631

Hanfland M, Syassen K, Fahy K, Louie SG, Cohen ML (1985) Pressure dependence of the first-order Raman mde of diamond. Phys Rev B 31:6896-6899

Hazen RM, Finger LW, Hemley RJ, Mao HK (1989) High-pressure crystal chemistry and amorphisation of a-quartz. Solid State Commun. 72:507-511

Heinemann S, Sharp TG, Seifert F, Rubie DC (1997) The cubic-tetragonal phase transition in the system majorite $Mg_4Si_4O_{12}$- pyrope$Mg_3Al_2Si_3O_{12}$, and garnet symmetry in the Earth's transition zone. Phys Chem Minerals 24:206-221

Heinz DL (1990) Thermal pressure in the laser-heated diamond anvil cell. Geophys Res Lett 17:1161-1164

Heinz DL, Jeanloz R (1987) Temperature measurements in laser-heated diamond cell. In: MH Manghnani, Y Syono (eds) High Pressure Research in Mineral Physics, p 113-116 Terra Scientific/American Geophysical Union, Tokyo/Washington, DC

Hemley RJ (1987) Pressure dependence of Raman spectra of SiO_2 polymorphs: α-quartz, coesite and stishovite. In: MH Manghnani, Y Syono (eds) High Pressure Research in Mineral Physics, p 347-360 Terra Scientific/American Geophysical Union, Tokyo/Washington, DC

Hemley RJ, Jackson MD, Gordon RG (1985) First-principles theory for the equations of state of minerals to high pressures and temperatures: application to MgO. Geophys Res Lett 12:247-250

Hemley RJ, Mao HK, Chao ECT (1986) Raman spectrum of natural and synthetic stishovite. Phys Chem Minerals 13:285-290

Hemley RJ, Mao HK, Bell PM, Mysen BO (1986) Raman spectroscopy of SiO_2 glass at high pressure. Phys Rev Lett 57:747-750

Hemley RJ, Bell PM, Mao HK (1987) Laser techniques in high-pressure geophysics. Science 237:605-611

Hemley RJ, Jephcoat AP, Mao HK, Ming LC, Manghnani MH (1988) Pressure-induced amorphisation of crystalline silica. Nature 334:52-54

Hemley RJ, Cohen RE, Yeganeh-Haeri A, Mao HK, Weidner DJ (1989) Raman spectroscopy and lattice dynamics of $MgSiO_3$ perovskite. In: A Navrotsky, DJ Weidner (eds) Perovskite: A Structure of Great Interest to Geophysics and Materials Science. p 35-44 American Geophysical Union, Washington, DC

Hemley RJ, Prewitt CT, Kingma KJ (1994) High-pressure behaviour of silica. *In:* PJ Heaney, CT Prewitt, GV Gibbs (eds) Silica—Physical Behavior, Geochemistry and Materials Applications. Rev Mineral 29:41-81

Hemley RJ, Mao HK, Goncharov AF, Hanfland M, Struzhkin VV (1996) Synchrotron infrared spectroscopy to 0.15 eV of H_2 and D_2 at megabar pressures. Phys Rev Lett 76:1667-1670

Hemley RJ, Mao HK, Shen G, Badro J, Gillet P, Hanfland M, Häusermann D (1997) X-ray imaging of stress and strain of diamond, iron, and tungsten at multimegabar pressures. Science 276:1242-1245

Hemley RJ, Goncharov AF, Lu R, Li M, Struzhkin VV, Mao HK (1998) High-pressure synchrotron infrared spectroscopy at the National Synchrotron Light Source. Il Nuovo Cimento D 20:539-551

Hercule S, Ingrin J (1997) Diffusion, incorporation-extraction and solubility of hydrogen in diopside. Am Mineral (in press)

Herzberg G (1950) Molecular Spectra and Molecular Structure I: Diatomic Molecules. Van Nostrand, London

Hofmeister AM (1987) Single-crystal absorption and reflection infrared of forsterite and fayalite. Phys Chem Minerals 14:499-513

Hofmeister AM (1991a) Calculation of bulk modulus and its presure derivatives from vibrational frequencies and mode Grüneisen parameters: solids with cubic symmetry and one nearest-neighbor distance. J Geophys Res 96:16181-16203

Hofmeister AM (1991b) Pressure derivatives of the bulk modulus. J Geophys Res 96:21893-21907

Hofmeister AM (1996) Thermodynamic properties of stishovite at mantle conditions determined from pressure variations of vibrational modes. *In:* Dyar D, McCammon CA, Schaefer MW (eds) Mineral Spectroscopy: A Tribute to Roger G. Burns, p 215-227 Geochemical Society, Houston

Hofmeister AM (1997a) Infrared reflectance spectra of fayalite, and absoprtion spectra from assorted olivines, including pressure and isotope effects. Phys Chem Minerals 24:535-546

Hofmeister AM (1997b) IR spectroscopy of alkali halides at very high pressures: calculation of equation of state and the response of bulk moduli to the B1-B2 phase transition. Phys Rev B 56:5835-5855

Hofmeister AM, Xu J, Mao HK, Bell PM, Hoering TC (1989) Thermodynamics of Fe-Mg olivines at mantle pressure: mid- and far-infrared spectroscopy at high pressure. Am Mineral 74:281-306

Hofmeister AM, Chopelas A (1991) Thermodynamic properties of pyrope and grossular from vibrational spectra. Am Mineral 76:880-891

Hofmeister AM, Chopelas A (1991) Vibrational spectroscopy of end-member silicate garnets. Phys Chem Minerals 17:503-526

Hofmeister AM, Ito E (1992) Thermodynamic properties of $MgSiO_3$ ilmenite from vibrational spectra. Phys Chem Minerals 18:423-432

Ingrin J, Hercule S, Charton T (1995) Diffusion of hydrogen in diopside: Results of dehydration experiments. J Geophys Res 100:15489-15499.

Inoue T (1994) Effect of water on melting phase relations and melt composition in the system Mg_2SiO_4-$MgSiO_3$-H_2O Up to 15 GPa. Phys Earth Planet Inter 85:237-263

Inoue T, Weidner DJ, Northrup PA, Parise JB (1998) Elastic properties of hydrous ringwoodite (γ–phase) in Mg_2SiO_4. Earth Planet Sci Lett 160:107-113

Inoue T, Yurimoto H, Kudoh Y (1995) Hydrous modified spinel, $Mg_{1.75}SiH_{0.5}O_4$: A new water reservoir in the mantle transition region. Geophys Res Lett 22:117-120

Isaak DG, Anderson OL, Goto T (1989) Elasticity of single-crystal forsterite measured to 1700 K. J Geophys Res 94:5895-5906

Isaak DG, Cohen RE, Mehl MJ (1990) Calculated elastic and thermal properties of MgO at high pressure and temperatures. J Geophys Res 95:7055-7067

Israel A, Vohra YK (1998) Growth of diamond anvils for high-pressure research by chemical vapor deposition, *In:* RM Wentzcovitch, RJ Hemley, WB Nellis, P Yu (eds) High Pressure Materials Research. p 179-184 Materials Research Society, Warrendale, PA

Ita J, Stixrude L (1992) Petrology, elasticity and composition of the mantle transition zone. J Geophys Res 97:6849-6866

Jackson I, Rigden SM (1996) Analysis of P-V-T data: Constraints on the thermoelastic properties of high-pressure minerals. Phys Earth Planet Inter 96:85-112

Jambon A (1994) Earth degassing and large-scale geochemical cycling of volatile elements. *In:* MR Carroll, JR Holloway (eds) Volatiles in Magmas. Rev Mineral 30:479-517

Jayaraman AJ (1983) Diamond anvil cell and high-pressure physical investigations. Rev Mod Phys 55:65-108

Jayaraman AJ, Wood DL, Maines RG (1987) High-pressure Raman study of the vibrational modes in $AlPO_4$ and SiO_2 (α-quartz). Phys Rev B 35:8316-8321

Jeanloz R (1980) Infrared spectra of olivine polymorphs: α–, β–phase and spinel. Phys Chem Minerals 5:327-341

Kanzaki M (1987) Ultrahigh-pressure phase relations in the system $Mg_4Si_4O_{12}$-$Mg_3Al_2Si_3O_{12}$. Phys Earth Planet Inter 49:168-175

Kanzaki M (1991) Stability of hydrous magnesium silicates in the mantle transition zone. Phys Earth Planet Inter 66:307-312

Karki BB, Warren SC, Stixrude L, Ackland GJ, Crain J (1997) Ab initio studies of high-pressure structural transformations in silica. Phys Rev B 55:3465-3472

Kieffer SW (1979c) Thermodynamics and lattice vibrations of minerals. 3. Lattice dynamics and an approximation for minerals with application to simple substances and framework silicates. Rev Geophys Space Phys 17:35-59

Kieffer SW (1985) Heat capacity and entropy: systematic relations to lattice vibrations. *In:* SW Kieffer, A Navrotsky (eds) Microscopic to Macroscopic. Rev Mineral 14:65-126

Kim M, Owen H, Carey PR (1993) High-performance Raman spectroscopic system based on a single spectrograph, CCD, notch filter, and a Kr+ laser ranging from the near-IR to near-UV region. Appl Spectrosc 47:1780-1783

Kingma KJ, Cohen RE, Hemley RJ, Mao HK (1995) Transformation of stishovite to a denser phase at lower-mantle pressures. Nature 374:243-245

Kingma KJ, Meade C, Hemley RJ, Mao HK, Veblen DR (1993a) Microstructural observations of α-quartz amorphization. Science 259:666-669

Kingma KJ, Hemley RJ, Mao HK, Veblen DR (1993b) New high-pressure transformation in a-quartz. Phys Rev Lett 25:3927-3930

Kohlstedt DL, Keppler H, Rubie DC (1996) Solubility of water in the α, β and γ phase of $(Mg,Fe)_2SiO_4$. Contrib Mineral Petrol 123:345-357

Kraft S, Knittle E, Williams Q (1991) Carbonate stability in the Earth's mantle: a vibrational spectroscopic study of aragonite and dolomite at high pressures and temperatures. J Geophys Res 96:17997-18010

Kruger MB, Williams Q, Jeanloz R (1989) Vibrational spectra of $Mg(OH)_2$ and $Ca(OH)_2$ under pressure. J Chem Phys 91:5910-5915

Kudoh Y, Takéuchi Y (1985) The crystal structure of forsterite Mg_2SiO_4 under pressure up to 149 kbar. Z Kristallogr 172:291-302

Le Toullec R, Pinceaux JP, Loubeyre P (1988) The membrane diamond anvil cell: a new device for generating continuous pressure and temperature variations. High Pressure Research 1:77-90

Lemos V, Camatgo F (1990) Effects of pressure on the Raman spectra of a 4:1 methanol-ethanol mixture. J.Raman Spectroscopy 21:123-126

Libowitzky E, Beran A (1995) OH defects in forsterite. Phys Chem Minerals 22:387-392

Liu LG (1992) Bulk moduli of $MgSiO_3$-perovskite. Phys Earth Planet Inter 72:12-20

Liu LG (1993) Volume, pressure and temperature dependences of vibrational frequencies. Phys Lett A 176:448-453

Liu LG, Mernagh TP, Lin CC, Xu J, Inoue T (1998) Raman spectra of hydrous β–Mg_2SiO_4 at various pressure and temperature. *In:* MH Manghnani, T Yagi (eds) Properties of Earth and planetary materials at high pressure and temperature. p 523-530 American Geophysical Union, Washington, DC

Liu LG, Mernagh TP, Lin CC, Irifune T (1997) Raman spectra of phase E at various pressures and temperatures with geophysical implications. Earth Planet Sci Lett 149:57-65

Liu LG, Mernagh TP, Hibberson WO (1997) Raman spectra of high pressure polymorphs of SiO_2 at various temperatures. Phys Chem Minerals 24:396-402

Liu LG, Mernagh TP, Irifune T (1994) High pressure Raman spectra of β–Mg_2SiO_4, γ–Mg_2SiO_4, $MgSiO_3$-ilmenite and $MgSiO_3$-perovskite. J Phys Chem Solids 55:185-193

Liu LG, Mernagh TP, Irifune T (1995) Raman spectra of pyrope and $MgSiO_3$-$10Al_2O_3$ garnet at various pressures and temperatures. High Temp High Press 26:363-374

Long DA (1977) Raman Spectroscopy. McGraw Hill, New York

Lu R, Goncharov AF, Hemley RJ, Mao HK (1998) Synchrotron infrared microspectroscopy: applications to hydrous minerals. Clay Minerals Society Workshop Lectures, Vol 9 (in press)

Lu R, Hofmeister AM, Wang YB (1994) Thermodynamic properties of ferromagnesium silicate perovskites from vibrational spectroscopy. J Geophys Res 99:11795-11804

Madon M, Price GD (1989) Infrared spectroscopy of the polymorphic series (enstatite, ilmenite, and perovskite) of $MgSiO_3$, $MgGeO_3$, and $MnGeO_3$. J Geophys Res 94:15687-15701

Manga M, Jeanloz R (1998) Temperature distribution in the laser-heated diamond cell. *In:* MH Manghnani, T Yagi (eds) Properties of Earth and planetary materials at high pressure and temperature. p 17-26 American Geophysical Union, Washington, DC

Manghnani MH, Vijayakumar V, Bass JD (1998) High-pressure Raman scattering study of majorite-garnet solid solutions in the system $Mg_4Si_4O_{12}$-$Mg_3Al_2Si_3O_{12}$. *In:* MH Manghnani, T Yagi (eds) Properties of Earth and Planetary Materials at High Pressure and Temperature, p 129-137 American Geophysical Union, Washington, DC

Mao H K, Hemley RJ (1991) Optical transitions in diamond at ultrahigh pressures. Nature 351:721-724

Mao HK, Hemley RJ, Chao ECT (1987) The application of micro-Raman spectroscopy to analysis and identification of minerals in thin section. Scan Micros 1:495-501

Mao HK, Shen G, Hemley RJ (1998) X-ray diffraction with a double hot-plate laser-heated diamond cell. *In:* MH Manghnani, T Yagi (eds) Properties of Earth and Planetary Materials at High Pressure and Temperature, p 27-34 American Geophysical Union, Washington, DC

Mao HK, Shu JF, Shen G, Hemley RJ, Li B, Singh AK (1998) Elasticity and rheology of iron above 200 GPa and the nature of the Earth's inner core. Nature (in press)

Mao HK, Xu J, Bell PM (1986) Calibration of the ruby pressure gauge to 800 kbar under quasi-hydrostatic conditions. J Geophys Res 91:4763-4767

Mao HK, Hemley RJ (1994) Ultrahigh-pressure transitions in solid hydrogen. Rev Mod Phys 66:671-692

Markgraf SA, Reeder RJ (1985) High-temperature structure refinements of calcite and magnesite. Am Mineral 70:590-600

Massey MJ, Baier U, Merlin R, Weber WH (1990) Effects of pressure and isotopic substitution on the Raman spectrum of α–Fe_2O_3: identification of two-magnon scattering. Phys Rev B 41:7822-7827

McMillan PF (1985) Vibrational spectroscopy in the mineral sciences. *In:* Kieffer SW, Navrotsky A (eds) Macroscopic to Microscopic: Atomic Environments to Thermodynamics. Rev Mineral 14:9-63

McMillan PF (1989) Raman spectroscopy in mineralogy and geochemistry. Ann Rev Earth Planet Sci 17:255-283

McMillan PF, Akaogi M (1987) The Raman spectra of β–(modified) spinel and γ–spinel Mg_2SiO_4. Am Mineral 72:361-364

McMillan PF, Ross NL (1987) Heat capacity calculations for Al_2O_3 corundum and $MgSiO_3$ ilmenite. Phys Chem Minerals 16:225-234

McMillan PF, Hess AC (1988) Symmetry, group theory and quantum mechanics. *In:* FC Hawthorne (ed) Spectroscopic Methods in Mineralogy and Geology. Rev Mineral 18:11-61

McMillan PF, Hofmeister AM (1988) Infrared and Raman spectroscopy. *In:* FC Hawthorne (ed) Spectroscopic Methods in Mineralogy and Geology. Rev Mineral 18:99-159

McMillan PF, Wolf GH (1995) Vibrational spectroscopy of silicate liquids. *In:* JF Stebbins, PF McMillan, DB Dingwell (eds) Structure, Dynamics and Properties of Silicate Melts. Rev Mineral 32:247-315

McMillan PF, Akaogi M, Ohtani E, Williams Q, Nieman R, Sato R (1989) Cation disorder in garnets along the $Mg_3Al_2Si_3O_{12}$-$Mg_4Si_4O_{12}$ join: an infrared, Raman and NMR study. Phys Chem Minerals 16:428-435

McMillan PF, Dubessy J, Hemley RJ (1996a) Applications to Earth Science and Environment. *In:* G Turrell, J Corset (eds) Raman Microscopy: Developments and Applications. p 289-365 Academic Press, New York

McMillan PF, Hemley RJ, Gillet P (1996b) Vibrational spectroscopy of mantle minerals. *In:* Dyar MD, McCammon CA, Schaefer MW (eds) Mineral Spectroscopy: A Tribute to Roger Burns, p 175-213 The Geochemical Society, Houston.

McSkimin HJ, Andreatch Jr P (1972) Elastic moduli of diamond as a function of pressure and temperature. J Appl Phys 43:2944-2948

Meade C, Jeanloz R, Hemley RJ (1992) Spectroscopic and X-Ray diffraction studies of metastable crystalline-amorphous transitions in $Ca(OH)_2$ and serpentine. *In:* Syono Y, Manghnani MH (eds) High-Pressure Research: Applications to Earth and Planetary Sciences, p 485-492 Terrapub/American Geophysical Union, Tokyo/Washington, DC

Meade C, Reffner JA, Ito E (1994) Synchrotron infrared absorbance measurements of hydrogen in $MgSiO_3$ perovskite. Science 264:1558-1560

Meade C, Jeanloz R (1988) Yield strength of MgO to 40 GPa. J Geophys Res 93:3261-3269

Meng Y, Weidner DJ, Fei Y (1993) Deviatoric stress in a quasi-hydrostatic diamond anvil cell: effect on the volume-based pressure calibration. Geophys Res Lett 20:1147-1150

Mernagh TP, Liu LG (1990) Pressure dependence of Raman spectra of garnet end-members pyrope, grossularite and almandite. J Raman Spectroscopy 21:305-309

Mishima O, Calvert LD, Whalley E (1984) "Melting ice I" at 77 K and 10 kbar: a new method for making amorphous solids. Nature 310:393-394

Moore RO, Gurney JJ (1985) Pyroxene solid solution in garnets included in diamond. Nature 318:553-555

Muinov M, Kanda H, Stishov SM (1994) Raman scattering in diamond at high pressure: isotopic effects. Phys Rev B 50:13860-13862

Nagel L, O'Keeffe M (1971) Pressure and stress-induced polymorphism of compounds with rutile structure. Mat Res Bull 6:1317-1320

Navrotsky A (1994) Physics and Chemistry of Earth Materials. Cambridge University Press

Olijnyk H (1992) Raman scattering in metallic Si and Ge up to 50 GPa. Phys Rev Lett 68:2232-2234

O'Neill B, Bass JD, Rossman GR, Geiger CA, Langer K (1991) Elastic properties of pyrope. Phys Chem Minerals 17:617-621

O'Neill B, Bass JD, Smyth JR, Vaughan MT (1989) Elasticity of pyrope-grossular-almandine garnet. J Geophys Res 94:17819-17824

Pacalo EG, Gasparik T (1990) Reversals of the orthoenstatite-clinoenstatite transition at high pressures and high temperatures. J Geophys Res 95:15853-15858

Palmer DC, Hemley RJ, Prewitt CT (1994) Raman spectroscopic study of high-pressure phase transitions in cristobalite. Phys Chem Minerals 21:481-488

Paterson MS (1982) The detemination of hydroxyl by infrared absorption in quartz, silicate glasses and similar materials. Bull Minéral 105:20-29

Parise JB, Leinenweber K, Weidner DJ, Tan K, Von Dreele RB (1994) Pressure-induced H bonding: neutron diffraction study of brucite, $Mg(OD)_2$, to 9.3 GPa. Am Mineral 79:193-196.

Pawley AR, McMillan PF, Holloway JR (1993) Hydrogen in stishovite, with implications for mantle water content. Science 261:1024-1026

Peercy PS, Morosin B (1973) Pressure and temperature dependence of the Raman active phonons in SnO_2. Phys Rev B 7:2779-2786

Piermarini GJ, Block S, Barnett JD (1973) Hydrostatic limits in liquids and solids to 100 kbar. J Appl Phys 44:5377-5382

Pine AS (1983) Brillouin scattering in semiconductors. In: Cardona M (ed) Light Scattering in Solids I. Topics in Applied Physics 8:253-273 Springer-Verlag, New York

Piriou B (1974) Etude des modes normaux par réflexion infrarouge. Ann Chim 9-17

Piriou B, Cabannes F (1968) Validité de la méthode de Kramers-Kronig et application à la dispersion infrarouge de la magnésie. Optica Acta 15:271-286

Poirier JP (1991) Introduction to the Physics of the Earth's Interior. Cambridge University Press, Cambridge

Prewitt CT, Finger LW (1992) Crystal chemistry of high-pressure hydrous magensium silicates. In: Y Syono, MH Manghnani (eds) High-pressure research: Application to Earth and planetary sciences, p 269-274 Terra Scientific/American Geophysical Union, Tokyo/Washington, DC

Rao KR, Chaplot SL, Choudhury N, Ghose S, Hastings JM, Corliss LM, Price DL (1988) Lattice dynamics and inelastic neutron scattering from forsterite, Mg_2SiO_4: phonon dispersion relation, density of states and specific heat. Phys Chem Minerals 16:83-97

Rauch M, Keppler H, Häfner W, Poe B, Wokaun A (1996) A pressure-induced phase trnasition in $MgSiO_3$-rich garnet revealed by Raman spectroscopy. Am Mineral 81:1289-1292

Richard G, Richet P (1990) Room temperature amorphization of fayalite and high pressure properties of Fe_2SiO_4 liquid. Geophys Res Lett 17:2093-2096

Reeber RR, Wang K (1996) Thermal expansion, molar volume and specific heat of diamond from 0 to 3000 K. J Elect Mat 25:63-67

Reffner J, Carr GL, Sutton S, Hemley RJ, Williams GP (1994) Infrared microspectroscopy at the NSLS. Synch Rad News 7:30-37

Reissland JA (1973) The Physics of Phonons. John Wiley and Sons, New York.

Reynard B, Takir F, Guyot F, Gwanmesia GD, Liebermann RC, Gillet P (1996) High temperature Raman spectroscopic and x-ray diffraction study of $\beta-Mg_2SiO_4$: some insights on its high temperature thermodynamic properties and on the $\beta-\alpha$ phase transformation mechanism and kinetics. Am Mineral 81:585-594

Reynard B, Guyot F (1994) High-temperature properties of geikielite ($MgTiO_3$-ilmenite) from high-temperature high-pressure Raman spectroscopy. Some implications for $MgSiO_3$-ilmenite. Phys Chem Minerals 21:441-450

Reynard B, Rubie D (1996) High pressure and high temperature Raman spectroscopic study of $MgSiO_3$-ilmenite. Am Mineral 81:1092-1096

Richet P (1988) Superheating, melting and vitrification of high-pressure minerals. Nature 331:56-58

Richet P, Fiquet G (1991) High-temperature heat capacity and pre-melting of minerals in the system MgO-$CaO-Al_2O_3-SiO_2$. J Geophys Res 96:445-456

Richet P, Gillet P, Ali Bouhfid M, Daniel I, Fiquet G (1993) A versatile heating stage for measuements up to 2700 K, with applications to phase relationships determinations, Raman spectroscopy and x-ray diffraction. J Appl Phys 73:5446-5451

Richet P, Ingrin J, Mysen BO, Courtial P, Gillet P (1994) Premelting effects in minerals: an experimental study. Earth Planet Sci Lett 121:589-600

Richet P, Gillet P (1997) Pressure induced amorphization of minerals: a review. Eur J Mineral 9:907-933

Ringwood AE (1991) Phase transformations and their bearing on the constitution and dynamics of the mantle. Geochim Cosmochim Acta 55:2083-2110

Ross NL (1997) The equation of state and high pressure behavior of magnesite. Am Mineral 82:682-688

Ross NL, McMillan PF (1984) The Raman spectrum of $MgSiO_3$ ilmenite. Am Mineral 69:719-721

Rossman GR (1988) Vibrational spectroscopy of hydrous components. *In:* FC Hawthorne (ed) Spectroscopic Methods in Mineralogy and Geology. Rev Mineral 18:193-206

Rossman GR (1996) Studies of OH in nominally anhydrous minerals. Phys Chem Minerals 23:299-304

Rossman GR, Smyth JR (1990) Hydroxyl contents of accessory minerals in mantle eclogites and related rocks. Am Mineral 75:775-780

Sakurai T, Sato T (1971) Temperature dependence of vibrational spectra in calcite by means of emissivity measurement. Phys Rev B 40:583-591

Salje EKH (1990) Phase transitions in ferroelastic and co-elastic crystals. Cambridge University Press

Sangster MJL, Peckham G, Saunderson DH (1970) Lattice dynamics of magnesium oxide. J Phys C 3:1026-1036

Schoen CL, Sharma SK, Helsley CE, Owen H (1993) Performance of a holographic supernotch filter. Appl Spectrosc 47:305-308

Schwoerer-Böhning M, Macrander AT, Abbamonte PM, Arms DA (1998) High resolution inelastic x-ray scattering spectrometer at the advanced photon source. Rev Sci Instr 69:3109-3112

Schwoerer-Böhning M, Macrander AT, Arms DA (1998) Phonon dispersion of diamond measured by inelastic X-ray scattering. Phys Rev Lett 80:5572-5575.

Schwoerer-Böhning M, Klotz S, Besson JM, Burkel E, Braden M, Pintschovius L (1996) The pressure dependence of the TA[110] phonon frequencies in the ordered alloy Fe_3Pt at pressures up to 7 GPa. Europhys Lett 33:679-682

Scott JF (1974) Soft-mode spectroscopy: Experimental studies of structural phase transitions. Rev Mod Phys 46:83-128

Serghiou GC, Hammack WS (1993) Pressure-induced amorphization of wollastonite ($CaSiO_3$) at room temperature. J Chem Phys 98:9830-9834

Sette F, Krisch MH, Masciovecchio C, Ruocco G, Monaco G (1998) Dynamics of glasses and glass-forming liquids studied by inelastic neutron scattering. Science 280:1550-1555

Sharma SK (1989) Applications of advanced Raman techniques in Earth sciences. *In:* HD Bist, JR Durig, JF Sullivan (eds) Raman Spectroscopy: Sixty Years On, Vibrational Spectra and Structure, p 513-568. Elsevier, Amsterdam.

Sherman WF (1980) Bond anharmonicities, Grüneisen parameters and pressure-induced frequency shifts. J Phys C 13:4601-4613

Singh AK, Mao HK, Shu JF, Hemley RJ (1998) Estimation of single-crystal elastic moduli from polycrystalline x-ray diffraction at high pressure: Application to FeO and iron. Phys Rev Lett 80:2157-2160

Sinogeikin SV, Bass JD, Kavner A, Jeanloz R (1997) Elasticity of natural majorite and ringwoodite from the Catherwood meteorite. Geophys Res Lett 24:3265-3268

Skinner BJ, Fahey JJ (1963) Observations on the inversion of stishovite to silica glass. J Geophys Res 68:5595-5604

Smyth JR (1987) $\beta-Mg_2SiO_4$: a potential host for water in the mantle? Am Mineral 72:1051-1055

Sotin C, Gillet P, Poirier JP (1985) Creep of high-pressure ice VI. *In:* J Klinger (ed) Ices in the Solar System. Reidel Publishing Company

Struzhkin VV, Goncharov AF, Hemley RJ, Mao HK (1997) Cascading Fermi resonances and the soft mode in dense ice. Phys Rev Lett 78:4446-4449

Teter DM, Hemley RJ, Kresse G, Hafner J (1998) High pressure polymorphism in silica. Phys Rev Lett 80:2145-2148

Thompson AB (1992) Water in the Earth's mantle. Nature 358:395-399

Tossell JA, Vaughan DJ (1992) Theoretical Geochemistry: Application of Quantum Mechanics in the Earth and Mineral Sciences. Oxford University Press, New York

Tsuchida Y, Yagi T (1989) A new, post-stishovite high-pressure polymorph of silica. Nature 340:217-220

Turrell G, Corset J (1996) Raman Microscopy. Developments and Applications. Academic Press, London

Wallace DC (1972) Thermodynamics of Crystals. John Wiley and Sons, New York

Wang SY, Sharma SK, Cooney TF (1993) Micro-Raman and infrared spectral study of forsterite under high pressure. Am Mineral 78:469-476

Wang Y, Weidner DJ, Liebermann RC, Zhao Y (1994) P-V-T equation of state of $(Mg,Fe)SiO_3$ perovskite: constraints on the composition of the lower mantle. Phys Earth Planet Inter 83:13-40

Wang Z, Rossman GR, Blake GA (1998) A new microsampling visible-infrared spectrometer based on optical parametric oscillator technology. Spectroscopy 13:44-47

Weidner DJ (1975) Elasticity of microcrystals. Geophys Res Lett 2:189-192

Weidner DJ, Carleton HR (1977) Elasticity of coesite. J Geophys Res 82:1334-1346

White WB (1974) The carbonate minerals. *In:* VC Farmer (ed) The Infrared Spectra of Minerals, p 87-110 Mineralogical Society, London

Williams Q (1992) A vibrational spectroscopic study of hydrogen in high pressure mineral assemblages. *In:* Y Syono, MH Manghnani (eds) High-pressure Research: Application to Earth and planetary sciences, p 289-296 Terrapub/American Geophysical Union, Tokyo/Washington, DC

Williams Q, Collerson B, Knittle E (1993) Vibrational spectra of magnesite ($MgCO_3$) and calcite-III at high pressures. Am Mineral 77:1158-1165

Williams Q, Hemley RJ, Kruger MB, Jeanloz R (1993) High-pressure infrared spectra of α–quartz, coesite, stishovite and silica glass. J Geophys Res 98:22157-22170

Williams Q, Jeanloz R, Akaogi M (1986) Infrared vibrational spectra of beta-phase Mg_2SiO_4 and Co_2SiO_4 to pressures of 27 GPa. Phys Chem Minerals 13:141-145

Williams Q, Jeanloz R, McMillan PF (1987) Vibrational spectrum of $MgSiO_3$ perovskite: zero pressure Raman and mid-infrared spectra to 27 GPa. J Geophys Res 92:8116-8128

Williams Q, Knittle E, Reichlin R, Martin S, Jeanloz R (1990) Structural and electronic properties of Fe_2SiO_4-fayalite at ultrahigh pressures: amorphization and gap closure. J Geophys Res 95:21549-21563

Wolf GH, McMillan PF (1995) Pressure effects on silicate melt structure and properties. *In:* JF Stebbins, PF McMillan, DB Dingwell (eds) Structure, Dynamics and Properties of Silicate Melts. Rev Mineral 32:505-561

Wolf GH, Wang SA, Herbst CA, Durben DJ, Oliver WF, Kang ZC, Halvorson K (1992) Pressure-induced collapse of the tetrahedral framework of crystalline and amorphous GeO_2. *In:* Y Syono, MH Manghnani (eds) High Pressure Research in Minerals Physics: Applications to Earth and Planetary Sciences. p 503-517 Terra Scientific/American Geophysical Union, Tokyo/Washington, DC

Wong PTT, Baudais FL, Moffatt DJ (1986) Hydrostatic pressure effects on the TO-LO splitting and softening of infrared-active phonons in α–quartz.

Yahagi Y, Yagi T, Yamawaki H, Aoki K (1994) Infrared absorption spectra of the high-pressure phases of cristoballite and their coordination. Solid State Comm 89:945-948

Yang HX, Ghose S (1994) In situ Fe-Mg order-disorder studies and thermodynamic properties of orthopyroxene $(Mg,Fe)_2Si_2O_6$. Am Mineral 79:633-643

Young TE, Green HW, Hofmeister AM, Walker D (1993) Infrared spectroscopic investigation of hydroxyl in β–Mg_2SiO_4 and coexisting olivine: implications for mantle evolution and dynamics. Phys Chem Minerals 19:409-422

Zaug J, Abramson E, Brown JM, Slutsky LJ (1992) Elastic constants, equations of state and thermal diffusivity at high pressure. *In:* Y Syono, MH Manghnani (eds) High Pressure Research in Mineral Physics: Applications to Earth and Planetary Sciences. p 157-166 Terra Scientific/American Geophysical Union, Tokyo/Washington, DC

Zaug J, Abramson EH, Brown JM, Slutsky LJ (1993) Sound velocities in olivine at Earth mantle pressure. Science 260:1487-1489

Zha CS, Duffy TS, Downs RT, Mao HK, Hemley RJ, Weidner DJ (1998) Single-crystal elasticity of the α and β polymorphs of Mg_2SiO_4 at high pressure. *In:* MH Manghnani, T Yagi (eds) Properties of Earth and Planetary Materials at High Pressure and Temperature, p 9-16 American Geophysical Union, Washington, DC

Zha CS, Duffy TS, Downs RT, Mao HK, Hemley RJ (1996) Sound velocity and elasticity of single crystal forsterite to 16 GPa. J Geophys Res 101:17535-17546

Zha CS, Duffy TS, Downs RT, Mao HK, Hemley RJ (1998) Brillouin scattering and x-ray diffraction of San Carlos olivine: direct pressure determination to 32 GPa. Earth Planet Sci Lett 159:25-33

Zha CS, Duffy TS, Mao HK, Downs RT, Hemley RJ, Weidner DJ (1997) Single-crystal elasticity of β–Mg_2SiO_4 to the pressure of the 410 km seismic discontinuity in the Earth's mantle. Earth Planet Sci Lett 147:E9-E15

Zhang J, Liebermann RC, Gasparik T, Herzberg CT, Fei Y (1993) Melting and subsolidus relations of SiO_2 at 9-14 GPa. J Geophys Res 98:19785-19793

Zhang L, Chopelas A (1994) Sound velocity of Al_2O_3 to 616 kbar. Phys Earth Planet Inter 87:77-83

Zouboulis ES, Grimsditch M (1991) Raman scattering in diamond up to 1900 K. Phys Rev B 43:12490-12493

Chapter 18

HIGH-PRESSURE ELECTRONIC
AND MAGNETIC PROPERTIES

Russell J. Hemley, Ho-kwang Mao, and Ronald E. Cohen

Geophysical Laboratory and Center for High Pressure Research
Carnegie Institution of Washington
5251 Broad Branch Road, NW
Washington, DC 20015

INTRODUCTION

Manifold materials make up the deep Earth: silicates and oxides of the deep fertile mantle, the transformed and processed subducted crust, plumes and volatile-containing partial melts, slag-like new materials at the core-mantle boundary, the convecting liquid metal of the outer core, the textured anisotropic inner core. The wide range of bonding states represented by these materials underlies their physical and chemical properties. At a fundamental level, these electronic and magnetic properties also directly influence large-scale global phenomena ranging from the initial differentiation of the planet, the formation and transmission of the Earth's magnetic field, the propagation of seismic waves, and the upwelling and downwelling of mass through the mantle. In comparison to common minerals and rocks, the many aspects of the behavior of these materials are not readily familiar owing to their distance, both physically and conceptually, from the near-surface environment of the planet. Yet, the study of the Earth as a whole cannot be considered complete without understanding and characterizing these materials at a fundamental level, as has been done for more familiar materials of the planet in evidence at the surface.

By far the most important consideration in pursuit of this knowledge is the effect on these materials of the extreme pressures and temperature of the deep interior, up to 363 GPa and in excess of 4500 K. In the 1920s, both Bridgman (1927) and Fowler (1927) suggested that at sufficiently high pressure the atoms that comprise ordinary materials may break down to form a dense, conducting assembly of electrons and nuclei. This idea was clarified within a few years with the development of the quantum mechanical theory of bonding in molecules and solids, and gave rise to Bernal's celebrated proposal that all materials must transform at very high pressure "to metallic or valence lattices" (Wigner and Huntington 1935, footnote 1). The proposition of the ultimate metallic state of dense matter is a natural consequence of the increasing electronic kinetic energy in compressed matter, as described in Chapter 19. Although its onset is material specific, this same effect of compression on electronic structure can have profound and sometime unexpected consequences for physical and chemical properties.

Over the range of pressures encountered within the Earth, rock-forming oxides and silicates compress by factors of two to three, and condensed volatiles, such as rare gases and molecular species can be compressed by well over an order of magnitude. With the large reduction in interatomic distances produced by these conditions, dramatic alterations of electronic and magnetic properties give rise to myriad transformations. Discrete electronic and magnetic transformations can occur, such as metallization and magnetic collapse (e.g. Knittle and Jeanloz 1986b, Cohen et al. 1997). More commonly, these changes are associated with, and drive, other transformations such as structural phase

0275-0279/98/0037-0018$05.00

transitions and chemical reactions. Materials considered volatiles under near-surface conditions can be structurally bound (as "valence lattices") in dense, high-pressure phases [e.g. hydrogen in ice—Goncharov et al. (1996), mantle silicates—Finger et al. (1989), and ferrous alloys—Badding et al. (1992)]. Noble gases become metallic (Goettel et al. 1989, Reichlin et al. 1989), but even before that, become refractory materials under pressure, with higher melting points than the dominant materials of the mantle and core (Jephcoat 1998). At still lower pressures, they form new classes of molecular compounds (Vos et al. 1992, Loubeyre et al. 1993). Elements such as Fe^{2+} and Mg^{2+}, which are commonly found in association, dissociate at depth (Mao et al. 1997a), whereas incompatible elements such as Fe and K may form alloys (Parker and Badding 1996). Indeed, under pressure the materials that reside in the deep Earth exhibit new phenomena, the study of which is not only a problem of first rank in the geosciences but stands at the forefront of modern condensed-matter physics (Hemley and Ashcroft 1998).

This chapter constitutes an introduction to studies of the electronic and magnetic properties of deep Earth materials and the techniques used to study them. General background on bonding and electronic structure of minerals can be found in various texts emphazing both the solid-state physics and chemistry points of view (Pauling 1960, Ashcroft and Mermin 1976, Harrison 1980, Kittel 1996). Because of the breadth of the field, the following constitutes only a brief overview of the subject. Techniques used for ambient pressure studies, including high-pressure mineral systems, have been described in earlier volumes of this series (e.g. Volumes 12 and 18). Here, we focus on *in-situ* high *P-T* techniques, applications, and results, paying special attention to very recent developments and prospects for the future.

We begin with a short review of fundamental properties, with emphasis on how these are altered under very high pressures and temperatures of the Earth's deep interior. This is followed by a discussion of different classes of electronic and magnetic excitations. We then provide a short description of various techniques that are currently used for high-pressure studies. Rather than being complete and tutorial, the aim of this section is to show the wide variety of methods that can be used to study these properties of minerals in situ under very high pressures. We conclude with a summary of selected examples as case studies, showing how the use of a combination of different techniques reveals insight into high *P-T* phenomena. The behavior of iron in different chemical environments and *P-T* regimes provides a unifying theme through many of these examples.

OVERVIEW OF FUNDAMENTALS

Bonding in deep Earth materials

The binding forces that hold atoms together are generally electrostatic in origin. The forces that keep atoms from collapsing into each other arise from increased kinetic energy as atoms are brought closer together, the Pauli exclusion principle that keeps electrons apart, the electrostatic repulsion between electrons, and ultimately (as the atoms are brought closer together) the electrostatic repulsion of the nuclei. Crystals can be characterized by the primary source of their binding energy. Following the conventional classifications set out by Pauling (1960), we briefly review for the purposes of the present chapter the following bonding types—ionic, covalent, metallic, van der Waals, and hydrogen bonding as found materials in the deep Earth.

Ionic. In ionic crystals, the primary source of binding is from the electrostatic attraction among ions. Prototypical examples are NaCl (Na^+Cl^-) and MgO ($Mg^{2+}O^{2-}$). This ionicity is driven by the increased stability of an ion when it has a filled shell of electrons. For example, oxygen has a nuclear charge $Z = 8$, which means that it would have two $1s$, two 2*s*, and four 2*p* electrons, but a filled *p* shell has six electrons. Thus, the O atom is highly reactive, and in the gas phase combines to form O_2 or other molecules with other atoms. In the simple picture of an oxide or silicate, two electrons are transferred from another atom to the O to form an O^{2-} anion from and a positively charged cation. This leads to a strong attractive electrostatic, or Madelung, interaction between the ions which greatly enhances the crystal stability.

Covalent. In a covalent bond, charge concentrates in the bonding region, increasing the potential and kinetic energy of interaction between electrons, but reducing the energy through the electron-nuclear interactions (since each electron is now on average close to both nuclei or atomic cores, which have a net positive charge to the electrons). Covalent bonds tend to be very directional since they are formed from linear combinations of directional orbitals on the two atoms. It is straightforward to develop models for such covalent materials, and such models have ranged from those based on empirical fitting parameters to those obtained directly from first principles (Stixrude et al., Chapter 19).

Metallic. The high-pressure behavior of metals, and in particular iron and iron alloys, is of paramount importance for understanding the nature of the Earth's core. Metals are distinguished by the fact that they easily conduct electricity due to partially occupied bands. The bonding in metals is often thought of as a positive ion core embedded in a sea of electrons, the negative electrons holding the ions together. This analogy can be misleading, since the 'sea' may be very anisotropic, and the metallic electrons may participate in covalent bonds.

Van der Waals. Another source of bonding is from dispersion, or van der Waals forces, fluctuating dipoles on separated atoms or molecules. Separated, non-overlapping charge densities do indeed have such an attractive force between them, and it varies as C_6/r^6 and higher order terms at large distances. These dispersion coefficients can be obtained from studying the scattering of atoms or molecules in the gas phase. Such dispersion interactions are also invoked to explain weak interlayer bonding in materials such as graphite and the weak binding of inert gases and closed shell molecules in minerals.

Hydrogen bonding. Hydrogen bonds are weaker interactions involving a hydrogen atom covalently bonded to an oxygen atom and weakly attracted to a neighboring oxygen. As such, the three atoms form a linear linkage with a hydrogen atom asymmetrically disposed between the oxygens, a configuration that is ubiquitous in hydrous materials. A range of degrees of hydrogen bonding at ambient pressure is evident in dense hydrous silicates that are candidate mantle phases (Prewitt and Downs, Chapter 9).

Many minerals exhibit combinations of the types of interactions outlined above. Most common is ionic and covalent bonding, both within the same bond and different degrees of each in different bonds present in the material. For example, silicates are considered about half-ionic and half-covalent (Stixrude et al., Chapter 19); transition metals such as Fe are both metallic and covalent. A molecular group that contains strong covalent bonds, can also be ionically bound to other ions in the crystal. An example is $CaCO_3$, where the C-O bonds are strong and directional, forming a planar triangle in the CO_3^{2-} carbonate group, which is bound ionically to Ca^{2+} ions.

Evolution of bonding with pressure and temperature

What happens to bonds, and more generally, the interactions between atoms on compression? As atoms are brought closer together, interatomic replusive forces become increasingly important, driving structures towards increased close packing. The attractive terms in molecular systems become less important, and the atoms begin to behave more like soft spheres. To first order, the chemistry and phase diagrams of many materials under pressure can be understood in terms of the packing of spheres of various sizes. However, interaction parameters developed to describe low-pressure phenomena fail to provide quantitative predictions at very high pressures. For example, equations of state of fluids and solids are often calculated from van der Waals interaction potentials for closed shell systems. However, a large number of high-pressure studies indicate that such equations of state are grossly inaccurate, even at the moderate crustal pressures (Hemley et al. 1989). The application of pressure in hydrogen-bonded systems causes a reduction in oxygen-oxygen distances, which generally enhances bonding between the hydrogen and the more distant oxygen atom. At mantle and core pressures, conventional hydrogen bonding is lost.

The behavior of ionic and covalent bonds may be more subtle. Although under pressure, ionicity might be expected to increase due to the increased difference of Madelung potentials between the anions and cations, in cases studied so far, ionicity generally changes little over the pressures found within the Earth (e.g. $CaSiO_3$; Stixrude et al. 1996). Covalency generally may increase as well, due to the increased hopping parameters with decreasing bond length, as is observed in solid Xe (see below). But it may give way to a new "valence lattice" (e.g. the upper stability limit in pressure for covalently bonded groups such as CO_3^{2-} in carbonate may be reached) or to a metallic modification. In metals themselves under pressure, the conductivity may increase or decrease; in some cases (e.g. carbon) pressure may drive a transition from a metallic phase (graphite) to an insulator (diamond). Moreover, these changes have an effect on magnetism, or the tendency of electron spins to align: theory and new experiments indicate that magnetism tends to decrease and disappear with increasing pressure.

These changes may be viewed in terms of thermodynamics by considering the free energy, $G = E + PV - TS$. The contributions from each type of bonding are in general strongly pressure, with a competition between density (i.e. PV term favors close packing and increases in coordination and hence a lower volume) and E (favors bond formation and minimization of repulsive interactions and the kinetic energy of the electrons). At high temperatures, the system will favor phases with higher entropy S. Thus these same considerations also apply to liquids at high P-T conditions; i.e. silicate and oxide melts as well as liquid iron alloys of the outer core. There is evidence that silicate melts undergo similar structural transformations such as coordination changes, although they are not abrupt but instead spread out over a range of pressures. In general, detailed knowledge of the electronic properties of liquids under extreme conditions is lacking, although considerable progress in theory is being made (Stixrude et al. Chapter 19). We conclude that the bonding in solid and liquid Earth materials under ambient conditions in general may not reflect their state under conditions of the deep Earth.

Electronic structure

Insight into these phenomena must come from examination of the changes in electronic structure induced by materials subjected to extreme P-T conditions. Because of the presence of long-range interactions in condensed phases, the bonding must in general be viewed from an extended point of view, though a local picture is often employed to simplify the problem. In a crystal (i.e. having periodic lattice), the states can be approximated by bands described in terms of functions (Bloch states) of the form $\phi(r) = u(r)\exp[-i\,\mathbf{k}\cdot r]$ and characterized by a quantum number $\mathbf{k}$, where $u(r)$ is a periodic function of position r (Kittel 1996, Stixrude et al., Chapter 19). The interaction between orbitals on different atoms gives rise to energy bands, which are dispersive as functions of $\mathbf{k}$, and the energy versus $\mathbf{k}$ is known as the band structure (Fig. 1). The core states, or deep levels for each atom, remain sharp delta function-like states, which may be raised or lowered in energy relative to their positions in isolated atoms. These core-level shifts are due largely to the screened Coulomb potential from the rest of the atoms in the crystal. The occupied

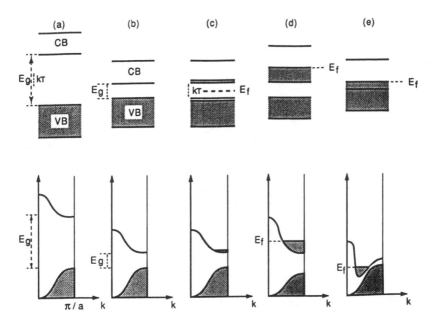

Figure 1. General types of band structures relevant to high-pressure mineralogy showing the valence bands (VB) and conduction bands (CB). The dashed line is the Fermi level (E_f). The stippled regions shows the filling of electrons. (a) Wide band-gap insulator (i.e. energy gap $E_g > kT$). (b) Intrinsic semiconductor (i.e. energy gap $E_g \sim kT$). (c) Intrinsic semiconductor with thermal excitations of electrons across the gap (e.g. $E_g \sim kT$). (d) Metal, showing the filling of electrons up to E_f (at $T = 0$ K). (e) Example of pressure-induced band overlap to form a metal (from Poirier 1991).

valence and empty conduction states no longer look like atomic states, but are broadened into energy bands. There are often intermediate states between the core levels and the valence states called semi-core states which are slightly broadened at low pressures, but which become broader and more different from atomic states with increasing pressure. The formation of ions is driven by the increased stability of an ion when it has a filled shell of electrons. The strong attractive electrostatic, or Madelung, interaction between the ions greatly enhances the crystal stability.

This picture now becomes a means for classifying different materials. Metals have partially occupied states at the Fermi level, the highest occupied energy level in a crystal, as shown in Figure 1. One can understand the insulating behavior of materials with fully filled bands by considering what happens when one applies an electric field. An electric field raises the potential at one part of the sample relative to another, and one would think that electrons would then flow down that potential gradient. However, in an insulator, the bands are filled, and due to the Pauli exclusion principle, nothing can happen without exciting electrons to states above the gap. This requires a large energy, so there is no current flow for small fields. Thus, the material has a finite susceptibility and is insulating. In a metal, the partially filled states at the Fermi level mean that current will flow for any applied field, and the susceptibility is infinite.

In a band picture of an insulator, the highest occupied levels form the valence band, designated E_v and the lowest energy empty levels form the bottom of the conduction band E_c. The difference is the band gap, $E_g = E_c - E_v$. Crystals with band gaps between occupied and unoccupied states should be insulators, and those with partially filled bands should be

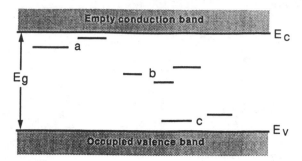

Figure 2. Localized levels states within the valence-conduction band gap E_g. (a) Shallow donor levels. (b) Deep levels. (c) Shallow acceptor levels (from Poirier 1991).

metals. Localized states may exist at energies between the valence and conduction bands (Fig. 2), and these can have major effects on optical and transport properties. If the gap is small or if the material has such intermediate states (e.g. by chemical doping such that electrons can be excited into the conduction bands or holes in the valence bands), the crystal is considered a semiconductor. In a non-magnetic system, each band holds two electrons. Thus, a crystal with an odd number of electrons in the unit cell should be a metal since it will have at least one partially filled band. Magnetic crystals that are insulators by virtue of local magnetic moments are known as Mott insulators (Mott 1990). These materials include transition metal-containing oxides present in the mantle (and considered core-components); such materials exhibit intriguing behavior as a function of pressure and temperature.

We should also consider the electronic structure of non-crystalline materials, including liquids. The existence of a band gap in crystals has been described in terms of the Brillouin zone and Bragg-like reflection of electrons. But a gap does not depend upon the periodicity of the atoms on a lattice. The electrons in the lowest energy states in the conduction band of a non-crystalline material (or disordered crystal) can be localized (Mott 1990). The difference between the localized and delocalized states is called the mobility edge. These may be important in describing the electronic properties of liquids and amorphous materials formed from deep Earth crystalline materials (e.g. metastably compressed fayalite or dense melts formed from iron oxides, as discussed below).

Magnetic properties

Magnetism arises from the presence of unpaired electrons. Electrons have magnetic moments, but if they are paired up in states as 'up' and 'down' pairs, there will be no net magnetism in the absence of a magnetic field. However, in open-shelled atoms there may be a net moment. According to Hund's rules, atoms (or isolated ions) will maximize their net magnetic moment, which lowers their total energy due to a decrease in electrostatic energy. In a crystal, interactions with other atoms, and formation of energy bands (hybrid crystalline electronic states) may lead to intermediate- or low-spin magnetic structures. As an example, consider Fe^{2+}, which has 6 d-electrons; two end-member situations can be considered (Fig. 3). In low-spin ferrous iron, the 6 d-electrons are paired up to fill 3 t_{2g} states, and there is no net-magnetic moment. In high-spin ferrous iron, five 'up' d-states are split in energy from five 'down' d-states by the exchange energy, and 5 d-electrons go into the 3 t_{2g} and 2 e_g lower-energy, majority spin states, and one electron into a high-energy t_{2g} state (Fig. 3). This gives a net magnetic moment component of $4\mu_B$ (bohr magnetons). The ideal magnetic moment component of Fe^{3+} is $5\mu_B$.

Such open-shell systems have often been discussed in terms of crystal field theory, which has been very successful in rationalizing optical spectra, crystal chemistry, and thermodynamic and magnetic properties of minerals, including their pressure dependencies.

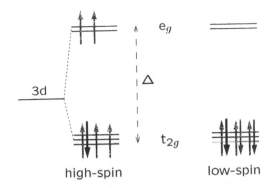

Figure 3. The lifting of the degenerate d-levels of ferric and ferrous iron in an octahedral environment. ('crystal field' splitting).

The principal assumption of crystal field theory is the existence of localized atomic-like d-orbitals. It has provided a way to rationalize the changes in energy levels in terms of local interactions of the transition metal cations with the coordinating atoms. The symmetry of the coordination polyhedron results in a lifting of the otherwise degenerate d-orbitals. The primary success of crystal field theory lies in the symmetry analysis of the splitting of the atomic-like states. These symmetry arguments are rigorous, but the origin of the splitting may not arise from a potential field represented as a point charge lattice, but rather to bonding hybridization. Furthermore, d-states are not pure atomic-like states but are dispersed across the Brillouin zone (energy varies with **k**).

It is not widely known that this alternative view has a sounder basis in quantum mechanics and more successful in explaining observed behavior. As shown by Mattheiss (1972), the origin of the 'crystal field' splitting can be traced to hybridization. In the language of quantum mechanics, the e_g-t_{2g} splitting is due to off-diagonal interactions, not shifts in the diagonal elements as would be the case for the conventional electrostatic crystal-field interaction. The splitting can be considered a ligand-field effect, but does not arise from changes in the electrostatic field at an atomic site. Rather, the splitting is due to d-d interactions between next-nearest neighbors and due to p-d and s-d interactions between neighboring oxygen ions. The d-d interactions lead to splittings that vary as $1/r^5$; although the p-d and s-d interactions follow a $1/r^7$ dependence (Harrison 1980); the metal-metal d-d splittings dominate. Thus, the $1/r^5$ dependence (Drickamer and Frank 1973) of the observed splittings cannot be taken as validation of the electrostatic crystal field model because the same dependence is given by hybridization effects.

In the ground state (i.e. at low temperatures), moments on individual ions or atoms can be oriented, giving rise to ferromagnetism if they are lined up in the same direction, or antiferromagnetism if they are oppositely aligned; this depends on the sign of the magnetic coupling J which is mediated by hybridization with other atoms in the crystal. At high temperatures the moments will disorder, but there are still local moments present. This paramagnetic state must be distinguished from the paramagnetic state that arises from loss of local moments (which we call magnetic collapse, or a high- to low-spin transition). Ferrimagnetism results when the magnetic moments are unequally distributed over different sublattices. A net spontaneous magnetic moment arises from the incomplete cancellation of aligned spins. An example is magnetite, Fe_3O_4, which consists of a sublattice of Fe^{2+} and one of Fe^{2+} and Fe^{3+}. For ferromagnetic and antiferromagnetic phases, the magnetic fields are aligned below the Curie and Néel temperatures, which gives rise to magnetic splittings.

The decrease in magnetism with pressure can be understood qualitatively from the increase in electronic bandwidths, which eventually become greater than the exchange splitting. This can be understood more quantitatively with the Stoner and extended Stoner models described in Chapter 19.

ELECTRONIC AND MAGNETIC EXCITATIONS

Excitations in a system are induced by external (e.g. electronic or magnetic) or internal (e.g. temperature) fields. The coupling of the response to the field is given by the frequency-dependent susceptibility. This can be characterized as the dielectric function $\varepsilon(\omega,\mathbf{k})$, where ω is the frequency and $\mathbf{k}$ is the wavevector (the dielectric function is also known as the relative permittivity). To relate this to refractive index n, a familiar property of minerals, we note that the dielectric function is complex,

$$\varepsilon(\omega) = \varepsilon_1(\omega) + i\,\varepsilon_2(\omega) \tag{1}$$

Alternatively, one can write the refractive index as a complex function,

$$N(\omega) = n(\omega) + i\,k(\omega) \tag{2}$$

where $\varepsilon = N^2$ by definition. It thus follows that $\varepsilon_1 = n^2 - k^2$ and $\varepsilon_2 = 2nk$. In general, the response of a crystal to an electric field is dependent on its orientation relative to the crystal axes (which gives rise to properties such as birefringence and optical activity; i.e. the dielectric function is a tensorial function). In the presence of an intense external electric field, the response can be a function of the magnitude of the field, and the susceptibility is a function of the field. This gives rise to non-linear optical response (e.g. multiphoton excitations) as described below. Electrons can be excited into extended, or itinerant, states (i.e. across the band gap), or the excitations may be local (i.e. forming a localized electron-hole pair, or exciton). Figure 4 shows the principal excitations and their approximate energy ranges considered here.

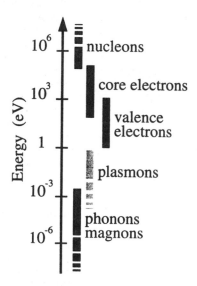

Figure 4. Approximate energy ranges of various types of excitations considered for high-pressure study.

Dielectric function

In general, there are several different contributions to the dielectric function. In materials that contain polar molecules, the molecules can align. In crystals containing non-polar atoms or ions, the particles can be displaced (ionic polarization) or the electronic charge distribution can be deformed (electronic polarization). The electronic contribution to the dielectric function of a material is formally determined by the sum of its electronic excitations (Ashcroft and Mermin 1976). Hence, measurement of the index of refraction of transparent minerals (i.e. below band gap or absorption edge) can be used to constrain the frequency of higher energy electronic excitations (e.g. in the ultraviolet spectrum) using various approximate formula of the type. In the simplest approximation, one can write,

$$n(\omega)^2 = F_1/(\omega_1^2 - \omega^2) + 1 \tag{3}$$

where ω_1 is the frequency of an effective high-frequency oscillator (electronic transition) that correlates with the energies of direct valence-conduction band transitions (Hemley et al. 1991, Eremets 1996) and ω is the frequency of light. Optical spectra of oxides and silicates from the near-IR to the vacuum UV are determined by electronic transitions involving the valence (bonding) band, the conduction (antibonding) band, and d-electron levels.

This approach has been useful in high-pressure studies when the window of the high-pressure cell precludes measurements above a critical photon energy (Hemley et al. 1991, Eremets 1996). For example, diamond anvils at zero pressure have an intrinsic absorption threshold of 5 eV (i.e. for pure type IIa diamonds). Moreover, type I diamonds contain nitrogen impurities that lower the effective threshold at zero pressure to below 4 eV (depending on the impurities). Although the band gap of diamond increases under pressure, the absorption threshold of diamond (and most likely the gap) decreases in energy under the stresses due to loading in the cell (Mao and Hemley 1991).

Piezoelectric crystals become polarized when subjected to a mechanical stress (compressive or tensile). Quartz is a common example. Piezoelectricity is important in the behavior of acoustic wave, but little is known about its pressure dependence in depp Earth materials. Although it has been suggested by analogy that silicate perovskites could exhibit ferroelectric properties, there is as yet no evidence for this behavior (see Hemley and Cohen 1992). The ultrasonic and Brillouin scattering techniques described in Chapters 15 and 17 can be used to investigate these phenomena at high pressure.

Electronic excitations

Extended excitations. Interband transitions are electronic excitations between bands, whereas intraband transitions are those within a band. An example of the latter is the damped frequency dependence of electronic excitations associated with itinerant electrons in a metal (which by definition has a partially filled band). This collective longitudinal excitation of the conducting electrons is known as the plasma oscillation. Plasma oscillations (plasmons) can be excited by inelastic scattering techniques at high pressures. Plasmons can also be excited in materials with a band gap (i.e. dielectrics, typically at higher energies). For example, in olivines the band gap varies from 7.8 eV in forsterite to 8.8 eV in fayalite (Nitsan and Shankland 1976).

Excitons. Electronic excitations may occur just below the band gap due to the creation of a bound electron-hole pair, called an exciton. The pair is characterized by a binding energy E_b, which is given by the $E_g - E_{ex}$, where E_{ex} is the excitation energy. In some respects, these excitations are the equivalent of (localized) transitions in molecular spectroscopy, but in a crystal they can move through it and characterized by a wavevector **k**. Continuing with the example of olivines at zero-pressure, a prominent absorption band due to an exciton is observed near the band gap (Nitsan and Shankland 1976). The absorption characteristics in this region in dense ferromagnesian silicates and oxides are important for assessing the heat transfer in the Earth, which is predominantly radiative at deep mantle temperatures.

'Crystal field' transitions. The color of many minerals arises from substitution of transition metals in the crystal structure. For example, the substitution of transition metals in olivines results in a strong absorption begin 1-2 eV below the band gap at zero-pressure. This has been ascribed to transitions involving the d-levels in the system (Nitsan and Shankland 1976). The development and application of crystal field theory to understand the behavior of open-shell systems ions, such as partially filled d-shell of transition elements in minerals, has a long history (Burns 1993). As discussed above, a more fundamental quantum mechanical picture shows that orbital hybridization of the central ion and its neighbors (e.g. oxygen atoms) must be considered in explaining spectral

changes. Nevertheless, the measurement of such spectra continues to find applications to a range of problems including crystalline site occupancy, coordination numbers in dense glasses, and radiative heat transfer (Keppler and Rubie 1993, Keppler et al. 1994, Shen et al. 1994).

Charge transfer. Charge transfer represents another important class of excitations in these materials. Anion-cation charge transfer involves transfer of electron density from an anion, viewed as a ligand (e.g. an oxide ion) to a metal (e.g. a transition metal). Many of these transitions in minerals occur in the ultraviolet region (Burns 1993). Intervalence charge transfer (e.g. metal-metal charge transfer) may occur between ions of different valence state, such as Fe^{+2} - Fe^{3+} or Fe^{2+} and Ti^{4+}. These transitions often give rise to bands in the visible spectrum which show marked shifts with pressure (Mao 1976).

Excitations by defects. Many defects are also associated with electronic properties or are electronic in origin. Electron-hole centers or defects can be produced by deformation or radiation, either naturally or in the laboratory. Examples of naturally occurring systems at ambient pressure that exhibit these properties are smoky quartz and numerous colored diamonds (see Rossman 1988). Pressure-induced defect luminescence can be found associated with pressure-induced phase transitions, as for example, in metastably compressed SiO_2 (Hemley 1987); this may be related to non-blackbody emission observed in shock-wave experiments (where it is thermally induced—Schmitt and Ahrens 1983). Another example is the intense visible luminescence associated with the B1-B2 (NaCl-CsCl) transition in CaO at 65 GPa (300 K) (Richet et al. 1989); luminescence appears to arise from electron-hole defects that form in the mixed phase sample during the transition (Hemley et al., to be published).

Electrical conductivity

Electrical conductivity is another measure of the response of the system to an applied electric field (static or frequency dependent). Electrical conductivity may be electronic or ionic; the former can be intrinsic or extrinsic. If there is a gap between the valence and conduction bands, the electronic conductivity can be thermally activated at finite temperature. This can also give rise to coupled distortions of the lattice as the electrons move through it; this is called a small polaron; see Kittel (1996). Naturally, electrical conductivity measurements provide a means for the direct identification of the insulator-metal transitions, induced by either pressure or temperature (or both; e.g. Knittle et al. 1986).

In general, electrical properties are also very sensitive to minor chemical impurities and defects in a given specimen (Tyburczy and Fisler 1995), such as Fe content, oxygen fugacity, and volatile content. The conductivity can be written as a complex quantity, giving real and an imaginary part (loss factor); this complex function is measured by impedance spectroscopy. Grain boundary conduction, for example, is observed to decrease under pressure as grain-grain contacts increase. Electric current can also be carried by ions (ionic conductivity). This may be especially noticeable at higher temperatures (e.g. premelting or sublattice melting; see Gillet et al., Chapter 17). Such behavior has been proposed for silicate perovskites by analogy to other perovskites as well as theoretical calculations (Hemley and Cohen 1992).

Magnetic excitations

Magnons are excited magnetic states which can be excited in light scattering via direct magnetic dipole coupling or an indirect, electric dipole coupling together with spin orbit interaction (Fleury and Loudon 1968). High-pressure measurements on mineral analogs have been performed to determine the metal-anion distance dependence of the so-called superexchange interaction J (Struzhkin et al. 1993b). This interaction parameter is a measure of the coupling of metal anions through a (normally) diamagnetic anion, such as the O between two Fe atoms. Magnon excitations may also be observed by infrared absorption.

OVERVIEW OF EXPERIMENTAL TECHNIQUES

An array of ambient-pressure techniques can now be used to probe electronic and magnetic excitations in minerals. Band structure is probed by photoemission, which involves interaction of electrons with the sample. Other such spectroscopies involving electron interactions include electron energy loss and auger (Cox 1987), each of which have been used for studying high-pressure phases quenched to ambient conditions. Here, we focus on techniques that have been applied to deep Earth materials in situ at high pressure. An excellent overview of many of the conventional mineralogical techniques applicable under ambient pressures is given in Volume 18 of this series (Hawthorne 1988). A number of these techniques can also be measured in situ in dynamic compression (e.g. shock-wave) experiments (Ahrens 1987).

Optical spectroscopies

Absorption and reflectivity. Optical spectroscopies have been one of the principle techniques used for in situ high-pressure investigations of minerals with high-pressure devices having optical access (e.g. diamond cells). A number of different types of absorption spectrometer systems, including double beam instruments, have been designed and built for high-pressure mineral studies. In optical experiments on samples in high-pressure cells using refractive optics (lenses), there is a need in general to correct for the index of refraction of the diamond because of chromatic aberrations (wavelength dependence of the focusing). This problem can also be overcome by the use of reflecting optics (e.g. mirror objectives). Versatile UV-visible-near IR absorption and reflectivity systems using conventional continuum light sources have been designed for high-pressure applications (Syassen and Sonnesnchein 1982, Hemley et al. 1998, Eremets 1996) (Fig. 5). Standard continuum sources and wide variety of laser sources that span the UV-visible and much of the IR regions are now available.

High-pressure optical spectra of a wide range of minerals (and analogs), have been measured over the years. These include measurements of pressure dependencies of band gaps, excitons, 'crystal field' spectra, impurity bands, and defects, as described above. We consider one example, the optical measurement of the refractive index. For the simple case of transparent minerals (e.g. wide band-gap insulators), the measurement in the visible range gives the real part of the dielectric function (i.e. off-resonance). Absorption (transmission) or reflectivity measurements with broad-band light of thin samples of comparable thickness to the wavelength of light give rise to patterns of constructive and destructive interference in the spectrum (Fig. 6). These fringes are straightforward to measure in a diamond cell. The spacing is determined by the refractive index and the thickness of the plate-like sample; each may be determined independently if the order of the fringe is known, and this can be found in several ways. Moreover, the wavelength dependence gives the dispersion of the index, which can be related to the higher energy electronic excitations, as described above.

Representative results are shown for MgO and diamond in Figure 6, which both show a decrease in refractive index with pressure. From this, we infer that the band gap of MgO increases with pressure, in agreement with theory (Mehl et al. 1988). Similar behavior is observed for hydrostatically compressed diamond (Eremets et al. 1992, Surh et al. 1992). In contrast, the band gaps of many other materials decrease under pressure (Hemley and Mao 1997) (see below for the case of Xe). Also, both theory and experiment indicate that

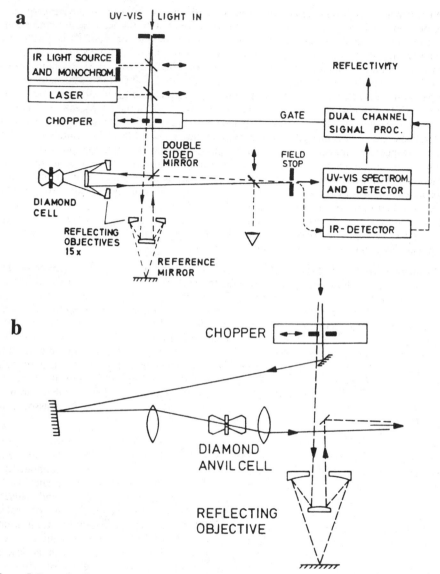

Figure 5. Example of a high-pressure UV-visible-near IR optical set up. Top: double beam instrument for reflectivity measurements. All reflecting microscope optics are used to remove chromatic aberrations and a grating spectrometer. Bottom: Modification of the double beam system for absorption measurements. One of the reflecting objectives is replaced by an achromatic lens. The system is capable of measurements from 0.5 to 5 eV; similar systems are in use in a number of high-pressure laboratories (from Syassen and Sonneschein 1982).

in diamond. the index can increase with non-hydrostatic stress. Brillouin scattering provides an independent measurement of the refractive index (Chapter 17); results for MgO are in good agreement with the interference fringe measurements (Zha et al. 1997). The relationship of this to changes in covalency and ionicity with pressure as predicted by theory is described in Chapter 19.

Synchrotron infrared spectroscopy. A recent major advance is the development and application of high-pressure spectroscopic techniques utilizing synchrotron radiation, specifically in the near- to far-infrared range. The technique has been shown to be particularly important for very small samples such as those in high-pressure cells. Synchrotron radiation sources have a very smooth and broad spectral range extending from the hard x-ray region (see Chapter 1) to very long wavelengths (low wavenumbers in principal well below 1 cm^{-1}); the infrared region can be measured with Fourier transform infrared (FTIR) techniques. The basic technique and applications to vibrational spectroscopy are described by Gillet et al. (Chapter 17). The technique can also be used to investigate electronic and magnetic excitations (Hemley et al. 1998). Optical studies under pressure, especially reflectivity and absorbance, provide information important for understanding fundamental changes in the electronic structure of materials close to insulator-metal transitions at very high pressures, including the appearance of interband and intraband excitations associated with these transitions. Measurements can also provide unique information on magnetic excitations, as shown in a very recent study of magnon excitations in $Sr_2CuCl_2O_2$ (Struzhkin et al. 1998), a perovskite-like material. This study showed how the pressure dependence of the IR and Raman modes can be used to distinguish between magnon and electronic excitations.

Luminescence. High-pressure luminescence measurements give in principle the same information as does absorption and reflectivity. Luminescence includes fluorescence and and phosphorescence, corresponding to shorter and longer-lived excited states (generally defined as less than or greater than one nanosecond, respectively). The transitions are typically governed by dipole selection rules. For the case of atomic-like transitions, such as localized excitations of ions in crystal fields, this can be understood in terms of the Laporte selection rules (i.e. $\Delta l = \pm 1$, where l is the angular momentum quantum number). A particularly important application in mineralogy is the use of this technique in constraining the proportion of Fe^{2+} and Fe^{3+} in various sites in silicates from d-orbital excitations (Rossman 1988). Charge transfer applications are described above. Other applications include excitations near and across band gaps, and excitations associated with defects.

Luminescence measurements are of obvious importance in the calibration of pressure. This includes excitation of Cr^{3+} in ruby by UV-visible lasers (e.g. He-Cd, Ar^+ ion) and measurement of the calibrated shift of the R_1 line in luminescence, currently calibrated to 180 GPa (Mao et al. 1978, Bell et al. 1986, Mao et al. 1986); see also Duffy and Wang (Chapter 14). The recent use of tunable lasers (e.g. titanium-sapphire) and pulsed excitations techniques have allowed the extension of the measurements of ruby luminescence to above 250 GPa (Goncharov et al. 1998). Recently, other luminescent materials, such as Sm-doped yttrium aluminum garnet (Sm-YAG), have been calibrated for this pressure range (Liu and Vohra 1996) as well as for high temperatures at lower pressures (Hess and Schiferl 1992). Other applications include the study of vibronic spectra, which are combined (coupled) electronic and vibrational excitations. Measurements at high pressure reveal the pressure dependence of both the electronic and vibrational levels in the system, which can be used to determine acoustic velocities (given assumptions about the dispersion of the modes and correct assignment of typically complex spectra (e.g. Zhang and Chopelas 1994); see also Gillet et al. (Chapter 17).

Optical light scattering

Raman scattering. Although most commonly used to study vibrational dynamics (Gillet et al., Chapter 17), laser light scattering spectroscopies can also be used to investigate high-pressure electronic and magnetic properties. Raman and Brillouin are inelastic light scattering techniques that measure transitions to an excited state from the

Figure 6 (opposite page). High-pressure refractive index measurements. (a) Schematic of the experimental configuration. The measurement is carried out by measuring in interference fringes produced by the platelet sample in the diamond cell. The ratio of reflected and incident light ($I_R(\lambda)$ and $I_0(\lambda)$ as a function of wavelength λ is given by the 'Fabry-Perot' equation, $I_R(\lambda)/I_0(\lambda) = 1 - 1/[1 + F \sin^2(2\pi\delta/\lambda)]$, where $F = 4R/(1-R)$, where R is the reflectivity and $\delta = 2nd$ and d is the sample thickness (Hemley et al. 1991). Reflection by several surfaces produces a multiple interference pattern. Measurements over this wide range of wavelengths provides a measure of the dispersion of the refractive index n. (b) Measured interference fringes for MgO at 9.5 GPa (300 K). (c) Pressure dependence of the refractive index as a function of pressure for MgO. Curve A: n at 2.1 eV, Curve B: n at 1.8 eV. Dashed line: extrapolation of previous measurements carried out to 0.7 GPa (Vedam and Schmidt 1966). A negative pressure derivative is found agreement with the earlier work, although the magnitude is much. (d) Pressure dependence of the refractive index of diamond, showing the different pressure dependencies as a function of wavelength (Eremets et al. 1992). The results of first-principles calculations of Surh et al. (1992) are also shown.

ground state (Stokes scattering) and to the ground state from a thermally excited state (anti-Stokes scattering). Low-frequency crystal field excitations of the appropriate symmetry can be probed, and the pressure dependence of such transitions have been studied in a few cases (Goncharov et al. 1994). The conduction electrons in metals can also be probed by Raman scattering, which in general gives rise to weaker broader features in the light scattering spectrum. High-pressure studies are complicated by the need the discriminate between the spurious background contamination from the true electronic contribution. Superconductivity in metals gives rise to structure in the electronic Raman spectrum, such as gap-like features at low frequency (Zhou et al. 1996).

Such studies include magnetic transitions in antiferromagnetic materials such as transition metal oxides predicted to undergo high-pressure insulator-metal transitions. These materials have magnetic anomalies that may be difficult to study directly at high pressure, but these anomalies may be associated with pronounced changes in magnetic light scattering spectra. These excitations have a strong temperature dependence because of the large temperature dependence of magnetic ordering. The high-pressure, low-temperature Raman magnon spectrum of CoO is shown in Figure 7 (Struzhkin et al. 1993a). A large number of excitations are observed, including several very strong bands, all of which can be described by the appropriate Hamiltonian. Measurements of the two-magnon excitations in Fe_2O_3 and NiO by high-pressure Raman scattering constrain the magnetic properties of these materials at high density (Massey et al. 1990a,b). The pressure dependence of higher-order effects, such as the coupling of the phonons with electron spin-pair excitations and the electronic structure of the material can also be probed (Massey et al. 1992). Although the higher-temperature behavior of such materials is of direct concern for the deep Earth (i.e. along the geotherm), detailed characterization of the low-temperature ground state is essential for understanding higher-temperature properties. The extremely low background signal in light-scattering experiments using ultrapure synthetic single-crystal diamond anvils (Goncharov et al. 1998) is enabling the extension of these light-scattering studies of electronic and magnetic excitations to a wider class of materials, including opaque minerals and metals (see also Chapter 17).

Brillouin and Rayleigh scattering. In principle, other optical light-scattering techniques can be used to investigate electronic and magnetic phenomena at high pressure. Brillouin involves the measurement of acoustic phonons; in metals, the measurements correspond to surface modes or plasmons. Brillouin scattering can also be used to determine the pressure dependence of the refractive index (dielectric function), usually in transparent materials. Rayleigh scattering is a quasi-elastic scattering spectroscopy. High-pressure studies of materials using this technique are difficult due to the typically strong background scattering near the excitation wavelength (laser line) by the anvils in the high-pressure cell. Consequently, few such studies have been reported.

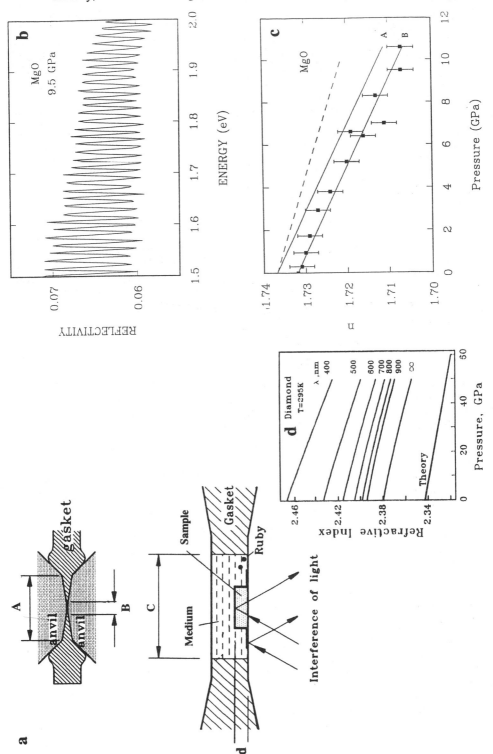

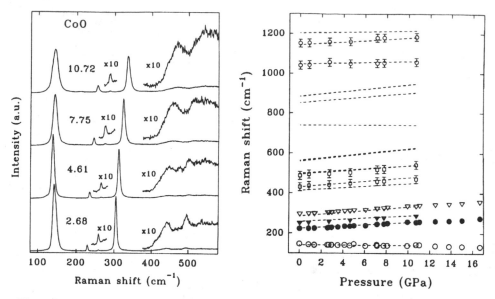

Figure 7. (a) High-pressure magnetic Raman scattering spectra of CoO measured at 20 K (Struzhkin et al. 1993a). The numbers refer to the pressure in gigapascals. (b) Pressure shifts of the magnetic excitations. The dashed lines correspond to the results of a fit to a quantum mechanical model (Hamiltonian), and show that all of the observed bands are magnetic excitations (see Struzhkin et al., 1993a). Under these conditions, CoO is an antiferromagnetic insulator (Néel temperature T_N = 293 K) and has a structure close to NaCl-type (nearly cubic, with a weak rhombohedral distortion).

Non-linear optical methods

One of the limitations of conventional one-photon absorption, reflectivity, and luminescence measurements is the fact that the absorption threshold of the anvils of the high-pressure cell precludes measurements on samples at higher energies (e.g. in the ultraviolet range), as mentioned above. Non-linear optical techniques can be used to access this region, however. For example, one can perform three-photon excitation of valence electrons to the conduction band or excitonic states in the vicinity of the conduction band, which is above the one-photon threshold of the anvil; this has been achieved in compressed salts in sapphire anvil cells (Lipp and Daniels 1991). It is now possible to apply and extend these techniques to study to high-pressure minerals. Another category of techniques in this class are four-wave mixing experiments, such as coherent anti-stokes Raman scattering. These rely on the third-order susceptibility and are enhanced when there are two-photon electronic resonances. Only a few high-pressure studies have been reported to date (Lipp and Daniels 1991, Reimann 1996).

Mössbauer spectroscopy

Mössbauer spectroscopy probes the recoil-less emission and resonant absorption of γ-rays by nuclei (McCammon 1995). The energy levels of the nuclei are sensitive to the electronic and magnetic fields present at the nuclei, and thus the oxidation state of an atom (e.g. ferrous versus ferric). Mössbauer spectroscopy reveals information on the hyperfine interactions and the local electronic and magnetic field at a nucleus. The hyperfine field is the magnetic field due to magnetic interactions. The technique has been used extensively in the mineralogy of deep Earth materials, including in situ high-pressure measurements (Fig. 8). A typical probe for mineral systems is based on the reaction, $^{57}Co + ^{0}\beta \rightarrow ^{57}Fe + \gamma$,

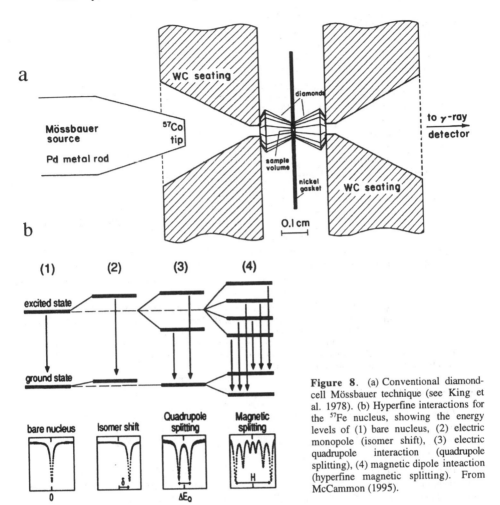

Figure 8. (a) Conventional diamond-cell Mössbauer technique (see King et al. 1978). (b) Hyperfine interactions for the ^{57}Fe nucleus, showing the energy levels of (1) bare nucleus, (2) electric monopole (isomer shift), (3) electric quadrupole interaction (quadrupole splitting), (4) magnetic dipole inteaction (hyperfine magnetic splitting). From McCammon (1995).

where ^{N+Z}X, and Z is the proton number, N the neutron number, and $Z + N$ is the mass. In a conventional Mössbauer experiment, one modulates the energy of the γ-rays by continuously vibrating from the parent source to introduce a Doppler shift of the radiation. The nuclear absorption levels are in general shifted by a change in chemical environment relative to the isotope in the source (e.g. ^{57}Fe, ^{119}Sn, ^{121}Sb, ^{153}Eu, ^{197}Au).

Mössbauer resonances compare the relative energy of the ground and excited states between nuclei in the source and sample (the so-called isomer shift). In a conventional experiment, the resonance corresponding to these isomer shifts corresponds to the source velocity at which maximum absorption appears. Three properties are commonly probed: the electric monopole interaction, the quadrupole splitting, and the magnetic splitting. The electric quadrupole splitting arises from the interaction beween the nuclear and quadrupolar moment and the local electric field gradient at the nucleus. This occurs for nuclei with spin quantum number $I > 1$ (i.e. for ^{57}Fe, $I = 3/2$). The interaction between magnetic dipole moments gives the magnetic splitting. Splittings associated with Curie and Néel transitions can also be observed in paramagnetic phases at low temperatures as a result of the slowing-down of relaxation times associated with re-orientation of the local magnetic fields.

Applications include determination of valence state, site occupancy, and local order. In studies of deep Earth materials at high pressure, each of these can be probed as a function of pressure (and temperature), including the measurement of changes across various transformations (crystallographic, electronic, and magnetic). Instrumentation consists of a radioactive source, vibrating drive, and detector (scintillator or proportional counter). Recent examples of high-pressure studies include a series of pyroxenes to 10 GPa (Zhang and Hafner 1992), measurements on $CaFeSi_2O_6$ to 68 GPa reveal evidence for a phase transition associated with a change in spin state, and extensions the technique into the megabar pressure range, as discussed below for FeO (Pasternak et al. 1997b).

Elastic x-ray and neutron scattering

Crystal structures are determined by the details of bonding and other interactions between the atoms in the material, so at one level structural methods, represent an essential means for studying bonding properties (see Downs and Prewitt, Chapter 9). Taking this correlation one step further, x-ray diffraction relies on the interaction between x-rays and electron distributions in materials, so these measurements give a detailed picture of the real space distribution of charge as determined by atomic scattering factors. This is a well-established technique in crystallography. It requires very high-quality crystals, corrections for absorption, and measurements to high q. Such studies have been attempted in some systems as a function of pressure, and changes in electron density associated with crossing the insulator-metal transition in iodine have been reported (Fujii 1996). Systematic measurements of this type for minerals under pressure apparently have not been undertaken. The availability of third-generation synchrotron radiation sources and the advent of a new generation of area detectors are important for extending these measurements to deep Earth materials at very high pressures. Complementary to the real-space distribution of the electrons is the momentum distribution, which can be obtained by high-pressure Compton scattering (see below).

Neutron scattering also provides information on magnetic and electronic properties. Magnetic scattering amplitude of neutrons by an atom depends on the direction of its magnetic moment (Furrer 1995). Thus, neutron diffraction can be used to map out the orientation of the magnetic moments in the unit cell of a crystal. Little work has yet been performed on minerals; nevertheless, experimental advances suggest potential applications, including high-pressure studies with large volume cells. Inelastic neutron scattering can be used to excite magnetic transitions; it requires high energies and the measurement of low-momentum transfer. Measurements of crystal-field transitions (at zero pressure) have been reported (Winkler et al. 1997). There has been considerable progress in studying pressure effects on electronic properties from inelastic phonon scattering. This includes studies of phonon anomalies in metals (e.g. to 10 GPa) that have been proposed to be coupled to changes in the Fermi surface (Klotz et al. 1998). Recent work has shown that is it possible to characterize the coupling of phonons with the high-spin/low-spin transitions (Schwoerer-Bohning et al. 1996).

High-pressure synchrotron x-ray spectroscopy

Numerous recent advances in synchrotron radiation techniques have resulted in significant developments in x-ray spectroscopy, as well as diffraction, of minerals and related materials under pressure. X-ray spectroscopies can be used as a probe of electronic structure; these techniques thus constitute an alternative to methods requiring use or measurement of charged particles interacting with a sample (e.g. electrons in photoemission). As such, x-ray spectroscopies can be used for in situ study of properties such as band structures; this is also an active area of theory (e.g. van Veenendaal and Carra 1997). Electronic spectroscopy can also be used to determine local structure.

Applications to mineralogy and geochemistry of Earth materials in the near-surface environment have been described previously (Brown et al. 1988). Despite these developments, until quite recently there have been very limited applications at high pressure. The development of high-pressure x-ray spectroscopy has been hindered in the past by insufficient synchrotron intensity and opaqueness of high-pressure vessels below 12 keV. Recently, third-generation sources have greatly boosted the intensities, and new high-pressure techniques such as the development of high-strength, but x-ray transparent Be gaskets have extended the window for measurements from 12 keV down to 4 keV. This extended x-ray energy range allows the study of pressure-induced effects on atomic coordination, crystal structures, and electronic properties of a wider class of minerals using a variety of x-ray spectroscopies. Recent experiments suggest numerous breakthroughs in the application of these new techniques to minerals at very high pressures (Mao et al. 1997b). The region below 4 keV (e.g. soft-x-ray and vacuum ultraviolet) is still opaque in all kinds of high-pressure devices, however, inelastic scattering techniques provide a means to access states that cannot be probed by direct (one-photon) processes due to the opaqueness of the anvil and gasket.

X-ray absorption spectroscopy (XAS). XAS involves measurement of the absorption associated with core level excitations to the continuum, typically by monitoring broad- band fluorescence. Different spectral regions in the vicinity of a core- level absorption edge are typically defined on the basis of the different information they contain (Fig. 9). The near edge x-ray absorption fine structure (XANES) provides information on the conduction band density of states. The extended x-ray absorption fine structure

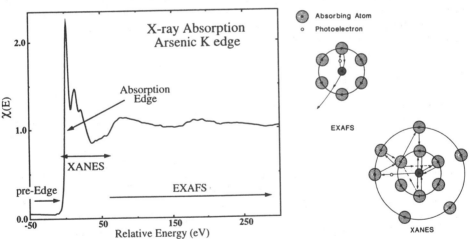

Figure 9. High-pressure x-ray absorption spectroscopy (XAS) of the arsenic K edge with different regions of spectroscopic interest indicated. The x-ray absorption near-edge structure (XANES) is also known as the multiple scattering region, and extends from a few eV below to about 50 eV above the edge. It corresponds to the energy region where the probability of transition of the core electron to the continuum or to the unpopulated shallow levels is highest. The region extending from about 30 eV to above 1000 eV is that of the extended x-ray absorption fine structure (EXAFS). The spectrum was measured at 8 GPa and room temperature at LURE, Orsay, France [from study by Badro et al. (1997)]. The transition probability for electrons drops and Kronig oscillations appear as a result of constructive interference of the ejected and scattered electron. Bottom: Schematic of the EXAFS and multiple-scattering XANES processes (from Brown et al. 1988).

(EXAFS) arises from multiple scattering of electrons in the environment of an atom, from which local structural information (interatomic distances and coordination) is obtained. Spin-dependent x-ray absorption spectroscopy can reveal magnetic properties even in the absence of long-range magnetic order (Hamalainen et al. 1992). High-pressure XAS has been used successfully for transition elements at moderate pressures with BC anvils (Wang et al. 1998), and for higher-Z materials (Itié et al.1992, Pasternak et al. 1997). EXAFS is now well developed for high-pressure studies (Itié et al. 1997), and can be measured by absorption or fluorescence. Previous high-pressure work with diamond anvils has been limited by the Bragg diffraction from the diamond and by the absorption edge of the diamond and the gasket. Because of the low energy of the Si edge, high-pressure studies of germanate analogs have been carried out (Itié et al. 1989). The development of more transparent gaskets, as outlined above, has helped to overcome these problems. EXAFS is particularly useful for study of pressure effects on disordered materials. A notable example is the observation of the coordination change in crystalline and amorphous GeO_2 (Itié et al. 1989), and the recent in situ study of the Ni coodination in $CaNiSi_2O_6$ glass to 40 GPa (Cormier et al. 1998). There is considerable promise in using EXAFS for local structural studies of liquids at high pressures and temperatures (Katayama et al. 1998). Mobilio and Meneghini (1998) provide a review of the variety of new synchrotron-based x-ray techniques for studying amorphous materials, including techniques such as anomalous scattering that can be used in conjunction with EXAFS.

X-ray emission spectroscopy (XES). In XES, deep-core electrons in the sample are excited by x-rays. The core-holes then decay through either radiative or non-radiative processes. For deep-core holes, the dominant decay channels are radiative processes, producing fluorescence, which is analyzed to provide information on the filled electronic states of the sample. The information provided by XES is complementary to that provided by XAS. Moreover, the final state of the fluorescent process is a one-hole state, similar to the final state of a photoemission process. Thus, the most important information provided by photoelectron spectroscopy, namely large chemical shifts in the core-level binding energies and the valence band density of states, is also available in XES. There have been significant developments in the ab initio treatment of such spectra, in particular, the electron-hole attraction, which requires going beyond a simple mean- field approach such as the local density approximation (Shirley 1998). This is needed for x-ray absorption and resonant inelastic x-ray scattering. Recent measurements of the effect of pressure on plasmon excitations in metals (e.g. the high-pressure phase of Ge) suggest the potential for investigating the electronic structure of core materials at megabar pressures (Mao et al. 1997b).

Recent examples of high-pressure XES include study of predicted high-spin/low-spin transitions in iron oxides and sulfides. A high-resolution Rowland circle spectrometer with a double crystal monochromator has been used (Fig. 10). A schematic of the K_β fluorescence is shown. In an atomic picture, the final state of the K_β is a hole state with $1s^23p^53d^5$. The measurement also reveal the XANES portion of the spectra, that is, including the pre-edge peak associated with transitions to unoccupied d–states (i.e. $3d$ in Fe). By measuring the absorption though selective monitoring of the fluorescence peaks, information on the contribution of different spins to the spectrum can be obtained.

X-ray magnetic circular dichroism (XMCD). For magnetic samples, x-ray magnetic circular dichroism (XMCD) measures the spin polarization of an electronic excitation by the use of circularly polarized x-rays. It is therefore in principle identical to conventional optical MDC with visible light (magneto-optical Kerr effect). In synchrotron radiation experiments, an x-ray phase plate changes the x-rays from linearly to circularly polarized. XMCD can be observed in both XANES and EXAFS. XMCD in XANES can

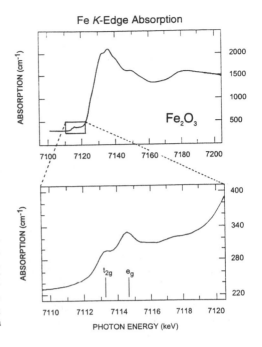

Figure 10. (a, right) Example of K edge absorption in Fe_2O_3 at zero pressure. The top figure shows the entire edge; the lower figure shows the 'crystal field' excitations evident in the lower energy region (C.C. Kao, private comm.). (b, below) Energy level diagram showing the excitations involved in the Fe^{2+} and Fe^{3+} K_β fluorescence measurements used to study high-spin/low-spin transitions in iron oxides and sulfides. The transition metal K_β emission results from the $3p \rightarrow 1s$ decay, and K_β emission line can be described to a good approximation by a two-step model (absorption followed by emission). In the final state, the $3p$ core-hole strongly interacts with the incomplete $3d$ shell. The consequence is the splitting of the K_β spectrum into two peaks, whose separation is controlled by the configuration interaction in the final state. The splitting usually exceeds the $3p$-$3d$ simple exchange interaction J. Furthermore, calculations show that satellite $K_\beta{}'$ at lower energy results predominantly from the $3p{\downarrow}3d{\uparrow}$ final state whereas the $K_{\beta1,3}$ line comes mainly from $3p{\uparrow}3d{\uparrow}$ final state with a small contribution of the $3p{\downarrow}3d{\uparrow}$ which appears in the spectrum as a low energy shoulder in the main peak.

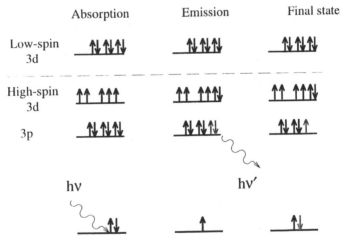

measure spin-resolved conduction band densities of states, whereas XMCD in EXAFS provides local magnetic structural information. The sign of the dichroism gives the ferromagnetic coupling between atoms in a metal.Very recently, high-pressure MCD measurements have been performed (Baudelet et al. 1997). For magnetic samples, magnetic circular dichroism is also observable in XES by exciting core-holes with circularly polarized x-rays. With the low end of the high-energy window extended down to 4 keV by the high-strength Be gasket, all elements above Ca ($Z = 20$) in principle can be studied at high pressure with a suitable choice of analyzer crystals. Crystal analyzers have been developed for inelastic x-ray scattering with sub-eV energy resolution of emission spectral lineshapes for studies of oxidation states, electronic energy level, spin states, and

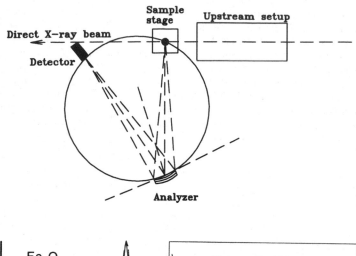

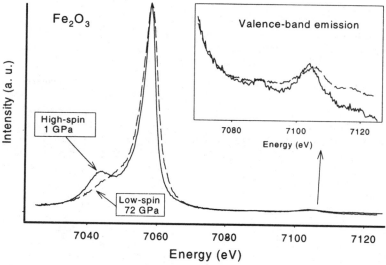

Figure 10, continued. (c) Schematic of a high-resolution Rowland-circle spectrometer for high-pressure synchrotron x-ray emission spectroscopy (Rueff et al. 1998; Badro et al., to be published). (d) K_β emission spectra showing the disappearance of the satellite peak at high pressure, indicative of a change in spin state in the sample. Smaller changes are observed in the weaker valence bands emission region, which is shown in the inset. The spectra were measured the NSLS, Brookhaven National Laboratory, and at the APS, Argonne National Laboratory.

trace element analysis. In particular, the important 3d and 4f emission bands can be studied to very high pressures.

Inelastic x-ray scattering spectroscopy (IXSS). IXSS measures the dynamical structure factor, and as such, it is similar to inelastic neutron scattering (Ghose 1988). Formally, $S(q, \omega)$ is related to the space-time Fourier transform of the density correlation function, which provides information on the electronic band structure and elementary excitations (such as phonons and plasmons) and in turn thermal, optical, magnetic, and transport properties of the material (Hill et al. 1996). IXSS has a number of advantages in comparison with other scattering probes, such as those that use (optical) light, electrons, and neutrons (Krisch et al. 1995, Kao et al. 1996). For example, light scattering can only probe zero-momentum transfer transitions; electron scattering suffers from multiple scattering effects and can only be used in high-vacuum conditions; neutron scattering needs very large samples. On the other hand, IXSS covers wide length (momentum) as well as temporal (energy) scales that are potentially important in high-pressure mineral studies. Low-energy excitations have been studied with ultrahigh resolution (~1.5 meV or 12 cm^{-1}) IXSS techniques (Ruocco et al. 1996, Sampoli et al. 1997). The resolution can be achieved is as good or better than what has been achieved by the best angle-resolved photoemission spectrometer. The potential for using this approach for measurements of vibrational densities of states is discussed in Chapter 17.

Another variation on these techniques is high-resolution resonant x-ray Raman scattering (Ederer and McGuire 1996), which is in principle like the more familiar Raman scattering at optical wavelengths. It can be viewed as an instantaneous absorption and emission process where the initial and final states are electronic or magnetic (rather than vibrational) states. Such resonant inelastic x-ray scattering, one detects electronic excitations of $\omega = \omega_1 - \omega_2$ and monitor momentum transfer $q = q_1 - q_2$. Examples include the study of the metal-antiferromagnetic insulator transitions (Isaacs et al. 1996), but the technique has not yet been used for high-pressure minerals. Studies at ambient pressure have shown that it is possible to measuring EXAFS-like spectra for K edges of low-Z elements by x-ray Raman; similar studies in high-pressure cells could provide probes of local structure (e.g. of Si) which cannot be studied by direct EXAFS techniques.

Compton scattering. Compton scattering is an inelastic scattering at high momentum transfer that probes electron momentum distributions. The details of the momentum distribution reveal information about bonding type (e.g. covalency) and band structure, particularly in combination with electronic structure theory (Isaacs et al. 1998). The availability of intense synchrotron radiation sources, particularly the high brightness at very high energies (~50 keV), together with the development of high-resolution spectrometers, are opening up new classes of Compton scattering experiments on small samples and at high pressures .

Nuclear resonance fluorescence spectroscopy (NRFS). As described above, Mössbauer spectroscopy has been used extensively in high-pressure mineralogy in laboratory studies with a radioactive parent source. High-pressure studies using a conventional Mössbauer source suffer from limited intensity for measurements on small samples, absorption by anvils, and background scattering. There have been major advances that exploit the temporal structure of synchrotron radiation to perform nuclear resonance spectroscopy in the time domain (Hastings et al. 1991). In this application, highly monochromatized x-rays from the synchrotron are used to excite narrow nuclear resonances, and the delayed photons are detected. Hyperfine splittings are reconstructed from the time-dependent intensity. The phonon densities of states around the nuclei can also be measured. Mössbauer spectroscopy has been used successfully to study Fe and Eu at second-generation synchrotron sources (Takano et al. 1991, Nasu 1996, Chefki et al. 1998). Very recently, the NRFS technique has been applied to the study of iron-bearing minerals under pressure, including silicate perovskites (Zhang 1997) (Fig.11). It has also been used to study magnetism up to megabar pressures (Lübbers et al. 1997).

Transport measurements

Electrical conductivity. There have been significant breakthroughs in the measurement of transport properties under pressure. Electrical conductivity has been measured under pressure in a wide variety of apparatus since the pioneering work of Bridgman (1949). The principal technique is the four-probe method, which has been

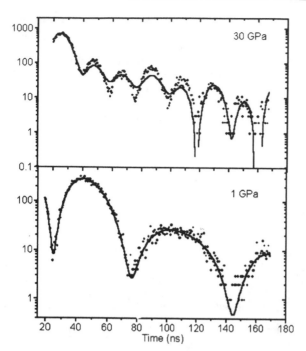

Figure 11. Nuclear resonance forward scattering spectra of $(Mg_{0.9}Fe_{0.1})SiO_3$ perovskite. The structure in the spectrum is assigned to different Fe^{2+} and Fe^{+3} sites and becomes better resolved at higher pressure (in the stability field of the perovskite) (from Zhang 1997). The high-pressure sample was contained in a He pressure-transmitting medium. This high-resolution technique utilizes the time-structure of the synchrotron source (see text). The spectra were measured at the ESRF, Grenoble, France.

extensively applied in diamond-cells (Mao and Bell 1977 1981) (Fig. 12). The technique has been widely used, including in conjunction with high P-T laser heating (e.g. Li and Jeanloz 1987) and resistive heating (e.g. Peyronneau and Poirier 1989) techniques. With the recent development of new laser-heating techniques, which provide reduced temperature gradients and large heating zones (Mao et al. 1998a), the accuracy of very high P-T electrical conductivity measurements is expected to be improved. Recently, the technique has been extended to above 200 GPa (from room temperture down to 0.05 K) (Eremets et al. 1998). New microlithographic techniques are allowing still smaller leads to be attached to or actually built into the surface of the anvils for achieving both higher pressure and higher accuracy (Fig. 12). The technique can also be coupled with a radiation field for carrying out photoconductivity measurements under pressure.

The complex impedance method commonly used under ambient pressures has been developed and applied at high pressure. Recently, it has been extended to large-volume (multi-anvil) devices (Poe et al. 1998). In this technique, the frequency of the current is varied over a wide range (from 0.01 Hz to 1 MHz); preliminary measurements on the $(Mg,Fe)_2SiO_4$ polymorphs to 20 GPa as a function of temperature indicate that the conductivity in the high-pressure phases (wadsleyite and ringwoodite) is a factor of 10^2 higher than in olivine (Poe et al. 1998), consistent with previous low-pressure studies. Another important example is conductivity measured along the Hugoniot in shock-wave experiments (Knittle et al. 1986), including the more recent reverberation approach (Weir et al. 1996).

Magnetic susceptibility. The magnetic susceptibility χ describes the response of a system to an applied magnetic field. Diamagnetic materials have negative χ. Paramagnetic and ferromagnetic materials have positive χ. For paramagnetic materials $\chi \sim 10^{-6}$ to 10^{-5} m^3/gauss and for ferromagnetic materials χ is several orders of magnitude higher. Typically, one reports the differential magnetic susceptibility, $\chi_\alpha = d\chi/dH$. For example, χ_α

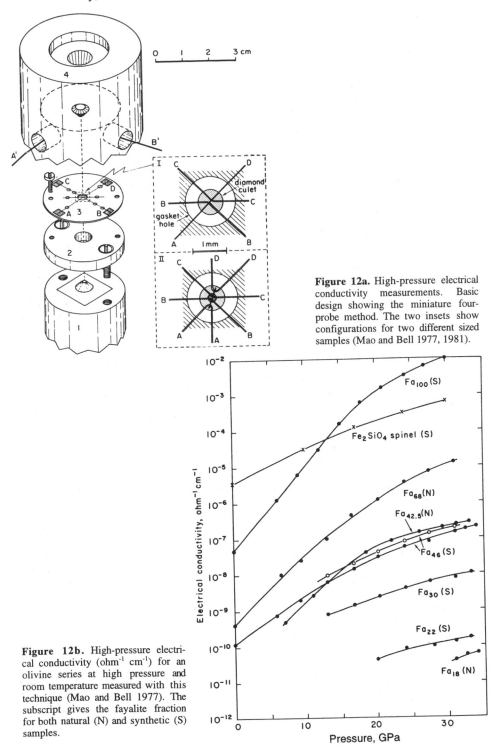

Figure 12a. High-pressure electrical conductivity measurements. Basic design showing the miniature four-probe method. The two insets show configurations for two different sized samples (Mao and Bell 1977, 1981).

Figure 12b. High-pressure electrical conductivity (ohm^{-1} cm^{-1}) for an olivine series at high pressure and room temperature measured with this technique (Mao and Bell 1977). The subscript gives the fayalite fraction for both natural (N) and synthetic (S) samples.

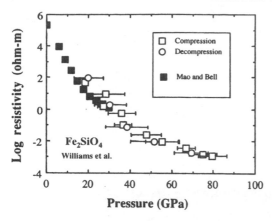

Figure 12c. Higher pressure measurements from Williams et al. (1990), plotted as resistivity (ohm-m) and compared to the earlier data of Mao and Bell (1977).

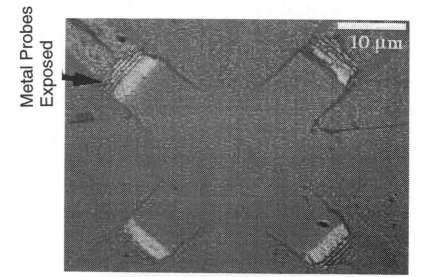

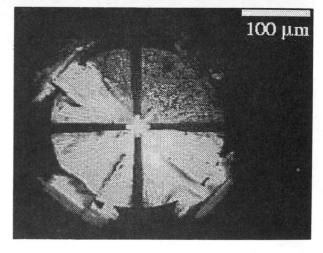

Figure 12d. Photomicrograph of leads deposited on the surface of diamond anvils using new microlithographic methods (Catledge et al. 1997). These new approaches have improved the accuracy and the pressure range of direct conductivity measurements under extreme conditions.

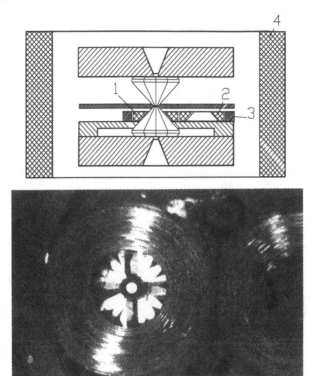

Figure 13. Coil assembly for high-pressure magnetic susceptibility measurements with the diamond cell. Top: Schematic showing the pick-up coil (1) compensation coil (2), excitation coil (3), and low-frequency modulation coil (4). Bottom: Photograph of the coil assembly and the lower diamond (gasket and sample removed). The low-frequency coil is crucial for increasing sensitivity of the technique (Timofeev 1992, Struzhkin et al. 1997).

= 1100 for Fe at $H = 0$. Several classes of high-pressure techniques have been developed. One involves measurement with SQUIDs (Webb et al. 1976). A second is an inductive technique developed by Tissen and Ponyatosvkii (1987) and later extended by Timofeev (1992). A schematic diagram of this method is shown in Figure 13. Originally applied to study superconductivity under pressure, it is currently being extended to investigate other pressure-induced magnetic transitions, including those in Fe (Timofeev et al., to be published).

Resonance methods

Electron paramagnetic resonance. Various techniques have been developed for EPR (ESR), which involves resonant microwave absorption between spin levels split in a magnetic field (Zeeman effect). This requires coupling of the microwave field, with the sample, which can be challenging at high pressure because of the small sample size (dimensions less than the wavelength of the radiation). Studies to ~8 GPa have been reported. Measurements on mantle silicates have been carried out at ambient pressures (e.g. Sur and Cooney 1989).

Nuclear magnetic resonance. Nuclear magnetic resonance is similar to EPR in that it too involves splitting of magnetic levels (Kirkpatrick 1988, Stebbins 1988). Because

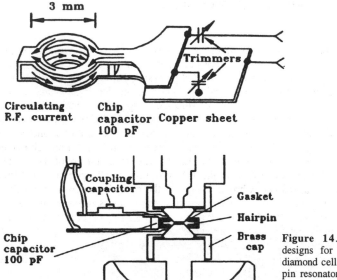

Figure 14. Example of one of several designs for high-pressure NMR in the diamond cell. This technique uses a hairpin resonator to direct the radio frequency signal into the sample. The trimmers provide fine frequency tuning. The cell is placed in a relatively large bore magnet (from Lee et al. 1992).

the magnetic moments of the nuclei are three orders of magnitude smaller than that of the electron, the level splitting is much lower in energy (radio frequency range, or ~100 MHz) at typical laboratory magnetic fields ($H \sim 5$-10 T). Proton NMR measurements have been carried out to 13 GPa (Ulug et al. 1991) and most recently to 17 GPa (I.F. Silvera, private comm.). A schematic of the technique developed by Ulug et al. is shown in Figure 14. NMR studies of high-pressure phases recovered at zero pressure have been carried out for some time. This includes ^{1}H and ^{29}Si NMR studies of quenched high-pressure hydrous phase (Phillips et al. 1997) and new magic angle spinning ^{29}Si relaxation techniques for characterizing naturally shocked samples (e.g. silica phases from Coconino sandstone) (Meyers et al. 1998).

A related measurement is nuclear quadrupolar resonance, which is similar to NMR but consists of measuring nuclear resonances in zero field. Measurements of the quadrupolar resonance ^{63}Cu in Cu_2O have been performed to 6 GPa, and in fact has been proposed as a pressure standard (Reyes et al. 1992). The numerous double-resonance techniques employed for years in condensed matter and chemical physics generally require larger sample volumes; the potential exists for employing these and other low-signal level techniques at very high pressures with the continued development of technology for increasing sample sizes under pressure (Chapter 1).

de Haas-van Alphen. This is a now standard technique for measuring Fermi surfaces of metals under ambient pressures but it generally requires large, perfect single crystals. High-pressure measurements have been performed in the past (Anderson et al. 1973). The advent of high-field magnets used in conjunction with new classes of diamond cells (Chapter 1) offers the possibility of extending these measurements to high pressures. Again, such studies of iron alloys under pressure could provide important constraints on the evolution of fundamental electronic properties of core materials with pressures. Other extensions of magnetic susceptibility methods are described above.

SELECTED EXAMPLES

Silicate perovskite

We now briefly summarize results of studies of the electronic and magnetic properties of some deep Earth materials, emphasizing the integration and application of different high-pressure techniques. The high-pressure behavior of $(Mg,Fe)SiO_3$ perovskite been the focus of considerable study because current studies indicate that it is the dominant phase of the lower mantle (Fig. 15). The primary aspects of the bonding in the material can be understood by examining the Mg endmember. First-principles calculations show that the material is highly ionic, with atoms having nearly spherical charge distributions, and it is a wide-band gap insulator (Fig. 15a,b). Both its ionic character and insulating character persist over the pressure range found within the Earth. The substitution of Fe in the structure introduces considerable absorption in the UV-visible-near IR spectra (Fig. 15c) (Keppler et al. 1994, Shen et al. 1994). A broad feature in the optical spectrum near 14900 cm^{-1} has been assigned to the $Fe^{2+} \rightarrow Fe^{3+}$ charge transfer transition (Keppler et al. 1994). The absorption through the near-IR and UV indicates $(Mg,Fe)SiO_3$ perovskite is a poor thermal insulator under lower mantle conditions where radiative heat transfer dominates. It has been proposed that the strong partitioning of Fe^{2+} relative to Mg^{2+} in magnesiowüstite is the result of stabilization of Fe^{2+} in the octahedral site of the oxide as compared to the (pseudo) dodecahedral site of the perovskite (Yagi et al. 1979, Burns 1993). The apparent crystal field stabilization has been used to estimate (or rationalize) partitioning between the two phases. Malavergne et al. (1997) find that partitioning results are consistent with observations for inclusions thought to have originated in the lower mantle (Kesson and FitzGerald 1992).

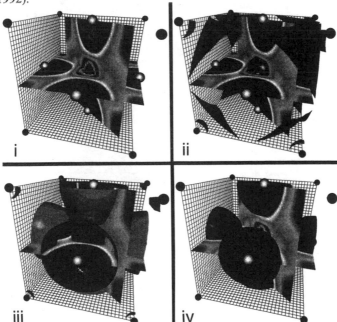

Figure 15a. Electronic properties of $(Mg,Fe)SiO_3$ perovskite. Calculated valence charge density for Mg end-member (ideal cubic structure). The valence charge density is the electron density associated with chemical bonding: i. Structure showing the Mg atoms at the corners, the oxygens at the centers of the faces, and the silicon at the center; ii. Iso-surface contours with 0.014 $e^-/bohr^3$; iii. 0.05 $e^-/bohr^3$; iv. 0.1 $e^-/bohr^3$ (from Hemley and Cohen 1992).

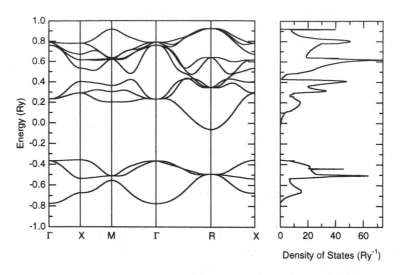

Figure 15b. Calculated band structure for the cubic $MgSiO_3$ perovskite (left) and the electron density of states showing the gap between the valence and conduction bands (from a study by Stixrude and Cohen 1993).

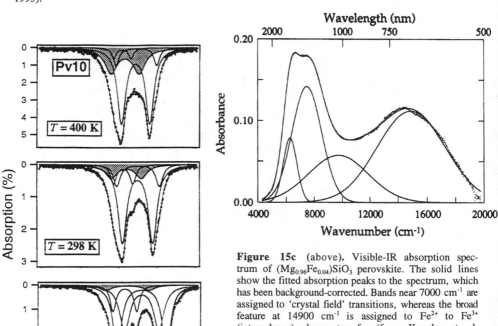

Figure 15c (above). Visible-IR absorption spectrum of $(Mg_{0.96}Fe_{0.04})SiO_3$ perovskite. The solid lines show the fitted absorption peaks to the spectrum, which has been background-corrected. Bands near 7000 cm^{-1} are assigned to 'crystal field' transitions, whereas the broad feature at 14900 cm^{-1} is assigned to Fe^{2+} to Fe^{3+} (intervalence) charge transfer (from Keppler et al. (1994). There is also evidence for higher energy features, as expected (e.g. approaching the band gap).

Figure 15d (left). ^{57}Fe Mössbauer spectra of $(Mg_{0.9}Fe_{0.1})SiO_3$ perovskite. The spectra were fitted with two Fe^{2+} doublets, an Fe^{3+} doublet, and a doublet with intermediate parameters indicative of thermally activated electron delocalization (from Fei et al. 1994).

Both crystallographic and Mössbauer studies indicate that Fe^{2+} in silicate perovskite resides in the (pseudo) dodecahedral site (see Hemley and Cohen 1992; Prewitt and Downs, Chapter 9). McCammon (1997) used Mössbauer measurements to show that a significant fraction of the iron in $(Mg,Fe,Al)SiO_3$ perovskite produced in multi-anvil experiments is Fe^{3+}. The incorporation of Fe^{3+} is strongly coupled to the Al^{3+}, but the extent to which this implies an oxidized lower mantle remains to be shown. Attempts have been made to determine the site occupancy of Fe^{2+} and Fe^{3+} from Mössbauer spectroscopy (McCammon 1998). Spectra of $(Mg_{0.95}Fe_{0.05})SiO_3$ reportedly synthesized at low f_{O_2} indicates that the Fe^{3+} goes in the octahedral site, whereas higher f_{O_2} conditions result in Fe^{3+} on both sites. Fei et al. (1994) reported evidence for a thermally activated electron delocalization from Mössbauer measurements on quenched samples (Fig. 15d). Recently, Zhang (1997) measured the Mossbauer effect in $(Mg,Fe)SiO_3$ perovskite (and clino-pyroxene) at high pressure by nuclear resonance forward scattering with synchrotron radiation (Fig. 11). Interestingly, the resolution of the spectrum improved considerably at high pressure (e.g. in the stability field of the perovskite). The results suggest that the thermally activated electron delocalization found in the quenched samples (Fei et al. 1994) may be suppressed at high pressure.

Measurements of the high P-T electrical conductivity of $(Mg,Fe)SiO_3$ perovskite (and magnesiowüstite) are essential for constraining the electrical conductivity profile of the deep mantle, and have been a major challenge in high-pressure mineralogy. Li and Jeanloz (1987) first measured the electrical conductivity of $(Mg,Fe)SiO_3$ perovskite at high pressures and temperatures by laser heating; subsequent measurements were carried out at lower maximum temperatures by Peyronneau and Poirier (1989) using a resistively heated cell. The latter results fit a hopping conductivity model. Poirier and Peyronneau (1998) interpret the electrical conduction in terms of a polaron hopping involving Fe^{2+} and Fe^{3+}, consistent with the Mössbauer results cited above (see also Duba et al. 1997). Katsura et al. (1998) reported that the temperature dependence of the conductivity differed significantly between samples measured at high pressure in its stability field and quenched to ambient pressure, confirming the need for in situ high-pressure measurements. As pointed out by Katsura et al. (1998) inconsistencies among the different experiments still remain; further work is needed to clarify the origin of these differences work (e.g. using new experimental techniques such as those identified above).

Fayalite

Fayalite undergoes dramatic pressure-induced changes in the optical and electrical properties on compression, and has been the subject of numerous high-pressure investigations. Fayalite transforms to a spinel structure at about 6 GPa (1273 K) and dissociates to a mixture of FeO and SiO_2 above about 22 GPa (Liu and Bassett 1986). Nevertheless, as is the case with many minerals, fayalite can persist metastably to much higher pressure in the absence of heating. Under such conditions, the zero-pressure absorption edge (Nitsan and Shankland 1976) shifts from the UV to the near IR with pressure such that samples became opaque to visible light above 15-18 GPa. As shown in Figure 12b, Mao and Bell (1977) found that the electrical resistivity concomitantly decreases by five orders of magnitude over this range, consistent with work of Smith and Langer (1982), Lacam (1983), and shock-wave study of (Mashimo et al. 1980). The low-temperature antiferromagnetic to paramagnetic transition was also tracked to 16 GPa by Mössbauer spectroscopy (Hayashi et al. 1987).

Subsequent optical, electrical, and x-ray measurements were extended to higher pressures, providing evidence that fayalite undergoes amorphization near 40 GPa and a further decrease in band gap and resistivity (Fig. 12c) (Williams et al. 1990). Near-IR reflectivity spectra measured down to 0.5 eV with the techniques described above (conventional source) to reported pressures of 225 GPa indicate that the metastable amorphous phase is not metallic (Williams et al. 1990). The results indicate that the transition observed in shock-wave compression experiments (McQueen and Marsh 1966) is likely due to disordering combined with significant changes in electronic properties in a metastable state (i.e. the equilibrium disproportionation does not occur). Although the possibility of having formed an amorphous phase gives some insights into the nature of the liquid silicate, the effects of temperature would be appreciable and therefore raise difficulties for direct comparisons between the low-temperature crystal and the liquid, especially as regards the electrical properties.

Wüstite

Wüstite displays a rich variety of the phenomena described in the preceding sections (Fig. 16). $Fe_{1-x}O$ [hereafter referred to as FeO] is non stoichiometric, and contains some ferric iron even in equilibrium with iron metal. The vacancies in wüstite form complex defect clusters (Hazen and Jeanloz 1984). With increasing pressure, more stoichiometric wüstite can be stabilized. In contrast to the behavior of MgO which remains in the rocksalt (B1) structure to at least 227 GPa (Duffy et al. 1995a), FeO exhibits complex polymorphism under pressure, and has been the subject much theoretical study (e.g. Sherman and Janssen 1995, Cohen et al. 1997). At room temperature and pressure, FeO is in the cubic rocksalt (B1) structure but undergoes a rhombohedral distortion at 17 GPa (Fig. 16a). As temperature is lowered at ambient pressure, the material passes the Néel temperature, becomes magnetically ordered, and also assumes a distorted rhombohedral structure. As pressure is increased, T_N increases (Zou et al. 1980), so that pressure promotes the rhombohedral phase. Isaak et al. (1993) showed that pressure promotes the rhombohedral distortion even in the absence of magnetism. Theory reveals the origin of the rhombohedral distortion with pressure. Analysis of the charge density as a function of rhombohedral angle (Isaak et al. 1993, Hemley and Cohen 1996) indicates that Fe-Fe bonding causes the rhombohedral strain, and the increase in Fe-Fe bonding with pressure is responsible for the increased distortion angle: the rhombohedral strain brings Fe ions closer together, and thus is favored with increasing Fe-Fe interactions.

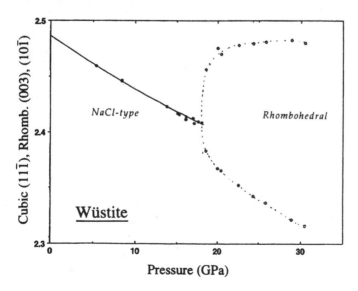

Figure 16a. Electronic and magnetic properties of wüstite. (a) Results of single crystal x-ray diffraction showing the splitting of the 111-class reflection of the NaCl-type (B1 structure) in the cubic – to - rhombohedral distortion at 17 GPa. The measurements were carried out in a He pressure-transmitting medium at room temperature (Shu et al. 1998).

Mössbauer experiments show evidence for a transition with increasing pressure to a low-spin phase (Fig. 16b), and also a transition to a non-magnetic phase with increasing temperature (Pasternak et al. 1997b). This is somewhat difficult to understand because the high-spin phase should be the high entropy phase, and thus increasing temperature should promote the high-spin magnetic phase, rather than the low-spin non-magnetic phase. It is possible that the transition with increasing temperature is a disordering of the local moments, rather than a loss of local moments, since temperature should promote disordering. Mössbauer measurements do not distinguish between loss of local moments and the disordering of moments. On the other hand, room-temperature K_β fluorescence spectra (Fig. 16c) measured to 133 GPa show no change in spin state (Badro et al., to be published). It will be important to supplement these x-ray and Mössbauer studies with electrical conductivity measurements and/or susceptibility measurements to find all of the boundaries of the expected complex phase diagrams for Mott insulating systems.

Further, a major transition occurs at high temperatures over this range of pressures (Fig. 16d) (Jeanloz and Ahrens 1980) and was found to be conducting (Knittle and Jeanloz 1986b); it is hexagonal and was identified from in situ diffraction studies to be the NiAs (B8) structure (Fei and Mao 1994). The stability fields for the three phases are shown in the phase diagram in Figure 16e. The structures of all three phases are closely related (Fig. 16f). Recent analysis of the diffraction data of Fei and Mao (1994) suggested the formation of a polytype or superlattice between B8 and anti-B8, with Fe in the As-site and O in the Ni-site (Mazin et al. 1998, Fang et al. 1998). These B8 and anti-B8 structures can be joined together smoothly, and the boundary between them is the rhombohedrally-distorted B1 structure (Mazin et al. 1998). This structure could form as a metastable intermediate phase; a unique, continuous structural transition between the phases may even occur.

Theoretical studies indicate that the high-spin/low-spin transitions in a number of structures are dominated by the local coordination. Thus, a transition metal ion in a smaller site will transform at lower pressures than one in a larger site. Thus ferric iron in solid solution substituting for Mg^{2+}, which is a smaller ion than Fe^{2+}, is likely to transform at lower pressures than in the pure ferrous iron compound. This suggests that magnesiowüstite will be low spin at even lower pressures than 100 GPa in the Earth. On the other hand, ferrous iron in the B8 or anti-B8 structures will remain high spin to much higher pressures. This also makes clear the distinction between magnetic collapse and metal-insulator transitions, since normal B8 FeO is predicted to be a high-spin metal, and anti-B8 is predicted to be a high-spin insulator.

Fe_2O_3 and Fe_3O_4

The results for FeO may be compared to similar studies of Fe_2O_3 and Fe_3O_4. Measurements on Fe_2O_3 demonstrate that a high-spin/low-spin transition may be coupled to phase transitions. A transition was first observed by shock-wave compression (Kondo et al. 1980, Rossman 1988). Reid and Ringwood (1969) predicted the existence of perovskite-type Fe_2O_3. Mössbauer measurements indicate a transition around 50 GPa

Figure 16 (cont'd, next page). (b) Mössbauer spectra of $Fe_{0.94}O$ as a function of pressure showing the evidence for a change in magnetic structure near 100 GPa (from Pasternak et al. 1997b). The inset shows the relative abundance of the low-spin component. (c) K_β emission showing an apparent persistence of the high spin state in this sample to 133 GPa; the results may be compared to the change observed in the equivalent spectrum measured for Fe_2O_3 shown in Fig. 10 (Badro et al., to be published). (d) High P-T resistivity of $Fe_{0.94}O$ from both diamond-cell and shock-wave experiments, latter compared to Fe (Knittle and Jeanloz 1986b, Knittle et al. 1986). (e) Phase diagram of FeO determined from a combination of studies: external resistance heating (open diamonds and triangles—Fei and Mao 1994), laser heating (solid symbols—Shen et al., to be published), and shock wave (open circle—Jeanloz and Ahrens 1980).

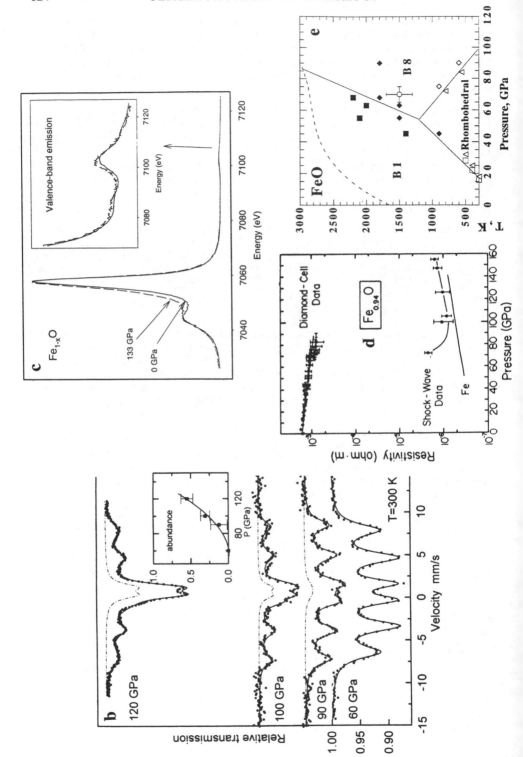

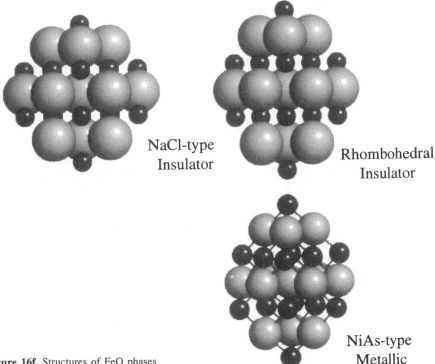

NaCl-type
Insulator

Rhombohedral
Insulator

NiAs-type
Metallic

Figure 16f. Structures of FeO phases
showing the similar bonding topologies.

(Syono et al. 1984, Suzuki et al. 1985). It was argued that the Fe^{3+} transforms from a high-spin state to a mixed-valence state above 50 GPa, with the iron atoms having different charges. Olsen et al. (1991) confirmed by x-ray diffraction the existence of a structural transition and identified the high-pressure phase as orthorhombic perovskite. Recent x-ray fluorescence experiments indicate that the material is low-spin in this phase (Fig. 10). High *P-T* diffraction studies show that an intermediate phase forms under laser heating at 34 GPa (Ma et al. 1998). Consistent with previous but less quantitative measurements of Endo and Ito (1982), Knittle and Jeanloz (1986a) reported that the resistivity of Fe_2O_3 dropped by about a factor of 10^3 at 48 GPa, close to the pressure of the transition observed by Mössbauer, x-ray diffraction, and shock-wave compression. Recent first-principles calculations carried out with the linear muffin-tin orbital (LMTO) and linearized augmented plane-wave (LAPW) methods indicate a high- to low-spin transition of the magnetic moment on Fe^{3+}, in agreement with recent experimental results (Gramsch and Cohen 1998). Coincident with the high- to low-spin transition is an insulator-to-metal transition that takes place with a predicted closing of the band gap at a calculated pressure of 28 GPa. These results for Fe_2O_3 may be compared with the behavior of Fe_3O_4, which contains both Fe^{+2} and Fe^{+3} and undergoes a phase transition near 20 GPa (Liu and Bassett 1986). Recently, the mineral has been examined using electrical resistivity techniques to 48 GPa, where compression-induced changes in electron hopping between Fe^{+2} and Fe^{+3} were found (Morris and Williams 1997).

Iron sulfides

The high-pressure behavior FeS provides another example of both interesting changes in electronic structure and how these changes are coupled to other alterations in physical properties (Fig. 17). A transition was first observed by x-ray diffraction at about 7 GPa. Subsequently, King et al. (1978) found by high-pressure Mössbauer measurements that the transition is associated with a loss of quadrupole splitting in the high-pressure phase (Fig. 17a); they suggested that the transition is electronic and due to either a charge transfer process or an insulator-metal transition. Fei et al. (1995) used the resistively heated diamond cell to study the P-T phase diagram with in situ synchrotron x-ray diffraction methods (see Prewitt and Downs, Chapter 9). The complex structure of FeS (III) was finally solved using high-resolution powder diffraction measurements together with a Reitveld refinement. The structure is monoclinic having space group $P2_1/a$ (Fei et al. 1998, Nelmes et al. 1998). Very recently, new x-ray emission techniques were used to study the K_β emission spectra, as described above. The results clearly show the K_β doublet structure induced by the magnetic hyperfine interactions is lost at the transition (Fig. b,c). These results, together with the x-ray diffraction data support that interpretation of an electronic transition. Notably, near-IR reflectivitiy measurements show no appreciable changes across the transition (Fig. 17d). This indicates that the higher pressure phase is not a high-carrier (high-conductivity) metal. Hence, the available data are most consistent with pressure-induced a high-spin/low-spin transition.

Iron

The high-pressure electronic and magnetic behavior of iron is of paramount importance for understanding the Earth's core. Three polymorphs of iron are well-established: the body-centered cubic phase (α –Fe) at low temperature and low pressure, the face-centered-cubic phase (γ –Fe) at high temperature, and the hexagonal-close-packed phase (ε -Fe) at high pressure (at least 300 GPa and 300 K, and 155 GPa and 2500 K; see Shen and Heinz, Chapter 12). Electrical resistivity increases dramatically across the α–ε phase transition (near 13 GPa). Meanwhile, the reflectivity of iron decreases markedly across the transition; measurements to 300 GPa showed that this low reflectivity continues to much higher pressure (Reichlin, Hemley, and Mao, unpublished).

The bcc phase is ferromagnetic. Theory shows that the magnetism stabilizes this structure, and that otherwise bcc iron would not form (see Stixrude and Brown, Chapter 8). In fact, there had been much discussion of bcc as the possible structure for iron in the Earth's inner core, but calculations showed that bcc iron is mechanically unstable at high pressure due to the loss of magnetism with pressure (Stixrude and Cohen 1995). On the other hand, the hcp iron is paramagnetic and is likely to be the predominant phase in the inner core. Magnetism may still play an important role in crystallization, rheology, and elasticity of iron at ultrahigh pressures. For instance, the observed elastic anisotropy of the inner core has been interpreted as a result of the magnetic field-induced orientation of iron crystals (Karato 1993, Stixrude and Brown, Chapter 8). The strong orientation dependence of lattice strain in hcp iron observed in the diamond cell may also be a result of magnetic interactions (Mao et al. 1998b).

Pressure effects on the valence band densities of states and magnetic properties of Fe can now be measured with the new synchrotron x-ray techniaues described above. The magnetic collapse in a related system, the RFe$_2$ Laves phases, was found using the new synchrotron nuclear resonance forward scattering technique (Lübbers et al. 1997). It will be of interest to extend this to iron and iron alloys at core pressures. In the case of Fe, large differences in density of states are predicted between bcc Fe and the two closed-packed phases (hcp and fcc). Spin-dependent K_β emission fine structure can be used to probe localized magnetic properties with XMCD. The element-specific nature of XES and XMCD will be particularly important in the study of iron, iron alloys, and other metals under pressure.

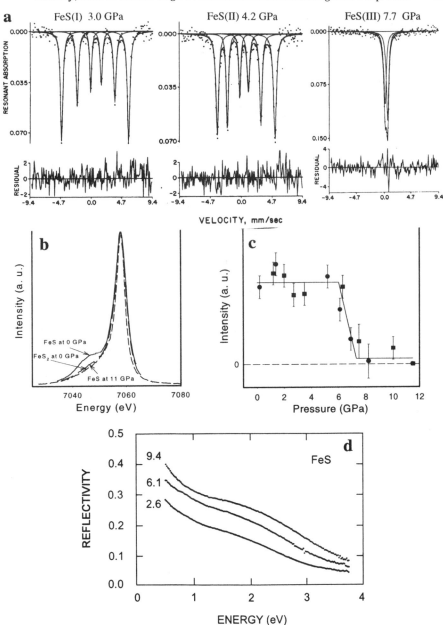

Figure 17. High-pressure transformation in FeS. (a) Mössbauer spectra of phases I, II, and III showing the collapse of the quadrupole splitting in the (room-temperature) high-pressure phase. (b) K_β' emission spectra of FeS (III) at 11 GPa as compared to FeS (I) (on decompression) and FeS_2 (low-spin configuration) (Rueff et al. 1998). (c) The amplitude of the K_β' satellite as a function of pressure. The solid line is a guide to the eye; the dashed line shows the zero intensity level. (d) Near infrared reflectivity as a function of pressure showing that the transition is not accompanied by a major change in reflectivity as expected for a transformation to a highly conducting metal (Hanfland, Hemley, and Mao, to be published). The pressures are listed in gigapascals on at the left.

Very recently, a novel synchrotron technique using both the nuclear Mössbauer effect and K electron emission has been developed for determination of phonon densities of state of iron-bearing materials (Seto et al. 1995, Sturhahn et al. 1995). Synchrotron x-rays are absorbed at the 14.414 keV nuclear resonance of ^{57}Fe, and subsequently the nucleus is deexcited by emission of K fluorescence. With the very narrow energy width (4.66 neV) and short life time (141 ns) of the ^{57}Fe nuclear resonance process, the electronic contribution is very efficiently separated from the pulse synchrotron radiation. Tuning the energy of the incident synchrotron radiation while monitoring the delayed K fluorescence yields gives a direct measure of the phonon density of state (Sturhahn et al. 1995). Such measurements at high pressures provide valuable information on kinetic energy, specific heat, mean force constants and other dynamical properties for iron and iron-containing minerals (see also Chapter 17).

Dense hydrous phases

As mentioned above, under pressure, volatiles can become structurally bound components, forming in the terminology of Bernal a "valence lattice". We consider just a few examples of the large effect of pressure on hydrogen speciation and bonding in deep Earth materials. The prototype system is H_2O ice, where recent experiments have shown that a symmetric hydrogen-bonded state of ice forms at 60 GPa, and persists to at least 210 GPa (Goncharov et al. 1996, see also Gillet et al. Chapter 17). This high-pressure phase is a non-molecular, partially ionic structure. The results indicate that the hydrogen-bond becomes symmetric at a critical oxygen-oxygen distance of 2.38-2.40 A, a result that is expected to be transferable to other systems, including hydrous silicates and oxides. The large change in vibrational frequencies in such symmetric hydrogen bonded phases has a significant effect on thermodynamic properties (see Chapter 17).

Dense hydrous silicates present different degrees of hydrogen bonding under ambient conditions. The pressure dependence of the OH stretching modes show a tendency toward increased hydrogen bonding, but decreased hydrogen bonding is also observed (Faust and Williams 1996, Hemley et al. 1998). Pressure-induced disordering of these minerals has been observed; this may be intimately associated with the behavior of the hydrogen through sublattice amorphization or 'melting' (e.g. Duffy et al. 1995b, Nguyen et al. 1997, Parise et al. 1998). These results point toward the importance of hydrogen-hydrogen repulsions at high material compression. Variations in hydrogen bonding may be a key factor in controlling polymorphism in hydrous phases at high pressure (Faust and Williams 1996). There is evidence that phase D (Prewitt and Downs, Chapter 9) breaks down at 50 GPa and high temperature (Shieh et al. 1998). However, the existence of still higher P-T hydrous phases is currently under study; these phases are likely to be unquenchable. Finally, an example of the change in bonding affinities is the formation of iron hydride at high pressure. This phase does not form under ambient conditions. It produces a dhcp structure at 3.5 GPa, which is stable to at least 60 GPa (Badding et al. 1992).

Trace elements: Xe

The low concentration of xenon in the atmosphere relative to cosmochemical abundances is a long-standing problem in geochemistry and has given rise to the proposal that the element may be sequestered at depth within the Earth (Ozima 1998). Because it is a highly compressible and closed shell material, Xe is also an excellent prototype system for studying pressure effects on physical properties. As a result, the material has been investigated both experimentally and theoretically to very high pressures (Fig. 18). Closing of the band gap has been studied using the optical techniques described. Near-IR absorption and reflectivity (Fig. 18a) show that solid becomes metallic at 130-150 GPa (Goettel et al. 1989, Reichlin et al. 1989). Calculations of the band structure (Fig. 18b) are in good agreement with the observations. These calculations further show the broadening of the bands arising with increased interatomic interactions with pressure, which is ultimately responsible for the metallization.

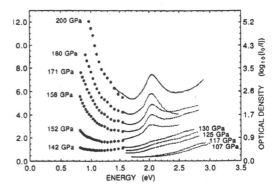

Figure 18a. High-pressure behavior of xenon. (a) Optical spectra showing the rise in IR absorption and the appearance of an interband transition indicative of metallization near 140 GPa (Goettel et al. 1989).

Prior to metallization, the material transforms from the fcc to the hcp structure (Jephcoat et al. 1987), with a metastable polytype structure forming under some conditions (perhaps not unlike that proposed for FeO described above). Recent studies have identified the equilibrium transition pressure for the direct fcc-hcp transition at 21 GPa (Caldwell et al. 1997). Experiments to 50 GPa and theoretical calculations to inner core pressures reported in this study show (surprisingly) no evidence for chemical affinity for Xe and Fe (Caldwell et al. 1997). The melting points of Xe, Kr, and Ar increase steeply with pressure and appear to exceed the melting point of iron at 30 GPa (Fig. 18c) (Jephcoat 1998). These results have led to the speculation that Xe may form as a free refractory phase in the deep Earth (Jephcoat 1998). Its solubility in mantle silicates and oxides under high *P-T* conditions remains to be investigated by direct in situ study.

CONCLUSIONS AND OUTLOOK

A range of new high-pressure techniques is revealing in dramatic fashion changes in bonding properties, as well as electronic and magnetic structure, induced by extreme pressures and temperatures in Earth materials. A number of factors control the nature of bonding in these materials under these conditions. Compression causes the mineral energetics to be increasingly controlled by repulsive interactions. In closed shell systems, which are characterized by largely spherical electron densities of the component atoms and ions, compression results in a tendency toward close packing at the expense of more open configurations dictated by directional bonding. At intermediate densities, those systems whose bonding properties may be described by orbital hybridization may be perturbed in complex ways as the extent of hybridization is changed by modification of interatomic distances. Under the most extreme conditions, electrons become increasingly destabilized in bonds and near ionic cores, and instead favor states of delocalization; however, the onset of this fundamental transition is strongly dependent on the elements that comprise the mineral. Finally, these trends in electronic properties a magnetic counterpart: magnetic collapse, with the high-spin states favored by Hund's rule giving way to states of low spin at high compression.

Recent work has shown that in general a variety of techniques is required to understand the evolution of mineral systems to very high pressures. This includes the use of newly developed theoretical methods, which are providing increasingly accurate

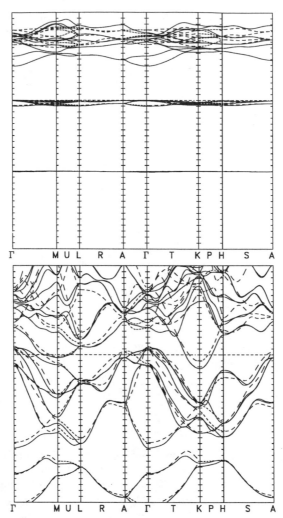

Figure 18b. High-pressure behavior of xenon. (b) Band structure of Xe (hcp structure) calculated using the LAPW method (see Chapter 19). At low pressure, Xe has almost atomic-like valence states, with very narrow band widths, as illustrated in the top figure for ambient pressure volume (500 bohr³/atom). With increasing pressure, the band broaden significantly and the material is metallic, as shown at the bottom for 250 GPa (100 bohr³/atom). One tick mark on the vertical axis corresponds to 1 eV.

predictions for energetic properties of these materials under extreme conditions. There is much to be learned about the origin of the behavior of transition-metal compounds and solid solutions at high pressures. The study of transitions in Mott insulators is a particularly exciting current problem, with important implications for condensed-matter theory as well. Future work should also focus on both defect properties and polyphase aggregates at high P-T conditions. That there can be profound effects of pressure on chemical properties is established; however, many phenomena are not yet understood nor are there guidelines for predicting chemical behavior such as element partitioning at deep mantle and core conditions. With the many recent developments in high-pressure experimental and theoretical techniques, the solution to these problems may be close at hand.

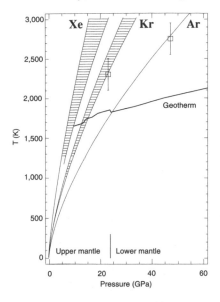

Figure 18c. High-pressure behavior of xenon. (c) Melting relations for heavy rare gas solids compared with the geotherm. The open symbols indicate the *P-T* region where the melting curves cross that of iron (Jephcoat 1998).

ACKNOWLEDGMENTS

We thank S. Gramsch, J. Badro, V.V. Struzhkin, A.F. Goncharov, and M. Somayazulu for very valuable assistance in the preparation of this chapter. This work was supported by NSF, NASA, DOE, and W.M. Keck Foundation.

REFERENCES

Ahrens TJ (1987) Shock wave techniques for geophysics and planetary physics. *In:* CG Sammis, TL Henyey (ed) Methods of Experimental Physics 24:185-235 Academic Press, San Diego

Anderson JR, Papaconstantopulos DA, McCaffrey JW, Schirber JE (1973) Self-consistent band structure of niobium at normal and reduced lattice spacing. Phys Rev B 7:5115-5121

Ashcroft NW, Mermin ND (1976) Solid State Physics. Holt Rinehart and Winston, New York

Badding JV, Mao HK, Hemley RJ (1992) High-pressure crystal structure and equation of state of iron hydride: Implications for the Earth's Core. *In:* Y Syono, MH Manghnani (ed) High Pressure Research in Mineral Physics: Application to Earth and Planetary Sciences. p 363-371 Terra Scientific, Washington, DC

Badro J, Gillet P, McMillan PF, Polian A, Itié JP (1997) A combined XAS and XRD study of the high-pressure behaviour of $GaAsO_4$ berlinite. Europhys Lett 40:533-538

Baudelet F, Odin S, Itié J-P, Polian A, Giorgetti C, Dartyge E, Pizzini S, Fontaine A, Kappler JP (1997) $Pt_{72}Fe_{28}$ invar metal studied by high pressure x-ray magnetic circular dichroism (XMCD). *In:* Crystallography at High Pressure Using Synchrotron Radiation: The Next Steps. ESRF, Grenoble, France

Bell PM, Xu J, Mao HK (1986) Static compression of gold and copper and calibration of the ruby pressure scale to pressures to 1.8 megabars. *In:* Y Gupta (ed) Shock Waves in Condensed Matter. p 125-130 Plenum, New York

Bridgman PW (1927) The breakdown of atoms at high pressures. Phys Rev 29:188-191

Bridgman PW (1949) The Physics of High Pressure. G. Bell and Sons, London

Brown GE Jr., Calas G, Waychunas GA, Petiau J (1988) X-ray absorption spectroscopy and its applications in mineralogy and geochemistry. *In:* FC Hawthorne (ed) Spectroscopic Methods in Mineralogy and Geology. Rev Mineral 18:431-512

Burns RG (1993) Mineralogical Applications of Crystal Field Theory, Second Edition. Cambridge University Press, Cambridge

Caldwell WA, Nguyen JH, Pfrommer BG, Mauri F, Louie SG, Jeanloz R (1997) Structure, bonding and geochemistry of xenon at high pressures. Science 277:930-933

Catledge SA, Vohra YK, Weir ST, Akella J (1997) Homoepitaxial diamond films on diamond anvils with metallic probes: the diamond/metal interface up to 74 GPa. J Phys Condens Matter 9:L67-L73

Chefki M, Abd-Elmeguid MM, Micklitz H, Huhnt C, Schlabitz W (1998) Pressure-induced transition of the sublattice magnetization in $EuCo_2P_2$: Change from local moment Eu(4f) to itinerant Co(3d) magnetism. Phys Rev Lett 80:802-805

Cohen RE, Mazin II, Isaak DG (1997) Magnetic collapse in transition metal oxides at high pressure: implications for the Earth. Science 275:654-657

Cormier L, Brown GE, Jr., Calas G, Galiosy L, Itié J-P, Polian A (1998) High-pressure x-ray absorption study of the local coordination environment of Ni in a $CaNiSi_2O_6$ glass. Geophys Res Lett:submitted

Cox PE (1987) The Electronic Structure and Chemistry of Solids. Oxford Science Publications, Oxford

Drickamer HG, Frank CW (1973) Electronic Transitions and the High Pressure Chemistry and Physics of Solids. Chapman and Hall, London

Duba A, Peyronneau, Visocekas F, Poirier J-P (1997) Electrical conductivity of magnesiowüstite/ perovskite produced by laser heating of synthetic olivine in the diamond anvil cell. J Geophys Res 102:27723-27728

Duffy TS, Hemley RJ, Mao HK (1995a) Equation of state and shear strength at multimegabar pressures: magnesium oxide to 227 GPa. Phys Rev Lett 74:1371-1374

Duffy TS, Meade C, Fei Y, Mao HK, Hemley RJ (1995b) High-pressure phase transition in brucite, $Mg(OH)_2$. Am Mineral 80:222-230

Ederer DL, McGuire JH (eds) (1996) Raman Emission by X-ray Scattering. World Scientific, Singapore

Endo S, Ito K (1982) Triple-stage high-pressure apparatus with sintered diamond anvils. *In:* A Akimóto, MH Manghnani (ed) High-Pressure Research in Geophysics p 3-12 Center for Academic Publishing, Tokyo

Eremets M (1996) High Pressure Experimental Methods. Oxford University Press, New York

Eremets MI, Shimizu K, Kobayashi TC, Amaya K (1998) Metallic CsI at pressures of up to 220 gigapascals. Science 281:1333-1335

Eremets MI, Struzhkin VV, Timofeev JA, Utjuzh AN, Shirokov AM (1992) Refractive index of diamond under pressure. *In:* AK Singh (ed) Recent Trends in High Pressure Research p 362-364 Oxford & IBH, New Delhi

Fang Z, Terakura K, Sawada H, Miyazaki T, Solovyev I (1998) Inverse versus normal NiAs structures as high-pressure phases of FeO and MnO. Phys Rev Lett 81:1027-1030

Faust J, Williams Q (1996) Infrared spectra of phase B at high pressures: hydroxyl bonding under compression. Geophys Res Lett 23:427-430

Fei Y, Mao HK (1994) In situ determination of the NiAs phase of FeO at high pressure and temperature. Science 266:1668-1680

Fei Y, Prewitt CT, Frost DJ, Parise JB, Brister K (1998) Structures of FeS polymorphs at high pressures and temperatures. *In:* M Nakahara (ed) Review of High-Pressure Science and Technology 7:55-58, Japan Society High-Pressure Science and Technology, Kyoto

Fei Y, Prewitt CT, Mao HK, Bertka CM (1995) Structure and density of FeS at high pressure and high temperature and the internal structure of Mars. Science 268:1892-1894

Fei Y, Virgo D, Mysen BO, Wang Y, Mao HK (1994) Temperature dependent electron delocalization in $(Mg,Fe)SiO_3$ perovskite. Am Mineral 79:826-837

Finger LW, Ko J, Hazen RM, Gasparik T, Hemley RJ, Prewitt CT, Weidner DJ (1989) Crystal chemistry of phase B and an anhydrous analogue: implications for water storage in the upper mantle. Nature 341:140-142

Fleury PA, Loudon R (1968) Scattering of light by one- and two-magnon excitations. Phys Rev 166:514-530

Fowler RH (1927) On dense matter. Mon Not R Astron Soc 87:114-122

Fujii Y (1996) Novel crystal physics under pressure. International Union of Crystallography XVII Congress and General Assembly, Seattle, Washington:C-1

Furrer A (eds) (1995) Magnetic Neutron Scattering. World Scientific, Singapore

Ghose S (1988) Inelastic neutron scattering. *In:* FC Hawthorne (ed) Spectroscopic Methods in Mineralogy and Geology. Rev Mineral 18:161-192

Goettel KA, Eggert JH, Silvera IF, Moss WC (1989) Optical evidence for the metallization of xenon at 132(5) GPa. Phys Rev Lett 62:665-668

Goncharov AF, Hemley RJ, Mao HK, Shu J (1998) New high-pressure excitations in parahydrogen. Phys Rev Lett, 80:101-104

Goncharov AF, Struzhkin VV, Ruf T, Syassen K (1994) High-pressure Raman study of the coupling of crystal field excitations to phonons in Nd-containing cuprates. Phys Rev B 50:13841-13844

Goncharov AF, Struzhkin VV, Somayazulu M, Hemley RJ, Mao HK (1996) Compression of H_2O ice to 210 GPa: Evidence for a symmetric hydrogen-bonded phase. Science 273:218-220

Gramsch S, Cohen RE (1998) First-principles LDA study of hematite: implications for the high-pressure behavior of Fe_2O_3. Abstracts of the International Union of Crystallography High Pressure Workshop, 14-16 Nov 1998, Argonne, IL:in press

Hamalainen K, Kao C-C, Hastings JB, Siddons DP, Berman LE, Stojanoff V, Cramer SP (1992) Spin-dependent x-ray absorption of MnO and MnF_2. Phys Rev B 46:14274-14277

Harrison WA (1980) Electronic Structure and the Properties of Solids: The Physics of the Chemical Bond. WH Freeman and Company, San Francisco

Hastings JB, Siddons DP, van Bürck U, Hollatz R, Bergmann U (1991) Mössbauer spectroscopy using synchrotron radiation. Phys Rev Lett 66:770-773

Hawthorne FC (eds) (1988) Spectroscopic Methods in Mineralogy and Geology. Rev Mineral 18.

Hayashi M, Tamura I, Shimomura O, Sawamoto H, Kawamura H (1987) Antiferromagnetic transition of fayalite under high pressure studied by Mössbauer spectroscopy. Phys Chem Minerals 14:341-344

Hazen RM, Jeanloz R (1984) Wüstite ($Fe_{1-x}O$): a review of its defect structure and physical properties. Rev Geophys Space Phys 22:37-46

Hemley RJ (1987) Pressure dependence of Raman spectra of SiO_2 polymorphs: α-quartz, coesite, and stishovite. *In:* MH Manghnani, Y Syono (ed) High-Pressure Research in Mineral Physics p 347-359 Terra Scientific, Tokyo/American Geophysical Union, Washington DC,

Hemley RJ, Ashcroft NW (1998) The revealing role of pressure in the condensed-matter sciences. Physics Today, 51:26-32

Hemley RJ, Cohen RE (1992) Silicate perovskite. Annu Rev Earth Planet Sci 20:553-600

Hemley RJ, Cohen RE (1996) Structure and bonding in the deep mantle and core. Phil Trans R Soc Lond A 354:1461-1479

Hemley RJ, Goncharov AF, Lu R, Li M, Struzhkin VV, Mao HK (1998) High-pressure synchrotron infrared spectroscopy at the National Synchrotron Light Source. Il Nuovo Cimento D 20:539-551

Hemley RJ, Hanfland M, Mao HK (1991) High-pressure dielectric measurements of hydrogen to 170 GPa. Nature 350:488-491

Hemley RJ, Mao HK (1997) Static high-pressure effects in solids. *In:* GL Trigg (ed) Encyclopedia of Applied Physics 18:555-572 VCH Publishers, New York

Hemley RJ, Zha CS, Jephcoat AP, Mao HK, Finger LW, Cox DE (1989) X-ray diffraction and equation of state of solid neon to 110 GPa. Phys Rev B 39:11820-11827

Hess NJ, Schiferl D (1992) Calibration of the pressure-induced frequency shift of Sm:YAG using the ruby and nitrogen vibron pressure scales from 6 to 900 K and 0 to 300 kbar. J Appl Phys 71:2082-2085

Hill JP, Kao CC, Caliebe WAC, Gibbs D, Hastings JB (1996) Inelastic x-ray scattering study of solid and liquid Li and Na. Phys Rev Lett 77:3665-3668

Isaacs ED, Platzman PM, Honig JM (1996) Inelastic x-ray scattering study of the metal-antiferromagnetic insulator transition in V_2O_3. Phys Rev Lett 76:4211-4214

Isaacs ED, Shukla A, Platzmann PM, Hamann DR, Barbiellini B, Tulk CA (1998) Evidene for covalency of the hydrogen bond in ice: a direct x-ray measurement. Phys Rev Lett:in press

Isaak DG, Cohen RE, Mehl MJ, Singh DJ (1993) Phase stability of wüstite at high pressure from first-principles LAPW calculations. Phys Rev B 47:7720-7731

Itié J, Baudelet F, Dartyge E, Fontaine A, Tolentino H, San-Miguel A (1992) X-ray absorption spectroscopy and high pressure. High Pressure Res 8:697-702

Itié J, Polian A, Calas G, Petiau J, Fontaine A, Tolentino H (1989) Pressure-induced coordination changes in crystalline and vitreous GeO_2. Phys Rev Lett 63:398-401

Itié JP, Polian A, Martinez D, Briois V, DiCicco A, Filipponi A, San Miguel A (1997) X-ray absorption spectroscopy under extreme conditions. J Phys IV France, Supp J Phys III 7:31- 38

Jeanloz R, Ahrens TJ (1980) Equations of state of FeO and CaO. Geophys J R AstronSoc 62:505-528

Jephcoat AP (1998) Rare-gas solids in the Earth's interior. Nature 393:355-358

Jephcoat AP, Mao HK, Finger LW, Cox DE, Hemley RJ, Zha C-S (1987) Pressure-induced structural phase transition in solid xenon. Phys Rev Lett 59:2670-2673

Kao CC, Caliebe WA, Hastings JB, Hämäläinen L, Krisch MH (1996) Inelastic x-ray scattering at the National Synchrotron Light Source. Rev Sci Instrum 67:1-5

Karato S (1993) Inner core anisotropy due to the magnetic field-induced preferred orientation of iron. Science 262:1708-1711

Katayama Y, Tsujii K, Oyanagi H, Shimomura O (1998) Extended x-ray absorption fine structure study on liquid selenium under pressure. J Non-Cryst Solids 232-234:93-98

Katsura T, Sato K, Ito E (1998) Electrical conductivity of silicate perovskite at lower-mantle conditions. Nature 395:493-495

Keppler H, McCammon CA, Rubie DC (1994) Crystal-field and charge-transfer spectra of $(Mg,Fe)SiO_3$ perovskite. Am Mineral 20:478-482

Keppler H, Rubie D (1993) Pressure-induced coordination changes of transition-metal ions in silicate melts. Science 364:54-56

Kesson SE, FitzGerald JD (1992) Partitioning of MgO, FeO, NiO, MnO, and Cr_2O_3 between magnesium silicate perovskite and magnesiowüstite: implications for the origin of inclusions in diamond and the composition of the lower mantle. Earth Planet Sci Lett 111:229-240

King H, Virgo D, Mao HK (1978) High-pressure phase transitions in FeS, using [57]Fe Mössbauer spectroscopy. Carnegie Inst Washington Yearb 77:830-835

Kirkpatrick RJ (1988) MAS NMR spectroscopy of minerals and glasses. *In:* FC Hawthorne (ed) Spectroscopic Methods in Mineralogy and Geology. Rev Mineral 18:341-403

Kittel C (1996) Introduction to Solid State Physics, 7th Edition. Wiley, New York

Klotz S, Besson JM, Hamel G, Nelmes RJ, Marshall WG, Loveday JS, Braden M (1998) Structures of FeS polymorphs at high pressures and temperatures. *In:* M Nakahara (ed) Review of High-Pressure Science and Technology 7:217-220 Japan Soc High-Pres Science and Technol, Kyoto

Knittle E, Jeanloz R (1986a) High-pressure electrical resistivity measurements of Fe_2O_3: comparison of static compression and shock-wave experiments to 61 GPa. Solid State Comm 58:129-131

Knittle E, Jeanloz R (1986b) High-pressure metallization of FeO and implications for the Earth's core. Geophys Res Lett 13:1541-1544

Knittle E, Jeanloz R, Mitchell AC, Nellis WJ (1986) Metallization of $Fe_{0.94}O$ at elevated pressures and temperatures observed by shock-wave electrical resistivity measurements. Solid State Comm 59:513-515

Kondo K, Mashimo T, Sawaoka A (1980) Electrical resistivity and phase transformation of hematite under shock. J Geophys Res 85:977-982

Krisch MH, Kao CC, Sette F, Caliebe WA, Hämäläinen L, Hastings JB (1995) Evidence for a quadrupolar excitation channel at the L_{III} edge of gadolinium by resonant inelastic x-ray scattering. Phys Rev Lett 74:4931-4934

Lacam A (1983) Pressure and composition dependence of the electrical conductivity of iron-rich synthetic olivines to 200 kbar. Phys Chem Minerals 9:127-132

Lee S-H, Conradi MS, Norberg RE (1992) Improved NMR resonator for diamond-anvil cells. Rev Sci Instrum 63:3674-3676

Li X, Jeanloz R (1987) Electrical conductivity of (Mg,Fe)SiO_3 perovskite and a perovskite-dominated assemblage at lower mantle conditions. Geophys Res Lett 14:1075-1078

Lipp M, Daniels WB (1991) Electronic structure measurements in KI at high pressure using three-photon spectroscopy. Phys Rev Lett 67:2810-2813

Liu J, Vohra YK (1996) Photoluminescence and x-ray diffraction studies on Sm-doped yttrium aluminum garnet to ultrahigh pressures of 338 GPa. J Appl Phys 79:7978-7982

Liu L-G, Bassett WA (1986) Elements, Oxides, and Silicates. Oxford University Press, New York

Loubeyre P, M J-L, RL T, L C-G (1993) High pressure measurements of the He-Ne binary phase diagram at 296K: Evidence for the stability of a stoichiometric Ne(He)$_2$ solid. Phys Rev Lett 70:178-181

Lübbers R, Hesse HJ, Grünsteudel HF, Rüffer R, Zukrowski J, Wortmann G (1997) Probing magnetism in the Mbar range: NFS high-pressure studies of RFe_2 Laves phases (R = Y, Gd, Sc). Highlights in X-ray Synchrotron Radiation Research, p. 38. ESRF, Grenoble, France

Ma Y, Mao HK, Hemley RJ, Shen G, Prewitt CT (1998) New high pressure phase of Fe_2O_3. Abstracts of the International Union of Crystallography High Pressure Workshop, Nov 14-16, 1998, Argonne, IL:in press

Malavergne V, Guyot F, Wang Y, Martinez I (1997) Partitioning of nickel, cobalt, and manganese between silicate perovskite and periclase: a test of crystal field theory at high pressure. Earth Planet Sci Lett 146:499-509

Mao HK (1976) Charge-transfer processes at high pressure. *In:* RGJ Strens (ed) The Physics and Chemistry of Rocks and Minerals p 573-581 John Wiley and Sons, New York

Mao HK, Bell PM (1972) Electrical conductivity and the red shift of absorption in olivine and spinel at high pressure. Science 176:403-406

Mao HK, Bell PM (1977) Techniques of electrical conductivity measurement to 300 kbar. *In:* MH Manghnani, S Akimoto (ed) High-Pressure Research: Applications to Geophysics p 493-502 Academic Press, Inc., NY,

Mao HK, Bell PM (1981) Electrical resistivity measurements of conductors in the diamond-window, high-pressure cell. Rev Sci Instrum 52:615-616

Mao H K, Hemley RJ (1991) Optical transitions in diamond at ultrahigh pressures. Nature 351:721-724

Mao HK, Bell PM, Shaner JW, Steinberg DJ (1978) Specific volume measurements of Cu, Mo, Pd, and Ag and calibration of the ruby R_1 fluorescence pressure gauge from 0.06 to 1 Mbar. J Appl Phys 49:3276-3283

Mao HK, Shen G, Hemley RJ (1997a) Multivariable dependence of Fe/Mg partitioning in the lower mantle. Science 278:2098-2100

Mao HK, Shen G, Hemley RJ, Duffy TS (1998a) X-ray diffraction with a double hot plate laser-heated diamond cell. *In:* MH Manghnani, T Yagi (ed) Properties of Earth and Planetary Materials at High Pressure and Temperature p 27-34 American Geophysical Union, Washington DC

Mao HK, Shu J, Shen G, Hemley RJ, Li B, Singh AK (1998b) Elasticity and rheology of iron above 200 GPa and the nature of the Earth's inner core. Nature:in press

Mao HK, Struzhkin VV, Hemley RJ, Kao CC (1997b) Synchrotron x-ray spectroscopy at ultrahigh pressures. Eos Trans Am Geophys Union 78:F774

Mao HK, Xu J, Bell PM (1986) Calibration of the ruby pressure gauge to 800 kbar under quasihydrostatic conditions. J Geophys Res 91:4673-4676

Mashimo T, Kondo KI, Sawaoka A, Syono Y, Takei H, Ahrens TJ (1980) Electrical conductivity measurements of fayalite under shock compression up to 56 GPa. J Geophys Res 85:1876-1881

Massey MJ, Baier U, Merlin R, Weber WH (1990a) Effects of pressure and isotopic substitution on the Raman spectrum of a-Fe_2O_3: identification of two-magnon scattering. Phys Rev B 41:7822-7827

Massey MJ, Chen NH, Allen JW, Merlin R (1990b) Pressure dependence of two-magnon Raman scattering in NiO. Phys Rev B 42:8776-8779

Massey MJ, Merlin R, Girvin SM (1992) Raman scattering in $FeBO_3$ at high pressures: phonon coupled to spin-pair fluctuations and magnetodeformation potentials. Phys Rev Lett 69:2299-2302

Mattheiss LF (1972) Electronic structure of the 3d transition-metal monoxides. I. Energy-band results. Phys Rev B 5:290-306

Mazin II, Fei Y, Downs R, Cohen RE (1998) Possible polytypism in FeO at high pressures. Am Mineral 83:451-457

McCammon CA (1995) Mössbauer spectroscopy of minerals. *In:* TJ Ahrens (ed) Mineral Physics and Crystallography: A Handbook of Physical Constants, AGU Reference Shelf 2, p 332-347 Am Geophys Union, Washington, DC

McCammon CA (1997) Perovskite as a possible sink for ferric iron in the lower mantle. Nature 387:694-696

McCammon CA (1998) The crystal chemistry of ferric iron in $Fe_{0.05}Mg_{0.95}SiO_3$ perovskite as determined by Mössbauer spectroscopy in the temperature range 80-293 K. Phys Chem Minerals 25:292-300

McQueen RG, Marsh SP (1966) Compressibility, elastic constants. *In:* SP Clark, Jr. (ed) Handbook of Physical Constants. Geol Soc Am Memoir 97:153

Mehl MJ, Cohen RE, Krakauer H (1988) Linearized augmented plane wave electronic structure calculations for MgO and CaO. J Geophys Res 93:8009-8022

Meyers SA, Cygan RT, Assink RA, Boslough MB (1998) ^{29}Si MAS NMR relaxation study of shocked Coconino sandstone from Meteor Crater, Arizona. Phys Chem Minerals 25:313-317

Mobilio S, Meneghini C (1998) Synchrotron radiation in the study of amorphous materials. J Non-Cryst Solids 232-234:25-37

Morris ER, Williams Q (1997) Electrical resistivity of Fe_3O_4 to 48 GPa: compression-induced electron hopping at mantle pressures. J Geophys Res 102:18139-18148

Mott NF (1990) Metal-Insulator Transitions. Taylor & Francis, New York

Nasu S (1996) High-pressure Mössbauer spectroscopy with nuclear forward scattering of synchrotron radiation. High Pressure Res 14:405-412

Nelmes RJ, McMahon MI, Belmonte SA, Allan DR, Gibbs MR, Parise JB (1998) Structures of FeS polymorphs at high pressures and temperatures. *In:* M Nakahara (ed) Review of High-Pressure Science and Technology 7:202-204 Japan Soceity for High-Pressure Science and Technology, Kyoto

Nguyen JH, Kruger MB, Jeanloz R (1997) Evidence for "partial" (sublattice) amorphization in $Co(OH)_2$. Phys Rev Lett 78:1836-1839

Nitsan U, Shankland TJ (1976) Optical properties of and electronic structure of mantle silicates. Geophys J R Astron Soc 45:59-87

Olsen JS, Cousins CSG, Gerward L, Jahns H (1991) A study of the crystal structure of Fe_2O_3 in the pressure range up to 65 GPa using synchrotron radiation. Phys Scripta 43:327-330

Ozima M (1998) Rare-gases in the mantle. Nature 393:304-305

Parise JB, Theroux B, Li R, Loveday JS, Marshall WG, Klotz S (1998) Pressure dependence of hydrogen bonding in metal deuteroxides: a neutron powder diffraction study of $Mn(OD)_2$ and β-$Co(OD)_2$. Phys Chem Minerals 25:130-137

Parker L, Badding JV (1996) Transition element-like chemistry for potassium under pressure. Science 273:95-97

Pasternak MP, Rozenberg GK, Milner AP, Amanowicz M, Zhou T, Schwarz U, Syassen K, Taylor RD, Hanfland M, Brister K (1997a) Pressure-induced concurrent transformation to an amorphous and crystalline phase in berlinite-type $FePO_4$. Phys Rev Lett 79:4409-4412

Pasternak MP, Taylor RD, Jeanloz R, Li X, Nguyen JH, McCammon CA (1997b) High pressure collapse of magnetism in $Fe_{0.94}O$: Mössbauer spectroscopy beyond 100 GPa. Phys Rev Lett 79:5046-5049

Pauling L (1960) The Nature of the Chemical Bond, 3rd Edition. Cornell University Press, Ithaca

Peyronneau J, Poirier JP (1989) Electrical conductivity of the Earth's lower mantle. Nature 342:537-539

Peyronneau J, Poirier JP (1998) Experimental determination of the electrical conductivity of the material of the Earth's lower mantle. *In:* M Manghnani, T Yagi (ed) Properties of Earth and Planetary Materials at High Pressure and Temperature p 77-87 American Geophysical Union, Washington DC

Phillips BL, Burnley PC, Worminghaus K, Navrotsky A (1997) 29Si and 1H NMR spectroscopy of high-pressure hydrous magnesium silicates. Phys Chem Minerals 24:179-190

Poe BT, Xu Y, Rubie DC (1998) Electrical conductivities of mantle minerals: in-situ high-pressure high-temperature complex impedance spectroscopy. *In:* M Nakahara (ed) Review of High-Pressure Science and Technology 7:22-24 Japan Society High-Pressure Science and Technology, Kyoto

Poirier JP (1991) Introduction to the Physics of the Earth's Interior. Cambridge University Press, New York

Raphael MP, Reeves ME, Skelton EF, Qadri SB, Kendziora C, Drews AR (1998) A new and improved method of measuring the pressure dependence of superconducting transition temperatures, $T_c(P)$. *In:* M Nakahara (ed) Review of High-Pressure Science and Technology 7:586-588 Japan Society High-Pressure Science Technology, Kyoto, Japan

Reichlin R, Brister K, McMahan AK, Ross M, Martin S, Vohra YK, Ruoff AL (1989) Evidence for the insulator-metal transition in xenon from optical, x-ray, and band-struture studies to 170 GPa. Phys Rev Lett 62:669-672

Reid AF, Ringwood AE (1969) High-pressure scandium oxide and its place in the molar volume relationships of dense structures of M_2X_3 and ABX_3 type. J Geophys Res:3238-3252

Reimann K (1996) Two- and three-photon spectroscopy of solids under high pressure. High Press Res 15:73-93

Reuff JP, Kao CC, Struzhkin VV, Badro J, Shu JF, Hemley RJ, Mao HK (1998) Pressure induced high-spin to low spin transition in FeS by x-ray emission spectroscopy. Phys Rev Lett:submitted

Reyes AP, Ahrens ET, Heffner RH, Hammel PC, Thompson JD (1992) Cuprous oxide manometer for high-pressure magnetic resonance experiments. Rev Sci Instrum 63:3120-3122

Richet P, Mao HK, Bell PM (1989) Static compression and equation of state of CaO to 1.35 Mbar. J Geophys Res 93:15279-15288

Rossman GR (1988) Optical spectroscopy. *In:* FC Hawthorne (ed) Spectroscopic Methods in Mineralogy and Geology. Rev Mineral 18:207-254

Ruocco G, Sette F, Bergmann U, Krisch M, Masciovecchio C, Mazzacurati V, Signorelli G, Verbeni R (1996) Equivalence of the sound velocity in water and ice at mesoscopic warelengths. Nature 379:521-523

Sampoli M, Ruocco G, Sette F (1997) Mixing of longitudinal and transverse dynamics in liquid water. Phys Rev Lett 79:1678-1681

Schmitt DR, Ahrens TJ (1983) Temperatures of shock-induced shear instabilities and their relationship to fusion curves. Geophys Res Lett 10:1077-1080

Schwoerer-Bohning M, Klotz S, Besson JM, Burkel E, Braden M, Pintschovius L (1996) The pressure dependence of the TA[110] phonon frequencies in the ordered alloy Fe_3Pt at pressures up to 7 GPa. Europhys Lett 33:679-682

Seto M, Yoda Y, Kikuta S, Zhang XW, Ando M (1995) Observation of nuclear resonant scattering accompanied by phonon excitation using synchrotron radiation. Phys Rev Lett 74:3828-3831

Shen G, Fei Y, Halenius U, Wang Y (1994) Optical absorption spectra of $(Mg,Fe)SiO_3$ silicate perovskites. Phys Chem Minerals 20:478-482

Sherman D, Jansen HJF (1995) First-principles prediction of the high-pressure phase transition and electronic structure of FeO: implications for the chemistry of the lower mantle and core. Geophys Res Lett 22:1001-1004

Shieh SR, Ming LC, Mao HK, Hemley RJ (1998) Decomposition of phase D in the lower mantle and the fate of dense hydrous silicates in subducting slabs. Earth Planet Sci Lett 159:13-23

Shirley EL (1998) Ab initio inclusion of electron-hole attraction: application to x-ray absorption and resonant inelastic x-ray scattering. Phys Rev Lett 80:794-797

Shu J, Mao HK, Hu J, Fei Y, Hemley RJ (1998) Single-crystal x-ray diffraction of wüstite to 30 GPa hydrostatic pressure. Neues Jahr Mineral Abh 172:309-323

Smith HG, Langer K (1982) Single crystal spectra of olivines in the range 40,000-5,000 cm^{-1} at pressures up to 200 kbar. Am Mineral 67:343-348

Stebbins JF (1988) NMR spectroscopy and dynamic processes in mineralogy and geochemistry. *In:* FC Hawthorne (ed) Spectroscopic Methods in Mineralogy and Geology. Rev Mineral 18:405-429

Stixrude L, Cohen RE (1993) Stability of orthorhombic $MgSiO_3$ perovskite in the lower mantle. Nature 364:613-616

Stixrude L, Cohen RE (1995) High pressure elasticity of iron and anisotropy of Earth's inner core. Science 267:1972-1975

Stixrude L, Cohen RE, Yu R, Krakauer H (1996) Prediction of phase transition in $CaSiO_3$ perovskite and implications for lower mantle structure. Am Mineral 81:1293-1296

Struzhkin VV, Goncharov AF, Hemley RJ, H. K. Mao, Moore SW, Graybeal JM, Sarrao J, Fisk Z (1998) Infrared and Raman magnon excitations $Sr_2CuCl_2O_2$ at high pressure. Phys Rev Lett:submitted

Struzhkin VV, Goncharov AF, Syassen K (1993a) Effect of pressure on magnetic excitations in CoO. Mat Sci Engineer A168:107-110

Struzhkin VV, Schwarz U, Wilhelm H, Syassen K (1993b) Effect of pressure on 2-magnon Raman scattering in K_2NiF_4. Mat Sci Engineer A168:103-106

Struzhkin VV, Timofeev YA, Hemley RJ, Mao HK (1997) Superconducting Tc and electron-phonon coupling in Nb to 132 GPa: Magnetic susceptibility at megabar pressures. Phys Rev Lett 79:4262-4265

Sturhahn W, Toellner TS, Alp EE, Zhang XW, Ando M, Yoda Y, Kikuta S, Seto M, Kimball C, Dabrowski B (1995) Phonon density of states measured by inelastic nuclear resonant scattering. Phys Rev Lett 74:3832-3835

Sur S, Cooney TS (1989) Electron paramagnetic resonance study of iron (III) and manganese (III) in the glassy and crystalline environments of synthetic fayalite and tephroite. Phys Chem Minerals 16:693-696

Surh MP, Louie SG, Cohen ML (1992) Band gaps in diamond under anisotropic stress. Phys Rev B 45:8239-8247

Suzuki T, Yagi T, Akimoto S, Ito A, Morimoto S, Syono Y (1985) X-ray diffraction and Mössbauer spectrum on the high pressure phase of Fe_2O_3. *In:* S Minomura (ed) Solid State Physics Under Pressure p 149-154 Terra Scientific, Tokyo

Syassen K, Sonnenschein (1982) Microoptical double beam system for reflectance and absorption measurements at high pressure. Rev Sci Instrum 53:644-650

Syono Y, Ito A, Morimoto S, Suzuki T, Yagi T, Akimoto S (1984) Mössbauer study on the high pressure phase of Fe_2O_3. Solid State Comm 50:97-100

Takano M, Nasu S, Abe T, Yamamoto K, Endo S, Takeda Y, Goodenough JB (1991) Pressure-induced high-spin to low spin transition in $CaFeO_3$. Phys Rev Lett 67:3267-3270

Timofeev YA (1992) Detection of superconductivity in high-pressure diamond anvil cell by magnetic susceptibility technique. Prib Tekh Eksper 5:186-189

Tissen VG, Ponyatovskii EG (1987) Behavior of Curie temperature of EuO at pressures up to 20 GPa. JETP Lett 46:287-289

Tyburczy JA, Fisler DK (1995) Electrical properties of minerals and melts. *In:* TJ Ahrens (ed) Mineral Physics and Crystallography: A Handbook of Physical Constants, AGU Reference Shelf 2 Am Geophys Union, Washington, DC

Ulug AM, Conradi MS, Norberg RE (1991) High pressure NMR: hydrogen at low temperatures. *In:* HD Hochheimer, RD Etters (ed) Frontiers of High Pressure Research p 131-141 Plenum, New York

van Veenendaal M, Carra P (1997) Excitons and resonant inelastic x-ray scattering in graphite. Phys Rev Lett 78:2839-2842

Vedam K, Schmidt EDD (1966) Variation of refractive index of MgO with pressure to 7 kbar. Phys Rev 146:548-554

Vos WL, Finger LW, Hemley RJ, Hu J-Z, Mao HK, Schouten JA (1992) High-pressure van der Waals compound in solid nitrogen-helium mixtures. Nature 358:46-48

Wang Y, Weidner DJ, Meng Y (1998) Advances in equation-of-state measurements in SAM-85. *In:* MH Manghnani, T Yagi (ed) Properties of the Earth and Planetary Materials at High Pressure and Temperature p 365-372 Am Geophys Union, Washington, DC

Weir S, Mitchell AC, Nellis WJ (1996) Metallization of fluid molecular hydrogen at 140 GPa (1.4 Mbar). Phys Rev Lett 76:1860-1863

Wigner E, Huntington HB (1935) On the possibility of a metallic modification of hydrogen. J Chem Phys 3:764-770

Williams Q, Knittle E, Reichlin R, Martin S, Jeanloz R (1990) Structural and electronic properties of Fe_2SiO_4-fayalite at ultrahigh pressures: amorphization and gap closure. J Geophys Res 95:21549-21563

Winkler B, Harris MJ, Eccleston RS, Knorr K (1997) Crystal field transitions in $Co_2[Al_4Si_5]O_{18}$ cordierite and $CoAl_2O_4$ spinel determined by neutron spectroscopy. Phys Chem Minerals 25:79-82

Yagi T, Mao HK, Bell PM (1979) Lattice parameters and specific volume for the perovskite phase of orthopyroxene. Carnegie Inst Washington Yearb 78:612-614

Zha CS, Mao HK, Hemley RJ, Duffy TS (1997) Elasticity measurement and equation of state of MgO to 60 GPa. Eos Trans Am Geophys Union 78:F752

Zhang L (1997) High pressure Mössbauer spectroscopy using synchrotron radiation on earth materials. *In:* Crystallography at High Pressure Using Synchrotron Radiation: The Next Steps. ESRF, Grenoble, France

Zhang L, Chopelas A (1994) Sound velocity of Al_2O_3 to 616 kbar. Phys Earth Planet Inter 87:77-83

Zhang L, Hafner SS (1992) High-pressure [57]Fe γ resonance and compressibility of $Ca(Fe,Mg)Si_2O_6$ clinopyroxenes. Am Mineral 77:462-473

Zhou T, Syassen K, Cardona M, Karpinski J, Kaldis E (1996) Electronic Raman scattering in $YBa_2Cu_4O_8$ at high pressure. Solid State Comm 99:669-673

Zou G, Mao HK, Bell PM, Virgo D (1980) High pressure experiments on the iron oxide wustite ($Fe_{1-x}O$). Carnegie Inst Wash Yearb 79:374-376

Chapter 19

THEORY OF MINERALS AT HIGH PRESSURE

Lars Stixrude

Department of Geological Sciences
425 E. University Ave.
University of Michigan
Ann Arbor, Michigan 48109

Ronald E. Cohen and Russell J. Hemley

Geophysical Laboratory and Center for High Pressure Research
Carnegie Institution of Washington
5251 Broad Branch Rd. NW
Washington, D.C. 20015

INTRODUCTION

Recent studies of minerals at very high pressures are yielding numerous surprises that present a considerable challenge to our understanding of mineral structure and bonding. In the context of earth sciences, the exploration of high pressure has opened up mineralogy to the investigation of the bulk of our planet that lies beneath the surface. The earth's interior is a unique environment in which the behavior of minerals often overturns our textbook intuition. Primarily because in situ experimental studies of minerals over the entire range of conditions encountered within the Earth are so new and just now being explored, the nature and evolution of the earth's interior is poorly understood, and difficult to predict with current theoretical understanding. In contrast, we have a better understanding in many ways of the interiors of distant stars. For example, we are able to calculate the structures and evolutionary history of stars with some certainty, an exercise that is not yet possible for the earth. The reason is that, because the pressure is so high, the electrons obey an almost trivial limiting behavior, the uniform electron gas (Ichimaru 1982). The underlying physics is that the kinetic energy of electrons increases with the charge density, ρ, as $\rho^{2/3}$ while the potential energy binding the electrons to the nuclei increases only as $\rho^{1/3}$; the kinetic energy dominates at high pressure and the electrons become unbound (see Bukowinski 1994 for an extended discussion).

The contrast with planetary interiors can be illustrated by expressing the pressure in terms of atomic units, one atomic unit (29.4 TPa) being comparable to the pressure required for complete ionization and the formation of a degenerate electron gas. The structure of planets are such that pressures are much less than unity. The behavior of planetary materials will be far from plasma-like, and therefore much more complex; the pressure-temperature domain is such that materials are best described as condensed matter. Using other language, although chemical effects are still dominant (i.e. properties controlled by the orbital structure of atoms), strong perturbations of these properties can introduce new and unsuspected complexity. A more useful pressure scale in the context of planetary interiors is formed from an energy typical of the spacing between electronic bands (1 eV) and a volume typically occupied by a valence electron in a mineral (20 Bohr^3 $\approx$ 3 Å^3). This pressure scale (~50 GPa) is characteristic of the earth's interior and of the bulk modulus of typical earth-forming constituents. On quite general grounds we must expect to find in planetary interiors not only significant compression and phase transitions, but also

0275-0279/98/0037-0019$05.00

electronic transitions (e.g. insulator to metal) and substantial changes in the mechanisms of bonding, all of which complicate our picture of planetary structure and evolution.

The recognition that minerals at high pressure are characterized by complex multi-phase behavior places tremendous demands on the required accuracy of theoretical methods. They must be general, applicable to essentially all classes of elements, and must not make any assumptions regarding the nature of bonding. This generality rules out most semi-empirical approaches that are based on our more traditional notions of mineral behavior, such as the otherwise powerful ionic model. Moreover, energies and volumes must be accurate to well within typical heats and volumes of solid-solid transformations, ruling out essentially all weak screening approaches that treat condensed matter in perturbative fashion beginning with the free electron gas. Indeed, early calculations using one such approach (Jensen 1938) incorrectly predicted that iron is substantially less dense than the earth's core (Birch 1952). On the other hand, the goal of theory is not only to provide accurate quantitative predictions of physical properties that can be measured experimentally (at least in principle); theory should provide important physical insight and understanding into phenomena that may not be apparent from measurements or from large-scale computations. Thus, the simpler approaches play an important role in providing this insight.

Over the past decade or so, theoretical methods have been developed that have sufficient generality and power to tackle the complex behavior we expect to find in high-pressure mineralogy. These first-principles methods are relatively new to the earth sciences literature and have not been widely reviewed. Moreover, unlike more venerable semi-empirical or ab initio models, they are based on a microscopic view of minerals that differs radically from our traditional intuition. This review explores some of the theoretical methods that have been applied in high-pressure mineralogy, but focuses on first-principles methods because of their relative historical and conceptual novelty. The following sections review in some detail modern first-principles methods based on density functional theory as they have been applied in the earth science literature, and the fundamental approximations upon which these methods are based. First-principles methods are contrasted with those ab initio and semi-empirical models that have also played an important role. We then discuss the derivation of observable quantities from density functional theory, and briefly review computational methods for solving the equations. Finally, we discuss applications of theory to understanding and predicting the behavior of minerals deep within the Earth, and explore some important unsolved problems and possible future directions.

THEORY

A wide range of theoretical methods have appeared in the earth sciences literature, and many have been applied to understanding the behavior of deep Earth materials. These methods differ vastly in the level of physics included and as a result in the quality and security of their predictions. Two extremes of the theoretical spectrum—semi-empirical to first principles—reflect two superficially different views of the microscopic world. The empirically-based ionic model, originally developed by Goldschmidt and Pauling (1960), forms the foundation of much of our understanding of mineral behavior. It continues to play an important conceptual role and forms the basis of most modern semi-empirical atomistic approaches. However, our traditional mineralogical intuition often fails us in the very high-pressure environment. The marked changes in electronic properties of materials that can occur under these extreme conditions require new approaches. First-principles methods view minerals in a very different way which, being more closely tied to the fundamental physics, permits understanding and prediction of unique high pressure behavior. This review focuses primarily on the first-principles approach as embodied in density functional theory by exploring its properties and capabilities and by contrasting it with semi-empirical and ab initio models that are more closely tied to our traditional views of mineral structure and bonding.

From the point of view of any first principles theory, solids are composed of nuclei and electrons; atoms and ions are constructs that play no primary role. This departure from our usual way of thinking about minerals is essential and has the following important consequences. We may expect our theory to be equally applicable to the entire range of conditions encountered in planets (and even stars), the entire range of bonding environments entailed by this enormous range of pressures and temperatures, and to all elements of the periodic table.

To illustrate this way of thinking about solids and to introduce some important concepts, we consider first the properties of the simplest system, the uniform electron gas with embedded nuclei. The total energy consists of the kinetic energy of the electrons, and three distinct contributions to the potential energy: (1) Coulomb interactions among nuclei and electrons (2) electron exchange and (3) electron correlation. The first contribution is straightforward and involves only sums over point charges and integrals over the electronic charge density. The latter two are corrections to the independent electron approximation which is invoked to render the full N-body problem (in which only Coulomb interactions appear) tractable.

Exchange and correlation account for local deviations from uniform charge that arise from the tendency of electrons to avoid each other. Correlation accounts for the mutual Coulomb repulsion, whereas exchange embodies the Pauli exclusion principle and the resulting tendency of electrons of the same spin to avoid each other. The net effect is that each electron can be thought of as digging a hole of reduced charge density about itself. Certain properties of the exchange-correlation hole are well understood; it is known for instance that its integrated charge must exactly balance that of the electron. Exchange and correlation reduce the total energy by reducing the Coulomb repulsion between electrons.

The total energy of our simple system is readily evaluated as a function of charge density; the equation of state then follows from differentiation. Assuming that the nuclei are in a close-packed arrangement, and including only the leading order high density contributions to exchange and correlation, the equation of state is (Hubbard 1984)

$$P = 0.176 r_s^5 \left[1 - \left(0.407 Z^{2/3} + 0.207 \right) r_s \right] \tag{1}$$

where P is the static (athermal) pressure, Z the nuclear charge, and the Wigner-Seitz radius

$$r_s = \left(\frac{3}{4\pi\rho} \right)^{1/3} \tag{2}$$

is a measure of the average spacing between electrons. The first term in Equation (1) is the kinetic contribution, the second due to the Coulomb attraction of the nuclei for the electrons and mutual repulsion of the electrons, and the third to exchange. Correlation, which is smaller than exchange at high density, has been neglected as has the mutual Coulomb repulsion of the nuclei.

Comparisons with the structure of planetary interiors reveal some fundamentally important aspects of planetary matter (Fig. 1). First, the net Coulomb attraction of the electrons by the nuclei plays an essential role at planetary densities—different mean nuclear charges account to first order for the difference in mean charge (and mass) density between Jupiter and earth. Second, screening has a first order effect on the equation of state,

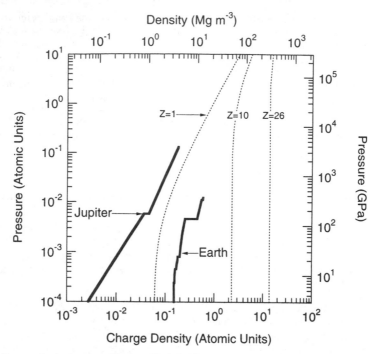

Figure 1. Pressure in the interior of Jupiter (Chabrier 1992) and earth (Dziewonski and Anderson 1981) as a function of mass density (top) and charge density (bottom). The charge density has been calculated from the observed mass density by assuming that the number of electrons is one-half the number of nucleons. Planetary structures are compared with limiting high density equations of state (Eqn. 1) for three values of the atomic number, Z.

accounting for the much lower densities of planets at a given pressure than predicted by Equation (1). In planetary matter, the charge density is substantially enhanced in the vicinity of the nucleus, reducing the ability of the point charges to attract the remaining (valence) electrons. Screening is weaker in the case of Jupiter because it contains dominantly lighter elements, and because the pressures are much greater. Nevertheless, for all the planets screening is so strong that it must be accounted for. The major part of the screening is from the tightly bound, essentially rigid core electrons. In the case of the terrestrial planets, the charge density near the nuclei is so much higher than in the interstitial region that this difference plays a central role in the design of modern computational methods.

First-principles level

We turn now from simple to real systems and at the same time from analytically expressible results to necessarily elaborate computations. Though the electronic structure will be non-trivial, we retain the charge density as a central concept. This is appealing because this quantity is experimentally observable; it is precisely what is measured by an x-ray diffraction experiment. While we will focus on density functional theory here, there are other first-principles methods that have been important in the earth science literature; among these is the periodic Hartree-Fock method (Dovesi et al. 1987). What all first-principles methods share is an approach that seeks to minimize approximations to the bare minimum. Some approximation is necessary since we are as yet incapable of solving the Schrödinger equation exactly for any mineralogical system.

Density functional theory. The general problem we are faced with in a non-uniform, non-degenerate electron system is formidable. Given a periodic potential set by the positions of the nuclei, we must solve the Schrödinger equation for the total wavefunction, $\Psi(r_1, r_2, ..., r_N)$ of a system of N interacting electrons, where N is on the order of Avagadro's number. Density functional theory (Hohenberg and Kohn 1964, Kohn and Sham 1965) is a powerful and in principle exact method of dealing with this problem in a tractable way (see Lundqvist and March 1987 for reviews).

The essence of this theory is the proof that the ground state properties of a material are a unique functional of the charge density $\rho(\vec{r})$. Among these properties are the ground state total energy

$$E = T + U[\rho(\vec{r})] + E_{xc}[\rho(\vec{r})] \tag{3}$$

and its derivatives (pressure, elastic constants, etc.) where T is the kinetic energy of a system of non-interacting electrons with the same charge density as the interacting system, U is the electrostatic (Coulomb) energy, including the electrostatic interaction between the nuclei, and E_{xc} is the exchange-correlation energy. A variational principle leads to a set of single-particle, Schrödinger- like, Kohn-Sham equations

$$\left[-\nabla^2 + V_{KS}\right]\psi_i = \varepsilon_i\psi_i \tag{4}$$

where ψ_i is now the wave function of a single electron, ε_i the corresponding eigenvalue and the effective potential

$$V_{KS}[\rho(\vec{r})] = \sum_{i=1}^{N}\frac{2Z_i}{|\vec{r} - \vec{R}_i|} + \int\frac{2\rho(\vec{r}')}{|\vec{r} - \vec{r}'|}d\vec{r}' + V_{xc}[\rho(\vec{r})] \tag{5}$$

where the first two terms are Coulomb potentials due to the nuclei and the other electrons, respectively, the last is the exchange-correlation potential, and the units are Rydberg atomic units: $\hbar^2/2m = 1$, $e^2 = 2$, energy in Ry, and length in Bohr.

The power of density functional theory is that it allows one to calculate, in principle, the exact many-body total energy of a system from a set of single-particle equations. The solution to the Kohn-Sham equations is that of the set of coupled generalized eigenvalue equations

$$H_{ij}(\vec{k})\psi_j(\vec{r}, \vec{k}) = \varepsilon_i(\vec{k})O_{ij}(\vec{k})\psi_j(\vec{r}, \vec{k}) \tag{6}$$

$$H_{ij}(\vec{k}) = \int\psi_i^*(\vec{r}, \vec{k})\left(-\nabla^2 + V_{KS}\right)\psi_j(\vec{r}, \vec{k})d\vec{r} \tag{7}$$

$$O_{ij}(\vec{k}) = \int\psi_i^*(\vec{r}, \vec{k})\psi_j(\vec{r}, \vec{k})d\vec{r} \tag{8}$$

where **H** and **O** are the Hamiltonian and overlap matrices, respectively and $\vec{k}$ is a vector in reciprocal space. Because the Kohn-Sham potential is a functional of the charge density, the equations must be solved self-consistently together with the definition of the charge density in terms of the wavefunctions

$$\rho(\vec{r}) = \int\sum_i n(E_F - \varepsilon_i)\psi_i^*(\vec{r}, \vec{k})\psi_i(\vec{r}, \vec{k})d\vec{k} \tag{9}$$

where n is the occupation number, and E_F is the Fermi energy.

Approximations. The Kohn-Sham equations are exact. That only approximate solutions have been possible to date is a limitation imposed only by our current ignorance of the exact exchange-correlation functional. If the exact exchange-correlation functional were known, we would be able to obtain exact solutions. All other terms in the Kohn-Sham equations are straightforward and readily evaluated. In addition to the essential approximation to the exchange-correlation potential, two other approximations are commonly invoked in some first principles calculations: the frozen-core approximation and the pseudopotential approximation. These three approximations are now discussed in detail:

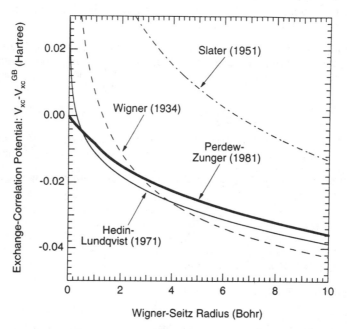

Figure 2. The difference between the exchange-correlation potential and its high-density limit (Eqns. 10,11) in (bold line) the local density approximation (LDA), and (other lines) three commonly used approximations to the LDA (Hedin and Lundqvist 1971, Slater 1951, Wigner 1934). For the Slater result, we use $\alpha = 2/3$ which yields the pure exchange potential.

The exchange-correlation potential. The exchange-correlation functional is known precisely only for simple systems such as the uniform electron gas (Fig. 2). The exchange portion is known analytically, as are the leading order contributions to correlation, in the limit of high density (Gell-Mann and Brueckner 1957)

$$V_{xc} = \frac{\partial}{\partial \rho} \left[\rho E_{xc} \right]$$ (10)

$$E_{xc} = -\frac{3}{4\pi} \left(\frac{9\pi}{4} \right)^{1/3} r_s^{-1} + A \ln r_s + B$$ (11)

where E_{xc} is the exchange-correlation energy, the first contribution to E_{xc} is exchange, and the constants, $A = (1-\ln 2)/\pi^2$ and $B = -0.046644$ (Perdew et al. 1996). At other densities, accurate values of the exchange-correlation potential are known from quantum Monte Carlo calculations (Ceperley and Alder 1980), which have been represented in a parametric form that obeys the high density limiting behavior (Eqns. 10, 11) (Perdew and Zunger 1981).

The precision of modern condensed-matter computations have made the accurate representation of the exchange-correlation potential of the uniform electron gas an important issue. In this context, one must be aware that approximate representations of V_{xc} have appeared frequently in the geophysical literature and are still in use. Of these, the Hedin-Lundqvist (1971) expression is most similar to the accurate Perdew-Zunger parameterization, that of Wigner (1934) the least. The Wigner approximation shows a much stronger dependence on density than the accurate potential, and leads to significant errors in density functional computations for solids. None of the commonly used approximate expressions satisfy the correct high-density limiting behavior (Eqns. 10, 11).

The charge density in real materials is highly non-uniform, and the exchange-correlation potential cannot be evaluated. Fortunately, simple approximations to the exchange-correlation potential have been very successful. The Local Density Approximation (LDA) is based on the uniform electron gas, taking into account nonuniformity to lowest order by setting V_{xc} at every point in the crystal to that of the uniform electron gas with a density equal to the local charge density (Lundqvist and March 1987).

The success of the LDA can be understood at a fundamental level in terms of the satisfaction of exact sum rules for the exchange-correlation hole (Gunnarsson and Lundqvist 1976). For example, the LDA correctly predicts an exchange-correlation hole of unit charge. Ultimately, the appropriateness of the LDA can be judged only by comparing its predictions to observation. Here, the LDA has been remarkably successful. The LDA has been shown to yield excellent agreement with experiment for a wide variety of insulators, metals, and semiconductors, for bulk, surface, and defect properties. The LDA also shows some important flaws; for example, it fails to predict the correct ground state of iron. It also shows failures in the transition metal oxides, predicting metallic ground states where insulating ground states are observed.

These failures have prompted the development of new exchange-correlation functionals. One shortcoming of the LDA may be its local character, that is, its inability to distinguish between electrons of different angular momenta or energy. Generalized gradient approximations (GGA) partially remedy this by including a dependence on local charge density gradients in addition to the density itself. Some care must be taken in constructing gradient approximations; a straightforward Taylor series expansion in the charge density gradient about the LDA result fails completely since it violates the sum rule for the exchange-correlation hole. The most widely used generalized gradient approximation satisfies the sum rules exactly (Perdew et al. 1996). This approximation and its forerunners have been shown to yield agreement with experimental data that is usually as good as LDA and often substantially better (Perdew et al. 1992). For example, GGA's correctly predict the bcc phase as the ground state of iron (Bagno et al. 1989). GGA is not a panacea however: for some materials, LDA results are in better agreement with experiment, e.g. 5d transition metals (Körling and Häglund 1992). Moreover, GGA does not alleviate the failure of LDA in the case of the transition metal oxides. The relationship of the GGA to the LDA can be expressed in terms of the enhancement factor

$$F(r_s, s) = \frac{V_{xc}^{GGA}(r_s, s)}{V_x(r_s)} \tag{12}$$

where V_x is the exchange potential and F is a function of the charge density and the non-dimensional charge density gradient $s = 24 / \pi^2 |\nabla \rho / \rho^{4/3}|$ (Fig. 3).

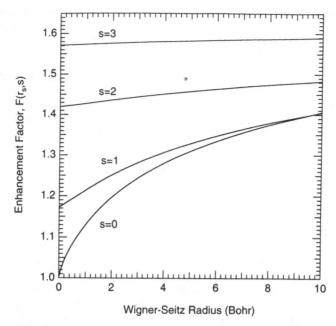

Figure 3. The effect of charge-density gradients on the exchange-correlation potential according to the generalized gradient approximation (GGA). The enhancement factor for zero gradient ($s = 0$) reflects the contribution of correlation to V_{xc}.

The frozen core approximation. The physical motivation for this approximation is the observation that only the valence electrons participate in bonding and in the response of the crystal to most perturbations of interest. Unless the perturbation is of very high energy (comparable to the binding energy of the core states), the tightly bound core states remain essentially unchanged. The frozen core approximation is satisfied to a high degree of accuracy for many applications, for example in the case of finite strains of magnitudes typically encountered in the earth's interior.

Within this approximation, the charge density of the core electrons is just that of the free atom, which can be found readily. We then need solve only for the valence electrons in Equation (4), often a considerable computational advantage. An important technical point is that, although in many cases the choice is obvious, there is no fundamentally sound way to decide *a priori* which electrons are core and which valence. Some care is required; for example, the 3p electrons in iron must be treated as valence electrons as they are found to deform substantially at pressures comparable to those in the earth's core (Stixrude et al. 1994).

The pseudopotential approximation. This approximation goes one step beyond the frozen core. It replaces the nucleus and the core electrons with a simpler object, the pseudopotential, that has the same scattering properties (Pickett 1989). The pseudopotential is chosen such that the valence wave function in the free atom is the same as the all electron solution beyond some cutoff radius, but nodeless within this radius. The advantages of the pseudopotential method are (1) spatial variations in the pseudopotential are much less rapid than the bare Coulomb potential of the nucleus and (2) one need solve only for the (pseudo-) wavefunctions of the valence electrons which show much less rapid spatial variation than the core electrons, or the valence electrons in the core region. This means that in the solution of the Kohn-Sham equations, potential and charge density can be represented by a particularly simple, complete and orthogonal set of basis functions (plane-waves) of manageable size. With this basis set, evaluation of total energies, stresses, and forces acting on the atoms is particularly efficient.

The pseudopotential is an approximation to the potential that the valence electrons "see" and its construction is non-unique; different pseudopotentials may yield significantly different predictions of bulk properties. Several different methods for constructing pseudopotentials have been developed (Lin et al. 1993, Troullier and Martins 1991, Vanderbilt 1990). Care must be taken to demonstrate the transferability of the pseudo-potentials generated by a particular method and to compare with all electron calculations where these are available. When these conditions are met, the error due to the pseudo-potential is generally small (few percent in volume for earth materials).

Ab initio models

Whereas first-principles methods seek to reduce approximations to a bare minimum, ab initio models construct an approximate treatment of some aspects of the relevant physics, such as the charge density or of the interactions between orbitals. The cost of additional approximation is often outweighed by the increase in computational simplicity and efficiency. For example, ab initio models have been widely used to explore transport properties or the properties of liquids which are very difficult (costly) to examine with fully first-principles approaches. Moreover, these models often yield insight that is sometimes difficult to extract from more complex and elaborate first principles calculations.

Gordon-Kim type approaches. This class of approaches, first introduced by Gordon and Kim (1972) bridges the gap between traditional ionic models of minerals and density functional theory. While based on the idea that materials are composed of closed-shell atoms or ions, it shows the power of density functional theory and the local density approximation even when these are not used self-consistently. Instead of solving for the charge density self-consistently with the potential, the total charge density is modeled by overlapping atomic or ionic charge densities and then the total energy is computed for that charge density using the LDA. This approximation leads to much faster computations because the charge density of isolated atoms or ions is easily calculated and self-consistency is not enforced. The method is less accurate than the self-consistent solution to the Kohn-Sham equations, even if the model density is good, because the local density form for the kinetic energy is not accurate enough in many cases (for example it does not give the proper shell structure for atoms). The Kohn-Sham approach does not make this approximation to the kinetic energy even in the LDA and the kinetic energy derives from the occupied orbitals. In subsequent work, the kinetic and correlation interactions were modified (leading to the term modified electron gas, or MEG) to give better results for atoms (Cohen and Gordon 1975, 1976). The MEG model generally gives equations of state that are too stiff for the rare gas solids, but is quite successful considering the simplicity of the model. These shortcomings of the simple Gordon-Kim model can be partially overcome in the case of rare-gas solids by allowing the atomic charge densities to respond to the embedding crystal potential, so that the atom is compressed with increasing pressure (LeSar 1988). However, the principal error arises from the simplicity of the principal interactions, particularly the Thomas-Fermi kinetic energy functional (scaled in the case of MEG) as opposed to the Kohn-Sham approach.

Care must be taken in the treatment of oxides within this approach because the O^{2-} ion is unstable in the free state. This difficulty is overcome by surrounding the ion with a Watson sphere which mimics the embedding crystal potential that stabilizes the ion in the crystal (Watson 1958). Thus O^{2-} is surrounded by a sphere of charge +2: when an electron moves far from the atom it sees an object of positive charge +1 behind it, the electron is bound and the configuration remains stable. The remaining question is how to choose the radius of the sphere. The original Gordon-Kim calculations of oxides used rigid ion potentials in which the ion was stabilized with a sphere whose radius was chosen so that the electrostatic potential in the sphere was equal to the Madelung potential at the site at a given volume (Muhlhausen and Gordon 1981). In subsequent calculations, an important contribution from the self-energy of the ion was included; this is crucial for predicting volume (i.e. pressure) dependent properties (e.g. Hemley et al. 1985, 1987). This approach was used to study a number of deep mantle minerals at high pressures.

The Potential Induced Breathing (PIB) model represented an extension of this approach (Boyer et al. 1985). It was also introduced to improve the accuracy of elasticity calculations, including the correct prediction of the deviation from the Cauchy conditions (accurate values for the shear and off-diagonal elastic constants, which for rigid ion potentials are equal; e.g. $c_{12}=c_{44}$ for cubic). In this model, the Watson-sphere radius is given by the Madelung potential as atoms are displaced or the lattice strained, $R_{wat}=Z_{wat}/P_{wat}$, giving a non-rigid ion, many-body, potential. The successes of this approach led to the development of the lattice dynamics of the PIB model (Cohen et al. 1987). Reasonable dispersion curves were obtained for the alkaline earth oxides. In the PIB model, the Watson-sphere radii are given by the Madelung potential, but the Madelung potential is not well behaved in the long-wave limit. A better procedure, though somewhat slower, is to optimize the total energy with respect to Watson sphere radii rather than to choose the radius using the Madelung potential (Wolf and Bukowinski 1988). This gives a Watson sphere radius close to that of PIB at zero pressure, but it changes more rapidly with compression than PIB due to the compression of the atom by short-range forces, in addition to the electrostatic crystal field. This model is known as the VIB, or variationally induced breathing, model. For example, anomalous behavior shown by the PIB model is absent in the VIB model. In the VIB model, the LO-TO splitting is the same as given by a rigid ion model, since all atomic deformations are spherical. There is no dipolar charge relaxation. In spite of the absence of atomic polarizability, the VIB model is very accurate and gives results that compare quite well with self-consistent results and experiment. Ab initio models have undergone further development by including the crystal potential in the atomic calculation, and a self-consistency cycle between the atomic densities and the crystal potential (Edwardson 1989, LeSar 1983). In the Self-Consistent Charge Deformation model (SCAD) (Boyer et al. 1997, Stokes et al. 1996) atomic densities are computed in the crystal potential, and states are occupied in order of energy, allowing charge flow between the atoms. The inclusion of non-spherical charge deformations has increased the accuracy of the models, but at the cost of increased complexity.

Tight binding. In its simplest parametric form, as originally formulated by Slater (1954) and extensively illustrated by Harrison (1989), the tight-binding method differs from those described so far in that the charge density does not appear explicitly. In this form of the method, the wavefunctions are constructed from basis functions consisting of atomic-like orbitals. For basis functions $\phi_{i\alpha}(\vec{r} - \vec{R}_i)$, where α labels the type of orbital (e.g. $s, p, d, ...$), and i labels the atom, the Hamiltonian and overlap matrices consist of elements

$$H_{i\alpha j\beta}(\vec{k}) = \sum_{l=0}^{\infty} exp[i\vec{k} \cdot \vec{R}_{ij}(l)]S_{\alpha\beta}[\hat{R}_{ij}(l)]h_{\alpha\beta}[R_{ij}(l)] \qquad (13)$$

$$O_{i\alpha j\beta}(\vec{k}) = \sum_{l=0}^{\infty} exp[i\vec{k} \cdot \vec{R}_{ij}(l)]S_{\alpha\beta}[\hat{R}_{ij}(l)]o_{\alpha\beta}[R_{ij}(l)] \qquad (14)$$

where $R_{ij}(l)$ is the distance between the i-th atom in the reference unit cell (labeled $l = 0$) and the j-th atom in the l-th unit cell, the $S_{\alpha\beta}$ are functions of direction only and, in the two-center approximation, the $h_{\alpha\beta}$ and $o_{\alpha\beta}$ are functions only of internuclear distance. Indices i and j run over all atoms in the unit cell, and l runs over all unit cells. Under the assumption that the basis set consists of functions with the symmetry of $s, p, d, ...$ atomic orbitals, the functions $S_{\alpha\beta}$ can be written in terms of spherical harmonics. The distance dependent functions, $h_{\alpha\beta}$ and $o_{\alpha\beta}$, are taken to be parametric functions of distance, with parameters chosen such that first principles results are reproduced. In this way, all explicit reference to the wavefunctions or charge density is eliminated. This simplifies the calculations tremendously, but renders the calculation non-self-consistent.

The non self-consistency of the tight binding approach has an important consequence which has not been widely recognized (Cohen et al. 1994). In general, the total energy can be written

$$E = \int \sum_i \varepsilon_i(\vec{k})d\vec{k} + F[\rho(\vec{r})] \qquad (15)$$

where the first term is a sum over the self-consistent eigenvalues, and the second term, a functional of the charge density, contains all non-band structure contributions to the energy. In non self-consistent calculations, the band structure now contains an arbitrary zero which must be fixed in order to calculate the total energy. The arbitrariness of the energy zero in the tight binding method can be exploited to recast the total energy as

$$E = \int \sum_i \varepsilon_i'(\vec{k})d\vec{k} \qquad (16)$$

where the new eigenvalues are shifted in energy such that $\varepsilon_i' = \varepsilon_i - F[\rho(\vec{r})]$. With this formulation, the total energy is given simply as a sum over the bands, eliminating the need for pair potential repulsive terms which are often included in other treatments. The parameters of the tight binding model are determined by fitting to accurate LAPW band structures and total energies. This approach has been very successful in describing the properties of a wide variety of monatomic systems such as iron, silicon, and xenon (Cohen et al. 1997b). The approach can be generalized to multicomponent systems including silicates but this has not yet been accomplished.

Semi-empirical atomistic models

The primary advantage of these highly approximate methods is that they are computationally fast, allowing one to examine much larger systems, or more complex physics than one could otherwise. In most cases they revert to our more traditional view of solids as being composed of ions rather than nuclei and electrons; the latter generally do not appear explicitly. This strictly precludes these methods from the study of systems in which electronic effects are important, such as those in which electronic (e.g. insulator-metal) transitions occur or in which magnetism is relevant. The atomistic picture also limits the transferability of semi-empirical models to a more or less narrow range of compounds or structures over which the bonding does not change substantially. Within the range of transferability, semi-empirical potentials may be made to fit experimental data with some accuracy and can lead to useful predictions of material behavior under conditions that have not yet been accessed experimentally, of previously unobserved behavior, or of similar materials.

Many of these models can be cast in the following form

$$E = \sum_{i<j} V_2(r_{ij}) + \sum_{i<j<k} V_3(r_{ij}, r_{ik}, r_{jk}) + \dots \qquad (17)$$

where E is the total energy of an atomic configuration, the sums are over the atoms in the system, and V_2 and V_3 are two-body (pair) and three-body potentials respectively. Higher order terms can be included although the sum may not converge rapidly for many systems. This approach is appealing in the case of ionic materials such as oxides and silicates because the largest part of the energy is the Madelung term, a sum of pair-wise interactions. The simplest semi-empirical model of an ionic solid consists of the Madelung term and pair-wise short-range repulsive forces. This widely studied model was originally applied to simple systems such as alkali halides but has more recently been used to investigate the behavior of oxides and silicates (Burnham 1990). However, short-range attractive and repulsive interactions, though energetically secondary, can strongly influence the structure and other properties of the system. These forces are often substantially more complex than the Coulomb potential and may require elaborate and non-unique functional forms and/or three-body or even higher-order terms for accurate representation.

DERIVATION OF OBSERVABLES

Total energy and band structure

For a given arrangement of nuclei (crystal structure) we may use any of the above methods to determine the total energy. First-principles and ab initio methods also yield the charge density, and the quasi-particle eigenvalue spectrum (electronic band structure). By examining the dependence of the total energy on perturbations to the volume V or shape of the crystal (described by the deviatoric strain tensor ε_{ij}) or to the positions of the atoms, the Helmholtz free energy F as a function of V, ε_{ij}, and temperature T can in principle be deduced. For example, the static pressure and the equation of state are simply given by the variation of the total energy with volume.

One may determine the elastic constants from total energy calculations. For small deviatoric strains under hydrostatic stress (Wallace 1972)

$$F(V, \varepsilon'_{ij}, T) = F_0(V) + F_{TH}(V, T) + \frac{1}{2} c_{ijkl}(V, T)\varepsilon'_{ij}\varepsilon'_{kl} \qquad (18)$$

where F_0 is the static (zero temperature) contribution, F_{TH} is due to the thermal excitation of electrons and phonons, and c_{ijkl} is the elastic constant tensor. This equation shows that combinations of elastic constants are related to the difference in total energy between a strained and unstrained lattice.

It is possible in principle to calculate thermal contributions to the thermodynamic and thermoelastic properties of crystals. Calculating thermal properties is much more difficult than calculating static properties, however. The reason is simple: the atomic vibrations induced by finite temperature break the symmetry of the crystal so that it is now periodic only in a time averaged sense. In the context of total energy calculations, our task is then to evaluate the partition function, an integral over all atomic configurations realized by a crystal at high temperature. While this is not difficult with semi-empirical or ab initio models, it is still essentially impossible with first-principles methods for most systems. More efficient ways of evaluating thermal free energies from first principles are required. Some future directions are indicated in the penultimate section.

Forces, stresses, and structures

The Hellman-Feynman theorem allows one to calculate first derivatives of the total energy directly in terms of the ground state wavefunctions. The application of this theorem allows one to determine the forces acting on every atom and the stresses acting on the lattice.

This is important for two related reasons. First, it allows one to determine ground state crystal structures very effectively. In first principles calculations, this has become possible only recently for relatively complex structures such as $MgSiO_3$ perovskite (Wentzcovitch et al. 1993). The key innovation has been the development of a structural optimization strategy based on a pseudo-Lagrangian that treats the components of the strain tensor and the atomic positions as dynamical variables (Wentzcovitch 1991). The optimization is performed at constant pressure. At each step of the dynamical trajectory, the Hellman-Feynman forces and stresses (Nielsen and Martin 1985) acting, respectively, on the nuclei and lattice parameters are evaluated and used to generate the next configuration. The optimization is complete when the forces on the nuclei vanish and the stress is hydrostatic and balances the applied pressure.

Second, once the ground state structure at a given pressure is determined, one can calculate the static elastic constants. This is done in a straightforward way by applying a deviatoric strain to the lattice and calculating the resulting stress tensor. The elastic constant, c_{ijkl} is then given by the ratio of stress σ_{ij} to strain ε_{ij}

$$\sigma_{ij} = c_{ijkl}\varepsilon_{kl} \tag{19}$$

Care must be taken to re-optimize the positions of the atoms in each strained configuration since vibrational modes typically couple with lattice strains in silicate structures. In addition, the elastic moduli under pressure (so-called effective elastic constants) may be defined in several ways; and attention must be given to the correct definition.

Linear response

This approach goes one step beyond the Hellman-Feynman theorem by computing changes in the total energy to second order. This is accomplished by computing the first order changes in the charge density in response to generalized perturbations. These perturbations may include the displacement of an atom, or the application of an electric field. The wavelength of the perturbation need not be commensurate with the unit cell, allowing one to investigate, for instance, phonon modes without resorting to supercells, a tremendous computational and conceptual advantage (Baroni et al. 1987).

In the context of mineralogy, the advantage of this approach is that properties that are related to second-derivatives of the total energy can be computed directly. Examples include elements of the dynamical matrix, the dielectric constant tensor and the Born effective charge tensor. With these quantities, the full phonon spectrum can be determined throughout the Brillouin zone (Lee and Gonze 1995, Stixrude et al. 1996). In addition to making contact with experimental observation of zone-center vibrational frequencies, these predictions allow one to investigate phase stability, and, to the extent that thermal properties are quasiharmonic, high temperature properties. The computation in polar substances is subtle and necessarily involves not only the calculation of force constants, but also that of the dielectric constant and Born effective charge tensors so that coupling to the macroscopic field at zone-center is properly accounted for (LO-TO splitting).

COMPUTATION

Methods

First-principles methods based on density functional theory solve Equations (6-8) by expanding the wavefunctions in a basis

$$\psi_i(\vec{r},\vec{k}) = \sum_{j=1}^{N} c_{ij}\phi_j(\vec{r},\vec{k}) \tag{20}$$

where N is the number of basis functions, ϕ_j and c_{ij} are the coefficients to be determined by solution of the Kohn-Sham equations.

The LAPW method is the current state-of-the-art in density functional theory computations. It makes no essential approximations beyond that to the exchange-correlation functional, allowing one to routinely solve for all electrons, core and valence. For example, it makes no approximations to the shape of the charge density or potential. The accurate representation of the potential and the core states means that the LAPW method is equally applicable to all elements of the periodic table, and over the entire range of densities of interest in planetary or astrophysical studies.

The LAPW differs from its forerunner, the APW method, in that in practical application, APW assumes a constant potential between the muffin-tin spheres that surround each nucleus (Bukowinski 1977, 1985). Because of its precise representation of the potential, the LAPW method is sometimes referred to as FLAPW, for "full-potential". LAPW shares the ability to precisely represent the full potential and charge density with the full-potential LMTO method (Söderlind et al. 1996). The FP-LMTO method is very similar to LAPW in its capabilities and level of accuracy, differing primarily in the details of the basis functions.

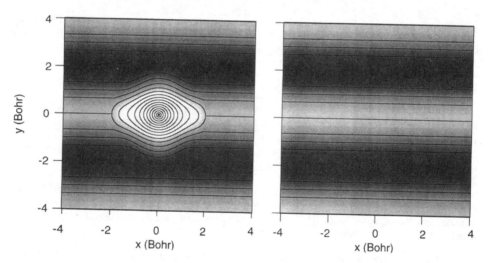

Figure 4. Left: A single LAPW basis function in the vicinity of a hydrogen nucleus located at the origin: $k = 0$, $\vec{G} = (0, \pi/2, 0)$, $l_{max} = 6$, $E_l(\mathrm{Ry}) = -(l+1)^{-2}$. Right: The plane wave $\vec{G} = (0, \pi/2, 0)$.

The accuracy and flexibility of the LAPW method derives from its basis which explicitly treats the first-order partitioning of space into near-nucleus regions, where the charge density and its spatial variability are large, and interstitial regions, where the charge density varies more slowly (Fig. 4) (Anderson 1975, Singh 1994, Wei and Krakauer 1985). These two regions are delimited by the construction of so-called muffin-tin spheres of radius R_{MT}^{α} centered on each nucleus α. A dual-basis set is constructed, consisting of plane-waves in the interstitial regions that are matched continuously to more rapidly varying functions inside the spheres. Within the muffin-tin spheres $r' < R_{MT}^{\alpha}$

$$\phi^{\vec{k}+\vec{G}}(\vec{r}) = \left[a_{lm}^{\alpha} u_l(E_l^{\alpha}, r') + b_{lm}^{\alpha} \dot{u}_l(E_l^{\alpha}, r') \right] Y_{lm}(\vec{r}/r) \qquad (21)$$

and for $r' > R_{MT}^{\alpha}$

$$\phi^{\vec{k}+\vec{G}}(\vec{r}) = exp\left[i\left(\vec{k} + \vec{G}\right) \bullet \vec{r} \right] \qquad (22)$$

where $\vec{r}' = \vec{r} - \vec{R}_{\alpha}$, $\vec{R}_{\alpha}$ are the positions of the nuclei, $\vec{k}$ is in the first Brillouin zone, $\vec{G}$ is a reciprocal lattice vector, u_l and $\dot{u}_l$ are, respectively, the solution to the radial part of the Schrödinger equation and its energy derivative for the spherically symmetric portion of the potential inside the muffin-tin sphere at energy E_l, and the coefficients a and b are determined by requiring continuity of the basis function and its first radial derivative on the muffin-tin sphere.

With this basis set, all-electron calculations for silicates or transition metals typically require on the order of 100 basis functions per atom. The primary disadvantage of the LAPW method is that the complexity of the basis functions makes it relatively intensive computationally. In practice, this limits the size of the system that can be studied. Even so, LAPW computations for structures as complex as that of $MgSiO_3$ perovskite (20 atoms in the unit cell) have been performed (Stixrude and Cohen 1993).

Basis sets consisting solely of plane waves, because of their analytical simplicity have significant advantages over the LAPW basis. However, all-electron calculations are virtually impossible with a plane wave basis set; the number of basis functions needed to represent the rapid spatial oscillations of the core region is much too large to be practical. For this reason, the plane wave basis is generally linked in practice to the pseudopotential approximation, in which the Fourier content of charge density and potential are limited by design.

Convergence

There are two primary convergence issues: the size of the basis, and the integrations over reciprocal space (e.g. Eqn. 9). Both LAPW and plane-wave basis sets have the property of smooth convergence; this means that convergence of the computations is readily assessed; quantities of interest vary smoothly as the basis set size is increased. In the LAPW method, the size of the basis set is described by the dimensionless quantity $R_{MT}K_{max}$, where K_{max} is the maximum wavenumber of the plane waves included in the basis set. In the pseudopotential method, the size of the basis is set by the maximum kinetic energy of the plane waves $E_{cut} = K_{max}^2$ in atomic units. Typical values for computations of silicates are $R_{MT}K_{max} = 7$ and $E_{cut} = 40$-80 Ry, depending on the pseudopotential that is used.

Sampling of the Brillouin zone is treated with the special points method, which has been shown to yield rapid convergence (Monkhurst and Pack 1976). For insulators, only a few points (1-10) are typically needed to achieve fully converged total energies; metals require denser sampling because of the often complex structure of the Fermi surface. The special points method constructs a uniform grid of k-points of specified resolution in the first Brillouin zone. The resulting set of k-points is divided into subgroups (stars) of symmetrically equivalent points. The Kohn-Sham equations are solved for only one member of each star, and the wavefunctions at other points in the star reconstructed with the appropriate symmetry operations, weighting the contribution of each star by its degeneracy.

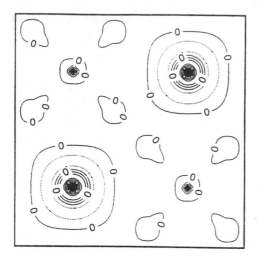 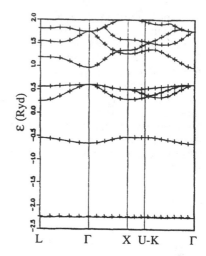

Figure 5 (left). Difference in charge density of MgO generated with overlapping ions using a Gordon-Kim model (potential-induced breathing, or PIB model) (Isaak et al. 1990) and calculated self-consistently using the linearized augmented plane wave (LAPW) method (Mehl et al. 1988). Contour interval: 0.005 e/Bohr^3 (see also Hemley and Cohen 1996).

Figure 6 (right). Band structure of MgO near zero pressure calculated using the LAPW method (lines) compared with that determined from the potential generated by the overlapping ion PIB charge density (Isaak et al. 1990, Mehl et al. 1988).

SELECTED APPLICATIONS

Bonding and electronic structure

Ab initio models give insight into bonding and electronic structure that are often not obvious from self-consistent computations alone. One example is the relationship among and meaning of ionicity, covalency, and band width. One might think for example that a purely ionic model would have atomic-like energy levels, and that band width would arise from hybridization or covalency. Figure 5 shows the difference in charge density of MgO computed with overlapping PIB ions and computed self-consistently using the LAPW method. The agreement is excellent. Furthermore, the bands that one finds using the crystal potential computed from the PIB charge density are in excellent agreement with the self-consistent band structure (Fig. 6). Thus, the PIB charge density is a good approximation to the self-consistent charge density for ionic materials such as MgO. A similar comparison should be made for rare-gas solids.

Thus, bands computed from the potential generated from overlapping ionic charge densities not only have width, but are in excellent agreement with self-consistent computations for ionic crystals. If one were to ask the origin of the band width in a tight-binding representation, one would find that the O $2p$ band width in MgO, for example, comes primarily from O-O $pp\sigma$ interactions (Kohan and Ceder 1996). We see that even a purely ionic charge density, generated by overlapping spherical ions, has a charge density that generates a potential, that when used in the KS equations implies a band width consistent with hybrid electronic states. There is a sort of duality in the description of ionic materials in that they can also be described from a charge density or tight-binding (or LCAO) perspective. In either case, there must be long-range Madelung terms in the total energy, that gives rise to LO-TO splitting in the lattice dynamics.

Not all ionic solids are formed from closed shell ions. For example, FeO (wüstite) and solid solutions between MgO and FeO (magnesiowüstite) behave like ionic solids, yet Fe^{2+} is a d_6 ion and is not closed shell. FeO belongs to a class of materials known as Mott insulators, which are discussed below. Such materials are very difficult to treat theoretically, and in spite of the importance of Fe in minerals, there is not yet a good method for obtaining first-principles results that are completely correct. Moreover, ab initio models such as PIB fail to give accurate predictions for the equation of state and other properties for these materials, even if one sphericalizes Fe, and treats it as ionic. Nevertheless, much can be learned about these materials from self-consistent computations as described below. Perhaps simple and accurate ab initio models can be developed for these materials, but it has not yet been done.

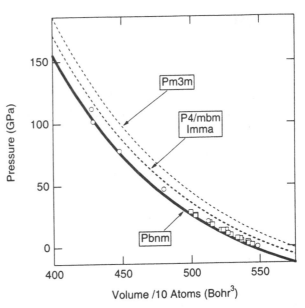

Figure 7. The equation of state of three different polytypes of $MgSiO_3$ perovskite as determined by LAPW calculations in the LDA approximation (Stixrude and Cohen 1993). The *Pbnm* structure is found to be most stable throughout the pressure regime of the earth's mantle. This is in agreement with experimental observations which so far have found no reproducible observations of other stable aristotypes. Circles and squares are experimental data from (Knittle and Jeanloz 1987) and (Mao et al. 1991), respectively.

Equation of state

The error due to the LDA in first principles calculations can be evaluated by comparing the results of LAPW calculations, which make no further essential approximations beyond the LDA, with experiment (Fig. 7). In investigations of silicates and oxides, it has been found that errors in volumes are typically 1-4% with theoretical volumes being uniformly smaller than experimental (Cohen 1991, 1992; Mehl et al. 1988, Stixrude and Cohen 1993). Part of this small difference is due to the higher temperatures of experiments (300 K) compared with the athermal calculations. This is a highly satisfactory level of agreement for a theory which is parameter free and independent of experiment.

All electron LDA computations of transition metals show errors of similar magnitude in the zero pressure volume; for the 3d and 4d metals, the calculations uniformly underestimate the experimental volumes, while for the 5d metals, the situation is more complex (Sigalas et al. 1992). For the heaviest materials, additional effects such as spin-orbit coupling, often neglected in computations, may become important and contribute to the discrepancy between theory and experiment. The generalized gradient approximation improves the agreement between theory and experimental equations of state for most materials including the 3d transition metals. In the case of iron, LAPW and FP-LMTO

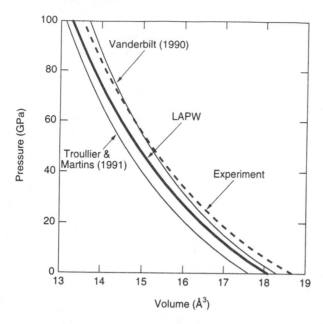

Figure 8. Equation of state of MgO from (dashed line) experiment (Duffy et al. 1995), (bold line) all-electron LDA (LAPW) calculations, (thin lines) pseudopotential calculations based on the indicated potentials (Troullier and Martins 1991, Vanderbilt 1990).

calculations differ from the experimentally measured room temperature equation of state by 3 % at zero pressure and by less than 1 % at core pressures (Sherman 1997, Söderlind et al. 1996, Stixrude et al. 1994); agreement with high temperature Hugoniot data is equally good (Stixrude et al. 1997).

Pseudopotential calculations make additional approximations that lead to additional errors (Fig. 8). These are small in magnitude and comparable in size to the LDA error. At this level of detail, different pseudopotentials yield results that differ from each other and from the all-electron LDA result from LAPW. Because the pseudopotential method is nearly as accurate as the much more elaborate LAPW method, it is often preferred for many applications, since its computational advantages allow much larger, and more complex systems to be studied.

Structure and compression mechanisms

The structure and compression mechanisms of a number of complex silicates have been studied with density functional theory at high pressure, including $MgSiO_3$ enstatite and perovskite, Mg_2SiO_4 forsterite, ringwoodite and inverse ringwoodite, and SiO_2 in the quartz, stishovite, $CaCl_2$ and columbite structures (Cohen 1991, 1992; Karki et al. 1997a, Karki et al. 1997d, Kiefer et al. 1998, Wentzcovitch et al. 1995a, Wentzcovitch et al. 1993, Wentzcovitch et al. 1995b, Wentzcovitch et al. 1998, Wentzcovitch and Stixrude 1997). Although simple ab initio models provide important predictions for high-pressure behavior of many deep earth materials, such as $MgSiO_3$ perovskite (Hemley et al. 1987, Cohen 1987b, Wolf and Bukowinski 1987), the need for an extended treatment relative to such simple approaches is readily seen: for example in the case of MgSiO,: the density, crystal structure parameters, and elasticity cannot all be explained from standard ionic models (i.e. assuming full charges on the ions) (Hemley and Cohen 1992). These investigations (1) provide an important test of the approximations upon which first-principles methods are based (2) illustrate in detail often not obtainable by experiment the nature of compression mechanisms and (3) provide a sensitive test of the hypothesis that

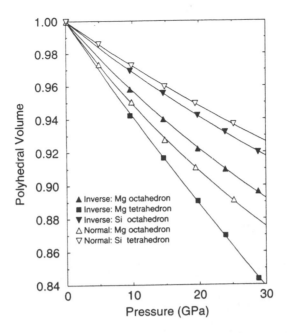

Figure 9. Compression of coordination polyhedra in normal and inverse Mg_2SiO_4 ringwoodite from LDA pseudopotential calculations (Kiefer et al. 1998).

some minerals undergo high-order symmetry-invariant phase transformations.

Using the method of Wentzcovitch (Wentzcovitch 1991, Wentzcovitch et al. 1993), the first principles optimization of complex crystal structures such as forsterite is an efficient procedure. Typically, on the order of 10-20 iterations are required for full structural convergence in this mineral with 7 internal degrees of freedom and three lattice parameters (Wentzcovitch and Stixrude 1997). In the case of ringwoodite, the results of first principles calculations show that, in the normal form, volume compression is mostly taken up by the MgO_6 octahedra which are much softer than the SiO_4 tetrahedra (Fig. 9). In the inverse form, the MgO_4 tetrahedra are the most compressible polyhedra, softer even than the MgO_6 octahedra in the normal form. This result can be understood by recognizing that, in the inverse structure, Si- and Mg-polyhedra combine to form relatively rigid octahedral layers, leaving most of the compression to the tetrahedral layers (Kiefer et al. 1998).

Phase stability

In many ways, phase stability provides the most stringent test of first-principles methods. The reason is that we are comparing total energies computed for two different structures, with different basis sets and Brillouin zones at the level of heats of transformations, generally a minuscule fraction of the total energy (less than one part per million).

First principles LDA total energy calculations of transformations in oxides and silicates have shown excellent agreement with experiment (Cohen 1991, 1992; Isaak et al. 1993, Karki et al. 1997b,d; Kingma et al. 1996, Mehl et al. 1988, Teter et al. 1998). These include the prediction, based on LAPW calculations, of the stishovite to $CaCl_2$ transition

(Cohen 1991, 1992) which was later found and confirmed experimentally (Kingma et al. 1995). Subsequent pseudopotential calculations are in excellent agreement with the earlier LAPW results showing that careful calculations yield consistent results even though computational methods may be quite different (Figs. 10 and 11). There is considerable interest in the possibility of still higher pressure phases that may be stable deep within the lower mantle. First principles calculations predict a phase transition near 96 GPa from the $CaCl_2$ structure to the columbite (α-PbO_2) structure (Karki et al. 1997d,e). This ultrahigh

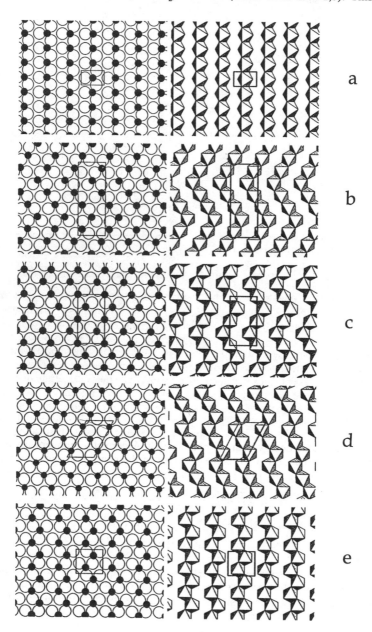

a

b

c

d

e

Figure 10. Opposite page: Representations of dense silica structures calculated by LDA (Teter et al. 1998). (a) $CaCl_2$, (b) SnO_2 (4 × 4) (c) $NaTiF_4$ (3 × 3), (d) $P2_1/c$ (3 × 2), and (1) a-PbO_2 structure types. The left-hand figures show one layer of the *ABAB* ... ... stacking of hcp oxygen anions (white) with one-half of the octahedral interstices filled with silicon ions (black). The right-hand figures show how these patterns form edge-sharing octahedral chains with various degrees of kinking (given in parentheses above).

Below, right: Enthalpies of the structures relative to that of stishovite as a function of pressure.

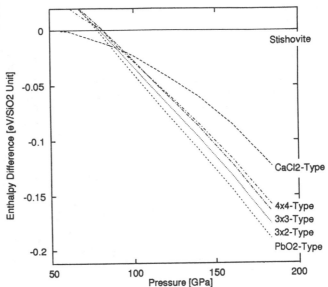

pressure structure is built of SiO_6 octahedra and differs from the lower pressure polymorphs primarily in the kinking of the chains of octahedra. Teter et al. (1998) found that there is a large number of closely related, and energetically competitive phases that are stable at pressures above 50 GPa (Fig. 10). These structures are all based on essentially close packing of the oxygen anions with different ordering of Si in the octahedral sites. The calculations predict that the columbite structure is stable above 80 GPa, in agreement with the calculations of Karki et al. (1997d,e), but that there should be extensive metastability of phases with similar bonding topology at these pressures; this appears to be observed experimentally (Hemley et al. 1994, Teter et al. 1998).

In the case of transition metals, the form of the exchange-correlation potential is critical. LDA fails to predict the correct ground state of iron, finding incorrectly that the hcp phase has a lower total energy than the bcc. The GGA correctly recovers the bcc ground state. Moreover, it accurately predicts the pressure of the phase transition from bcc to hcp near 11 GPa (Stixrude et al. 1994). This is an important result because the energetics are particularly subtle in the case of this transition since it involves a ferromagnetic and a non-magnetic phase. These calculations find that the hcp phase is the stable low temperature phase of iron at pressures beyond 11 GPa, in excellent agreement with experiment (Mao et al. 1990). First principles calculations show that it is highly unlikely that bcc will reappear as a stable phase at extreme pressures and temperatures, independent of the form of the exchange-correlation potential. The reason is that bcc is found to undergo an elastic instability with respect to a tetragonal strain at high pressure (Fig. 12). At pressures beyond 150 GPa, the bcc structure will spontaneously distort (Stixrude and Cohen 1995). The inner core is likely composed of a close-packed structure, either hcp or a similar hexagonal or nearly hexagonal phase, or fcc.

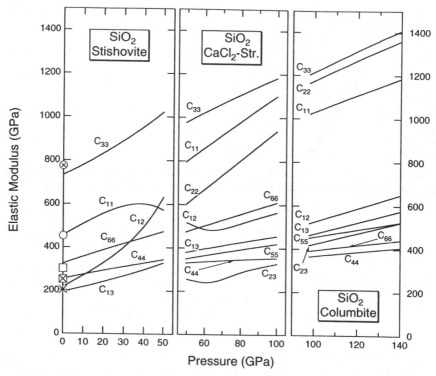

Figure 11. Elastic constants of three high-pressure phases of silica from LDA pseudopotential calculations (Karki et al. 1997c). The pressure of the phase transition from stishovite to the $CaCl_2$ structure is in excellent agreement with experiment (Kingma et al. 1995); while the predicted phase transition to the columbite structure is consistent with some of the diffraction data reported by Kingma et al. (1996) (see also Teter et al. 1998). Experimentally measured elastic constants of stishovite are indicated by the symbols (Weidner et al. 1982).

Figure 12. First principles LAPW calculations of the energy vs. c/a ratio of iron within GGA (solid curves) and LDA (dashed) approximations at two volumes: 70 $Bohr^3$ (~10 GPa, upper curves) and 50 $Bohr^3$ (~200 GPa, lower curves). The c/a ratios of the bcc and fcc structures are indicated. The change in the curvature of the energy surface at bcc is an elastic instability (Stixrude et al. 1994).

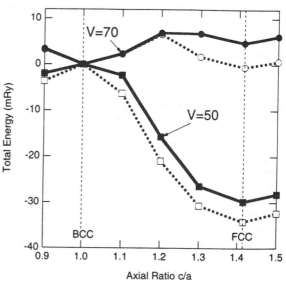

Another approach towards phase stability is the investigation of the dynamical stability of lattices. One example of this approach is that of Stixrude et al. (1996) who performed linear response computations with the LAPW method on $CaSiO_3$ perovskite in the cubic perovskite structure (Fig. 13). Unlike $MgSiO_3$, which is orthorhombic, experimental studies have found $CaSiO_3$ to be cubic. Unexpectedly, however, cubic $CaSiO_3$ was found in the theoretical calculations to be unstable at the M- and R- points on the Brillouin zone boundary. The unstable vibrational modes correspond to rotations of the SiO_6 octahedra which lower the symmetry. Frozen phonon calculations confirmed this result. The results suggest strongly that $CaSiO_3$ is not cubic but tetragonal or lower symmetry. The small strain predicted theoretically is consistent with the expected precision of x-ray studies that were interpreted as showing a cubic structure for $CaSiO_3$.

Ab initio models can also be used to study the stability of a variety of structures to search for possible phase transitions. This was done successfully for Al_2O_3, where the PIB model showed a high pressure elastic instability (Cohen 1987a). Detailed PIB computations for different structures showed a phase transition at high pressures to the Rh_2O_3 II structure (Cynn et al. 1990), and the transition was confirmed and pressure computed accurately using the LAPW method with the PIB structural parameters (Marton and Cohen 1994). Pseudopotential calculations are in good agreement (Thompson et al. 1996). These computations predicted a phase transition near 90 GPa, in excellent agreement with later experiments (Funamori and Jeanloz 1997). The pseudopotential calculations predict a further transition to a perovskite phase at pressures above those found in the lower mantle (i.e. at 223 GPa).

Elastic moduli

The elastic moduli are of central importance in studies of the earth's interior since they govern the passage of seismic waves, our primary source of information on the structure of the subsurface. Despite their importance, density functional calculations of the elastic constants of earth materials have appeared only recently. The key development has been that of an efficient structural optimization scheme (Wentzcovitch et al. 1993), and calculation of stresses from the Hellman-Feynman theorem. Elastic constants are determined by calculating the stress generated by deviatoric strains applied to the equilibrium structure as described above. It is straightforward to demonstrate that one is within the linear regime by performing the calculation at a variety of values of the strain magnitude and extrapolating to the limit of zero strain (Karki et al. 1997b). These calculations show that strains of the order of 1 % are appropriate for silicates and oxides. By carefully choosing the symmetry of the applied strain, it is possible to calculate all elements of the elastic constant tensor with a small number of different strains. For example, the three elastic constants of a cubic mineral can be determined from a single strain; four different strains have been used for orthorhombic minerals (9 independent elastic constants) (Silva et al. 1997).

The full elastic constant tensors of a number of silicates and oxides have been determined with the plane-wave pseudopotential method, including that of MgO periclase, $MgSiO_3$ perovskite, Mg_2SiO_4 forsterite and ringwoodite, and SiO_2 in the stishovite, $CaCl_2$, and columbite structures (Karki et al. 1997a,b,c; Kiefer et al. 1997, Silva et al. 1997, Wentzcovitch et al. 1993). Once the elastic constant tensor is determined, it is straightforward to calculate the elastic wave (seismic) velocities in any direction, the elastic anisotropy, and the seismic-wave velocities of isotropic aggregates. Results for silica (Fig. 11) show several interesting features: (1) the phase transition from stishovite to the $CaCl_2$ structure is associated with an elastic instability ($c_{11}-c_{12}->0$) that originates from the coupled octahedral rotation and basal plane shear that relates the two structures; (2) this elastic

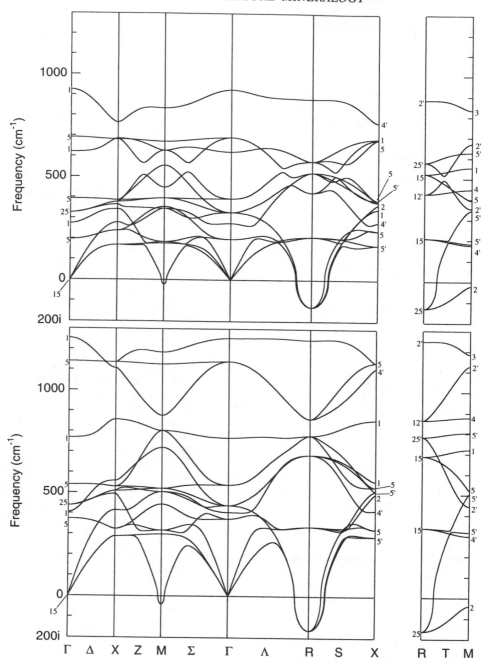

Figure 13. Phonon spectrum of the cubic phase at unit cell volume $V = 310$ Bohr^3 ($P = $ -8 GPa) (a) compared with that at $V = 240$ Bohr^3 ($P = 80$ GPa) (b) (Stixrude et al. 1996). All zone-center modes (Γ-point) are stable. However, zone-boundary modes at the M- and R-points are unstable, as shown by the existence of imaginary frequencies. The final panel (M-R) shows that the edges of the cubic Brillouin zone are everywhere unstable. The magnitude of the imaginary eigenfrequencies increases with compression. Symmetry designations are those of Cowley (1964) as corrected by Boyer and Hardy (1981).

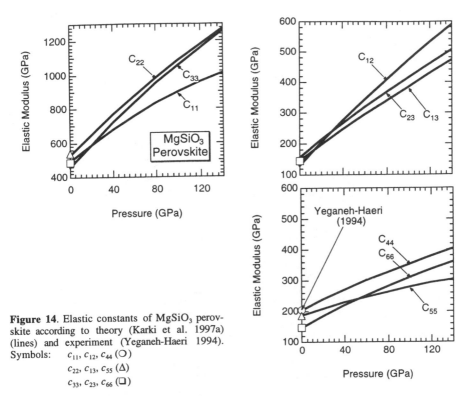

Figure 14. Elastic constants of MgSiO$_3$ perovskite according to theory (Karki et al. 1997a) (lines) and experiment (Yeganeh-Haeri 1994). Symbols: c_{11}, c_{12}, c_{44} (O)
c_{22}, c_{13}, c_{55} (Δ)
c_{33}, c_{23}, c_{66} ($\square$)

instability is accompanied by a very large (60 %) change in shear wave velocity; so large that it may be seismologically observable if as little as a few percent of free silica exists in the earth's mantle, and (3) all phases of silica are highly anisotropic elastically with the anisotropy diverging in the neighborhood of the phase transition.

Extensive computations have been performed for MgSiO$_3$, although it is difficult to study because there are 20 atoms in the primitive unit cell. MgSiO$_3$ appears to be more ionic than stishovite (Hemley and Cohen 1992, Stixrude and Cohen 1993), but the fully charged ionic model is not as successful as it is for simple minerals such as MgO. In the case of MgO, ab initio ionic models find proper violations of the Cauchy relations in alkali halides and alkaline earth oxides are found which agree reasonably with experiment (Mehl et al. 1986). For perovskite the predictions of single crystal elastic constants for MgSiO$_3$ (Cohen 1987b) were in reasonable agreement with later experiments (Yeganeh-Haeri 1994), and high-pressure elastic constants are still not available experimentally. More recently, large scale computations of elastic constants have been performed for MgSiO$_3$ using a plane wave basis with pseudopotentials (Karki et al. 1997a) (Fig. 14). Agreement with available experiments is excellent, and these computations give predictions of the elasticity of perovskite for pressures throughout the mantle. These results will be benchmarks for future computations of thermoelasticity at mantle temperatures.

Magnetic collapse

This phenomenon may occur in a wide variety of transition metal bearing minerals at high pressure and has been predicted on the basis of first principles calculations in FeO at pressures above 100 GPa (Isaak et al. 1993) (Fig. 15). The predicted magnetic collapse is a

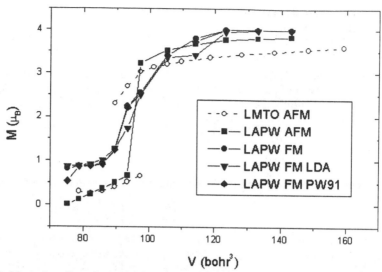

Figure 15. Magnetic collapse in FeO shown as a function of volume predicted from different levels of theory. FM and AFM correspond to ferromagnetic and antiferromagnetic calculations, and results are shown for both LAPW (linearized augmented plane wave) and LMTO (linearized muffin-tin orbital) methods. The generalized gradient approximation (GGA) results are labeled PW91. The high-spin/low spin transition is discontinuous in AFM and continuous in FM (Cohen et al. 1998).

type of high-spin low-spin transition, which has been observed in other materials such as NiI_2 and MnS . High-spin low-spin transitions can be either continuous higher-order phase transitions, or first-order phase transitions. Recent Mössbauer measurements indicated a 100 GPa transition in FeO (Pasternak et al. 1997), although recent x-ray K_β spectra show no transition to at least 130 GPa (see Hemley, Mao, and Cohen, this volume). More work is needed to understand these reported differences.

Magnetic collapse can be understood in terms of pressure-induced changes in the band structure with the Stoner model, discussed below. This picture contrasts with the conventional local atomic view which focuses only on the Brillouin zone center. In the conventional view, a high-spin to low-spin transition occurs when the crystal-field splitting between d orbitals of e_g and t_{2g} character exceeds the exchange splitting between up- and down-spin states (Burns 1993). Furthermore, the conventional view is that the e_g-t_{2g} splitting is due to the Coulomb potential from the transition metal atom's coordination polyhedron, and that the splitting increases as r_s^{-5}. Although reported crystal-field splittings vary approximately as r_s^{-5} in many cases, this view does not appear to be correct. In fact, the on-site contributions to the crystal field-splitting that arise from the electrostatic interaction with the surrounding O^{2-} ions have the wrong sign, and would give e_g lower than t_{2g} by a small amount, rather than t_{2g} being lower, as observed. Instead, the main contribution to the e_g-t_{2g} splitting is due to d-d hybridization between transition metal ions, which also varies as r_s^{-5}. These d-d interactions operate only at the Brillouin zone-center (Γ-point), also the only point where the e_g - t_{2g} symmetry is appropriate. At other points in the Brillouin zone, the band width is due to hybridization with O 2p states (Mattheiss 1972a,b). Optical spectroscopy measures not only the states at Γ, but rather is sensitive to vertical transitions throughout the zone, and thus to the entire bandwidth. The d-p interactions lead to a band width that behaves approximately as r_s^{-7}. Thus, the band width and apparent crystal field splitting should vary between r_s^{-5} and r_s^{-7} with increasing pressure. The distinction between the present picture and the conventional view can be seen most easily in terms of the behavior of the Hamiltonian. In the conventional view, it is the on-site, diagonal, elements that vary rapidly with pressure, giving rise to increased splitting, whereas band theory says that it is the off-diagonal, covalent, contributions which change the band widths with pressure.

The decrease in magnetism with increasing pressure can be understood in a qualitative sense in that as pressure increases, band widths increase, and eventually become greater than the exchange splitting. This can be examined quantitatively with the Stoner model (Cohen et al. 1997a). In the Stoner model, the effect of magnetism on the total energy is

$$\Delta E = \frac{-M^2 I}{2} + \frac{M^2}{2N(0)} \tag{23}$$

where M is the magnetic moment, the Stoner Integral I is an atomic property, and $N(0)$ is the density of states at the Fermi level (or top of the valence band) in the non-magnetic state. The first-term is the exchange energy due to magnetization and the second is the change in the band energy with magnetic moment. Minimizing ΔE with respect to M leads to the Stoner criterion for a stable magnetic state, $IN(0) > 1$. As pressure is increased I, being a property of the atom, remains constant but $N(0)$ decreases as band widths widen with increasing hybridization. In the absence of new bands crossing the Fermi level or changes in band topology at the Fermi level or top of the valence band, magnetism will decrease and disappear with increasing pressure.

Mott insulators

Band theory is known to fail in the case of transition metal oxides in the sense that it often predicts metallic ground states when these are observed to be insulating. The cause of this failure may be similar to the failure to predict accurate band gaps in other materials, but it is also believed that there is a more specific failure of LDA-like theories for the transition metal oxides, which is that LDA is local, as discussed above, and does not distinguish between electrons of different angular momentum. The Hohenberg-Kohn theorem says that such orbital dependent potentials should not be necessary to find the energy and ground-state charge density of a system, but the exact functional that would give this behavior is unknown, and would likely be extremely complex in order to give the proper charge density of transition metal oxide compounds, especially those that involve orbital ordering. The main problem with LDA-like theories is believed to be the mean-field treatment of the local Coulomb repulsion U, which is a measure of the increase in energy when an electron is added to an atom.

A promising method for realistic computations of real Mott insulators, but which needs further investigation is the LDA+U model (Anisimov et al. 1993, Anisimov et al. 1991, Mazin and Anisimov 1997). In this method an orbital dependent potential is added to simulate the effect of U. The parameter U can be estimated by computing the change in energy with orbital occupancy in constrained LDA calculations. LDA+U appears to be an excellent approximation at zero pressure where the band width (W) is small, but it is unclear whether it will give reasonable results at high pressures, where U/W is much smaller. The parameter U/W decreases with pressure since U is relatively insensitive to pressure, but the band width increases rapidly with pressure due to increased hybridization. One explanation for the success of band theory in the case of the high pressure transitions in transition metal oxides discussed above is that U/W is small under at high pressure, making band theory more applicable and predictive. Transitions predicted for the other transition metal oxides should be re-examined.

High-temperature and transport properties

The effect of temperature is directly linked to the dynamics of a crystal: thermal contributions can be written as a sum over lattice vibrations (phonons). Because dynamics break the symmetry of the primitive unit cell, determination of dynamics from first principles is a substantial computational challenge. This can be accomplished either by determining the dynamical matrix and diagonalizing to find the phonon frequencies (e.g. via linear response, discussed above), or by molecular dynamics, a brute force approach which simply solves Newton's equations of motion for a periodic array of nuclei. Molecular dynamics has the advantage of also being able to treat transport properties such as chemical diffusivity, and liquids. Nearly all previous studies of vibrations, high temperature properties, or transport properties have been based on semi-empirical or ab initio methods. Molecular dynamics simulations using the pseudopotential method have now been performed on deep earth materials for the first time in a study of liquid iron at core conditions (Wijs et al. 1998). The first-principles investigation of other solid and fluid earth materials by this technique represents an exciting future direction.

The biggest advantage of fast, ab initio methods such as PIB or VIB is that one can perform lattice dynamics and long molecular dynamics simulations on systems of reasonable size, and study a wide variety of thermodynamic and transport properties. For example, Isaak et al. (1990) performed lattice dynamics on MgO as a function of lattice strain, going beyond the normal quasiharmonic approximation, and studied the effects of temperature and pressure on elasticity and the equation of state. In that study, the dynamical matrix was found throughout the Brillouin zone in order to obtain the free energy, and this was repeated for different lattice strains. In spite of increased computational power in the last eight years, no such study has yet been done self-consistently. The Isaak et al. study gave what are still one of the few estimates of cross derivatives of pressure and temperature on elasticity that are available. The results were then used to help understand the increase in seismic parameter $d\ln V_s/d\ln V_p$ with depth in the Earth, a quantity that has been difficult to constrain experimentally.

Going beyond lattice dynamics, Inbar and Cohen (1995) determined the thermal equation of state of MgO using molecular dynamics and the PIB model. Such studies are just becoming possible using self-consistent methods, and still have not been performed. Using molecular dynamics and the VIB model, it is also possible to study complex phenomena, such as thermal conductivity (Cohen 1998) and diffusion (Ita and Cohen 1997, 1998). The diffusivity of O in MgO, for example, is obtained in agreement with measurements within experimental error.

Because high temperature properties are difficult to obtain self-consistently, Wasserman et al. (1996) applied the particle-in-a-cell model to compute thermodynamic properties of iron at high pressures and temperatures in conjunction with a new fast and accurate tight-binding model (discussed above in the tight binding section) that allows computations for large unit cells (Cohen et al. 1994). They obtained excellent agreement with shock compression data up to ultrahigh pressures and temperatures (Stixrude et al. 1997). They also performed molecular dynamics using the tight-binding model for iron liquid at outer core conditions (Fig. 16), and obtained an estimate of the viscosity of liquid iron in excellent agreement with subsequent pseudopotential calculations (Wijs et al. 1998). These results are geophysically significant because the viscosity of the outer core had been uncertain by 13 orders of magnitude.

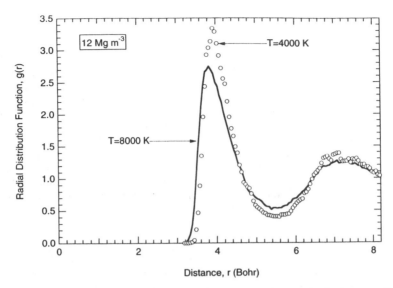

Figure 16. Radial distribution function of liquid iron at two temperatures at a density corresponding to the bottom of the earth's outer core. Calculations were performed via molecular dynamics based on the tight-binding total energy method of Cohen et al. (Cohen et al. 1994). The Stokes-Einstein viscosity calculated from these simulations is 5($\pm$3) centipoise, and is weakly dependent on temperature over the range shown.

CONCLUSIONS AND OUTLOOK

Modern first-principles methods are now capable of realistic predictions of many experimentally observable properties of minerals such as the equation of state, phase stability, crystal structure, and elasticity. Parameter-free and completely independent of experiment, these methods have been shown to reproduce observations even of subtle features such as phase transitions and the elastic anisotropy with good accuracy. First-principles methods including density functional theory represent the ideal complement to the experimental approach towards studying the behavior of earth materials under extreme conditions. This review has only given an indication of the realm of application of modern condensed-matter theory to the study of minerals at high pressure. We have reviewed only a subset of the important calculations that have been performed. One foresees accelerated progress on a number of fronts in this challenging field resulting from the continued interplay of theory and experiment.

ACKNOWLEDGMENTS

This work supported by the National Science Foundation under grants EAR-9628199, EAR-9614790, and EAR-9729194 (LPS); EAR-9305060, EAR-9870328, and EAR-9614363 (REC); EAR-9710475, EAR-9526763 and EAR-9706624 (RJH). The computations at the Geophysical Laboratory were performed on a Cray J90/16-4096, which is supported by NSF grant EAR-9512627, the Keck Foundation, and the Carnegie Institution of Washington.

REFERENCES

Anderson OK (1975) Linear methods in band theory. Phys Rev B 12:3060-3083

Anisimov VI, Solovyev IV, Korotin MA, Czyzyk MT, Sawatzky GA (1993) Density functional theory and NiO photoemission spectra. Phys Rev B 48:16929-16934

Anisimov VI, Zaanen J, Andersen OK (1991) Band theory and Mott insulators: Hubbard U instead of Stoner U. Phys Rev B 44:943-954

Bagno P, Jepsen O, Gunnarson O (1989) Ground-state properties of third-row elements with nonlocal density functionals. Phys Rev B 40:1997-2000

Baroni S, Giannozzi P, Testa A (1987) Green-function approach to linear response in solids. Phys Rev Lett 58:1861-1864

Birch F (1952) Elasticity and constitution of the earth's interior. J Geophys Res 57:227-286

Boyer LL, Hardy JR (1981) Theoretical study of the structural phase transition in $RbCaF_3$. Phys Rev B 24:2577-2591

Boyer LL et al. (1985) Beyond the rigid ion approximation with sphercally symmetric ions. Phys Rev Lett 54:1940-1943

Boyer LL, Stokes HT, Mehl MJ (1997) Application of a Kohn-Sham-like formulation of the self-consistent atomic deformation model. Ferroelec 194:173-186

Bukowinski MST (1977) A theoretical equation of state for the inner core. Phys Earth Planetary Inter 14:333-344

Bukowinski MST (1985) First principles equations of state of MgO and CaO. Geophys Res Lett 12:536-539

Bukowinski MST (1994) Quantum geophysics. Ann Rev Earth Planet Sci 22:167-205

Burnham CW (1990) The ionic model: Perceptions and realities in mineralogy. Am Mineral 75:443-463

Burns RG (1993) Mineralogical Applications of Crystal Field Theory. Cambridge University Press, Cambridge

Ceperley DM, Alder BJ (1980) Ground state of the electron gas by a stochastic method. Phys Rev Lett 45:566-569

Chabrier G (1992) The molecular-metallic transition of hydrogen and the structure of Jupiter and Saturn. Astrophys J 391:817-826

Cohen AJ, Gordon RG (1975) Theory of the lattice energy, equilibrium structure, elastic constants, and pressure-induced phase transformations in alkali-halides. Phys Rev B 12:3228-3241

Cohen AJ, Gordon RG (1976) Modified electron-gas study of the stability, elastic properties, and high-pressure behavior of MgO and CaO crystals. Phys Rev B 14:4593-4605

Cohen RE (1987a) Calculation of elasticity and high pressure instabilities in corundum and stishovite with the potential induced breathing model. Geophys Res Lett 14:37-40

Cohen RE (1987b) Elasticity and equation of state of $MgSiO_3$ perovskite. Geophys Res Lett 14:1053-1056

Cohen RE (1991) Bonding and elasticity of stishovite SiO_2 at high pressure: linearized augmented plane wave calculations. Am Mineral 76:733-742

Cohen RE (1992) First-principles predictions of elasticity and phase transitions in high pressure SiO_2 and geophysical implications. In: Y. Syono and M.H. Manghnani (eds) High-Pressure Research: Applications to Earth and Planetary Sciences, p 425-431 Terrapub, Tokyo

Cohen, R.E (1998) Thermal conductivity of MgO at high pressures. In: M Nakahara (ed) Review of High-Pressure Science and Technology 7:160-162, Japan Soc. High-Pres Science Technol., Kyoto, Japan

Cohen RE, Boyer LL, Mehl MJ (1987) Lattice dynamics of the potential induced breathing model: First principles phonon dispersion in the alkaline earth oxides. Phys Rev B 35:5749-5760

Cohen RE, Mazin II, Isaak DG (1997a) Magnetic collapse in transition metal oxides at high pressure: Implications for the Earth. Science 275:654-657

Cohen RE, Mehl MJ, Papaconstantopoulos DA (1994) Tight-binding total-energy method for transition and noble metals. Phys Rev B 50:14694-14697

Cohen RE, Stixrude L, Wasserman E (1997b) Tight-binding computations of elastic anisotropy of Fe, Xe, and Si under compression. Phys Rev B 56:8575-8589

Cohen RE, Fei Y, Downs R, Mazin II, Isaak DG (1998) Magnetic collapse and the behavior of transition metal oxides: FeO at high pressures. In: Wentzcovitch R, Hemley RJ, Nellis WJ, Yu P (eds) High-Pressure Materials Research. Proceedings of the Fall 1997 Materials Research Society Meeting, Vol 499 Pittsburgh, PA in press

Cowley R (1964) Lattice dynamics and phase transitions of strontium titanate. Phys Rev 134:A981-A997

Cynn H, Isaak DG, Cohen RE, Nicol MF, Anderson OL (1990) A high pressure phase transition of corundum predicted by the Potential Induced Breathing model. Am Mineral 75:439

Dovesi R, Pisani C, Roetti C, Silvi B (1987) The electronic structure of α-quartz: A periodic Hartree-Fock calculation. J Chem Phys 86:6967-6971

Duffy TS, Hemley RJ, Mao HK (1995) Equation of state and shear-strength at multimegabar pressures—magnesium oxide to 227 GPa. Phys Rev Lett 74:1371-1374

Dziewonski AM, Anderson DL (1981) Preliminary reference earth model. Phys Earth Planet Inter 25:297-356

Edwardson PJ (1989) Corridors-between-adjacent-sites model for the four phases of $KNbO_3$. Phys Rev Lett 63:55-59

Funamori N, Jeanloz R (1997) High pressure transformation of Al_2O_3. Science 278:1109-1111

Gell-Mann M, Brueckner KA (1957) Correlation energy of an electron gas at high density. Physical Review 106:364-368

Gordon RG, Kim YS (1972) Theory for the forces between closed-shell atoms and molecules. J Chem.Phys. 56:3122-3133

Gunnarsson O, Lundqvist BI (1976) Exchange and correlation in atoms, molecules, and solids by the spin-density-functional formalism. Phys Rev B 13:4274-4298

Harrison WA (1989) Electronic structure and the properties of solids. Dover, New York

Hedin L, Lundqvist BI (1971) Explicit local exchange-correlation potentials. J Phys C:Solid State Phys 4:2064-2083

Hemley RJ, Cohen RE (1992) Silicate perovskite. Ann Rev Earth Planet Sci 20:553-600

Hemley RJ, Jackson MD, Gordon RG (1985) First-principles theory for the equations of state of minerals to high pressures and temperatures: application to MgO. Geophys Res Lett 12:247-250

Hemley RJ, Jackson MD, Gordon RG (1987) Theoretical study of the structure, lattice dynamics, and equations of state of perovskite-type $MgSiO_3$. Phys Chem Minerals 14:2-12

Hemley RJ, Prewitt CT, Kingma KJ (1994) High-pressure behavior of silica. Rev Mineral 29:41-81

Hohenberg P, Kohn W (1964) Inhomogeneous electron gas. Phys Rev 136:B864-B871

Hubbard WB (1984) Planetary Interiors. Van Nostrand Reinhold, New York

Ichimaru S (1982) Strongly coupled plasmas: high-density classical plasmas and degenerate electron liquids. Rev Mod Phys 54:1017-1059

Inbar I, Cohen RE (1995) High pressure effects on thermal properties of MgO. Geophys Res Lett 22:1533-1536

Isaak DG, Cohen RE, Mehl MJ (1990) Calculated elastic and thermal properties of MgO at high pressure and Temperatures. J Geophys Res 95:7055-7067

Isaak DG, Cohen RE, Mehl MJ, Singh DJ (1993) Phase stability of wüstite at high-pressure from 1st-principles linearized augmented plane-wave calculations. Phys Rev B 47:7720-7731

Ita J, Cohen RE (1997) Effects of pressure on diffusion and vacancy formation in MgO from non-empirical free-energy integrations. Phys Rev Lett 79:3198-3201

Ita J, Cohen RE (1998) Diffusion in MgO at high pressure: Implications for lower mantle rheology. Geophys Res Lett 25:1095-1098

Jensen H (1938) Das Druck-Dichte Diagramm der Elemente bei höheren Drucken am Temperaturnullpunkt. Zeits Phys 111:373-385

Karki BB, Stixrude L, Clark SJ, Warren MC, Ackland GJ, Crain J (1997a) Elastic Properties of ortho-rhombic $MgSiO_3$ perovskite at lower mantle pressures. Am Mineral 82:635-638

Karki BB, Stixrude L, Clark SJ, Warren MC, Ackland GJ, Crain J (1997b) Structure and elasticity of MgO at high pressure. Am Mineral 82:51-60

Karki BB, Stixrude L, Warren MC, Ackland GJ, Crain J (1997c) Ab initio elasticity of three high-pressure polymorphs of silica. Geophys Res Lett 24:3269-3272

Karki BB, Warren MC, Stixrude L, Ackland GJ, Crain J (1997d) Ab initio studies of high-pressure structural transformations in silica. Phys Rev B 55:3465-3471

Karki BB, Warren MC, Stixrude L, Ackland GJ, Crain J (1997e) Ab initio studies of high-pressure structural transformations in silica (erratum, vol 55, pg 3465, 1997). Phys Rev B 56:2884-2884

Kiefer B, Stixrude L, Wentzcovitch R (1998) Ab initio investigation of structure and compression in normal and inverse ringwoodite. Am Mineral (in press)

Kiefer B, Stixrude L, Wentzcovitch RM (1997) Elastic constants and anisotropy of Mg_2SiO_4 spinel at high pressure. Geophys Res Lett 24:2841-2844

Kingma K, Mao HK, Hemley RJ (1996) Synchrotron x-ray diffraction of SiO_2 to multimegabar pressures. High Press Res 14:363-374

Kingma KJ, Cohen RE, Hemley RJ, Mao HK (1995) Transformation of silica to a denser phase at lower-mantle pressures. Science 374:243-245

Knittle E, Jeanloz R (1987) Synthesis and equation of state of $(Mg,Fe)SiO_3$ perovskite to over 100 gigapascals. Science 235:668-670

Kohan AF, Ceder G (1996) Tight-binding calculation of formation energies in multicomponent oxides: application to the MgO-CaO phase diagram. Phys Rev B 54:805-811

Kohn W, Sham LJ (1965) Self-consistent equations including exchange and correlation effects. Phys Review 140:A1133-A1138

Körling M, Häglund J (1992) Cohesive and electronic properties of transition metals: The generalized gradient approximation. Phys Rev B 45:13293-13297

Lee C, Gonze X (1995) Ab-initio calculation of the thermodynamic properties and atomic temperature factors of SiO_2 alpha-quartz and stishovite. Phys Rev B 51:8610-8613

LeSar R (1983) Phys Rev B 28:6812

LeSar R (1988) Equations of state of dense helium. Phys Rev Lett 61:2121-2124

Lin JS, Qteish Q, Payne MC, Heine V (1993) Optimised and transferable non-local separable ab-initio pseudopotentials. Phys Rev B 47:4174-4180

Lundqvist S, March NH (1987) Theory of the Inhomogeneous Electron Gas. Plenum Press, London

Mao HK et al. (1991) Effect of pressure, temperature, and composition on lattice parameters and density of $(Fe,Mg)SiO_3$ perovskites to 30 GPa. J Geophys Res 96:8069-8079

Mao HK, Wu Y, Chen LC, Shu JF, Jephcoat AP (1990) Static compression of iron to 300 GPa and $Fe_{0.8}Ni_{0.2}$ alloy to 260 GPa: Implications for composition of the core. J Geophys Res 95:21737-21742

Marton FC, Cohen RE (1994) Prediction of a high pressure phase transition in Al_2O_3. Am Mineral 79:789-792

Mattheiss LF (1972a) Electronic structure of the 3d transition -metal monoxides. II. Interpretation. Phys Rev B 5:306-315

Mattheiss LF (1972b) Electronic structure of the 3d transition-metal monoxides. I. Energy-band results. Phys Rev B 5:290-306

Mazin II, Anisimov VI (1997) Insulating gap in FeO: Correlations and covalency. Phys Rev B 55: 12822-12825

Mehl MJ, Cohen RE, Krakauer H (1988) Linearized augmented plane wave electronic structure calculations for MgO and CaO. J Geophy Res 93:8009-8022

Mehl MJ, Hemley RJ, Boyer LL (1986) Potential induced breathing model for the elastic moduli and high-pressure behavior of the cubic alkaline-earth oxides. Phys Rev B 33:8685-8696

Monkhurst HJ, Pack JD (1976) Special points for Brillouin-zone integrations. Phys Rev B 13:5188-5192

Muhlhausen C, Gordon RG (1981) Electron-gas theory of ionic crystals, including many-body effects. Phys Rev B 23:900-923

Nielsen OH, Martin R (1985) Quantum mechanical theory of stress and force. Phys Rev B 32:3780-3791

Pasternak MP et al. (1997) High pressure collapse of magnetism in $Fe_{0.94}O$: Mössbauer spectroscopy beyond 100 GPa. Phys Rev Lett 79:5046-5049

Pauling L (1960) The Nature of the Chemical Bond. Cornell University Press, Ithaca, New York

Perdew JP, Burke K, Ernzerhof M (1996) Generalized gradient approximation made simple. Phys Rev Lett 77:3865-3868

Perdew JP et al. (1992) Atoms, molecules, solids, and surfaces—application of the generalized gradient approximation for exchange and correlation. Phys Rev B 46:6671-6687

Perdew JP, Zunger A (1981) Self-interaction correction to density-functional approximations for many-electron systems. Phys Rev B 23:5048-5079

Pickett WE (1989) Pseudopotentials in condensed matter systems. Comput Phys Rep 9:114-197

Sherman DM (1997) The composition of the earth's core: constraints on S and Si vs. temperature. Earth Planet Sci Lett 153:149-155

Sigalas M, Papaconstantopoulos DA, Bacalis NC (1992) Total energy and band structure of the 3d, 4d, and 5d metals. Phys Rev B 45:5777-5783

Silva Cd, Stixrude L, Wentzcovitch RM (1997) Elastic constants and anisotropy of forsterite at high pressure. Geophys Res Lett 24:1963-1966

Singh DJ (1994) Planewaves, Pseudopotentials, and the LAPW Method. Kluwer Academic, Norwall, Massachusetts

Slater JC (1951) A simplification of the Hartree-Fock method. Phys Rev 81:385-390

Slater JC, Koster GF (1954) Simplified LCAO method for the periodic potential problem. Phys Rev 94:1498-1524

Söderlind P, Moriarty JA, Willis JM (1996) First-principles theory of iron up to earth-core pressures: structural, vibrational, and elastic properties. Phys Rev B 53:14063-14072

Stixrude L, Cohen RE (1993) Stability of orthorhombic $MgSiO_3$-perovskite in the Earth's lower mantle. Nature 364:613-616

Stixrude L, Cohen RE (1995) Constraints on the crystalline structure of the inner core: Mechanical instability of BCC iron at high pressure. Geophys Res Lett 22:125-128

Stixrude L, Cohen RE, Singh DJ (1994) Iron at high pressure: linearized augmented plane wave calculations in the generalized gradient approximation. Phys Rev B 50:6442-6445

Stixrude L, Cohen RE, Yu RC, Krakauer H (1996) Prediction of phase transition in CaSiO$_3$ perovskite and implications for lower mantle structure. Am Mineral 81:1293-1296

Stixrude L, Wasserman E, Cohen RE (1997) Composition and temperature of earth's inner core. J Geophys Res 102:24729-24739

Stokes HT, Boyer LL, Mehl MJ (1996) Spherical self-consistent atomic deformation model for first-principles energy calculations in ionic crystalline solids. Phys Rev B 54:7729-7736

Teter DM, Hemley RJ, Kresse G, Hafner J (1998) High pressure polymorphism in silica. Phys Rev Lett 80:2145-2148

Thompson KT, Wentzcovitch RM, Bukowinski MST (1996) Polymorphs of alumina predicted by first principles: Putting pressure on the ruby scale. Science 274:1880-1882

Troullier N, Martins JL (1991) Efficient pseudopotentials for plane-wave calculations. Phys Rev B 43:1993-2003

Vanderbilt D (1990) Soft self-consistent pseudopotentials in a generalized eigenvalue formalism. Phys Rev B 41:7892-7895

Wallace DC (1972) Thermodynamics of Crystals. John Wiley & Sons, New York

Wasserman E, Stixrude L, Cohen RE (1996) Thermal properties of iron at high pressures and temperatures. Phys Rev B 53:8296-8309

Watson RE (1958) Analytic Hartree-Fock solutions for O^{2-}. Phys Rev 111:1108-1110

Wei SH, Krakauer H (1985) Local-density-functional calculations of the pressure-induced metallization of BaSe and BaTe. Phys Rev Lett 55:1200-1203

Weidner DJ, Bass JD, Ringwood AE, Sinclair W (1982) The single-crystal elastic moduli of stishovite. J Geophys Res 87:4740-4746

Wentzcovitch RM (1991) Invariant molecular dynamics approach to structural phase transitions. Phys Rev B 44:2358-2361

Wentzcovitch RM, Hugh-Jones DA, Angel RJ, Price GD (1995a) Ab initio study of MgSiO$_3$ *C2/c* enstatite. Phys Chem Minerals 22:453-460

Wentzcovitch RM, Martins JL, Price GD (1993) Ab initio molecular dynamics with variable cell shape: application to MgSiO$_3$ perovskite. Phys Rev Lett 70:3947-3950

Wentzcovitch RM, Ross NL, Price GD (1995b) Ab initio study of MgSiO$_3$ and CaSiO$_3$ perovskites at lower-mantle pressures. Phys Earth Planet Inter 90:101-112

Wentzcovitch RM, Silva Cd, Chelikowsky JR, Binggeli N (1998) A new phase and pressure induced amorphization in silica. Phys Rev Lett 80:2149-2152

Wentzcovitch RM, Stixrude L (1997) Crystal chemistry of forsterite: a first principles study. Am Mineral 82:663-671

Wigner E (1934) On the interaction of electrons in metals. Phys Rev 46:1002-1011

Wijs GAd et al. (1998) The viscosity of liquid iron at the physical conditions of the earth's core. Nature 392:805-807

Wolf GH, Bukowinski MST (1987) Theoretical study of the structural properties and equations of state of MgSiO$_3$ and CaSiO$_3$ perovskites: implications for lower mantle composition. In: M.H. Manghnani and Y. Syono (eds) High Pressure Research in Mineral Physics, p 313-331 Terrapub, Tokyo

Wolf GH, Bukowinski MST (1988) Variational stabilization of the ionic charge densities in the electron-gas theory of crystals: Applications to MgO and CaO. Phys Chem Minerals 15:209-220

Yeganeh-Haeri A (1994) Synthesis and re-investigation of the elastic properties of single-crystal magnesium silicate perovskite. Phys Earth Planet Inter 87:111-121